Principles of
Human Nutrition

Principles of Human Nutrition

Martin Eastwood

University of Edinburgh, Western
General Hospital, Edinburgh, UK

CHAPMAN & HALL

London · Weinheim · New York · Tokyo · Melbourne · Madras

Published by Chapman & Hall, 2–6 Boundary Row, London SE1 8HN, UK

Chapman & Hall, 2–6 Boundary Row, London SE1 8HN, UK

Chapman & Hall Gmbh, Pappelallee 3, 69469 Weinheim, Germany

Chapman & Hall USA, 115 Fifth Avenue, New York, NY 10003, USA

Chapman & Hall Japan, ITP-Japan, Kyowa Building, 3F, 2-2-1 Hirakawacho, Chiyoda-ku, Tokyo 102, Japan

Chapman & Hall Australia, 102 Dodds Street, South Melbourne, Victoria 3205, Australia
Chapman & Hall India, R. Seshadri, 32 Second Main Road, CIT East, Madras 600 035, India

First edition 1997

© 1997 Martin Eastwood

Typeset in 10/12 Sabon by Photoprint, Torquay, Devon.

Printed in Great Britain by The Alden Press, Osney Mead, Oxford

ISBN 0 412 57650 3

A catalogue record for this book is available from the British Library

Library of Congress Catalog Card Number: 96–84974

This book is dedicated to Jenny for all the reasons she knows and without whom it would not have been possible.

Contents

Contents

Acknowledgements

Whilst the responsibility for this book is entirely mine, there are many people who have given so much help.

David Brock; Jennifer Browne; Maureen Cole; Jenny Eastwood; Anne Jenkinson; David Kritchevsky; Janet Lambert; Margaret McAllister; Ian Penman; Fiona Phillips; Gillian Poole; Deborah Pryde; Afolabi Sawyerr.

I am also grateful to Nigel Balmforth at Chapman & Hall for his gentle, persuasive support during the writing of this book, and to Alison Conneller throughout the production process.

Chapter 1

An introduction to human nutrition

1.1 Introduction

- An adequate provision of all nutrients in the correct proportions is a prerequisite for health.
- Analysis of medical and nutritional data requires the use of modern mathematical concepts and techniques.
- Nutrients are necessary for growth of the whole body, cellular and chemical structure and repair and the provision of energy.
- The body copes better with an excess than with a deficiency of nutrients, with the exception of alcohol.
- The body metabolizes dietary nutrients, whether in insufficient, adequate or excessive amounts, in a manner determined by genetic make-up.
- World population size is an important determinant of availability of nutrition and hence health.
- There are threats to population nutrition by adverse environmental changes.

This book looks at nutrition as an exciting science, which extends into all of biology either as a source or a recipient of nutrition, as a sociological force, as a branch of medicine or as an art as in cooking.

Nutrition – the science and the art – studies the provision and fate of nutrients in the body. The biochemistry of the metabolism of ingested nutrients is dictated in each individual by species, gender and inherited constitution. All living creatures require a range of dietary chemicals for growth and metabolism. These chemicals are obtained from the accumulated metabolism of a range of organisms. The individual requires for good health an adequate total calorie as well as essential nutrient intake to provide for the needs of a genetically determined constitution which dictates protein and enzyme structure and hence metabolism.

Modern analysis was, until the 1980s, based on principles defined by Euclid. His geometry was a science of properties and relations of magnitudes in space, as lines, surfaces and solids. Observations in nature may have to be considerably distorted or 'smoothed out' to be described within this framework.

Medical and nutritional investigation is concerned with the quantification of recorded data. In nutritional studies this is complicated by individual and haphazard human eating patterns and the metabolic response to irregular food intake. Traditional mathematical models have been superseded by the modern concept of Fractals and Fractal analysis. Here the whole pattern is represented in each of the tiny parts.

The term **Fractal** was introduced by a modern mathematician, Benoit Mandelbrot. The word Fractal is derived from the Latin, 'fractus' meaning irregular or fragmented. Fractals are the description of a type of distribution of points in space but also taking distribution in time into account. Fractal analysis is therefore most relevant to the complex patterns of many medical and nutritional observations.

1.1.1 Food utilization

The food is broken down to chemicals of molecular size which are readily absorbed and utilized by the body. The process of absorption is dictated by the nutrient needs of the body and bioavailability. In general, energy providing nutrients have a high bioavailability value, whereas for micronutrients this value is lower and more variable. Some nutrients, e.g. divalent cations, calcium and magnesium, are only absorbed in an amount necessary for the needs of the body, as an excess of these can be toxic.

Waste products of metabolism are excreted in breath (CO_2), urine (in general, water-soluble compounds of molecular weight less than 300 Da) and bile (in general, fat-soluble, molecular weight more than 300 Da). Accumulation of metabolic waste products has disadvantageous effects on growth, metabolism and well-being. Nutrients will contribute to bodily needs in several ways:

1. Provision of energy.
2. Creation of structure.
3. The provision of essential small molecular substances which the body cannot synthesize.

Some nutrients are sources of carbon and nitrogen which pass into the metabolic pool, to meet the body's general needs, e.g. carbohydrates, fats and amino acids. Carbohydrates and lipids are necessary fuels for metabolic activity, to a variable extent for structure and in some instances in the synthesis of hormones. The whole range of amino acids is relevant for adequate structural growth. Amino acids may also be utilized at times of nutritional deprivation as a source of energy.

An important aspect of nutrition is the availability of dietary sources of nutrients. Causes of dietary deficiencies range from a lack of all nutrients (famine), to absence or omission of individual food items from the diet for social, economic, cultural, religious or personal reasons. Nutrients may not be absorbed from the intestine in some illnesses. A deficiency or excess of overall calorie intake or of individual nutrients may result in nutritional disorders.

1.1.2 Metabolism of nutrients

The metabolism of nutrients by enzymes is dictated by the individual's gene structure and the induction of enzymes and in turn by species and gender. These distinctions are complex, subtle and only partially understood.

The nutrient needs and subsequent metabolism by the individual will be influenced by growth in the young and in pregnancy, and modified by disease, drugs, alcohol and tobacco. Towards the end of life there are important changes in the effectiveness of absorption and utilization of the nutrients consumed.

1.1.3 Essential nutrients

Other nutrients are essential in that these molecules cannot be synthesized within the body and can only be provided by the diet. Such essential nutrients provide for metabolic processes, vitamins, e.g. ascorbic acid, trace elements, e.g. selenium, and for structure (e.g. proteins, essential amino acids, vitamins and trace elements).

The science of nutrition is devoted to defining requirements for essential nutrients, amino acids, essential fatty acids, vitamins and trace elements. Recommended daily amounts (RDA) are dependent upon diverse factors such as growth, pregnancy and illness and are only carefully determined approximations. Implicit in the requirement for essential dietary constituents is that the human race is not independent of the environment. Thus, man is part of a food chain as a recipient or producer of food.

1.1.4 Nutritional deficiency and excess

It is not possible to live for more than 2–3 minutes without oxygen. However, life can continue without water for between 2 and 7 days depending upon the ambient temperature and the amount of exercise being taken. Survival without any food at all, but with water, may be for 60–120 days, depending upon the body stores. Females and those with considerable subcutaneous fat survive for longer than slightly built males.

There are individual responses to nutritional deficiency and excess. Though in general weight increase in association with overall excessive eating and weight loss is associated with inadequate dietary intake. The failure to provide the essential amino acids, fats, vitamins and trace elements leads to specific lesions which may progress to morbidity and death. There is no nutritional explanation for the apparent synthesis of essential vitamins by some individuals. When scurvy was a problem in the Royal Navy the fleet would come into land every 2 months to take on board provisions specifically to reduce the prevalence of scurvy. However, on the long sea voyages some individuals died quite quickly of scurvy and others appeared to be unaffected. Similarly, the different types of beri-beri suggest individual metabolic responses to thiamin deficiency.

In general the body copes better with an excess than with a deficiency of nutrients, with the exception of alcohol. Consequently there is an inclination to eat somewhat more than is required. The body copes less well with an excess of dietary fatty or fat-soluble compounds than an excess of water-soluble dietary components. Fatty nutrients, e.g. lipids, are stored and, if the storage load becomes excessive, then the body is disadvantaged. Water-soluble dietary excesses may be excreted metabolically modified or unchanged in the urine. Excess dietary protein and lipid intakes may be metabolically modified to structural or storage tissues or possibly be excreted in bile and urine. The variable pathways whereby these processes occur will be determined by the range of variants of the same enzyme (isoenzymes) which form the metabolic enzyme structure of the individual.

1.1.5 Social, population and environmental influences on nutrition

The reliable provision of food requires an organized society. A society that is disorganized through war, epidemics of infections or natural disaster is less able to produce or deliver food than a well-structured stable society with a sufficiency of healthy workers. It is important that food is grown which is appropriate for the particular population's social, cultural and religious beliefs. The influences on nutrition (Figure 1.1) include:

- food availability and intake
- acceptable taste which meets cultural and social mores
- sufficient but not excessive suitable nutrients and chemicals which will vary with age, gender, growth and health
- the provision of a sufficiency of oxygen and water
- a nutrient intake which meets the requirements and constraints set by the individual's genetic constitution
- a ready loss of breakdown products of metabolism

It has been suggested that diet may affect behaviour. In some ancient cultures certain foods were thought to have magical qualities capable of giving special powers of strength, courage, health, happiness and well-being. It is possible that some food constituents may affect the synthesis of brain neurotransmitters and thus modify brain functions. It is therefore important to integrate dietary effects on brain chemicals into our wider understanding of human behaviour.

Until there is an understanding of such nutritional and metabolic mechanisms confused advice may be disseminated. Pathology which is provoked by the metabolic response of even a small proportion of the population may erroneously be applied to the population as a whole.

The world population will increase by between 2.5 and 3.5 billion by the year 2025. There are even projections that the world population will be greater than 8 billion in the year 2025, that is, it will have grown four-fold in a century. To provide adequate contraception for the world's fertile female population would cost US$ 9 billion annually, the amount the world spends on arms every 56 hours.

While the world population grew from 1.49 billion in 1890 to 2.5 billion in 1950 and to 5.32 billion in 1990, the energy consumption grew from 1 terawatt in 1890 to 3.3 terawatts in 1950 to 13.7 terawatts in 1990. Energy consumption grows as the population grows and as poor people become richer. Each individual in the developing world uses 0.28 kilowatts (kW) a year while those in the developed world use 3.2 kilowatts and those in the United States use 9 kW. If the poor of the world were to increase their use to a modest amount of energy, say 2.3 kW a year, this could mean a seven-fold increase in world energy consumption. This, combined with a doubling of their numbers by 2050, could mean a 14-fold increase in energy consumption. At the present rate of consumption there is sufficient oil and gas to last 50 years. Energy costs in industrialized countries are low, but energy is wasted. Energy costs in the developing world are steadily rising; cheap biomass fuels, e.g. wood and dung and crop wastes, are insufficient for needs. Over 2 billion people suffer from acute scarcity of fuel wood and lack other energy supplies. Deforestation and the growing use of tree, crop and dried dung for fuel, removes these vital soil replenishers. These changes in energy requirements and utilization could have serious effects on the global carbon availability summarized in Table 1.1.

By the year 2030 the level of the sea will have risen 18 cm due to global warming. Global warming causes the drying of wells, a decrease in fuel which will be less easily obtained, fewer raw materials for industry, less fodder for livestock and even a decrease in availability of medicinal plants. More time spent on collecting fuel and

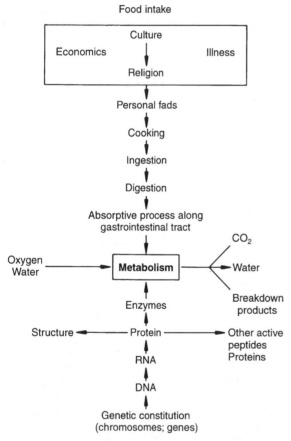

FIGURE 1.1 Metabolism represents a relationship between food intake and the enzymes characteristic of an individual dependent upon genetic constitution. Also important are oxygen and water intake and the ability to excrete carbon dioxide, water and metabolic breakdown products.

TABLE 1.1 Global carbon availability

Balance	Source	Amount
Gained	Fossil fuels	5 billion tonnes
	Plant respiration	50 million tonnes
	Deforestation	2 billion tonnes
	Biological and chemical absorption	50 billion tonnes
Lost	Plant photosynthesis	100 billion tonnes
	Biological and chemical absorption to ocean	104 billion tonnes

water and fewer livestock would result in less manure for fuel or fertilizer, money would be spent on fuel and farm yields would fall. The result overall would be lower income, poorer nutrition and poorer health.

A definition of health should include the concept of sustainability or the ability of the ecosystem to support life in quantity and quality. The converse of sustainability is entrapment. Entrapment leads to dependence on outside aid, forced migration, starvation or civil war.

Environmental damage is in itself serious and nutritionists must be aware of the potential human nutritional consequences of this damage.

Depletion of the productivity of land

Some 25 billion tons of top soil are lost annually from the world's crop lands. As a result, there will be an increase in the world's area of deserts.

Deforestation

Over 11 million hectares of tropical forest are felled each year, leading to soil erosion, with resulting floods and development of desert. Subsequent flood control measures can impede migration of fish and fishing practices.

Availability of soil micronutrients

Oil pollution or a change in soil pH alters the bioavailability of micronutrients to plants and hence the whole food chain.

Use of chemical fertilizers

Considerable progress has been made in agriculture with the so-called Green Revolution crops. High-yielding staple crops provide a consistent and predictable calorie intake in a wide number of countries. However, the fertilizers used to maintain the plants are quality sources of nitrogen, but do not contain the rich array of micronutrients, e.g. sulphur, manganese, zinc and copper provided by the traditional manures.

The trace element content of the soil declines and this is reflected in plant growth. There is now a serious possibility that a new nutritional bottle neck is developing wherein the constraint to growth is not macronutrients, but micronutrients such as zinc.

The reduction of crop rotation practices and other modern agricultural techniques may result in zinc deficiency in soil and hence in plants. Goitre has also been attributed to the use of fertilizers which impair the bioavailability of iodine.

Industrial pollution

Large stretches of rivers contain dangerous heavy metals and chemicals in concentrations high enough to kill fish and endanger human life. Despite legal restrictions, factories and sewerage works dump dangerous substances, e.g. mercury, cadmium, insecticides, herbicides and chemicals such as chloroform, into the sea, lakes and rivers. Some contamination also comes from old dumps and old mine workings. The food chain is adversely affected by this pollution.

	Third World	Industrialized World
Population	Increasing	Static
Age distribution	Young	Elderly
Diseases	AIDS Tuberculosis Hunger War	Coronary heart disease Cancer
Food	Insufficient	Too much
Agriculture	Marginal, removing trees	Intensive Pesticides Herbicides
Fuel	Timber, loss of forests	Fossil fuels Air pollution
Pollution	People	By-products of industry, transport, packaging

FIGURE 1.2 Population characteristics: some differences between the Third World and the Industrialized World.

Smoking

There is considerable overlap between tobacco smoking and conditions in which bad nutrition is believed to play an aetiological role.

Deprivation

The definition and measurement of deprivation is important because of the relationship between deprivation, poor nutrition and ill health.

One measurement scale of deprivation is the Jarman underprivileged area 8 (UPA 8) score. The variables include unemployment, overcrowding, no car, low owner occupation, social class, poor skills, pensioners living alone, single parent, lack of amenities, ethnic status, children under 5 and having moved within 1 year. Unemployment is an important single element in the definition of deprivation. Hospital admissions correlate highly with all measures of deprivation.

Some differences in population characteristics between the Third World and Western civilizations are shown in Figure 1.2

Further reading

Campbell, D.A., Radford, J.M.C. and Burton, P. (1991) Unemployment rates: an alternative to the Jarman index. *British Medical Journal*, 303, 750–5.

Godlee, F. (1991) Strategy for a healthy environment. *British Medical Journal*, 303, 836–8.

Gopalan, C. (1992) *Bulletin of Nutrition Foundation of India*, 13, 1.

Leader (1991) Fractals and medicine. *Lancet*, 338, 1425–6.

Lowrey, S. (1991) Housing. *British Medical Journal*, 303, 838–40.

Smith, R. (1993) Overpopulation and overconsumption. *British Medical Journal*, 306, 1285–6.

Weaver, L.T. and Beckerleg, S. (1993) Is health a sustainable state? A village study in The Gambia. *Lancet*, 341, 1327–8.

1.2 History of food

- Over the centuries travel has introduced migrants, armies and traders to new foods and modes of cooking.
- Revolutions and industrialization have altered cooking and eating habits.
- Knowledge of nutritional concepts has recently altered dietary intake.
- The advent of supermarkets and the importation of foods from all over the world has further changed dietary intake.

1.2.1 History of diet around the world

Ancient Egypt

Grain, bread and porridge have been the basis of food since the beginning of nutritional history. In Egypt, bread was made as flat cakes from toasted grains of barley, wheat or millet. The meal mixed with water in a paste was either dried in the sun or baked on the flat stones of the hearth. Primitive grains required toasting before the hard outer husks could be removed. However, the Egyptians found a strain of wheat which could be threshed without being toasted, and consequently dough could be made from a raw flour. Subsequently the Egyptians found out how to produce beer and used this knowledge in bread-making to create sponginess in the dough when baked and so leavened bread was produced.

Egyptian bread is said to have had a sour taste, suggesting they used lees of beer or a sour dough for their leaven. Sour dough is a piece of fermented dough saved from a previous baking and is used to start the fermenting process of the new dough.

In 13th century BC Egypt, it was usual for people to eat two meals a day, a light morning meal and a more substantial evening meal consisting of several dishes.

Ancient Greece

The Greeks made unleavened bread from coarse wheat which they favoured over barley meal, baking their bread in hot ashes and later in bread ovens. Increasingly they used flour which had been finely sieved, removing most of the chaff. There were a wide range of breads available to the ancient Greeks and also to the Romans. For leaven, like the Egyptians, both the Greeks and the Romans used sour dough.

The Roman Empire

The earliest Romans were a rural people and ate a thick porridge of barley or beans and green vegetables with flat barley bread, hard-baked in ashes. Cheeses were made from goats' milk, meat was a rarity and fish hardly used. Food was flavoured with garlic, parsnip, olives and olive oil. When Rome became all-powerful the poor still ate thick grain soups of millet and coarse bread together with a little turnip or a few beans. Raw olives or goats' cheese or figs were delicacies and occasionally there was cooked pork or meat balls which were produced in cook shops which were to be found throughout the cities. For long periods in Rome bread was given without payment.

The wealthy few had food from all over the world – spices from India, south-east Asia and China, wheat from Egypt, ham from Gaul and wine from Greece. Ginger came from China through central Asia; cloves from Indonesia by sea to Ceylon and then on by sea and land to Alexandria.

Pepper was a very important element in Roman cooking and was brought overland from India through Egypt. A meal might consist of hors d'oeuvres: a salad of mallow leaves, lettuces, leeks and mints and fish dishes garnished with sliced eggs, rue and tuna fish. The main course included a kid, meat-balls and beans together with chicken and a ham. The meal finished with a dessert of ripe apples and vintage wines. The salads were dressed with wine, oil and vinegar and liquamen, the sauce made from fermented salted anchovies. Hams were boiled with dried figs and bay leaves and baked with honey in a pastry coating. Chickens were roasted or boiled in a variety of spiced sauces. Roman manors had their own oyster tank to ensure a fresh supply. The Roman physician, Galen, taught his followers that it was harmful to eat fruit with a meal.

By the second century BC cooking had become an art in Rome. The Romans ate three meals a day. Breakfast between 8:00 and 10:00 am, bread and cheese and a glass of water. Lunch, eaten at noon with little ceremony, was usually bread with cold meat, vegetables and fruit with a little wine. Dinner was at about 8 o'clock in the evening in winter and 9 o'clock in the summer. Most food was eaten with the fingers, which were rinsed from time-to-time though the diners were provided with knives, toothpicks, spoons and napkins.

The requirement for grain necessitated the importation of wheat on a large scale from Egypt, Sicily and North Africa. Special docks and lighthouses were built for these grain ships. Rotary mills powered by animals were used and sophisticated ovens with systems of draughts and chimneys controlled the heat. The flour was milled into various grades ranging from the finest white to coarse flour considered suitable only for slaves.

The Roman army stretched from the Scottish border through to Egypt and as far East as the edge of the Black Sea. The soldiers were issued with daily rations of grain or bread, meat, wine and oil and the cost was deducted from their pay. Cheese, vegetables and salt were included in their basic rations. When the legions were on the march they carried a scythe to cut crops, a metal cooking pot, a mess tin and 3 days' emergency food of hard tack, dried cooked grain which could be eaten without further cooking, salted pork and sour wine. Meat was boiled or grilled and the soldiers were issued with spits as standard equipment. A quern to grind flour and a portable oven were carried for every 10 men. During peace time a soldier was expected to grind his own grain and bake his own bread. Soldiers supplemented their diet by hunting. In war time troops foraged from the enemy countryside. Soldiers made their own cheeses from the milk of animals kept at the forts.

Roman army officers lived on fresh meat and imported edible snails, olives, vintage wine, pepper, fish sauce, hams and oysters. The Roman army was huge, 300 000 men in the first century AD. This required substantial organization in supplying the garrison and moving food around the Empire. The cost of such provision of food eventually became an intolerable burden on the Roman taxation system.

Mosaic dietary laws

Jews have a diet established under Mosaic law. These laws defined clean and unclean food. Blood was seen as the life-force and was forbidden, so all food animals had their blood drained as in kosher ritual slaughter. Fish that swam with their fins and had scales were accepted as clean; shell-fish which have neither fins or scales but swim in water were unclean. Cows, sheep and goats which chew the cud and have cloven hooves were clean; the pig, which could not live on grass, was difficult to herd and had little stamina for the nomadic way of life, was unclean. The modification to that diet depended on the dispersion of the Jewish race either to the North of Europe or along the southern shores of the Mediterranean Sea. Eastern European Jews adopted central European cooking, e.g. 'gefilte fish', poached fish cakes made with the flesh scraped from the skin and bones and mixed with onions, seasoning and breadcrumbs bound with egg. The Jews of the Mediterranean lands used fish, fruit, nuts and vegetables.

India

Indian philosophers stressed the importance of food for the uplifting of the soul and the health of the body. They suggested spices such as cloves and cinnamon were warming; coriander and cumin cooling. They also believed in pure and impure foods; rice and honey were considered purer than other foods.

The Indus valley was the centre of civilization around 2000 BC when it was invaded by Aryan invaders from Iran and Afghanistan. The economy of these Aryan warrior nomads was based on cattle. They lived primarily on meat and milk products. Barley was the staple grain, ground into flour and cakes and eaten with butter. Crushed toasted barley was mixed into a gruel with curds, clarified butter and milk. Food was eaten by hand. Thick gruels were licked off the fingers and thin ones drunk from bowls or cups made of clay. The inhabitants of the Indus valley were driven south into India. Aryans gradually adopted the use of rice, wheat and beans and learned the use of spices, including tumeric, long peppers (pepper from vines similar to black pepper), sour oranges and sesame, but continued to use clarified butter for cooking. Rice was cooked with mung beans into a thick gruel or khicri. Raw ginger was eaten after meals to aid digestion. The juice of the soma plant was mixed with rice or curds.

The early Aryans were tribal and their society was divided into castes. The Brahmins were the highest caste with strict rules of purity, particularly concerning food. Food could be polluted by being touched by lower-caste people which was believed to affect its purity. To protect themselves from pollution, strict laws governing the toilet and behaviour of cooks were developed. Brahmins could only eat food prepared by other Brahmins. There were huge blood sacrifices, particularly of cattle, and until the end of the fifth century the meat was eaten after the sacrifice. As society developed, effigies of horses and cows made of dough were substituted and a ban on killing cattle was introduced.

Beliefs were very austere and did not allow the eating of meat, fish or eggs, and the people were consequently strict vegetarians. Other Hindus, also strict vegetarians, do not eat rank smelling foods and ban onions and garlic from their kitchens. There may be a separate side kitchen in which onions and garlic may be cooked for dishes not considered ritually pure.

The Buddhist diet was a compromise between the diets of the Hindus and Brahmins. Buddhists were not forbidden meat, merely not allowed to kill for food. Vegetarianism was, however, encouraged for all people. The cow was sacred and no longer killed for food or sacrifice. In the temple in modern India Hindu gods are offered vegetarian dishes.

Until the 15th century the rulers of northern India came into contact with and were influenced by Persian culture. Many of the dishes had a Persian origin, particularly the samusak pastries. Samosas made of meat, hashed and cooked with almonds, walnuts, pistachios, onions and spices were presented. The word samusak comes from the Persian word sanbusa, a triangle, and is similar to the samosas sold as snack foods in modern times. Meals would start with sherbet and bread in the form of thin round cakes. Roasted meat was cut into a large sheet which was divided into six pieces, one piece being placed before each man. Round dough cakes made with ghee were stuffed with a mixture of flour, almonds, honey and sesame oil. On the top of each dough cake was a brick-shaped sweet cake made of flour, sugar and ghee. Meat was served in large porcelain bowls, and cooked with ghee, onions and green ginger. In addition, rice cooked in ghee was served with chicken.

Northern India was invaded by the Mughals at the end of the 15th century. They came from Uzbekistan in central Asia. Their cuisine was similar to that of the Persians, with a variety of grains, green vegetables and meats, which were oily, sweet and spicy.

China

In the China of the last millennium BC grains were cooked whole as flour milling did not come into general use until about the first century AD. An Imperial Court banquet would include roast turtle and fresh fish, bamboo shoots and reed tips. Meanwhile, the common people lived on a diet of beans and grains flavoured with sour or bitter herbs. Salt and sour plums were the earliest seasoning used. Around the second century BC fermented salted soya beans became popular and were produced on a commercial scale. By the fifth century Chinese cooks had a choice of herbs and pickled meats for bitter flavours – honey and maltose from grains for sweet flavours, and prickly ash, mustard or ginger for hot flavours.

The Chinese knelt on mats or flat cushions to eat. The food was laid out either on the floor or on tables. Chopsticks were beginning to be widely used, but soup stews were eaten with spoons and grains were eaten with the fingers. Soup stews were very popular. For the rich there were beef soups seasoned with sour plums, pickled meat sauce and vinegar. Other soups contained venison, salted fish, bamboo shoots and rice, beef, dog or turnip. The poor had soups made of vegetables and grain without meat.

In the Imperial courts of China 960–1280 AD Chinese cooking reached great heights. Seasoning was important: sesame, anise, ginger, black pepper, onion, salt, cardamoms and vinegar. Other seasonings such as orange peel, soy sauce, peppermint, cinnamon and liquorice were used. Rice was brought in from Vietnam at the beginning of the 11th century with resulting nutritional advantage. China became one of the richest countries and as a result of trading new foods and culinary skills were available. From India came the refining of sugar and black pepper; from Persia came coriander and pastries.

Stir-fry became a central cooking method. Woks came into common use and wheat flour doughs and pastries were mastered. Bean curd was discovered and became increasingly popular and noodles became generally used. Sweet sauces made from fermented flour were used to flavour stir-fry dishes. Raw meat and fish were both great delicacies, being flavours which were intense and natural. Preserved pickled meats and fish were very popular. Sparrows were pickled with fermented rice and barbecued.

As the Chinese Empire developed during the 1600 hundred years from the time of Christ, a very complicated system of cooking was developed by the northern Chinese aristocracy. Vinegar, fermented bean paste and soy sauces were developed. In the winter meat and fish were preserved by pickling or made into fermented sauces. Vegetables were pickled with salts. At festivals, popular foods were gruels and packets of grain wrapped in cucumber leaves. A small bear was steamed with onion, ginger, orange peel and salt after being marinated in fermented bean sauce. The fat from a boiled pig was skimmed off for separate use. Discs of boiled dough were made from wheat flour and other doughs from a balm of white rice and wine were left to simmer by the fire. Dishes were spiced with ginger, rice wine, prickly ash pepper and fermented bean sausages, as well as the bitter bark of magnolia.

A Chinese proverb says: 'you are what you eat'. Chinese attitudes to food and health were dependent on this. The Chinese held to the humoural belief that the universe and everything in it was composed of four elements: fire, air, earth and water, and four qualities, heat, cold, moisture and dryness. Treatment for bodily disorders was based on the strength and interaction of these elements and qualities. The basic division of the Chinese beliefs was between the bright, dry, warm male principle, Yang, and the cold, dark, moist female principle Ying (Yang and Ying).

The human body is a reproduction of the cosmos. To be healthy was a reflection of the general harmony among the various virtues, while illness is a sign of disharmony of heat, cold, moisture and dryness. These elements were controlled by food which also had the four qualities. Fundamental to the Chinese theories on nutrition was the idea that food and cures come from the same source. Cooling foods such as green vegetables, and fruits treat fevers and rashes, while heating foods such as liver and chicken could treat debility and weakness.

In China, soup was used as a tonic to maintain health. A healthy soup proposed for the liver was made of dog meat, sour plums, Chinese leeks and hemp, while for the lungs a soup of yellow millet, chicken, peach and onion was recommended. Chicken soup was meant to be particularly good after childbirth. (It is also interesting that the Jewish general belief is that chicken soup treats all ailments.)

A goat's heart marinated in rose water and barbecued with safflower was a prescription for tachycardia, and a goat's leg and cardamom for strength. Tisanes made from ginger were recommended for general strengthening. Chestnuts or salted bamboo shoots and sesame were served several times a day and the women of the house frequently drank strengthening gruels. A bean gruel made with a little salt and ginger was regar-

ded as good both for kidneys and as a cure for vomiting.

Buddhism came to China from India at about the time of Christ. With it came preaching against the killing of animals for food. As Buddhism progressed in China the prohibitions of meat eating were not accepted generally, except in Buddhist monasteries, famous for the excellence of their cuisine where the cooking followed strict Buddhist regulations. However, they also insisted on five colours: red, green, yellow, black and white; five flavours, bitter, salt, sweet, hot and sour; and five styles of cooking, raw, simmered, barbecued, fried and steamed. All were represented in a temple meal. Dishes were contrived to create the illusion of eating meat; flour and water pastes were made to resemble animal barbecues. Gluten was used for both stir-fry dishes and barbecues.

In Europe during the Dark Ages trade with the East stopped and pepper disappeared from northern Europe. By the time of the Norman conquest of England eastern spices and pepper were once again being traded by Arabs living in Spain and sending trading ships to India, south-east Asia and China. This trade was then developed by Venice, which eventually became dominant in trading, bringing food from the East to the markets of the north of Europe.

When sweet potatoes were introduced into China they were an immediate success because they grew readily in poor soils and adverse weather conditions. They were also attractive because they were sweet, in an area where sugar and sweeteners were expensive. By the beginning of the 19th century sweet potatoes were a staple food for half of the population of northern China. Similarly, chillies arrived in China about 1700 and were introduced into Schezuan yunnan cooking. Maize was added to the Chinese diet around the middle of the 16th century, but was never as popular as rice and wheat flour became more popular.

Japan

Buddhism moved to Japan from China and Korea during the seventh century. When Buddhism came to Japan the emperor forbade the eating of any meat except by the sick. Within 100 years both chicken and fish were exempted from this rule. A stricter interpretation of Buddhist law came in the 12th century with the spread of Zen Buddhism from China. The attempts to introduce a rigid vegetarian diet were only partially successful.

Until the mid-19th century the Japanese were somewhat reluctant to kill four-legged animals, particularly cattle, for food. However, most modern Japanese eat beef. Zen Buddhism developed and formalized the tea ceremony which had been current in China some 700 years previously.

Zen belief in restraint and simplicity was expressed in the tea ceremony with its strict rules of formal behaviour. These were undertaken in the Zen temples. Only foods in season could be used. Two main styles of tea ceremony cooking developed: one at the Daaitokuji temple near Kyoto in the 14th century and the other in the 16th century at the Obakusn temple near Tokyo. At Daaitokuji the meal was prepared in individual servings. The Obakusn meal was based on Chinese vegetarian cooking and retained the Chinese practice of serving all food in large dishes in the centre of the table, from which diners could help themselves. Such a meal started with green tea and a sweet cake served according to the formal tea ceremony style, followed by a plate of cold hors d'oeuvres and a clear soup with bean curd and ginko nuts. Next came a number of different foods cooked by simmering: bean curd balls, aubergine, rolls of thin bean curd sheets, mushrooms, bamboo, lotus roots, chillies, ginko nuts and pine needles. After this arrived a steamed dish followed by a dish of braised vegetables, a deep-fried dish, a salad of chrysanthemum leaves with a walnut dressing, a vegetable stew, fruit and finally rice cooked with a little green tea. Such a feast is known as Lohan's delight, Lohan being a Buddhist saint.

Ancient Persia

The Persian Court in the sixth century AD regarded the rearing of animals and birds for the table as being very important. Wild asses were fattened with clover and barley and then cooked with yoghurt and spices. Chickens were reared on hemp, oil and olives and after being killed were hung for 2 days by the feet and then by the neck before they were cooked. Other dishes included milk-fed kids and calves and fat beef cooked in a broth of spinach, flour and vinegar. Hares and pheasants were made into ragoutes. In the summer the Persians ate nut and almond pastries made with gazelle fat and fried in nut oil. Foods imported from Europe and Asia were available to the Persians. Fresh coconut was served with sugar and dates and stuffed with nuts. They also ate sweet preserves of lemons, quinces, Chinese ginger and chestnuts and drank sweet wine.

Islam

Mohammed preached that food was a gift from God, 'So eat of what God has given you, lawful and good, and give thanks to God's favour if Him it is you serve'. Pork or any animal found dead, blood or animals killed as an offering to a pagan god, fish without scales (including shell-fish), alcohol and fermented liquids were all forbidden, or halan. Carnivorous animals and birds were forbidden. Permitted foods were halal. Animals killed for food have to be slaughtered by an approved butcher who must say 'In the name of God, God is most great' and cut the animal's throat to allow the blood to drain out. Animals who die by disease, strangulation or beating are not acceptable. This practice is still followed by modern Moslems.

The month-long fast of Ramadan is in memory of the prophet's revelations and is for the health of the soul. Fasting is a way of reaping spiritual rewards. Nothing is eaten or drunk during daylight hours in the month of Ramadan. Each evening after sunset the fast is broken with three dates and water, followed, after final sunset, by prayer and a meal. All Moslems must follow Ramadan after the age of responsibility, 12 years in girls, 15 years in boys. Exceptions are the elderly in poor health, pregnant and nursing women, menstruating women, the sick, travellers and labourers. The meal is of an ordinary size, not extra quantity to fill the stomach after fasting. In Saudi Arabia today people eat a meal of bread, milk or sour milk together with a braised or stewed meat dish.

The lifestyle of the Arab Califs who ruled Egypt, Iran and the eastern Mediterranean in the 13th century AD was influenced by the Persian traditions, including cooking. Trade with the East brought a wide range of foods to the Arab world and was of a high level of sophistication and luxury. Cleanliness was all-important, in particular the cleaning and preparation of the food and hand washing. Meat, usually lamb, was cooked with fruit such as oranges, lemons, pomegranates, red currants, apples and apricots. Fresh vegetables such as carrot, onion, aubergine, spinach and leeks appeared in many meat dishes. Meat was fried in the rendered down fat from sheep's tails. Some meats were fried before boiling and almonds and other nuts were used to make gravies. Spices such as ginger, cinnamon, pepper and caraway from China and India as well as local spices, cumin, coriander, were used with meat dishes. Rice was a luxury and was mixed with meat into a pilau-style dish.

The New World

Many of the foods eaten in Europe, and which are regarded as Mediterranean foods, came originally from central or meso America. The early hunters in Central America lived on mammoth or barbecued bison and relied on gathering seasonally available plants. Alternative sources of protein were gophers, squirrels, rabbits and mice. These hunter–gatherers had necessarily to be nomadic.

In the warm, well-watered central valleys there was an abundance of fish and fowl. The new concept of returning seeds to the ground for harvesting was the beginning of agriculture. The avocado pear and some kind of squash were the first to be cultivated. Between the period of 5000 and 3000 BC maize and beans were cultivated, but at that stage, provided only 10% of the total diet. It is interesting that even then, Mexican food was already heavily spiced with chilli.

Villages in this area date from approximately 3000 BC and a basic triad of plants, maize, beans and squash were grown. The early maize was only

the size of a strawberry. The diet was supplemented with deer and fish from lagoons and neighbouring forests. Cannibalism may also have been practised.

During the Olmec period of middle America from 1500 BC to 100 BC the basic crop was maize, which even today accounts for 90% of the inhabitants' diet. It was possible to obtain two crops of maize per year. The agriculture was based on slash and burn; that is, a patch of forest was felled during the short dry season, the wood burnt, seeds sown using a simple digging stick and the crops harvested. This was repeated until the ground became arid when it was then allowed to rejuvenate over the next 5 years. Such a system supported both the Olmec and the Mayan civilizations. It sustained only a limited number of people, however, and both civilizations collapsed when it was no longer possible to grow sufficient food to sustain the populations. In the Olmec period the growth of food was complicated by the flooding of large rivers. At this stage, the dog and the turkey had been domesticated and served as sources of food. Limitations for these civilizations were that the plough had not been invented and there were no draught animals, the beast of burden being man. The cities of meso America had a carefully planned supply and control of water, with main aqueducts carrying water to the city.

After the land had been cleared, weeds would grow very readily and the great cities which developed would disappear quite quickly after the agricultural land had been exhausted and the population moved on to new and fertile land.

By the time of the Spanish conquest, the Aztecs grew maize as the main crop, chillies as seasoning with additional squashes and beans. The latter provided nitrogen and protein in the diet. There were no dietary dairy products and very little meat. Chia was used to make a kind of porridge. Maguey was an all-purpose plant; the spikes served as needles, the fibre was used for making cloth and the juice from the heart of the plant was used to make the alcoholic beverage 'pulque' which is still drunk in Mexico. During the Aztec period pulque was used for ritual intoxication and may have been used to sedate people waiting to be sacrificed. Central to Aztec agriculture was human sacrifice, as it was believed that the safe-guarding of the crop cycle depended on such rituals. At its most intense 50 000 people a year were sacrificed in order to ensure the rising of the sun and continued crops. The Spanish brought many of the foods grown in central America back to Spain. They were then grown in Spain and are now associated with what is called 'the Medi-terranean diet'.

Other imports from the New World included nasturtiums from the West Indies, used for their flowers and leaves in salads. Potatoes, turkeys and chocolate came from central America. Potatoes also came to Europe from the mountainous parts of South America. Initially, they were a curiosity in English cooking and in mainland Europe pota-toes were not eaten to any great extent. In Ireland, by the middle of the 17th century they were an established staple food. Turkeys were introduced into England about 1524, having been imported from Spain. They replaced swans, peacocks and bustards as festive foods, being relatively cheap, readily stuffed, roasted and baked in pies. Fruit and vegetables coming from central America were sweet potatoes, peanuts, maize and chillies.

Dietary influences of the Crusades

After the Crusades, spices and new foods were brought into northern Europe. Sugar was unknown in Europe until the 11th century when it came to Europe, first from the Middle East and then from Spain, but by the 17th century, sugar plantations, run by slave labour in the Caribbean and Brazil, enabled Europe to indulge its fast-growing taste for sugar. Spiced sugar comfits were nibbled in the long fasting hours of Lent as a medicinal aid.

Cooking practices began to change. Crusaders learnt to cook meats in almond milk and to fry meat first without boiling. Cooking meat with fruit, which is a Middle Eastern custom, began to be adopted in European dishes. Rice grown by the Arabs in Spain was imported into French and English cooking. The Arab custom of highly col-oured dishes was also imported into London and Paris; foods were dyed green with parsley, yellow

with egg yolks and red with sandalwood, cinna-mon or alkenet.

Rice, oranges, figs, dates, raisins, spinach, almonds and pomegranates were all imported. Dishes were made with rose hips, shredded almond, chicken, red wine, sugar and strong pep-per and thickened with rice flour. Other recipes included eels seasoned with ginger, cinnamon, cloves, cardamom, galingale, long peppers and saffron. Galingale is a rhizome belonging to the ginger family. Cloves, cardamom, nutmegs, mace and rose water were all Arab ingredients which were imported after the Crusades. Blancmange, made with rice sweetened with sugar and flav-oured with almonds, is a Middle Eastern dish. The Arabs who were the mainstay of culture in the European Dark Ages maintained much of the Greek and Indian philosophies and science. They applied these philosophies to food, whereby foods were classified and used to balance the humours in man.

In contrast, the early Christians believed that only Christ had healing powers. Illness was a punishment for wrong-doings, to be treated by fasting and prayer. However, by the time of the Crusades, some foods were seen as having medici-nal properties. This all emanated from the Salerno School near Naples founded by Benedictine monks. Knights returning from the crusades often stopped there and were cured. Even 400 years later, the English physician Andrew Boorde was influenced by the Salerno school of teaching. Foods were still regarded as hot or cold, dry or wet, according to the humoural theory. Fruits, milk products and red meat were all to be eaten with caution. Specific foods were believed to be suitable or unsuitable for different diseases and there were even different diets for different types of men. 'Sanguine men, who are hot and moist, should be careful in eating meat, but not eat fruit.' Such people had to be careful with their food or they would become fat and gross. 'Phlegmatic men are cold and moist and should not eat white meat, herbs or fruit, but only onions, garlic, pep-per, ginger and hot and dry meats. Choleric men are hot and dry and should avoid hot spices and wine. Melancholic men are cold and dry and should not eat fried or salted meats and drink only light wines.'

1.2.2 History of European diet

In late mediaeval Europe feast alternated with famine. Large trenches of hand-baked coarse bread were cut into oblongs to serve as plates. There were jugs of water and wine on side tables. Diners were provided with a broad knife and spoon and rinsed and dried their hands after taking their place at the table for large and important meals. Most meals were taken by hand from the serving plates. Fine white bread was trimmed into finger-shaped sops and used to mop up liquid, including wine. Potage or soups were eaten with spoons from shared bowls and mopped up with sops. Meats and other foods were sliced and placed on the bread trenches. The meat slices were held with the fingers, and before being eaten, were dipped in a sauce the consistency of mustard. At the end of each course the softened trenches of bread were collected to be given to the poor. These were replaced by a soteley which consisted of coloured scenes sculpted from marzipan made with ground almonds and sugar. These were often decorated with banners which might depict the four seasons or the Christmas story.

At large banquets there would be a boar's head with gilded tusks, a heron, sturgeon and a pie made with cream, eggs, dates, prunes and sugar. The next course might include venison served in spice, wheat gruel, stuffed suckling pig and peacocks, skinned roasted and served in their plumage. The third course had more roast birds, quinces in syrup, grilled pork rissoles, custard tarts and pies of dried fruit and eggs.

The essence of mediaeval cookery lay in mixture. The quantities of spices used were quite significant. For example, one 15th century house used five pounds (2.2 kg) of pepper, two-and-a-half pounds (1.1 kg) of ginger, three pounds (1.4 kg) of cinnamon and one-and-a-quarter pounds (0.6 kg) each of mace and cloves in a year. Sugar was expensive and regarded in a similar fashion in cooking as a spice. Raisins, dates and saffron were introduced to northern Europe from the Middle East. Pastry and the stylish shaping of pies came from Persia along with recipes for traditional Chinese pastries.

The diet of ordinary people was very different from the nobility at court. Such people lived on cheeses, curds, cream and oatcake. Others, more fortunate, ate two or three meals a day and enjoyed wine or beer, pork or meat, cheese, dried beans and bread, with the occasional chicken, eggs, pepper, cumin, salt, vinegar and sufficient vegetables.

Potage, a porridge-like soup thickened with cereal or bread was popular in England. A porridge made of boiled ground wheat moistened with milk and covered with saffron was served with venison at the court of Richard II. The poor, when they could afford meat, made a potage of dried beans boiled in bacon stock mashed and served with bacon.

Throughout Europe peasants lived on a similar diet of bread, cheese and pork which was usually salted. In northern Europe peasants ate more rye or black bread than in the south. A Lenten bread was made of barley and oats. In times of shortage, bread was made of oats, peas or beans. The leaven would probably have been sour dough. The bread was heavy, hard-crusted and coarse. In England it was usually baked on a hearth, stone or in a pot buried in the fire embers. In France, peasants were forced to use bread ovens belonging to their landlords for which they paid with a portion of bread dough. Rats and mice polluted the stored grain, weevils burrowed into the dried beans, bacon was rancid, cheese was mouldy and wine sour. Bread was made from rye infected with ergot fungus which brought with it the terrible consequences of induced abortions. There was never enough fodder to keep more than the few animals needed for breeding, alive through the winter. In the autumn animals would be killed for their meat, salted or smoked, and preserved for the winter and spring. Turnips, beans and peas were dried. During times of famine the poor would only have cabbages and turnips without bread or salt.

In the Middle Ages, towns were relatively small. Within the city were private gardens and around the city walls were fields and vegetable gardens. The offal from various meat and fish was dumped anywhere with resulting pollution of the streams.

In the Christian calendar there were 200 fast days a year, when meat, milk and eggs were all forbidden, and only fish or vegetables were allowed. There were particular privations during Lent at the end of winter when food stocks were already low. The one meal a day allowed for the 6 weeks of Lent offered a daily diet of salted fish, stock fish or red salted herrings with mustard sauce. Fresh fish was also permissible; consequently some monasteries had their own fish pools. The only soup allowed was dried peas boiled in water and flavoured with fried onions. More prosperous individuals could use dried fruit such as currants, figs and dates and sweetmeats of crystallized ginger or candied violets.

Cooking techniques changed during Lent. Milk made from ground almonds replaced cows' milk for poaching and stewing. Oil was used rather than butter or lard. Sea-birds were sometimes allowed in religious houses because they were a form of water creature. Fast or fish days continued in England until after the Reformation and it was not until the mid-17th century that statutory fish days were abandoned in England. The long fasts of Advent and Lent ended with Christmas and Easter. Christmas in England was celebrated not only with new wine from Gascogny, but also certain days of feasting and entertainment. The end of Lent was celebrated with a great feast on Easter Sunday, meals on that day traditionally included a lamb or a kid in many European countries. In mediaeval times it was usual to give gifts of hard-boiled eggs painted with vegetable dyes on Easter Sunday. This practice has continued into the present times.

The 16th and 17th centuries

By the 16th century came the development of printing and cookery books. During this period English merchants became extremely rich, with fortunes made from trade to India and the West Indies. There was little ability to store food, so the diet reflected seasonal availability. Some of the cookery books discussed the curative properties of specific diets. A recipe for whitening and retaining teeth recommended rose water, sage, marjoram, alum and cinnamon. Rosemary was believed to have almost miraculous powers and was used to treat colds, toothache, aching feet, bad breath, sweating, lack of appetite, gout, consumption and madness. Bed-wetting could be cured by eating fried mice.

In England in 1603 breakfasts of cold meat, cheese and egg were eaten between 6:00 and 7:00 am. Dinner, the main meal of the day, was between 11:00 am and 12:00 noon, and a light supper was taken at about 6:00 pm. By the late 18th century fashionable people in London were eating as late as 7:00 pm, although this was not the pattern in the country. Breakfast was a meal of cold meats and ale, eaten about 9:00 or 10:00 am in towns, and supper had become a late-night snack. Afternoon tea was taken between breakfast and dinner with tea and bread and butter or buttered toast. Dinner plates of pottery or pewter replaced the mediaeval trench of bread, and forks were slowly introduced. Nevertheless, monarchs such as Louis XIV of France always ate with their fingers. The dishes, both sweet and savoury, were laid out on the table in geometric patterns. The third course of fish and confection was similarly presented.

Around the periphery of London there were small, intensive market gardens, manured by excrement. Fruit and vegetables were grown for the population of the capital and were sold from barrows or in markets, for example Covent Garden.

In the 17th century increasing trade among European countries led to new diets. Trade with India and China resulted in tea being imported. A variety of beverages, such as **tea, coffee** and **chocolate** were introduced into Europe.

The 1000-year-old Chinese habit of drinking **tea** was a novelty in Europe. The Portuguese brought tea to Europe in the middle of the 16th century. The Dutch brought small quantities to France and by the mid-17th century large amounts were introduced into England by the

East India Company. The first teas imported were green teas. The leaves were picked, rolled and steamed to prevent further fermentation. Initially very weak tea was made and drunk with sugar but no milk. By the end of the 17th century bohea or black tea, which is a stronger and less astringent tea was available. Cream or milk was then added to the tea to counteract the acidic effects of the tannin. Tea became popular throughout all social classes and was drunk throughout the day. Tea replaced beer as the drink for many English farm labourers.

Coffee was introduced into Europe during the 17th century. The coffee shrub is native to Ethiopia. By the 16th century coffee was drunk throughout the Moslem world. It was introduced to Europe through Venice by the beginning of the 17th century. The first coffee shop opened in Oxford in 1650. Many of the early coffee shops became associated with dining clubs and were important centres of political and social life.

Drinking chocolate became popular throughout Europe during the 17th century. Chocolate was initially imported by the Spanish from Mexico where it had been drunk by the Mayans and Aztecs. When first introduced into Spain it was a secret maintained by monasteries. The Spanish added sugar, cinnamon and vanilla in place of the chillies in the Aztec recipe. By the middle of the 17th century drinking chocolate was available throughout Europe though it was not until the mid-18th century that chocolate bars were produced.

Tea, coffee and chocolate replaced the previously widely drunk mulled wine. As they were rather bitter in flavour, sugar was taken to improve the flavour.

France was the most important European wine producer in the mediaeval era, sustaining a tradition which had existed since Roman times. Wines were exported to England from the Weine basin and from Gascogny. The new wines from Bordeaux arrived in England just in time for Christmas. Mediaeval wines were lighter than the modern ones and were at their best after about 4 months. They were stored in wooden barrels. The Parisians preferred light white wines, whereas the English liked red wines. By the end of the 16th century wines from the warm south were recognized as having a higher alcohol content than those grown in the north. These stronger wines could be kept for several years and would improve with keeping.

The favourite wine during the second half of the 17th century was champagne. Champagne wines were first developed under Henry IV of France at the beginning of the century. It was only slowly that their capacity to form a sparkling wine was appreciated, largely under the guidance of Dom Perignon from the Abbey of Hautvillers near Rheims. Champagne was the most popular wine at the court of Charles II in London.

Claret was imported in barrels from Bordeaux, usually drunk warmed with spices as mulled wine. Younger wines were more expensive than older wines because wines kept in barrels tended to deteriorate after a year or two. Maturing of wine in bottles later became more common and with this development the concept of wine improving with age.

By the mid-17th century the great tradition of French cooking had begun. This was a cuisine based on a series of techniques; basic preparation, bouillon and roux, the use of bouquet garni, egg whites for clearing consommeé and stuffings made with mushrooms and other vegetables. Pieces of meat and mutton were slowly cooked. Eighteenth century French and subsequent cooking preferred the infusion of carefully selected flavours. The previous menus containing multi-flavoured sauces and exotic game birds were thus replaced.

From the 17th century with the increasing availability of sugar, puddings were slowly established as a regular feature of a meal and served with other savoury dishes in the second course. There were a wide range of fruit pies, fool's cream, syllabubs and fritters. The English pudding was transformed by the introduction of the boiling cloth. Before this innovation boiled puddings – both the sweet and spicy versions of sausage and savoury puddings – were cooked in animal intestines. These had to be fresh, so boiled puddings could only be made during the seasonal autumn slaughter of livestock for the winter. However, once boiled puddings were cooked wrapped in cloth, they could be enjoyed at any time of the year. Quaking or shaking puddings of cream, breadcrumbs, sugar and eggs flavoured with spices were cooked in well-floured bags in simmering water. Apple puddings of apple, sugar and butter were wrapped in pastry skin and boiled in a cloth. The boiled sweet pudding became the national dish, consisting of flour, suet, milk and eggs, and was usually boiled in the same utensil as the meat of the day, the square of bacon, cabbage or other green vegetables in one net, the potatoes in another and the roly poly in a cloth. Roly poly pudding with dried fruit was served as a first course, with a similar purpose to the Yorkshire pudding, to reduce the appetite for the more expensive later courses. Plum puddings with dried fruit were increasingly developed in the 17th and 18th centuries, with fruit, sugar and spices such as cinnamon, nutmeg, ginger, cloves and maize.

The 18th and 19th centuries

Butcher's meat required specialized killing and cutting which often took place in remote towns, whereas rabbits, chickens and pigs were killed locally. During the Industrial Revolution in Britain, workers crowded into the rapidly growing towns to man the new machines. Women worked in factories and were also responsible for the preparation and cooking of family meals, without the training they had previously received from their totally domesticated and now distant mothers. There was a resulting decline in nutrition.

In the 18th and 19th centuries rural life was very much dominated by the availability of seasonal fresh fruits and game from the rural activities of fishing and hunting. Fishing towns supplied cod, lobster, sole, skate and whiting. Fresh vegetables were grown in the garden, with strawberries and raspberries in June, and peaches, nectarines, plums and pears in September. Asparagus and cucumber were also grown.

The farmworkers had a very simple diet. There was bacon from the family pig, kept in a sty at the back of the cottage, eaten with fresh vegetables, bread and home-made lard flavoured with rosemary. However, many of the farmworkers came close to starvation, particularly during times of poor harvest. They lived on potatoes, and in Scotland on oatmeal, milk and sometimes fresh herrings. Potatoes were increasingly popular as they were easy to cook and provided the essence of a hot meal. They also had the advantage of lasting three-quarters of the year. Fresh meat was a luxury, eaten only on Sundays. A pot roast was made by placing the meat with a little lard or other fat in a covered iron saucepan kept over the fire. In the north oatmeal and tea were provided for workers. Rural workers in the south did not necessarily have gardens provided and lived almost exclusively on bread with little salt, bacon or cheese.

The development of restaurants followed the French Revolution when chefs, who had lost their aristocratic employers, opened restaurants, resulting in a more general, intense and continuing interest in recipes and food. There was a defining of cooking styles, with precise cooking instructions for the preparation of purées, essences, sauces and garnishes with a perfect balance between well-chosen flavours. This led to an almost total domination by French cooking of food in Europe. It has been suggested that, by the mid-19th century, the urban British middle-class, unlike the French, had lost contact with their own country origins and consequently an understanding of the origins of their foods. The British cooking pattern was plain, leaving foods to taste of themselves, whereas the French haute cuisine depended on the cook adding to their flavour.

The introduction of spices into Britain came from exposure to India. The recipe for Worcester sauce was brought back from India and curries were introduced through the East India Company. The first recipes for mulligatawny soup, using curry powder, came at the beginning of the 19th century. Such soups were thickened with barley, bread or split peas. The British community living in India combined Indian and British foods, one such being the development and increasing popularity in the 18th century of chutneys made with tamarinds, mangoes, limes, and aubergines.

Aubergine being a vegetable which was unknown in Britain at that time, although known in southern Europe.

In the industrial areas there was considerable starvation. During the 1840s the diet of the majority was stodgy and monotonous, and for many, deficient in both quantity and nutriment. Badly housed parents lost many of their children and those who survived were undernourished, rachitic and sometimes deformed. Meat was often eaten only two or three times a week, with the main or even sole food source being bread and potatoes.

At the beginning of the 19th century the British soldier's daily ration was one pound of bread (450 g) and a quarter of a pound (110 g) of meat. Much of the meat for the military, including the Navy, was reared in the Scottish Highlands. The cattle were driven along roads by drovers to Smithfield in London. This trade was central to the Highland economy. In the army barracks there were two coppers for each company, one for meat and the other for vegetables, so the food could only be boiled. There were no canteens. There were two meals, one at 7:30 am and the other at 12:30 pm. On overseas service soldiers were provided with salt pork or salt beef or dried biscuits.

There were considerable problems of storage during this period with resulting mass adulteration and upgrading of food. Bakers bleached inferior grades of flour with alum to make bread appear white. Flour was diluted with ground peas and beans, beer was adulterated with acids, milk was thickened with arrowroot, the skins of Gloucester cheese were coloured with red lead, and old port crusting was imitated by lining the bottle with a layer of super tartrate of potash. Hedgerow clippings were used to adulterate tea. Leaves of blackthorn, ash and elder were boiled, dried and coloured on copper plates. Ground coffee was diluted with chicory and toasted corn. Flour was mixed with chalk, pipe clay, powdered flints and potato flour. Second-hand tea leaves were sold by servants to merchants. The tea leaves were then mixed with gum and dried with black lead before selling them as fresh tea leaves. It was only by the

1870s that parliament legislated against food adulteration.

The introduction of the railways enabled food to be carried rapidly around the country. By the end of the 18th century, fresh salmon in ice could be brought from Scotland to London, initially by road and later by the railway. In the 1860s and 1870s the development of the railway system into the Mid-West of the United States and the cattle lands of South America opened up new fertile sources of food. The railway and rapid ship movements meant that grain and cattle could be brought from North America and South America to the industrial areas of Europe. Later came the introduction of reaping machines and self binders requiring fewer workers which increased cheap wheat and other crop production.

Canned meats and vegetables were used by the Royal Navy in the Napoleonic Wars, but this was only partially successful. By the late 19th century canned Californian pineapples and peaches were available in Britain.

The long period of urban malnutrition became apparent during the Boer war when nearly 40% of the British volunteers had to be rejected because of being physically impaired by inadequate diet. The result of this and other findings, for example the Rowntree Report on poverty, resulted in parliamentary Acts providing free school meals for children of poor families and pensions for the elderly. By the middle of the 19th century, advances in knowledge of nutrition enabled adequate nutritional provision to be made for developing schoolchildren.

The 20th century

The diet of the British working classes at the beginning of the 20th century was dominated by bread, sugar, lard, cheese, bacon and condensed milk. The meat was brought chilled from Argentina or frozen from New Zealand.

In Europe, factories were attracting workers from the land so that as the available agricultural production was reduced the diet was augmented by overseas food. Novel methods were found to keep foods on long journeys. Cattle in Argentina were slaughtered solely for their skins for leather.

The development of a meat concentrate, e.g. Bovril, made the proteins available for use by distant populations. The introduction of refrigerated ships enabled meat to be carried long distances in prime condition. Later in the 20th century, food could be carried by refrigerated lorries, so that lettuces, strawberries and melons could be brought in good condition from France to Britain. Successful canning was another important development. Canned salmon and canned peaches became the traditional Sunday tea for many people during the 1939–1945 war.

The long working hours in factories resulted in eating problems, because the traditional time for the working mens' main meal was the middle of the day. Good employers provided canteens where workers could eat at that time. At the Cadburys' works canteen a meal of roast beef and two vegetables was available at midday. For the majority of workers canteens were not available until the 1940s. Often food was taken in to work to be eaten at midday. This might be a pie, or a basin of meat and vegetable stew or cold sandwiches, and tea. Coal miners would take in a bottle of cold tea and a tin of sandwiches. There were no set meal breaks and they ate as they worked. Other workers relied on stalls in market places close to the factory. When possible, men returned to their homes for their midday meal. However, changes in working patterns altered eating patterns throughout the world. At the same time the women, when not working but having virtually annual pregnancies throughout their childbearing years, were not even given these facilities

1.2.3 History of eating patterns in Scotland

An example of the development of eating patterns in an industrial society can be found in Scotland. In 1949, Kitchin and Passmore described three distinct eras of nutrition in the general Scottish population. The first era was that of a self-supporting agricultural community. The diet was in the Viking tradition, which included rye, wholemeal bread, oat and barley porridge; fish (especially herring); boiled meat and broths of sheep,

lamb, goat, ox, calf and pig; cheese, butter and cream; beer and mead; and among the wealthy, wine. The most common vegetables were cabbage and onions; apples, berries and hazelnuts were also popular.

The second era was the age of the Industrial Revolution. As industry expanded and the population increased, so food requirements exceeded home food production, necessitating the import of food from overseas. Significant differences in health appeared between the urban and rural populations, in part because of the better quality of the countryman's diet. During the 19th century, 10% of the population were too poor to buy sufficient food for themselves. Such malnutrition, in addition to bad sanitation, inadequate overcrowded housing, insufficient land for farming, and a host of acute and chronic infectious diseases, led to rickets, poor stature, and high maternal and infant mortality.

A study in 1903 by Patton, Dunlop and Inglis of the diet of the labouring classes in Edinburgh, identified a large proportion of poorly developed and undersized children and adults. Two groups were studied; families with assured and adequate incomes, and the poor, who were unable to buy the necessities of life, either because of inadequate income or because of employment that was only casual. Some families living in poor housing on small irregular wages refused to take part in the study. The range of expenditure on food varied from 2.5 to 9.5 pence per man per day. The daily nutrient intake per man varied from 1100 to 4800 kilocalories (kcal). The wife and children of the poor families lived on tea, bread and potatoes, the tradition being that the man ate butcher meat daily. The larder was replenished with small quantities of food bought each day. Alcohol was a great problem, affecting work record and hence income. In contrast, the families of workmen receiving regular wages, ate meals consisting of bread, potatoes, oatmeal, eggs, beef, mutton, ham, butter, herrings, cod, sugar, rice and barley. Fresh vegetables were confined to potatoes, cabbage and peas.

In Edinburgh before the First World War, over 12% of the milk contained tubercle bacilli. The consequence was that the bovine, milk-borne form of tuberculosis occurred in 32.4% of tuberculosis patients aged under 5 years, 29% aged 5–16, and 2.9% of adults.

The third era was that of state planning. During and after the Second World War, the Government was obliged to control and provide an adequate food supply for the whole population, so that malnutrition could, despite the obstacles to the importation of food, be avoided as far as possible. This was the golden age of the nutritionist. Good nutritional practice, school meals and works canteens were available to the entire population. The consequences were striking. During the period 1941–1945 boys 13 years old in Glasgow grew to be on average $\frac{3}{4}$ of an inch (2 cm) taller and 3 pounds (1.4 kg) heavier than those in the same age group in 1935–1939, and $3\frac{1}{2}$ inches (9 cm) and $12\frac{1}{2}$ pounds (5.7 kg) heavier than those of 1910–1914.

Since then, a fourth era has emerged, the era of the supermarket and a free market. Increasing overall prosperity, more efficient home farming practices and ready availability of food from all over the world means that many have the choice of a wide range of foods from an efficient wholesale and retail system. The constraints now are personal and communal preferences, and financial limitations. There have been considerable increases in height, weight and lifespan. However, the population now has new and unprecedented opportunities for inappropriate or excessive indulgences. In general, however, the human race copes better with nutritional excesses than with insufficiency.

An important consequence of the advent of this fourth era is that Scotland's nutrition is dominated by a common range of choice on supermarket counters, selected by a few supermarket purchasing managers. Despite diligence in ensuring that our food is clean and wholesome, there are widespread concerns about chemical additives ('E numbers'), bacterial and viral infections, pesticide residues, and the perils of excessive intake of proteins or saturated fatty acids. Food – long recognized as a vector of infectious disease – is now seen to have a role in the aetiology of non-infectious disease such as coronary heart disease, maturity-onset diabetes and obesity.

Quality of life is important. Well-being and protection from stress are not inappropriate ambitions. The French have a reverence for food which has nothing to do with their concern for longevity. They are fortunate in being blessed with both exciting food and longevity. A rigidly controlled diet may not result in longer life, it may only seem longer. The value of survival and quality of life will vary with circumstances and may be judged only by the individual or possibly the community.

Further reading

Burnett, J. (1989) *Plenty and Want – A History of Food in England from 1815 to the Present Day*, Routledge, London.

Davies, N. (1982) *The Ancient Kingdoms of Mexico*, Penguin Press, London.

Fieldhouse, P. (1986) *Food and Nutrition, Custom and Culture*, Chapman & Hall, London.

Kitchin, A.H. and Passmore, R. (1949) *The Scotsman's Food: an Historical Introduction to Modern Food Administration*, E. & S. Livingstone, Edinburgh.

Leeming, M. (1991) *The History of Food*, BBC Books, London.

Patton, N.D., Dunlop, J.C. and Inglis, E.A. (1903) *Study of the Diet of the Labouring Classes in Edinburgh*, Otto Schulze, Edinburgh.

Tannabill, R. *Food in History*, Penguin.

Tansey, G. (1995) *Food System Guide*, Earthscan.

1.3 Domestic water

- Water regulations require the provision of water wholesome at the time of supply.
- Inspection of water treatment ensures compliance with regulations.
- Permitted quantities are defined for substances which may occur in domestic water.
- The organisms responsible and the effects of microbiological contaminants of water are discussed.

1.3.1 Introduction

Water occurs in most foods and as free fluid. In organized societies water – free from bacteria and other pollutants – is made available through a water system.

1.3.2 Water standards

In the United Kingdom the Water Act 1989 places a duty on water companies to supply water that is wholesome at the time of supply. Wholesomeness is defined by reference to standards and other requirements of the water supply (water quality) and to Regulation 1989 which incorporates the relevant requirements of EC drinking water directives (80/778/EEC).

In assessing the adequacy of water treatment, inspectors assess the:

- disinfection of water supplies
- provision of water-treatment facilities
- use of chemicals, products and materials of construction
- action and studies which have been undertaken to reduce lead (plumbosolvency) and parasites
- colour, turbidity, odour, taste, nitrate, aluminium, iron and manganese and polycyclic aromatic hydrocarbons (PAH) content of water

There must also be sampling programmes for the microbiological quality of the water leaving the treatment works with analysis for pesticides and bacteria, particularly faecal coliforms. Analyses are made for a wide range of pesticides, atrazine, simazine, chlortoluron, dichlorprop, dichlorvos, isoproturon, ioxynil, methyl chlorphenoxybutyric acid (MCPB), necoprop, propyzimide, (2,4-dichlorophenoxyacetic acid (2,4-D), fenpropimorph, tetrachloromethane, phosphorus, antimony, arsenic and selenium.

The disinfection of the water supply involves the reaction of the disinfectant with organic contaminants. These are known as disinfection by-products. Chlorine is very widely used as a disinfectant in England and Wales and this produces 'chlorination by-products' including trihalomethanes (THM) such as chloroform. There is no strong evidence that these are harmful.

TABLE 1.2 Permitted quantities of substances in domestic water at time of supply

Substance	Standard ($\mu g/l$)
Polycyclic aromatic hydrocarbons (PAH)	
(Six specific PAH)	0.2
Benzo 3–4 pyrene	0.01 (annual average)
Pesticides	
Individual substances	0.1
Total pesticides	0.5
Aluminium	200
Lead	50
Nitrates	50
Trihalomethanes (THM)	100 (average over 3 months)

Most water distribution mains are constructed of cast or ductile iron. Before 1970 these were given an internal anti-corrosion coating of coal tar pitch which may contain up to 50% of PAH. This PAH may leach into the water supply in solution or in suspension, though surveys show that PAH is not in fact present in the majority of samples taken. The medical consequences of ingestion of PAH are not known.

The regulations set standards for permitted quantities of many substances. Those for PAH, pesticides, aluminium, lead, nitrates and trihalomethanes (THM) are shown in Table 1.2. The main source of nitrate in ground water is leaching from agricultural farmland.

1.3.3 Contaminants

Cryptosporidium is a protozoan parasite which is a cause of diarrhoea in humans. Blue–green algae occur naturally in all inland waters and concentration can alter rapidly in response to various climatic factors and the availability of essential nutrients. Sometimes these blue–green algae may release toxins, though it is not known whether these enter the water supply.

Viruses are important causes of ill-defined outbreaks of sickness and diarrhoea. These viruses usually enter the water cycle through faecal contamination. Techniques for monitoring viruses in water supplies are complex and expensive and there is no present provision for their identification in domestic water supplies.

Fresh water which is contaminated from industrial and domestic effluent and agricultural fertilizer run-off is at risk of allowing the growth of blooms or scums of toxic cyanobacteria. Microbiological contamination of the water supply, occurring through after-growth within the distribution system, is very difficult to eradicate.

Summary

1. Most countries oblige water companies to supply water which is wholesome. Wholesomeness is defined by reference standards which include the freedom of the water from bacterial contamination and the presence of chemicals, products and materials only at concentrations which do not present health risks. It is also demanded that the water looks and smells clean.
2. Problems arise for the water supply from contamination by faecal material, contamination by pesticides and fertilizers and also from the piping which carries the water, e.g. the pipe coating or lead.
3. Other problems are protozoa and algae which are present in reservoirs, lakes and other sources of water. The implications of such contamination have yet to be determined.

1.4 Food chain

- The food chain is the process whereby nitrogen and carbon are synthesized to complex and essential nutrients by microorganisms, seaplankton and plants. These are eaten by ruminants, vegetarian animals and fish and in turn by carnivores and humans.
- The food chain may be contaminated by pollutants, chemicals or bacteria, most seriously affecting seeds, storage organs and growing organisms.
- Additives may be introduced into food to enhance appearance, quality, acceptability or shelf-life.
- Non-nutrients in the food chain may be infective, chemical or synthetic residues and may be industrial or agricultural in origin.
- Toxins may be from plants, microbes or mushrooms.
- There are many significant water-borne toxins.

1.4.1 Introduction

This section focuses on the biological and evolutionary, rather than the producer to consumer, food chain.

There is a complex chain of events which precedes food being eaten by humans. Plants absorb inorganic chemicals whether they be nutrients or toxicants from the aqueous phase of the soil. Through ingestion and digestion of plants by animals such chemicals can be transferred from one organism to another. Decaying organic matter, whether animal or plant, will return inorganic chemicals to the soil solution which may be removed once again by the action of microorganisms. Rain water dissolves soil matter and may carry inorganic chemicals from the land to the sea through streams and rivers. The chemicals may return to the terrestrial system through precipitation into stream and river sediments (Figure 1.3). The chain may, however, be distorted or contaminated by a pollutant, a chemical or by infection. Mining, smelting, coal mining and coal utilization, dredging of water ways, chemical manufacturing systems, sewage sludge, fertilization, irrigation and traffic patterns may alter these nat-

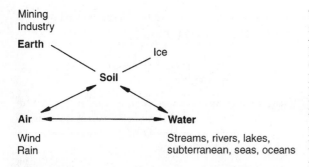

FIGURE 1.3 Physical vectors of water-soluble chemicals through soil, water and air.

TABLE 1.3 Factors which may disrupt the human food chain

Poor weather conditions
Changes in soil fertility and composition
Pests or bacteria
Pollution of water
Drought
War and social disruption
Incompetent food transport systems
Poor farming and fishing equipment or practices
Absence of healthy labour
Disincentives to efficient farming and fishing
Overfarming and fishing
Poverty
Ignorance
Illness
Cultural and religious non-acceptance of products

ural cycles. These utilities, situated in localized areas, may be traced as sources of pollution entering the human food chain.

The food chain achieves conversion of nitrogen and carbon to complex and essential nutrients by microorganisms, sea plankton and plant growth and storage organs and seeds. The products are utilized by ruminants, vegetarian animals and fish which in turn may be eaten by insects, birds, animals, fish and crustacea. All of these, according to the prevailing local culture may be eaten by humans. Economics, religion and social advancement determine what is eaten and the manner and form in which it is eaten. Important in the choice of what is eaten by humans is the attractiveness of the food or prepared dish. This may depend upon the expected appearance, smell, taste, feel and sound, colour, shape, texture, flavour and consistency.

There is the potential for disruption of the food chain by the factors listed in Table 1.3.

1.4.2 Food additives

The use of food additives is controlled by regulations of the Ministry of Agriculture, Fisheries and Food, the Food Advisory Council (FAC) and many other advisory committees on nutrition. European Union Committees and the Codex Alimentarus are of increasing relevance. These policy-making bodies must be informed by scientific studies (e.g. toxicity studies). There are, however, inevitable tensions between regulations and the need to produce food economically.

A food additive (Table 1.4) may be regarded as a substance, either synthetic or natural, which is normally not consumed as food itself, but is deliberately added, usually in small amounts, and is intended to remain in food for some desirable function or effect during manufacturing, production, processing, storage or packaging or in the finished product. Food additives enhance the appearance, quality, consumer acceptability and flavour of foods, extend shelf-life and allow mass production, mass distribution and ready availability. They facilitate and standardize preparation, enhance nutritional value and provide for the specific dietary requirements. Food additives may be used to improve colour. Azo dyes are colour-fast and are commonly used as colouring agents.

Food additives, e.g. salts and spices have been used throughout recorded food history. Vitamin C has been added to foods to prevent scurvy, and more recently vitamin D to improve bone structure, iodine to reduce goitre and fluoride in water to reduce caries.

The functions of nutritional additives include enrichment and fortification. Enrichment replaces nutrients lost during processing through oxidation or reaction with food components, e.g. ascorbic acid and thiamine, which are lost through chemical and enzymatic degradation. Fortification is the addition during manufacture of substances to

TABLE 1.4 'E' numbers

Colourings and Additives	E number
Yellow	E100–E110
Red	E120–E129
Blue	E131–E133
Green	E140–E142
Brown and black	E150–E155
Plant extracts	E160–E163
Other colourings	E170–E180
Preservatives	E200–E297
Antioxidants	E300
Emulsifiers, stabilizers and auxiliary compounds	E322–E495
Acids, bases and related materials	E500–E529
Anti-caking agents and auxiliary additives	E530–E585
Flavour enhancers and sweeteners	E620–E640
Glazing agents and auxiliary additives	E900–E914
Flour treatments, improvers and bleaching agents	E920–E928
Packing gases	E941–E948
Sweeteners	E950–E967

improve dietary intake, e.g. vitamins in margarine.

There are also negative effects of food additives. Anti-oxidants, spices, gums and waxes may result in dermatitis. Tartrazine, the anti-oxidants BHA and BHT and sacramen may cause urticaria. There is an increased risk of tartrazine sensitivity in aspirin-intolerant patients. Benzoic acid and sodium benzoate, which are used to suppress the growth of yeast and bacteria in food, can cause skin and asthmatic reactions, particularly in patients already suffering from asthma. Sodium monoglutamate can produce the 'Chinese restaurant syndrome', bronchospasm, flushing and headache.

Excluded from this are substances Generally Regarded As Safe (GRAS). These include pesticide chemicals and their residues before food processing, colour additives and chemicals which have been sanctioned for use over a prolonged period.

1.4.3 Food preservatives

Preservatives are added to foods which readily perish and deteriorate due to bacterial coloniza-

tion. The prevention of botulism is perhaps the most significant example.

1.4.4 Processing aids

Despite seasonal variations in availability of natural products, the food industry ensures that products are uniform in appearance. Butter naturally varies in yellow colour throughout the year and its composition, e.g. milk with variable milk sugar, proteins and fat content, may have to be adjusted to achieve a consistent product.

1.4.5 Non-nutrients in the food chain

Infective

The most common dietary contaminants of meat products are food-borne microbes or fungi arising from improper sanitation, inadequate refrigeration during storage or insufficient cooking.

Chemical

These include antibiotics, pesticides, metals, industrial chemicals and fertilizers. Nitrates are used as an additive in preserving cured meats and are naturally present in celery, beetroot and lettuce accumulating from nitrogenous fertilizers and also occurring in drinking water following leaching from farms. Microbial nitrate reducing activity is found in the mouth where oral bacteria reduce nitrate recirculated in the saliva to nitrite. If the nitrate content of the water drunk by babies is increased then nitrate reduction and subsequent absorption of nitrites can result in methaemoglobin formation. N-nitrosation occurs at low pH in the stomach from salivary nitrate with nitrosamine formation and the potential for carcinogenic change. Most nitrosamines are metabolized in tissues, but *N*-nitrosoproline a stable nitrosated amine is excreted in urine and indicates N-nitrosamine formation. High dietary fat reduces *N*-nitrosoproline excretion. There is also an endogenous non-microbial synthesis system for nitrate which is derived from dietary protein and can be a significant source of nitrates.

Synthetic residues in animal carcasses

Such residues may arise from veterinary prescription of drugs, the ingestion of feed or water contaminated by pesticide residues or contact with other pollutants in the environment, e.g. insecticides.

The most common types of drugs used in animal production are:

- antimicrobial drugs which are used to control and prevent diseases and promote growth. Antimicrobial drugs are used in the treatment of mastitis, enteritis, pneumonia and to eradicate parasites. Feeding supplements of low levels of the antibiotics chlortetracycline and oxytetracycline improves growth rates and feed efficiency. This increases the rate of weight gain, hence reducing the time to achieve a required weight. Monensin affects the rumen microbes but not the metabolism of the animal and improves daily weight gain by approximately 16%.
- parasiticides and pesticides to control helminths, e.g. round and tape worms, liver fluke, coccidia, insects (flies, lice, mange, mites, grubs, etc.) and other parasites.
- hormones which may be used to stimulate growth, encourage the development of lean meat, increase feed efficiency or prevent or terminate pregnancy in female cattle. Anabolic implants are growth stimulants and vary in use with type, age and sex of the animal. Calves implanted during nursing show an average enhanced increase in daily weight gain of 9%. Cattle implanted from weaning to finishing average a 15% increase in daily weight gain. Finishing cattle average an 11% increase in daily weight gain. Melengestrol acetate is a synthetic progesterone which is a contraceptive for heifers, improves average weight gain by 11% and the feed efficiency conversion by 11%.
- feed additives, vitamins and minerals: such products leak into the food chain as a result of incorrect dosage or route of administration and the species and site of administration. Lack of information about a drug's turnover time

resulting in the sale of treated animals without allowing sufficient withdrawal time, premature slaughter of treated animals, contamination of feed milling equipment and bins, or exposure to water containing antibiotics used to flush alleys and waste bins may also result in additives used in animal production appearing inappropriately in food.

The enteric absorption of ingested drugs varies with the dose level, the drug formulation and the addition of an adjuvant (this is a compound added to lengthen the period of release). Human exposure is reduced as the animals may metabolize, excrete and dilute these compounds. Some metabolites may have a toxicity which is different from that of the parent compounds and such toxicity may vary from species to species.

Industrial

Phthalate esters are widely used in industry and their annual production is of the order of millions of pounds in weight. These spread to the environment as pesticide carriers and insect repellents. They are also used as plasticizer additives and slowly leach from the plastic product over a prolonged period into the environment. They are relatively modest in their toxicity, though they may leach from plastic bags to stored blood. Milk, cheese, lard and other fatty substances can extract phthalate plasticizers from plastic tubing used in processing and from PVC-based plastic containers and film. Concentrations are generally low, in single figures per million, though the concentration is increased in fatty foods (Figures 1.4 and 1.5). The levels resulting in acute toxicity are likely to be of 50–1000 mg/kg body weight, or may reach even higher levels. The long-term effects are not known.

There is widespread distribution of these chemicals in the marine and freshwater systems. This, combined with the high lipid solubility of these compounds, raises the problem of food chain accumulation in fish, seafood and water fowl. These chemicals are quite difficult to identify using standard chemical analytical methodology.

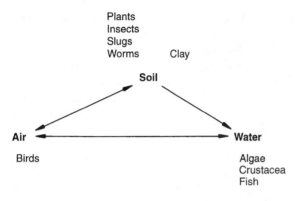

FIGURE 1.4 Physical vectors of fat-soluble chemicals tend to be concentrated in the lipids of plants, birds and animals, though may be present in clays.

Agricultural

DDT is no longer used in many agricultural systems because of its toxicity but, despite this, it remains in the environment. There is a phenomenon of *biomagnification* which shows that concentrations of such chemicals can increase by 1000 to 1 000 000 times as they pass to the top of the food chain causing serious deformities or even death (Figure 1.6). Phytoplankton may have a DDT concentration of 0.0025 ppm which compares with lake trout at 4.83 ppm. Zooplankton may have a DDT concentration of 0.123 ppm and herring gull eggs 124 ppm. As phytoplankton collect nutrients for plant growth they also accumu-

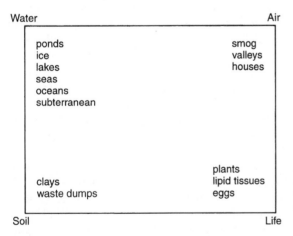

FIGURE 1.5 Concentrations of chemicals in static and mobile phases: water, air, soil and life.

late non-polar contaminants which may be in the water in very low concentrations.

1.4.6 Toxins

See section 9.3.4, page 306.

Plant toxins

Plants may serve as poisons. Hemlock, which is a member of the parsley family, is a deadly poison. Most chemicals which are natural poisons in plants are known as 'secondary' products, produced by the plant only to form part of the defence mechanism of the plant against herbivores and pathogens. Plants do also produce certain 'primary' products such as amino acids and

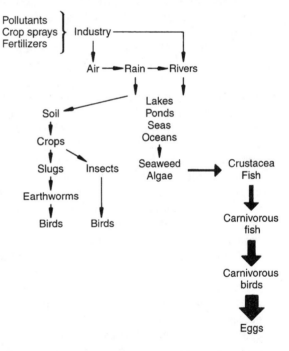

FIGURE 1.6 Food chain effect and biomagnification factor. Biomagnification factor is a sequence of concentration within organisms compared with the immediate soil or water environment in which the organisms live. The biomagnification, for example of DDT, may be of the order of 10^3 to 10^6 between algae and fish-eating birds; and between the soil and birds that eat earthworms. At all of these stages other creatures can eat intermediate stages in the chain and so concentrate chemicals in their own tissues.

amines which may enter the human food chain and cause poisoning. Humans have, through trial and error, learned to avoid certain poisonous plants as foods. Some plant poisons are minimized in their toxicity by cooking.

There is a system within cells (p-glycoproteins) which is found especially in the mucosa of the gastrointestinal tract, oesophagus, stomach and colon which is protective against toxic substances in plants, bacteria and fungi by pumping toxic substances from the cell. Drugs commonly used in cancer chemotherapy – against which resistance develops – have a common origin in that they are all products of plants, fungi and bacteria. Resistance appears to be related to the amount of p-glycoprotein 170 in the tumour cell. The multi drug resistance-1 gene associated with p-glycoprotein 170 is amplified in patients receiving drugs of plant or bacterial origin. This perhaps accounts for why the oesophagus, stomach and colon are so resistant to cancer chemotherapy.

Proteinaceous substances, *lectins*, are found throughout plants in seeds, bulbs, bark and leaves acting as protection against unwanted predators. They are resistant to proteolysis and bind strongly to the brush border of the small intestinal epithelium resulting in increased growth of the small intestine. Bacterial colonization of the small intestine can then occur. Lectins bind to specific sugars of the epithelial membrane glycoconjugates.

Toxicants are grouped together by chemical classification, e.g. alkaloids, cyanogenic glycosides, amino acids, or by their effect, e.g. carcinogens, oestrogens.

Alkaloids

Alkaloids contain nitrogen as part of a heterocyclic ring structure and are often bitter to taste. They are present in about 25% of all plant species, there are possibly 6000 different alkaloids. Alkaloids are important causes of liver, lung and heart damage, neurological disorder and birth defects in livestock to which alkaloids present a much greater risk than to humans because of the plants and therefore toxins consumed as food. Some do, however, pass into the human food

chain including solanum alkaloids and pyrrolizidine alkaloids.

Solanum alkaloids The solanum alkaloids include sugar-based alkaloids (glycoalkaloids) found in potatoes, apples, egg plants, roots and leaves of tomatoes and sugar beet roots. These alkaloids have anticholinesterase properties. The poisoning effects include influenza-like symptoms, headache, nausea, fatigue, vomiting, abdominal pain and diarrhoea. The alkaloid concentration in potatoes is in the order of 2–13 mg solanine/100 g fresh weight, predominantly in the potato skin, and particularly marked in the green parts of the potato after exposure to light, resulting in a dangerous concentration of 80–100 mg/100 g fresh weight (20 mg/100 g is the generally accepted upper limit of safety for solanine). Cooking the potato does not reduce solanine concentrations, as the solanine is stable at increased temperature.

Pyrrolizidine alkaloids Some 3% of the flowering plants in the world – about 6000 species – contain pyrrolizidine alkaloids (senecio alkaloids) with about 250 different chemical structures. Pyrrolizidine alkaloid poisoning generally presents as hepatic disease with chronic liver failure and periportal fibrosis resulting in cirrhosis and veno-occlusive disease as the central veins of the liver are blocked by connective tissue. Pyrrolizidine poisoning usually occurs from contact with weeds contaminating food crops, including wheat or corn, harvested with the grain. The problem may be compounded by plants of Senecio and Crotalaira being used as folk medicine. They may also be taken in the form of herbal teas, e.g. gordolobo yerba. In Japan, flower stalks of *Petasites japonicus* –which are used as a cough medicine – also contain the pyrrolizidine alkaloid, petasitenine.

Aconites, the dried root stalks of plants in the Aconitium family are used as herbal medicines to treat rheumatism, neuralgia and cardiac complaints. They are prepared by boiling or soaking in water which hydrolyses the toxic alkaloids into less toxic derivatives. To avoid severe poisoning a limit of 1.5–3.0 g of cured aconitum root stalk is recommended.

Cyanogenic glycosides
These are synthesized by plants and contain sugar molecules and α-hydroxynitriles (cyanohydrins). When these are degraded by plant enzymes, nitrile groups are eliminated as hydrogen cyanide which is toxic; 50–60 mg is an average fatal dose in humans. Cyanogenic glycosides have been detected in 110 plant families and over 2000 plant species. These include the plants identified in Table 1.5.

Cyanogenic glycoside poisoning is most seen in populations eating lima beans and cassava, though the concentration in lima varies markedly from the American white bean to the Puerto Rican small black. In cassava, however, cooking removes or destroys cyanogenic glycoside and the enzymes that cause the liberation of hydrogen cyanide. Linseeds, traditionally used as laxatives, are cyanogenic but toxic effects are unknown at traditional levels of intake, though altered processing and increased intake can lead to unexpected toxicity. The cyanide content varies from batch to batch (4–12 mmol/kg) and contains the cyanogenic glucoside as cassava. Poisoning from cyanogenic glycosides can also occur from the ingestion of bitter almonds, choke, cherry seeds and drinking tea made from peach leaves.

Amino acids and amines

Amino acids There are more than 200 amino acids but only 25 to 30 are universally incorporated into proteins or occur as intermediates in metabolism. Some uncommon amino acids interfere with metabolism and consequently are toxic. The lathyrogens are toxic amino acids which cause lathyrism. This results from the ingestion of seeds of the lathyrus species including chickling vetch, the flat-podded vetch and the Spanish vetchling. Lathyrism is characterized by muscular weakness and paralysis of the lower limbs and may be fatal. It is particularly prevalent in times of famine when there is reduced choice in food.

Another uncommon plant amino acid which is poisonous is 3-methylenecyclopropylpropionic acid (hypoglycin A). This is found in the fruit of the tropical tree, *Blighia sapida*. This plant is poisonous when the fruit is unripe or when the inadequately cooked fruit is eaten. The consequences are severe vomiting, coma, acute hypoglycaemia and death within 12 hours. The toxicity is due to interference with the oxidation of fatty acids and, as a result, undernourished individuals are particularly vulnerable.

There are also selenium-containing amino acids, selenomethionine and selenocystine. Excess ingestion of the selenium amino acids results in dermatitis, fatigue, dizziness and hair loss. Such amino acid poisoning results from the consumption of the nuts from the monkey nut tree, *Lecythis olloria*, found in Central and South America.

Amines These include serotonin, noradrenaline, tyramine, tryptamine and dopamine. These are very active pharmacologically, are potent vasoconstrictors and are found in bananas, plantains, pineapples, avocados, tomatoes and plums. While levels of these amines normally eaten are readily detoxified, excessive consumption of these plants may defeat the natural detoxification systems. Patients receiving the monoamine oxidase inhibitor group of antidepressants must observe dietary restrictions, avoiding amine-rich diets.

Glucosinolates
These are produced by the brassica plant family, including cabbage, kale, Brussel sprouts, cauliflower, broccoli, turnips, garden cress, water cress,

TABLE 1.5 Hydrogen cyanide (HCN) levels liberated from food crops containing cyanogenic glycosides

Food	Hydrogen cyanide (HCN) yield (mg/100 g)
Bitter almond seed	290
Peach seed	160
Cassava leaves	104
Apricot seed	60
Lima bean	
Puerto Rico small black	400
American White	10

radishes and horseradish, rape seed, brown, black and white mustard. There are more than 70 different glucosinolates and most of the crucifer plants contain several of these families. Glucosinolates are responsible for the pungent flavours of horseradish and mustard and the characteristic flavour of turnip, cabbage and other related plants. They have been shown to interfere with thyroid function in experimental animals. Toxic effects arise from metabolic products formed from the action of the enzymes thioglucosidases, which break the glucosinolates into glucose, organic nitriles, isothiocyanates and thiocyanate ions. It is these latter two substances which modify thyroid function. It is, however, only at very high ingestion rates of raw plants, e.g. 500 g of raw cabbage daily for 2 weeks, that any effect on the thyroid gland is noticed.

Specific toxic effects

Carcinogens

- **Safrole** is a carcinogen found in several oils, including oil of sassafras, camphor and nutmeg. Safrole has been found in 53 plant species and in 10 plant families, and produces liver cancer if sufficient is added to a rat diet. Safrole represents 75% of the weight of oil of sassafras, which was previously used as a flavouring agent in root beer. Black pepper contains small amounts of safrole and larger amounts of piprine, which has been shown to be carcinogenic to mice.
- **Furanocoumarins** are carcinogenic chemicals produced by celery, parsley and parsnip. However, the concentration in these plants is low but may increase in diseased plants. The most common furanocoumarins are psoralen, bergapten (5-methoxypsoralen) and zanthotoxin (8-methoxypsoralen). Cycasin is found in cycads which are important sources of starch for tropical and sub-tropical populations. Such compounds are capable of producing liver, kidney, intestinal and lung cancers in rats.
- **Pyrrolizidine alkaloids** have caused cancer in rodents, and human cancers have been reported

from the use of herbal remedies containing these alkaloids.
- **Stevioside** is a very sweet glycoside from *Stevia rebaudiana*. Steviol is the aglycone resulting from bacterial hydrolysis, and has potential for carcinogenesis.
- **Cycasin** is a glycoside present in the nuts of *Cycas circinalis*. This is harmless as the glycone, but the aglycone methylazoxymethanol is carcinogenic. The β-glucosidase which splits off the glycone is found in tissues and in colonic bacteria.

Oestrogens At least 50 plants are known to contain chemicals which have oestrogenic activity, including carrots, soy beans, wheat, rice, oats, barley, potatoes, apples, cherry, plums and wheatgerm. Oestrogens are also present in vegetable oils such as cotton seed, sunflower, corn, linseed, olive and coconut oils. The oestrogenic activity rests in isoflavones, coumestans or resorcyclic acid lactones. It is doubtful if physiological effects would be elicited in humans by normal consumption of foods containing these weakly oestrogenic chemicals

Mutagens The cooking and processing of meat and fish at high temperatures results in heterocyclic amines with mutagenic and carcinogenic potential as judged by the Ames' test.

Flavonoids Flavonoids are polyphenolic glycosides which occur in edible plants, e.g. citrus fruits, berries, root vegetables, cereals, pulses, tea and coffee. They are hydrolysed by bacteria in the saliva and intestine to quercetin, kaempferol and myricetin. These aglycones show mutagenicity on the Ames' test. However, there is further bacterial degradation of the aglycone and rapid excretion.

Miscellaneous toxic agents

Beans The broad-bean, or fava bean, can produce acute haemolytic anaemia (favism) which is prevalent in Mediterranean countries, China and Bulgaria. The disease is characterized by nausea, dyspnoea, fever and chills and occurs 5–24 hours after broad-bean ingestion. Individuals who are susceptible to favism are deficient in glucose 6-phosphate dehydrogenase which also results in resistance to malaria.

Lentils Red lentils (*Lensculinaris*) are pulses which produce modest crops. Similar pulses from *Vicia sativa* are sometimes substituted in the diet of some populations. Cultivars of *V. sativa* may contain two neurotoxins, L-β-cyanoalanine and γ-L-glutamyl derivatives at a concentration of approximately 0.1%, a level capable of being toxic to animals. Most, if not all, of the neurotoxins are lost if the seeds are soaked and soaking and cooking water is discarded. *V. sativa* also contains pyrilidine glucoside. *V. sativa* and *V. faba* can cause favism which is a haemolytic anaemia in individuals who lack glucose 6-phosphate dehydrogenase in their red cells. This condition is common in people of Mediterranean origin.

Myristicin This is a potent hallucinogenic chemical produced by dill, celery, parsley, parsnip, mint and also nutmeg. It is said that as little as 500 mg of raw nutmeg may produce psychoactive symptoms while 5–15 mg of powdered nutmeg may result in euphoria, hallucinations and a dream-like feeling, followed by abdominal pain, depression and stupor.

Oxalates Oxalates may be produced endogenously by the metabolism of ascorbic acid or the amino acid glycine, which is also derived from the families Polygonaceae, Chenopodiacea, Portulacaceae and Fidoidaceae. Spinach contains 0.3–1.2%, rhubarb 0.2–1.3%, beet leaves 0.3–0.9%, tea 0.3–2% and cocoa 0.5–0.9%.

Gossypol This is the yellow colouring of cotton, *Gossypium*. It is found in the pigment glands of the leaves, stems, roots and seeds and may form 20–30% of the weight of the gland. When ingested, the result is depressed appetite and loss of body weight, cardiac irregularity and circulatory failure or pulmonary oedema. A major source of gossypol in the diet can be cotton seed oil, which may be found in salad oil, margarine and

shortening. Gossypol has also been used in China with a 99% effectiveness as a male anti-fertility agent.

Diterpenoids The honey from wild rhododendrons (*Rhododendron luteum* and *Rhododendron ponticum*) may be poisonous due to the nectar containing toxic diterpenoids (grayanotoxins).

Toxicity and hazards

In considering toxins of plant origin, the concept of **toxicity**, the ability to produce a harmful effect must be distinguished from **hazard**, the capacity of the chemical to produce toxicity under the circumstances of exposure.

Microbial toxins

Microbial toxins are poisonous metabolites produced by bacteria, filamentous fungi, mushrooms and algae. Bacterial fungi grow commonly in food, competing for nutrients with animals and humans, as mushrooms and algae are themselves a form of food supply. Bacteria and fungi may proliferate within a food as it is stored or may proliferate on entering the gastrointestinal tract.

Bacterial toxins
Most illnesses caused by pathogenic microorganisms result from proliferation of pathogenic microorganisms in the host, usually in the gastrointestinal tract. In other cases, poisoning arises from the ingestion of toxins. Food-borne diseases usually cause gastrointestinal disturbance. Toxins produced by bacteria have different modes of action and have individual toxic characteristics (Table 1.6).

Some toxins produce the same type of cellular disorder and therefore the toxin is named after the specific action. Toxins that cause enteric disorders, e.g. cholera, salmonellosis and *E. coli* are called **enterotoxins**. Others, such as tetanus and botulinum, are **neurotoxins**. Bacterial toxins are

TABLE 1.6 Relative potency of toxins

Toxin	Lethal dose to mice (μg/kg body weight)
Bacterial toxins	0.00003–1
Animal venoms	10–100
Algal toxins	10–1000
Mushroom toxins	>1000
Mycotoxins	1000–10 000

usually either **endotoxins** or **exotoxins**. Endotoxins are released upon the disintegration or death of bacterial cells in the body. These produce specific toxic effects either in specific tissues or to the whole body. They are toxic when there is massive bacterial infection in the body and their effects may be due to an over-reaction of the host immune system. Exotoxins are special proteins excreted by toxigenic bacteria in various foods and are generally extremely poisonous and fatal, e.g. botulism, cholera and gastroenteritis.

There are five sources of bacteria which cause food-borne illness:

- faecal matter or urine of infected humans or animals
- nasal and throat discharges of sick individuals or asymptomatic carriers
- infections on body surfaces of food-handlers
- infected soils, muds, surface waters and dust
- sea water, marine materials and marine life.

Food is only the final link in a chain of infections and as a suitable medium may determine the degree to which a product is infected. For example cholera-causing organisms do not thrive in acid foods. Botulism is caused by food contaminated by strains of *Clostridium botulinum*. This highly lethal organism is the most poisonous known to man.

Mycotoxins Mycotoxins are highly poisonous compounds, of low molecular weight, produced by moulds or fungi which are contaminants of fruit and agricultural products. If mould growth occurs on any food there is the possibility of mycotoxin production which may persist long after the mould has disappeared. A large number of commonly consumed foods therefore may potentially contain mycotoxins. These include wheat, flour, bread, corn meal and popcorn, which may contain aflatoxin, ochratoxin, serigmatocystin, patulin, penicillic acid, deoxynivalenol or zearalenone. Peanuts and pecans may contain aflatoxins, ochratoxin, patulin and strigmatocystin. Apples and apple products may contain patulin. Growth of microorganisms in food will be influenced by moisture content, relative humidity, temperature, food composition, presence of competing microorganisms and fungal strain. The critical moisture content varies with the commodity. Storage fungi are primarily aspergilli and some penicillia. Variations in the moisture content of stored materials in different areas throughout storage bins (hot spots) allow fungal growth and toxic development. Cereal grains are a good substrate for toxin production, whereas seeds that are high in protein, soy beans, peanuts and cotton seeds support certain toxins but not others. Other materials in the commodity, e.g. zinc, can also affect the ability of fungal growth and toxin production, e.g. aflatoxin and soya beans.

Significant food-borne mycotoxins These include aflatoxins, ochratoxin A, citrinin, patulin, penicillic acid, zearalenone, trichothecenes and alternaria toxins.

- **Aflatoxins** are produced by some strains of *Aspergillus flavus* and *A. parasiticus*. There are six main aflatoxins: B_1, B_2, G_1, G_2, M_1, M_2. Aflatoxin B_1 is a principal member of the aflatoxin family. Aflatoxin M is a metabolite of B_1, it is found in the milk of dairy cattle which have ingested mouldy feed and readily produces cancer of the liver. It is stable in raw milk and processed milk products, and is unaffected by pasteurization or processing into cheese or yoghurt. The widespread use of milk and milk products by children make this toxin of importance. Foods commonly contaminated with aflatoxins include peanuts, peanut oil, corn and beans. Aflatoxin B can kill poultry and domestic animals. Aflatoxins are potent liver toxins and carcinogenic in animals but, it appears from studies in the United States in otherwise fit people, not in humans. Other studies have shown that aflatoxin contamination of food correlates with the incidence of liver cancer in high-risk areas, such as south-east Asia and tropical Africa where malnourishment and viral hepatitis are endemic. The mortality rate from liver cancer among individuals infected with hepatitis B and who are HBsHg-positive is 10 times higher than in individuals who are HBsHg-positive but eat small amounts of these infected materials. It has been suggested that 50% of liver cancer cases in Shanghai are related to aflatoxin exposure.

- **Ochratoxins** are produced by *Aspergillus ochraceus* and other *Aspergilla* species. They can contaminate corn, pork, barley, wheat, oats, peanuts, green coffee and beans. Ochratoxin A, which frequently occurs in wheat and barley, can cause kidney damage in rats, dogs and pigs and may cause kidney disease in humans. Only 2–7% of the ochratoxin A in barley is transmitted to beer during processing. Over 80% of ochratoxin A is destroyed on roasting of coffee and variable losses of ochratoxin A occur when baking with toxin-contaminated flour.

- **Citrinin** is a yellow compound produced by *Penicillium* and *Aspergiculla* species and is a contaminant of yellow peanut kernels from damaged pods. It may be strongly nephrotoxic though in general is less toxic than ochratoxin A.

- **Patulin** is toxic to bacteria, mammalian cell cultures, higher plants and animals. Patulin is produced by a dozen *Penicillium* and *Aspergiculla* species. It is a principal cause of apple rot and a common pathogen on many fruits and vegetables. It is a contaminant of fruit juices world-wide, particularly apple juice. It is unstable in the presence of sulphydryl compounds and sulphur dioxide. When fruit juices are left to ferment more than 99% of the patulin is destroyed.

- **Zearalenone** is an oestrogenic compound which causes vulvovaginitis and oestrogenic responses in pigs. It is produced by *Fusarium* species and is found in moist corn in autumn and winter, is not very toxic and has not been implicated in human disease.

Mushroom toxins

There are thousands of mushroom species. In the United States there may be more than 5000. There are many species with very similar appearance and yet quite different tissues and cellular structure. Edible or poisonous species may differ quite radically in different environmental growth conditions. The poisonous *Amanita muscarina* comes in three colours, dark red, yellowish orange and white. These vary in intensity with age or exposure to sun and rain. The orange-capped toxic variety can be confused with the edible Acesarea. A characteristic feature of a species may be changed by mechanical damage which makes for easy mistakes in identity. There are many individual responses to the toxins which have resulted in conflicting reports regarding edibility in the literature. Responses vary with the number of mushrooms eaten, the preparation, length of cooking, age and the health of the individual as well as the amount of toxin present in the mushroom. There is no simple test for the toxicity of fungi.

Types of mushroom poisoning
These include: cytotoxic, haemotoxic, neurotoxic, hallucinogenic, gastrointestinal, disulfiram-like activity and carcinogenic.

Cytotoxic mushrooms The most important toxins in this group are amatoxins and phallotoxins. Amatoxins are 10–20 times more toxic than phallotoxins and there is no known antidote. Phallotoxins are hepatotoxic whereas amatoxins are both strongly hepatotoxic and nephrotoxic. Fungi containing amanitin include the species *Amanita galerina* and *A. conocybee*. Among the most poisonous of mushrooms is *Amanita phalloides*: this has a large cap which is greenish brownish in tone; the smell is of raw potato; the taste, reported by survivors, is said to be quite good. The *Galerina* are small brown to buff mushrooms with moist, sticky caps, found on logs buried deep in moss and are equal in toxicity to *A. phalloides*.

Haemotoxic mushrooms The Ear mushroom causes inhibition of blood clotting when eaten in sufficient amounts.

Neurotoxic mushrooms *Amanita muscarina* and *A. pantherina* are important examples of neurotoxic mushrooms which cause increased salivation, lacrimation, sweating and severe gastrointestinal disturbances.

Hallucinogenic mushrooms Two mushrooms, *Psilocybe* and *Panaeolus* may cause euphoria and excitement as well as muscle incoordination and weakness of arms and legs. Panic reactions may follow psychedelic visions of intense bright-coloured patterns and are associated with an inability to distinguish between fantasy and reality.

Gastrointestinal toxic mushrooms These cause abdominal cramp, intense abdominal pain, nausea, vomiting and diarrhoea which may be incapacitating and may even cause death in children. *A. agaricus* includes the widely available mushrooms sold in supermarkets as well as a number of phenol-smelling yellow stainers. These mushrooms are variably edible or toxic, suggesting the existence of local mildly toxic forms.

Disulfiram-like mushrooms These induce hypersensitivity to ethanol and when eaten with alcohol can cause severe flushing of the face, palpitations, tachycardia, nausea and vomiting. Disulfiram may be offered to alcoholics as an external deterrent to drinking.

Carcinogenic mushrooms Laboratory assays have identified carcinogenic and mutagenic properties in mushrooms. These include the false morel gyromitrin and two common types, the common commercial supermarket mushroom, *Agaricus*

bisporus and the Japanese forest mushroom, *Cortinellus shiitake*. The toxins responsible are unstable to heat.

1.4.7 Significant water-borne pathogens

Cyanobacteria

Cyanobacteria are found world-wide and were described as long ago as the 12th century when dogs died after ingesting neurotoxins from benthic which is attached to the sediment cyanobacteria. Cyanobacteria are members of the genera *Microcystis, Anabaena, Aphanizomenon* and *Oscillatoria*. They contain lipopolysaccharide endotoxins or potent agpatotoxins (microcystins) and neurotoxins, e.g. anatoxins, saxitoxins. There are few epidemiological studies which have assessed the dangers to human health from freshwater cyanobacteria, though exposure to the cyanobacterial blooms has led to skin reactions, conjunctivitis, rhinitis, vomiting, diarrhoea and atypical pneumonia. Cyanobacteria can form blue–green, milky blue, green, reddish or dark brown blooms and scums to ponds and lakes.

Among the algal toxins are those which can cause serious problems with the liver, the skin and the central nervous system.

Saxitoxin

Most seafoods are safe and unlikely to cause illness, though in regions where reef fish live there is the danger of ciguatera, a paralytic shell food poisoning and neurotoxic shell-fish poisoning. The food industry takes elaborate precautions to eliminate such contamination. However, many holiday makers may be endangered by harvesting molluscs and fishing without being aware of pollution in the water.

Saxitoxin, is associated with paralytic shell-fish poisoning which results from the ingestion of molluscs (mussels, clams, oysters, scallops) contaminated with the neurotoxins of the dinoflagellates, *Gonyaolax catenella*. The dinoflagellates are one cell algae-like organisms which grow in water and may fluoresce and light the sea at night. They also produce red tides or blooms so that the colour of the water may change to yellow, red, brown or green, depending on the nature of the pigments present. The amount of poison in shell-fish depends on the number of dinoflagellates in the water and the amount of water filtered by the shell-fish. The molluscs are themselves unaffected by the toxin which is, however, readily released when the mollusc is eaten. The dinoflagellate synthesizes the saxitoxin, a substance highly neurotoxic in humans in a dose of 0.5–0.9 mg. It acts by preventing sodium ions from passing through the membranes of nerve cells and prevents neurotransmission. The saxitoxin is stable to heating and alterations in pH. Symptoms begin rapidly after eating shell-fish, appearing as numbness in the lips, tongue and finger tips. This is followed by numbness in the legs, arms and neck, accompanied by general muscular incoordination. Respiratory paralysis occurs within 2–12 hours.

Ciguatoxin

Ciguatera is a food poisoning following the eating of contaminated fish, which have concentrated the poison while eating dinoflagellates, e.g. *Gambierdiscus toxicus*. Herbivorous fish eat algae on rocks, and accumulate toxins in their tissues. Though the dinoflagellate poison is 40 000 times more toxic than cyanide, the fish is not affected by the toxin. Over 400 fish species may harbour ciguatoxin. The fish that are affected are fish that swim at the bottom of the water near reefs and are found between 35°N to 35°S latitudes and include barracuda, snapper, red snapper, jack, amber jack, grouper, chinamen fish, parrot fish, surgeon fish, moray eel, sea bass and shark. Toxic species vary from area to area. The symptoms of ciguatera include nausea, numbness and itching, aching jaws and teeth and reversal of hot and cold sensations. The fish responsible for ciguatera vary in the degree of poisonousness which may alter over a period of one to several years. There are no physical or chemical tests for the presence of ciguatoxin only the awareness of having eaten the fish.

Brevetoxin

Brevetoxin is a marine biotoxin from the dinoflagellate, *Gymnodinium breve*, blue-green algae abundant in the Gulf of Mexico. This causes red tides which kill many fish and infect bivalve molluscs. The growth of these dinoflagellates depends on salinity, wind and water temperature, maximum growth being at between 17.5 and 18°C. Two heat-stable neurotoxins have been isolated from *Gy. breve*. The symptoms are those of neurotoxic fish poisoning but with paralysis not having been reported.

Aplysiatoxin

An algal toxin which causes 'swimmers itch' has been isolated from the widely distributed blue-green alga, *Lyngbya majuscula*. This dermatitis is uncommon because the algal toxin readily decomposes when exposed to strong sunlight, though their ingestion may contribute to the risk of developing cancer.

Summary

1. In the food chain, nitrogen and carbon are converted to complex and essential nutrients by microorganisms, sea plankton and plant growth and storage organs and seeds. There is a complex chain of events that precedes food being eaten by humans, which progresses from the inorganic chemicals in the soil to plants, which are ingested and digested by animals. When carnivores eat herbivores there is a transfer of chemicals from plants through animals which may be concentrated in one particular animal species, organ or egg.
2. Food additives are substances, either synthetic or natural, which are normally not regarded as food, but added in small amounts to influence the desirability of food or to preserve food. The use of additives is carefully controlled by the food authorities. Additives are also used for

enrichment and fortification where nutrients are lost through processing or where there is seen to be a shortage of a nutrient within the diet of a population.

3. Preservatives are added to foods to prevent them from deteriorating as a result of bacterial colonization. Other additives are used to ensure that the products are uniform in appearance.

4. Food-borne microbes or fungi are other non-nutrients in the human food chain. Antibiotics, pesticides, metals, industrial chemicals and fertilizers can enter the food chain and hence human food. Some of the synthetic residues arise from veterinary prescription of drugs, ingestion of feed or water contaminated by pesticide residues and thereafter eaten by animals which are eaten by humans. Similarly feed additives, vitamins and minerals which enhance growth and health of animals may subsequently be ingested by humans.

5. Industrial contaminants may leach out into the soil and be taken up by plants and enter the human food cycle.

6. Many plants, plant leaves, stalks and roots contain poisonous chemicals which may well occur to protect the plant from being eaten. Humans through the centuries have had to learn which plants are free of poisons. Some plant poisons, including alkaloids, are valuable in medicine in small amounts but in larger amounts are poisonous. Cyanogenic glycosides are found in lima beans and cassava. Cooking reduces the toxicity of these compounds which liberate hydrogen cyanide. Some amino acids and amines are poisonous, by interfering with metabolism. Glucosinolates interfere with thyroid function and have other toxic effects on metabolism. There are low levels of carcinogens in oils from a wide number of plants, including celery, parsley and parsnip. Other plants, including carrots, soya beans, wheat, barley, potatoes and apples contain oestrogens. The bean can be toxic to certain populations causing favism, a form of haemolytic anaemia. Myristicin, produced by dill, celery, parsley, parsnip and mint can be hallucinogenic.

7. Toxins can be produced by bacteria, fungi, mushrooms and algae. Bacterial pathogens are important causes of food-borne illness. Mycotoxins are highly poisonous compounds produced by moulds or fungi, sometimes within food during storage. The most important of these are aflatoxins, which are carcinogenic in susceptible populations. Among mushrooms there are a number of very poisonous species, although the response varies with the amount eaten, cooking and the health of the person, as well as the amount of toxin present in the mushroom. Mushrooms may be cytotoxic, or have haemotoxic, neurotoxic, hallucinogenic or gastrointestinal toxic effects.

8. A wide number of pathogens occur in the water and include cyanobacteria in drinking water. Saxitoxin and ciguatoxin are toxins produced by shell-fish and fish infected by dinoflagellates.

Further reading

Coates, M.E. and Walker, R. (1992) Interrelationships between the gastrointestinal microflora and non-nutrient components of the diet. *Nutrition Research Reviews*, 5, 85–96.

Elder, G.H., Hunter, P.R. and Codd, G.A. (1993) Hazardous freshwater cyanobacteria (blue-green algae). *Lancet*, 341, 1519–20.

European Parliament and Council Directive No. 95/2/EC (1995) *Food additives other than colours and sweeteners*. Official Journal of the European Communities, L61/1.

Jalpas, J.S. (1994) Some aspects of cancer medicine. *Journal of the Royal College of Physicians of London*, 28, 136–42.

D'Mello, J.P.F., Duffus, C.M. and Duffus, J.H. (1991) *Toxic substances in crop plants*. Royal Society of Chemistry, Cambridge.

Rosling, H. (1993) Cyanide exposure from linseed. *Lancet*, 341, 177.

Winter, C.K., Seiber, J.N. and Muckton, C.F. (eds) (1990) *Chemicals in the Human Food Chain*, Van Nostrand Reinhold, New York.

Chapter 2

Genetics

2.1 Introduction to genetics

- Genotype is the genetic constitution of an organism.
- Two parents contribute to the total inheritance of the offspring.
- A gene is a store of genetic information held in the form of DNA which determines the synthesis of a specific peptide chain.
- The difference between different individuals depends on their genetic make-up which is expressed through the synthesis of proteins and isoenzymes.
- Mendelian laws describe the inheritance of a specific characteristic.
- The common phenomenon of polymorphism is exemplified in humans by different forms of haemoglobin resulting in sickle cell disease and thalassaemia.
- Predisposition to and survival from disease may be related to genetic constitution.

In order to understand the genetic basis of metabolism, it is necessary to be familiar with the principles of genetics, population genetics and the molecular basis of health and disease.

Human genetics is the science which looks at the inherited variations in humans, a study of the mechanisms of evolution, and the process of change in gene frequency. Medical genetics is the application of these principles to medical practice.

Every individual has a specific potential for survival and reproduction which is dictated by genetically determined characteristics which influence metabolism, fecundity, birth, growth and death rate.

2.1.1 Principles of genetics

A **gene** is the store of genetic information held in the form of deoxyribonucleic acid (DNA). The gene – which may exist in several forms – determines the synthesis of particular peptide chains. The genetic constitution is the **genotype**. The particular form of a gene is an **allele** and determines metabolism. How that gene is expressed enzymatically and the resultant characteristic is the phenotype. The two parents obviously contribute to the total heritage of the offspring, less directly the four grandparents and the influence of each great-grandparent is even less conspicuous. If the inheritance is of a genetic defect – particularly if the defect carries the potential for morbidity or premature mortality – then the interest of the medical geneticist is aroused. However, such characteristics do not appear equally in all offspring. A characteristic or phenotype from one parent may be dominant to the other parent's recessive phenotypic contribution. Phenotypes conceal a great variety of recessive genotypes which may become apparent in subsequent generations.

The genetic process was clarified by Mendel through his studies on the growing of peas. He identified that where genes from the parent generation are both dominant or both recessive then the offspring will be homozygous for that gene, whether dominant or recessive. If there is a mix of dominant and recessive genes the resulting individual is heterozygous. Thus, homozygous is where there are two identical genes at a locus of a chromosome; heterozygous is when these two genes are different. Genes from each parent are in pairs which separate into single units, pass to the next generation independently of each other, and are never present together in the next generation. This is Mendel's first law (Figure 2.1). The second law is that pairs of genes pass to the next generation as though they were independent of one another.

Epistasis is when one gene eliminates the phenotypic effect of another. The epistatic relation is independent of dominance. Dominance and epistasis are terms referring to the expression of a gene, not to any intrinsic property. A gene is said to have a high penetrance if the carriers are clearly distinguishable from non-carriers. A gene is said to be pleiotrophic if it has a range of phenotypic manifestation.

All genes are likely to be pleiotrophic dominant in respect to one aspect of their expression and recessive in another. A favoured gene will increase in frequency and expression and a favourable characteristic will eventually become dominant. However, when the circumstances change then that gene may become disadvantageous.

An *allele* is one of several forms of a gene occupying a given position or locus on a chromosome. Different alleles of the same gene are called multiple alleles. Gene frequency in the population is the frequency of one kind of allele on a set of chromosomes. Autosomes are all the chromosomes except the sex chromosomes. A normal cell is diploid, that is, it has two copies of each autosome on each pair of chromosome.

After conception, when two sets of chromosomes pair, the pairs of alleles determine the characteristics of the developing individual. Some inherited characteristics probably are dependent upon several alleles; that is, there is a cumulative influence of several genes. Height and weight are dependent upon a number of allelic and other, including nutritional, influences.

Genes close together on a chromosome tend to remain close to each other during the sexual development of the next generation. Females have two X chromosomes and males only one. This is relevant in the transmission of those characteristics carried on the X and Y chromosomes.

During cell division there is paired exchange between closely associated chromatids. A *chiasma* (Figure 2.2) is the site where chromatids are broken at corresponding points and which join in a

Inheritance

Mendel's laws

1. Each body cell has paired chromosomes. Genes are duplicated, e.g.

AA	Aa	aa
Homozygous	Heterozygous	Homozygous

; or

2. Each ovum or sperm (gamete) has one chromosome; the genes are single

A   A   A   a   a   a

When the ovum is fertilized, then the result may be

	Homozygous		Homozygous	
Parents	AA	AA	aa	aa
Offspring	AA	AA	aa	aa
Parents	One parent AA One parent Aa		Parents both Aa	
Offspring	AA	AA	AA	Aa
	Aa	Aa	Aa	aa

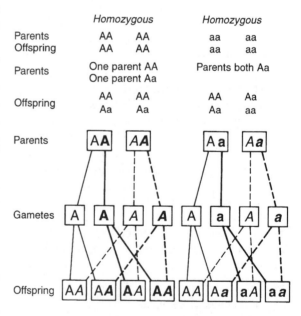

FIGURE 2.1 Mendel's first law states that genes are units which segregate. Members of the same pair of genes are never present in the same fertilized ovum (gamete) but always separate and pass to different gametes. Mendel's second law states that genes separate independently. Members of different pairs of genes move to gametes independently of each other.

crossover manner producing new chromatids. This process is called breakage and reunion and leads to recombination. Translocation occurs when part of one chromosome breaks and becomes attached to a different chromosome.

The possibility of a chiasma between two points on a chromosome is proportional to the distance separating the two points. Thus, as the distance between two points increases so does the probability of a crossover. The closer genes are to one another, the tighter they are linked and less likely they are to be split.

Major genes have defined functions which determine specific characteristics, as distinct from polygenes which control several apparently different functions. Major genes function at a single locus which may exist in different allelomorphic states. Alleles may be:

- completely dominant
- incompletely or partially dominant; the phenotype of the heterozygote is intermediate between that of the two homozygotes
- codominant and contribute equally to the phenotype

According to Mendel's second law, in the second generation alleles segregate so that equal numbers

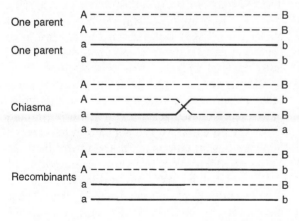

FIGURE 2.2 Chiasma formation. The two identical chromosomes from one parent are shown by a broken line, the others are shown by an unbroken line. These join to form a bivalent in the meiosis. Crossing over occurs between two of the chromosomes; both are broken at the same site and the opposite ends are joined. The result is two recombinant chromosomes.

of each of the four types of gamete are produced. Two general types of offspring are produced:

- two parental types
- a recombinant type where the dominant of one parent and the recessive of the other parent are combined

The phenotypic variation of a measurable characteristic of a population usually takes the form of a unimodal frequency distribution which can be described as mean and variance. If there is inbreeding then the variability decreases as the genetic uniformity increases.

2.1.2 Population genetics

A **species** consists of a set of individuals who can actually or potentially interbreed and may adapt but are reproductively separate from other species. A population is an inter-breeding group of individuals.

Evolution is a process of adjustment by selection of existing genetic possibilities. Such mutations or changes in genotype in a group may take place at the edge of a defined and settled geographical area. This area where change takes place may be adjacent to an environment where new characteristics are needed to survive.

Any mutation must be consistent with the viability of the organism if it is not to be a lethal mutation. A nonsense mutation in a gene is one which prevents the protein specific to the gene from being synthesized. Other mutations, called suppressers, may allow the nonsense mutation to be overcome so that the protein can be synthesized.

As fitness to survive the rigours of the environment increases so genetic variability is reduced; that is, inbreeding results in fewer heterozygotic and more homozygotic individuals. There are systematic effects in which the size and direction of the change are determined including:

- immigration into and migration from a population; the effect will be dependent on the proportion of the immigrant and total populations
- unique mutations which may provide novel forms essential for evolution

- selection and differential contribution by different types to the succeeding generations

A micro evolution consists of a change in the population either to fit into a new environment or over time in a changing environment. If a population is large, mating is random, and there are no migrations or selective mutations, then the population will develop an equilibrium of gene and genotype frequency after one or two generations. If a rare allele is favoured it increases in frequency slowly at a 5% increment until equilibrium is achieved.

The size of the breeding population is important relative to the total number of adults. The effective population size may be affected by social prohibitions or an unequal sex ratio, e.g. when large numbers of men are killed during war. Non-recurrent events, such as rare mutations, hybridization and selective incidents may also influence genetic variability.

Selection may occur through differential survival affecting any stage of the developmental cycle. The newly formed fertilized egg, the zygote, develops into a baby, an infant and then an adult. Gametes (sperms or ova) are formed, mating takes place to ensure that the gametes unite and to produce the zygote of the next generation.

Evidence from field and laboratory studies suggests that polymorphism is a very common phenomenon. Different alleles coexist in a population creating genetic polymorphism, the occurrence together in the same locality of two or more discontinuous forms of a species, in such proportions that the rarest of these cannot be maintained merely by recurrent mutation. Polymorphism is said to be transient if it occurs because one gene is in the process of replacing another. It is stable if the gene frequency is at, or moving towards a stable equilibrium point. A good example in nature is melanism in the peppered moth where the coal black mutant becomes dominant to the usual pale speckled moth. The black pigmentation helps the moth survive in dirt of industrial areas.

In humans, sickle cell disease is an example of stable polymorphism. There are numerous types of haemoglobins in humans. Most are rare but haemoglobin S, haemoglobin C and haemoglobin E are common in some parts of the world, being three alleles on the same locus. The sickle cell condition is caused by haemoglobin S. The abnormal forms differ from normal in the amino acid sequence of the haemoglobin and may be distinguished by electrophoresis. These abnormal forms are disadvantageous because they are less effective in carrying oxygen than haemoglobin A. Under conditions of low oxygen tension the red cells of the homozygote crenate, the distorted cells are inefficient in carrying oxygen and also occlude blood vessels. The survival rate of the SS homozygote is only 20% of the average for all genotypes. On the other hand, those heterozygous individuals who are a mix of haemoglobin A and haemoglobin S, i.e. haemoglobin AS type, have shown greater resistance to malaria caused by *Plasmodium falciparum*. The fertility of AS women is greater than of A women because of a reduced abortion rate.

Another abnormal haemoglobin type is thalassaemia, controlled at a different locus, and occurring primarily in Mediterranean races. This condition is associated with a deficiency of red cell glucose 6-phosphate dehydrogenase. The homozygote is non-viable and carriers may suffer from haemolytic disease at birth, when treated with certain drugs and after eating certain raw beans. The red cells of heterozygous individuals carrying glucose 6-phosphate dehydrogenase deficient alleles are resistant to malarial sporozoites. Such gene frequency differences occur in over 50% of individuals in malarial areas

All selection involves either differences in mortality or differences in reproduction so that gradually there are changes in the polymorphism and hence the potential for the population to increase. Spontaneous mutations occur at a rate of between $1 : 10^5$ to $10 : 10^7$ per locus per generation

J.B.S. Haldane looked at the evidence for selection in survival by human birth weights. Those babies with an optimal birth weight between 7.5 pounds (3.4 kg) and 8.5 pounds (3.9 kg) had a survival rate of 98.5%. The majority of those who died belonged to weights which were outside that range. The reasons for babies having a particular weight range are numerous, but nevertheless there is this selection process at the beginning of life.

Change in gene frequency results in the elimination of unfit types under intense selection. It is possible for the population to be severely reduced until a favoured dominant phenotype becomes the most abundant. Haldane calculated that the number of deaths to secure the substitution of one generation to another may be 30 times the population's size in one generation. A large number of individuals may be lost as a result of genetic heterogeneity. Haldane argued that because of the cost of evolution very few characters may be selected at one time unless they are controlled by the same genes. The rate of evolution is limited by the cost of the change. Deleterious mutations in the human population may act on the foetus before birth. The zygote may not be implanted or an embryo may be aborted at an early stage. Once a selective force eliminates a fraction of the potential population, the next selective force acts on the individuals that remain, and so on, until all selection has been completed. The mode of evolution is an elementary process of change in gene frequency. Evolution is an immensely complicated process and takes place by differential contribution of different individuals in a species to the succeeding generation.

These effects are clearly seen in humans. There are distinct attributes in isolated populations, for example, the Aborigines of Australia who have been isolated for many generations. This is in contrast to the pooling in gene type which is taking place at the present time through migration to the United States, Canada, white Australia and New Zealand.

In Britain there have been a series of epidemics over the centuries which must be selective in defining the gene population. In the Middle Ages the plague was rampant; in the 19th century, tuberculosis, and in the 20th century coronary artery disease and possibly cancer are the main killer diseases. The next generations may be peculiarly vulnerable to other conditions, which may include viral diseases. The population which survives a particular epidemic may be at an advantage genetically though those very genetic characteristics which enable survival in one stress situation may create a subsequently more vulnerable population. The genetic make-up of a person is tested when that person meets a particular environment or situation. There may be advantages to that genetic constitution or it may result in predisposition to a particular disease. A constitution which may in the heterozygous state be protective may in the homozygous form be dangerous, as discussed in relation to sickle cell anaemia and thalassaemia.

This present generation is presented with dietary choice, clean food, good sanitation, attractive accommodation and relative freedom from lethal infectious diseases. In addition, the population is exposed to tobacco smoking and sedentary occupations. This total environmental change must identify the strengths and vulnerability of the genetic stock. The controversy remains whether a condition or state of development is dictated by nature or nurture, though nurture certainly presents challenges to nature.

Summary

1. Human genetics is a science that looks at the inherited variations in humans, the study of the mechanism of evolution and the process of change in gene frequency.
2. A gene is a store of genetic information held in the form of deoxyribonucleic acid (DNA). The gene determines the synthesis of particular peptide chains. The genetic constitution is a genotype. The particular form of a gene is an allele and determines metabolism.
3. The two parents contribute directly to the total inheritance of the offspring and they in turn are affected by their own parents.
4. Inheritance of a genetic characteristic is determined by laws described by Mendel.
5. Mendel's first law states that genes from each parent are in pairs which separate into single units, are passed to the next generation independently of each other, and are never present

together in the next generation. The second law is that pairs of genes pass to the next generation as though they were independent of one another.

6. Genes are said to be dominant or recessive. An allele is one of several forms of a gene occupying a given position or locus on a chromosome. Alleles may be varyingly dominant and this affects the characteristics of the next generation. If there is a mix of dominant and recessive genes, the resultant individual is heterozygous. Homozygous is where there are two identical genes in a chromosome. This means that a particular characteristic will be readily expressed.

7. A species consists of a set of individuals who may interbreed. Evolution is a process of adjustment by selection of existing genetic populations. Such changes in a group may take place at the edge of a defined and settled geographical area and may relate to an adaptation to an environment where new characteristics are needed to survive. Mutations may influence the viability of the organism. Immigration and movement within the population may affect the genetic pool.

8. Such polymorphism may express itself in the physiology of humans and will affect the way in which the types of enzymes are synthesized and consequently metabolism and disease.

Further reading

Brock, D.J.H. (1993) *Molecular Genetics for the Clinician*, Cambridge University Press, Cambridge.

Connor, J.M. and Ferguson-Smith, M.A. (1993) *Essential Medical Genetics*, 4th edn, Blackwell Scientific Publications, Oxford.

Cook, L.M. (1976) *Population Genetics. Outline Studies in Biology*, Chapman & Hall, London.

Emery, A.E.H and Rimoin, D.L. (1990) *Principles and Practice of Medical Genetics*, 2nd edn, Churchill Livingstone, Edinburgh.

Weatherall, D.J. (1991) *The New Genetics and Clinical Practice*, Oxford University Press, Oxford.

2.2 Molecular basis of health and disease

- Those conditions where the characteristics of the genotype are responsible for a particular pathology are the province of medical genetics.
- Mendelian disorders are a single-gene defect or a single-locus disorder. Other genetically determined disorders are more complex.
- Chromosomal disorders may be the result of the loss, addition or abnormal arrangement of one or more of the chromosomes.
- The mapping of genes on chromosomes will facilitate the description and interpretation of the genetic and biochemical basis of metabolism and metabolic abnormalities.
- It is difficult to identify the molecular biological basis of disease with poor or excessive nutrition.
- There is a genetic component in the development of each individual's metabolic pathways.

2.2.1 Introduction

Medical genetics has enabled the scientist and clinician to study the aetiology of non-infective disease at the cellular and molecular level. Many conditions are now recognized as belonging to a group of conditions called genetic disease, wherein characteristics of the genetic material are either entirely or partially responsible for the pathology.

2.2.2 Genetic disorders

Mendelian disorders

These are single-gene defects or single-locus disorders (Figure 2.3). They result from a mutant allele or a pair of mutant alleles at a single locus. Such changes can be inherited or arise *de novo* through a mutation. When an allele is dominant or recessive alleles may result respectively in auto-somal dominant or recessive conditions of. Modern geneticists suggest that these conditions are not attributable to dominant or recessive genes but rather that the consequent phenotypes are domin-

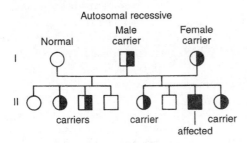

Autosomal dominant

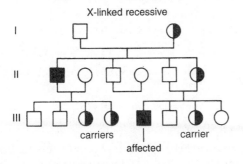

Defect is transmitted to half of offspring, whether male or female

FIGURE 2.3 Families with Mendelian single-gene defects. **Autosomal dominant**: half the offspring, whether male or female, are affected in the third generation. **Autosomal recessive**: when a carrier marries a carrier then two of the children are carriers and one child is affected. In contrast when a carrier marries a non-carrier then two of the children are carriers. **X-linked recessive**: the characteristic is carried by the mother, in the third generation the daughters but not the sons of the affected male are carriers. Among the children of the female carrier, half the sons are affected and half the daughters are carriers.

ant or recessive. The inheritance is either auto-somal dominant, autosomal recessive, autosomal codominant or X-linked inheritance dominant or recessive.

Autosomal dominant conditions
In these disorders:

- both homozygotes and heterozygotes manifest the condition
- affected patients have an affected parent
- the risk is 1 in 2 to each child of one affected and one unaffected parent
- both sexes are equally affected
- both sexes are equally likely to transmit the condition

Variable expression, non-penetrance or mutation may result in mildness or severity of the condition. Even a dominant disease shows variable expression. Non-penetrance may result in a person with no signs of the condition carrying the genes from an affected parent and producing an affected child, for example familial combined hyperlipidaemia, Huntington's disease or polyposis coli.

Autosomal recessive conditions
These diseases are determined by a single autoso-mal locus. The condition is manifest only in peo-ple who are homozygous for the abnormal allele (aa). The parents of affected children are pheno-typically normal carriers Aa. Children of two parents heterozygous for a particular phenotype will be 25% normal, 50% also heterozygotes and 25% will show clinical expression of the condi-tion. Each child has a 1 : 4 risk of being affected. The distinctive features are:

- phenotypically normal parents may have one or more affected children
- unless an affected person mates with a carrier, all the children are unaffected
- both sexes are equally affected
- the condition can be measured biochemically, e.g. the haemoglobinopathies, cystic fibrosis, phenylketonuria or sickle cell anaemia

Autosomal codominant
This is the simplest form of Mendelian inherit-ance. The characteristics are determined by a

single genetic locus with two alleles (alternative forms of a gene Aa) located on one of the autosomes (any chromosomes except the sex chromosome X or Y). In the heterozygote Aa both alleles are expressed. The homozygote form is AA or aa each of which shows a different phenotype. Many biochemical variants, e.g. isoenzymes (different types of the same enzyme) are codominant. Examples are:

- blood groups ABO, Rhesus
- red cell enzymes acid phosphatase, adenylate kinase
- cell surface antigens, human leukocyte antigen systems (HLA)

X-linked inheritance
The mother provides the X chromosome and the father the Y chromosome to the son. The specific features are:

- The disease affects mainly males.
- Affected males have unaffected parents but may have affected maternal uncles.
- The disease is transmitted by carrier women who are usually asymptomatic, half of the sons of a carrier are affected and half the daughters are carriers.
- The children of an affected male are unaffected but all of his daughters are carriers.
- The daughters of an unaffected man and a carrier woman are affected.

The variability of X-linked conditions may be because of the suppression of one of the female X chromosomes, achieved, in part, by methylation of a dinucleotide. The choice of which X chromosome is inactivated (lyonized) is random but persists through the life of that cell line. Some genes in the tip of the short arm of the X chromosome escape inactivation and are expressed in the normal manner, e.g. Duchenne muscular dystrophy, or haemophilia A and B.

X-linked dominant inheritance is not common. Affected males have all normal sons but affected daughters. These include X-linked hypophosphataemic rickets which occurs only in females but is believed to be lethal in males.

Chromosome disorders

These are the result of the loss, gain or abnormal arrangement of one or more of the chromosomes and are in general the result of a numerical or structural mutation in the parent's germ cell. There are 23 chromosomes in the human germ cell and 46 chromosomes in the diploid cell. Polyploidy is when multiples of 23 occur and are most frequently observed in spontaneously aborted infants, with triploidy 69 and tetraploidy 92 chromosomes. Where a single chromosome is lost or gained then the result is aneuploidy. This usually results when a chromosome fails to separate during division. The resulting child may have a mosaic of normal and aneuploidy chromosome complements. The consequence of the development of aneuploidy is either trisomy, an extra chromosome or monosomy, loss of a chromosome, e.g.:

- Down's syndrome chromosome 21
- sex chromosome 47, XXY Klinefelter's syndrome
- 45, X Turner's syndrome

If there is chromosomal breakage then translocation may occur with complex genetic consequences.

Multifactorial disorders

The aetiology is complicated but such disorders occur when a phenotypic characteristic is revealed by a particular environmental situation. Such associations are demonstrated by twin, sibling or family studies. This genetic predisposition is important in the understanding of conditions which are often attributed entirely to an outside influence, e.g. nutrition. There has to be a predisposition highlighted by an environmental precipitant, for example the response to alcohol by men and women and different racial groups. Flushing occurs with acetaldehyde in the Mongol races, but only when drinking alcohol. Other examples are insulin-dependent diabetes, hypertension, colonic cancer and manic depressive disorders.

Somatic genetic disorders

These are mutations in somatic cells which may result in tumours. Alterations in large groups of genes may be involved. A gene often observed to be affected in tumour formation is the p53 gene, a tumour suppressor gene found on the short arm of chromosome 17. A series of cancers have been associated with alterations in regions of the chromosomes within the malignant cell nucleus.

Mitochondrial disorders

An extreme form of non-Mendelian inheritance is when the genotype of only one parent is inherited and the other is permanently lost. This contrasts with Mendelian genetics when reciprocal crosses show the contribution of both parents to be equally inherited. Usually it is the mother whose genotype is preferentially or solely inherited. This maternal inheritance can be shown where the inheritance of the genes of the mitochondria is contributed entirely by the ovum and not by the sperm. Mitochondrial DNA appears to mutate more rapidly than nuclear DNA in mammals.

2.2.3 Human genome mapping

Molecular biologists have embarked on a description of the complete linkage map of the human genome. The plan is to map the genes and their position on the chromosome and the gene abnormality for the most common genetic disorders. The human genome contains 3×10^9 base pairs (bp) of DNA packed into 23 chromosomes. Any disease that affects more than one person per 1×10^9 is almost certainly described somewhere in the genetic literature. The recessive nature of some human cancers has been demonstrated using recombinant DNA techniques and other genes associated with cancer have been localized to chromosomal regions.

Mutations and germ-line mosaics

Mutation rates can be calculated on the basis of a balance existing between the rates at which identified genes are eliminated by natural selection and created by mutation. Most mutations for autosomal recessives occur in normal homozygotes and may be transmitted by phenotypically normal carriers over many generations before appearing in an affected person.

It is easier to identify a population with a disadvantaged gene through the disease expression than to study populations who are vulnerable or invulnerable to environmental factors such as a deficiency or excess of dietary constituents, tobacco smoking or alcohol. Some progress has been made in identifying factors determining an individual response to alcohol.

A genetic basis to a condition, physical attribute or disease may be determined by:

1. **The DNA sequence.** The delta-globin gene was discovered on chromosome 11 during DNA sequencing studies. The nucleotide sequence suggests that the transcribed protein is a member of the β-globin family.
2. **A definable protein abnormality.** α-1 antitrypsin was discovered as a serum protease inhibitor. Absence of this enzyme was shown to be associated with pulmonary emphysema and cirrhosis of the liver. This has been mapped to chromosome 14.
3. **A disease entity.** Cystic fibrosis has been shown by its familial occurrence pattern to be a genetic disease. Linkage analysis has implicated a gene on chromosome 7.

2.2.4 Interpreting patterns of disease inheritance

The interpretation of human pedigree patterns is complex and beyond the scope of this book (see McKusick, 1987 for further details).

Mendelian disorders are caused by a mutation at a single genetic locus. Gene mapping attempts to locate precisely the disease loci on a chromosome and then to search for cloned sequences within the gene to identify how this differs from the advantageous gene.

Genetic markers are Mendelian characters used to follow a small section of a chromosome through a pedigree. Ideally, a marker should have a known chromosomal location, be highly polymorphic, show codominant inheritance and be reasonably measurable in the blood. Genetic heterogeneity implies clinical similarity produced by different genes. In practice, a number of gene mechanisms may be involved.

Locus heterogeneity

This term is applied when an apparently single clinical disease is caused by either of two or more separately located genes. Predisposition to breast cancer is an example, e.g. p53 gene on chromosome 17p. However, only 40% of breast cancers can be accounted for by this mechanism.

Intra-locus heterogeneity

Different mutations or deletions within a single gene may cause different phenotypes.

Intra-family heterogeneity

This is a situation wherein the disease manifestations and clinical course are very variable even within a family with the same inherited gene defect. This variability may be due to the action of two or more modifying genes.

Anticipation

This refers to the severity of the disease increasing with successive generations.

Genomic imprinting

This is when a deletion on a chromosome results in a different clinical consequence in males and females.

Inter-family heterogeneity

Between families there is a great variation in disease phenotype, but the disease within families is remarkably constant. Any heterogeneity in such conditions is due to intra-locus differences rather than being due to two separate but closely linked genes.

It is likely that differential methylation which is sex-specific at the gamete level causes this variation in gene expression. The Barr body which characterizes the female X chromosome is associated with this methylation.

Haplotypes and linkage disequilibrium

A disease mutation can occur when there is a particular coincidence of individual genetic attributes, e.g. female, blood group, HLA-A type. This would suggest a predictable point on the chromosome for the mutation to occur.

2.2.5 Nutrition and the molecular basis of disease

It can be seen how very difficult it is to identify the molecular biological basis of many of the diseases which have been attributed to poor or excessive nutrition. The metabolic process by which each nutrient is converted to energy or to structure will be individual and dependent upon the efficiency of the isoenzyme complexes in the metabolic pathways. This individual metabolic response will be genetically determined. For example, some individuals may be able to eat an excess of fat and dispose of the fat in a safe manner, others may drink large amounts of alcohol without any mishap. The digestive enzymes and any isoenzyme differences are critical only when there is an excess or deficiency of a nutrient when the important differences, and hence vulnerabilities, of individuals will be exposed.

Summary

1. A group of conditions called genetic diseases are associated with alterations in the genetic material which may be entirely or partially responsible for the pathology.

2. Mendelian disorders are those in which there is a single-gene defect or single-locus disorder. These result from a mutant allele or a pair of mutant alleles at a single locus. Inheritance is either autosomal dominant, autosomal recessive, autosomal codominant–dominant or X-linked inheritance, dominant or recessive.

3. In autosomal dominant conditions, both homozygotes and heterozygotes manifest the condition and include familial combined hyperlipidaemia, Huntington's disease and polyposis coli.

4. Autosomal recessive conditions are determined by a single autosomal locus and are manifest in individuals who are homozygous for the abnormal allele. Each child of two parents heterozygous for a particular phenotype, will have a 1 : 4 risk of clinical expression of the condition, such as the haemoglobinopathies, cystic fibrosis, phenylketonuria and sickle cell anaemia.

5. Autosomal codominant conditions are determined by a single genetic locus with two alleles on one of the autosomes. Many biochemical variants, e.g. isoenzymes, are codominant, for example blood groups A, B, O and Rhesus.

6. X-linked inheritance: these are conditions transmitted by females, mostly affecting males and include X-linked hypophosphataemic rickets.

7. Chromosome disorders result from the loss, gain or abnormal arrangement of one or more of the chromosomes and include Down's syndrome chromosome 21, sex chromosome, 47, XXY, Klinefelter's syndrome, and 45, X Turner's syndrome.

8. Multifactorial disorders occur when a phenotypic characteristic is revealed by a particular environmental situation, e.g. metabolism of alcohol and the excess acetaldehyde which occurs in the Mongol races but only when alcohol is consumed.

9. Molecular biologists have embarked on a description of the complete linkage map of the human genome which will identify a genetic basis for many conditions.

10. The metabolic process by which each nutrient is metabolized will be individual and dependent upon the efficiency of the genetically determined isoenzyme complexes in the metabolic pathway.

Further reading

Davis, K.E. and Read, A.P. (1988) *Molecular Basis of Inherited Disease*, IRL Press, Oxford.

Emery, E.A.H. (1986) *Methodology in Medical Genetics*, 2nd edn, Churchill Livingstone, Edinburgh.

Evans, D.G.R and Harris, R. (1992) Heterogeneity in genetic conditions. *Quarterly Journal of Medicine*, 84, 563–5.

McKusick, V.A. (1987) *Mendelian Inheritance in Man*, 8th edn, John Hopkins' University Press, Baltimore.

2.3 Genetic basis of metabolism

- Within the nucleus are genes which determine the synthesis of specific proteins.
- DNA is a polymer of deoxyribonucleotides, helical in shape covalently linked to a triphosphate ester of a nitrogenous base.
- DNA is transcribed to RNA which is a template for the synthesis of protein.
- Proteins synthesized in the cell are sorted and distributed into different cell compartments where their function is specific. The function of each cell protein is genetically determined by enzymatic modification of amino acid residues.
- Cellular metabolic function depends upon the cell being divided into compartments.

- Metabolism in part depends upon chemicals moving between compartments by specific transport mechanisms.

2.3.1 Introduction

The cell, the 'triumph of evolution' is the fundamental unit of living organisms. Structure and function within cells are compartmentalized (Figure 2.4). Chemicals move between compartments by specific transport mechanisms. Cells can interact with their environment by receipt of chemical signals at the external surface and transduction of these signals within the cell, and by secretion of synthetic products from the cell.

2.3.2 Cell compartments

The cell compartments, shown in Table 2.1, include the nucleus, cytosol, mitochondria, endoplasmic reticulum (ER), Golgi apparatus, lysosome and peroxisome.

Nucleus

The nucleus contains the genes for the synthesis of cellular proteins. These genes are composed of deoxyribonucleic acid (DNA), and the total number of genes is referred to as the genome. The DNA is complexed with protein and forms exceptionally long continuous strands called chromosomes. Each diploid human cell contains 23 pairs of chromosomes. The nucleus is surrounded by a double membrane. The layers of the membrane contain pores which are presumed to be involved in the transfer of material to the cytosol.

The DNA in the nucleus is replicated in total when cells divide. Exact copies of each chromosome are distributed identically between the two daughter cells of each division. Transcription is another form of DNA copying. Small sections of the genome are copied selectively and each segmental copy (transcript) is composed of ribonucleic acid (RNA) not DNA.

Cytosol

The interior of the cell (cytoplasm) has an aqueous phase – the cytosol – in which many of the enzymes catalysing metabolic reactions occur. Some enzymes, however, are membrane-bound in

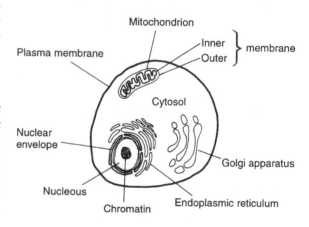

FIGURE 2.4 The cell consists of a number of compartments, each separated from the cytosol by a membrane.

TABLE 2.1 The compartments of the cell

Compartment	Boundary	Cell volume (%)	Function
Nucleus	Nuclear envelope	5	Gene transcription
Cytosol	Plasma membrane	6	Protein synthesis
Mitochondria	Mitochondrial envelope	25	Energy production
Endoplasmic reticulum	Folded membrane	10	Protein modification
Golgi apparatus	Membrane stacks	5	Protein sorting
Lysosome	Closed membrane	< 1	Protein degradation
Peroxisome	Closed membrane	< 1	Oxidation reactions

the various organelles found within the cytoplasm.

Mitochondria

There may be about 1000 mitochondria per cell. They are ovoid organelles with a double membrane, the inner membrane being convoluted. Many of the enzymes bound to the mitochondrial membranes function to transport and oxidatively metabolize in order to produce energy that is stored in the phosphate bonds of adenosine triphosphate (ATP).

DNA is transcribed into messenger RNA (mRNA) within the mitochondria. The mRNA transcripts are then translated into protein in the cytosol. The mitochondria possess genes capable of encoding some, but not all, of the proteins they contain. All mitochondrial DNA is inherited from the female parent since the section of the male gamete which enters the female ovum does not contain any mitochondria.

Endoplasmic reticulum

These are membrane sheets of rough endoplasmic reticulum, so-called because of its electron micrograph appearance which results from the binding of ribosomes. Rough endoplasmic reticulum is involved in protein synthesis.

Ribosomes

These organelles are composed of ribosomal RNA (rRNA) and protein. The rRNA is encoded by genes in the nucleus. The sequence of the rRNA genes is highly conserved within species and differences between rRNA genes can be used as a basis for taxonomic classification and for estimating the evolutionary proximity of different species.

In human cells the ribosome is composed of two subunits, both containing rRNA and protein. These are designated on the basis of centrifugation characteristics as 40S and 60S.

Golgi apparatus

This is a stack of membrane cisternae, often found close to the nucleus, which is involved in the sorting of protein for transport both within and outside the cell.

2.3.3 The biochemical basis of gene transcription

The strands of DNA are polymers of deoxyribonucleotides; RNA of ribonucleotides. Each unit is composed of the sugar deoxyribose (DNA) or ribose (RNA), covalently linked to a triphosphate ester of a nitrogenous base. The bases are adenine and guanine (purines) and cytosine and thymine (DNA) or uracil (RNA) (pyrimidines) (Figure 2.5). The corresponding deoxyribonucleotides are called deoxyadenosine triphosphate (dATP), deoxyguanosine triphosphate (dGTP), deoxycytidine triphosphate (dCTP) and deoxythymidine triphosphate (dTTP). For RNA the ribonucleotides are ATP, GTP, CTP and UTP.

The polymer is composed of phosphodiester bonds linking the 3' carbon of one deoxyribose ring with the 5' carbon of the next (Figure 2.6).

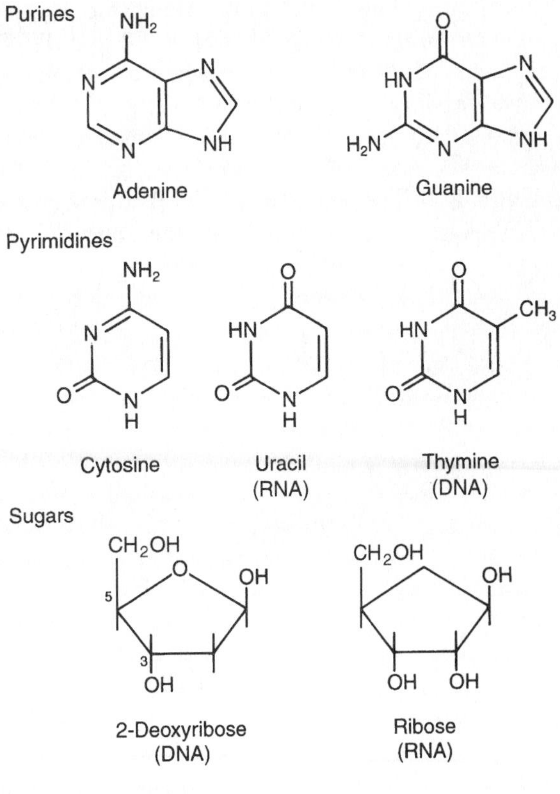

FIGURE 2.5 Purines, pyrimidines and sugars present in the nucleotides of DNA and RNA.

The polymer is helical in shape, with the base moieties protruding into the centre of the helix. Two single helices of DNA interact to form a duplex strand – the double helix (Figure 2.7). The size and shape of the bases in each helix determine the ability of the two strands to interact. Because of the constraints of the helical backbone and the space within the centre of the helix, the strands will only associate if adenine and guanine on one strand lie opposite and pair with thymine and cytosine on the other. Each of the associating strands is complementary to the other, the sequence of one strand determining the sequence of the other. The base pairs, the interacting purine–pyrimidine bases on opposite strands, are non-covalently linked by hydrogen bonding (Figure 2.8).

This complementation is the basis for the replication of DNA and the transcription of RNA. One DNA strand acts as a template for the elaboration of a complementary second strand, by the polymerization of dATP, dGTP, dCTP and dTTP to form DNA or by the polymerization of the corresponding ribonucleotides ATP, GTP, CTP and UTP to form RNA. UTP is uracil triphosphate – uracil pairs opposite adenine in RNA (Figure 2.8). The replication of DNA thus requires a DNA template, a supply of the four deoxy-

ribonucleotides and an enzyme, DNA polymerase, capable of catalysing the polymerization. Similarly, the transcription of DNA to complementary RNA requires a DNA template, the four ribonucleotides ATP, GTP, CTP and UTP and an enzyme, RNA polymerase.

The double helix has further structure imposed upon it. DNA molecules are made more compact by the coiling of the double helix around protein – particularly the histone proteins. In this coiled, double-helical state, the sequence of bases inside the helix is inaccessible for transcription. The chromosomes are highly compacted structures of DNA coiled with proteins. The DNA strands in human cells are linear. Genes are lengths of linear DNA that are transcribed into RNA. Not all of

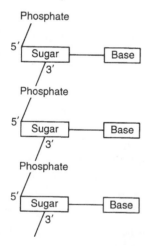

FIGURE 2.6   Nucleic acid structure. The 5' phosphate end is at the top and the 3' hydroxyl group is at the bottom of the molecule.

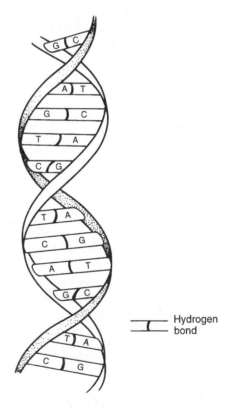

FIGURE 2.7 Structure of DNA. The two bands are the sugar–phosphate backbones of the two strands which run in opposite directions. The four nucleotide bases C, A, T and G, are linked by hydrogen bonds.

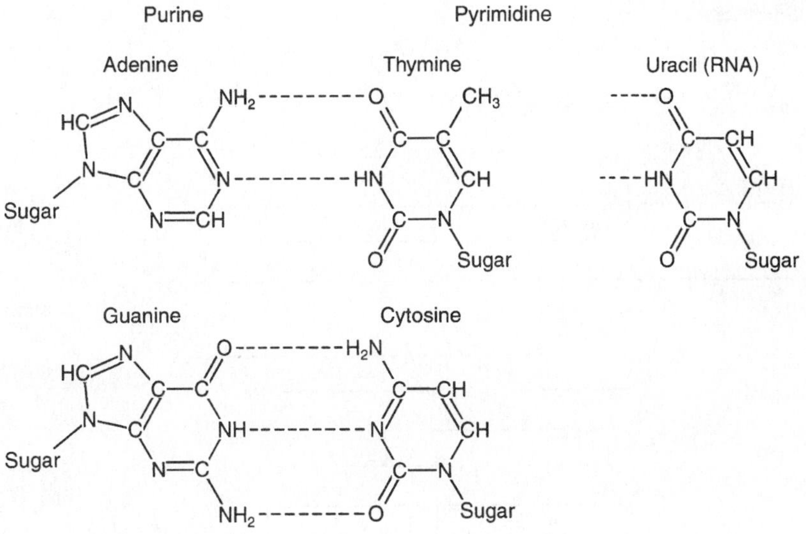

FIGURE 2.8   Base pairings in DNA. Purines always face pyrimidines as complementary A–T (adenosine and thymine) and G–C (guanine and cytosine) base pairings.

the DNA in the chromosomes forms genes. Much of the DNA makes up sequences between genes and has no known function. Other sequences, although they are not transcribed, act as regulatory elements by facilitating the binding of functional proteins to the DNA strand (promoters and enhancers). Even within the genes, only some of the DNA supplies sequence information for the synthesis of protein, the expressed regions or exons. Intervening regions between exons (introns) are transcribed into RNA initially, but are removed (spliced out) before the finished mRNA leaves the nucleus. Gene transcription requires unwinding of the coil and separation of the strands of duplex DNA in the region of the gene to be transcribed.

Gene transcription, with its requirements for conformational change in the DNA strand, specific DNA-dependent RNA polymerase and DNA regulatory elements is necessarily a complex process. The functions of the cell depend on the proteins it contains. Cell growth and differentiation, as well as metabolism, depend on the synthesis of certain proteins at certain times in the life cycle of the cell, in a concerted manner. Disordered gene transcription, or the synthesis of non-functional

proteins because of DNA mutations may result in disaster for the organism. Figure 2.9 summarizes the control of gene transcription.

2.3.4 Protein synthesis and distribution

After leaving the nucleus, mRNA enters the cytoplasm and becomes available to act as a template for the synthesis of protein, i.e. the code of bases in RNA is translated into a colinear sequence of structural units of protein – the amino acids. In order to act as template for protein synthesis, the mRNA must first bind to specific sites on the ribosome (Figure 2.10).

The **genetic code** is well characterized (Figure 2.11). Three consecutive bases (a codon) in the RNA molecule encode the addition of a specific amino acid to the elongating polypeptide chain, depending on their sequence. This specificity occurs because of the recognition of each codon by a separate class of RNA molecules called transfer RNA (tRNA). There are specific tRNAs for each amino acid. They function as bivalent adaptors allowing only one specific amino acid to bind to an attachment site at one end of the molecule, and recognizing the codon for the amino acid by means of an 'anticodon' triplet of bases at the

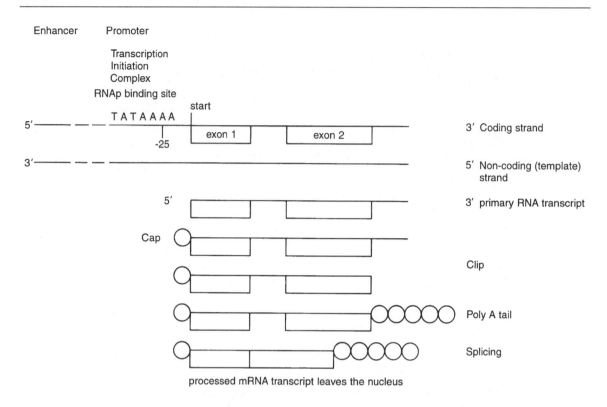

FIGURE 2.9 The transcription–initiation complex of DNA-dependent RNA polymerase (RNAp) and other transcription factors assembles and binds to the coding DNA strand. The region of the binding site for this complex is called the promoter region. There are a number of conserved sequences in the promoter region, among them the 'TATA box', centred at 25 bases upstream from the transcription start site. Other regulatory sequences, 'enhancers', may be found at sites distant (and either upstream or downstream) from the start site. The primary RNA transcript is modified before leaving the nucleus. A nucleotide cap is added at the 5' end. The sequences downstream of the last exon are clipped off and a poly-adenosine tail is added. Finally the sequences in between exons are spliced out.

other end. The process of protein biosynthesis is summarized in Figure 2.12.

Protein sorting or trafficking

The process of distribution of proteins synthesized in the cell is known as protein sorting or trafficking. All cellular proteins are synthesized on rough endoplasmic reticulum. Some are then secreted from the cell, others are distributed selectively to various organelles, and yet others remain within the cytosol. It is clear that there must be mechanisms regulating the distribution of these proteins since their disposal is specific.

The developing understanding of the processes of protein secretion and receptor-mediated endocytosis is providing insight into the mechanisms of selective protein distribution within the cell.

2.3.5 Human genetic variation

The human genome contains long segments of non-coding, non-regulatory DNA of unknown function and perhaps 50 000 genes. The average human gene is approximately 10 000 bases long. In addition to the genes, the regions of promoter and enhancer sequences are of considerable importance. When one considers the vast length

Elongating polypeptide chain

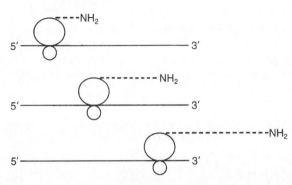

FIGURE 2.10 The 40S (smaller) ribosomal subunit binds to the 5' nucleotide cap of the mRNA. The ribosome then moves along the mRNA molecule scanning for a start codon (sequence AUG, coding for methionine). Several functional proteins (initiation factors and elongation factors) regulate the translocation of the ribosome along the mRNA molecule and the addition of amino acids (carried by transfer RNA molecules) to the polypeptide chain.

of coding and regulatory DNA within the genome, the number of proteins coded for, and the variety of alterations which can occur (of which point mutation is but one example), it becomes clear that there is enormous scope for genetic variation between individuals.

The function of each cell protein, including, of course, the enzymes, is determined by its amino acid sequence and enzymatic modification of some of the amino acid residues occurring after translation. In turn, the sequence of each protein is determined ultimately by the sequence of purine and pyrimidine bases in its gene. The activity and amount of all cell proteins are thus controlled principally by regulation of gene transcription. Although homology of certain DNA sequences is sufficient to assign a species to organisms, within any species there are variations. Nowhere is this more evident to us than when we inspect each other. Clearly, differences in our environment account for some of the differences in the appearance and behaviour of individuals. Many differences, however, have a genetic basis. This point can be illustrated by consideration of certain proteins.

Selective protein distribution

Proteins destined for secretion are synthesized with a 'leader sequence' at the N-terminus (initially translated end). This leader sequence directs the protein to be translocated to the interior of the endoplasmic reticulum (ER), and is later cleaved from the protein before secretion. Within the ER secretory proteins become glycosylated. Vesicles containing the entrapped glycoprotein bud off the ER, move through the cytosplasm, become coated with other specific cytosolic proteins and fuse with the Golgi membranes. Here, further modification takes place and then further vesicles containing the secretory proteins bud off from the other side of the Golgi stack, move through the cytoplasm, fuse with the plasma membrane of the cell and release their contents to the exterior. Similarly, substances endocytosed at the plasma membrane are entrapped by vesicles which fuse with endosomes (another type of vesicle which buds from the Golgi). Separation of the internalized ligand and the cell surface receptor takes place in the endosome and fusion with lysosomes may occur later.

Endocytotic vesicles budding from the Golgi carry secretory proteins for initial cell storage and later regulated secretion and in doing so become coated in a protein called clathrin. In contrast, vesicles budding from the Golgi carrying secreted proteins become coated in coatomer proteins, but not clathrin. Clathrin-coated endocytotic vesicles also contain a protein complex called AP-2 in the coat, whereas clathrin-coated vesicles formed at the Golgi contain a distinct complex, AP-1. The exclusive association of AP-2 with endocytotic vesicles is probably determined by its interaction with the cytoplasmic tails of receptor proteins embedded in the plasma membrane of the cell. It is conceivable that other transient coat constituents of vesicles and the sequences of proteins contained within them may play a part in determining their cellular targets.

Haemoglobin is the oxygen-carrying protein of the blood. Transport of oxygen to tissue cells is perhaps the most fundamental requirement for metabolism. The transport of water and fuel (food) is subsidiary only in the sense that cell death occurs more rapidly from deficiency of oxygen than from deficiency of water or fuel. Haemoglobin is an abundant protein and is readily accessible for study. It is not surprising therefore that haemoglobin was among the first proteins to be studied and that it is one of the best characterized human proteins.

The example of haemoglobin illustrates how a single point mutation within an exon (the segment of a gene represented in the RNA product) can have significant effects on the function of an expressed protein.

Similar mutations occur in the genes for other proteins. Many severe diseases are caused by alterations or premature termination of the amino acid sequence of enzymes in metabolic pathways which lead to loss of catalytic function, e.g. phenylketonuria, galactosaemia and the 'storage diseases'.

Thus, genetically based alterations in the synthesis and sequence of one protein may give rise to variably severe consequences. It is highly likely that such mechanisms underly the differences of handling of some nutrients between individuals.

The uniqueness of the DNA sequences of individuals can, in fact, be demonstrated using restriction fragment length polymorphism or 'genetic finger-printing', a technique which has found well-known forensic application.

UUU UUC } Phe	UCU UCC UCA UCG } Ser	UAU UAC } Tyr	UGU UGC } Cys
UUA UUG } Leu		UAA UAG } TERM	UGA } TERM UGG } Trp
CUU CUC CUA CUG } Leu	CCU CCC CCA CCG } Pro	CAU CAC } His CAA CAG } Gln	CGU CGC CGA CGG } Arg
AUU AUC AUA } Ile AUG Met	ACU ACC ACA ACG } Thr	AAU AAC } Asn AAA AAG } Lys	AGU AGC } Ser AGA AGG } Arg
GUU GUC GUA GUG } Val	GCU GCC GCA GCG } Ala	GAU GAC } Asp GAA GAG } Glu	GGU GGC GGA GGG } Gly

All the triplet codons have meaning: 61 of the codons represent amino acids. All the amino acids except tryptophan and methionine have more than one codon. Three codons cause termination (TERM). The order of bases in a codon is written in the same way from 5' to 3'.

Ala	Alanine	Gln	Glutamine	Leu	Leucine	Ser	Serine
Arg	Arginine	Glu	Glutamic	Lys	Lysine	Thr	Threonine
Asn	Asparagine	Gly	Glycine	Met	Methionine	Trp	Tryptophan
Asp	Aspartic	His	Histidine	Phe	Phenylalanine	Tyr	Tyrosine
Cys	Cysteine	Ile	Isoleucine	Pro	Proline	Val	Valine

FIGURE 2.11 The genetic code. The codons are shown in the messenger RNA format.

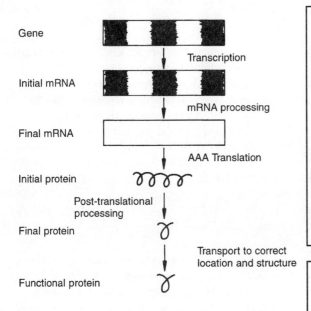

Gene

Transcription

Initial mRNA

mRNA processing

Final mRNA

AAA Translation

Initial protein

Post-translational processing

Final protein

Transport to correct location and structure

Functional protein

FIGURE 2.12 An overall picture of protein biosynthesis, from gene to functional protein. AAA is amino acid sequence.

Haemoglobin

Adult human haemoglobin is a tetramer. It comprises two alpha- and two beta-chains. Each of the four polypeptide chains has an attachment site for haem, a ring molecule which binds oxygen. Important changes in the shape of the whole protein occur when oxygen and other small molecules bind to it. Sickle cell anaemia, a disease characterized by the abnormal shape of red blood cells in situations of low oxygen concentration, is caused by the alteration of just one amino acid (position 6) of the 146 in the β-chain. This change occurs because of a single base alteration in one codon of the beta-chain gene. This is not the only mutation in the haemoglobin gene. Several hundred mutant haemoglobins have been identified. The consequences of each mutation depend upon the effect on the functions of the globin chains (particularly oxygen binding and dissociation) and their interaction with each other in the tetramer.

The enzyme UDP-glucuronyl transferase catalyses the addition of glucuronic acid to bilirubin. Hereditary absence of this enzyme causes a severe disease with jaundice and kernicterus, which is usually fatal in infancy – Crigler–Najjar syndrome type 1. Crigler–Najjar syndrome type 2 is associated with severe deficiency of the same enzyme activity, but the jaundice is less severe and kernicterus can be prevented. Gilbert's syndrome is also associated with deficiency of UDP-glucuronyl transferase, but it is an entirely benign, very common condition usually diagnosed incidentally.

Extracting DNA from cells

It is simple to extract and purify the DNA from human leukocytes. Blood is collected, and the leukocytes are centrifuged into a band between the red blood cells and the plasma. Detergents are applied to disrupt external and internal cell membranes and proteins – including those bound to DNA – can be removed by precipitation with solvents such as phenol. DNA can be precipitated by the addition of ethanol, and then re-suspended in buffer. The DNA can be digested into fragments by the application of specific restriction endonucleases. These enzymes, mainly derived from bacteria, cleave DNA at particular sequences of bases. The fact that the size of the resulting fragments is variable from one individual to another implies that the cleavage sites are differently arranged between individuals, and thus that their DNA sequences are not identical.

2.3.6 Enzymes and isoenzymes

An ability to adapt to a changing environment is of paramount importance to all organisms. One such environmental change is the availability and type of nutrients and their processing by the metabolic system. Such an adaptation will depend upon a responsive regulatory system, which includes enzymatic activity. It is not clear how an individual's genetic make-up may influence the body in relation to nutrient intake. The array of enzymes provided by the genetic configuration of the individual will be individual to that person. Hence, the metabolic pathways in normal, abundant and deficient dietary states will be dictated by the enzyme types, cellular distribution and activity. Such differences in enzyme activity and response result from the isoenzymes or enzyme separation in the cell or body. Although an individual metabolic pathway must consist of a sequence of enzymatically catalysed reactions, there will be different rates of activity, according to the body's enzymes or isoenzyme profile. An isoenzyme – sometimes called an isozyme – is a member of a group of enzymes which are structurally similar, and which catalyse the same reaction, but with differing rates of activity. These isoenzymes may differ by only an amino acid or by different aggregation of subunits of polypeptides.

Inter-individual differences in enzyme profile

An example is the hot flushing following ethanol ingestion experienced by Mongol races (see Chapter 6). It is due to an enzymatic difference between Caucasians and Mongol races in the type and activity of the enzyme glyceraldehyde dehydrogenase. A deficiency in the enzyme's activity results in accumulation of blood glyceraldehyde, causing flushing.

Another enzymatic difference between individuals is the slow and fast acetylation of certain drugs. Acetylation of fat-soluble chemicals is a liver enzymatic activity which facilitates the biliary excretion of water-insoluble xenobiotics and other extraneous compounds. Caffeine is eliminated in the urine after hepatic acetylation and hence becomes more soluble in water. Susceptibility to wakefulness after drinking coffee may be dictated by slow acetylator activity.

Before the discovery of isoenzymes, differences in metabolic activity between tissues and to a lesser extent between individuals were accounted for by variations in the amounts of enzyme and substrate available, cell permeability, compartmentalization and hormone effects. Some of these are important, but equally important are the qualitative aspects of tissue enzymology.

Enzymes are, fundamentally, **catalysts** of biochemical reactions. Their activity can be regulated by three mechanisms.

1. Enzyme molecule synthesis and degradation. In general, enzyme synthesis is constant, i.e. it is a zero-order reaction. Degradation is a first-order reaction, i.e. the degradation rate is a percentage of the available enzyme pool. When synthesis and degradation rates are equal, the steady-state exists. Changes in steady-state result in changes in enzyme activity and are dependent upon the half-life of the enzyme. Such change is slow and takes place over 10 minutes to several days.

2. Conversion from an inactive to an active form. This is a rapid mechanism, e.g. by phosphorylation or dephosphorylation. Glycogen synthetase and pyruvic dehydrogenase are active in the dephosphorylated form, whereas phosphorylase, an enzyme involved in glycogen breakdown, is active only when phosphorylated. A single enzyme may be responsible for the initiation of a major metabolic change (see 'One enzyme initiates a major metabolic change'). Many enzymes, especially proteases, are synthesized as inactive 'pro-' forms, known as zymogens. Trypsinogen, chymotrypsinogen and pepsinogen are examples. These are converted, in the appropriate circumstances, to the active form. This is achieved by proteolytic cleavage at a peptide bond. This reaction may be very specific. Trypsinogen is cleaved by enterokinase, a small intestinal brush border enzyme. Trypsin in turn activates the other pancreatic zymogens. Similar systems occur in blood coagulation, fibrinolysis, hormone action and the complement system.

One enzyme initiates a major metabolic change

For example, a hormonal trigger for glycogen degradation activates adenyl cyclase at the cellular membrane. Cyclic AMP is synthesized from ATP and stimulates kinase which catalyses the activation of phosphorylase kinase. Phosphorylase kinase catalyses the phosphorylation of phosphorylase b to the active a form. The active enzyme then acts on glycogen to form glucose-1-phosphate which then leads to the formation of glucose.

3. **Changes in concentration of metabolic intermediates.** Changes in concentrations of substrates, cofactors, activators and inhibitors provide the fine and immediate control of enzyme activity. Enzyme activity is related to substrate and cofactor concentration, according to the Michaelis–Menten equation.

$$V = \frac{V_m \times [S]}{K_m + [S]}$$

where V is the velocity of the reaction and V_m is the maximal velocity, K_m is the Michaelis constant, i.e. the concentration of substrate at which the velocity is half V_m, and $[S]$ is the concentration of substrate or cofactor. The Michaelis constant indicates the physiological concentration at which the enzyme will function. This identifies the possibility of catalytic effectiveness at a tissue substrate concentration.

An alternative enzymatic relationship between velocity and substrate concentration is a sigmoidal curve. This occurs as a result of co-operative binding of substrate to some enzymes of multiple sub-units. The first molecule attaches to a subunit and causes a change in the shape of the subunit, which in turn facilitates the binding of substrate to a second subunit, and so on. The best-known example of this allosteric interaction is the binding of oxygen by the subunits of haemoglobin. Activation and inhibition of enzyme activity are important. Activation may originate in the cell or be mediated from an extraneous source which activates a receptor. An inhibitor may compete with a substrate for the binding site. Alternatively, the inhibitor or activator may be attached at another site and elicit allosteric interactions either to reduce or augment enzyme activity. The effector may also alter the rate of release of product from the enzyme, thereby altering K_m of the reaction.

Table 2.1 highlights metabolically significant variables for which isoenzymes have been implicated.

Isoenzyme and metabolic reversibility

Compartmentalization of pathways is an essential component of metabolism. Isoenzymes are separated into different cellular compartments. Metabolism is a series of discrete unidirectional chemical reactions catalysed by polyisoenzyme complexes. The same reaction may occur in different directions in different compartments of a cell,

TABLE 2.2 Metabolic variables and the isoenzymes implicated

Variable	Isoenzyme
K_m	Hexokinase
	Pyruvate kinase
	Glutaminase
	Creatine kinase
Substrate and cofactor	Aldolase
	Alcohol dehydrogenase
	Isocitrate dehydrogenase
Allosteric properties	Hexokinase
	Pyruvate kinase
	Aspartate kinase
	Glutaminase
	Fructose bisphosphatase
Subcellular localization	Isocitrate dehydrogenase
	Adenylate kinase
Dietary and hormonal control	Hexokinase
	Tyrosine aminotransferase
	Pyruvate kinase
	Arginase

or in two different cells within the same organism. Each will be catalysed by a different isoenzyme, with different K_m values and at least one will require the input of energy from a different reaction, or the efficient removal of the products of the thermodynamically unfavourable reaction.

Isoenzyme compositions in tissues

The liver and muscle have different isoenzyme compositions. There are distinct muscle-type phosphofructokinase, aldolase, enolase and pyruvate kinase enzymes which are not present in liver and vice versa. Glycogen phosphorylase is an other example. Glucokinase is present in the liver and not present in the muscle. These differences reflect the different metabolic requirements of the tissues.

Of the 10 enzymes involved in the sequential metabolism of glucose to pyruvate nine have isoenzymes. Enzyme multiplicity may arise as a result of various factors.

Genetic factors

Multiple alleles at a single genetic locus The heterozygous individual with two different allelic variants (one on the maternally derived chromosome and one on the paternally derived chromosome) will produce two different types of enzyme subunits. If the enzyme is composed of multiple subunits, an individual heterozygous for the genes of some or all of the subunits will be capable of assembling a greater variety of multimers.

Multiple genetic loci The organism may produce one protein with a given enzyme function in one tissue, and a different protein which catalyses the same reaction in a different tissue. Gene expression varies from tissue to tissue and at varying times in the overall development, from foetus to adult and even with ageing in the adult. Multiple gene loci produce differences in isoenzyme profile.

Secondary or post-translational alterations in isoenzymes

Enzyme subunits can be modified to produce a range of composite enzymes from the same gene complex. Only part of the enzyme subunit may be involved.

Aldolase is encoded at three genetic loci. In muscle, aldolase A has two subunits, A alpha and A beta. The transition from A alpha to A beta is by slow deamination of an asparagine residue near the carboxyl terminus. The post-translational process may be tissue-specific, creating differences in tissue isoenzymes. For example, pyruvate kinase is a tetrameric enzyme. Its activity is inhibited when the enzyme is phosphorylated. The predominant isoform in the liver is designated L as it is susceptible to phosphorylation. Two other isoforms are less susceptible to phosphorylation. In this way hormone action inhibits consumption of glucose by the liver (by phosphorylation of the L isoenzyme) when the blood glucose level is low and the substrate is more urgently required by other tissues such as brain and muscle.

Apparent multiplicity

Artefacts or apparent isoenzymes from the same enzyme or proenzyme can be created under differing conditions of extraction and storage conditions. Only permanent forms are considered to be true isoenzymes.

Summary

1. Metabolism is in part dependent on the structure of the cell. Structure and functions are distributed between cells and chemicals move between compartments by specific transport

mechanisms. The cell compartments include the nucleus, cytosol, mitochondria, endoplasmic reticulum, Golgi apparatus, lysosome and peroxisome. The nucleus contains the genes for the synthesis of cellular proteins. Cytosol is an aqueous phase containing many of the enzymes catalysing metabolic reactions. The mitochondria include the enzymes which function to transport oxidatively metabolized nutrients. The endoplasmic reticulum is involved in protein synthesis.

2. The strands of DNA within the nucleus are polymers of deoxyribonucleotides; RNA polymers of ribonucleotides. Each unit consists of deoxyribose covalently linked to bases, adenine and guanine (purines) and cytosine and thymine (DNA); uracil (RNA) (pyrimidines). The function of the DNA and RNA is dependent upon the interacting purine and pyrimidine bases. The reading of

the code of the DNA and RNA is in triplets of any three of these bases. The order of any three of these bases determines the amino acid sequence of proteins produced by the DNA molecule.

3. Not all of the DNA in the chromosome forms genes; much of it makes up sequences between genes and has, as yet, no known function. Other sequences, while not transcribed to proteins, act as regulatory elements by facilitating the binding of functional proteins to the DNA strand.

4. The regions which are to be transcribed into proteins are initially transcribed into mRNA. mRNA from the nucleus enters the cytoplasm and becomes available to act as a template for the synthesis of protein, i.e. the triplets of bases in RNA are translated into a colinear sequence of structural units of amino acids in protein.

5. Proteins synthesized in the cell are subsequently sorted or trafficked to other parts of the cell or other organs.

6. The human genome contains long segments of non-coding, non-regulatory DNA of unknown function and some 50 000 genes. The average human gene is approximately 10 000 bases long. The activity and amount of all cell proteins are controlled principally by the regulation of gene transcription.

7. Isoenzymes (isozymes) are proteins which are identical in all respects of the prime function but differ in the efficiency of the function by virtue of small but important amino acid variations from the other isoenzymes in that family. The range of enzymes provided by the genetic configuration is individual to each person. Hence, the metabolic pathways in normal, abundant and deficient dietary states will be dictated by the enzyme amounts and activity, in turn, dependent on the isoenzymes or enzyme separation in the cell or body.

8. Enzyme activity may be regulated by three mechanisms: enzyme synthesis and degradation; conversion from an inactive to an active form; and changes in concentration of metabolic intermediates. The activity of an enzyme is described by Michaelis–Menten equations.

Further reading

Freedland, R.A. and Briggs, S. (1980) *A Biochemical Approach to Nutrition*. Chapman & Hall, London.

Lewin, B. (1990) *Genes IV*, Oxford University Press, Oxford.

Rider, C.C. and Taylor, C.B. (1980) *Isoenzymes*, Chapman & Hall, London.

Rosenthal, N. (1994) Regulation of gene expression. *New England Journal of Medicine*, **331**, 931–3.

Stryer, L. (1988) *Biochemistry*, 3rd edn, W.H. Freeman & Co., New York.

Ureta, T. (1978) The role of isozymes in metabolism: a model of metabolic pathways as the basis for the biological role of isozymes. *Current Topics in Cellular Regulation*, **13**, 233–58.

Zubay, G. (1993) *Biochemistry*, W.C. Brown, Iowa, USA.

2.4 Genome methodology

- Genetic loci may be identified by tracing inherited family characteristics or propensity to certain diseases.
- Specific probes allow the identification of particular gene sequences which facilitate further studies of chromosomes and gene sequences.
- Studies of chromosomes and gene sequences use techniques including polymerase chain reactions (PCR).

2.4.1 Introduction

Isolating individual genes from the genome

Genes are small and occupy only a small part of a genome. A typical mammalian genome is 10^9 base pairs. A gene of 10 000 bp is only 0.0001% of the total nuclear DNA.

Specific probes are required to identify particular gene sequences. Usually an mRNA which represents a particular protein is used as the probe. A highly labelled radioactive probe of RNA or DNA may be used whose hybridization with a gene is assayed by autoradiography. Another technique only requires knowledge of a small sequence of the protein and in this method short oligonucleotides are synthesized which correspond to this protein. A variety of oligonucleotides can be synthesized corresponding to possible alternative codons especially of the third base. The triplet code can be used to trace the oligonucleotide sequences. Using this oligonucleotide synthetic technique a single strand of DNA or genome DNA can be produced.

2.4.2 Gene mapping

To isolate genes where no protein product is known it is necessary to map the identified genetic locus. Physical methods map the loci in cytogenetic terms. The nomenclature of chromosome regions is based on the bands and sub-bands of suitably stained chromosomes demonstrated by cytogeneticists. For example, Xp21 means sub-band 1 of band 2 of the short arm (p = short arm; q = long arm) for the X chromosome. Linkage methods give genetic distances in centi-Morgans (cM)

Physical methods of gene mapping

In situ *hybridization*

This is a direct method wherein a cloned DNA fragment is hybridized directly to a spread of chromosomes in the metaphase stage of cell division. This phase of chromosome division is accumulated within a cell system during growth by a mitotic inhibitor such as colchicine.

In situ *hybridization*

The accumulated cells are fixed and spread on a slide. The chromosomes are treated with trypsin, stained with Giemsa and the resulting dark and light bands of individual chromosomes can be identified.

Chromosomes are denatured on a slide to leave DNA single-stranded but the chromosome morphology remains intact. The slide is then exposed to a labelled single-stranded DNA probe which hybridizes to any matching sequence on the denatured chromosome. The results are analysed by examining a large number of cells and logging the chromosomal location of each silver grain in the emulsion. A histogram of grain count plotted against chromosomal position will show a strong peak at the location of the matching sequence.

Somatic cell hybrids

This is a widely used technique. If mouse and human cells are fused by treatment with polyethylene glycol the resulting mixtures are unstable and tend to shed the human chromosomes in a more or less random way. Eventually stable cell lines containing small sets of mouse chromosomes plus a few human chromosomes remain.

Once a collection of well-characterized hybrids has been prepared, the presence of other human gene products can be correlated with the presence of particular human chromosomes. If the original human cell contains a translocated or double-deleted chromosome then the hybrids can map a DNA fragment relative to the translocation break point.

However, most chromosomal abnormalities in both familial and sporadic neoplasia affect many genes and many systems. Sometimes a specific disease is associated with a small chromosomal abnormality. Prader–Willi syndrome and retinoblastoma are examples. Other cancers are being mapped and cloned. Chromosomal breaks as well as deletions can cause disease. When chromosomes are involved in translocations or inversions, the break point may disrupt a gene and move it into an environment where its expression is inappropriate. Many tumour types are associated with specific chromosome break points.

Gene mapping linkage analysis

Linkage studies Two genetic loci are linked if they segregate together in pedigrees more than by random chance. Loci are linked because they are next to each other on the same chromosome. Linkage analysis is the same in all species. The aim is always to identify and count recombinants in suitable crosses. In a family with an autosomal dominant disease the pedigree may have a restriction length polymorphism type. Human linkage studies rely on exceptional families and statistics. The problem in humans is that families are small and often difficult to locate and information must be combined from several families.

Analysis of linkage studies The analysis of a linkage study requires the writing of a table showing the 'lod scores' for a disease and a marker at a range of recombination values.

'Lod scores'

For any linkage study data the lod score is the ratio of the probability that the loci are linked to the probability that the data could have arisen from unlinked loci. Such scores are usually calculated for recombination fractions 0.0, 0.05, 0.10 up to 0.50 (independent assortment). A lod of zero means the assumption of linkage or no linkage is equally valid. A positive lod score favours linkage and a negative lod is evidence against linkage at the given recombination fraction. The thresholds of significance are 3 and −2. This means the odds must be 1000 : 1 in favour of linkage before evidence is accepted. A lod score of 3.0 means that the overall probability of linkage is 95%. A plot of lod score against recombination fraction takes several forms.

Multilocus linkage analysis In this analysis a set of families with an identified disease type are studied for a series of markers or from a candidate region of a chromosome. Lod scores can be calculated for each marker. Standard lod score analysis cannot combine data for more than two loci. It is important to fix the gene closest to the framework of the markers so that the closest flanking markers can be used as 'handles' in attempts to clone the gene. The modern computer can identify the unknown locus, e.g. a known disease and a fixed framework of chromosomal marker loci and calculate the overall likelihood of these being connected in an aetiological relationship. Here, instead of a lod score there is a location score.

Location score analysis is a very powerful tool and becomes more informative as the gene locations become apparent. When enough negative information accumulates then it is possible to do exclusion mapping though biological markers must be accurately identified.

Sib pair analysis This is a simplified form of linkage analysis used for mapping recessive characters. If one parent has marker alleles ab and the other cd, then there is a 1 : 4 chance that two children will have the same types. However, if two children have the same recessive conditions, they will of necessity inherit the same alleles of markers close to the gene responsible for the disease. Sib pair analysis studies affected sibs and looks for markers which are shared more frequently than 1 : 4. This is the manner in which the cystic fibrosis locus was mapped to chromosome 7.

Tracking disease genes using linked markers The prerequisite for such an exercise is a Mendelian disease with a known map location and one or two links in restriction fragmentation length polymorphism (RFLP, see below). Such analyses require family studies and are only applicable if the pedigree is suitable. It does not require any knowledge of the molecular pathology and therefore is applicable to diseases where the gene has not yet been isolated and heterogeneous mutation diseases. It requires:

- that a marker is found closely linked to the disease locus and for which the person at risk of transmitting the disease is heterozygous
- study of other members of the family in order to calculate which of the marker alleles is on the chromosome carrying the disease allele
- use of the marker to discover whether the pathological chromosome or its normal homologue was passed on to the person requiring the diagnosis.

2.4.3 Direct tests for disease genes

Gene deletions

Among many mutational changes the gene is rendered inoperative due either to a physical deletion or partial deletion. These changes are not detectable cytogenetically but are demonstrated by a failure of a probe to hybridize to the DNA.

Direct detection of point mutations

Probes may be several hundred or several thousand nucleotides long and hybridize to any sequence showing 95% or greater concordance. This does not happen if there is a small degree of mismatching. The general method is to use very short probes. This enables the singular elements within a genome to be identified.

In order to map the nucleic acid sequence at a molecular level, the DNA molecule is broken at defined points using restriction enzymes.

Ribonuclease A cleavage
Single base mutations in both cloned and genomic DNA sequences can be detected by cleavage of mismatches in RNA : DNA duplexes with ribonuclease A. A single-stranded RNA probe is synthesized from a cloned DNA fragment applicable to the mutated region. This probe is then hybridized to its complementary sequence in cloned and genomic DNA sequences or amplified by the polymerase chain reaction (PCR). The resulting RNA : DNA duplex can then be used to identify mismatch positions.

The lengths of nucleotides produced can be accurately analysed as short sequences of double-stranded DNA. Each restriction enzyme has a target in a duplex DNA of about 4–6 base pairs long. The enzyme cuts the DNA at every point at which its target sequences occur. When the DNA molecule is cut with suitable restriction enzymes to distinct fragments, these are separated by molecular weight by gel electrophoresis. Restriction mapping uses overlapping fragments created by different enzymes, each splitting the DNA at a different point. A difference in restriction maps between two individuals is called a restriction fragmentation length polymorphism (RFLP). Restriction markers identify genetic loci and hence mutations. This allows the relevant genetic loci to be placed on a gene map even if the abnormal gene or protein is not known at that time. From such basic information diagnostic tests or even isolation of the gene can be developed.

Every individual has a unique constellation of restriction sites; such a combination of specific regions is called a **haplotype**. The use of DNA restriction analysis to identify individuals has been nicknamed 'DNA finger-printing'. One way of using RFLP is the lod score. RFLP can also be used to place genes on a genetic map and is used to map genomes and in particular the human genome.

The use of yeast strains with known genetic translocation mapping has enabled chromosomal regions to be identified and mapped using hybridization techniques.

It is also possible to compare the nucleotide sequence of a gene with the amino acid sequence of a protein and determine either the amino acid sequence of the protein or the nucleotide sequence corresponding to that protein primary sequence, i.e. whether they are colinear. The restriction map of DNA will exactly match the amino acid map of the protein. There are, however, extra regions in the DNA not representing protein.

Messenger RNA always includes a nucleotide sequence that corresponds exactly with the protein product. The gene may include additional sequences that lie within the coding region interrupting the sequences that represent the protein, exons represented in the messenger RNA and introns missing from the messenger RNA.

DNA hybridization

The hydrogen bonds which hold the double helix of DNA together can be disrupted by heat or high salt concentration so that the strands are separated (denaturation). Restoring the double helix restores the original properties of the DNA. This technique can be used to isolate DNA segments. The nucleic acid sequence is identified by a recombination technique to a complementary nucleic acid sequence by a zipper-like effect. When nucleic acids from different sources but similar sequences join together, they anneal with each other. This phenomenon is called hybridization, e.g. between parts or lengths of DNA and RNA. The ability of two nucleic acid preparations to hybridize constitutes a precise test for complementary sequences since only complementary sequences can form a duplex structure.

It is now a routine procedure to obtain the DNA corresponding to any particular gene. Cloning a fragment of DNA allows indefinite amounts to be produced from even a single original molecule. A clone is a number of cells or molecules all identical to an original ancestral cell or molecule.

By genetic analysis the chromosome carrying a genetic trait is identified. The gene is tracked to a region of the chromosome by genetic characterization of individuals with grossly abnormal chromosomes. The search continues at a molecular level for the gene within the region that can be associated with the disease. Once a particularly susceptible gene is isolated then it is possible to search for a certain sequence in that gene which is associated with a particular allele in individuals with that condition. Hybrid DNA molecules are constructed by using restriction enzymes to cleave the DNA into short nucleotide sequences.

The cloning of DNA uses the ability of bacterial plasmids and phages to continue to live after additional sequences of DNA have been incorporated into their genomes. Such an insertion generates a hybrid or chimeric plasmid or phage. These chimeras can thereafter be replicated in the bacteria. Copies of the original foreign fragment can be retrieved from the continuing generations of bacteria.

Cloning of specific genes requires the identification or characterization of particular regions or sequences of the genome. A probe is required to react with the target DNA. The phage or plasmid is called a cloning vector. A probe is needed which would react with the target DNA, e.g. a known protein product with a messenger RNA coding for the protein. It may be possible to identify messenger RNA in the cytoplasm that represents a particular unknown gene product. This is important in diseases where the abnormal gene product is not known.

Another technique is called *blunt-end ligation* which relies on the ability of the T4 DNA ligase to join two blunt-ended DNA molecules together, i.e. they lack any protruding single strands. When DNA has been cleaved with restriction enzymes, cutting both strands at the same position, blunt-end ligation can be used to join the fragments directly together.

Cloning of specific genes

The cloning vector DNA is cleaved at appropriate sites. It is possible to construct hybrid molecules that can amplify the amount of material or to express a particular sequence. Critical to such a system is that the inclusion of the new sequence does not upset any essential function. A restriction enzyme is used with a single target site in a non-essential part of the DNA vector. Plasmid genomes are circular so that included DNA fits into the circle. These are perpetuated indefinitely and then isolated according to size by gel electrophoresis. Long non-circular genes can also be used. Non-essential DNA can be replaced by foreign DNA. The foreign DNA fragment is joined to a cloning vector by a reaction between the ends of the fragment and vector. Restriction enzymes make staggered cuts in the DNA to produce short complementary single-stranded sticky ends. Specific enzymes cleave each of the two strands of duplex DNA at different sites. The DNA fragments on either side fall apart and produce a complementary single-strand region. Complementary ends are annealed by base pairing. When two quite different molecules are cleaved with specific enzymes, identical sticky ends are generated which can be joined, though other areas may open up and the plasmid may rejoin.

The purpose of cloning is to amplify the inserted DNA. If the mode of expression of the foreign DNA is being studied then the foreign DNA must be inserted into the plasmid in an appropriate orientation.

In order to obtain and study a particular gene the order of events is reversed. The genome is cloned first. Clones containing a particular sequence are selected. Vectors carrying DNA from the genome itself are called genomic or chromosomal DNA clones. Cloning an entire genome, as opposed to specific fragments, is called a shotgun experiment. The genome is broken into fragments, the fragments are put into a cloning vector to generate a population of chimeric vectors. A set of such cloned fragments is used to form a genome library. As new probes are found they can be tested against the library collection of grouped fragments. An alternative with mRNA is to prepare single-strain DNAs from the entire messenger RNA populations and to produce a single-strand DNA library. These can be stored and tested when a new probe is available.

A particular genome clone can be selected from the library by colony hybridization. Bacterial colonies carrying chimeric vectors are lysed on a nitrocellulose filter. The DNA is denatured *in situ*, fixed and hybridized with a radioactively labelled probe. A genome clone needs only to carry some of the probe's sequence to react with it.

Chromosome walking

A clone may be isolated which is believed to contain the known gene or a region that is of interest. Sections of the chromosome are hybridized with clones from the reference library. It is possible to study lengths of hundreds of kb and regions of the genome by systematically moving along the chromosome. This is known as 'chromosome walking'.

Copying messenger RNA onto DNA

To isolate a particular mRNA, two cell types are needed, one that expresses the RNA and one that does not. This technique uses subtractive hybridization. The mRNA of the target cell line is used as substrate to prepare a set of single-strand DNA molecules corresponding to all the expressed genes. This is then hybridized with all of the mRNA of another closely related cell. The sequences that are common to both cell types are removed. After discarding all the DNA sequences that hybridize for the other mRNA, those that are left are regarded as peculiar to that cell and can be characterized This has been used to isolate clones in the T-cell receptor but not in the closely related B lymphocyte.

In order to identify a DNA sequence which represents a particular protein, the responsible mRNA is used as the starting point. Reverse transcription allows synthesis of duplex DNA from the mRNA. This is particularly easy if the mRNA carries a poly(A) tail at the 3' end.

Synthesis of duplex DNA

A primer is annealed to the poly(dA). The enzyme engages in the usual 5'–3' elongation, adding deoxynucleotides one at a time. The product of the reaction is a hybrid molecule consisting of a template RNA strand paired with a complementary DNA strand. The original mRNA is degraded by alkali which does not affect DNA. The product is a single-stranded DNA that is complementary to the mRNA. The hairpin at the 3' end of the single DNA provides a primer for the next step, *E. coli* DNA polymerase I converts this single-stranded DNA into a DNA duplex.

Digested strands of DNA are separated on gel electrophoresis. DNA is denatured to give single-stranded fragments which are transferred from agarose gel to a nitrocellulose filter where the fragments are immobilized. This system is known as **Southern blotting**. The DNA fragments are separated and those corresponding to a particular probe are isolated directly from a digest of the clone DNA. DNA immobilized on nitrocellulose can be hybridized *in situ* with a radioactive probe. Only those fragments complementary to a particular probe will hybridize and can be identified by autoradiography. **Northern blotting** is used for RNA, and **Western blotting** for proteins.

An alternative procedure is dot blotting wherein cloned DNA fragments are spotted next to one another on a filter. The filter is hybridized with a solution containing the radiolabelled probes. The radioactive intensity of the dot corresponds to the extent to which the RNA is represented in the clone.

Polymerase chain reaction (PCR)

This reaction requires a knowledge of the sequences on either side of the target region and allows such a region between two defined sites to be amplified. PCR allows sequences of interest to be selectively amplified against a background of an excess of irrelevant DNA. A target sequence of up to 1 kb can be amplified 10^5 to 10^6-fold. A prerequisite is a unique flanking sequence so that specific oligonucleotide primers can be used.

At the moment direct detection is limited to well-understood diseases which are relatively homogeneous at the molecular level. A preparation of DNA, often just an extract of the whole genome, is denatured. The single-stranded preparation is annealed with two short primer sequences that are complementary to sites on the opposite strands on either side of the target region. DNA polymerase is used to synthesize a single strand from the 3'-OH end of each primer. The entire cycle can be repeated by denaturing the preparation and starting again. The number of copies of the target sequence grows exponentially, doubling with each cycle. More recently production of DNA polymerase from a thermophilic bacterium has meant that the same enzyme remains active through the heating steps required for the denaturation–renaturation cycles, allowing a sequence to be amplified up to 4×10^6 times in 25 cycles. The length of the target sequence is determined by the distance between the two primer sites up to 2 kb. If a replication event causes an error then this will also will be amplified. This technique enables the identification of individual alleles in a genome and is a very powerful tool in modern molecular biology.

Polymerase chain reaction

Heat-stable DNA polymerase is used and the reaction is run by putting a tube containing all the ingredients through perhaps 30 temperature cycles. After gel electrophoresis, genomic target sequences can be demonstrated and a search made for the presence or absence of target sequences.

Summary

1. Specific probes are required to identify particular gene sequences; usually an mRNA which represents a particular protein is used as a probe.
2. To isolate genes where no product protein is known, the method is to map the identified genetic locus. This may be facilitated by physical methods which include *in situ* hybridization, somatic cell hybridization and gene mapping linkage analysis. Many of these are based on family studies where a set of families with an identified genetic characteristic are studied for a series of markers or for a candidate region of a chromosome.
3. Among many mutational changes, a gene may lose function due to physical deletion or partial deletion. These are demonstrated by the failure of a probe to hybridize to that particular DNA. In order to map the nucleic acid sequence at a molecular level, the DNA molecule is broken at defined points using specific restriction enzymes. It is also possible to compare the nucleotide sequence of a gene with the amino acid sequence of a protein and determine the amino acid sequence of the protein or the nucleotide sequence corresponding to that protein primary sequence. The technique of cloning a number of cells or molecules allows a genetic trait to be identified. The gene is tracked to a region of the chromosome by genetic characterization of individuals with grossly abnormal chromosomes. The cloning of specific genes requires the identification or characterization of particular regions or sequences of the genome.
4. Digested strands of DNA or RNA or protein can be separated on gel electrophoresis; this constitutes the 'blotting' method of identification of RNA and DNA or protein.
5. The polymerase chain reaction (PCR) allows a region between two defined sites in the gene to be amplified. This is a powerful tool in allowing sufficient amounts of DNA to be developed for further analytical work.

Further reading

Lewin, B. (1990) *Genes IV*, Oxford University Press, Oxford.

Trayhurn, P. and Chester, J.K. (1996) Workshop on molecular biological techniques in nutritional science. *Proceedings of the Nutrition Society*, 55, 573–618.

Chapter 3

Nutritional status

3.1 Outline

Nutritional status and the ability of an individual to respond to deficiency and excess of available food are determined by genetic inheritance and isoenzyme constitution. The nutrition of the mother prior to and during pregnancy and that of the individual throughout intrauterine development and from birth to the time of assessment as well as age, sex, culture, geographical and socio-economic factors may each be significant determinants.

Adequate nutrition cannot be assumed merely by the absence of clinical features of deficiency disease. Indeed some degree of malnutrition has been identified in a significant proportion of the hospital populations of affluent countries.

3.2 Optimal requirements of nutrients

- Good nutrition and hence good health requires that all nutrients are provided in adequate amounts.
- The science of nutrition identifies the amounts of each nutrient to be taken in an optimal diet for an individual with defined needs.
- Dietary guidelines have been determined for individuals with differing physiological needs, growth, pregnancy, lactation, middle-aged to elderly or ill.
- Specific nutrient requirements and recommendations are discussed.

3.2.1 Introduction

Optimal dietary requirements must of necessity be that intake which is most likely to result in good health or to ensure that the individual is in a favoured condition to be able to achieve good health. This includes:

- optimal body function
- freedom from infection
- resistance to disease and ability to cope with diseases
- effect of treatment regimes
- longevity
- quality of life

Appropriate nutrition requires that all nutrients, carbohydrates, lipids, proteins, minerals, vitamins and water are taken in adequate amounts and in the correct proportions. This is essential for normal organ development and function, reproduction, repair of body tissues and combating stress and disease. The nutrient intake should be appropriate for sustained activity and effective physical work. Many nutrients require the presence of other nutrients if they are to fulfil their activity within the body. It therefore behoves nutritionists to make recommendations to the community.

The definition of needs is very exacting. The important question is; need for what? The needs for a growing child who is physically active and laying down tissues of a wide range of types, neural including brain, muscle, enzyme systems, liver tissue, bone and connective tissue, are in contrast to the mature exercising young adult in the physical prime of life. The pregnant or lactating woman's needs will be of a different qualitative and quantitative nature. The teenage mother who is growing as well as sustaining a growing infant has particular needs. As the person ages the body requires much more care and maintenance as all activities reduce in intensity. Needs are also altered by stress, illness and trauma, and complicated by culture, religious dictates and economy.

The fulfilment of nutritional needs is modulated by agricultural economics, particularly in developing countries. The recognition of this basic premise has been essential for the survival of every major civilization – Roman, Persian, Indian and Mayan. The amount recommended will always exceed needs dictated by the inefficiency of the system, the degree of losses by digestion, absorption and metabolic transformations that accompany utilization.

3.2.2 Dietary reference values and measurements

A number of measurements are used to compare dietary reference values for food energy and nutrients.

3.2.3 Nutritional recommendations

Recommended nutritional allowances (RDAs) reflect needs of a population, established by physiological and metabolic measurements, but give no indication of the metabolism and disposition of the chemical. **Dietary guidelines** interpret these into practical statements, for real-life nutrition. Recommended Nutrient and Dietary Allowances indicate the requirements of individual nutrients for defined populations groups, e.g. babies, toddlers,

References terms used for nutritional studies

BMR (Basal Metabolic Rate): the rate at which the body uses energy when the body is at complete rest. Values depend on age, sex and body weight. For a 65 kg man BMR is approximately 7.56 MJ/day. For a 55 kg woman, BMR is about 5.98 MJ/day.

PAL (Physical Activity Level): a multiple of BMR; the ratio of overall daily energy expenditure to BMR. The values range from 1.4 for a person with light energy expenditure in work with non-active leisure pursuits to 1.9 for a man in energy-demanding work whose leisure time pursuits are also energy-demanding.

PAR (Physical Activity Ratio): PAL and PAR are discussed on pages 73 and 88.

Terms relating to energy and nutrient intakes:

RDI: Recommended Daily Intake of nutrients for the United Kingdom, 1969.

RDA Recommended Daily Amounts of food energy and nutrients for groups of people in the United Kingdom, 1979.

EAR Estimated Average Requirement for a group of people for energy, protein, vitamins or minerals. Half will usually need more than the EAR and half less.

LRNI Lower Reference Nutrient Intake for protein, vitamins or minerals. An amount of the nutrient that is enough for only a few people in a group who have low needs.

RNI Reference Nutrient Intake for protein, vitamins or minerals. An amount of the nutrient that is enough or more than enough for about 97% in a group. If the average intake of a group is the RNI, then the risk of a deficiency in the group is extremely small. The value is equivalent to RDA or RDI.

Safe intake. The term used to indicate intake or range of intakes of a nutrient for which there is not enough information to estimate RNI, EAR or LRNI. It is an amount which is enough for almost everyone, but not so large as to cause undesirable effects.

DRV Dietary Reference Value; a term used to cover LRNI, EAR, RNI and safe intake.

pregnant and lactating women. Nutritional and dietary guidelines recommend intakes of food, milk, meat and vegetables, etc.

A recommended dietary allowance takes the view of the body as a biochemical system, a machine, and considers the needs of healthy normal individuals and populations.

3.2.4 Development of nutritional and dietary advice

Every parent endeavours to instruct children on what to eat. All recorded literature contains dietary advice, for example in the Old Testament and the Koran there are important landmarks in clean food policies. W.O. Atwater was the first director of the Office of Experimental Stations in the US Department of Agriculture. In 1894, he published a table of food composition and dietary standards for the US population. Atwater's dietary standards were intended to represent the average needs of humans for protein and total calories, fat and carbohydrate in order to provide a balance to energy utilization. Minerals and vitamins had yet to be discovered. At the beginning of this century Caroline Hunt classified food into five groups:

- milk and meat
- cereals
- vegetables and fruit
- fats and fat foods
- sugars and sugary foods

The amounts of foods were listed in familiar household units. Such information was then extended into a series of guides for the average family.

The recession of the 1930s changed the utility of such guides and cost-effective guides became necessary. The food planners recognized that cereal foods, potatoes and dry beans supply energy and nutrients in a cheap form. Hazel Stiebeling, the developer of these guides, drew attention to the importance of a balance between 'protective foods', e.g. calcium, vegetables and fruit, or 'nutrient-dense', and 'high-energy and protein foods'.

The first RDAs were defined in the USA in 1941, and listed specific recommendations for calories and protein, iron, calcium, vitamin A and D, thiamin, riboflavin, niacin and ascorbic acid. The need for health education to underline these concepts was also identified. In 1943 the 'basic seven' food guide was issued. These were foods to be eaten on a regular basis:

- green and yellow vegetables
- oranges, tomatoes and grapefruit
- potatoes and other vegetables and fruits
- milk and milk products
- meat, poultry, fish, eggs, dried peas and beans
- bread, flour and cereals
- butter and fortified margarine.

This guide develops these recommendations into the concept of exchanges. When fruit is scarce then vegetables may be an alternative. These recommendations were expanded in 1946 into a National Food Guide with recommended servings. By 1958 a simpler system had been written with a basic four groups. The recommendation was for a daily minimum of two helpings from four food groups:

- milk and milk products
- meat, fish, poultry, eggs, dry beans and nuts
- four servings of fruit and vegetables
- four servings of grain products

These were to be regarded as a basic diet, the expectation was that other foods would also be eaten.

In 1977, *Dietary Goals* for the US was written by a US Senate Select Committee. Quantitative goals were laid down for protein, carbohydrate, fat, fatty acids, cholesterol, sugars and sodium. These recommendations aroused controversy. Nevertheless, since then the public's desire for an authoritative, consistent and achievable food guide has engendered a series of reports. The most recent ones combine food recommendation with graphics which illustrate dietary intake in a food guide pyramid, the recommended guide to daily food choices (with servings rigidly defined by the US Food and Drugs Administration) being:

- LEVEL 1: bread, cereal, rice and pasta 6–11 servings
- LEVEL 2: vegetable group 3–5 servings; fruit group 2–4 servings
- LEVEL 3: milk, yogurt and cheese group 2–3 servings, meat, poultry, fish, dry beans, eggs and nut group 2–3 servings
- LEVEL 4: fats, oils and sugars to be used sparingly

The diet should be reduced in salt, fat and saturated fat and alcohol intake should be moderate. The guide recommended a substantial intake of vegetables, fruit and grain products. RDAs are issued separately from these recommendations.

The estimation of nutrient requirements has proved to be very difficult. The British Panel on Dietary Reference Values of the Committee on Medical Aspects of Food Policy used five criteria to define intakes (see page 72). Alternative approaches are to define the intake of a nutrient which prevents impairment of well-being or allows storage to take place.

The criteria used in establishing standards of RDA and RDI are reflected in the consequent recommendations. These criteria may be based on traditional descriptions or perceived consequences of long term shortcomings in intake. Should avoidance of disease or total body saturation be the criterion? A recommended intake for each known nutrient is desirable for each group in the population.

Nations and Communities have their own tables of reference of daily intakes or recommended daily allowances (RDIs and RDAs). In the UK these tables are based on a mixed omnivorous diet and are prepared by committees of experts. While RDI is 'recommended daily intake' and RDA 'recommended daily allowance', in practice they are equivalent. The RDIs are used in a variety of ways, including the assessment of dietary surveys.

RDIs (RDAs) are standards by which nutrients in the food eaten by different sections of the community can be assessed. Every year the British National Food Survey examines the diets of differ-

ent populations by region. In this way shortcomings in a diet can be identified. RDIs tend to be generous in their recommendations, being appropriate for 97% of the population. When planning diets for groups in institutions, e.g. old peoples' homes, prisons and armed forces, it is important that the overall diet meets the RDI. In planning food supplies International Agencies use RDIs to plan long-term aid for developing regions and in calculating food supplies for famine relief.

3.2.5 Dietary standards

Nutritional labelling

Manufactured, canned and packed foods should show the amount of important nutrients printed on the label, in absolute amounts and as a proportion of the reference intake. This is an area of increasing importance in nutrition and is central to the transfer of knowledge to the consumer. The US Food and Drug Administration (FDA) has issued regulations on such labelling. All terms are rigidly defined, and include serving size 30 g for cookies (biscuits), low fat < 3 g per serving. The European Union recommends a labelling policy. The information is given as nutrient per 100 g.

Raw food, meals served in restaurants, cafeterias and aeroplanes and small businesses are as yet exempt.

Nutrient density

This is the amount of nutrient expressed in relation to the energy content of that food and the RDI. The nutrient value of any food may be expressed with its content of nutrients and of energy, each related to RDI. This is the nutrient density and for any one nutrient of food is:

$$\frac{\text{Nutrient (nutrient in 100 g)/(energy in 100 g)}}{\text{RDI of nutrient/(RDI of energy)}}$$

Such figures indicate whether or not that particular item of food is a good or a poor source of a particular nutrient. In defining RDIs the committees responsible perform four tasks:

1. They determine whether the compound is an essential nutrient for humans.
2. They assess whether deficiency of the nutrient may lead to a defined clinical deficiency disease and whether replenishing that nutrient leads to the reversal of the defined clinical deficiency condition; there are, however, different definitions of deficiency.
3. They assess the effect of biological availability, for example haem iron is more readily absorbed than iron in vegetable foods.
4. They consider separately RDIs for energy from those of nutrients. For example, the RDIs for nutrients and energy are given in separate tables in the United States. The energy requirement for an individual will vary with physical activity, leisure and stage of life. RDIs for energy therefore represent an average for the population, whereas protein, vitamins and minerals supply the needs of the majority of the population and are well below the upper 'safe' limits.

Variations and definitions of deficiency may be illustrated by attitudes to vitamin C deficiency. The United States Food and Nutrition Board requires that normal tissue be saturated with vitamin C, while in Britain the Department of Health considers that the value should be one which is compatible with being free from the ill-effects of deficiency. Consequently, the RDI for ascorbic acid is 70 mg in the USA and 40 mg in the UK. A requirement of the RDI is safety, so the estimate is that the RDI is safe for 95% of the population. RDIs however do not take allowance for stresses such as infection, injuries or other illnesses.

The recommendations take into account availability of food, e.g. protein which will be set differently for developed and developing countries.

Geographical differences in RDIs

The RDIs of 15 European countries and the WHO/FAO vary enormously. The protein recommendation ranges from 60 to 120 g, calcium 500 to 1500 mg, ascorbic acid 30 to 100 mg. It has been suggested that there might be a need for two standards of requirements.

Utilization of recommendations

The community physiological requirement or 'safe level' is useful for evaluating diets and for defining and reversing unsatisfactory low intake of one or more nutrients.

Nutrient units

Liu and Guthrie (1994) have suggested the concept of 'nutrient unit'.

$$\text{Nutrient unit (Nu)} = \frac{\text{amount of the nutrient}}{\text{RDI for the nutrient}} \times 100$$

The Nu for any individual excluding pregnant and lactating women would be 100. Therefore any food that contains say 4% of the RDI of a nutrient would have 4 Nu of that nutrient present. There are specific requirements for different age and sex groups, for example for iron. In many situations the Nu will vary very little from 100. The Nu for calcium in one cup of milk becomes (288 mg/ 900 mg) × 100 = 33 or 33% of the RDI (USA RDI). Therefore, an individual can readily see that they need three cups of milk to achieve their daily requirement of calcium, or that one cup will provide them with one-third of their daily requirement.

The reference daily intake or 'desirable range' is intended for teaching, for menu-planning by housewives, dieticians or caterers and perhaps for agricultural economic planning. The RDAs reflect the current scientific judgement on the amount of each nutrient that is needed for the maintenance of health.

The United Kingdom panel on DRVs uses 'dietary reference values' (DRVs). These are estimates of reference values and not recommendations for intakes. They may be used as yardsticks for the assessment of dietary surveys and food supply statistics, guidance on appropriate dietary composition and meal provision, or for food-labelling purposes. The values are closely related to the biological parameters used to derive the figures. The requirement for a nutrient varies from one person to another, and may alter with the composition and nature of the diet as a whole. There is no absolute requirement for fat, sugars or starches, though there are for essential fatty acids. Estimates of requirements may be made from:

- the intake of a nutrient to maintain a given circulating concentration or degree of enzyme saturation or tissue concentration
- the intake of a nutrient by individuals or by groups which is associated with the absence of any signs of deficiency disease
- the intake of a nutrient needed to maintain balance, noting that the period over which such balance needs to be measured differs for different nutrients and between individuals
- the intakes of a nutrient needed to cure clinical signs of deficiency
- the intake of a nutrient associated with an appropriate biological mark of nutritional adequacy

There is no single criterion to define the requirement for all nutrients. The notional mean requirement is the estimated average requirement (EAR) (Figure 3.1). The reference nutritional intake (RNI) is two notional standard deviations above the EAR, while the lower reference nutrient intake (LRNI) is two notional standard deviations below the mean. At very high levels of consumption there may be evidence of undesirable effects. However, it is difficult to establish DRV with great confidence. There are considerable shortcomings in many of the estimations. There are differences for age and sex and it is appropriate to establish DRVs for all such groups of the population. DRVs during pregnancy will vary with the age and activity of the mother. For most nutrients no increment for pregnancy is necessary in the mature Western mother; however, the growing teenage mother or marginally sustained mother does have extra and very real needs to increase dietary intake. DRVs have been set only for infants fed with artificial feed as it is assumed that human breast milk will automatically provide all the baby's requirements.

Dietary energy can also be expressed as estimated average requirements (EAR) for different age and gender groups. EAR for energy are multiples of the basal metabolic rate, i.e. the BMR multiplied by a factor determined by physical activity level (PAL). The physical activity level is the ratio of overall daily energy expenditure to BMR. A PAL of 1.9 would reflect a very active

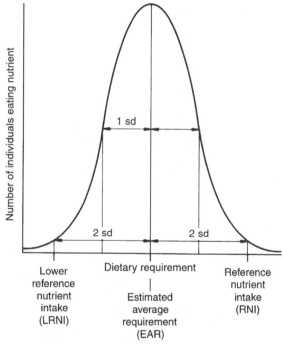

1 sd = 1 standard deviation;
68.3% of observations lie within 1 sd
2 sd = 2 standard deviations;
95.45% of observations lie within 2 sd

FIGURE 3.1 Estimated average requirements (EAR); lower reference nutrient intake (LRNI) and reference nutritional intake (RNI). LRNI and RNI are 2 standard deviations above and below the EAR.

work pattern. The physical activity level of 1.4 is a minimum of activity at work and leisure. A physical activity level of 1.5 should be used for individuals aged 60 years. Old people have low levels of energy expenditure and intake with a consequent risk of nutritional deficiency.

The 1985 report of the FAO/WHO/UNO Expert Consultation

The energy requirement of an individual is the level of energy uptake for food which will balance energy expenditure when the individual has a body size and composition and level of physical activity consistent with long-term good health, and that will allow for the maintenance of economically necessary and socially desirable physical activity. In women who are pregnant or lactating, the energy requirement includes the energy needs associated with the deposition of tissue or the secretion of milk at rates consistent with good health. The level of energy intake should recognize:

- the basal metabolic rate
- growth in the case of the young
- physical activity which is divided into occupational and discretional activities
- the thermic effect of food

The maintenance component was taken to be 1.4 × BMR for both men and women while the other components are expressed as a function of BMR. The average daily requirements of an elderly person might be approximately 1.55–1.75 × BMR.

3.2.6 Specific nutritional requirements

Protein

Protein is relatively constant at 10–12% of energy intake and diminishes in parallel with the age-related fall in energy. This may be a factor in muscle loss with age. A safe intake of protein should not be lower than 0.75 g/kg/day (WHO/FAO/UNO, 1985). Dietary protein provides nitrogen in an organic form for the renewal of amino

---

Energy cost

The physical activity ratio (PAR) is a multiple of BMR, an estimate based on the duration and type of physical activity. It is 1.2 × BMR for relaxation and 4 × BMR for gentle walking. Total energy activity for a day (the sum of the constituent PARs) is calculated by dividing the day into periods of activity, e.g. bed (8 hours) working day (7–8 hours for 5 days) with the remainder being variable. In this variable period PAR may vary over short or prolonged periods of activity. A PAR of 2 would apply to gentle domestic activities, 3 for gentle walking or village cricket, 4 for heavy housework, golf and DIY jobs or 7 for exercise provoking breathlessness and sweating. Energy cost per hour may be expressed in millijoules (MJ) and, for a sample of activities, is: lying in bed 0.2 MJ, watching television 0.3 MJ, housework 0.7 MJ and walking 0.9 MJ.

---

acids for their various functions including proteins in cell walls, plasma proteins, muscles, enzymes and collagen. The amino acids of protein can be deaminated and may act as an energy source in their own right.

Dietary amino acid requirements

Proteins differ in their biological quality, dependent upon the amounts and proportions of essential amino acids. A protein which is rich in all of the essential amino acids would score higher on the scale of biological quality than a protein deficient in one or more essential amino acids.

It is necessary to define protein and specific amino acid needs in diets both with abundant and also deficient amounts of nutrients, including protein and specific amino acids. The protein requirement of all age groups should be based on the recommendations in the report of the FAO/WHO/UNO Expert Consultation, where the values were based on estimates and the amounts of high-quality egg or milk protein required for nitrogen (N) equilibrium as measured in nitrogen balance studies (Table 3.1). The estimated average intake of protein increases from 10.6 g/day at 4–6 months to 14.8 g/day at 4–6 six years and 22.8 g/day at 7–10 years. In the male, protein requirement increases from 33.8 g/day in the 11–14-year-old male to 42.6 g/day in the 50+ year-old male. In the female, corresponding values were 33 g/day for 11–14-year-olds to 37 g/day for the over-50s. Athletes may require more dietary protein depending on the muscle power required in the sport. It was suggested that an addition should be made of 6 g/day for pregnancy and 11 g/day for lactation during the first 6 months and then 8 g/day required after 6 months as the protein content of the breast milk falls after this. For infants and children additions were made for growth and in pregnancy and lactation additions were made to account for the growth of the foetus and to allow adequate breast milk production. There is relatively little change with age in the requirement for protein for maintenance, values falling from 120 mg (N)/kg/day at 1 year to 96 mg (N)/kg/day for adults. This assumed during growth an efficiency of dietary utilization of 70%.

In the case of the elderly the recommended nitrogen intake is the same as for younger adults, 0.75 g protein/kg/day. Daily protein intakes in the United Kingdom have tended to increase to figures of 84 g for men and 64 g for women. There has been concern that excessive intakes of protein may be associated with health risks.

TABLE 3.1 Estimated average intake of protein by age, sex, pregnancy + lactation.

	male	female
	g/day	
4–6 m	10.6	10.6
4–6 y	14.8	14.8
7–10 y	22.8	22.8
11–14 y	33.8	33
50 + y	42.6	37
pregnancy		+ 6
lactation 0–6 m		+ 11
6 m +		+ 8

Dietary allowances have been defined as operational or factorial:

- **Operational**: this defines the amount of each amino acid which has to be eaten to keep bodily functions within normal or identified limits e.g. maintenance of weight, amino acid concentrations and excretion of nitrogen in the urine. These are practical day to day nutritional issues. In human nutrition, outcome may be described rather than defined as normal, optimal, maximal and engendering well-being. The amount of individual amino acids necessary to maintain protein balance may be different from that required for maximum rates of protein turnover, dependent upon age, sex and metabolic status and special conditions. The requirements of a baby, or teenage girl who is pregnant, or an endurance athlete or an elderly person recovering from a stroke will thus be different.
- **Factorial**: these are based upon an understanding of the underlying biology of the amino acid's metabolism. The rate and amounts required for and in the metabolism of each amino acid matches the amino acid's biological importance. An ideal is when the intake of the amino acid under study supports the requirements of a range of processes. This approach assumes a knowledge of the overall metabolism of the amino acid, alone and in the context of variable needs and amount of accompanying amino acids. Net protein turnover in the body is measured during growth, in the mother and foetus during pregnancy and the loss through secretion of protein, amino acids and non-protein nitrogen in milk. Loss of nitrogen in urine is also measured. This approach defines a minimum or obligatory need which by definition is less than the recommended dietary intake. In a factorial analysis of amino acid requirements the pattern of necessary amino acids needed is set by the pattern of different amino acid utilization, for protein synthesis and other pathways.

The effect of gastrointestinal protein secretion becomes important in the compromised nitrogen balance. In such circumstances with mucoproteins and enzymes being lost to the body, losses of amino acids may be specific. This depends on the manner in which nitrogen is lost. If the nitrogen loss is due to amino catabolism and hence in the form of urea, then the loss includes all amino acids. If the loss is as specific proteins then it may be accentuated and specific if there is a increase of particular amino acids in that protein. The amino acid composition of enzyme and mucous secretions that escape digestion in the small intestine and are degraded in the colon is different from that of proteins degraded metabolically. The non-essential amino acid : essential amino acid ratio (2.3 : 1) in small intestinal protein secretions compares with 1 : 1 of the body proteins. Secreted intestinal mucus is particularly rich in threonine and cysteine. Continued loss of such enriched proteins from the gastrointestinal tract is significant, especially at times of dietary protein shortage.

It is difficult to achieve consensus between operational and factorial approaches. In the simplest factorial model the minimum amino acid needs are divided between net protein production and maintenance.

Estimation of the biological value of a protein

Nitrogen excretion The nutritional value of a protein is measured by first establishing the rate of nitrogen excretion on a protein-free diet. Thereafter, known amounts of the protein being tested are added to the diet and the effect on nitrogen excretion measured.

$$\text{Biological value of a protein} = \frac{\text{Retained nitrogen}}{\text{Absorbed nitrogen}} \times 100$$

If the protein provides all the needs for protein synthesis at a rate equal to protein turnover, then the biological value will be 100. The standard is whole chicken egg protein.

If half the nitrogen fed is lost by excretion then twice the required intake of that protein is necessary to achieve equilibrium.

Amino acid content An alternative approach is to measure the amino acids in the protein. The figure can be compared with that of the egg protein standard.

$$\text{Amino acid score} = \frac{\text{mg of amino acid in 1 g of test protein}}{\text{mg of amino acid in 1 g of reference protein}}$$

Not all amino acids measured chemically are biologically relevant. There may be losses during cooking, lysine being vulnerable to cross-linking with other amino acids. The **protein efficiency ratio** (PER) is calculated as the weight gain per weight of protein eaten by young rats.

Following an enriched protein diet, the enzymes which degrade amino acids are activated especially in the liver. The priority is protein synthesis. This is reflected in the K_m of the catabolic and synthesising enzymes, with values for protein-synthesizing enzymes being close to 10^{-3} mM and those for catabolizing enzymes closer to 1 mM. Therefore, the catabolizing enzymes only come into play during conditions of abundance of amino acids.

The **biological value** of protein is calculated with reference to growth in the young. The majority of individuals, however, are neither young nor growing. It may be argued that the middle-aged and elderly are replacing tissue and their needs are probably similar to those in growth.

Sugar

Sugar or sucrose is a readily available source of energy. The sucrose intake varies from country to country. It has been suggested that in the UK, the average intake of non-milk extrinsic sugar (i.e. other than lactose in milk) should not exceed 60 g/day or 10% of the total dietary intake. This nomenclature of non-milk sugars refers primarily to sucrose.

> In 1987, the daily sucrose intake in the UK was 104 g per person, providing 14% of the body energy. Honey and glucose, at 16 g per person per day, provided 2%, and lactose at 23 g per person per day, a further 3%. Sucrose intake as packaged sugar was on average 26 g per person per day; total sugars were 95 g per person per day – 18% of the body intake. Breast or bottle-fed infants obtained 40% of energy from lactose. For pre-school children 25–30% of daily food energy was provided by intake of sugars. Older children and adults tend to take less sugar, 17–25%.

Starches

It has been suggested that starches should provide the balance of dietary energy not provided by the whole protein, fat and non-milk extrinsic sugars; that is, on average 37% of total dietary energy for the adult population. The same principle should be applied to children over 2 years old.

Such calculations are for starch which is digestible in the small intestine. The calculations do not include resistant starch.

Fats

Adults
Dietary expert committees have recommended a reduction of the fat content of the diet to 30–35% of energy. Unsaturated fatty acids should be increased. Saturated fatty acids should provide only 10% of food energy.

The properties of *trans* fatty acids are not sufficiently understood to make recommendations as to the influence of these in the diet, though the UK COMA committee suggested not more than 2% of total fat intake.

A ratio of unsaturated to saturated fatty acids of more than 0.45, even approaching 1.0 and a mixture of n-6 and n-3 polyunsaturated fats and mono-oleic acid, should be a target. The n-3 *cis* polyunsaturated fatty acids should provide 0.2% of total energy intake and n-6 1.0%. *Cis*-mono-unsaturated fatty acids should provide 12% of daily total energy intake. The target is a plasma cholesterol concentration of less than 5.2 mmol/l (200 mg%).

A Report of the British Nutrition Foundation Task Force in 1992 suggests that this is achievable by eating fewer meat products, dairy products and baked foods and by eating more oily fish (herring, mackerel, sardine and salmon) fruit, vegetable and wholemeal cereals. Saturated fat should be replaced by starch in the form of bread, presumably without butter, cereals, fruit and vegetables.

> Intakes of essential fatty acids in Great Britain have been suggested to be 11.7 g/day of n-6 fatty acids and 1.6 g of n-3 fatty acids. The main sources of n-6 fatty acids are vegetables, fruits and nuts (3.0 g), cereal products (2.6 g) and vegetable oils (2.4 g). The main sources of n-3 fatty acids are vegetables (0.4 g), meat and meat products (0.3 g), cereal products (0.3 g) and fat spreads (0.3 g). Most adult western diets provide 15–18 g of essential fatty acids and in general healthy individuals have body reserves of 100–500 g in adipose tissue.

Babies
Human milk is rich in linoleic acid (C18 : 2), and is important for the development of infants. The linoleic acid content of milk varies in amount (3–12%), being dependent upon maternal dietary intake and possibly smoking habit. The average linoleic acid content of human breast milk is of the order of 377 mg/kg baby body weight. α-Linolenic acid (C18 : 3) provides 0.4% of human milk giving an intake of 40 mg/kg body weight and DHA (22 : 6) as 0.2%, 20 mg/kg body

weight. Infant formula feeds do not meet these expectations.

Pregnant and lactating women
Their requirements are similar to other adult recommendations.

n-6/n-3 Polyunsaturated fatty acid ratios
In human milk the ratio of n-6/n-3 polyunsaturated fatty acids is of the order of 11 : 1. In practice, the important ratio is of the essential fatty acids, linoleic acid (n-6) to α-linolenic acid (n-3) (see text describing fatty acid nutrients). These two compete for the enzyme 6-desaturase; the ideal ratio is not known, but in adults a ratio of 5 : 1 has been recommended.

Monounsaturated fatty acids
An intake of total monounsaturated fats 15% (oleic acid 12%) of total energy intake has been

recommended for adults. All these dietary proposals should be accompanied by and protected by an adequate intake of antioxidants, e.g. vitamin E, selenium, ascorbic acid and β-carotene. Vitamin E intake is linked to unsaturated fatty acids, especially linoleic acid intake, and is contained in similar sources. The recommendation is 0.4 mg vitamin E per g linoleic acid.

Non-starch polysaccharides (fibre)
Children of less than 2 years old should not take foods rich in non-starch polysaccharides at the expense of more energy-rich foods which they require for adequate growth. It is not at the present time possible to define a non-starch polysaccharide intake, though it has been suggested that an adult diet should contain an average for the population of 18 g non-starch polysaccharides from a variety of food whose constituents contain it as a naturally integrated component.

Summary

1. Appropriate nutrition requires that all nutrients, carbohydrates, lipids, proteins, minerals, vitamins and water are eaten in adequate amounts and correct proportions. This is essential for normal organ development and function, reproduction, repair of body tissues and combating stress and disease. The nutrient intake should be appropriate for sustained activity and effective physical work.
2. Dietary reference values for food energy and nutrients include BMR (Basal Metabolic Rate), PAL (Physical Activity Level) and PAR (Physical Activity Ratio). Terms relating to energy and nutrient intakes include and have included RDI (Recommended Daily Intake), RDA (Recommended Daily Amounts) of food energy and nutrients, EAR (Estimated Average Requirement) for a group of people, LRNI (Lower Reference Nutrient Intake), RNI (Reference Nutrient Intake) for protein, vitamins or minerals, Safe Intake and DRV (Dietary Reference Value).
3. Dietary guidelines interpret these into practical statements, for real-life nutrition. Recommended nutrient and dietary allowances indicate the requirements of individual nutrients for defined population groups, e.g. babies, toddlers, pregnant and lactating women. Nutritional and dietary guidelines recommend intakes of food, milk, meat and vegetables, etc.
4. The energy requirement of an individual is the level of energy uptake for food which will balance energy expenditure when the individual has a body size and composition and level of physical activity consistent with long-term good health, and allow for economically necessary and socially desirable physical activity. In women who are pregnant or lactating, the energy requirement includes the deposition of tissue or the secretion of milk at rates consistent with good health.
5. Recommended dietary protein intake is 10–12% of energy intake and diminishes in parallel with the age-related fall in energy. This may be a factor in muscle loss with age. A safe intake of protein should not be lower than 0.75 g/kg/day. Proteins differ in their biological quality,

dependent upon the amounts and proportions of essential amino acids. A protein which is rich in all of the essential amino acids would score higher on the scale of biological quality than a protein deficient in one or more essential amino acids. It is necessary to define protein and specific amino acid needs in diets with both abundant and also deficient amounts of nutrients, including protein and specific amino acids.

6. If the protein provides all the needs for protein synthesis at a rate equal to protein turnover, then the biological value will be 100. The standard is whole chicken egg protein. An alternative approach is to measure the amino acids in the protein. The figure can then be compared with that of the egg protein standard, the amino acid score

7. The sucrose intake varies from country to country. In the UK, the average intake of non-milk extrinsic sugar (i.e. other than lactose in milk) should not exceed 60 g/day or 10% of the total dietary intake.

8. Starches should provide the balance of dietary energy not provided by the whole protein, fat and non-milk extrinsic sugars; that is, on average 37% of total dietary energy for the adult population. The same principle should be applied to children over 2 years old.

9. The fat content of the diet should be 30–35% of the dietary energy content. Unsaturated fatty acids should replace saturated fatty acids and provide 10% of food energy. A ratio of unsaturated to saturated fatty acids should be more than 0.45, even approaching 1.0; a mixture of n-6 and n-3 polyunsaturated fats and mono-oleic acid should be a target. Human milk is rich in linoleic acid (C18:2, n-6), and is important for the development of infants.

Further reading

Department of Health (1991) Report on health and social subjects. No. 41. Dietary Reference Values for Food Energy and Nutrients for the United Kingdom. Report of the panel on Dietary Reference Values of the Committee on Medical Aspects of Food Policy. HMSO, London, pp. 2–3.

Food and Agriculture Organization/World Health Organization/United Nations (1985) *Energy and Protein Requirements*. Technical Report Series, no. 724. WHO, Geneva.

Hexted, D.M. (1975) Dietary Standards. *Journal of the American Dietetic Association*, 66, 13–21.

Liu, J.-Z. and Guthrie, H.A. (1994) Nutrient labelling – a tool for nutritional education. *Nutrition Today*, 17, 16–21.

Reeds, P.J. (1990) Amino acid needs and protein storing patterns. *Proceedings of the Nutrition Society*, 49, 489–97.

Unsaturated Fatty acids: Nutritional and Physiological Significance. Report of British Nutrition Foundation Task Force (1992) Chapman & Hall, London.

Welsh, S., Davis, C. and Shaw, A. (1992) A brief history of food guides in the United States. *Nutrition Today*, 27, 6–11.

3.3 Evaluation of dietary intake

• Dietary intake is measured to assess both total nutrient intake and intake of individual constituents.
• Methods have been developed which can be used for individuals or for population surveys.

3.3.1 Introduction

Data which identify food consumption are collected for a variety of reasons:

• to estimate the adequacy of dietary intake of the population

• to investigate the relationship between diet and health and nutritional status
• to evaluate nutritional education, intervention and food fortification programmes

Information is obtained on food and dietary intake by individuals or groups of individuals either by methods of measuring food intake or by methods of converting food intake to nutrient intake.

3.3.2 Methods of measuring food intake

Central to any nutritional epidemiological study is a correct measurement of dietary intake. The absoluteness of the results can be checked in two ways. One is to include markers of internal validity, which could include asking the subject for the same information in different ways, and to look for consistency. Alternatively one can use an absolute method, e.g. a weighed record.

Error must be estimated so that allowance can be made in the final analyses. Histories of what people have eaten and drunk may be written, recorded by word, or visually. In other instances, trained field workers are necessary. Much of this methodology depends on the sophistication of the society that is being studied.

Prospective records of food intake are more accurate than recollections of past intake. The weighed dietary record – which has been regarded as the most accurate method – contains inherent problems. Weighing is an intrusion before eating and shortcuts may distort results in the weighed eating pattern.

Common errors associated with the collection of data on food intake include:

- omission of data on inter meal snacks, parts of meals, entire meals or entire days either deliberately by the subject or in error
- recording of wrong amounts of foods
- incorrect identification of food
- recording of data on the wrong subject's form

Errors may also be associated with the entry of data including: reading errors, wrong identification numbers, transcription errors, omission or double entry of data or lines or segment of data being transposed.

Description, weighed and estimated records

Subjects can be taught to describe and give an estimate of the previously weighed food before eating and to record any left-overs. In such exercises it is important that the recipes are given, and an information pool of average recipes is made available, comparable with those of manufactured foods. Such records, either weighed or described estimates, are quite different in accuracy from precise weighing which is necessary if food composition tables are not available. In such a detailed process raw ingredients, cooked foods, meals or snacks and the individual portions must all be weighed. Aliquots must be obtained for chemical analysis.

Measurement of intake

A number of methods are available for estimating food intake – written statements, portable tape recorders and even plastic models of food and photographs. Some methods include a combined tape recording and an electronic scale. The detailed manual coding of food records is being replaced by 'menu' computer programs based on food groups. Systems include placing the food on the scale and pressing the appropriate key; nutrient intakes are then calculated directly by the computer, using the weight of food and the database. All such records require a high degree of cooperation from the individuals involved and validation is crucial to ascertain whether subjects are over- or under-reporting.

A major source of error is the estimation of portion size and the imprecision may be of the order of 50%. The coefficient of variation may be of the order of 50% for foods, but less than 20% for nutrients. The error of estimation of the weight of food is probably random and the error will be minimized by increasing the number of observations.

Diet histories

All the methods rely on asking people to remember accurately the frequency and quantities of food eaten on previous occasions.

The length of time of a study may be important and it has been suggested that 3–4 days may be the optimum period rather than the traditional 7 days. In some circumstances 2-day or even 1-day records may be sufficient. The period of observation necessary varies for different nutrients. The number of days necessary to classify 80% of subjects correctly into the extreme thirds of the distribution also varies. The 7-day record is probably sufficient to classify the distribution for nutrient intake of a population for energy and energy-yielding nutrients into three major groups. Longer periods are required for alcohol, some vitamins, minerals and cholesterol.

Household data

Aggregate data based on surveys of groups of people rather than individuals can be used for community surveys of nutrient intake. These studies are usually based on the household and give indications of the food consumption pattern between communities. Regional and socioeconomic differences can be highlighted. However, such studies do not extend to the individual in those areas. In such community surveys, four methods of measurement are used: (i) food accounts; (ii) inventories; (iii) household recall; and (iv) list recall.

The **food account** method is based on the household. The individual who is responsible for buying the food and preparing it keeps a record of the quantities of food entering the house, including purchases, food from the garden, gifts, payment in kind and other sources. Such a survey assumes that there is no change in the average level of food stocks, but it is obvious that there will be variation within that period in the amount of food held within the house's foodstore. The Household Food Consumption and Expenditure Survey (the National Food Survey) has been completed on an annual basis in Great Britain for half a century. It began in 1940 to examine the nutritional quality of the diet of urban working-class households, to assess the value of the wartime food policy. By 1950, all sections of the population were surveyed and hence these reports give an account of the British food habits since 1950.

The **inventory** method requires the respondents to keep a record of all food coming into the house. A larder inventory is carried out at the beginning and end of the survey. The advantage of this method is that it gives a direct measure of the amount of food and nutrient available for consumption within a single household. The method, as well as identification of total purchases, is used in the National Food Survey.

In the **household record** method, the foods available for consumption, raw or processed, are weighed or estimated in household measures, allowing for preparation waste, etc. Food eaten by visitors is calculated and subtracted from the total. Allowance is made for food waste either by collecting the waste or estimating loss. This technique is best suited for populations in which most of the diet is home-produced rather than preprocessed. It is the ideal method for unsophisticated societies.

The **list recall** method requires recall of the amount and cost of food obtained for household use over a given period, from 24 hours up to the more usual 1 week. In this way the estimate of food costs and net household consumption of both foods and nutrients can be obtained. This method is important where most food is purchased rather than home-produced.

Population studies

In a series of population studies it can be shown that the dietary pattern can be grouped basically on food sources of macronutrients. The information on which these are based will be from production, import and export data.

Countries differ in the types of carbohydrate staples available, primarily wheat, potatoes or rice. There are major differences in the sources of proteins, animal or vegetable and fish sources and fats, predominantly animal fats, tropical oils, coconut palm and palm kernel oils or other vegetable fats. Countries where populations have more wheat or potatoes available tend to have more animal fats and protein available including beef, pork, milk, eggs and butter. Food sources of fats and proteins are similar in these countries but total fat availability is lower in Central and South American nations in comparison with North America, European and Mediterranean countries.

Central and South American countries have more carbohydrate and vegetable proteins, particularly rice, beans and grains. In Asia, rice is a predominant source of starch and populations in these countries have more vegetable proteins, particularly cassava, sweet potatoes, beans, nuts and grains.

The diet of Tanzania, Thailand and Sri Lanka consists of fats predominantly from animal sources and tropical oils, proteins derived from fish and vegetable sources and carbohydrates from rice, fruit, vegetables and sugar. Those in Cyprus and Finland are characterized by fats derived from animal and vegetable fats, protein from animal products excluding fish, and carbohydrates from wheat, fruit, vegetables and sugar (Figures 3.2 and 3.3).

Utilization of extensive food nutrition surveillances Seven-day measurements of food intake have been obtained with measures, for example, of height and weight of school-children. Anthropometric and blood pressure measurements have also been obtained.

Individual diet histories
The diet history, like the questionnaire method, is a repeatable and relatively valid method. It covers significant periods of time and so compensates for the potential distortions due to week to week

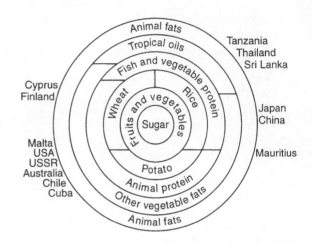

FIGURE 3.2 The dietary patterns of a number of countries listed by food sources of macronutrients that are predominant in their respective food supply. (From Posner, B. (1994) *Nutrition Reviews,* 52, 201–7, with permission.)

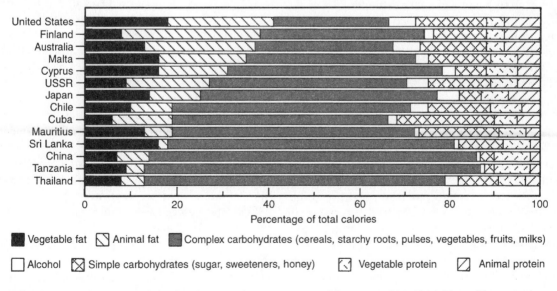

FIGURE 3.3 FAO macronutrient data by animal structure/vegetable source 1984–1986. Vegetables, proteins and carbohydrates in the diet in a number of countries. (From Posner, B. (1994) *Nutrition Reviews,* 52, 201–7, with permission.)

variations in diet. This is of particular importance for some nutrients, e.g. vitamins D and A.

Interviews are best based on prepared protocols which allow for individual foods, set meals and variations in intake over time. Interviewer training is very important. Disadvantages are reliance on memory, interviewer bias and subject bias in response to being interviewed.

Diet history usually consists of:

- detailed interview to measure amounts and frequency (usually per month) of a wide variety of foods
- cross-check food frequency list
- 24-hour recall
- 3-day record (optional)

It is important that at least the first three elements are used.

24-hour recall The individuals interviewed are asked a systematic series of questions to ensure recollection and description of all food and drink consumed in the 24 hours before the interview with emphasis on food consumption meal-by-meal and looking for day-to-day and seasonal variation. Implicit in this method is that the interviewer is very familiar with the food habits of the local population. The 24-hour recall is quick and easy to administer. The limitation is that it does not take account of day-to-day variation and differences between the weekday and weekend intake. Furthermore, individuals with low intakes tend to report higher than accurate intakes, and those with high intakes tend to report lower. There is consequently a reversion to the mean.

The advantages of retrospective methods are that they are quick, cheap and independent of motivation and literacy. The disadvantages are that they rely on memory, observer bias is possible and there is no measure of day-to-day variation in the diet. These methods are also of limited value in children under the age of 12 unless seen with their parents and also individuals may wish to imply that they eat a better diet than they actually do. There are regional and cultural differences and sauces and spices in some diets may alter the micronutrient intake

Questionnaires

The administration of questionnaires is a widely used method for assessing diet. It is easy and cheap. Questionnaires may be self-administered or interviewer-administered. The benefit of the former is that interviewer bias is eliminated. However, the questions must be very simple and unambiguous and require the subjects to be able to read and write. Questions may be left unanswered and there may be a low response rate. An interviewer-administered questionnaire ensures answers to more complex questions, completion of all questions and an explanation of problems. Questions are usually of the frequency and amount type (FAQ). Individuals are asked how often they usually consume an item of food or drink. Closed questionnaires have defined questions, demand an easily classified answer whereas open questionnaires ask for comment. It can be appropriate for questions or questionnaires to be closed rather than open, in that the interviewer asks precisely the same questions of all subjects. Questionnaires range from simple to very comprehensive and may include from 9 to 190 food items. The foods in the questionnaire should be the minimum number which includes the major sources of nutrients for the majority of subjects in that population.

In measuring diet by questionnaire in any epidemiological study, it is important to measure diet at the correct time, preferably at the time when the condition associated with the diet is being induced. Such questionnaires are complicated and can be misleading through the omission of questions relating to the inclusion of other elements which may be present in or absent from the food. These may include additives, contaminants, chemicals, natural toxins and the nutritional effects of the method of food preparation/cooking.

Frequency categories should always be continuous, for example 'never', 'less than once per month', 'one or two times per month', 'three or four times per month', or 'more than three or four times per month'. This should be followed by a listing of the number of days per week the item is consumed. If the person answering the question is unable to find an appropriate answer then the sensitivity of the results will be reduced. It is important to define the contribution of 'within-subject error'.

Questionnaires do require a significant amount of development and validation but can have the advantage of being precoded for computer analysis.

Accuracy of food frequency questionnaires
Both diet records and diet histories have a number of practical limitations. Quantitative food frequency questionnaires can be used to estimate habitual intake of food items in epidemiological studies. Such questionnaires comprise questions on the frequency of consumption which can be grouped into lists of foods with similar nutrient composition expressed as amount/quantities per day, per week or per month. Portion sizes are given in everyday units such as slices, cups and spoonfuls. Most foods should appear in the same position in a hierarchy of measurements, e.g. the same quartile, when compared with more accurate methods, namely weighed-food records. Sugar, cream products and vegetables are the foods most likely to be unreported, and bread, potatoes and fish over-reported, contributing to the inaccuracies of the questionnaire method of obtaining nutritional information.

Dietary studies are most successful in societies which are simple in structure, closed communities, self-sufficient farming communities, long-distance seafarers or the very young. It is much more difficult to assess accurately the nutritional intake of a complex society with constant access to food.

3.3.3 Converting food intake to nutrient intake

The conversion of food consumption to nutrient intake requires computer software, a nutrient database and an appropriate program. A program should ideally accept data from dietary histories so that information on each food in a meal eaten is recorded, and converted by the computer into constituent nutrients. There may be differences in the nutrient value imposed by the bioavailability of that food. The bioavailability of a nutrient is the proportion of that nutrient that is available for utilization by the body.

Chemical analysis

These techniques all have disadvantages in free-living populations. Analysis of food may be necessary when the chemistry of a particular food is not available in a composition of food table or when no information is available on which foods are an important source of a nutrient or other food components of interest.

The **duplicate portion** technique is the most accurate and is the benchmark for other techniques. Its accuracy is dependent on obtaining a sample identical to the food consumed by the subject under study. Such a study is complicated by inter-meal snacks, or if insufficient food is prepared for a true duplicate to be obtained.

With the **aliquot sampling** techniques, the weights of all foods eaten and drinks taken are obtained and an aliquot, say 10%, is collected for analysis. However errors can arise because of sampling techniques.

The **equivalent composite** technique is where the weights of all foods eaten and beverages drunk are recorded. At the end of the survey a sample of raw foods equivalent to the mean daily amounts of food eaten by an individual are analysed.

More recently, the **doubly-labelled water method** has been used which does not affect the way of life of the subjects. This method measures energy expenditure which should equal intake in subjects maintaining weight. A carefully weighed oral dose of $^2H_2{}^{18}O$ is taken and urine samples are then collected over the next 15 days. Carbon dioxide production is measured as the difference in the water pool ($^2H_2$) and the bicarbonate and water pool (^{18}O). Changes in body weight and the water pool can be used to correct measured energy intake in relation to energy expenditure from the doubly-labelled water method.

An alternative, less expensive method of measuring protein intake is the **24-hour urinary nitrogen method**. The completeness of the urine samples is estimated by the PABA (*p*-aminobenzoic acid) check method. PABA is given orally at the beginning of the collection, and should reappear in the voided urine, indicating completeness of collection. Complete 24-hour urine collections are important to validate studies of dietary methods in nutritional epidemiology.

Body weight is also an important measurement from which estimates of basal metabolic rate may be calculated.

Food composition tables

National epidemiological studies may look at nutrient intake as well as food consumption data. To be relevant to nutritional epidemiologists food composition tables must encompass the following:

- Foods included in the table: these must be comprehensive and appropriate for the population studied. It is important that a food is described unequivocally and what is meant by a particular food is clearly apparent. This is easy for raw fruit and vegetables but becomes more difficult for cooked dishes where the recipe may vary.

   The number of nutrients included in the table for each food must be sufficient for the study in question.

- Method of expression of amounts of nutrients.
- Nutritionally appropriate methods used for the estimation of each nutrient. It is important that there is standardization of methods and these should include replicate analyses and analysis of reference materials provided by central bodies.

Analysis of individual foods

The disadvantage of using food composition tables and nutrient databases is that each value is the average of a limited number of samples analysed for each food. Sampling errors are large, especially for mixed food dishes and meals. These add to the total error and variation in results from dietary intake studies. Differences in water content are a main cause of variation in the chemical measurement of nutrients. Therefore, foods containing a large amount of water are always subject to large variation. The least variable nutrient is probably protein. Estimates for energy, protein, fat, carbohydrate and mono- and disaccharides are probably accurate within 10%. The vitamin and mineral content is much more speculative, and less accurate.

Summary

1. Food consumption data are collected to identify whether the population is eating sufficiently well, to look at the relationship between diet, health and nutritional status, and to look at nutritional education, intervention and food fortification programmes.
2. A correct measurement of dietary intake is central to any nutritional epidemiological study. It is important that the validity of the results is checked and error must be estimated so that allowances can be made in the final analyses.
3. Prospective records of food intake are more accurate than recollections of past intake.
4. A number of techniques have been described including description, weighed and estimated records, measurement of intake, diet history, household data and inventory methods. Other methods include household records, list-recall methods and population studies. The 7-day measurement of food intake has been obtained with measures of anthropometric blood constituents. Individual diet histories and 24-hour recall are important assessments of nutritional intake. The accuracy of food frequency questionnaires can be assessed.
5. The conversion of food consumption to nutrient intake is a complex process which requires chemical analyses of the various food constituents, the writing of food consumption tables and analysis of individual foods.

Further reading

Bingham, S.A. and Nelson, M. (1991) Assessment of food consumption and nutrient intake, in *Design Concepts in Nutritional Epidemiology* (eds B.M. Margetts and M. Nelson), Oxford Medical Publications, Oxford University Press, Oxford.

Posner, R M., Franz, N. and Quatromoni, P. (1994) Nutrition and the global risk for chronic diseases – The Inter Health Nutrition Initiative. *Nutrition Reviews,* 52, 201–7.

Schofield, C. and James, W.P.T. (1985) Basal metabolic rate. *Human Nutrition: Clinical Nutrition,* **39C** (Suppl. I), 1–96.

West, C.E. and van Staveren, W.A. (1991) Food consumption, nutrient intake, and the use of food consumption tables, in *Design concepts in Nutritional Epidemiology* (eds B.M. Margetts and M. Nelson), Oxford Medical Publications, Oxford University Press, Oxford.

3.4 Measurement of energy

- Energy balance is dependent on nutrient intake and energy expenditure.
- Basal metabolic rate is the consumption of energy at complete rest.
- Energy metabolism is affected by heat loss which is proportional to body surface area.
- Energy expenditure can be measured precisely, at rest and during physical activity, and varies with age, sex and specific states, e.g. growth, pregnancy and lactation.
- Methods of measurement of energy expenditure vary in complexity, cost and suitability to the duration of the investigation.

3.4.1 Introduction

Energy is continuously required for cell repair and growth , but only intermittently for work, though food intake to provide this energy is intermittent. There is loss of nutrient energy when food is converted to mechanical energy; about 65% is dissipated as heat.

Twelve people sitting talking in a room produce heat at 60 kJ per minute, equivalent to a 1 kw electric fire.

Only 25–35% of nutrient energy is used for mechanical work and less than 10% is for basic physiological activity, e.g. cardiac and respiratory contractions. At complete rest and without physical work (basal metabolism), energy is still required for the activity of the internal organs and to maintain body temperature. This is called basal metabolic rate (BMR).

3.4.2 Energy expenditure at rest

BMR is calculated as body weight/surface area in metres^2. The surface area of the body is used to enable comparison of measurements of BMR in individuals of varying size. The surface area is calculated from height and weight using nomograms. BMR is more closely related to lean body mass than to surface area. During sleep the overall metabolic rate approximates to the BMR.

Metabolic rate calculated from body weight alone can be used as an index of total energy intake for groups of individuals, again assuming the ratio of total energy expenditure to basal metabolic rate is 1.6 on average though not in sedentary individuals (Table 3.2) . Rates of work or energy expenditure are calculated in watts (1 W = 1 kJ/s).

Changes in energy metabolism are due to the effect of and interactions between the environmental and biological factors which influence metabolism. The **Brody–Kleiber metabolic equation** attempts to predict the basal metabolism of all mammals regardless of size from their physical measurements, e.g. surface area. The argument for the equation is:

- it is theoretically possible to define a mammal of unit mass from which the basal metabolism of all mammals can be calculated by a simple mathematical formula

- a change in body mass has the same energetic effect regardless of species or individual differences in structure or body composition, i.e. $P = aM^b$, where P = basal metabolism, a = mass coefficient, M = body mass and b = mass exponent. b is equal to 0.75 and a to 70 kcal/$M^{0.75}$ (mass in kilograms for mammals). Heat loss is proportional to body surface area.

Body mass alters during life. It is therefore important to differentiate the quantitative from qualitative changes in mass. During neonatal growth there is a quantitative increase in mass associated with qualitative changes in body composition and form. In young adult life, body composition and form remain fairly constant while mass continues to increase. In middle age body mass may increase due to increasing fat and retention of water, both of which are metabolically somewhat inactive. In old age there may be a decline in body mass.

TABLE 3.2 Equations for estimated basal metabolic rate (BMR) from body weight

Age (years)	BMR (MJ/24 hours)
Under 3	
m	BMR = 0.249 wt − 0.127
f	BMR = 0.244 wt − 0.130
3–10	
m	BMR = 0.095 wt + 2.110
f	BMR = 0.085 wt + 2.033
10–18	
m	BMR = 0.074 wt + 2.754
f	BMR = 0.056 wt + 2.898
18–30	
m	BMR = 0.063 wt + 2.754
f	BMR = 0.062 wt + 2.036
30–59	
m	BMR = 0.049 wt + 3.653
f	BMR = 0.034 wt + 3.538
Over 60	
m	BMR = 0.049 wt + 2.459
f	BMR = 0.038 wt + 2.755

wt = body weight in kg

The BMR of a 70-kg man is approximately 60–75% of the total daily expenditure, i.e. 6.3 MJ (1500 kcal)/day. This maintains normal body functions, temperature and the sympathetic nervous system. Resting metabolic rate (RMR) is measured in an individual at rest at a temperature which makes no demands on energy production over a period of at least 8–12 hours after the last meal. The BMR is measured in the morning upon wakening after 12–18 hours of rest. It is somewhat lower than RMR. Determinants of metabolic rate which are invariable are: age, sex and genetic constitution. The variable elements are: the diet that antedated the test, body composition and weight, temperature, hormones, smoking, drugs and stress. Differences in RMR due to body composition can be corrected by relating the figures to fat-free mass. The reduction in metabolic rate in adults with age is a function of loss of lean tissue, and also the rate of cellular metabolism declines. Women have a lower resting metabolic rate because of less lean tissue, though the measurement may be higher when measured as fat-free mass. The RMR in women falls before ovulation and rises by about 5% after ovulation. Insulin, thyroid hormones and adrenaline also influence resting metabolic rate.

The actual metabolic mass of an animal results from a combination of processes reflecting the various functional and structural transformations that occur during its lifetime. The metabolic activity over a lifetime dictates the functional mass and composition of an animal. This metabolic activity varies during growth, pregnancy, lactation, illness, injury or senescence, but may be more constant during maturity.

3.4.3 Regulation of energy balance

In the regulation of energy balance, nutrient intake and energy (E) expenditure are related in the formula:

$$\delta E = E_{in} - E_{out}$$

Inappropriately high intakes or low expenditure produce energy excesses, increase fat storage and result in a gain in body weight. E_{in} is the energy available for metabolism of the foods and E_{out} is formed from two components:

$$E_{out} = E_{exer} + E_{ther}$$

where E_{exer} is the energy available for metabolism of the foods lost from the body in urine and faeces, and E_{ther} is heat production (thermogenesis).

The thermic effect of physical exercise will vary according to the intensity of work performed and the duration of activity. The 70-kg man requiring a maintenance energy intake of 10.5 MJ (2500 kcal)/day will require 3.2 MJ (750 kcal)/day or 30% of these energy requirements for muscular activity. Clearly the thermic effect of physical energy will be the most variable of all the components of E_{ther}. The metabolic efficiency of physical work is approximately 30%.

Adaptive thermogenesis is believed to account for no more than 10–15% of total energy expenditure, but may be important in the long term. This may be due to a change in resting metabolic rate due to adaptation to environmental stress, e.g. temperature, food intake, emotional stress and other factors. During under-nutrition there is progressive decline in the resting metabolic rate. During over-nutrition there is an increase in resting metabolic rate of the order of 10–15%. These are in part due to sympathetic nervous system activity, adrenaline, thyroid hormones and insulin.

Thermogenesis is the increase above BMR caused by the thermic effect of food intake. It is a by-product of cellular and body maintenance, the thermic effect of food, the thermic effect of physical exercise, exercise heat production, and the phenomenon of adaptive thermogenesis. The thermic effect of food is an increase in energy expenditure over the RMR following a meal. Heat is produced in response to an alteration in metabolic efficiency associated with changes in environmental conditions. The relative contribution of each to the total energy expenditure can be calculated. The effect of diet is complex and not well understood. All elements of diet are thermogenic. This is in part due to the energy required for digestion, absorption, transport, metabolism and storage of the ingested food. There may be other influences on the sympathetic nervous system by dietary carbohydrates. The thermic effect of food is said to be approximately 10% of calorie intake, though the effect of specific nutrients may vary. A complex network of dietary and hormonal factors act to regulate diet-induced thermogenesis in humans.

There is a genetic element to this measurement. In determining the response to food, the thermic effect of food is complex and not consistent, for example between obese and lean subjects. Part of the difference may be due to insulin resistance associated with obesity. It has been suggested that exercise plays a role in energy balance, both by expending energy and regulating food intake. Aerobic fitness and the timing and size of a meal are determinants of the metabolic response to exercise, and account for some of the differences between lean and obese subjects.

The sympathetic nervous system and the indirect effect of adrenaline and noradrenaline may be involved in some of the changes. The type of nutrients in food, the substrates which result from the ingestion of that food, and the signal triggered by that food, each play a part in the thermogenesis. The signal may be altered by intake and may produce differences in sympathetic activity. Undoubtedly fasting suppresses and sucrose stimulates sympathetic activity. In addition to insulin mediating glucose metabolism, thyroid hormones are also important.

The effect on energy metabolism following carbohydrate over-feeding may result from energy-requiring processes which are quite different from those from an individual mixed meal. This may result from differences in the metabolic fate of the carbohydrate. A proportion of ingested glucose if in excess is converted into lipid rather than oxidized or converted into glycogen. Lipogenesis from glucose is relatively inefficient. Part of the increase in energy expenditure may be attributable to the metabolic change, as well as to activation of the sympathetic nervous system. It is possible that the thermogenic effect of fat is mediated by free fatty acids or the hormones which are stimulated by such fatty acids.

It is probable that proteins produce a larger and more sustained thermic response than carbohydrate or fat. This may well reflect the energy cost of the synthesis of tissue proteins. Skeletal muscle is involved in more than half the total protein turnover and the fasting state fall in muscle synthesis can account for most of the change in whole body turnover. This has implications for energy expenditure. Metabolic rate may reflect changes in protein synthesis.

3.4.4 Energy requirements of humans

The total daily energy expenditure is the sum of the BMR, thermic effect of food eaten, and the energy expended in physical activity. The total energy expenditure is expressed as a multiple of BMR and is affected by the physical activity level (PAL); therefore it is necessary to know the sex, age and body weight to calculate the BMR, and the intensity of the various activities of work and leisure. The energy expended in physical activity varies.

Energy expended in physical activity

For example, 1.2 × BMR for sitting, if no physical activity takes place, and 4 × BMR for walking on the level at an average pace. The duration and intensity of physical activity is important. There is an effect of occupation, non-occupational activities and intensity of activity. This enables an estimate of daily energy expenditure of groups, so that physical activity level is characterized by description of life-style.

Daily activity expenditure equals

BMR × [time in bed] + [time at work × PAR] + [non-occupational time × PAR],

where PAR is the physical activity ratio.

Energy expenditure can be described as light, moderate or very hard:

- **Light work** is less than 170 W (2.5 kcal/min); it includes golf, assembly work, gymnastic exercises, brick laying, painting.
- **Moderate work** is 350–500 W (5–7.4 kcal/min); it includes general labouring with a pick and shovel, agricultural work, ballroom dancing and tennis.
- **Very hard work** is 650–800 W (10–12.5 kcal/min); it includes lumber work, furnace stoking, cross country running and hill climbing.

Young children

Energy requirements decline from 1.7 MJ (405 kcal)/kg/day for boys at 3 years to 1.2 MJ (290 kcal)/kg/day at 9 years; for girls, 1.6 MJ (385 kcal)/kg/day at 3 years declines to 1.1 MJ (255 kcal)/kg/day at 9 years.

Older children, adolescents and adults

In this age range a man weighing 74 kg, with an estimated average requirement for energy and a physical activity level of 1.4 uses approximately 10.6 MJ/day (2550 kcal/day). For a woman weighing 60 kg, comparable figures would be 8 MJ/day (1900 kcal/day).

Pregnancy

Maintenance of a normal pregnancy requires energy for increases in tissue mass and metabolic activity. The increased tissue mass includes the uterus, foetus and placenta, increased blood volume and an increase of 2–2.4 kg of fat which provides an energy reserve for lactation. The energy cost of the changes in tissue mass during the whole of pregnancy is on average 40 000 kcal (167 MJ) in women of 60 kg non-pregnant body weight. The total increase in BMR over the duration of the pregnancy is of the order of 30 000 kcal (126 MJ), giving an overall total of about 70 000 kcal (293 MJ). However, it is unusual for the mother to take this increment in calories and there is no apparent risk to the mother or the foetus for failing to meet these additional requirements. It has been suggested that the increment in EAR for pregnancy should be 0.8 MJ/day (200 kcal/day) above the pre-pregnant EAR only during the last trimester.

Nonetheless, women who are underweight at the beginning of pregnancy and women who do not reduce activity may need more nourishment. Growing teenage mothers also need extra calories for their own growth as well as that of their baby. Other women may not need extra because of a decline in physical activity.

Lactation

Women who are exclusively breast-feeding until the baby is 3–4 months old have different requirements from those who are supplementing breast-feeding. The gross energy content of breast milk is 280 kJ/100 g (67 kcal/100 g). The energy cost is in the order of 2.7 MJ (650 kcal) per day. It has been suggested that the additional energy requirement is of the order of 1.9–2.4 MJ (450–570 kcal)/day.

The elderly

The energy expenditure of the elderly is reduced, but in general so is physical activity. There is large variation between individuals. An active 70-year-old may have as high an energy expenditure as a sedentary 40-year-old. There is also a decline in the basal metabolic rate, which is particularly affected by the reduction in the fat-free mass. In ageing, weight may remain steady but fat replaces lean tissue. In the old and very old (over 75 years) sickness and disability and a reduction in body weight (including lean mass) alter energy requirements. In estimating energy requirements a standard value for physical activity level in the elderly of 1.5 BMR is useful, regardless of sex or whether the individual is housebound, is in an institution or living at home.

3.4.5 Measurements of energy expenditure

Dietary survey

See page 78, Evaluation of dietary intake.

Heart rate monitoring

This is a simple and inexpensive method for measuring total energy expenditure of large groups of people. This method is a socially acceptable and sufficiently accurate method for estimating habitual total energy expenditure in free-living children to whom other methods are not acceptable. There is a significant decline with age in energy expenditure during physical activity and also a decline in duration of physical activity.

> Younger children (7–9 years) spend more time (470 min/day) engaged in physical activity than older children (12–15 years) (280 min/day). Boys spend approximately one-third more time (460 min/day) in physical activity than girls (320 min/day).

Calorimetry

This is considered the best method for measuring energy expenditure over a long period of time. The measurements is of heat loss and not heat production. Heat storage may complicate the result, and this is not a useful method for short-term measurements.

Direct calorimetry
This measures heat produced and the excretion of water and carbon dioxide. This is the basis of the Atwater and Rosa respiratory chamber.

Indirect calorimetry
This is the most widely used technique. Heat production (metabolic rate) is determined from oxygen consumption and carbon dioxide production. If the urinary excretion of nitrogen is also known, then the type and rate of fuel oxidation within the body can be calculated. Such measurements of oxygen consumption and CO_2 production may be derived from the respiration chamber or from portable respirometers. Oxygen consumption is proportional to energy and heat production. The ratio of excreted carbon dioxide to oxygen is peculiar to each major nutrient, i.e. the respiratory quotient for glucose oxidation $= 1$, for animal fat $= 0.7$ and for protein $= 0.8$. Measuring the urinary nitrogen allows an estimate of the protein being metabolized. Oxygen consumption can be measured over long periods

in respiration chambers or into closed circuit systems, e.g. Benedict Roth spirometer, Douglas bag or the Max Planck respirometer.

Doubly labelled water method

This method (see p. 83) is generally accepted as the most accurate technique for measuring energy expenditure over a period of time. It interferes minimally with free-living subjects, but is expensive and requires that intake equals expenditure.

The total energy expenditure includes the energy cost of basal metabolism, physical activity, thermogenesis and the cost of synthesizing new tissues. The total energy requirement of children can be calculated by adding estimates of the energy value of new tissue deposited during normal growth to the estimates of total energy expenditure obtained by the doubly labelled water method. Previous estimates of infant intake contained problems because of the difficulties estimating dietary intake.

Summary

1. Energy is continuously required for cell repair and growth, but only intermittently for work, though food intake to provide this energy is intermittent.
2. At complete rest and without physical work (basal metabolism), energy is still required for the activity of the internal organs and to maintain body temperature. This is called basal metabolic rate (BMR). Basal metabolic rate = weight/surface area in metres^2. BMR in a 70-kg man is approximately 60–75% of the total daily expenditure, i.e. 1500 kcal/day. Determinants of metabolic rate which are invariable are: age, sex and genetic constitution. The variable elements are: the diet that antedated the test, body composition and weight, temperature, hormones, smoking, drugs and stress.
3. Heat loss is proportional to body surface area.
4. In the regulation of energy balance, nutrient intake and energy expenditure are related in the formula: $\delta E = E_{in} - E_{out}$. Inappropriately high intakes or low expenditure produce energy excesses, increase fat storage and result in a gain in body weight. E_{in} is the energy available for metabolism of the foods and E_{out} is formed from two components: $E_{out} = E_{exer} + E_{ther}$, where E_{exer} is the energy available for metabolism of the foods lost from the body in urine and stools, and E_{ther} is heat production (thermogenesis).
5. The energy expenditure for light work is less than 170 W (2.5 kcal/min) (e.g. golf, assembly work, gymnastic exercises, brick laying, painting). That for moderate work is 350–500 W (5–7.4 kcal/min) (e.g. general labouring with a pick and shovel, agricultural work, ballroom dancing and tennis). That for very hard work is 650–800 W (10–12.5 kcal/min) (e.g. lumber work, furnace stoking, cross-country running, hill climbing).
6. The energy requirements of children vary with age. Pregnant and lactating women have increased energy needs.
7. Measurements of energy expenditure include heart rate monitoring, direct and indirect calorimetry and the doubly labelled water method.

Further reading

Heusner, A.A. (1985) Body size and energy metabolism. *Annual Review of Nutrition*, 5, 267–93.

Leibel, R.L., Edens, N. K. and Fried, S.K. (1989) Physiologic basis for the control of body fat distribution in humans. *Annual Review of Nutrition*, 9, 17–43.

Livingstone, M.B.E. *et al.* (1992) Daily energy expenditure in free-living children: comparison of heart rate monitoring with the doubly labelled water method. *American Journal of Clinical Nutrition*, 46, 343–52.

Prentice, A.M., Coward, W.A., Davis, H., Murgatroyd, P., Black, A., Goldberg, G., Ashford, S., Sawyer, M. and Whitehead, R.G. (1985) Unexpectedly low levels of energy expenditure in healthy women. *Lancet*, i, 1419–22.

Suarez, P.K. (1996) Upper limits to mass specific metabolic rates. *Annual Review of Physiology*, 58, 583–605.

Woo, R., Daniels-Kush, R. and Horton, E.S. (1985) Regulation of energy balance. *Annual Review of Nutrition*, 5, 411–33.

3.5 Body composition

- Body composition can be measured as fat mass and fat-free or lean body mass.
- Lean body mass consists of viscera, muscles, organs, blood and bones.
- Body composition can be assessed by clinical examination, anthropomorphic or biochemical measurements.

3.5.1 Introduction

The body is not of constant composition, and is an assembly of different organs of differing composition. Body composition is affected by nutritional status. This may be assessed by clinical examination.

Conventional two-compartment models of body composition separate body weight into fat mass and fat-free or lean body mass, the latter including viscera, muscles, organs, blood and bones.

The human body can be regarded as being in three compartments:

1. Cell mass (55% of the total weight) which is the active tissue performing the work of the body
2. Extracellular support tissue (30% of the total weight) which supports the cell mass. This includes: (i) blood, plasma and lymph and extracellular fluid; and (ii) minerals and protein fibres in the skeleton and the connective tissue including collagen.
3. Energy stores (15% of the total body weight), predominantly held in adipose tissue, subcutaneously and around organs and some glycogen. Most of the protein but only 1 kg of fat are essential cell components, the residual fat is storage.

The relative proportions will vary with a number of factors: sex, age, stage of development, physical fitness, hormonal status, pregnancy and even mood. In the embryo the proportion of water is higher than that in the mature animal:

- 28-week foetus   88% water
- new-born baby   75% water
- 2-month baby   65% water
- 4 months to adult 60% water

The proportion of fat increases somewhat with age. In the elderly the amount of water slowly declines by small amounts.

3.5.2 Measuring composition of the major body compartments

The dilution principle

This measurement is based on adding a known amount of readily measurable substance Q into a body compartment of volume V which allows free and even distribution throughout the compartment. After a defined period the concentration of Q (C) can be measured and then the volume can be calculated, using the equation

$$V = Q/C$$

The requirements for the accuracy of such a measure are that Q readily diffuses throughout the compartment and that there is no metabolism of Q or binding to the walls or macromolecular component of that compartment.

Clinical findings in poor nutrition

Hair: lack of lustre; sparse and thin; straight in Negroes; depigmented

Face: depigmented; nasolabial sebaceous gland dysfunction; 'moon face'

Eyes: Bitot's spots (bilateral desquamated thickened conjunctival epithelium); conjunctival and corneal xerosis (dry, thickened, wrinkled, pigmented); keratomalacia (corneal softening)

Lips: angular stomatitis; angular scars; cheilosis

Tongue: smooth or red and swollen; atrophic papillae

Teeth: edentulous, caries; mottled enamel

Gums: spongy bleeding gums

Glands: thyroid and/or parotid enlargement

Skin: xerosis; follicular hyperkeratosis (hair follicles plugged with keratin); petechiae (small subcutaneous bleeds); pellagrous dermatosis (symmetrical erythema exposed to light and mechanical irritation; dry and scaling); flaky dermatosis; scrotal and vulval dermatitis

Nails: koilynichia

Subcutaneous tissues: oedema of dependent tissues e.g. legs and sacral; loss of subcutaneous fat

Skeletal system: craniotabes (deformed skull bones); frontal and parietal bone bossing; epiphyseal enlargement; beading of ribs; persistent open anterior fontanelle; deformities of thorax; tender bones; failure to grow to expected height

Muscle/nervous system: muscle wasting; sensory loss; loss of ankle and knee jerks; loss of proprioception (position sense); loss of vibration sense; calf tenderness

Gastrointestinal: enlarged liver; diarrhoea

Haematological: anaemia

Cardiovascular: cardiac enlargement; tachycardia

Infections: tuberculosis; undue sensitivity to infection

Sexual development: correct stages of puberty:
Females: breast development, onset of menstruation, amenorrhoea, pubic hair development, fertility, size of babies.
Males: genital (penis) size, pubic hair

Emotional: listless and apathetic; mental confusion

(Modified from Davidson and Passmore, 1986).

Total body water

This can be measured by chemical methods, e.g. urea, antipyrine or ethanol. Alternatively, isotopic methods can be used, e.g. deuterium or tritium. The dilution is measured on a blood sample. The result is ideally expressed as the 'tritium' or 'antipyrine' space. This alerts the reader to the type of measurement, which may be different with each chemical. The usual result is 40 litres and accounts for 50–65% of body weight.

Extracellular water

A number of substances have been used, including inulin, sucrose, sodium thiocyanate, sodium thiosulphate and isotopic bromide ion. Thiocyanate enters red cells so a correction requires to be made. The extracellular compartment usually accounts for 18–24% of the body weight.

Cell water and cell mass

If the total body water and the extracellular water have been measured then the cell water can be calculated by difference. Approximately 70% of the total body cell mass is water, the cell mass will weigh 36 kg and represent 55% of the total body weight. Only half of the normal body weight is made of cells which are active in metabolism.

An alternative method is to use ^{40}K, the natural isotope of potassium which can be measured using a whole-body counter.

Body fat

The proportion of fat in the body can be measured by underwater weighing using Archimedes' principle, which states that the density $d = $ mass/volume. The density of fat is 0.90, while that of the whole body is about 1.10. So, if x is the percentage of fat in the body

$$\frac{100}{d_{body}} = \frac{100 - x}{1.10} + \frac{x}{0.90}$$

$$x = \frac{495}{d_{body}} - 450$$

The density of the body can be measured by first weighing in air and then in water. If M is the mass of the body, and V the volume, then $d = M/V$. The measurement has to take into account the buoyancy of the lungs, which is measured by a nitrogen washout method.

The body volume is measured by displacement. The difference in weight in air and submerged in water is the volume of the displaced water. The volume corresponding to this mass of water is calculated by dividing by the density of water at the time of the underwater weighing. A drawback to this method is the inability to measure the density of the lean body mass.

Dilution principle for fat stores

Gases which are more soluble in fat than water and which dissolve rapidly in body fat, e.g. ^{85}Kr or cyclopropane have been tried but lack precision.

Skeleton

The methods for assessing the size and composition of the bony skeleton are limited, but include:

- **biochemistry:** plasma calcium; urinary calcium; plasma alkaline phosphatase; plasma vitamin D; urinary hydroxyproline (collagen); calcium and phosphorus balance
- **radiology:** bone age; isotope bone scans; computed tomography (CT) scanning
- **bone biopsy**

3.5.3 Estimation of body composition

The method used to estimate body composition will depend upon the interest of the analyst. The anatomist will be interested in the size of organs, physiologists in the components of cells, membranes and extracellular compartments, nutritionists in fat, protein, water and mineral content and the butcher in meat, fat and bone. A straightforward way to estimate any of these is carcass analysis. The results and the repeatability of such measurements are very

limited. Non-destructive methods enable observations to be made over a period of time, and are more acceptable in human studies.

> The normal composition of a 65-kg man is:
> - protein   11 kg   (17%)
> - fat   9 kg   (13.8%)
> - carbohydrate   1 kg   (1.5%)
> - water   40 kg   (61.6%)
> - minerals   4 kg   (6.1%)
>
> Of these, a proportion is storage tissue, predominantly fat. The fat store may range from 9 kg in this example to a significant proportion of the overall body weight in the obese. Available carbohydrate stores are meagre at 200 g. Approximately 2 kg of protein could be retrievable as energy but with consequences to muscle mass. Only 10% of the body water can be lost without profound consequences for well-being.

Physical measurements
Height and weight
Possibly the most effective diagnostic examination is to look at the patient and to decide whether the patient is underweight, normal or overweight.

Height and weight are important indices of growth in children and adults. Life assurance company tables give acceptable weights for heights for adults. The measurement of children is more importantly recorded on height and weight velocity charts; that is, the changes over time are recorded. Up to the age of 3 children's length is measured lying on a flat surface. Over this age then the standing height is recorded. Weight should be naked weight. Notes on the development of puberty are recorded on 'distance' charts, which note the advent of sexual and pubic characteristics.

Approximate measurements of muscle mass and body mass are made by the mid-arm circumference and triceps skin-fold thickness. The lower 10% of the range indicates significant underdevelopment. In the upper 10% of the range individuals can be regarded as obese.

Weight has to be judged in the context of the height of the individual. This relationship is contained in the Quetelet index or Body Mass Index (BMI). This is used as an obesity index, using weight (kg) and height (metres) as weight/height^2.

- very obese   > 40
- obese   30–40
- plump   25–30
- acceptable   < 25
- underweight   < 20

Weight for age is expressed as a percentile of a reference population. In the first percentile are the lowest 10%, i.e. 90% are greater in that particular measurement. The 90% percentile means that these are the highest 10% in that measurement. Many malnourished children in developing countries come in the first percentile of these very Western-based charts, so percentage of the median weight of the reference population is used. The indices are age-dependent (60% weight for age is severe malnutrition in the infant but only moderate in the schoolchild).

Z-scores, that is standard deviation scores, have merit in the severely compromised child. The Z-score for weight for age is calculated by subtracting the median weight of the reference population at the child's age from the child's weight and dividing by the standard deviation (SD) of the weight of the reference population at that age.

The reference population commonly referred to has been taken from a North American population. Many of these children are overweight; therefore the weight for age and weight for height are skewed above the median. Different SDs are used above and below the median. Height for age is normally distributed and therefore there is only one set of SD required. New percentile charts are being produced.

Z-scores are comparable across ages; therefore an SD deviation is equivalent, e.g. − 4.0. The implications will however vary from age group to age group. The WHO report suggests that the Z-score should be used universally in developing countries, the danger point where intervention is advisable being a figure where Z = less than − 4.0.

Body shape

Subcutaneous fat As the majority of fat in the body is stored subcutaneously then subcutaneous fat can be measured by skin-fold callipers. There are various measures of skin-fold thickness (Figure 3.4):

- Triceps: at a point equidistant from the tip of the acromion and the olecranon process
- Subscapular: just below the tip of the inferior angle of the scapular
- Biceps: at the mid-point of the muscle with the arm hanging vertically
- Supra-iliac: over the iliac crest in the mid-axillary line.

Durnin and Womersley (1974) made measurements of skin-fold thickness and body density using underwater weighing and produced tables which enable calculations of percentage of body fat weight from skin-fold measurements.

A line drawn between the L_4–L_5 lumbar disc space and the umbilicus is the clinical dividing point between the depots of fat. Skin-folds (subscapular, triceps, abdominal, thigh) and body circumferences (waist, i.e. the narrowest region between the bottom of the rib cage and anterior

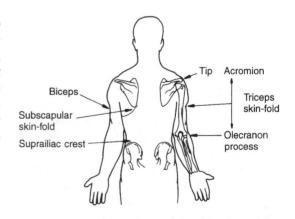

FIGURE 3.4   Sites of measurement of skin-fold thickness.

superior iliac spine; abdomen at umbilicus; pelvis at pubic symphysis; thigh; arm) have been used. In addition, X-rays, ultrasound and magnetic resonance imaging have been used. The L_4–L_5 landmark produces optimal male–female separation for upper and lower body fat. Men have 53% of their fat in the upper body, while women have 46% in the upper body. There is wide individual variability in the relative amount of adipose tissue in the intra-abdominal visceral depots and men have more of their adipose tissue in this compartment. Men have 79% of their fat in the subcutaneous region, women 92%; men 21% in the visceral area and women 8%.

Cardiovascular risk appears to be related to relative amounts of visceral adipose tissue, and not to the absolute amount of visceral tissue.

> It is possible that total adiposity correlates directly with insulin production and the waist : height circumference ratio correlates inversely with the fraction of insulin extracted by the liver. This may be due to increased concentrations of hepatic free fatty acids which may interfere with hepatic catabolism of insulin and provide substrate for over-synthesis of very low density lipoproteins.

Excess fat around the waist and abdomen (central obesity) is assessed by the waist : hip ratio. The level of the measurement varies immensely and it is important to record how the measurement was taken. The waist is variously defined as the minimum circumference of the mid-section of the body where the body narrows, seen from the front, or the measurement midway between lower rib margin and iliac crest. The hip circumference might be defined as the maximum circumference of the midsection, taken at the maximum extension of the stomach in front, the measurement at the iliac crest or the maximal diameter at the buttock or trochanteric region. A waist : hip ratio of greater than 1.0 rather than waist measurement alone is a predictor of a greater risk of diabetes and cardiovascular disease than excess fat around the hips and thighs. This is because there is a significant correlation between waist : hip ratio and intra-abdominal subcutaneous fat. This correlation does not extend to body mass index.

This pattern of fat distribution is strongly associated with total body weight, except in the stocky southern Italian populations. Individuals with predominantly central android or apple distribution of fat have higher rates of atherosclerotic heart disease, stroke, hypertension, hyperlipidaemia and diabetes than individuals whose adipose tissue is distributed in a peripheral pattern, gynoid or pear. The age of onset of obesity may also dictate distribution of fat. There are sex differences in the way in which body fat is mobilized during weight loss. Men appear to lose weight in the abdominal area more readily than women, while women lose fat more readily from their hips.

Head and chest circumferences
In young children, head circumference is considered an index of brain development and is related to chest measurement (3rd to 4th rib level on the sternum insertion).

Muscle function
Hand dynamometry records the strength of squeeze. Unfortunately this method is dependent upon attention and motivation and culture, e.g. males may want to show how strong they are.

Biochemical and other tests

There are precise and important laboratory methods for measuring nutritional status for vitamins and for electrolytes and trace elements; these are summarized in Chapters 7 and 8, respectively.

Electrolyte composition
The electrolyte composition of the extracellular and intracellular fluids is quite different (Table 3.3). In the intracellular compartment, the major cation is potassium, whereas in the extracellular compartment the major cation is sodium. Considerable energy utilization is required to maintain

TABLE 3.3 Normal distribution of ions in intracellular and extracellular fluids

	Intracellular mEq\l	Extracellular mEq\l
Cations		
Na	10	145
K	150	5
Ca	2	2
Mg	15	2
Total	177	154
Anions		
Cl	10	100
bicarbonate	10	27
sulphate	15	1
organic acids		5
phospate	142	2
Proteins		19
Total	177	154

the electrolyte concentrations. Cellular activity in muscle, nerve or secretory cell requires changes in ionic concentrations. The control of the cell milieu interior is essential to life.

More general biochemical measurements are the plasma albumin and proteins. Albumin has a half-life of 20 days and is therefore a long-term indicator of nutritional status. A plasma albumin of under 25 g/l suggests severe malnutrition. However, plasma albumin may fall in other conditions, e.g. renal disease. Plasma proteins include transferrin and retinol binding protein, complement component (C3) and fibronectin which have shorter half-lives and therefore reflect a more recent nutritional status. There is as yet no consensus as to which of these plasma proteins is the best to use for routine assessment.

Nitrogen loss can be measured in urine from the total urea in a 24-hour urine collection, using the equation:

$$\text{Total nitrogen loss} = \text{24 h urinary urea} \times 0.035 + 2\,\text{g}$$

2 g being inevitable loss, e.g. skin, hair, intestinal cells.

Metabolic measurements include plasma cortisol, glucose, interleukin-6 and albumin. Indices of inflammation include white cell count, especially the neutrophil count and C-reactive protein.

Estimating lipid-free body mass
The estimation of body composition depends on the body being regarded as lipid-free body mass. The lipid-free mass can therefore be calculated from one measurement, e.g. water or potassium. It is assumed that this measurement is a constant figure. During growth this will however alter in protein, water, glycogen, bone and electrolyte composition.

Water
This can be estimated using isotopes of hydrogen or oxygen. There is some metabolism of water, so the method over-estimates by 4% with hydrogen-labelled and 1% with oxygen-labelled water.

Potassium
The stable isotope of potassium, ^{40}K, is used. This has a relatively high natural specific activity. Low background counts limit the precision of this method in large animals. An alternative method, which has the disadvantage of using radioactive ^{42}K, measures the dilution of an administered dose in the body potassium.

Neutron-activation analysis
This method allows the measurement in a non-destructive manner of the nine elements which account for 99% of body mass. The whole body N can be measured with an accuracy better than Kjeldahl analysis of the carcass. Similar accuracy is obtained for oxygen, sodium, phosphorus, chloride and calcium.

The total heat of combustion of body tissues can be calculated from the protein, lipid and glycogen content. These can be distinguished from the N and O content; the C : H ratios are relatively constant. Differences provide the potential for estimating energy stores.

Bioelectrical impedance and body composition
This method has the appearance of simplicity. Simple bioelectrical impedance analysers measure

a series of segmental resistances or impedances, the size of which are determined by skeletal dimensions. The method is cheap, easy to use and non-traumatic for the subjects, but the physical realities of the system are complex.

Imaging

The development of imaging techniques based upon ultrasound, X-ray, computed tomography (CT) and nuclear magnetic resonance imaging (MRI) are used to estimate body composition. Indeed, there is a good correspondence between cross-sectional adipose estimations by MRI images and carcass analysis

---

Magnetic resonance imaging (MRI)

Certain nuclei, e.g. H, possess spin and a magnetic moment. In a magnetic field the magnetic moment processes about the applied field direction at a frequency proportional to the intensity of the magnetic moment. By irradiating with electromagnetic energy at the resonant frequency, the procession angle can be increased. This energy is lost in relaxation of the spin-lattice. This alteration in spin energy can be received in a receiver coil and the different amplitudes of spin relaxations induced in the system can be recorded. The relaxation will vary in different chemical structures, e.g. it is prolonged in free water and shorter in water held in hydrated proteins. The major sources of protons of mobile molecules are the H nuclei in water and the CH_2 of acylglycerol. Both proton density and relaxation time can be used to facilitate tissue discrimination.

Summary

1. The body is not of constant composition, and is an assembly of different organs of differing composition. Body composition is affected by nutritional status. This may be assessed by clinical examination.
2. Conventional two-compartment models of body composition separate body weight into fat mass and fat-free or lean body mass, the latter including viscera, muscles, organs, blood and bones.
3. Measurements of body composition include the use of the dilutional principles for total, extracellular and cellular water, body fat and a range of measurements for skeletal mass. Height and weight, body shape, subcutaneous fat, anthropometric measures and muscle function tests are also utilized. Biochemical estimations include electrolytes in extracellular and intracellular compartments, plasma proteins and urinary nitrogen.
4. Lipid-free body mass can be calculated from isotope-labelled water, potassium, neutron activation analysis, bioelectrical impedance, imaging or nuclear magnetic resonance imaging (MRI).

Further reading

Bishop, C.W., Bowen, P.E. and Ritchley, S.I. (1981) Norms for nutritional assessment of American adults by upper arm anthropometry. *American Journal of Clinical Nutrition,* **34**, 2530–9.

Davidson and Passmore (1986) In *Human Nutrition asnd Dietetics* (eds R. Passmore and M.A. Eastwood), 8th edn. Churchill Livingstone, Edinburgh.

Durnin, J.G.V.A and Womersley, J. (1974) Body fat assessed from total body density and its estimation from skin fold thickness measurements on 481 men and women aged 16 to 72 years. *British Journal of Nutrition,* **32**, 77–97.

Fuller, M.F., Fowler, P.A., McNeill, G. and Foster, M.A. (1990) Body composition: the precision and accuracy of new methods and their suitability for longitudinal studies. *Proceedings of the Nutrition Society,* **49**, 423–36.

McLaren, D.S. (1981) *A Colour Atlas of Nutritional Disorders,* Wolfe Medical Publications Ltd, London.

Shann, F. (1993) Nutritional indices: Z, centile or percent? *Lancet,* **341**, 526–7.

Smith, D.N. (1993) Bioelectrical impedance and body composition. *Lancet,* **341**, 569–70.

WHO Working Group (1986) Use and interpretation of anthropometric indications of nutritional status. *Bulletin of the World Health Organization,* **64**, 629–41.

Chapter 4

Nutritional epidemiology

4.1 Outlines of nutritional epidemiology

- Epidemiology is the relationship between possible causes and the distribution and frequency of diseases in human populations.
- Epidemiology has been an important tool in the identification of nutritional needs.
- Epidemiological studies require protocols to study carefully defined populations.
- Such studies require precise statistical evaluation.
- The interpretation of these results must make biological sense.
- Meta-analysis is an important statistical tool which enables relevant trials to be analysed together.

4.1.1 Introduction

Epidemiology is the study of the relationship between possible determining causes and distribution of the frequency of diseases in human populations. Epidemiological research has been one of the mainstays of our understanding of nutrition.

There are three objects of epidemiology:

- to describe the distribution and extent of disease problems in human populations
- to elucidate the aetiology of diseases
- to provide the information necessary to manage and plan services for the prevention, control and treatment of disease

What is being studied must be carefully defined and the interpretation of the results of the study be within the framework of the original definition.

Nutritional epidemiology is in addition to the study of the nutritional determinants of disease, the measurement of nutritional status in relation to environment and dietary intake. Most clinical measurements of nutritional status are used to identify deficiencies in overall intake or intake of individual nutrients.

4.1.2 Historical nutritional studies

The great nutritional studies of the early part of this century led to the discovery of dietary deficiencies in the form of beri-beri, goitre, pellagra, rickets, scurvy and xerophthalmia. These spectacular discoveries led to the concept of single dietary factors in the aetiology of disease. There is an understandable desire to identify simple and singular causes of contemporary diseases. However, it is now apparent that the system is complex, with groups vulnerable in addition to nutritional factors to multiple other predisposing factors, for example environmental infection, activity and individual genetic variation. Furthermore, only a proportion of individuals who are at risk develop such nutritional problems.

Reaching rapid conclusions based on epidemiological associations all too frequently mars the reputation of nutrition as a science, and delays identification of disease processes. Many such associations are unlikely to be causal; instead, these are markers for the net effect of many variables that influence morbidity and mortality.

4.2 Epidemiological research methodology

A systematic approach to epidemiological research involves:

- identifying the area of study
- developing a specific study question
- writing a research protocol
- completing the research described by the protocol
- analysis of the data with emphasis on (i) problems with the protocol; and (ii) statistics
- correct interpretation of the results examining the relationship between exposure and outcome

The best epidemiological studies include large numbers of variables carefully and independently measured. Examples of such studies are the Framingham study; the British Cancer Register; and the US Nurses Study.

Variables which are easy to identify but are complicated in interpretation are: smoking; sex; alcohol intake; social class; coffee consumption; and national differences.

Epidemiological studies can be seen as being experimental or observational. In experimental investigations a particular set of conditions are established for subjects by the investigator. In observational studies the investigator observes and measures but has no control over the actions of the subjects. Examples of experimental studies are **community trials** and clinical or **field trials.**

Community trials use populations whose disease status is unknown but most of whom will be healthy, whereas clinical or field trials work with subjects who already have a disease, that is therapeutic trials, or studies of subjects who are known to be free from the disease at the time of the study (prevention studies). In both instances, the general design is to divide the population into a treatment or exposure group and a control group and to compare the effect of a treatment relative to a control regime given at the same time under the same conditions. This design enables the effects of the treatment to be established. It is important that there is perfect randomization between the control and the treatment groups. Both populations should be studied for an equivalent and biologically relevant length of time unaffected by the commitment threshold of the subjects and investigators. The observer and the subject should be blind to the treatment regime but there should be maintenance and assessment of compliance. Intelligent and understanding professional statistical advice is essential. It is mandatory that this advice be taken before the trial rather than as an after-thought or in response to the comments of the editor of a journal.

4.2.1 Types of nutritional studies

Sources of information for nutritional studies are given in Table 4.1. Each source has its own value and sphere of usefulness.

Ecological studies

Observational studies include ecological studies. This is the study of groups who may be identified over a particular period or at a determined time or

TABLE 4.1 Sources of information for nutritional studies

Survey of sickness
Sickness absence statistics
Notification of diseases
Gereneral practitioner statistics
Hospital out-patient statistics
Hospital in-patient statistics (public or private)
Mortality statistics
Post-mortem statistics

place, (country, province or city), or with similar sociodemographic features.

Cohort studies

Cohort studies are similar to field trials, with the exception that exposure is not randomly assigned by the investigator. Here individuals who are exposed and those who are not exposed are compared. Cohort studies measure exposure accurately and the outcome is not influenced by measured exposure. There is less likely to be information bias; a measure of risk is obtained. However, such studies are expensive and time consuming.

Case control studies

Individuals with an outcome or disease (subjects) are compared with those who do not suffer from, or do not yet manifest the disease (controls). Factors thought to be involved in the condition are measured in both groups and thus the risk factor can be calculated. The definition of the population from which cases and controls are taken and the interval from the time of exposure or removal from exposure may influence the results. In case control studies a range of exposure levels must be included, allowing a gradient of risk to be established.

Case control studies are not expensive, can be completed quickly and multiple risk factors can be examined at the same time. They are particularly appropriate for rare diseases. Inaccuracies may, however, arise in that the data are obtained retrospectively, exposure may not be accurately identified or the disease may in fact influence exposure.

Cross-sectional studies

Such studies measure exposure and disease state at the same time.

Selection of population

It is difficult to define a normal population to research. There is a wide range of gene pools which any population will represent. In selecting a population it is important to pose the question and then to decide what useful information can be obtained from that population which must be readily available and which can then be defined in terms of age, sex, race and habitation.

4.2.2 Sampling

Sampling is required in order to obtain manageable numbers of individuals who are truly representative of the population to be studied. A system of random sampling is used:

- **Simple random sampling** is where the individuals are numbered sequentially and then the randomly numbered individuals are included in the study.
- **Systematic sampling** is a variation of simple random sampling.
- **Stratified sampling** is where measurements are made in each sampling unit, which may differ in magnitude from one sub-group to another, e.g. weight, body size, age or sex.
- **Cluster sampling** is where there are small units which may be sampled, e.g. households.
- **Multi-stage sampling** combines the different forms of sampling, working from larger to progressively smaller sampling frames.

The concept of 'intention to treat' must be used; that is, that everyone who is recruited into a trial should be included in the results. Those who fail to complete a trial (non-responders) pose a problem. They are a group different from those who do complete the trial (responders) and this may make the sample unrepresentative of the required target population. Thus, those who remain in the study may not be representative of those of the population as a whole. It is important to describe the non-responders and to define whether or not they are representative of the population as a whole.

Epidemiological definitions

Validity: in order for a study to be valid the findings must be representative and reproducible – this is **external** validity. An alternative interpretation of validity identifies whether the measure utilized actually measures outcome – this is **internal** validity. The standard against which the validity is judged may not, however, be absolute.

Sensitivity measures the proportion of truly affected subjects who are appropriately classified.

Specificity measures the proportion of truly unexposed individuals who have been correctly classified.

Repeatability, precision reliability or reproducibility is the consistency with which exposure is measured; similar results must be obtained if the experiment or analysis is repeated. *Confidence intervals*: 95%, 99% or greater confidence limits must be established.

Bias is any trend in the collection, analysis, interpretation, publication or review of data which leads to a conclusion which is systematically different from the truth. Bias can include selection bias and information bias, recall bias and interviewer bias.

Confounding factors are variables other than that being studied which may influence the outcome. For a variable to be a confounder it must be associated with but not causally dependent on the exposure of interest. Where possible, confounders can be controlled for in selection of subjects, by randomization, restriction or matching. **Randomization** is where subjects are allocated to different treatment exposures to ensure that any differences which are identified do not occur by chance. In contrast, **restriction** is achieved by including only

---

those subjects in whom the variable under study is the same. **Matching** ensures that potentially confounding factors are identically distributed in each group in the study, e.g. sex is a confounder and therefore it is necessary to study men and women separately. Other confounders are age, pregnancy, lactation, growth, activity, both physical and emotional, and stress and these should be equivalent in both groups.

A combination of conditions may be required for a disease or condition to occur. Where time is an important prerequisite for the development of a condition, the time and duration of exposure must be identified. This is known as the **induction period**. The **latent period** is the time between the induction of the disease and the clinical manifestation which permits diagnosis. It is helpful to the theory if there is dissociation between exposure and outcome across a range of dose–response relationships.

4.2.3 Measurement of outcome

Having determined that an outcome can be measured with sufficient validity, sensitivity and specificity, it is then important to evaluate: (i) the role of chance; and (ii) hypothesis testing.

This requires the null hypothesis to be tested, which is the state in which there is no relationship between exposure and outcome. An alternative hypothesis is that a low level of association is present.

Bradford Hill suggested nine points which may suggest causality:

1. strength of association
2. consistency
3. specificity
4. relationship in time
5. biological gradient
6. biological plausibility
7. coherence of evidence
8. experimental evidence
9. analogy

He set out the conditions which had to be met before it could be concluded that an association observed in a case-controlled study could be interpreted as indicating cause and effect:

- the alienation of bias in the selection of patients and controls, in the way that patients and controls reported their histories and in the way that interviewers recorded data
- an existence of an appropriate time relation between exposure to the suspected agent and the development of the disease
- the lack of any other distinction between the affected patients and their controls that could account for the observed association or of any common factor that could lead both to the specific exposure and the development of the disease

It is debatable whether these criteria have been met for the place of dietary fibre in the aetiology of the diseases which have been allegedly associated with fibre deficiency. The principal reason for the persistence of the association is that the concept makes good sense, rather than that the case has been made by good science.

Outcome measurements are of three types:

1. The observation of the event, for example the first clinical sign of a disease (incidence).
2. The recording of presence or absence of disease at a particular time (prevalence), the accuracy of which may be influenced by migration or mortality.
3. The measurement of level of disease on a metric scale which may similarly be subject to inaccuracies.

Nutritional epidemiological studies frequently study dietary intakes which represent a continuum of exposure. There may be a dose–response relationship between intake and the consequent disease or condition.

Proper estimates of exposure and consequent relationship requires:

- the probability distribution and measurement errors

- the distribution of true exposure and confounders in the population studied

It may well be that both distributions are multivariate. Therefore, the variability of each component in the measurements has to be examined, but so also has the interrelationship. This requires multivariate analysis.

4.2.4 Meta-analysis in epidemiology

Many small trials give inconsistent results. If, however, the effect were to identify the ability to modify the risk of serious disability then the accumulation of such small studies might be important. Meta-analysis is a precise statistical tool in which the results of all the relevant trials which have been undertaken can be analysed together. It is important that patients are included on an 'intention to treat' basis. All outcomes can be recorded in the same way and all trials and all patients can be included, irrespective of their inclusion in published results. It is not clear, however, whether epidemiologists can use meta-analysis for identification of hazards to health.

Epidemiological data for nutrition may not be categorized as readily as clinical trial data. Consequently, there is a subjective judgement entering into the decision to include or not include a particular study or item within a study. All substantial studies should be considered for inclusion, whether or not they have been published, and thus avoid the effect of publication bias. This is most likely to occur in a situation where there is a discouragement or a reluctance to publish results which fail to show any positive effects. Consequently, in conducting a meta-analysis it is important to include unpublished studies where the cases and controls have been properly selected. A weakness of meta-analysis is the criteria for exclusion of studies. Do the results of such excluded studies materially differ from the chosen studies? If so, in what way? Any differences should be reflected in the overall conclusions.

Summary

1. Epidemiology is a study of the relationship between possible determining factors and the distribution of the frequency of disease in human populations. Nutritional epidemiology requires the measurement of nutritional status in relation to environment and dietary intake.
2. The great epidemiology studies of this century led to the discovery of beri-beri, goitre, pellagra, rickets, scurvy and xerophthalmia.
3. The science of epidemiology demands a precise identification of the area of study, a specific study questionnaire, a research protocol, and effective analysis and interpretation of the resulting data.
4. Epidemiological studies can be experimental or observational. In experimental investigations a particular set of conditions are established for subjects by the investigator. In observational studies, the investigator observes and measures but has no control of the actions of the subjects. Studies may be community trials, or clinical or field trials.
5. Sampling is required to obtain a manageable number of individuals who are truly representative of the population to be studied. A system of random sampling is used.
6. An outcome must be measured with appropriate validity, sensitivity and specificity. It is also important to evaluate the role of chance and also to test a hypothesis. Bradford Hill has identified nine points which indicate causality between exposure and outcome.
7. Meta-analysis is a precise statistical tool in which the results of all relevant trials, even those which are small and may have given inconsistent results, can be analysed together.

Further reading

Doll, R. (1994) The use of meta-analysis in epidemiology. Diet and cancer of the breast and colon. *Nutrition Reviews*, 52, 233–7.

Editorial (1990) Volunteering for research. *Lancet*, 340, 823–4.

Margetts, B.M. and Nelson, M. (1991) *Design Concepts in Nutritional Epidemiology*. Oxford Medical Publications, Oxford University Press, Oxford.

Chapter 5

Basic nutrients

5.1 Nutrients and chirality

- The bonding of groups to the four valencies of carbon gives two possible molecular structures, the one being the non-superimposable mirror image of the other.
- Isomers that are mirror images of each other are called enantiomers; they have identical chemical properties except when metabolized by enzymes.

5.1.1 Introduction

Pasteur discovered the phenomenon of optical isomers of organic compounds. This led him to state that 'the forces of nature are not symmetrical'. He discovered that sodium ammonium tartrate could be divided into optically inactive and active crystals. These molecules were identical in physical properties except for the direction of rotation of polarized light. They had identical chemical properties except towards enzymes.

5.1.2 Enantiomers

The two optical isomers are mirror images of each other, a phenomenon called **chirality** by Kelvin from the analogy of the mirror-image relationship between the left and right hand. Isomers that are mirror images of each other are called **enantiomers**. A mixture of equal parts of enantiomers is called a **racemic** modification and is optically inactive, the two enantiomers cancelling each other.

When an enantiomer is reacted with a reagent that is itself optically active then the result will be dependent upon which isomer is involved in the reaction; the other will unaffected. The dominant product of the reaction results from the best stereochemical fit being selected.

The relationship between chirality and optical activity depends on the valencies of carbon being directed towards the vertices of a regular tetrahedron (Figure 5.1). The bonding of four different groups to the central atom gives two possible molecular structures, one being the non-superimposable mirror image of the other. Several chiral centres make for a series of optical compromises, which give rise to a series of asymmetrical molecules.

5.1.3 Chirality and nutritional biochemistry

This phenomenon of chirality is important in nutritional biochemistry, since it affects many biochemical, pharmacological and metabolic processes. Every amino acid, with the exception of glycine, is optically active. All of these amino acids have the same configuration about the carbon atom carrying the alpha (α) amino group. There are two enantiomers of amino acids, L- and D-. The L-α-amino acid is the naturally occurring form throughout the animal kingdom. The D-form is found in bacteria. When D-amino acids are presented to the body they are deaminated and lose their amino acid identity. Proteins and hence enzymes – being made of optically active amino acids – become optically active reagents. Enzymes are therefore very sensitive to the distinctions of chirality.

The L-form of the α-amino acid is metabolized in the body at a rate which is 20 times faster than the D-enantiomers. This difference is dependent upon the ease with which enzymes interact with the shape of the substrate. Polypeptides composed of the naturally occurring L-amino acids are stabile compared with the corresponding D-enantiomers.

Monosaccharides are polyhydroxylated aldehydes, aldoses or ketones. The middle carbon of glyceraldehyde has four different substitutions which result in two enantiomers, a D- or L-form, each with optical activity. The C4, C5 and C6 sugars can be formed (in a theoretical sense) through the condensation of acetalde-

hyde to either glyceraldehyde or dihydroxyacetone. This gives a series of optically active monosaccharides; these differences are important biologically.

Some hormones may exist as isomers; a racemic mixture of a hormone may result in important differences in activity between the isomers, e.g. adrenaline.

> The difference in the smells of lemons and oranges is solely one of chirality, the smell of lemons is the L- and the smell of oranges the D-isomer of the same chemical (Figure 5.2). L-asparagine tastes bitter and the D-form tastes sweet. Another chemical, carvone, tastes of caraway when in the L-form and of spearmint in the D-form.

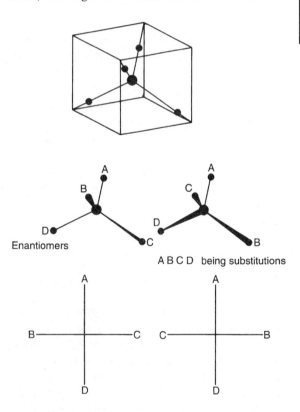

FIGURE 5.1 The two optical isomers are mirror images of each other. The bonding of four different groups to the central atom gives two possible molecular structures, one being the non-superimposable mirror image of the other.

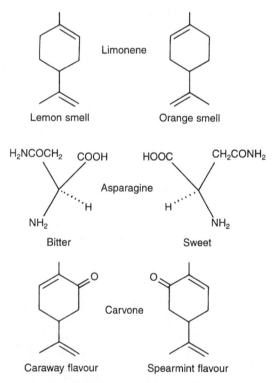

FIGURE 5.2 Examples of the effect of chirality on physical attributes, smell, taste and flavour.

Summary

1. Isomers are compounds which have the same molecular formula but different chemical structure.
2. Some molecules, because of their shape, are said to possess 'handedness' or chirality. Chirality exists when isomers differ in the arrangement of some of their atoms in space.
3. The molecules are mirror images of each other, and are called enantiomers. They may differ biologically, especially in the manner they are attacked by enzymes.

Further reading

Solomons, T.W.G. (1992) *Organic Chemistry*, 5th edn, John Wiley & Sons, New York.

5.2 Amino acids

- There are 20 L-α-amino acids which are important in human nutrition.
- There are eight essential amino acids which cannot be synthesized by the human body.
- Amino acids, through peptide linkage, form peptides and proteins, which have a wide range of biological functions.
- Amino acids which are hydrophobic are found within protein structure, whereas amino acids which are more hydrophilic are found on the surface of the protein.
- Proteins differ in their biological value or quality depending upon the proportions of constituent essential and non-essential amino acids.
- Dietary allowances for proteins are estimated by their total and relative essential and non-essential amino acid content.

5.2.1 Introduction

There are 20 amino acids which are important in human nutrition. These are listed in Table 5.1. The amino acids have a variety of side chains which provide a range of chemical characteristics. The side chain determines the properties of the amino acid which are classified by the chemistry of these side chains (R group). The carbon to which the carboxyl is attached is the α-carbon. Amino acids have four different groups around the α-carbon resulting in optically active L- or D-isomers or enantiomers. The L-forms are conjugated into proteins and biological systems. The D-form is found in the walls of bacteria and in some antibiotics. D-amino acids are transported very slowly across membranes compared with L-forms of the same amino acid.

L-α-amino acids can form peptides or protein structure by forming a peptide linkage between the amino groups and carboxyl groups, –NHCO–. The individual properties of the amino acids with their varied size, shape and side chains each contribute to very complex and functional protein structure.

The specific folding characteristics of a protein is the consequence of the amino acid sequence in the chain. Hydrophobic residues are in the interior of the protein and hydrophilic on the outside.

At pH 7 amino acids may be negatively charged (aspartic acid and glutamic acid) or positively charged (lysine and arginine). The charge on others (histidine) is dependent upon the pH. Asparagine and glutamine are polar, while being neutral. As a generalization, amino acids that have

TABLE 5.1 Amino acids. The mass is the molecular weight minus that of water. E indicates essential amino acid; (E) indicates facultatively essential (see p. 111). Occurrence indicates frequency of occurrence of each amino acid residue in the primary structure of 207 unrelated proteins of known sequence. Relative hydrophobicity (kcal/mol) compared with glycine; the larger the value, the more hydrophobic the molecule.

Amino acid	Abbreviation	Mass (Da)	Volume (Å^3)	Occurrence in proteins (%)	Relative hydrophobicity
Polar R groups					
asparagine	Asn	114	117	4.4	12.1
glutamine	Gln	128	144	3.9	11.8
serine	Ser	87	89	7.1	7.5
threonine	Thr E	101	116	6.0	7.3
tyrosine	Tyr (E)	163	194	3.5	8.5
Non-polar R groups					
alanine	Ala	71	140	9.0	0.45
cysteine	Cys (E)	103	109	2.8	3.6
glycine	Gly	57	60	7.5	0
isoleucine	Ile E	113	167	4.6	0.24
leucine	Leu E	113	167	7.5	0.11
methionine	Met E	131	163	1.7	3.8
phenylaline	Phe E	147	190	3.5	3.15
proline	Pro	97	123	4.6	n.a.
tryptophan	Trp E	186	228	1.1	8.3
valine	Val E	99	140	6.9	0.40
Charged R groups					
aspartic acid	Asp	115	111	5.5	13.3
glutamic acid	Glu	129	138	6.2	12.6
lysine	Lys E	128	169	7.0	11.9
arginine	Arg	156	173	4.7	22.3
histidine	His	137	153	2.1	12.6

side chains which are hydrophilic, i.e. water-soluble, are found on the surface of proteins. Amino acids which are hydrophobic are found within the protein (phenylalanine, leucine, isoleucine, valine, methionine). Glycine, alanine, serine, threonine and cysteine may be placed in the protein in a less-defined position. Proline and cysteine are important in dictating peptide and protein structure. Proline's side chain forms part of the main structure and its shape means that the direction of the main chain is altered by the formation of a bend. Cysteine can form a covalent –S–S– disulphide bridge with another sulphur-containing residue and this creates stable linkages.

5.2.2 Classification of amino acids

Amino acids can be grouped in a number of ways. They can be classified chemically based on structure (Figure 5.3) or nutritionally (Figure 5.4).

Chemical classification

- Mono-amino, mono-carboxylic amino acids: glycine (Gly), alanine (Ala), valine (Val), leucine (Leu), isoleucine (Ileu)
- Hydroxy-amino acids: serine (Ser), threonine (Thr)
- Basic amino acids: lysine (Lys), arginine (Arg), histidine (His)
- Acidic amino acids and amides: aspartic acid (Asp), glutamic acid (Glu), asparagine (Asn), glutamine (Gln)
- Sulphur-containing amino acids: cysteine (Cys), methionine (Met)
- Aromatic amino acids: phenylalanine (Phe), tyrosine (Tyr), tryptophan (Trp)
- Imino acid: proline (Pro)

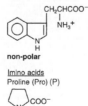

FIGURE 5.3 Amino acids are classified by the substitution in the side chain (R group). The carbon to which the carboxyl is attached is the alpha (α)-carbon. The side chain determines the properties of the amino acids. The amino acids are optically active L or D isomers or enantiomers. This is important in their biological actions.

Functional classification

This separates the amino acids into four groups. The polarity or non-polarity suggests the manner that the amino acid will be incorporated into proteins, polar on the outside, non-polar in the interior of the protein. Amino acids may be charged or uncharged according to the pH of the environment.

Amino acids may undergo further reactions:

- Enzymatic **acetylation** and **methylation**, usually of lysine: this suppresses positive charges forming on the amino group.
- **Phosphorylation**: the addition of a phosphate group to the hydroxyl group of serine or tyrosine and occasionally threonine.
- **Glycosylation**: the addition of a carbohydrate to an amino acid. A sugar may be attached to the amino group of asparagine to form an N-linked oligosaccharide. Occasionally a sugar

FIGURE 5.4 Amino acids important in human nutrition showing full name and abbreviated name. The amino acids are listed as overall structure; non-polar or polar, basic or acidic. This gives a clue as to the function of the amino acids in proteins. Hydrophilic amino acids are found on the surface of proteins. Amino acids which are hydrophobic are found within the protein.

may link to the hydroxyl group of serine or threonine to form an O-linked oligosaccharide.

The side chain is the major factor in determining the transport system that is utilized by an amino acid. Among neutral amino acids bulk and lipophilic properties of the side chain are all important.

5.2.3 Amino acid structure

Polar amino acids

These have no charge (see Figure 5.3); examples are serine (Ser) and threonine (Thr). The side chains are small, aliphatic with a hydroxyl group which gives either hydrophobic or hydrophilic properties. The hydroxyls are somewhat polar acting as hydrogen donators or acceptors. Threonine has a centre of symmetry of which only one isomer occurs naturally.

Asparagine (Asn) and **glutamine** (Gln) are the amide forms of aspartic and glutamic acid. They occur as amino acids in their own right and are incorporated into proteins. The amide groups are labile at extremes of pH and hydrolyse to aspartic acid and glutamic acid. When present as an amino terminal group in a peptide the glutamine may cyclize spontaneously to pyrolidone carboxylic acid.

Tyrosine (Tyr) is an aromatic (cyclic, benzene ring structure) amino acid residue which contributes to the ultra-violet absorption and fluorescent characteristics of proteins. The hydroxyl groups of tyrosine are relatively reactive and are readily nitrated and iodinated.

Non-polar amino acids

These are hydrophobic and interact with one another.

Glycine (Gly) is the simplest amino acid, has no side chain, is symmetrical and therefore does not exist in the D- or L-form. The presence of glycine in a polypeptide chain gives the chain flexibility.

Alanine (Ala), **isoleucine** (Ile), **leucine** (Leu) and **valine** (Val) are aliphatic residues which form a somewhat homogeneous class, having an inert side chain and hydrophobic properties. The molecular surfaces and shapes are varied, giving a wide potential for forming a variety of polypeptide structures. Isoleucine has an extra centre of asymmetry, but in nature and therefore in the diet, only one isomer is involved in protein synthesis.

Methionine (Met) and **cysteine** (Cys) are sulphur-containing amino acids. The long alkyl side chains of methionine produce a hydrophobic molecule which has a somewhat reactive sulphur grouping in the thioether group which is especially vulnerable to oxidation. The thiol group of cysteine is extremely reactive, ionizing at a slightly alkaline pH. Covalent disulphide bonds between cysteine residues occur in some proteins as cystine. Cystine is incorporated into proteins as cysteine, the disulphide bond developing later.

Phenylalanine (Phe) and **tryptophan** (Trp) are aromatic amino acid residues which contribute to the ultra-violet absorption and fluorescent characteristics of proteins. The aromatic ring of phenylalanine is very hydrophobic and chemically inert.

Proline (Pro) is a cyclic imino acid which has an aliphatic side chain, without functional groups and bonded covalently to the nitrogen of the peptide group. In polypeptide structures the proline readily forms bends in the protein chain. Proline has no amide groups for hydrogen bonding; consequently the protein structure is made rigid by the constraints placed on rotation by the cyclic five-membered ring. The 4-hydroxyproline derivative is important in the protein collagen.

Acidic amino acids

These owe their acidity to an extra carboxyl group.

Aspartic acid (Asp) and **glutamic acid** (Glu): the side chains differ in having one or two $-CH_2-$ groups, each with a terminal carboxyl group. These are ionized at pH > 5 and are polar, and hence can chelate cations. The protonated forms may act as a hydrogen donor or acceptor but rarely at neutral pH in proteins. The carboxyl groups are reactive and form esters with other amino acids and can be reduced to alcohols.

Basic amino acids

These include amino or imidazole groups.

Lysine (Lys) and **arginine** (Arg): the side chain of lysine is a hydrophobic chain of four methylene groups ($-CH_2-$) and an amino group, which is polar charged at biological pHs. The non-ionized amino group readily undergoes acylation, alkylation, arylation and deamination reactions. The arginine side chain has three hydrophobic methylene groups and a basic κ-guaninido group which is ionized.

Histidine (His): the imidazole side chain has special properties which are important in reactions requiring a catalyst. The tertiary amine has nucleophilic reactivity, important for enzymatic reactions. The imidazole group is free of steric hindrance. This is a strong base at neutral pH. In the non-ionized form one N is an electrophile and the other a nucleophile, respectively donor and acceptor for hydrogen bonding.

5.2.4 Essential and non-essential amino acids in nutrition

Humans can synthesize some amino acids (non-essential) from glucose and ammonia, via the Krebs' cycle or from free amino acids by transamination or reductive amination. However, amino groups are not transferred freely between all amino acids; there are nine amino acids (see Table 5.1) for which humans have no amination capability and therefore no synthetic ability. These are **essential** amino acids which must therefore be provided in the diet and which were defined originally by Rose as those amino acids which must be included in the diet to ensure optimal growth.

There are few hints from chemical data which indicate why some amino acids are essential and others readily synthesized. Some essential amino acids are clustered in the family of amino acids with apolar R groups. Neither molecular volume, nor – as will be seen in data given elsewhere – does the genetic code or triple codon for the

essential amino acids give any indication as to why an amino acid is essential.

Two other amino acids, tyrosine and cysteine, are **facultatively essential**. They are synthesized from essential amino acids and become essential only if there is a deficiency of their precursor essential amino acid.

Facultatively essential amino acids

Tyrosine is synthesized from phenylalanine, while cysteine synthesis requires methionine (a sulphur provider) as a precursor, the remainder of the molecule being derived from serine. Phenylalanine and methionine are both essential amino acids. If their presence in the diet is at or below minimum requirement levels, then tyrosine and cysteine in turn become essential.

Some amino acids may be regarded as conditionally essential, requiring preformed carbon side chains and substituted groups from other amino acids. Glycine, serine and cysteine may well function as an inter-related group, with the need for adequate provision of each. The requirements of the nitrogen cycles, e.g. glutamate cycle, may well increase the requirements for those particular amino acids involved in these cycles.

The following factors may complicate amino acid requirements:

- A lack of a primary amino acid may limit the utilization of other amino acids leading to equal loss of carbon and nitrogen.
- There might be a lack of non-essential amino acids; this could result in the deamination of essential amino acids to provide nitrogen.
- Lack of conditional essential amino acids, which leads to problems in the balance of nitrogen and carbon substrates and the need for specific if not essential amino acids.

Amino acid regulation of protein turnover

Amino acid requirements include the maintenance of protein turnover above a certain limit, regardless of net protein retention. It is not yet clear whether protein turnover is regulated by the rate of supply of any amino acid or whether it is restricted to a limited number of amino acids. The physiological state of the subject may be a variable: age, growth, pregnancy or health.

Gastrointestinal protein loss

Faecal nitrogen losses have important effects on dietary nitrogen needs. Nitrogen fixation by gut flora results in a faecal amino acid pattern which is different from that of the dietary amino acids. There is substantial secretion of protein into the intestinal tract. Only a proportion is recycled in the small intestine. A large proportion of the protein passing through the ileum is of endogenous origin.

5.2.5 Non-protein pathways of amino acid metabolism

It is not strictly correct to equate nitrogen balance with protein balance. Some items in the diet, especially human milk, contain significant amounts of non-protein nitrogen and amino acids which are necessary for the synthesis of nitrogenous compounds not linked to overall protein metabolism. Haem and creatine are not recycled and are a drain on stores of methionine and glycine. An adult man requires 1 g glycine/day to support creatine and haem synthesis. On a protein-free diet, such a man would have to mobilize 1.5 g glycine from body stores.

Two aromatic amino acids (tyrosine and tryptophan) are important in hormone synthesis, tyrosine forming thyroxine and the catecholamines, adrenaline and noradrenaline, and tryptophan forming serotonin.

Summary

1. There are 20 amino acids which are important in human nutrition. The amino acids have a variety of side chains which provide a range of biological properties. The amino acids are classified by the nature of their side chain. The L-forms are conjugated into proteins and biological systems. The D-form is found in the walls of bacteria and in some antibiotics.
2. Humans can synthesize most of the amino acids from glucose and ammonia. These are called non-essential amino acids and can be synthesized via Krebs' cycle or from free amino acids by transamination or reductive amination. There are eight amino acids – isoleucine, leucine, lysine, methionine, phenylalanine, threonine, tryptophan and valine – which humans cannot synthesize. Hence, it is essential that these amino acids are provided in the diet.
3. Tyrosine and cysteine are facultatively essential. They are synthesized from essential amino acids. They only become essential if there is a deficiency of their precursor essential amino acid.
4. Some amino acids are conditionally essential, requiring preformed carbon side chains and substituted groups from other amino acids, e.g. glycine, serine and cysteine may well function as a inter-related group, with the need for adequate provision of each. The requirements of the nitrogen cycles, e.g. glutamate cycle, may well increase the requirements for glutamate.
5. Proteins differ in their biological quality, depending upon the proportions of essential and non-essential amino acids. A protein which is rich in all of the essential amino acids would score higher on the scale of biological quality than a protein with no essential amino acids.

Further reading

Carpenter, K.J. (1994) *Protein and Energy*, Cambridge University Press, Cambridge.

Creighton, T.E. (1984) *Protein Structures and Molecular Principles*. W.H. Freeman & Co., New York.

Food and Agriculture Organization/World Health Organization/United Nations (1985) *Energy and Protein Requirements*. Technical Report Series, no. 724. WHO, Geneva.

Monod, J., Changeux, J.P. and Jacob, F. (1963) Allosteric proteins and molecular control systems. *Journal of Molecular Biology*, 6, 306–29.

Perutz, M. (1990) *Mechanisms of Cooperativity and Allosteric Regulation*, Oxford University Press, Oxford.

Reeds, P.J. (1990) Amino acid needs and protein storing patterns. *Proceedings of the Nutrition Society*, 49, 489–97.

Zubay, G. (1993) *Biochemistry*, 3rd edn, Wm Brown, Iowa, USA.

5.3 Protein

- Proteins are high molecular weight polyamides.
- Proteins consist of a sequence of amino acids which form the primary structure.
- When the primary structure is folded a secondary structure results by free rotation around bonds. The tertiary structure of a protein is the final format of the protein. All formats are dictated by the sequence of amino acids.
- Some protein molecules are complexes of more than one polypeptide chain. These sub-units form a quaternary protein structure.
- Proteins are important in cellular and tissue structure and are biologically active.
- Enzymes are proteins. Some belong to a family of isoenzymes which may catalyse the same reaction but with varying efficiency.
- An individual's enzyme make-up will be determined by the types of isoenzymes, which in turn are dependent on genetic make-up.
- The pattern of isoenzymes dictates an individual's metabolism.

5.3.1 Introduction

Proteins are high molecular weight polyamides, consisting of one or more chains of amino acids which then fold into a form that imparts a particular function. Proteins vary in size (from 1000 to 1 000 000 Da) and length, with some extending to 2000 amino acid residues. The average protein has a molecular weight of between 24 000 and 37 500 Da, which is 200–280 amino acid residues long. Proteins differ in their amino acid sequence rather than possessing a unique amino acid content. With 20 amino acids available – each with a singular contribution to make to the structure – the permutations are enormous.

The amount of each of the amino acids in a protein, and their place in the primary structure, decides the shape and biological properties of that protein. Specific groupings of amino acids in a section of the protein result in biological properties which are individual to that protein, e.g. in an enzyme active centre.

Collagen consists of one-third glycine and one-quarter proline, the remainder the usual spectrum of amino acids. Similarities in proteins reflect common ancestry of structural genes. The distribution of amino acids within the body of the protein overall is generally of the frequency expected from tables of amino acids. Though methionine is found at the amino-terminal end more frequently than would be expected, a tyrosine at the carboxy-terminal end is extremely rare.

Proteins are present in all living tissues and are the principal material of skin, muscle, tendons, nerves and blood; they form enzymes, antibodies and even have a supporting role in molecular biology.

The enzymatic activity of purely protein enzymes, i.e. those which do not involve co-enzymes, are dependent upon the chemical properties of the functional groups of the side chain of nine amino acids:

- imidazole ring of histidine
- carboxyl groups of glutamate and aspartate
- hydroxyl groups of serine, threonine and tyrosine
- amino groups of lysine
- guanidinium group of arginine
- sulphydryl group of cysteine

The groups act as general acids and bases and catalyse proton and group transfer reactions.

Metals, e.g. cobalt, iron, manganese, copper, zinc and molybdenum function as cofactors in enzyme reactions. They are points of positive charge, interact with two or more ligands and exist in two or more valency states.

5.3.2 Protein structure

Primary structure

Proteins consist of a sequence of amino acids, singular to that protein, the sequence of which is called the **primary structure**. Amino acids form peptides and proteins by linkage through a covalent peptide bond. Short chains up to 20 amino acids in length are peptides. The condensation and formation of a peptide bond between the carboxyl group of one amino acid and the amino group of another (Figure 5.5) determines the eventual conformation of the protein. Subsequently the polypeptide chain can take up secondary or tertiary protein structure.

A polypeptide chain has an amino-terminal or N-terminal end and a carboxy-terminal or C-terminal end. In describing protein structure there is a convention whereby the N-terminal amino acid residue (having the free amino group) is written on the left side and the C-terminal amino acid

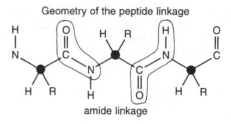

FIGURE 5.5 Peptides are joined by amide linkages; the geometry of such an amide linkage is shown.

residue (having the free carboxyl group) at the right end.

Resonance

This is important in the peptide bond. The structure is a compromise between two resonating hybrids. In one the C–N is a single bond with no overlap between the lone electron pair of the nitrogen and the carbonyl carbon. The other structure has a double bond between the amide nitrogen and the carbonyl carbon. All of the atoms directly connected to the C and N are held in a straight line, the amide plane. This amide link limits the number of orientations possible in the peptide chain and is important in determining the tertiary structure of proteins.

The frequency of association of amino acids and amino acid sequences with particular structural forms allows the predictions of how a primary structure will conform in its final tertiary protein.

Secondary structure

When the primary structure is folded then a secondary structure, made possible by free rotation around bonds, results. A sequence of amino acids will bend in a particular configuration which is determined by the charges on the amino acids in that sequence.

Extended polypeptide chains tend to develop a slight right-handed twist. All cross-over connections between proteins are right-handed. β-sheets of globular proteins (see p. 116) equally are always twisted in a right-handed format. Such twisted sheets form the backbone of protein structures. Proline is an important amino acid which creates an angle within a polypeptide chain.

A major factor in the conformation of proteins is the presence of non-covalent bonds (Figure 5.6). These are usually created in an aqueous solution, the water forming a shell of ionic charges around and between molecules. The non-covalent bonds between different side chains in different regions of the molecules especially proteins include:

- **Van der Waals forces**: these are attractions

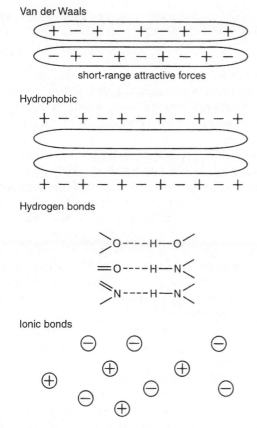

FIGURE 5.6   Non-covalent bonds.

between two adjacent atoms. They are weak reactions but are important in large molecular aggregations.

- **Hydrophobic interactions**: these occur between amino acids with apolar side chains in response to the presence of water in soluble areas of the peptide chain. The hydrophobic amino acids tend to join together, to avoid contact with the aqueous environment and to form hydrophobic regions within the protein. The most hydrophobic of the amino acids, tryptophan, phenylalanine and tyrosine, are characterized by bulky aromatic side chains. Leucine, valine and methionine are less strongly hydrophobic.

- **Hydrogen bonds**: these are weak electrostatic bonds that involve an H atom being shared between two other atoms (both electronegative) such as O and N. Groups in peptide side chains can be donors or H acceptors. One of each is required to form an H-bond. Liquid water may also interact with the NH and CO groups.

- **Ionic interactions**: these occur between oppositely charged side chains. This interaction is modified by water. The basic amino acids, lysine, arginine and histidine, may interact with the acidic aspartic acid and glutamic acid. A basic amino acid may also react with negatively charged phosphate groups and nucleic acid groups. C and N terminal COO^- and N^+ can interact with each other and side chains.

The secondary structure describes various functions:

- **Alpha-helix**. In this formation, polypeptides form right-handed helical spirals, which are long rods in which each residue's carbonyl group forms a hydrogen bond with the amide –NH group four amino acids along the chain (Figure 5.7). The stability of the chain is dependent upon stable hydrogen bonds and tight packing when the chain folds. There is no free space in the centre of the spiral. The side chains project out from the helix. Amino acids vary in their readiness to adopt this configuration. Once the first spiral is established, consecutive spiral forms readily follow, through co-operative folding. Glutamic acid, methionine and alanine are associated with α-helix formation, whereas glycine, proline and asparagine are less frequently involved in such secondary motifs.

- **Beta-sheets.** In this structure, two or more fully extended polypeptide chains, forming flat sheets, are brought together side by side. Hydrogen bonds form between the NH and carbonyl groups of adjacent chains (Figure 5.7). Amino acids with a branched or bulky side chain or an aromatic ring (valine, isoleucine, tyrosine, leucine, phenylalanine, tryptophan) are commonly found in such conformations, unlike glutamic acid, aspartic acid and proline. Proline has no NH group and therefore cannot hydrogen bond.

- **Reverse** or **beta-bends.** These are found in globular proteins. In such structures reverses in chain direction occur. These reverse turns give proteins their globular structure and special geometry. Such a system is created by a carbonyl hydrogen bonding with an amide NH three positions along the chain. The amino acids proline, glycine, asparagine, aspartic acid and serine are particularly associated with such conformations. Glycine, being small, can readily form a flexible hinge. Proline has a more restricted and structurally prescribed formation which facilitates bend formation.

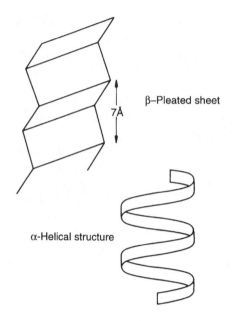

FIGURE 5.7 Secondary structures; alpha (α) helix and flat beta (β)-pleated sheets.

Tertiary structure

The tertiary structure of a protein is the formation assumed by the chain and is the structure with the lowest free energy and is therefore the most stable. The folding structure of a protein is determined primarily by the amino acid sequence. This process of folding takes place in the endoplasmic reticulum. The complicated formation and enormous range of shapes – and hence function – of protein arises from the variation in surface localized charges and shape. The sum of the charges will give the protein a net electrovalent charge in an electrical field.

The distribution of charges on the amino acids is such as to permit a particular tertiary folded structure. The folding takes place in water, along with dissolved cations and anions which create an important environment for this folding process. Cross linkage through –S–S– bonds or hydrogen bonds make for the development of the tertiary structure of proteins.

The affinity of the peptides for water is called **hydrophilicity**, the converse being **hydrophobicity**. Proteins are in contact with water on their surface, i.e. the hydrophilic surface. The normal environment of proteins is water with a relatively high ionic strength produced by a variety of salts. Some proteins are not stable in water but require a modest ionic solution for stability. These salts compete with water in ionic electrostatic interactions, and also increase the solubility of the peptide bond in water.

The hydrophobic sequences are buried in the interior of the protein, in a relatively water-free environment. Most proteins are a mixture of hydrophobic and hydrophilic domains. Hydrophobic characteristics arise from a sequence of hydrophobic amino acids. Water molecules are generally excluded from the protein interiors. On the occasions when there is water in the protein interior, water forms an integral part of the protein structure. There are hydrophobic cavities in the protein, isolated from the aqueous bulk solvent. The water molecules attach by hydrogen bonds to polar groups of the protein. Such buried water molecules fill holes and pair with internal protein polar groups.

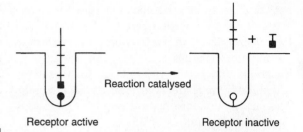

FIGURE 5.8 Enzyme active sites, with receptor seen active and after the reaction has been catalysed. The substrate enters the receptor site and attaches to an amino acid or other polar site. After the reaction the two products disengage and move from the site. The enzyme active site may be activated or inactivated by phosphorylation which changes the bonding characteristics to a favourable or unfavourable state.

Ligands, active sites and receptors

The binding of small molecules to proteins may be important in forming and protecting or changing the structure of a molecule with consequences for function and structure. Proteins may contain a cofactor such as iron or zinc. When a small molecule binds to a protein it is called a **ligand**. Such ligand arrangements are found with enzymes in which there is an active site with a catalytic reaction. When a small molecule binds to a specific site on the protein, there is a change in the conformation in such a way that activity of the protein is altered.

An **active site** (Figure 5.8) is an area of the protein structure which facilitates the binding of a specific ligand through adsorption. If the protein is an enzyme then this region is an active site. If the region is somewhat larger or involves transport or hormonal processes then this region is called a **receptor**. Implicit in the concept of active site is that the ligand is altered chemically in a catalytic process. A receptor will receive and pass on the ligand or a message will be activated, for example across a cell membrane. The formation and function of the active site is dependent on the positioning of specific amino acids in a particular formation. Reactive groups function as electrophiles or nucleophiles. Amino acid nucleophiles in the active site of enzymes include hydroxyl groups of serine and tyrosine, carboxyl groups of aspartate and glutamate and sulphydryl groups of cysteine.

Water is a poor solvent for non-polar substances. Such substances have to force a cavity into which they can nestle; this is an energy-consuming exercise. Once the cavity is formed, the solute will then rearrange its bonds, so that the free energy of the system is minimized, and stability returns. Such areas can occur in hydrophobic domains within the protein tertiary chain. Non-polar molecules tend to aggregate in water, so that the surface area of the cavity decreases.

The tertiary structure of a protein in general is found in two conformations:

- **Fibrous proteins** which have elongated structures with the polypeptide chains extending to long strands. This is based on an α-helix or β-sheet. Such proteins provide structure in the cell or tissue, e.g. connective tissue.
- **Globular proteins** have a tertiary structure which is partially a helical secondary structure. Most enzymes come into this format, as do those proteins involved in gene expression and regulation.

Quaternary structure

Some protein molecules are complexes of more than one polypeptide chain. Each chain with its tertiary structure forms a larger protein molecule. The resultant combination of sub-units gives the quaternary protein structure.

Allosteric proteins are proteins consisting of more than one polypeptide chain. This quaternary

structure exists in a number of alternative conformations and may therefore have different biological properties in each conformation. These
allosteric proteins play an important role in both
metabolic and genetic regulation. Such proteins
are able to rotate freely around covalent bonds.
The change in conformation requires alterations
of non-covalent bonds.

5.3.3 Proteins and evolution

The same protein can be found in a variety of
widely different organisms, from bacteria to
plants and humans, and to some extent proteins
may help to identify phylogenetic relationships.
Proteins which have different functions in the same
organism may have similar structure or sequences.
The respiratory coenzyme cytochrome c has been
found in more than 80 different species, which
suggests a phylogenetic relationship. Similar usage
of a protein by a variety of species include the
globins, the A and B chains of haemoglobin and
myoglobin. Different proteins may evolve at different rates over time. The globin gene may be
traced back 600 million years to an ancestral
invertebrate. Gene evolution appears to owe
much to gene duplication; one gene is conventional and retains the status quo, the other gene
may vary and can be amplified in importance with
changes in the environment.

5.3.4 Allosteric proteins and enzymes

Different allosteric protein and hence enzyme
shapes are dictated by the DNA of the gene which
encodes for that enzyme. The enzyme may be part
of a family of isoenzymes, all of which catalyse
the same reaction but with varying efficiency
because of subtle differences in structure. The
metabolic characteristics of an individual are
determined by the translation by RNA into proteins and isoenzymes which will be of varying
potency and efficiency (Figure 5.9).

Cooperative substrate binding and modifications of enzymatic activity by metabolites may
arise in proteins with two or more structures in
equilibrium. Such proteins are made up of several
sub-units, arranged symmetrically. The structures

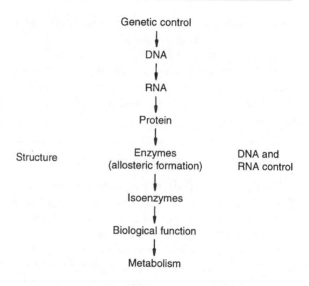

FIGURE 5.9 Allosterism and enzyme activity.

differ in the arrangement of the sub-units and the
number and energy of the bonds between them. In
one situation, the sub-units may be held together
by strong bonds that would resist the tertiary
change needed for substrate binding. In the other
state these bonds would be relaxed. In the transition between them the symmetry of the molecule
would be preserved.

Allosteric enzymes have the biological advantage that no direct interaction occurs between the
substrate of the protein and the regulatory metabolite that controls its activity because control
would be due entirely to a change in structure
induced in the protein when it binds its specific
effector. Such allosteric enzymes include glycogen
phosphorylase and glutamine synthetase.

Glycogen phosphorylase

Glucose phosphorylase is the key enzyme in the
control of glycolysis – the mobilization of energy
from glycogen. This is a complex allosteric protein that is subject to activation and inhibition.
Phosphorylase kinase is activated by phosphorylation of a pair of serine residues and inhibited by
hydrolysis of the serine phosphate bonds. In muscle, phosphorylase kinase is activated by the
release of calcium ions from the sarcoplasmic

reticulum, which also stimulates muscle contraction. When activated phosphorylase kinase catalyses the stepwise phosphorylation of glycogen with the release of glucose-1-phosphate.

Glycogen phosphorylase structure

The enzyme is a dimer of two identical sub-units, each a single polypeptide chain of 842 amino acid residues to which a pyridoxal phosphate is attached to Lysine680. Its unphosphorylated form, phosphorylase B, is inactive but is activated by cyclic AMP when it reaches 80% of the activity of the phosphorylated form. This form, known as phosphorylase A, exhibits near-maximal activity without cAMP. Each of the two forms is subject to regulation by other factors. Phosphorylase B is partially activated by the weak effector inosine monophosphate (IMP) and phosphorylase A by glucose. Phosphorylase is a complex structure, each of its sub-units consist of two domains of a core of pleated β-sheets flanked by α-helices. The N-terminal domain includes the serine phosphate, the activating AMP and inhibiting glucose 6-phosphate (G6P) binding sites, the glycogen storage site and a small part of the catalytic site. The C-terminal domain complements the catalytic site and also contains the neighbouring site where the inhibitory nucleosides and purines bind. In the enzyme dimer the two sub-units are joined end-to-end in a small contact making up no more than 7% of the surface area in the phosphorylase B and 10% in A. One side of the dimer is convex with a radius of curvature matching that of the glycogen particle. It contains the entrance to the catalytic tunnel and the glycogen storage site. The side that faces away from the glycogen particle contains a regulatory phosphorylation site and the overlapping AMP and G-6-P binding sites.

Most of the binding sites for substrates and effectors are widely separated. Despite this, binding of ligands to any of the sites can affect all the others. Change in phosphorylation of Serine14 induces a change from the weakly activated B to the inhibited A structure at the sub-unit boundary. Such a phosphorylation results in the burial and ordering of the amino-terminal 16 residues and the exposure and disorder of the carboxy-terminal five residues in phosphorylase A and the reversal in phosphorylase B. This is accompanied by changes in hydrogen bonding. The dominant interaction is responsible for the allosteric transition. The tense structure (T) state is inactive and the active state is the relaxed state (R). The transition from the T to the R structure consists of rotation of one sub-unit relative to the other by 1°. Allosteric effects are provided by tower helices which tilt and slide relative to each other close to the active centre. These helices rigidly block access to the catalytic site, a polypeptide loop in the T structure. Displacement allows substrates as well as additional cation side chains, to move into the catalytic site. This converts the coenzyme phosphate from the mono to the dianionic form, which is necessary for catalytic activity.

Phosphofructokinase

This is a further controller of glycolysis in the cells. As in haemoglobin, cooperativity and feedback inhibition arise from a transition between two alternative quaternary structures in which one pair of rigidly linked sub-units rotates relative to the other and rearranges the relative bonds between them. The effector sites lie at the sub-unit boundaries where the allosteric transitions take place. In phosphofructokinase the catalytic sites span the sub-unit boundary where the allosteric transitions take place and are directly affected by the position of the boundary. The alterations in function are the result of changes in tertiary structure, comparable with a set of levers.

Summary

1. Proteins are high molecular weight polyamides, consisting of one or more chains of amino acids which then fold into a form that imparts a particular function.

2. The amount of each of the amino acids in a protein, and their place in the primary structure, decides the shape and biological properties of that protein. Specific groupings of amino acids in one section of the protein may result in biological properties which are individual to that protein, e.g. in an enzyme active centre.

3. Proteins consist of a sequence of amino acids which is singular to that protein. The sequence is called the primary structure. Amino acids form peptides and proteins by linkage through a covalent peptide bond.

4. When the primary structure is folded then a secondary structure results, made possible by free rotation around bonds.

5. A major factor in the conformation of proteins are non-covalent bonds. The non-covalent bonds between different side chains in different regions of the molecules especially proteins include Van der Waals, hydrophobic interactions, hydrogen bonds and ionic bonds. The secondary structure describes α-helix or β-sheet formation.

6. The tertiary structure of a protein is the formation assumed by the chain. It is the structure with the lowest free energy and therefore the most stable. It is often found as fibrous proteins. Such proteins provide structure in the cell or tissue. Globular proteins have a helical secondary structure. Most enzymes and proteins involved in gene expression and regulation fit into this format.

7. A small molecule which binds to a protein is called a ligand. Such ligand arrangements are found with enzymes in which there is an active site with a catalytic reaction.

8. Some protein molecules are complexes of more than one polypeptide chain and form a larger protein molecule, the quaternary protein structure, e.g. allosteric proteins. There are a number of alternative conformations of this quaternary structure which have different biological properties, an important principle in metabolic and genetic regulation.

9. Different allosteric protein and enzyme shapes are dictated by the DNA of the gene encoding that enzyme. The metabolic characteristics of an individual are determined by the translation of RNA into different isoenzymes.

Further reading

Creighton, T.E. (1984) *Protein Structures and Molecular Principles*, W.H. Freeman & Co., New York.

Dean, P.M. (ed.) (1995) *Molecular Similarity in Drug Design*, Blackie Academic and Professional, London.

Doolittle, R.F. (1995) The multiplicity of domains in proteins. *Annual Review of Biochemistry*, **64**, 287–314.

Food and Agriculture Organization/World Health Organization/United Nations (1985) *Energy and Protein Requirements*. Technical Report Series, no. 724, WHO, Geneva.

Monod, J., Changeux, J.P. and Jacob, F. (1963) Allosteric proteins and molecular control systems. *Journal of Molecular Biology*, **6**, 306–29.

Perutz, M. (1990) *Mechanisms of Cooperativity and Allosteric Regulation*, Oxford University Press, Oxford.

Reeds, P.J. (1990) Amino acid needs and protein storing patterns. *Proceedings of the Nutrition Society*, **49**, 489–97.

Zubay, G. (1991) *Biochemistry*, 3rd edn, W.C. Brown, Iowa, USA.

5.4 Fats

- Dietary fat consists of triglycerides, phospholipids and sterols.
- Triglycerides consist of glycerol and three fatty acids.

- Glycerol is the alcohol component of glycerides, phospholipids and wax. The distribution of individual fatty acids as esters of glycerol carbons 1, 2 and 3 is important.
- Dietary fatty acids usually have the *cis*-formation.
- The major classes of fatty acids are distinguished by the distribution and number of double bonds.
- The four groups of fatty acids, n-3, n-6, n-7 and n-9, cannot interconvert.
- Essential dietary fatty acids linoleic (n-6) and α-linolenic (n-3) are not synthesized by humans and are essential to the diet.
- Linoleic and α-linolenic acid acids can be further extended to longer chain polyunsaturated fatty acids.
- *Trans* unsaturated fatty acids in general behave and are metabolized as saturated fatty acids.
- Cholesterol belongs to the family of steroid alcohols important in cell wall and membrane structure and is a precursor of bile acids, of adrenal and gonadal hormones and vitamin D.
- Lipids are structural, provide energy stores or have a metabolic role.

5.4.1 Introduction

Fat is a heterogeneous mixture of lipids, predominantly triglycerides but also includes phospholipids and sterols. Triacylglycerols (triglycerides) are the principal dietary lipids which are stored in fat stores in humans and consist of esters of fatty acids, both saturated and unsaturated and glycerol. Cholesterol, cholesterol ester and phospholipids are important in the structure of cell membranes, mitochondria, lysosomes and the endoplasmic reticulum. Lipids are important in providing insulation against the cold.

The principal dietary sources of fat are dairy products, meat, margarine and other fats, biscuits, cakes and pastries. Plant storage fats are present in nuts, cereal grains and fruits such as the avocado. Other dietary lipids include cooking fats, salad oils and mayonnaises. Eggs are a source of lipid, predominantly as saturated and mono-unsaturated fatty acids, lipoprotein, triacylglycerols, cholesterols and phospholipids. Many foods contain structural fats, phospholipids and glycolipids, cholesterol and plant sterols. Eating brain as a food provides animal sphingolipids to the diet. Dairy products contain milk fat globule membranes. Green leafy vegetables contain galactolipids and there are membrane lipids in cereal, grains, vegetables and fruit.

5.4.2 Major lipid categories

Fatty acids (see Table 5.2)

Saturated fatty acids

Most saturated fatty acids are straight-chain structures with an even number of carbon atoms. Saturated fatty acids have a basic formula $CH_3(CH_2)_nCOOH$ where n can be any even number from 2 upwards. Almost without exception dietary fatty acids are formed as even numbers of carbons in an unbranched chain. In naturally occurring lipids the number of carbons in the chain ranges from two to more than 30, the most common fatty acid being palmitic, n = 14 and stearic acid, n = 16. The generic structure of a straight-chain saturated fatty acid is $CH_3-(CH_2-CH_2-)_nCOOH$.

Unsaturated fatty acids

Of nutritional and biological importance are the fatty acids with double bonds, i.e. unsaturated. Removal of hydrogen bonds results in ethylenic double bonds which are unsaturated. The hydrogen atoms on either side of the double bond in the fatty acid molecules are of *cis* geometrical configuration or *trans* configuration (Figure 5.10). These are **stereoisomers** in that the two forms differ in the arrangement of their atoms in space; *cis-* means that the hydrogens are on the same side

TABLE 5.2 Name, structure and occurrence of the most common fatty acids. (Reproduced from Geigy Scientific Tables, Ciba-Gergy Ltd., with permission.)

	Name	Structure	Remarks
Saturated straight chain acids			
C-1	Formic acid (methanoic acid)	H·COOH	Occurs in human urine and many plant materials
C-2	Acetic acid (ethanoic acid)	CH_3·COOH	Present in most biological materials. Formed from ethanol by many species of aerobic bacteria and from pentoses by some anaerobic species
C-3	Propionic acid (propanoic acid)	CH_3·CH_2·COOH	Formed by bacterial decomposition of carbohydrates
C-4	n-Butyric acid (butanoic acid)	CH_3·$(CH_2)_2$·COOH	Occurs in traces in many fats
C-5	n-Valeric acid (pentanoic acid)	CH_3·$(CH_2)_3$·COOH	
C-6	Caproic acid (hexoic acid, hexanoic acid	CH_3·$(CH_2)_4$·COOH	
C-8	Caprylic acid (octanoic acid)	CH·$(CH_2)_6$·COOH	Component of many fats
C-10	Capric acid (decanoic acid)	CH_3·$(CH_2)_8$·COOH	Component of many animal and vegetable fats
C-12	Lauric acid (dodecanoic acid	CH_3·$(CH_2)_{10}$·COOH	Major component of vegetable fats (esp. laurel). In smaller quantities in depot fat of animals, milk fat, fishliver oils
C-14	Myristic acid (tetradecanoic acid)	CH_3·$(CH_2)_{12}$·COOH	Component of almost all animal fats (1–5%) and vegetable fats, esp. milk fat, fish oils, palm oil, nutmegs
C-16	Palmitic acid (hexadecanoic acid)	CH_3·$(CH_2)_{14}$·COOH	Widely distributed in nature. Present in almost all fats
C-17	Margaric acid (heptadecanoic acid)	CH_3·$(CH_2)_{15}$·COOH	Occurs in traces in mutton fat
C-18	Stearic acid (octadecanoic acid)	CH_3·$(CH_2)_{16}$·COOH	Found abundantly in important edible fats. Also occurs in vegetable fats
C-20	Arachidic acid (eicosanoic acid)	CH_3·$(CH_2)_{18}$·COOH	Occurs in traces in many seed and animal fats
	Heneicosanoic acid	CH_3·$(CH_2)_{19}$·COOH	
C-22	Behenic acid (docosanoic acid)	CH_3·$(CH_2)_{20}$·COOH	Present in traces in animal fats and seed fats. Constitutes 50% of the spleen cerebrosides in GAUCHER's disease

C-24	Lignoceric acid (tetracosanoic acid)	$CH_3 \cdot (CH_2)_{22} \cdot COOH$	Component of sphingomyelins and of kerasin (spleen cerebroside in GAUCHER's disease). Also found in some vegetable fats and bacterial and insect waxes
C-26	Cerotic acid (hexacosanoic acid)	$CH_3 \cdot (CH_2)_{24} \cdot COOH$	Occurs free and combined. In Chinese wax (cetyl ester), beeswax, wool fat
C-28	Montanic acid (octacosanoic acid)	$CH_3 \cdot (CH_2)_{26} \cdot COOH$	Component of montan wax, beeswax, Chinese wax
C-30	Melissic acid (triacontanoic acid)	$CH_3 \cdot (CH_2)_{28} \cdot COOH$	Occurs in beeswax

Mono-unsaturated fatty acids

C-10	cis-Δ^9-Decenoic acid	$CH_2 = CH(CH_2)_7 \cdot COOH$	Occurs in butter and milk fats and sperm head oil
C-12	Linderic acid (cis-Δ^9-dodecenoic acid)	$CH_3 \cdot (CH_2)_6 CH = CH(CH_2)_2 COOH$	Occurs in various seed oils, e.g. Lindera obtusiloba
C-12	Lauroleic acid (cis-Δ^9-dodecenoic acid)	$CH_3(CH_2)_5 CH = CH(CH_2)_3 COOH$	Occurs in sperm head oil and blubber
C-14	Myristoleic acid (cis-Δ^9-tetradecenoic acid)	$CH_3(CH_2)_3 CH = CH(CH_2)_7 COOH$	Occurs in milk fat and depot fat of many animals
C-16	Palmitoleic acid (cis-Δ^9-hexadecenoic acid)	$CH_3(CH_2)_5 CH = CH(CH_2)_7 COOH$	Widely distributed marine oils depot and milk fats, animals, vegetable oils and fats
C-18	cis-Δ^6-Octadecenoic acid (petroselinic acid)	$CH_3 \cdot (CH_2)_{10} \cdot CH = CH \cdot (CH_2)_4 \cdot COOH$	Occurs in seeds of aromatic plants (parsley, celery, etc.) and in some umbellate fats
C-18	Oleic acid (cis-Δ^9-octadecenoic acid)	$CH \cdot (CH_2)_7 \cdot COOH$ $\|\|$ $CH \cdot (CH_2)_7 \cdot CH_3$	Most abundant of the unsaturated fatty acids. Present in nearly all natural fats (one-third of fatty acids of cow's milk; phosphatides). Occurs in traces in human urine.
C-18	Elaidic acid (trans-Δ^9-octadecenoic acid)	$CH_3 \cdot (CH_2)_7 \cdot CH$ $\|\|$ $CH \cdot (CH_2)_7 \cdot COOH$	Formed by isomerization of oleic acid
C-18	trans-Vaccenic acid (trans-Δ^9-octa-decenoic acid)	$CH_3 \cdot (CH_2)_5 \cdot CH$ $\|\|$ $CH \cdot (CH_2)_9 \cdot COOH$	Occurs in many animal fats and vegetable oils
C-18	Δ^{12}-Octadecenoic acid	$CH_3 \cdot (CH_2)_4 \cdot CH = CH \cdot (CH_2)_{10} \cdot COOH$	Occurs in partially hydrogenated peanut oil
C-20	Gadoleic acid (Δ^9-eicosenoic acid)	$CH_3 \cdot (CH_2)_9 \cdot CH = CH \cdot (CH_2)_7 \cdot COOH$	Cis and trans forms. In many fish and marine animal oils, in vegetable oils, in brain phosphatides
C-22	Cetoleic acid (Δ^{11}-docosenoic acid)	$CH_3 \cdot (CH_2)_9 \cdot CH = CH \cdot (CH_2)_9 \cdot COOH$	Occurs in various marine oils

	Name	Structure	Remarks
C-22	Erucic acid (cis-Δ^{13}-docosenoic acid)	$CH\cdot(CH_2)_{11}\cdot COOH$ $\parallel$ $CH\cdot(CH_2)_7\cdot CH_3$	Occurs in seed oils, esp. rapeseed oil
C-22	Brassidic acid ($trans$-Δ^{13}-docosenoic acid)	$CH_3\cdot(CH_2)_7\cdot CH$ $\parallel$ $CH\cdot(CH_2)_{11}\cdot COOH$	Formed by isomerization of erucic acid
C-24	Selacholeic acid (nervonic acid, cis-Δ^{15}-tetra-cosenoic acid)	$CH\cdot(CH_2)_{13}\cdot COOH$ $\parallel$ $CH\cdot(CH_2)_7\cdot CH_3$	Occurs in shark and ray liver oils, in brain cerebrosides (nervone) and sphingomyelins
C-26	Ximenic acid (Δ^{17}-hexa-cosenoic acid)	$CH_3\cdot(CH_2)_7\cdot CH = CH\cdot(CH_2)_{15}\cdot COOH$	Occurs in *Ximenia americana* (tallow-wood). A hexacosenoic acid is found with nervonic acid in brain cerebrosides

Polyunsaturated fatty acids

	Name	Structure	Remarks
C-6	Sorbic acid ($\Delta^{2,4}$-hexadienoic acid)	$CH_3\cdot CH = CH\cdot CH = CH\cdot COOH$	Occurs as lactone in oil of unripe mountain ash berries
C-18	Linoleic acid (cis-cis-$\Delta^{9,12}$-octa-decadienoic acid)	$CH_3\cdot(CH_2)_4\cdot CH$ $\parallel$ $CH\cdot CH_2\cdot CH$ $\parallel$ $CH\cdot(CH_2)_7\cdot COOH$	Widely distributed in plant, esp. in linseed, hemp and conttonseed oils. Also in lipids of animals (component of phosphatides, etc.). Essential dietary component
C-18	Hiragonic acid ($\Delta^{6,10,14}$-hexadecatrienoic acid)	$CH\cdot(CH_2)_2\cdot CH = CH\cdot CH_3$ $\parallel$ $CH\cdot(CH_2)_2\cdot CH = CH\cdot(CH_2)_4\cdot COOH$	Occurs in sardine oil
C-18	Linolenic acid ($\Delta^{9,12,15}$)	$CH\cdot CH_2\cdot CH = CH\cdot CH_2\cdot CH_3$ $\parallel$ $CH\cdot CH_2\cdot CH = CH\cdot(CH_2)_7\cdot COOH$	Occurs in many vegetable oils, esp. drying oils such as linseed oil. Also in traces in animal fats (phosphatides). Essential dietary component
C-20	Timnodonic acid ($\Delta^{4,8,12,15,18}$-eicosapentaenoic acid)	$CH_3\cdot CH = CH\cdot CH_2\cdot CH = CH$ $\mid$ $CH\cdot(CH_2)_2\cdot CH = CH\cdot CH_2$ $\parallel$ $CH\cdot(CH_2)_2\cdot CH = CH\cdot(CH_2)_2\cdot COOH$	Occurs in sardine oil, cod-liver oil, pilot whale oil and oil from *Squalus sucklei* (spiny dog fish)
C-20	Arachidonic acid ($\Delta^{5,8,11,14}$-eicosatetraenoic acid)	$CH_3\cdot(CH_2)_4\cdot CH = CH\cdot CH_2\cdot CH$ $\parallel$ $CH\cdot CH_2\cdot CH = CH\cdot CH_2\cdot CH$ $\parallel$ $CH\cdot(CH_2)_3\cdot COOH$	Occurs in animal lipids (liver, phosphatides). Synthesized in animals from dietary linoleic acid

C-22	Clupanodonic acid ($\Delta^{4,8,12,15,19}$-doco-sapentaenoic acid)	$CH_3 \cdot CH_2 \cdot CH = CH \cdot (CH_2)_2 \cdot CH$ $| $ $CH \cdot (CH_2)_2 \cdot CH = CH \cdot CH_2 \cdot CH$ $\|$ $CH \cdot (CH_2)_2 \cdot CH = CH \cdot (CH_2)_2 \cdot COOH$	Occurs in fish oils
C-24	Nisinic acid ($\Delta^{4,8,12,15,18,21}$-tetracosahexaenoic acid)	$CH_3 \cdot CH_2 \cdot CH = CH \cdot CH_2 \cdot CH = CH \cdot CH_2$ $CH \cdot (CH_2)_2 \cdot CH = CH \cdot CH_2 \cdot CH = CH$ $\|$ $CH \cdot (CH_2)_2 \cdot CH = CH \cdot (CH_2)_2 \cdot COOH$	Occurs in tunny oil

Branched chain fatty acids

C-4	Isobutyric acid (2-methyl-propanoic acid)	$\begin{matrix} CH_3 \\ CH_3 \end{matrix} \!\!\!> CH \cdot COOH$	Occurs free in carob beans (*Ceratonia siliqua*), as ethyl ester in croton oil; also in faeces and as product of enzymic breakdown of proteins. Intermediate in metabolism of valine
C-5	Isovaleric acid (3-methylbutanoic acid)	$\begin{matrix} CH_3 \\ CH_3 \end{matrix} \!\!\!> CH \cdot CH_2 \cdot COOH$	Occurs in root of valerian, tobacco leaves, volatile oils, depot fat of dolphins and porpoises, as glyceride in human faeces. Formed from leucine in bacterial degradation of proteins. Intermediate in metabolism of leucine
C-5	Tiglic acid (*cis*-2-methyl-Δ^2-butenoic acid)	$CH_3 \cdot CH = C \cdot COOH$ $\qquad\quad \|$ $\qquad\quad CH_3$	Occurs in croton oil (glyceride); in Roman cumin oil (esters), in geranium oils. Intermediate in metabolism of isoleucine

Hydroxy fatty acids

C-18	Ricinoleic acid (*cis*-12-hydroxy-Δ^9-octadecenoic acid)	$CH \cdot CH_2 \cdot CH(OH) \cdot (CH_2)_5 \cdot CH_3$ $\|$ $CH \cdot (CH_2)_7 \cdot COOH$	As glyceride, chief constituent of castor oil
C-23	2-Hydroxytricosanoic acid	$CH_3 \cdot (CH_2)_{20} \cdot CH(OH) \cdot COOH$	Component of normal brain cerebrosides to extent of about 7% of total fatty acids
C-24	Cerebronic acid (phrenosinic acid, 2-hydroxytetracosanoic acid)	$CH_3 \cdot (CH_2)_{21} \cdot CH(OH) \cdot COOH$	Component of cerebroside phrenosin (cerebron). About 15% of total fatty acids of brain cerebrosides
C-24	2-Hydroxynervonic acid (2-hydroxy-Δ^{15}-tetracosenoic acid)	$CH_3 \cdot (CH_2)_7 \cdot CH = CH \cdot (CH_2)_{12} \cdot CH(OH) \cdot COOH$	Component of cerebroside hydroxynervone (of which the isomeric Δ^{17}-acid is also a component). About 12% of total fatty acids of brain cerebrosides

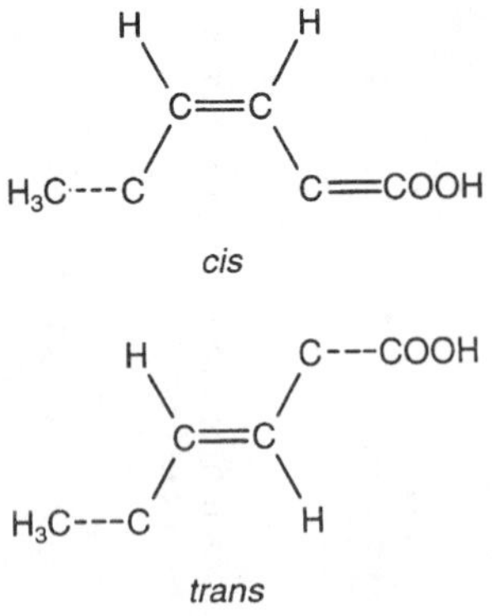

FIGURE 5.10 Geometric isomerism in unsaturated fatty acids; *cis* and *trans*.

(the most common configuration for fatty acids in nature), whereas in *trans* the hydrogens are on the opposite side. This results in differing physical properties and response to enzymatic attack.

Monoenoic unsaturated fatty acids These contain one unsaturated double bond. The more common have an even number of carbon atoms with chain length of 16–22 carbons and a double position with the *cis*- position, often in the delta 9 position. A double bond causes restriction in the movement of the acyl chain at that point. The *cis* configuration introduces a kink into the average molecular shape. This means that the *cis* form is less stable thermodynamically and has a lower melting point than the *trans* form.

Polyunsaturated fatty acids These are derived from monoenoic fatty acids, the position of the second double bond being dictated by the synthetic processes. Mammalian enzymes may only remove hydrogen atoms between an existing double bond and the carboxyl group. Further desaturation has to be preceded by chain elongation. Unsaturated fats with one or more *cis* double bond are more common in natural lipids than are *trans*. The *cis* bond affects the linearity of the chain of methylene groups in that the molecule folds back on itself. Polyunsaturated fats are susceptible to oxidation but are protected by natural antioxidants, e.g. vitamin E.

Nomenclature of polyunsaturated fatty acids

There are several nomenclatures for the polyunsaturated fatty acids. The nomenclature may be based on the saturated parent acid, number of carbon atoms and position of the double bonds. The differences are dictated by whether the numbering is taken from the methyl or carboxyl end. Numbering from the carboxyl (COOH) end is the chemical nomenclature and called the Geneva system. Numbering is from the carbon 1 carboxyl group (Figure 5.11). Examples:

oleic acid; *cis*-9-octadecaenoic acid
elaidic acid; *trans*-9-octadecaenoic acid

A short-hand system indicates the number of double bonds: stearic acid C 18 : 0 and oleic or elaidic acid C 18 : 1. The position of the double bond is given by *cis*-9-, 18 : 1 (oleic acid), or *trans*-9, 18 : 1 (elaidic acid).

An important aspect of unsaturated fatty acids is the opportunity for isomerism, which may be either **positional** or **geometric**. Positional isomerism occurs when double bonds are located at different positions in the carbon chain. A 16-carbon mono unsaturated fatty acid may have positional isomeric forms with double bonds at C7 and C9. These are called Δ7 and Δ9, the position of unsaturation is numbered with reference to the first of the pair of carbon atoms between the double bond.

Linoleic acid can be written as *cis* (Δ-)9, *cis* (Δ-)12-18 : 2 or (*cis,cis*)9,12-octadecadienoic acid, showing that it is an 18-carbon fatty acid with *cis* double bonds 9 and 12 carbons from the carboxyl end.

The alternative nutritional or biological system uses the prefix n – or historically ω. In this system the numbering is from the terminal methyl end. The main dietary unsaturated fatty acids families are n-3, n-6, n-7, n-9. This numbering system is determined by the position of the first double bond from the methyl carbon atom (Figure 5.11). This double bond determines the number of double bonds which can be inserted.

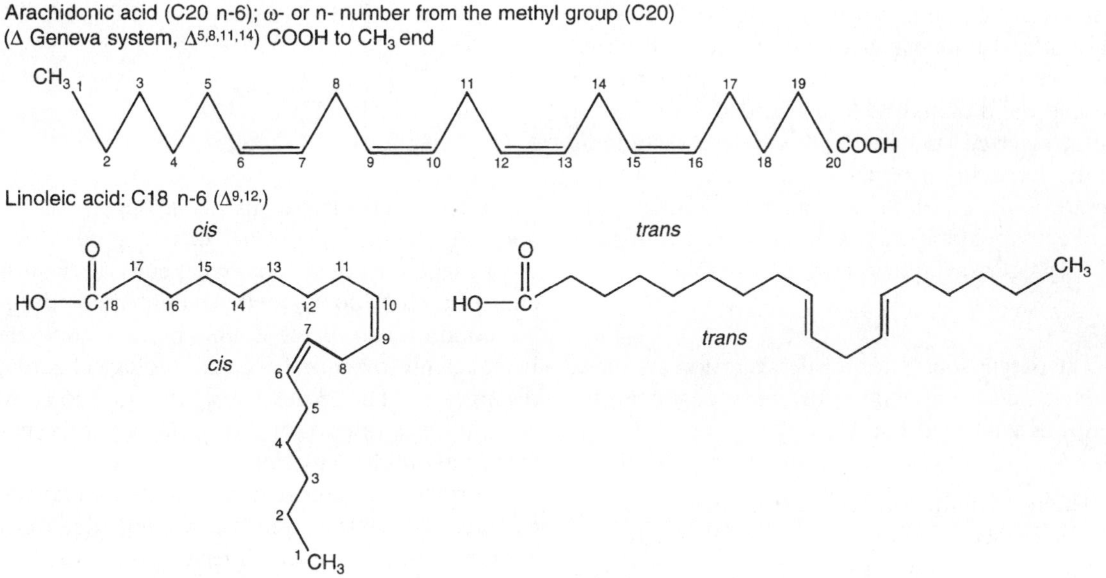

Arachidonic acid (C20 n-6); ω- or n- number from the methyl group (C20)
(Δ Geneva system, $\Delta^{5,8,11,14}$) COOH to CH_3 end

Linoleic acid: C18 n-6 ($\Delta^{9,12,}$)

FIGURE 5.11   Structure of polyunsaturated fats and numbering systems with the ω- or n- system used in nutrition and delta (D) or Geneva system of chemistry. Examples are given of arachidonic acid, C20 n-6, and linoleic acid, C18 n-6. The chemical numbering with the Geneva system is from the carboxyl group so that the carboxyl group is 1 and the double bonds are indicated by the proximal carbon of the double bond, e.g. C20: (the position of a *cis* or *trans* is indicated. The ω- or n- numbering is from the methyl end.

Essential fatty acids

The essential fatty acids and their longer-chain molecular products are necessary for the maintenance of growth, good health and reproduction. They are important in biological membranes and affect the permeability of the membrane to water and sugars and metal ions.

Deficiency of essential fatty acids in adults is rare but has been seen in children fed virtually fat-free diets. The skin abnormalities were those previously seen in experimental animals with dermatosis of the skin, increased water permeability, increased sebum secretion and decreased epithelial hypoplasia.

Humans are unable to desaturate fatty acids towards the methyl end of the fatty acid chain, a failure which produces three distinct non-interconvertible families of fatty acid. Polyunsaturation is undertaken by three desaturases, Δ4, 5 and 6 which introduce double bonds between carbon atoms 4–5, 5–6 and 6–7.

Fatty acid deficiency

When animals are made deficient in fatty acids the body weight decreases, the heart enlarges, and there is decreased capillary resistance. Cholesterol accumulates in the lungs and endocrine organs alter, in that the thyroid gland shrinks and abnormalities in reproduction have been described in both females and males. The essential fatty acid activity of any one fatty acid is measured by restoration of normality to rats deprived of essential fatty acids. This consists of growth and restoration of water permeability and normal skin.

Animals are unable to insert double bonds into fatty acids at carbon position 12 and 15; therefore linoleic and linolenic acids cannot be synthesized and are essential. The fatty acids which have a double bond at n-3, n-6, n-7 and n-9 cannot be interconverted in animal tissues. Arachidonic acid, the main product of the elongation and desaturation of linoleic acid, has essential fatty acid activity but is only essential when insufficient of its precursor linoleic acid is available.

The parent substrates for desaturation are oleic, linoleic and α-linolenic acids. They give rise to a series of fatty acid families:

- The **n-9 family**, originating from oleic acid n-9; a non-essential fatty acid family. Oleic acid (*cis*-9,C-18 : 1) is the parent fatty acid. Oleic acid can increase the chain length to become eicosatrienoic acid.
  C-18 : 2, 6, 9
  C-20 : 2, 8, 11
  C-20 : 3, 5, 8, 11
  C-22 : 3, 7, 10, 13
  C-22 : 4, 7, 10, 13

- The **n-6 family**, originating from linoleic acid n-6; linoleic acid: *cis, cis*-9,12 (C-18 : 2) is the parent fatty acid, with the n-6 numbering from the methyl end. Linoleic acid can extend to gamma-linolenic and arachidonic (C-20 : 4, n-6)
  C-18 : 3, 9, 12, 15
  C-18 : 4, 6, 9, 12, 15
  C-20 : 4, 8, 11, 14, 17
  C-20 : 5, 8, 11, 14, 17
  C-20 : 5, 7, 10, 13, 16, 19
  C-22 : 4, 7, 10, 13, 16, 19

- The **n-3 family**, originating from α-linolenic acid n-3; α-linolenic acid: *cis, cis, cis*-9, 12,15 (C-18 : 3) is the parent fatty acid. α-Linolenic acid (C-18 : 3, n-3) can add methylene groups to increase the chain length to eicosapentaenoic acid.

C-18 : 3, 9, 12, 15
C-18 : 4, 6, 9, 12, 15
C-20 : 4, 8, 11, 14, 17
C-20 : 5, 8, 11, 14, 17
C-20 : 5, 7, 10, 13, 16, 19
C-22 : 4, 7, 10, 13, 16, 19

The essential fatty acids all belong to the n-3 and n-6 groups, but not n-7 or n-9.

Essential fatty acid activity depends on the presence of a *cis*-9, *cis*-12 methylene-interrupted double bond system. If the double bond is converted from *cis* into *trans* this essential biological activity disappears. The process of desaturation and elongation is important in the tissue synthesis of some polyunsaturated fats.

The parent unsaturated fatty acids are extended by alternate desaturation, i.e. the introduction of a double bond and chain-lengthening reactions. The enzymes involved in this are limited in number and fatty acids from each family compete for the enzymes. Linoleic acid and linolenic acid are the preferred substrates.

If there is a deficiency of essential fatty acids then non-essential fatty acids are metabolized preferentially. The end-products of non-essential fatty acids extension cannot function in cell membranes or in eicosanoid precursors. The ratio of eicosatrienoic acid (all *cis*-5,8,11; C-20 : 3 from oleic acid) to arachidonic acid (all *cis*-5,8,11,14; C-20 : 4 from linoleic acid) in the plasma (the triene : tetraene ratio) is a biochemical index of essential fatty acid deficiency. The ratio should be between 0.2–0.4. It is possible that *trans* fatty acids may influence the metabolism of essential fatty acids by inhibiting desaturases.

All essential fatty acids are polyunsaturated fats. All polyunsaturated fatty acids are not necessarily essential. There is competition between n-6 (essential) and n-9 (non-essential) and also between n-6 fatty acids. Linoleic acid is essential at a dietary level of about 1% of energy. The n-3 fatty acids (parent α-linolenic acid) are important components of brain and retinal lipid tissue. These fatty acids cannot be synthesized in the animal and their dietary provision may be particularly important in early life.

Desaturation of fatty acids; the basis of essentiality

The initial desaturation introduces a double bond at position 6 into the first member of each fatty acid family. The desaturation enzyme is common to all of these processes. The order of affinity of the substrate for the $\Delta 6$ desaturase is 18:3 < 18:2 < 18:1. Linoleic acid is converted into arachidonic acid by this enzyme. If the absolute amount of linoleic acid is low or absorption is abnormal, then deficiency problems arise. Alternatively, deficiency may result from genetic disorders, e.g. a lack of specific desaturase enzymes. These include rare diseases such as Reye's syndrome and the Prader–Willi syndrome. There may also be reduced $\Delta 6$ desaturase activity as a result of normal biological variation. Gamma-linolenic acid bypasses this enzyme system. If the diet contains small amounts of linoleic acid but a massive amount of other fatty acids, the $\Delta 6$ desaturase system may be overwhelmed. If there is insufficient linoleic acid, then oleic acid is extended to a 20-carbon atom structure cis-5,8,11-eicosatrienose instead of arachidonic acid.

Dietary fatty acids

Dietary lipids will contain saturated, mono, unsaturated and polyunsaturated fatty acids. The amount and type will differ from food source to food source. Mammalian animal storage fats are predominantly saturated (e.g. palmitic acid) and mono unsaturated (e.g. oleic acid) fatty acids. Ruminant fats contain monounsaturated (e.g. stearic acid) fatty acids because of the rumen desaturases. Milk fat contains a high proportion of saturated fatty acids with a chain length of 12 carbon atoms or less (C4–C8). Milk from cows, sheep and goats is relatively rich in short- and medium-chain fatty acids and *trans* and branched-chain fatty acids. The proportions are seasonal, being dependent upon the feed. Human milk is somewhat richer in oleic acid than ruminant milk, and contains 7% of linoleic acid.

The unsaturated C16, C18, C20 and C22 fatty acids predominate in the lipids of freshwater plant or animal life. The important saturated fatty acid is palmitic acid, usually present as 10–18% of total fatty acids. The oils of fish and marine animals are rich in polyunsaturated fatty acids of the n-3 family, including C20 and C22 with up to six double bonds. Freshwater plant and animal fats contain unsaturated C16, C18, C20 and C22 fatty acids. The only important saturated fatty acid is palmitic acid.

Fatty acids in vegetable and fish oils

Vegetable oils	Fish oils
n-6	n-3
18 : 2 linoleic	18 : 3 linolenic
18 : 3	18 : 4
20 : 3	20 : 4
20 : 4 arachidonic eicosapentanoenoic acid	20 : 5
22 : 4	22 : 5
22 : 5	22 : 6

In land animals the unsaturated oleic acid and the saturated palmitic acid predominate and are found in hard fats with a low melting point. Dietary fats contain small amounts of short-chain C4 and C8 fatty acids. In plant seeds, oleic acid and palmitic acid are predominant with linoleic acid a minor component. A diet rich in linoleic (n-6) and α-linolenic acid (n-3 family) can be obtained by eating vegetable seed oils. Arachidonic acid (n-6) is not present in vegetable oils but is synthesized from linoleic acid, meat being a good dietary source. Erucic acid (C22 : 1) is the principal fatty acid in the oil of the rape plant seed, an important temperate crop despite the erucic acid being toxic to the myocardium. A new variety of rape seed, cambra, contains only 2% erucic acid.

An important effect on the food fatty acid composition is industrial processing, particularly catalytic hydrogenation which improves the stability and physical properties of the fat by reducing the overall fatty acid unsaturation. The total number of double bonds in the fatty acid molecule is reduced and the double bonds are shifted along the hydrocarbon chain. A proportion of the double bonds present in the original oil in the *cis* configuration are isomerized to the *trans* configuration.

Some seed oils have a significant fatty acid content of *trans* unsaturation, though this is not common. All green plants contain small amounts of *trans*-3-hexadecenoic acid. Ruminant fat may also contain isomeric fatty acids. Fatty acids with *trans* bonds may also be monounsaturated or polyunsaturated, with *cis* and *trans* bonds within the same molecules. The British diet contains on average 6–7 g of *trans* unsaturated fatty acids, half of which comes from industrial hydrogenation and the remainder from ruminant products.

Bacteria are a source of branched-chain fatty acids which are usually saturated, though these can also occur in animal fats. The presence of a branched chain lowers the melting point. Butter fats, bacterial and skin lipids contain significant amounts. Chain branch and substitution is noted by prefix br-16 : 0 for branched chain, hexadecaenoic acid or ho-16 : 0 for hydroxypalmitic acid.

Trans *fatty acids*

The biohydrogenation of fats occurs commonly in the rumen of ruminants. Polyunsaturated fatty acids of plant origin, e.g. linoleic and linolenic acid can undergo partial or complete hydrogenation by anaerobic rumen bacteria. *Trans* fatty acids provide some 6% of the dietary fat intake in the British diet containing milk, butter and other dairy products, margarine, meat and meat products, fish, eggs and cereal-based foods. These important sources of *trans* fatty acids provide approximately half of the *trans* unsaturated fatty acids in the diet.

Trans fatty acids, in addition to occurring naturally, can be formed during the partial hydrogenation of a *cis* unsaturated fatty acid. This can occur either biologically or industrially.

Important sources are the industrially hydrogenated fish and vegetables, or fish and vegetable oils, used in the manufacture of margarine, frying oils, shortenings and specialty products. Dairy product *trans* fatty acids are mainly C14 to C18, particularly vaccenic acid (*trans*-11, C18 : 1) and elaidic acid (*trans*-9, C18 : 1). *Trans* fatty acids produced industrially are more complex and more variable in type. Vegetable oils producing *trans* acid include soya bean, rape seed, sunflower seed and palm kernel. Soya bean oil may contain 30–50% of *trans* fatty acids of the C18 series, particularly *cis, trans*- and *trans, cis*-dienoic acid. Hydrogenated fish oils may yield a longer chain *trans* monoenoic acids C18 to C22. *Trans* fatty acids do not in general possess essential fatty acid activity; a small amount in the diet is unlikely to cause any pathological problems.

The *cis* double bonds can isomerize to become fatty acids with one or more *trans* bonds as well as acids with both *cis* and *trans* bonds. The double bond system *trans* may be separated by one or more methylene groups or may be placed between adjacent pairs of carbon atoms. The centres of unsaturation are said to be **conjugated**.

Cyclic fatty acids

These are uncommon but are important metabolic inhibitors and are found in many bacteria, plants and fungi.

Oxyacids

These are major components of surface waxes, cutin and suberin of plants.

Glycerol triesters

The glycerol triesters are the esters of glycerol, substituted at the three alcohols to form three distinct classes of lipid; triacylglycerol, phospholipids and waxes. The principal substitutions in the glycerol esters are fatty acids.

Glycerol (1,2,3-propanetriol) is the alcohol present in the natural triester glycerides, phospholipids and waxes. The formula of glycerol is:

$HOCH_2CHOHCH_2OH$
1   CH_2OH
2   CHOH
3   CH_2OH

In triacylglycerols of vegetable origin the C2 position is esterified to a C18 unsaturated fatty acids; in pig triacylglycerols the C2 position is occupied by palmitic acid. In rodent fats the distribution is random. Many other natural triacylglycerols have an unsaturated fatty acid in the C2 position. The C1 position is usually occupied by a saturated fatty acid. Over 50% of the triacylglycerol fatty acids are unsaturated and 16 : 0 fatty acid is the most common saturated fatty acid.

Triacylglycerols
These are esters of glycerol and fatty acids (Figure 5.12). The distribution of fatty acids in the C1, C2 or C3 position varies. The numbering C1, C2, C3 is based on a conventional L-configuration, when this notation is used the letters sn (stereospecific numbering) are used. This distribution of fatty acids gives a wide range of fatty acid ester triacylglycerols from different sources. The distribution appears to be dependent upon complex and perhaps partially understood biological rules.

The type and quantity of dietary fatty acids are important in determining the proportions of fatty acids in the ester. On average, the fatty acid composition of triacylglycerides is approximately 20% palmitic acid, 7% palmitoleic, 50% oleic and 10% linoleic.

Medium-chain triglycerides are prepared from coconut oil and contain C8 : 0 and C10 : 0 fatty acids. They are important in clinical nutrition as they are readily hydrolysed by pancreatic lipase and pass to the liver in the portal vein as fatty acids. This provides a ready form of energy to patients with steatorrhoea.

Physical properties of triacylglycerides The physical properties of acyl lipids are affected by the individual fatty acids and the melting point is important. Cell membranes are not able to function when the fatty acids are crystalline. For mammals the critical temperature is 37°C and poikilotherms anything from − 10° to over 100°. *Trans* fatty acids in triacylglycerols pack together more closely than those of *cis* isomers and this affects the melting point. A *trans* bond has little effect on the conformation of the chain. This makes its physical properties much closer to that of a saturated fatty acid. Oleic acid with one *cis* bond in the chain has a melting point of 13°C, whereas elaidic acid, the *trans* isomer of oleic acid, has a melting point of 44°C.

Phospholipids
These are derivatives of phosphatidic acid, esterified with phosphoric acid on the C3 position and with fatty acids on the C1 (usually saturated) and C2 (unsaturated) positions. Phosphatidyl inositols are esters of a cyclic derivative of glucose. The cardiolipins (phosphatidylglycerol and diphosphatidylglycerols) are esters of phosphatides with an additional glycerol and are predominantly linoleic acid esters. Sphingomyelins contain a dihydroxy amine sphingosine.

Wax esters
These are esters of long-chain fatty acids with long-chain fatty alcohols, formula R'COOR". These waxes are found in some bacteria, the oil of the sperm whale, the flesh oils of several deep sea fishes and in zooplankton. They are poorly hydrolysed by the pancreatic lipase of the human digestive system and are of little nutritive value.

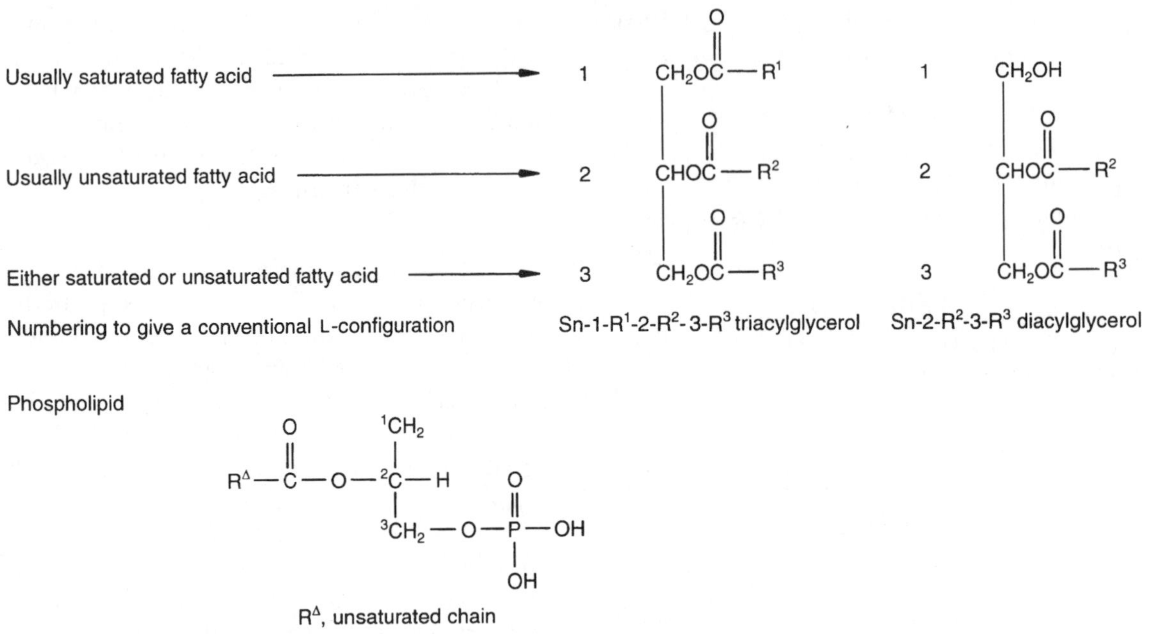

FIGURE 5.12   Triacylglycerol. Structure of acylglycerols and phospholipids.

Cholesterol and sterols

The sterol ring system is widespread and is an important hydrophobic structure. They contain alcohols and form esters with fatty acids.

Cholesterol (Figure 5.13) is found in all animal tissues and is a structural component of cell walls and membranes, precursor of bile acids, adrenal and gonadal hormones and vitamin D. It also can accumulate in atheromatous lesions of arterial walls.

Cholesterol belongs to the family of steroidal alcohols containing between 27 and 30 carbon atoms. All possess the 17-carbon cyclopentano-phenanthrene ring. All contain a 3-β-hydroxyl group and an endocyclic double bond usually in the 5,6 position, with a side chain which has varying degrees of saturation and unsaturation. Cholesterol exists as the free sterol or as esters with fatty acids. The more common sterols are found in higher animals and plants, with the exception of ergosterol, found in yeasts. Sterols may however be found in fungi, algae and marine invertebrates.

Other sterols include:

- Coprostanol (5-β-cholestanol), 5-β-cholestan-3β-ol
- Stigmasterol (24-α-ethylcholesta-5 : 22-dien-3β-ol) is widely distributed in plants but only calabar and soya beans contain sufficient to serve as sources
- Ergosterol (24-β-methylcholesta-5 : 7 : 22-trien-3β-ol) occurs in yeast, together with 5-α : 6-dihydroergosterol. Ergosterol contains one more carbon atom than cholesterol.
- Lumisterol (9-β : 10-α-ergosterol), pyroergocalciferol (9-α : 10-α-ergosterol), isopyroergocalciferol (9-β : 10-β-ergosterol)
- Ergocalciferol (vitamin D_2) 5,6-*cis*-ergocalciferol and 5 : 6-*cis*-cholecalciferol (vitamin D_3) the natural vitamin D of fish oils, are both derived from 7-dehydrocholesterol by ultraviolet irradiation

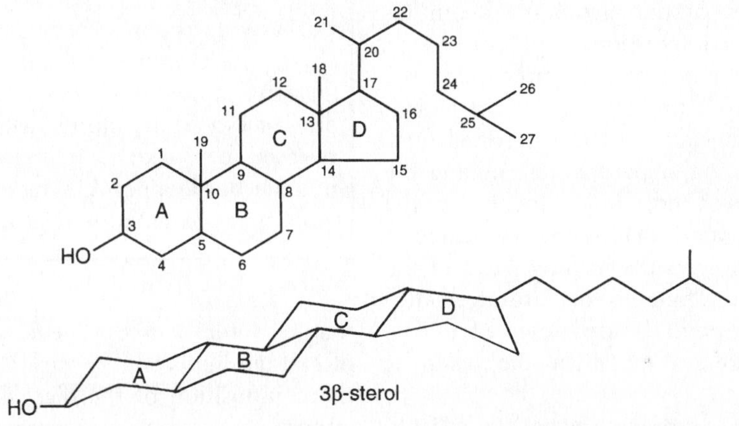

FIGURE 5.13   Structure of cholesterol showing the numbering system and the flat, geometric structure of the molecule.

- Brassicasterol occurs in rape-seed oil and is 24-β-methylcholesta-5 : 22-dien-3-βol
- Campasterol (24-α-methylcholest-5-en-3-β-ol) is obtained from rape-seed oil, soya-bean oil and wheat germ oil
- Spinasterol is a 7 : 22 diene found in spinach or alfalfa
- Sitosterol; the five sitosterols are the most widely distributed plant sterols. β-sitosterol (24-α-ethylcholest-5-en-3β-ol) is the principal sterol of cottonseed and calycanthus oil and is also found in soya bean oil, wheat-germ oil, corn oil, rye germ oil, cinchona wax and crepe rubber
- Fucosterol and sargasterol are found in fresh-water green and brown algae. Fucosterol is 24-ethylidenecholest-5-en-3β-ol and sargasterol is the 20-α-methyl isomer
- Mycosterols are sterols originating in fungi and include zymosterol (5-α-cholesta-8 : 25-dien-3β-ol) ; acosterol and fecosterol are minor sterols with double bonds at the 8 : 9 position and episterol has the double bond in the 7 : 8 position. Marine sterols vary from species to species; gastropods have cholesterol as the principal sterol whereas pelecypods (bivalves) contain a complex of C28 and C29 sterols. These include chalinasterol [24 (28) methylene-cholesterol] which is found in sponges, oysters, clams and sea anemones.

5.4.3 Functions of lipids

These can be described as structural, storage and metabolic, although individual lipids may have several different roles at different times, or even at one and the same time.

Structural lipids

Lipids play an integral part in biological membranes. The importance of lipids in such barriers lies in their ability to exclude water and other molecules. These are important at surfaces and in membranes, functioning as barriers between one environment and another.

All living cells contain a membrane providing a structure in which many metabolic reactions take place. In mammals the lipids involved are the glycerophospholipids and unesterified free cholesterol, while in plants the glycosylglycerides are predominant, especially in the chloroplasts. The importance of the compounds involved in membranes is the possession of hydrophilic groups close to hydrophobic groupings. These are called amphiphilic. The composition of lipid molecules on either side of many membranes is quite different. This is called membrane asymmetry. The physical properties of the membrane are influenced by the lipid composition.

In lipids which are esters of fatty acids the hydrocarbon chain of the fatty acid is the hydrophobic moiety. The nature of the esterified fatty acid chain plays a major part in determining the physical properties of those lipids. Increase in the number of double bonds determines the degree of unsaturation and lowers the melting point of the acyl chains. Within families of saturated fatty acids the melting point is also lowered as the chain length decreases or when the chain is branched.

The membranes of storage organs, e.g. adipose tissue, the liver, muscle and kidney mitochondria, contain n-6 fatty acids, particularly arachidonic acid. Nervous tissue, the retina and the reproductive organs have a substantial content of longer chain fatty acid with five or six double bonds, particularly of the n-3 family. The biosynthesis of eicosanoids (which includes prostaglandins) requires polyunsaturated fatty acids of the n-3, n-6 family.

Structural lipids eaten as animal and plant flesh are phospholipids and glycolipids. They are, by the nature of their role in the cell, enriched in polyunsaturated fatty acids.

Storage lipids

In humans the general reservoir of lipids is adipose tissue. Milk fat is an energy store for the benefit of the newborn and the egg-yolk lipids for the developing chick embryo. Lipids are in a dynamic state and constantly being broken down or removed from the tissue or replaced. Lipids transported as lipoproteins are taken into tissues where they may be stored as energy reserves in the adipose tissues, incorporated into the structural lipids of membranes or oxidized to supply energy, depending on the requirement of the body.

Food fats are the adipose tissues of animals, fish and seed oils. Storage and barrier function depend on the bulk properties of lipids. The high energy densities of triacylglycerols makes them ideal as

Carbohydrate stores provide only half as much energy per gramme as lipid stores and contain an equal amount of water. When lipid stores are metabolized more metabolic water is released per gramme compared with carbohydrate but less per ATP molecule (0.12 water molecules compared with 0.132).

long-term fuel sources. The fatty acid composition of storage lipids is very variable and depends on the composition of the diet. Triacylglycerols are important concentrated water-insoluble energy stores.

The storage fats of monogastric animals are influenced by the fat content of the diet. The adipose tissue of ruminants is determined by the bacteria of the rumen with the production of saturated and monounsaturated acids, though unusual fatty acids, both *trans* and branched, are also produced. If the diet of pigs and poultry is derived from carbohydrates, then the subsequent synthesis of fat results in predominantly saturated and monounsaturated fats. The inclusion of vegetable oils such as soya bean oil in the diet can increase the linoleic acid content of the carcass.

Fish oils are derived from two principal sources, either lean fish whose reserve oils are stored as triacylglycerols in the liver, e.g. cod, or fatty fish, e.g. mackerel, where the oils are stored in the flesh. The types of fatty acids are dependent upon season and diet but are characterized by polyunsaturated fatty acids, n-3 type, of 20 or more carbons in length.

Storage triacylglycerols used commercially in seed oils are fleshy fruit, e.g. the exocarp of the olive or seed endosperm, e.g. rape. The fatty acids vary with seed source, for example palmitic acid in palm oil and oleic acid in olive oil.

Lean meats contain arachidonic acid (20 : 4, n-6). The storage fats in meat are largely saturated and monounsaturated fatty acids. Plant leaf fatty acids are predominantly palmitic acid, palmitoleic acid, oleic acid, linoleic acid and α-linolenic acid.

Only 12 oil-bearing seeds are important commercially, and only two of these are used for industrial purposes, linseed oil (rich in α-linolenic acid) and castor oil (ricinoleic acid, a laxative). The fatty acid in rape-seed oil is now predominantly oleic acid. The lupin or evening primrose contains γ-linolenic acid.

Summary

1. Dietary fat is a heterogeneous mixture of lipids, predominantly triglycerides but also includes phospholipids and sterols.
2. Fatty acids have a basic formula $CH_3-(CH_2)_n-COOH$ where n can be any even number from 2 upwards. Unsaturated fatty acids are important nutritionally and biologically. There are several nomenclatures for the unsaturated fatty acids which may be based on the saturated parent acid, the number of carbon atoms and position of the double bonds. The differences are dictated by whether the numbering is taken from the methyl or carboxyl end.
3. The four main dietary unsaturated fatty acid families are n-3, n-6, n-7 and n-9. This numbering system is determined by the position of the first double bond from the methyl carbon atom.
4. Glycerol (1,2,3-propanetriol) is the alcohol present in the natural triester glycerides, phospholipids and waxes.
5. Cholesterol is a structural component of cell walls and membranes, precursor of bile acids, of adrenal and gonadal hormones and vitamin D. It can also accumulate in atheromatous lesions of arterial walls. Cholesterol belongs to the family of steroidal alcohols containing between 27 and 30 carbon atoms and exists as the free sterol or as esters with fatty acids.
6. The more common sterols are found in higher animals and plants, with the exception of ergosterol, found in yeasts. Sterols are also found in fungi, algae and marine invertebrates.
7. The essential fatty acids and their longer-chain molecular products are necessary for the maintenance of growth, good health and reproduction. The essential fatty acids all belong to the n-3 and n-6 groups, linoleic acid (n-6) and α-linolenic acid (n-3). The fatty acids which have a double bond at, n-3, n-6, cannot be synthesized by humans.
8. The functions of lipids are structural, storage and metabolic, although individual lipids may have several different roles at different times, or even at one and the same time.
9. Structural lipids are important at surfaces and in membranes, functioning as barriers between one environment and another. Such barriers are able to exclude water and other molecules.
10. Food fats are the storage fats of animals and plants. The high-energy densities of triacylglycerols make them ideal as long-term fuel sources. The fatty acid composition of storage lipids is very variable and depends on the composition of the diet.

Further reading

Gurr, M.I. and Harwood, J.L. (1991) *Lipids*, Chapman & Hall, London.

Shoppee, C.W. (1964) *Chemistry of the Steroids*, Butterworths, London.

Trans Fatty Acids (1987) Report of the British Nutrition Foundation, London.

Willett, W.C., Stampfer, M.J., Manson, J.E., Colditz, G.A., Speizer, F.E., Rosner, B.A., Sampson, L.A. and Hennekens, C.H. (1993) Intake of *trans* fatty acids and risk of coronary heart disease among women. *Lancet*, **341**, 581–5.

5.5 Carbohydrates

- Carbohydrates consist of monosaccharides, disaccharides and polysaccharides.
- Monosaccharides are either polyhydroxyaldehydes (aldoses) or polyhydroxyketones (ketoses). These have chiral properties.
- Glucose, a monosaccharide is the most important dietary energy source.
- Disaccharides of nutritional importance are sucrose, lactose and maltose.

5.5.1 Introduction

Carbohydrates are an important source of energy in all human diets. The amount in the diet varies according to the economy, the range being 40–80% of calorie intake. Carbohydrates are synthesized from carbon dioxide and water. The primary structures are monosaccharide sugars (Figure 5.14), which may dimerize to disaccharides, e.g. sucrose, or polymerize extensively to form polysaccharides, e.g. cellulose, pectin and hemicellulose. In the human, carbohydrates are central to energy utilization as glucose and to storage as starch, a polymeric carbohydrate.

5.5.2 Chemistry of simple natural carbohydrates and their derivatives

Monosaccharides
α-D-Glucose
Fructose
Mannose
Xylose
Galactose
Disaccharides
Sucrose
Lactose
Maltose
Derivatives of mono- and disaccharides
Hydrogenated derivatives (or sugar alcohols)
Sorbitol
Mannitol
Polymers of glucose
Dextrose

Monosaccharides

Monosaccharides chemically are either polyhydroxyaldehydes (aldoses) or polyhydroxyketones (ketoses). Carbohydrates have centres of asymmetry, called **chiral centres**, which are optically active. Optically active describes the two forms of the molecule termed as dextrorotatory or levorotatory, depending on the direction that they rotate plane-polarized light (Figure 5.15). The symbol d or (+) is used for dextrorotatory rotation and l or (−) refers to levorotatory rotation. Such rotation can be tested by making a solution of the compound and measuring the rotation of the plane of polarized light through the solution. Where there is a mixture of equal amounts of the d and l form this is known as a **racemic mixture**.

An understanding of the stereochemistry of sugars is based on configurational properties. For common sugars the prefix D or L is used for the area of asymmetry most remote from the aldehyde or ketone end of the molecule. The reference compound is glyceraldehyde (Figure 5.16). C4, C5 and C6 sugars are derived from glyceraldehyde or dihydroxyacetone by the stepwise addition of formaldehyde. Aldehydes can form hydroxyl compounds with the carbonyl group. If a molecule of water is added, the result is an aldehyde hydrate. If a molecule of alcohol is added the product is a **hemiacetal**; adding a second alcohol molecule produces an **acetal**. Sugars often form intramolecular hemiacetals; this they do readily in water so the resulting compounds form a five- or six-membered ring.

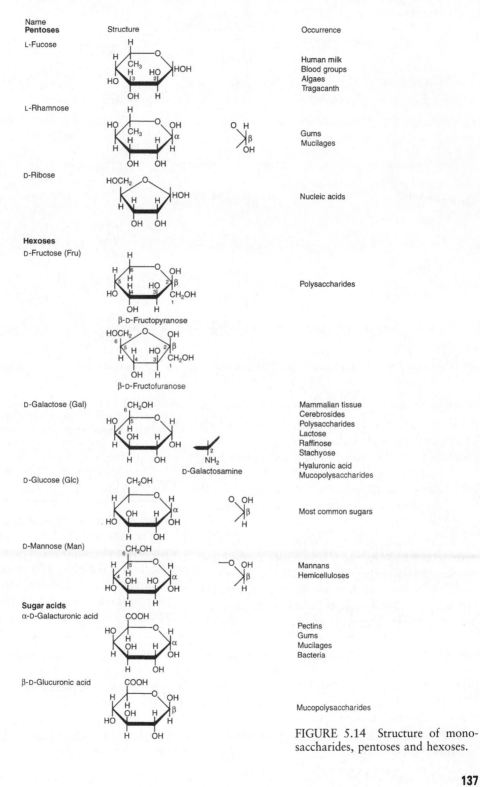

Name	Structure	Occurrence

Pentoses

L-Fucose

Human milk
Blood groups
Algaes
Tragacanth

L-Rhamnose

Gums
Mucilages

D-Ribose

Nucleic acids

Hexoses

D-Fructose (Fru)

Polysaccharides

β-D-Fructopyranose

β-D-Fructofuranose

D-Galactose (Gal)

Mammalian tissue
Cerebrosides
Polysaccharides
Lactose
Raffinose
Stachyose
Hyaluronic acid
Mucopolysaccharides

D-Galactosamine

D-Glucose (Glc)

Most common sugars

D-Mannose (Man)

Mannans
Hemicelluloses

Sugar acids

α-D-Galacturonic acid

Pectins
Gums
Mucilages
Bacteria

β-D-Glucuronic acid

Mucopolysaccharides

FIGURE 5.14 Structure of mono-saccharides, pentoses and hexoses.

Polyhydroxyaldehyde
(aldoses)

CHO
|
[CHOH]$_n$
|
CH$_2$OH

Polyhydroxyketose
(ketoses)

CH$_2$OH
|
CO
|
[CHOH]$_n$
|
CH$_2$OH

Aldose, e.g. glucose

C^{1}HO
|
HC^{2}OH
|
HOC^{3}H
|
HC^{4}OH
|
HC^{5}OH
|
C^{6}H$_2$OH

Ketose, e.g. fructose

C^{1}H$_2$OH
|
C^{2}O
|
HOC^{3}H
|
HC^{4}OH
|
HC^{5}OH
|
C^{6}H$_2$OH

FIGURE 5.15 The designation of dextrorotatory and laevorotatory is based on the structure of polyhydroxy-aldehyde (aldoses) and polyhydroxyketone (ketoses).

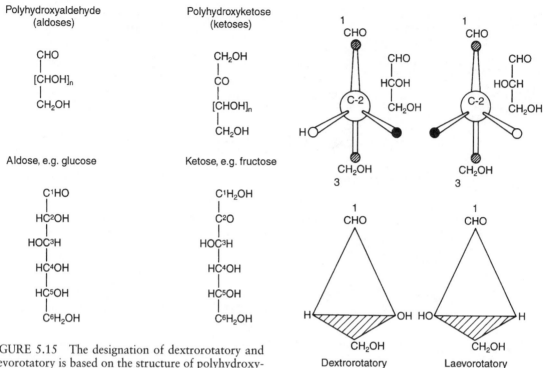

Dextrorotatory   Laevorotatory

FIGURE 5.16 The structure of carbohydrates is based on glyceraldehyde. This gives the dextrorotatory or laevorotatory properties. Dextrorotatory D- is called D series; this applies if the secondary alcohol furthest away from the principal function, e.g. aldehyde, keto, carboxyl group has the same configuration as D-glyceraldehyde.

The α-designation for the D-series means that the aldehydal C-1 hydroxyl group is on the same side of the structure as the ring oxygen and the β indicates the reverse. For the L-series the opposite is the case.

Glucose exists in two different forms: α-D-glucose and β-D-glucose. The hexoses formed are numbered as shown in Figure 5.17.

When the sugar is dissolved in water the hemi-acetal is in equilibrium with the straight-chain hydrated form. The straight chain form which usually represents only a small fraction of the total can convert to either hemiacetal, α or β. The conversion of one stereoisomer to another in solution is referred to as mutarotation. Hemiacetals with five-membered rings are called **furanoses** and hemiacetals with six-membered rings are called **pyranoses** (Figure 5.18). In general the pyranose form dominates.

Substitution at the equatorial position is easier because there is less chance of steric hindrance with other substitutions. The most stable conformation is the chair form which results in the maximum number of substitutions in the equatorial position. The furanose ring is non-planar and can exist in more than one conformation.

Another way of showing the structure is the Haworth projection. The alternative and most frequently used conformations are the 'chair' and 'boat' forms in which the atomic groups of the ring carbons are either perpendicular to the plane of the ring, i.e. axial, or parallel to the plane of the ring, i.e. equatorial (Figure 5.19).

Cyclic structures

α-D-(+)-Glucose

```
H—C—OH
    |
H—C—OH
    |
HO—C—H      O
    |
H—C—OH
    |
H—C
    |
  CH₂OH
```

β-D-(+)-Glucose

```
HO—C—H
    |
H—C—OH
    |
HO—C—H      O
    |
H—C—OH
    |
H—C
    |
  CH₂OH
```

Haworth rings

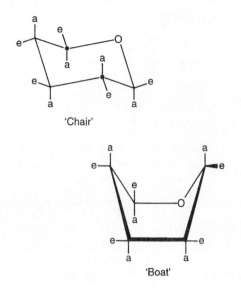

FIGURE 5.19 The 'chair' and 'boat' structure of hexoses. a-orientated substitutions are axial; e-orientated substitutions are equatorial.

Chair structures

(axial)   (equatorial)

FIGURE 5.17 Glucose exists in two isomeric forms: anomers which differ in the configuration of C-1. These are represented by a cyclic structure, the Haworth rings, or the chair structures.

Glucose is readily methylated in the α or β position. Such derivatives of glucose are called **glucosides** and of galactose, **galactosides**. The most stable conformation is one where the bulkiest group –CH₂OH occupies an equatorial position (Figure 5.20). Of all D-aldohexoses, β-D-(+)-glucose is able to conform to a shape in which every bulky group is in the equatorial position. This may explain why β-D-(+)-glucose is the most common organic chemical in nature.

Glycosidic linkage

A **glycoside** is an acetal formed by interaction of an alcohol with the carbonyl of a carbohydrate; the link between the sugar and an alcohol is a glycosidic bond (Figure 5.21). A glycoside can be formed with an aliphatic alcohol, phenols and hydroxycarboxylic acids, as well as with another sugar. Glucose can form a glycoside with the alcohols of glucose in the α and β conformation. Monosaccharides can link through glycosidic bonds to form disaccharides and larger carbohydrate structures. Thus carbohydrate acetyls are formed, i.e. OR where R is a sugar.

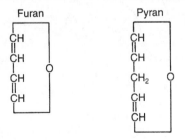

FIGURE 5.18 Carbohydrates: the general structures of the ring form are based on the C4 furanoses and the C5 pyranoses. The carbohydrate structures are analogous to the heterocyclic compounds furan and pyran.

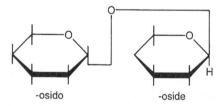

Bulky groups equatorial

Bulky groups axial

FIGURE 5.20 Glucose showing the two geometric forms. When the bulky groups (CH₂OH) are **equatorial** (in the plane of the molecule) the configuration is stable. When such groups are **axial** (at right-angles) to the plane of the molecule, the molecule is less stable.

-osido   -ose

If the linkage is glycosidic on one side only, then suffixes -osido and -ose are used

-osido   -oside

If the linkage is glycosidic on both sides, then the suffixes -osido and -oside are used

FIGURE 5.21 Oligosaccharides are joined by glycosidic linkages.

Disaccharides

Maltose has the structure 4-O-(α-D-glucopyranosyl)-D-glucopyranose. Both halves of the molecule contain the six-membered pyranose ring. Mal-

tose is a disaccharide joined by a glycosidic bond between the C-1 carbonyl of one D-(+)-glucose molecule and the C-4 (providing the –OH as an alcohol) of a second D-(+)-glucose molecule with an α-(1–4) glycosidic linkage (Figure 5.22). The C-1 carbon determines the configuration of the linkage, as this is the active carbon and hence α. The second D-(+)-glucose is an α-D-glucosopyranosyl group. Maltose has a free aldehyde group and is therefore a reducing sugar.

Cellobiose is a disaccharide formed from two D-(+)-glucose molecules. These are two pyranose rings and a glycosidic linkage to the –OH on C-4. The D-glucose units are connected by a β rather than an α linkage. (+)-Cellobiose is a 4-O-(β-D-glucopyranosyl)-D-glucopyranose.

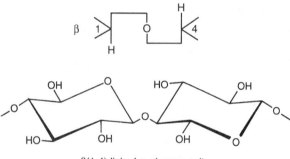

β(1-4)-linked D-glucose units

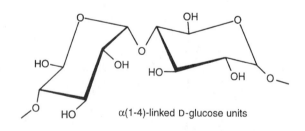

α(1-4)-linked D-glucose units

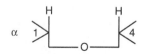

FIGURE 5.22 1–4-linked glucose units; β- and α-linkages.

Reducing sugars

Carbohydrates that can reduce Fehling's, Benedict's or Tollen's solution (complexed cupric solutions) are known as reducing sugars. All monosaccharides, whether aldose or ketose, are reducing sugars. Most disaccharides are reducing sugars with the exception of sucrose. The reason why sucrose is a non-reducing sugar is that it does not contain a free aldehyde or ketone group, being a β-D-fructoside and an α-D-glucoside. In contrast, lactose – which has reducing properties – is a substituted D-glucose to which a D-galactosyl unit is added. The D-(+)-glucose unit has a free aldehyde group for oxidation.

Lactose is a reducing sugar, a disaccharide of D-(+)-glucose and D-(+)-galactose joined through a β linkage. Lactose is a substituted D-glucose in which a D-galactosyl unit is attached to one of the oxygens, a galactoside not a glucoside. The glycosidic linkage involves a –OH at C-4. Lactose is 4-O-(β-D-galactopyranosyl)-D-glucopyranose.

Sucrose is a non-reducing sugar consisting of a D-glucose and a D-fructose unit joined by a glycosidic linkage between C-1 of glucose and C-2 of fructose. Sucrose is a β-D-fructoside and an α-D-glucoside, a D-glucopyranose and a D-fructofuranose. Sucrose may be regarded as α-D-glucopyranosyl-β-D-fructofuranoside or β-D-fructofuranosyl-α-D-glucopyranoside.

The structures of the most common disaccharides are shown in Figure 5.23.

5.5.3 Types and forms of sugar

Sugars are a source of sweetness to food. Sugars which are naturally present or added, add to the flavour of certain foods which are not particularly sweet in themselves, e.g. milk, fresh vegetables, sauces, canned vegetables, mayonnaise, breakfast cereals, canned and packet soups. Sugars are also important in soft drinks.

Sugars contribute to existing flavours and generate new flavours. Sugars can be used by yeast and bacteria as a substrate in fermented products. Sugars can also have a preservative action in jam and also contribute to the 'shelf-life' of cakes.

There are different ways in which sugar occurs in food:

- From natural constituents of such foods as fruit, milk, vegetables and honey.
- As an essential ingredient of prepared foods, preserved sweets, chocolates and cakes.
- An optional additive such as sucrose added to tea, coffee and other foods.

The attraction of sucrose is in the amount that can be taken, whether as a concentrate or diluted in the bulk of food.

Cane sugar

Sugar cane is grown in tropical countries between 25° north and south of the equator. The crop takes some 9–18 months to grow to a height of 3–5 metres, and ripens during the cooler and drier parts of the year. After harvesting, an average of four to five ratoons grow. This is a very energy-efficient crop, three times as efficient as potatoes and seven times as energy-efficient as wheat. Four end-products of sucrose manufacturing are obtained:

- Raw sugar, subsequently converted to refined syrup and sugar.
- Cane molasses – the non-crystallizable residue from the production of raw sugar which is used for cattle feed, fermented to alcohol or converted into rum or fuel. It can also be used for growing yeast and may be cleaned up as treacle.
- Bagasse – cane fibre used for fuel, paper and board.
- Filter press mud used as fertilizer.

Raw sugar is converted into refined sugar by removing the colour and residual impurities.

Beet sugar

Beet sugar is a temperate crop which is very energy-efficient in growth. The sucrose is extracted from the roots after slicing the beet; the sucrose is extracted by diffusion. The end products are:

- sucrose – usually refined white and ready for consumption
- beet molasses – normally fermented and used for cattle feed
- beet pulp – used for animal feed

There are other sources of sugar: gur or gaggary, which is made in India, West Africa and Pakistan, is a crude mixture of sucrose, crystals and syrups obtained by extracting and concentrating sugar cane juice in an open pan over a fire. The juice is concentrated and solidified in coconut shells or wooden moulds. Variants on this include panella and khandsari sugar.

Other speciality sugars

Demerara sugar is a golden-brown sugar produced in Demerara, Guyana. This contains residual molasses which gives flavour and slight moisture. London Demerara is a brand name for white sugar with added molasses. Muscovados or Barbados sugar contains molasses to give a treacle flavour.

Castor sugar has smaller crystals than ordinary granulated sugar, and icing sugar is a finely powdered sugar produced by granulating sugar through a hammer mill.

Lump sugar is a soft crystalline mixture of refined and partially refined cane sugar.

Transformed sugar is a patented process (Tate

Name	Structure	Occurrence
Lactose (4′-[β-D-galactopyranosido]-D-glucopyranose)	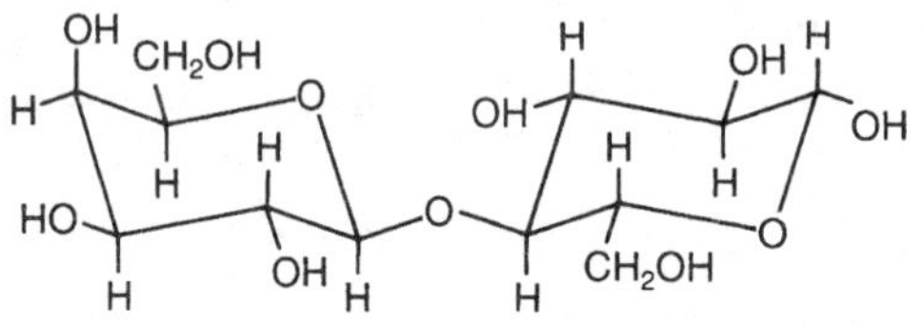	Constituent of mammalian milk (4-8%). Only faintly sweet
Maltose (4′-[α-D-glucopyranosido]-β-D-glucopyranose)	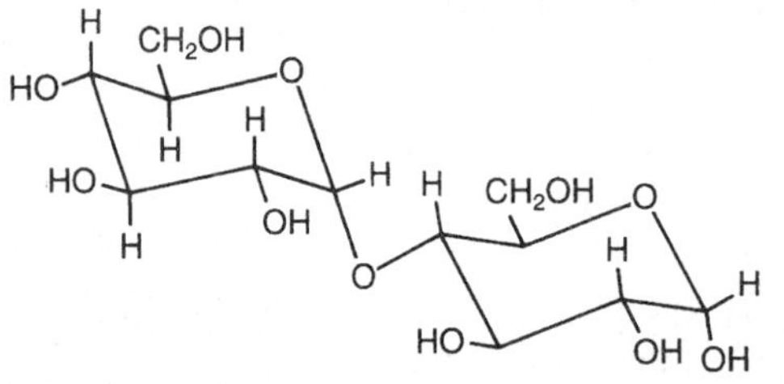	Break-down product of starch and glycogen arising in the course of digestion. Found free in some plants (barley) and in honey
Sucrose (saccharose, cane sugar, beet sugar, α-D-glucopyranosido-β-D-fructofuranoside)	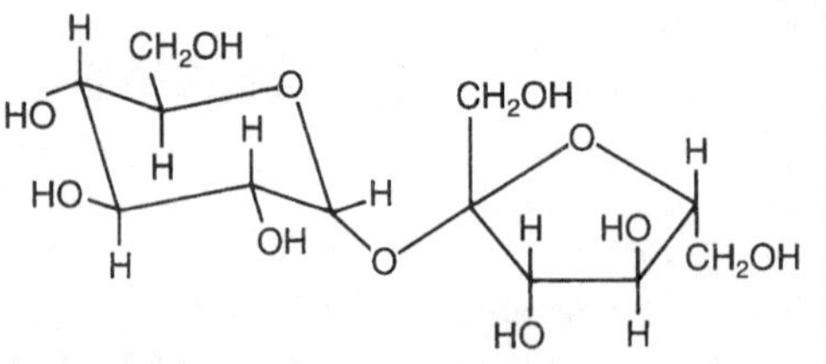	Almost universal in the vegetable kingdom

FIGURE 5.23 Disaccharide structures. (Reproduced from Geigy Scientific Tables, Ciba-Geigy Ltd, with permission.)

& Lyle) which has reduced bulk, is non-sticky and used in dry mixing and vending packs.

Other sugars

Glucose is made from a high-conversion glucose syrup. The amount of added glucose consumed in the diet is small. Fructose and maltose may be obtained from high fructose and high maltose syrups. Lactose is extracted from whey, a by-product of cheese manufacture. Most lactose is consumed as milk.

Syrups and mixtures

During the course of the manufacture of sucrose, a number of sucrose syrups are produced:

Molasses: in the production of sucrose, impurities and lime are removed and a clear solution obtained which is concentrated. Sugar crystals and a mother liquor called molasses are produced which, when partially refined and blended, becomes treacle. Golden syrup is prepared from a mixture of by-products of sugar cane refining, which are further refined, decolourized and concentrated.

Invert syrup: this is a mixture of glucose and fructose in equal parts. When (+) sucrose is hydrolysed by dilute aqueous acid or by the action of the enzyme invertase (from yeast) the result is equal amounts of D-(+)-glucose and D-(+)-fructose. This hydrolysis is accompanied by a change in the sign of optical rotation from positive to negative, the reaction is called the **inversion** of (+)-sucrose. The levorotatory mixture of D-(+)-glucose and D-(+)-fructose is called invert sugar. It has a chemical composition like that of honey, but without any of the congeners which give honey its particular flavour. Invert syrup is sweeter than sucrose syrup and is used in the food industry, especially in brewing. It is derived from warming a solution of sucrose with mild acid, for example, citric acid. Invert syrup differs from sucrose syrup, in that it is sweeter on a molar basis, less viscous, more water absorbent and more readily fermented.

Glucose syrups: these are made from starch from either maize, potato, wheat or cassava. These contribute 16% of the sugars and syrups

consumed. The starch is hydrolysed by acid or by enzymes. Partially hydrolysed products are called maltodextrins which are several glucose units in length. These are spray-dried powders which readily retrograde and have poor storage properties in syrup form. Maltodextrins provide bulk without sweetness. This is further treated with enzymes which produce glucose and other sugars. Other enzymes can be used, which produce maltose rather than glucose as the main end-product.

High fructose syrups: these are made from glucose syrups in which 94% of the dissolved sugars is glucose. These are high-conversion glucose syrups. To convert the glucose to fructose, the syrup is diluted and treated with an enzyme which converts the glucose into a mixture of glucose and fructose in the ratio of 52 : 42. It is possible to separate this by chromatography into high-glucose and high-fructose fractions. A high-fructose fraction has the benefit of greater sweetness in relation to bulk.

The world production of sugar is approximately 100 million tons, 55% from sugar cane and 45% from sugar beet. Sugar cane is a very important crop in many tropical and sub-tropical countries.

5.5.4 Sugars in prepared foods

Bulk

In sweets, jams and cakes, sugars are an important part of the food and provide the characteristic taste, texture and flavour.

Texture

Sugars form an important part of the characteristics of cakes and sweets and effect crispness, crunchiness, a feeling of softness, lightness, chewiness, etc. In cakes made with sugar, there is an entrapment of air in the mixture when sugar is creamed with fat. When the mixture is baked, the gelling of starch is delayed and this increases the temperature at which the egg coagulates. In this way, trapped air expands and the mixture rises fully before the heat sets the system. The texture of biscuits results from the particular formulations and baking conditions used. The sugars are usually solid and crystalline and this makes the texture close and crunchy.

Preservation

High concentrations of sugars prevent the growth of bacteria and moulds, in part because of inhibition of the availability of water to the bacteria. Sugar-reduced products may have to be kept refrigerated or have a very reduced life after opening. The traditional way of preserving is either sugar or salt. Jams and marmalade contain 60% dissolved solids, most of which is as sugars, partly from the fruit and partly added. Most jams contain 70% sugar.

Browning reaction

Sugars can undergo two major types of browning reaction. If sugar is heated, browning occurs, called caramelization. There is also a heat-induced reaction between sugar and the amino acids of the proteins in food, known as the 'Maillard reaction', though the nutritional effects of this reaction are unknown.

5.5.5 Sugars in specific food

- Cereals and cereal foods: there may be little or no added sugars in some cereal foods, for example bread. Sweet biscuits contain 25% sugar and some breakfast cereals require sugar in order to give particular flavours.
- Milk and milk products: sugars make up approximately 5% of milk solids in whole milk, 5% in skimmed milk and 52% in dried skimmed milk. Lactose is substantially less sweet than other sugars. Sucrose is added to sweetened condensed milk, primarily to act as a preservative. Ice-cream contains 14% of added sugars; soft ice-cream substantially less. These sugars provide bulk, sweeteners and texture. Fruit yoghurt contains approximately 9% of added sucrose.
- Fruits: most edible fruits are rich in sugars, e.g. 9% (86% of the dry weight) in eating apples; 64% (82% dry weight) in raisins.
- Canned fruit: the liquid or syrup can be sucrose syrup or sugar-rich fruit juices. The strength of the syrup depends on the fruit and sourness. The amount of sugar is added to help preserving characteristics of the product. The high sugar content in canned fruit helps to conserve colour, flavour and texture as well as sweetness.
- Vegetables: the energy content of vegetables comes from sugars which range from 2–9% of the weight. The amount of sugar declines with boiling.
- Preserves, sweets and chocolates: sugars are major components in jam (69%), milk chocolate (54%) and are also essential to the flavouring of caramel and toffee.
- Beverages: sugars are added to soft drinks to make them sweet. The added sugars are traditionally sucrose and glucose syrup, but fructose and fructose syrup are also used.
- Alcoholic drinks: sugars may be present in sherries and beers either as small amounts, 1–3%, or as major ingredients as in sweet sherry (7%) and liqueurs (up to 30%).
- Cakes, pastries and puddings: these are rich in sucrose or glucose syrup and fructose syrup. They constitute 16% of shortbread, 31% of sponge cake and 12% of blancmange.
- Processed meat products: small amounts of sucrose are added to corned beef and luncheon meat to make the products softer.

Summary

1. Monosaccharides, chemically, are either polyhydroxyaldehydes (aldoses) or polyhydroxyketones (ketoses); they possess chiral centres and are optically active. The two forms of the molecule are described as dextrorotatory (d or +) or levorotatory (l or −) depending on the direction that they rotate plane-polarized light. A mixture of equal amounts of the d and l forms is known as a racemic mixture.

2. Monosaccharides include α-D-glucose, fructose, mannose, xylose and galactose. Monosaccharides form intramolecular hemiacetals, so the resulting compounds form a five (furanose) or six-membered (pyranose) ring. Carbohydrates that can reduce Fehling's, Benedicts's or Tollen's solution (complexed cupric solutions) are known as reducing sugars.
3. Monosaccharides can link through glycosidic bonds to form disaccharides and larger carbohydrate structures. Disaccharides include sucrose (glucose and fructose); lactose (glucose and galactose); maltose (glucose and glucose).
4. The commercial interest in sucrose results in its being available in a series of forms. There are many commercially advantageous effects of sugar in prepared foods.

Further reading

Dobbing, J. (ed.) (1989) *Dietary Starches and Sugars in Man: A Comparison.* ILSI Human Nutrition Reviews, Springer-Verlag, London.

Eriksson, C. (ed.) (1981) *Maillard Reaction in Food.* Pergamon Press, Sweden.

Report of the British Nutrition Foundation's Task Force (1987) *Sugars and Syrups.* British Nutrition Foundation, London.

Zubay, G. (1993) *Biochemistry*, 3rd edn, Wm C. Brown Publishers, Dubuque, Iowa, USA.

5.6 Polysaccharides

- Polysaccharides are polymers of monosaccharides joined by glycosidic bonds.
- Starch and glycogen are long-chain molecules with a straight-chain backbone and branching side chains.
- Polysaccharides are important energy stores and form starch and cellulose in plants and glycogen in humans.
- The granules of starch in the plant may gelatinize during cooking and subsequently undergo retrogradation
- The physical and chemical formats of starch affect digestion whether in the small intestine or colon.
- Starches and other polysaccharides interact with other food components.

5.6.1 Introduction

In biology most carbohydrates exist as high molecular weight polymers. Polysaccharides are simple sugars connected by glycosidic bonds. The sugars involved are D-glucose, D-mannose, D- and L-galactose, D-xylose, L-arabinose, D-glucuronic acid, D-galacturonic acid, D-mannuronic acid, D-glucosamine, D-galactosamine and neuraminic acid.

Cellulose is a major structural polysaccharide forming approximately half of the carbon found in plants (see p. 153). Cellulose consists of polymeric glucose with a glycosidic linkage of the β- (1–4) type. D-Glucose in the chair form of a pyranose ring exists in a rigid format. Such a linkage structure is characteristic of cellulose with an extended polysaccharide chain. Cellulose has a molecular weight of 50 000 Da or greater, forming chains which join together as bundles, with a diameter of 100–250 Å with approximately 2000 chains in such a bundle.

5.6.2 Starch and glycogen

The two major polysaccharides for energy storage are starch in plant cells and glycogen in animal cells.

Starch and glycogen

Both are α-(1–4) homopolymers with occasional α-(1–6) linkages at branch points. They differ in their chain lengths and branching patterns. Glycogen is highly branched with an α(1–6) linkage occurring every 8–10 glucose units along the backbone with short side chains of approximately 8–12 glucose units each (see Figure 5.26). Starch occurs both as unbranched amylose and as branched amylopectin (Figure 5.25). Amylopectin has α(1–6) branches but these occur at every 12–25 glucose residues and with longer side chains, 20–25 glucose units long.

Starch

Starch is a homopolysaccharide, being formed solely of glucose units and is a very important source of energy in human nutrition. Starch is the main food reserve in plants. During photosynthesis starch is deposited in special parts of the plant, for example tubers (potato), kernels (maize) and grains (wheat). Not only is starch widely used in traditional human consumption, new technologies are constantly being applied to develop novel uses. Starch is used to include fat replacers based on maltodextrins, and sweets based on polyols and non-digestible starch derivatives. The world production of starch is of the order of between 20–25 million tonnes per year, 10–20 million tonnes of which are used for food purposes. In contrast, the world production of sucrose is of the order of 100 million tonnes per year, of which 99% is used in foods.

Starch granule structure

Starch consists of glucose units which may join together in different ratios of amylose and amylopectin. The shape, size and composition of the granules are dependent on the amylose : amylopectin ratio.

An α(1–4)-linked unit in a polyglucose results in a natural turning of the chain and forms a **helix**. Such a helix, when it winds round the molecule of iodine, results in the characteristic blue colour of the amylose–iodine complex. An α(1–4) configuration is much more susceptible to degrading enzymes which are less frequently found for the β(1–4) linkages and structure.

Amylose This is a linear unit with glucoses joined by α(1–4) linkages. It is insoluble in cold water. Because of the tetrahedral structure of the carbon atom and the α(1–4) linkage, the natural form of starch is a helical structure with a turn every six glucose units. Amylose consists of between 200 and 2000 glucose units, depending on the source. The helical structure can form complexes with ions wherein the ion enters the core of the helix. When this occurs with iodine the brown iodine reacts with the white starch to give a blue colour.

Amylopectin This is a branched structure with glucose joined by α(1–6) linkages at the branch point, and α(1-4) linkages in the linear sections (Figure 5.24). The polymer contains between 10 000 and 100 000 glucose units, but no helical structure exists. There is no blue colour formation with iodine, a red-brown colour being produced. Amylopectin is insoluble in cold water but is soluble in hot water. It does not readily become retrograded, and on cooling may form a loose gel.

5.6.3 Physical and chemical properties of starch

Starch is deposited in the plant in the form of microscopic granules which are insoluble in cold water. Heating in excess water results in a soluble preparation. In dilute solutions ($< 1\%$), a precipitate forms; at high concentrations a gel is produced, the production of which can exclude water from the starch–water matrix, a process called **syneresis**.

Gelatinization

As a starch granule is heated in water the water is absorbed, the granule structure being altered by the loss of crystallization of amylopectin. This is followed by swelling, hydration and solubilization. The viscosity increases until the viscous solution of starch paste is formed, a process called **gelatinization**. This is characterized by an increase in consistency that appears to result from the creation of a starch network that binds the swollen granules together. There are also changes in crystalline melting, loss of birefringence and starch

Name	Structure	Occurrence
Amylopectin (α-amylose)	Highly-branched molecule composed of several hundred unit-chains, each of which comprises 20-26 α(1-4)-linked glucose residues; the unit-chains are interlinked by glycosidic bonds from the reducing group to C-6 of a glucose residue in an adjacent chain: 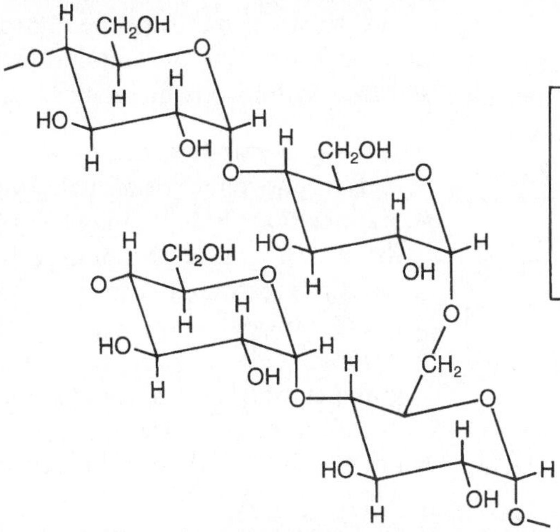 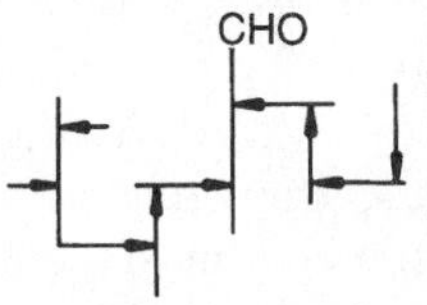	Main constituent of starch (usually ca. 80%)
Amylose (β-amylose)	Essentially a linear chain of α (1-4)-linked glucose residues: 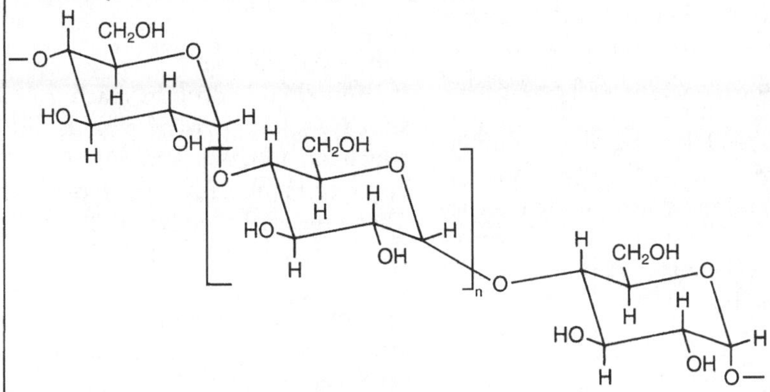	Constituent of starch (ca. 20%). Absent in some starches, e.g. that of 'waxy' maize (corn)

FIGURE 5.24 Structure of starch: amylopectin and amylose. (Partly reproduced from Geigy Scientific Tables, Ciba-Geigy Ltd, with permission.)

solubilization. The point of initial gelatinization and the range over which it occurs is governed by starch concentration, method of observation, granule type and heterogeneities within the granule population under observation.

> The gelatinization temperature is a characteristic of the starch source, and varies between 72°C for waxy corn and maize starch and 85°C for wheat starch. Gelatinized starch is susceptible to amylase digestion. On cooling, this gelatinized starch undergoes **retrogradation**.

Retrogradation

Starch retrogradation is a process which occurs when the molecules comprising gelatinized starch begin to reassociate in an ordered structure. In its initial phases two or more starch chains may form a simple junction point, which may then develop into a more extensively ordered region. Retrogradation occurs in two steps:

1. Aggregation of the polysaccharide chains with phase separation into polymer-rich and polymer-deficient phases (gelation).
2. Slow crystallization of the macromolecules in the polymer-rich phase.

The length of time taken for these phases depends on the concentration and the chemistry of the starch. The whole process may require from 2 to 30 days. Retrogradation is more favourably achieved at low temperatures. Crystallinity is particularly marked in 50–60% gels and is not found in more dilute or concentrated gels. Ultimately, under favourable conditions, a crystalline structure appears. Retrograded starch is not hydrolysed by amylase and belongs to the family of resistant starches (see Table 5.3).

Starch paste
Pure amylose has a lower hot viscosity that pure amylopectin and therefore the amylopectin : amylose ratio is important in dictating the viscosity of the starch. Starch paste has properties of clarity, viscosity, texture, stability and taste which depend on the degree of gelatinization. The starch granules do not swell and there is no development of viscosity below 65°C. Starch paste may be affected by production processes that affect the intra-granule hydrogen bond which is important to maintain the integrity of the granule during gelatinization. Processes such as prolonged heating, exposure to high temperature, too much stirring during and after cooking, or exposure to specific pH conditions are all factors which reduce starch paste function. Freezing and thawing also have profound effects on starch paste and increase retrogradation. Modified starches are produced which provide stability under processing conditions and freeze/thaw conditions. Such granular starch modifications include:

- Cross-link starches: this may result from exposure to phosphate or adipate.
- Stabilization: stability may be made a feature of the starch granule by adding large molecular weight substituent groups, e.g. acetate and hydroxypropyl. This weakens the inter- and intramolecular hydrogen bonds within the granules and causes a reduction in the gelatinization temperature of the starch.
- Pre-cooked starches: these develop viscosity in water without the application of heat. Pre-cooking of starch occurs by extrusion of native granules at low water content.

Converted starches
If an internal hydrolysis reaction is applied to the starch granule, it is possible to produce starches which at 15–20% concentration give the same viscosity as the native starch of 5% concentration. There are improved textural properties and these starches give a firm gel.

Hydrophobic formation Substitution of native starches with cyclic dicarboxylic acid anhydrides produces hydrophobic starches which can be used in beverages, emulsions, clouding agents, flavour encapsulation, vitamin protection, salad dressings and creams.

Enzymatic cyclization using the enzymes cyclodextrin glucanotransferase can result in cyclodextrinase where several glucose monomers are joined together into a circular cone. A hydrophilic rim occurs at the top and bottom and a hydrophobic region in the middle and this may bind and hold molecules in the cavity. The size of the cyclodextrin can be determined by the substrate. Such a system can retain odours and flavours with consequent manufacturing advantages.

The phosphate content of native starch, that is, chemically linked phosphorus, can also affect the physical properties. The amount of phosphorus may be very small, varying from 0.03% in waxy corn or maize starch, to 0.06% in potato starch.

Chemical and enzymatic hydrolysis of starch

The process occurs in two stages, liquefaction and saccharinification.

1. Liquefaction: gelatinization and reduction in viscosity are required. This allows enzyme accessibility and also reduction in viscosity and hence allows high concentrations of starch to be used.
2. Saccharinification: the hydrolysis reaction can be stopped at a pre-determined point and results in a mixture of starch-hydrolysis products of defined molecular weight. Enzymes, including β-amylase and amyloglucosidase, are used. The product is called dextrose equivalent (DE) depending on the degree of hydrolysis of the starch and is a measure of the dextrose content. Starch is 0 DE and dextrose 100 DE.

DE, however, is a somewhat imprecise measurement in that it does not provide information as to the composition or properties of the product.

Hydrolysis products of starch

- Maltodextrins: these are starch hydrolysis products with a DE of 20.
- Dried glucose syrups: these are starch hydrolyses with a DE > 20 and when spray-dried are called dried glucose syrups. They are usually found in the ratio range of 30 to 60 DE. Above 60 DE the dried material is hydroscopic and difficult to keep dry.
- Maltose syrups: when β-amylase is used in the hydrolysis process, maltose is the main product.
- Hydrolysates: these are very high DE glucose syrups, 80 to 98.

5.6.4 Structure of starch in food

Starchy foods are usually heated in order to become palatable. In addition, they are often dried or frozen to increase ease of preparation in the domestic kitchen, to improve palatability, and to extend shelf-life. All preparations have an effect on starch's functional and nutritional properties (Tables 5.3 and 5.4).

Bread has a moisture content of 35–40% which is close to the optimum conditions for starch crystallization. Amylose retrogrades rapidly and within hours of cooking. On the other hand the

TABLE 5.3 *In vitro* nutritional classification of starch

Starch	Occurrence	Digestion in small intestine
Rapidly digestible starch	Freshly cooked starchy food	Rapid
Slowly digestible starch	Most raw cereals	Slow but complete
Resistant starch		
1. Physically inaccessible starch	Partly milled grain and seeds	Resistant
2. Resistant starch granules	Raw potato and banana	Resistant
3. Retrograded starch	Cooled, cooked potato, bread and cornflakes	Resistant

(From Englyst and Kingman, 1990)

TABLE 5.4 *In vitro* digestibility of starch in a variety of foods

| | % RDS | % SDS | % RS | | |
			RS$_1$	RS$_2$	RS$_3$
Flour, white	38	59	–	3	t
Shortbread	56	43	–	–	t
Bread, white	94	4	–	–	2
Bread, wholemeal	90	8	–	–	2
Spaghetti, white	55	36	8	–	1
Biscuits, made with 50% raw banana flour	34	27	–	38	t
Biscuits, made with 50% raw potato flour	36	29	–	35	t
Peas, chick, canned	56	24	5	–	14
Beans, dried, freshly cooked	37	45	11	t	6
Beans, red kidney, canned	60	25	–	–	15

The values are expressed as a percentage of the total starch present in the food. RDS = rapidly digestible starch; SDS = slowly digestible starch; RS = resistant starch; t = trace. (Englyst and Kingman (1990))

process of staling is a slow process which involves a change in the physical form of amylopectin. In addition, water moves through the bread from the crumb to the crust. There is also a decrease in the proportion of soluble starch and also in the starch's ability to swell in cold water. However, none of these affects the susceptibility of these starches to α-amylase hydrolysis.

Resistant starch

This may occur during food processing, in the baking of bread, cooling and storage. Amylose forms a strong retrograded starch stable to 120°C while amylopectin retrograde starch can be disrupted by gentle heating. When stale bread is reheated the retrograded amylopectin – which is the cause of the staling – reverts to the native form. The amount of resistant starch in foods is influenced by the water content, pH, temperature and duration of heating, and the number of heating and cooling cycles, freezing and drying.

In bread, resistant starches form immediately after baking and remain stable during storage. In contrast, resistant starch production in boiled potatoes is time-dependent; freshly cooked potatoes contain < 2% resistant starch, but this increases during cooling to about 3%. Reheating may reduce the degree of retrogradation, but on cooling the amount of resistant starch increases further. Similar changes may be found with wheat starch.

There is a rough correlation between the amy-lose content and the yield of resistant starch after cooking and drying. Waxy starch forms very little resistant starch, whereas high-amylose starch forms over 30% of resistant starch.

The processing conditions can affect the amount of gelatinization and the level and type of retrogradation of starch. Quick-cooking rice has a porous structure and a grain which hydrates very rapidly. Non-sticky grains result from controlled retrogradation of gelatinized starch. A par-boiling process modifies the physical, chemical and textural characteristics of the rice. Such a change reduces the leaching of vitamins, proteins and starch from the grain into the cooking water. The rice grains are steeped and steam-dried, resulting in gelatinization of the starch granules which are held in a proteinaceous matrix without damaging the starch granules. On the other hand, the retrogradation of starch in cooked rice may result in a tough and rubbery product.

Another processing technique which affects gelatinization and the general physicochemical properties of rice is extruder cookers. It is possible for the amylose and amylopectin to be separated by the physical process with an effect on stickiness, expansion, carbohydrate solubility and digestibility. The starch granules may be disrupted; there may be partial starch breakdown and increased levels of resistant starch, but these effects depend on the amylose : amylopectin ratio in the original preparation.

5.6.5 Interaction of starches and sugars with other food compounds

Starch/protein interactions

Protein is associated with starch molecules on the granule surface. Such aggregations alter the granule's physical properties. These interactions are important in bread and pasta products. Bread is a colloidal system which is physically unstable. There is formation of a coherent viscoelastic mass, which is dough. Protein in the dough forms a matrix in which the starch granules are separated at high concentrations. These starch granules gelatinize in part during baking and the glutinous matrix coagulates to form bread with its characteristic structure. The degree of gelatinization of starch is very dependent on the moisture content and the temperature and also on the type of wheat products used.

Pasta is made from durum semolina which is very hard due to a high protein content. During extrusion protein forms a protective coating around the starch granule. There is partial starch gelatinization and the coagulated protein forms a fibrillar structure. In contrast, protein is not important in preventing starch leaching. The starch granules are stored in the cells. The granules gelatinize and swell until the cells are filled. It is only when the cells are broken, when the starch is freed that a sticky product is produced.

Sugar/protein interactions

Sugars in the dough alter the texture of the resulting bread by delaying the gelatinization of starch and also the denaturation of proteins. Such reactions include the typical Maillard reactions, the products of which are formed when wheat cereal products are toasted. Aldopentoses are more reactive than aldohexoses which in turn are more reactive than disaccharides. The important reaction is with lysine so that hexoses and pentoses remove 42–66% of lysine respectively. The loss of lysine is very dependent upon the protein.

The **Maillard reaction** takes place when reducing sugars, e.g. glucose, fructose or lactose, are subjected to heat in the presence of protein. The reaction is an initial condensation between an amino acid group and a carbonyl group to form a Schiff base and water. After this there is cyclization and isomerization in acidic conditions (Amadori rearrangements). The result is a 1-amino-1-deoxy-2-ketose derivative.

Other amino acids which may be involved in Maillard reactions are tryptophan, arginine and histidine. Sucrose does not react, but may be degraded by heat, producing molecules which can react with free amino acids. This is particularly marked in dry heating. It can also result in loss of nutritional qualities of the protein in baked food. Such losses are particularly marked in biscuits, which are dry-heated.

Starch/lipid interactions

Fatty acids and monoglycerides can form inclusion compounds with amylose. This occurs when the hydrocarbon of the lipid is found within the helical cavity of amylose. Such association is found much less commonly with lecithin and not at all for acylglycerols. This is because of steric hindrance preventing the acylglycerols from entering the starch helix. The quantity of lipids also affects the ability of starch to retrograde. Amylopectin is also able to form complexes with fatty acids but to a much lesser degree than amylose.

Cereal starches form inclusion complexes with monoacyl lipids. The latter are present in the cereal complex with 15–20% of the total amylose. Monoglycerols are used to prevent retrogradation of amylose. They also have important effects on the physical properties and cooking properties of starch.

Starch/sugar/mineral interactions

Cations in general have little affinity for starch and glucose syrups at neutral pHs.

Summary

1. In biology most carbohydrates exist as high molecular weight polymers. These polysaccharides are simple sugars connected by glycosidic bonds.
2. Cellulose is a major structural polysaccharide and accounts for approximately half of the carbon found in plants. Cellulose consists of polymeric glucose with a glycosidic linkage of the β-(1–4) type.
3. The two major polysaccharides for energy storage are starch in plant cells and glycogen in animal cells. They differ in their chain lengths and branching patterns. Glycogen is highly branched with an α(1–6) linkage occurring every 8 to 10 glucose units along the backbone with short side chains of approximately 8 to 12 glucose units each. Starch occurs both as unbranched amylose and as branched amylopectin.
4. The chemical and physical changes associated with gelatinization and retrogradation have important consequences for food production and intestinal absorption.
5. Starch is hydrolysed by the enzyme amylase. Starches which are not hydrolysed by amylase are resistant starches.
6. Starch, proteins or lipids may interact to form aggregates.

Further reading

Annison, G. and Topping, D.L. (1994) Nutritional role of resistant starch. *Annual Review of Nutrition*, **14**, 297–320.

Atwell, W.A., Hood, L.F., Lineback, D.R., Varriano-Marstoatone, and Zobell, H.F. (1988) The terminology and methodology associated with basic starch phenomena. **Cereal Food World**, 33, 306–11.

Dobbing, J. (ed.) (1989) *Dietary Starches and Sugars in Man: A Comparison*, ILSI Human Nutrition Reviews, Springer-Verlag, London.

Englyst, H.N. and Cummings, J.H. (1986) Digestion of polysaccharides of potato in the small intestine of man. *American Journal of Clinical Nutrition*, **45**, 423–31.

Englyst, H.N. and Kingman, S.M. (1990) Dietary fiber and resistant starch, a nutritional classification of plant polysaccharides, in *Dietary Fiber* (eds D. Kritchevsky, C. Bonfield and J.W. Anderson), Plenum Press, New York.

Report of the British Nutrition Foundation Task Force (1990) *Complex carbohydrates in Foods*, Chapman & Hall, London.

Southgate, D.A.T. (1976) *Determination of Food Carbohydrates*, Applied Sciences Publishers, London.

5.7 Dietary fibre

- Dietary fibre is a collective term for dietary complex carbohydrates (non-starch polysaccharides) which form the structure of plant cell walls.
- Dietary fibre can be regarded as a sponge which, as it passes along the gastrointestinal tract, alters absorption in the duodenum, bile acid metabolism, caecal fermentation and faeces weight.
- The origin of dietary fibre, and/or how it is extracted, can alter its effect on nutrition and gastrointestinal function.

5.7.1 Introduction

Dietary fibre is a member of a family of dietary complex carbohydrates which have individual and diverse actions. The chemistry of individual polymers cannot, however, identify or predict the biological action in the gastrointestinal tract. Each fibre is peculiar in its biological action, and is affected by extraction, physical format and processing. One way of classifying dietary polysaccharide is to include polymers which exceed 20 sugar residues as dietary complex carbohydrates. Another name for dietary fibre is **non-starch polysaccharides** (NSP).

All of these materials are polymeric carbohydrates and lignins which together form the plant cell wall and which are not digested as they pass through the upper gastrointestinal tract.

Considerable anatomical differences exist between and within economically important plant groups. These differences occur in the cell wall and could be important in determining the diversity of actions in the gastrointestinal tract of sundry dietary fibres.

5.7.2 Distribution of fibre in plants

Dietary fibre in naturally occurring foods consists of plant cell walls, the structure of which differs not only among plant species but also during normal development within one species or even a single cell. The composition of the cell wall is dependent on the plant species, the tissue type, the maturity of the plant organ at harvesting, and to some extent post-harvest storage conditions. Non-carbohydrate components of plant cell walls may also influence the nutritional potential of the plant fibres, e.g. proteins, lignin, phenolic esters, cutin and waxy materials and suberin.

Parenchymatous tissues are the most important source of vegetable fibre. The vascular bundles and parchment layers of cabbage leaves, runner beans, pods, asparagus stems and carrot roots are relatively immature and only slightly lignified on harvesting and hence digestion. The soft fruits such as strawberries and raspberries contain very little dietary fibre but abundant amounts of water. Lignified tissues are of greater significance in cereal sources such as wheat bran and oat products. Cereals contain very little pectic substances but there is substantial arabinoxylan in wheat and β-glucan in barley and oats. The distribution of polysaccharides within the plant tissue also varies. Much of the βglucan in oats is concentrated in the cells of the outermost layer of the seeds, whereas the β-glucans in barley are more evenly distributed.

5.7.3 The chemistry of fibre

An important aspect of the plant cell wall is the interlocking of water-soluble polysaccharides to form biological barriers which are water-resistant. Many of the constituents of the plant cell wall, hemicelluloses and pectins, are soluble in water after extraction. This solubility, which is unmasked by extraction processes, contrasts with the insolubility of the complex polysaccharide of the intact cell wall.

The backbone of the plant cell wall, cellulose, is a polymer of linear β(1–4)-linked glucose molecules, several thousand molecules in length (Figure 5.25). Cellulose occurs largely in a crystalline form in microfibrils, coated with a monolayer of more complex hemicellulosic polymers held tightly by hydrogen bonds. These are embedded in a gel of pectin polysaccharides. The cellulose microfibrils are coated with a layer of xyloglucans bound by hydrogen bonds, and this enables the insoluble cellulose to be dispersed within the wall matrix.

Hemicelluloses

Substitution of the hydroxyl group at C6 with xylose, as in xyloglucans, makes it more soluble in alkali and water. Hemicelluloses provide part of the true rigidity of the cell wall. The important hemicelluloses are xyloglucans, xylans and β-glucans. Xyloglucan is a linear (1–4)-β-D-glucan chain substituted with xylosyl units which may be further substituted to form galactosyl-(1–2)-β-xylosyl or fucosyl-(1–2)-D-galactosyl-(1–2)β-D-xylosyl units.

The **pectins** may act as biological glue, cementing cells together. The precise function of pectins within the cell wall is unclear, but they are closely associated with calcium. Most pectins are probably derived from the primary cell wall and appear to be soluble only after calcium ions are removed. The principal cross-linkage is provided by the helical (1–4)-β-D-galactosyluronic groups from adjacent polysaccharides (Figure 5.25) and

Name	Structure	Occurrence
Dextrans	$\alpha(1\text{-}6)$-Linked glucose residues in branched or straight chains, for instance: 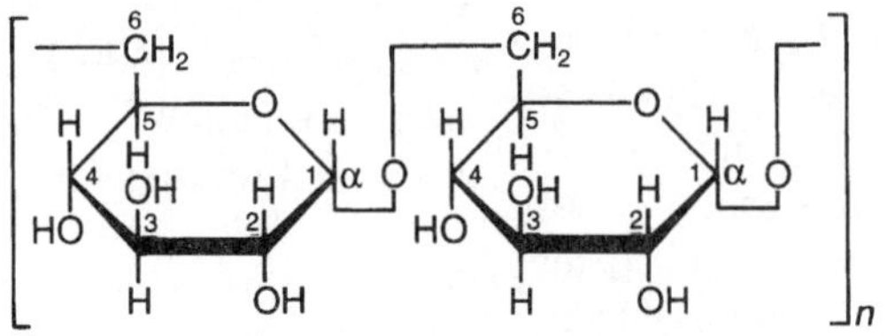	
Glycogen (liver starch)	Highly-branched molecule resembling amylopectin and consisting of unit chains of $\alpha(1\text{-}4)$-linked glucose residues interlinked by $\alpha(1\text{-}6)$-glycosidic bonds: 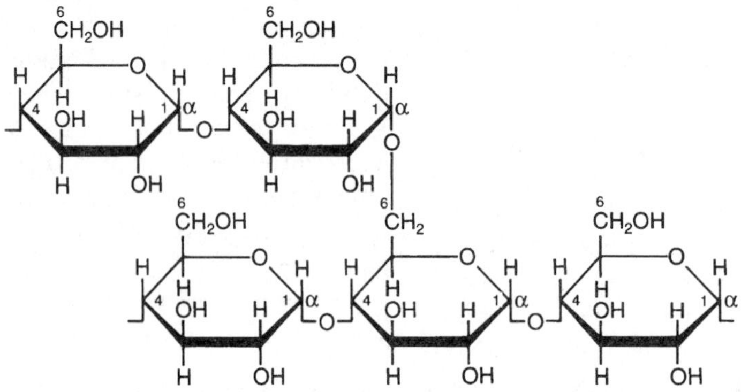	Reserve carbohydrate of animal tissues. Converted in muscle to lactic acid during glycolysis. Also present in yeast
Pectic acid (pectins)	Linear chain of $\alpha(1\text{-}4)$-linked D-galacturonic acid residues: 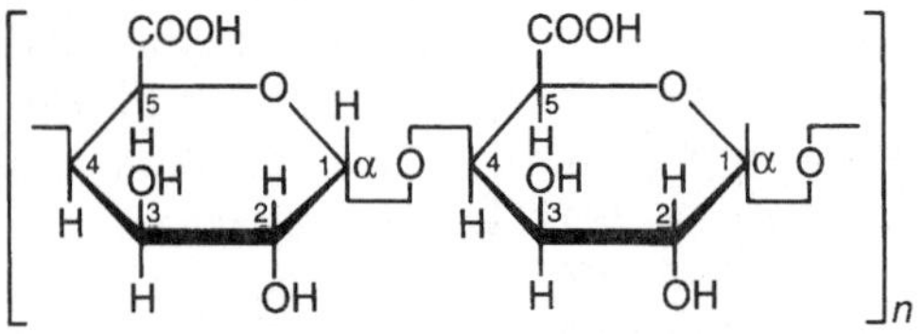	Important cell-wall constituent of plants. Occurs as Ca salt or methylester
Cellulose	Linear chain of $\beta(1\text{-}4)$-linked glucose residues: 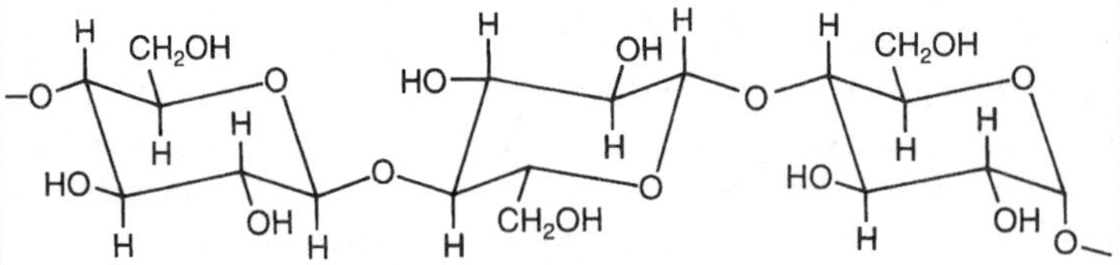	Chief structural polysaccharide of plants. Also found in algae, bacterial membranes, and as tunicin in some lower animals. Not digested by man

FIGURE 5.25 Structure of dextrans, glycogen, pectin and cellulose. (Partly reproduced from Geigy Scientific Tables, Ciba-Geigy Ltd, with permission.)

condensation with calcium converts soluble pectin into rigid 'egg-box' structures. The extent of calcium cross-bridging or esterification through aromatic linkages and even degree of branching and size of neutral sugar side chains influence gel flexibility, cell wall porosity and interaction with hemicellulosic polymers.

The glycoproteins within the cell wall can provide extensive cross-linkages across the different polysaccharide components of the cell wall and may act to form a network with the cellulose microfibrils within a hemicellulose pectin gel.

5.7.4 Gums and mucilages

The plant gums and mucilages are a poorly defined group of exudates from plants. Mucilages are gums dissolved in the juices of the plant. Plant gums are a heterogeneous group of complex, branched heteropolysaccharides. They are obtained from plants often occurring as a reaction to injury. Mucilages are associated with seeds and are very water-soluble. They are functionally important in seeds in dry areas where the mucilages bind water so that after rain the plant can rapidly grow and pass through the cycle to produce a new seed and so survive the next drought.

Some plant polysaccharides are harvested from the plant and purified before inclusion as processed food components or used therapeutically. These are more often soluble polysaccharides and include the following:

- Guar gum (mol. wt. 0.25×10^6), a linear non-ionic galactomannan.
- Gum karaya (mol. wt. 4.7×10^6), a cylindrical complex polysaccharide partially acetylated and highly branched with interior galacturono-rhamnose chains to which are attached galactose and rhamnose end groups. Glucuronic acid is also a component.
- Gum arabic (mol. wt. $0.5–1.5 \times 10^6$), with a complex acidic heteropolysaccharide based on a highly branched array of galactose, arabinose, rhamnose and glucuronic acid. Uronic acid residues tend to occur on the periphery of an essentially globular structure.
- Gum tragacanth (mol. wt. $0.5–1 \times 10^6$), a complex gum with two major components, bassorin and tragacanthin composed of arabinose, fucose, galactose, glucose, xylose and galacturonic acid.

These gums and mucilages are used as additives by food manufacturers, and as such may contribute less than 2% of a food.

5.7.5 Physical properties of dietary fibre

An important function of insoluble fibres is to increase viscosity in the intestinal contents. Other polymeric components of the diet (proteins, gelatinized starch) and mucous glycoproteins liberated from the epithelia, particulate material present in chyme (such as insoluble fibre or hydrated plant tissues), will also contribute to a lesser extent to overall viscosity. The viscosity of these substances in the intestine is sensitive to changes in ionic concentration due to intestinal secretion or absorption of aqueous fluids.

Vegetables undergo structural change during cooking and mastication, e.g. cellular disintegration. The cells after cooking are ruptured and the cell contents lost. The grinding of foods before cooking and ingestion may also have pronounced effects on fibre action. Cell walls may be disrupted and the reduced particle size of some fibre preparations such as wheat bran may decrease the biological efficacy. The effect of other cooking processes, e.g. Maillard reactions, etc. are not known.

5.7.6 Quantitative measurement of fibre

There are two approaches to the analysis of dietary fibre – gravimetric and gas–liquid chromatography (GLC). Gravimetric methods measure by weighing an insoluble residue after chemical and enzymatic solubilization of non-fibre constituents. The remaining protein is assayed and subtracted from the original weight. The GLC methods involve the enzymatic breakdown of starch and the separation of the low-molecular weight sugars, acid hydrolysis to free sugars, conversion to alditol acetates and finally separation and measurement of neutral monomers with GLC, together with determination of uronic acid and lignin. The GLC methods enable the nature of the carbohydrate to be determined in more detail.

5.7.7 Actions of fibre along the gastrointestinal tract

The four major effects of dietary fibre on gastro-intestinal activity (Figure 5.26) relate to:

- the rate of gastrointestinal absorption
- sterol metabolism
- caecal fermentation
- faeces weight

Rate of intestinal absorption

There are two main components to the role of dietary fibre in the upper gastrointestinal tract, to prolong gastric emptying time, and to retard the absorption of nutrients. Both are dependent on the physical form of the fibre, and in particular its viscosity.

The inclusion of viscous polysaccharides in carbohydrate meals reduces the postprandial blood glucose level concentrations in humans. There

appears to be no correlation between the rate of gastric emptying and postprandial concentrations of blood glucose.

Diets which contain a substantial amount of complex carbohydrate content tend to be bulky, and require longer times to eat. In one experiment the time taken to eat a whole apple took longer (17 minutes) than that required for purée (6 minutes) or apple juice (1.5 minutes) in equicaloric amounts.

Gastric emptying is affected by the physical nature of the gastric contents as well as the chemistry of the components. Isolated viscous fibres tend to slow the gastric emptying rate of liquids and disruptible solids. Gastric emptying for a fibre on its own may be quite different from that when ingested along with other dietary constituents such as fat and protein. The gastric emptying time for different fibre sources is variable.

Rates of release of nutrients from dietary fibre in the intestine are influenced by factors such as the intactness of tissue histology, degree of ripeness, and the effects of processing and cooking.

There is no evidence to suggest that viscous polysaccharides inhibit transport across the small intestinal epithelium. It is more likely that their viscous properties inhibit the access of nutrients to the epithelium. Two mechanisms bring nutrients into contact with the epithelium. Intestinal contractions create turbulence and convection currents which mix the luminal contents bringing material from the centre of the lumen close to the epithelium. Nutrients then have to diffuse across the thin, relatively unstirred layer of fluid lying adjacent to the epithelium. Increasing the viscosity of the luminal contents may impair both convection and diffusion of the nutrients across the unstirred layer. In the case of isolated polysaccharides such as guar gum, the slowing of nutrient absorption appears to be a function of viscosity.

In the case of whole plant material the influence on absorption appears to be due to the inaccessibility of nutrients within the cellular matrix of the plant. The effects on absorption can be minimized by grinding the food before cooking or by thorough chewing; both processes open the cellular structure.

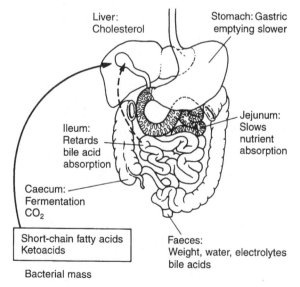

FIGURE 5.26 Action of dietary fibre (non-starch polysaccharides) along the gastrointestinal tract.

Inadequate mixing of luminal contents by soluble polysaccharides due to increased viscosity may also slow the movement of digestive enzymes to their substrates.

Viscous polysaccharides tend to delay small bowel transit, possibly due to resistance to the propulsive contractions of the intestine. In rats most of this delay is secondary to alterations in ileal motility, with transit through the upper small intestine little affected.

Complex carbohydrates – particularly those that possess uronic and phenolic acid groups or sulphated residues such as pectins and alginates – may bind magnesium, calcium, zinc and iron. However, there are other constituents of plant cells, e.g. phytates, silicates and oxalates which also chelate divalent cations. The binding of minerals may be reduced by acid, protein, ascorbate and citrate.

The reduction in absorption of minerals and vitamins could, in theory, have adverse nutritional consequences, particularly in populations eating diets inherently deficient in these nutrients; that is, in developing countries or fastidious health food-conscious communities where diets may be marginal in micronutrients but high in fibre. Children are particularly vulnerable to such conditions. Customary Western diets contain levels of minerals and vitamins in excess of daily requirements. Mineral balance studies have indicated that for people on nutritionally adequate diets, the ingestion of mixed high-fibre diets or dietary supplementation with viscous polysaccharides is unlikely to cause mineral deficiencies.

The ingestion of dietary fibre may similarly affect drug absorption in two ways; by reducing gastric emptying or inhibiting mixing in the small intestine. Viscous polysaccharides delay the absorption of paracetamol. Quite separately, if a drug enters the enterohepatic circulation any bacterial metabolism of the drug may be altered by coincidental fermentation of fibre and thus the half-life of a drug may be increased or decreased. An example of this is digoxin, which has a narrow therapeutic range and is passively absorbed in the small intestine and so is affected by gastric emptying or decreased small intestinal absorption.

Digoxin is also reduced in the colon to an inactive metabolite, which then will be absorbed.

Alteration in sterol metabolism

Dietary fibre has been shown to have an effect on sterol metabolism. This effect is not simple and it is possible that dietary fibre displaces fat from the diet, or that polyunsaturated fats frequently eaten in conjunction with the fibre may also be important. The direct effect of fibre on sterol metabolism may be through one of several mechanisms:

- altered lipid absorption
- altered bile acid metabolism in the caecum
- reduced bile acid absorption in the caecum
- indirectly via short-chain fatty acids, especially propionic acid, resulting from fibre fermentation

An important action of some fibres is to reduce the reabsorption of bile acids in the ileum and hence the amount and type of bile acid and fats reaching the colon. Bile acids may be trapped within the lumen of the ileum either because of a high luminal viscosity or because they bind to the polysaccharide structure. A reduction in the ileal reabsorption of bile acid has several direct effects. The enterohepatic circulation of bile acids may be affected. In the caecum, bile acids are deconjugated and 7α-dehydroxylated. In this less water-soluble form bile acids are adsorbed to dietary fibre in a way that is affected by pH and is mediated through hydrophobic bonds, thereby increasing the loss of bile acid in the faeces. The consequence of this is that the enterohepatic pool is initially reduced. This may be renewed by increased synthesis of bile acids from cholesterol, reducing body cholesterol.

Other fibres, e.g. gum arabic, are associated with a significant decrease in serum cholesterol without increasing faecal bile acid excretion.

The fibres that are most effective in influencing sterol metabolism (e.g. pectin) are fermented in the colon, as demonstrated by an increased breath hydrogen production. It is unlikely that the physiological effect is due entirely to adsorption to fibre in the colon. This is in contrast to the important sequestrating effect of fibre in the ileum. There might be an alteration in the end-products of bile acid bacterial metabolism which are absorbed from the colon and return to the liver in the portal vein, modulating either the synthesis of cholesterol or its catabolism to bile acids. Alternatively, it is probable that bacteria bind bile acids in the colon after the initial deconjugation and dehydroxylation.

The precise relationship between the proportions of ileal and caecal absorption of bile acids is difficult to estimate. This is a variable which is dependent in part on the amount and type of fibre. A further complication is the bacterial colonization of the ileum simulating caecal bacterial activity. Approximately 25% of the body pool of cholic acid and 50% of chenodeoxycholic acid pass into the caecum either to be absorbed or excreted in faeces.

Substrate for caecal fermentation

The colon may be regarded as two organs, the right side a fermenter, the left side affecting continence. The right side of the colon is involved in nutrient salvage so that dietary fibre, resistant starch, fat and protein are utilized by bacteria and the end-products absorbed for use by the body.

The colonic flora is a complex ecosystem consisting largely of anaerobic bacteria which outnumber the facultative organisms at least 100 : 1. The colonic flora of a single individual consist of more than 400 bacterial species. The total bacterial count in faeces is 10^{10} to 10^{12} colony forming units/ml. Despite the complexity of the ecosystem the microflora population is remarkably stable. Although wide variations in the microflora are found between individuals, studies in a single subject show the microflora to be stable over prolonged periods. There is some attraction in identifying individual bacteria. It is, however, more profitable for physiological and nutritional studies to regard the caecal bacterial complex as an important organ in its own right, complementary to the liver in the enterohepatic circulation.

The caecal bacterial flora are dependent upon dietary and endogenous sources for nutrition. There are variations in the amounts of substances passing through the intestine from the ileum, with an inverse relationship between caecal bacterial metabolism and upper intestinal nutrient absorption. Dietary fibre has an influence on bacterial mass and enzyme activity. The consensus view is that while the caecal bacterial mass may increase as a result of an increased fibre content in the diet, the types of bacteria do not alter.

The process whereby a compound is bacterially dissimilated in the caecum under anaerobic conditions is complex and varied, leading to partial or complete decomposition with the end products being:

- absorbed from the colon to be utilized as nutrients
- absorbed and re-excreted in the enterohepatic circulation
- excreted in the faeces

In addition to this the colon is part of the excretion system provided by the liver and biliary tree, i.e. the enterohepatic circulation. Poorly water-soluble chemicals of molecular weight of about 300–400 Da, are excreted in the bile with enhanced water-soluble properties through chemical conjugation with glucuronide, sulphate, acetate, etc. or made physically soluble by the detergent properties of bile acid. These may be endogenous, e.g. bile acids, bilirubin and hormones, or exogenous, e.g. drugs, food additives, pesticides. They pass unabsorbed through the small intestine. In the caecum these biliary excretion products, and also unabsorbed dietary constituents, e.g. resistant starch, fat and proteins and mucopolysaccharides secreted by the intestinal mucosa, are fermented by the bacterial enzymes. The fermentation process of biliary excretion products removes substitutions which have enhanced water-solubility and enabled biliary excretion to occur. The bacterial metabolic products are less water-soluble. Some of the end-products of the fermentation of biliary excretion compounds are reabsorbed, metabolically altered, reconjugated in the liver and excreted in bile; hence an enterohepatic circulation is established.

The effects of dietary fibre in the colon may be summarized in terms of:

- susceptibility to bacterial fermentation
- ability to increase bacterial mass
- ability to increase bacterial saccharolytic enzyme activity
- water-holding capacity of the fibre residue after fermentation

Enlargement of the caecum is a common finding when some dietary fibres are fed and this is now believed to be part of a normal physiological adjustment. Such an increase may be due to a number of factors, prolonged caecal residence of the fibre, increased bacterial mass, or increased bacterial end-products.

The **fermentation** of fibre yields hydrogen, methane and short-chain fatty acids. Hydrogen is readily measured in the breath and has a diurnal variation with its nadir at mid-day and an increase in the afternoon. Diverse sources of fibre influence the evolution of hydrogen in different ways. Disaccharides generate hydrogen more rapidly than trisaccharides which in turn evolve hydrogen more quickly than oligosaccharides. More complex carbohydrates may not be fermented as rapidly and may require induction of specific enzymes before they can be utilized.

Methane-producing organisms are said to be strict anaerobes. There are wide diversities in the proportion of individuals who exhale methane in their breath. The range within different healthy adult populations is from 33 to 80%. The breath methane status of an individual remains stable throughout the day and over prolonged periods, yet faeces from healthy individuals – regardless of breath methane excretion status – will always produce methane. This suggests that all individuals produce methane but that there must be a critical level produced if methane is to spill over into the breath.

It has been demonstrated that considerable methane excretion only takes place when sulphate-reducing bacteria are not active. The metabolic end-product of dissimilatory sulphate reduction is thought to be toxic to methanogenic bacteria. When sulphate is present, sulphate-reducing bacteria have a higher substrate affinity for hydrogen than do methanogenic bacteria.

Some non-absorbed carbohydrates, e.g. pectin, gum arabic, oligosaccharides and resistant starch, are fermented to short-chain fatty acids (chiefly acetic, propionic and *n*-butyric), carbon dioxide, hydrogen and methane. The production of short-chain fatty acids has several possible actions on the gut mucosa. All of the short-chain fatty acids are readily absorbed by the colonic mucosa, but only acetic acid reaches the systemic circulation in appreciable amounts. Butyric acid is metabolized before it reaches the portal blood, propionic acid is metabolized in the liver. Butyric acid appears to be used as a fuel by the colonic mucosa and *in vitro* studies of isolated cells have indicated that the short-chain fatty acids and butyric acid in particular are the preferred energy sources of colonic cells. Short-chain fatty acids are potent stimulants of cellular proliferation not only in the colon but also in the small intestine.

Short-chain fatty acids are the predominant anions in the human faeces. They are derived by fermentation of complex carbohydrates, resistant starch and non-starch polysaccharides.

> The caecal fermentation of 40–50 g of complex polysaccharides will yield 400–500 mmol total short-chain fatty acids, 240–300 mmol acetate, and 80–100 mmol of both propionate and butyrate. Almost all of these short-chain fatty acids will be absorbed from the colon. This means that faecal short-chain fatty acid estimations do not reflect caecal and colonic fermentation, only the efficiency of absorption, the ability of the fibre residue to sequestrate short-chain fatty acids, and the continued fermentation of fibre around the colon, which presumably will continue until the substrate is exhausted.

The absorption of short-chain fatty acids from the colon in humans is concentration-dependent, and associated with bicarbonate secretion. Bicarbonate appears consistently in the colonic lumen during short-chain fatty acid absorption, a process independent of the chloride–bicarbonate exchange. It is possible that there is an acetate–bicarbonate exchange at the cell surface, but the precise mechanism is not understood. There is a stimulatory effect of short-chain fatty acids on sodium absorption from the colonic lumen. This may be related to the recycling of hydrogen ions. The unionized short-chain fatty acid crosses into the cell where it dissociates and hydrogen ion is moved back into the lumen in exchange for sodium. Thus, short-chain fatty acids provide a powerful stimulant to sodium and water absorption.

The fermentation also produces other carbon fermentation products which when absorbed are metabolized into the glycerol of hepatic glycerides and amino acids.

The presence of bacteria in the colon produces an 'organ' of intense, mainly reductive, metabolic activity. This is in contrast to the liver, which is oxidative. The range of metabolic transformations that the intestinal flora perform on ingested compounds is wide. The major enzymes involved in these activities include azoreductase, nitrate reductase, nitroreductase, β-glucosidase, β-glucuronidase, and methylmercury-demethylase. The action of fibre on the activity of these enzymes may be species-dependent and animal studies are not always predictive of the action in humans.

Faeces weight

Faeces are complex and consist of 75% water; bacteria make a large contribution to the dry weight, the residue being unfermented fibre and excreted compounds. There is a wide range of individual and mean faeces weights. In a study in Edinburgh the individual variation was between 19 and 280 g over 24 hours. The amount of faeces excreted varies quite markedly from individual to individual and by any one individual over a period of time, though why such variation occurs is unknown. Of the dietary constituents, only dietary fibre influenced faeces weight.

The most important mechanism whereby dietary fibre increases faecal weight is through the water-holding capacity of unfermented fibre. However, fibre may influence faecal output by another mechanism. Colonic microbial growth may be stimulated by ingestion of such fermentable fibre sources as apple, guar or pectin. This is an uncertain route, as there is not always an increase in faeces weight as a result of eating these fibres. There may also be an added osmotic effect of products of bacterial fermentation on faecal mass, though this is not as yet a well-defined contribution.

One of the major functions of the colon is to absorb water and produce a plasticine-type of faeces which can be voided readily and at will from the rectum. The ileum contains a viscous fluid, the viscosity being created by mucus and water-soluble fibres whose molecular weight, degree of cross-linkages and aggregation will affect the viscosity. If the viscosity increases to a certain point, peculiar to the constituent macromolecules, then a sol or hydrated carbohydrate complex will result. The sol will be coherent and homogeneous.

The concentration of ileal effluent in the caecum and colon is the result of the absorption of water. This might be expected to create a gel. Faeces are not a gel, however, but a plasticine-like material, heterogeneous without viscosity and made up of water, bacteria, lipids, sterols, mucus and fibre. In the caecum there is therefore a marked physical change, in part as a result of bacterial activity, in part by the presence of bacteria themselves. Such a plasticine-like structure is lost in watery diarrhoea. The mechanism of this change, whether physiological or pathological, is unknown but some of the steps involved are described below.

In the colon water is distributed in three ways:

1. Free water which can be absorbed from the colon.
2. Water that is incorporated into bacterial mass.
3. Water that is bound by fibre.

Faeces weight is dictated by:

- the time available for water absorption to take place through the colonic mucosa

- the incorporation of water into the residue of fibre after fermentation of the fibre
- the bacterial mass

Wheat bran added to the diet increases faecal weight in a predictable linear manner and decreases intestinal transit time. The increment in faecal weight is independent of initial weight. Wholemeal bread, unless of a very coarse nature, has little or no effect on faecal weight. The particle size of the fibre is all-important, coarse wheat bran being more effective than fine wheat bran. The greater the water-holding capacity of the bran, the greater the effect on faecal weight. The effect of the water-binding by wheat bran is such that in addition to an increase in faecal weight, other faecal constituents, namely bile acids – which in absolute amounts do not increase – are diluted by faecal water and hence their concentration decreases. The increment in faecal weight per gram of wheat bran varies in different populations. For control subjects, an increase in wet faecal weight, while depending on the particle size of the bran, is generally of the order of 3–5 g/g fibre. However, in individuals with irritable bowel syndrome and symptomatic diverticulosis, the increment is of the order of 1–2 g wet weight/g fibre. This suggests that there is a difference in the handling of the fibre in the intestine in these situations.

Bacteria are an important component of the faecal mass. It is not known what percentage are living and what percentage are dead and as such are being voided. The fermentation of some fibres results in an increase in the bacterial content and hence faecal weight. Other fibres, of which pectin is an important example, are fermented without any such effect. It is possible that some fibres which increase stool weight in association with an increased bacterial mass do so because of an increase in excreted bacteria adherent to unfermented fibre.

The degree to which free water is absorbed from the colon will be affected by a number of factors; these are poorly understood. An increase in the short-chain fatty acid concentration of faeces appears to be related to an increased output of faecal water, which may suggest that under some circumstances short-chain fatty acid absorption is less efficient and in part determines faecal output. This is in sympathy with the view of Hellendoorn (1978) who suggested an important role for fibre fermentation products on stool weight and transit time. The demonstration that short-chain fatty acids were absorbed rapidly in the colon suggested that they play no part in determining faecal output. However, it would appear that there is continued fermentation of some complex carbohydrates, e.g. ispaghula in the distal colon. Under these circumstances the faecal short-chain fatty acids may influence faecal water osmolality, absorption and stool weight.

The effect of fibre in the colon may be summarized as:

$$\text{Faeces weight} = W_f(1 + H_f) + W_b(1 + H_b) + W_m(1 + H_m)$$

where W_f, W_b and W_m are, respectively, the dry weights of fibre remaining after fermentation in the colon, bacteria present in the faeces, and osmotically-active metabolites and other substances in the colonic contents which could reduce the amount of free water absorbed, and H_f, H_b *and* H_m denote their respective 'water-holding capacities' (i.e. the weight of water resistant to absorption from the colon, per unit dry weight of each faecal constituent).

Summary

1. Dietary fibre is a term for the family of dietary complex carbohydrates and lignins in plant cell walls. Alternative descriptions include non-starch polysaccharides (NSP). These plant cell wall polymeric carbohydrates and lignins are not digested in the upper gastrointestinal tract.
2. These complex carbohydrates have individual and diverse actions along the gastrointestinal tract. The chemistry cannot, however, identify or predict the biological action of individual fibres

in the gastrointestinal tract. Each fibre is peculiar in its biological action, and is affected by extraction, physical format and processing.

3. In the upper gastrointestinal tract the physical properties of the dietary fibre (NSP) are important in slowing the rate of absorption of nutrients.
4. Some dietary fibres may alter sterol turnover, usually by increasing faecal bile acid excretion.
5. Some dietary fibres increase caecal bacterial growth and metabolism.
6. Some dietary fibres increase faecal weight through a combination of the water-holding capacity of the fibre not fermented by bacteria, by bacterial growth, and the osmotic effect of bacterial fermentation products in the colonic lumen.

Further reading

Hellendoorn, E.W. (1978) Fermentation as the principal cause of the physiological activity of indigestible food residue, in *Topics in Dietary Fiber Research* (ed. G.A. Speller), Plenum Press, New York, pp. 127–216.

Kritchevsky, D. and Bonfield, C. (1994) *Dietary Fibre in Health and Disease*, Eagan Press, St Paul Minnesota, USA.

Kritchevsky, D., Bonfield, C. and Anderson, J.W. (eds) (1990) *Dietary Fiber*, Plenum Press, New York.

Report of the British Nutrition Foundations Task Force (1990) *Complex Carbohydrates in Foods*, Chapman & Hall, London.

Schweitzer, T.F. and Edwards, C.A. (eds) (1992) *Dietary Fibre – a Component of Food*, Springer-Verlag, London.

Southgate, D.A.T. (1976) *Determination of Food Carbohydrates*, Applied Sciences Publishers, London.

Trowell, H., Burkitt, D. and Heaton, K. (1985) *Dietary Fibre, Fibre-Depleted Foods and Disease*, Academic Press, London.

Chapter 6

Alcohol

6.1 Alcohol as a nutrient

- Alcohol is a non-essential nutrient but is significant socially
- The metabolism of alcohol is dependent on genetic predisposition and gender
- Alcohol has a substantial nutritional value
- Alcohol in excess has profound physical and emotional consequences

6.1.1 Introduction

Alcohol is an interesting dietary component to discuss in detail. This simple molecule can be used as a model to underline the different modalities of nutrition, manufacturing, social, genetic predisposition and gender differences in metabolism, nutritional value and the consequences of excess.

Yeast is the main organism responsible, through the fermentation of carbohydrates, for the production of alcohol. The popularity of alcohol results from its effects on mood and an induced sense of well-being. The ingestion of alcohol is a social activity; a meeting, occasion or function is better enjoyed in the relaxing ambience coincidental with the consumption of alcohol.

6.1.2 Socioeconomic and sociomedical aspects

Alcohol is used in widely diverse cultures. Fermented drinks include Kava made from the pepper plant which is widely used in the Pacific for ceremonial occasions. Mild exposure leads to a fatigued retreat into suspended animation, chronic exposure to biochemical disturbances and disease.

In Britain, alcohol has traditionally been drunk in the home and in public houses or clubs. Increasingly alcohol is enjoyed as part of a restaurant meal. Men drink alcohol more than women in a ratio of approximately 2 : 1 depending on the venue. Some 3% of British adults drink every day, 34% weekly, 52% monthly and less than 10% less frequently than monthly.

In Western society, there is an ambivalent attitude to alcohol. On the one hand it is a social facilitator, on the other hand its excessive use is associated with most challenging sociomedical problems. Great skill is employed in developing different forms of alcoholic drinks – wines, fortified wines, beers and spirits. Much of the Western social scene depends to some extent on the use and in part the cultured acclaim of the most perfect of these drinks. The comfortable feeling created by modest amounts of ethanol dispels potential anxieties in most of the population including clinicians. On the other hand, alcohol leads to drunkenness, violence, motor vehicle driving offences, hooliganism, social degradation, impotence, marital disharmony, disease and brain atrophy.

Both acute and chronic ethanol intake can impair work performance and decision taking. There is an increased risk and severity of accidents associated with alcohol intake.

In England and Wales there may be 28 000 excess deaths directly attributable to alcohol abuse. In Britain it was estimated in 1986 that alcoholism cost the country £1.7 billion. At least an average of 8 working days each year per head of population are lost through alcoholism. Most alcoholics work, and they have a three-fold increased risk of accidents over non-drinkers. Between 3–5% of staff in an average company, independent of seniority, will have an alcohol-related problem. Alcohol accounts for 5% of the energy intake in the United States where the cost of the sick alcoholic amounts to over $100 billion a year. This does not include the distress to the family, society and workplace.

6.1.3 Alcohol intake

Assessment and recommendations

It is not uncommon for doctors to neglect to take an accurate ethanol intake history, and patient self-reporting notoriously underestimates intake. In assessing alcohol intake an empirical unit has been defined equivalent to 8–10 g of ethanol. A unit is found in: one single measure of spirits, one glass of wine, a measure of fortified wine, $\frac{1}{2}$ pint of beer or lager, or one small glass of sherry.

The safe limits of alcohol intake are different for men and women (Table 6.1) though there are, as discussed below, individual differences.

Probably the best method for identifying alcohol abuse is the Michigan Alcohol Screening Test (MAST). This is a 10-question modification (devised by Pokorny *et al.*, 1972) from the original 25 weighted questions described by Selzer.

TABLE 6.1 Comparison of the effects of alcohol consumption in men and women

Effect	Weekly consumption (units)	
	Men	Women
Safe	< 20	< 13
Risk increases with consumption	21–50	14–35
Hazardous	< 51	< 36

Alternative methods of identifying excessive alcohol use include random measurements of blood and breath ethanol concentrations and serum γ-glutamyltransferase activity. This enzyme activity is useful diagnostically but may revert to normal after about 5 years of ethanol excess.

6.1.4 Morbidity and mortality

In many studies of ethanol-associated deaths, there is a suggestion of a U- or J-shaped relation between ethanol consumption and mortality, the death rate in abstainers being higher than in those consuming 1–10 units a week. This finding may be spurious and reflect an undue number of previously heavy drinkers now mortally compromised by ethanol, but abstaining in their terminal state. Life-long total abstainers from alcohol form a small section of the population. Other health hazards such as cigarette smoking may be more common in the abstaining group or, alternatively moderate drinking may be good for health, for example having the same effect on platelets as aspirin, with similar benefits to the ageing vasculature. Or perhaps there is a spectrum of distinct populations, those who do not drink, those who drink in moderation, and those who drink excessively. Crucial to the critical evaluation and interpretation of morbidity and mortality patterns in alcoholics, moderate drinkers and abstainers is a knowledge of disease which antedated the drinking.

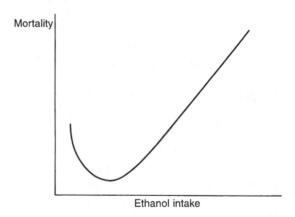

FIGURE 6.1 The U- or J-shaped relationship between mortality and ethanol intake.

Daily admissions to a general hospital in London included 12% of alcohol-related admissions, 7% of those admitted to a general surgical ward, and 26% admitted to an overnight observation ward. Many of the admissions were young males.

In a 15-year study of young Swedish Army conscripts, there was a strong association between death rate and alcohol intake. With an ethanol intake greater than 250 g per week (25–30 units), the risk of death rose to three times that of those with a moderate intake of 1–100 g per week (10–12 units). One-third of deaths were due to violence or suicide. In this group of 49 464 men there was a linear relationship between alcohol intake and mortality.

The accumulated evidence is that in contrast to the J-shaped relationship between alcohol and mortality described in some studies (Figure 6.1) there is a linear relationship between alcohol intake and morbidity.

6.2 Alcoholism

Alcoholism is suggested to occur more frequently than by chance in certain families. Family influences, common environment, pharmacogenetic and psychopharmacological factors and various psychiatric abnormalities have all been implicated. The mental disturbances described include depression, anxiety states and anti-social personality disorders.

Population genetic studies include 'normal' drinking groups, population studies of alcohol metabolism and population genetic studies in twins and adopted children. A study in Copenhagen showed a four-fold increase in the incidence of alcoholism among male adoptees, adopted soon after birth. Although 5% of all persons consuming alcohol proceed to alcohol abuse, only one-fifth of these develop cirrhosis. Such a selective process has led to speculation that genetic vulnerabilities and additional infections (e.g. viral) may be relevant.

The separation of genetics and environment in studies of alcoholism is almost impossible. The shortcomings of studies include contact in youth of adoptees with alcoholic parents, studies having too few subjects, and problems with the source of subjects. There is an interplay between predisposing genetic factors and biologically controlled protective factors, for example an innate or acquired aversion to alcohol sensitivity symptoms, a genetically based susceptibility to alcohol abuse or no access to alcohol despite the subject's predilection. Changes in ethanol and acetaldehyde metabolism as a result of genetically determined variations in the enzymes involved are responsible for individual and racial differences in alcohol drinking habits and acute and chronic reactions to alcohol. In addition, there is varied vulnerability to organ damage after chronic alcohol abuse.

Table 6.2 indicates the genetic and environmental determinants of the effects of differing levels of alcohol consumption.

TABLE 6.2 Genetic and environmental determinants of effects of differing levels of alcohol consumption

Consumption level	Determinant	
	Genetic	Environmental
Safe	Personality Psychiatric predisposition	National and family attitudes to alcohol
Liability to dependence	Pharmacogenetic e.g. acetaldehyde concentrations Personality	Cultural patterns of drinking Continuous or bout drinking
Liability to disability	End-organ vulnerability HLA8 antigen–dependent	Reaction of employers, spouse and family Vulnerability to accidents

6.3 Alcohol metabolism

An understanding of the metabolism of ethanol is important in an understanding of the role that it plays in the aetiology of a number of conditions. In order to achieve a blood alcohol concentration of 0.1%, the alcohol intake has to approach 20% of the individual's resting energy expenditure. Ethanol cannot be stored and can be regarded metabolically as a water-soluble carbohydrate and also as a substrate for ATP production.

Ethanol is readily converted into fat, but not into carbohydrate when consumed with normal amounts of food. The calories provided by alcohol should be regarded as lipid. In large doses the intake of ethanol has pharmacological effects, which may result in emaciation. The effect of alcohol is a direct one, and not due to coincidental malnutrition.

Between 2–10% of ethanol absorbed is excreted unchanged by the kidneys and lungs. The remainder is oxidized in the liver and other tissues. Ethanol creates striking imbalances in the metabolic processes in the liver.

> ### Microsomal ethanol oxidizing system
>
> There is an accessory pathway which comes into play at high ethanol concentrations, the microsomal ethanol oxidizing system (MEOS). This is dependent upon a specific cytochrome P_{450} activity. While the cytochrome alcohol oxidase is a P_{450}IIEI system, it is possible that other isoenzymes have yet to be identified MEOS has a relatively high K_m for ethanol (8–10 mM compared with 0.2–2mM for ADH), hence the concentration differential. A third pathway, peroxisomal catalase, is a possible but unproven pathway of alcohol metabolism in humans.

6.3.1 Alcohol dehydrogenase

Ethanol is metabolized through acetyl-CoA pathways (Figure 6.2), the first stage in fatty acid synthesis. The initial step involves the enzyme alcohol dehydrogenase (ADH) which converts ethanol to acetaldehyde by oxidation. ADH is a cytosolic zinc-containing enzyme which is a dimeric protein of two sub-units, each of molecular weight 40 kDa. The enzyme has a broad substrate specificity which includes the dehydrogenation of steroids, oxidation of glycols in the metabolism of noradrenaline and ω-oxidation of fatty acids.

Liver ADH exists in multiple forms, based on the association of different sub-units in various permutations (Levels 1 to 4). The divergence at each of the four levels corresponds with a number of gene duplications.

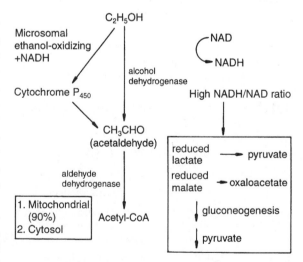

FIGURE 6.2   Ethanol metabolism.

ADH – Levels 1–4

At the first level there is a variation of protein units and their arrangements. There are a minimum of five genes known to be related to the sub-units of class I, II and III which occur in Level 3. The affinity for ethanol diminishes from I to III. These differences in substrate specificity result from differences in structure in one segment of the otherwise unaltered enzyme protein chain-folds. The segment surrounds the second zinc atom. This segment is associated with differences in the quaternary structure (dimer or tetramer) of the enzyme and with the presence or absence of the zinc atom itself. Lack of this segment coincides with the tetrameric formation and its presence with the dimeric quaternary structure.

The enzymes vary in their inducibility with steroid hormones, Class I being inducible and abundant in the liver. Class III is constitutive and also found in placenta, testis and brain. The genome structures of Classes I and II are very similar in intron positions; both have nine exons with identical borders, the introns being similar to those in plant ADH. The minor differences between the major human ethanol-oxidizing isoenzymes, especially Class I, correlate with single positions at the active site mainly for residues 48 and 93.

Ethnic variations in ADH activity

The different ADH isoenzyme spectra have no major effects on an individual's physiological reaction to alcohol. The difference in ADH alleles is quite different between Caucasians, Japanese, Chinese, native Americans, Black Americans and Brazilians. While 5–10% of the English, 9–14% of the German and 20% of the Swiss populations have the 'atypical' phenotype, of ADH_2, this variant is found in 85% of Japanese, Chinese and other mongoloid races. The black American population has the variant ADH Indianapolis in 25%

of the population. The role of this alcohol dehydrogenase in alcohol excess is unclear. The system of enzymes includes isoenzymes and variants, only some of which can oxidize ethanol.

Alcohol oxidation

In the ADH-mediated oxidation of ethanol, hydrogen is transferred from the substrate to nicotinamide adenine dinucleotide (NAD) forming the reduced form, NADH. The result is an excess of reducing equivalents in the hepatic cytosol as free NADH, as the system for the removal of NADH is overwhelmed. The acetaldehyde produced is converted to acetate by aldehyde dehydrogenase which also converts NAD to NADH. The rate-limiting factor is the re-oxidation of NADH. The availability of ADH only becomes relevant in severe protein malnutrition. The NADH/NAD ratio is important in liver metabolism during ethanol oxidation. All known pathways of ethanol oxidation produce acetaldehyde, which in turn is converted to acetate. There are endogenous sources of acetaldehyde, e.g. as a result of the activity of:

- deoxypentosephosphate aldolases
- pyruvate dehydrogenase
- phosphorylphosphoethanolamine phosphorylase
- cleavage of threonine to acetaldehyde and glycine by a threonine aldolase

More than 90% of acetaldehyde oxidation is in the liver, though the genes for these enzymes are found in other tissues. There are several aldehyde dehydrogenases (ALDH) both in the cytoplasm and mitochondria of most species, though most of those with a low K_m activity, $ALDH_2$, are in the mitochondria. The current understanding of the multiplicity of aldehyde dehydrogenase enzymes is not as great as that of ADH. The tetrameric liver aldehyde dehydrogenases (NAD^+-specific) and dimeric forms utilize either NAD or NADPH. The second duplicatory level is dependent upon their intracellular position, either cytosolic or mitochondrial. Mitochondrial metabolism of acetaldehyde is reduced with chronic alcohol consumption.

The acetaldehyde may bind covalently to liver microsomal proteins. This binding increases with long-term alcohol consumption and MEOS activity. A stable adduct is formed with the ethanol-inducible microsomal P_{450}IIIEI, serum albumin and haemoglobin. Acetaldehyde may also act as a neoantigen with the generation of circulating antibodies against acetaldehyde-altered proteins and complement-binding acetaldehyde adducts which may contribute to the perpetuation or exaggeration of liver disease.

ALDH isoenzymes

Many human ALDH isoenzymes have been identified. Human livers contain four major isoenzymes. A single ALDH locus governs the synthesis of cytosolic $ALDH_1$ and a single $ALDH_2$ governs the mitochondrial $ALDH_2$. The two non-allelic genes $ALDH_{3a}$ and $ALDH_{3b}$ are involved in the synthesis of the $ALDH_3$ isoenzymes. $ALDH_1$ and $ALDH_2$ are basically similar with a 65% degree of homology, but the sequences of the exon which includes the sequence for the signal peptide (17 amino residues) of $ALDH_2$ are not homologous.

ALDH variants

There are individual and ethnic differences in alcohol metabolism and clearance rates which may be additional factors in the production of acetaldehyde and flushing reactions. The molecular difference between the two variants of ALDH is in a single amino acid substitution Glu→Lys in position 14 from the N-terminal end. This enzyme variant results in a high blood acetaldehyde after ethanol ingestion, followed by flushing and uncomfortable feelings. This may limit the ability to consume large amounts of alcohol, and possibly a propensity to alcoholic liver disease.

Approximately 50% of Orientals, including the Japanese, have no hepatic $ALDH_2$ but have an inactive aldehydase variant. The American Indians, a sub-set of the mongoloid races, exhibit post-ethanol ingestion flushing and yet alcoholism is widely prevalent in the Red Indian population. The role of acetaldehyde in flushing is not certain though one mechanism may be through the release of histamine. Post-ethanol ingestion flushing is found in 5–10% of Caucasians but is not associated with the ALDH variant.

6.3.2 Alcohol and xenobiotics

Interactions occur between xenobiotics and ethanol following acute ethanol consumption which affects the metabolism of the former. This results in an increased blood clearance of warfarin, phenytoin, tolbutamide, propranolol, rifampin and pentobarbital. In chronic exposure to ethanol there are complex effects with drugs because of competition for common metabolic processes involving P_{450}, interference with the supply of NADPH, and inhibition of glucuronidation. All of these effects may influence metabolism of drugs including methadone, morphine and tranquillizers. Other drug processes such as acetylation and sulphation are unaffected. Nevertheless, this alteration in the habitual hepatic metabolic processes produced by ethanol results in a vulnerability to drugs and other chemicals with conversion to toxic metabolites, particularly those utilizing the cytochrome P_{450} system, e.g. carbon tetrachloride. Similar effects are shown with halothane, isoniazid and phenylbutazone. Ethanol also affects the metabolism of steroids causing, for example, a decrease in blood testosterone concentrations. The metabolism of vitamin A and D is also altered.

6.4 Alcohol and the body

6.4.1 Effects on individual organs and tissues

Mouth

Clinical effects include enlargement of the parotid glands and alterations in parotid secretions.

Oesophagus

There is an increased incidence of oesophageal cancer, reflux oesophagitis, Barrett's oesophagus and dysmotility in the oesophagus. Retching after an excess of alcohol can lead to a Mallory–Weiss tear at the lower end of the oesophagus with consequent haematemesis and other complications.

Stomach

Metabolic effects

A significant fraction of the alcohol ingested at social levels is oxidized in non-hepatic tissues, primarily in the stomach. ADH in the stomach and intestine belongs to Class I and is almost exclusively an ADH_3 isoenzyme. This gastric process may account for 20% of rat ethanol metabolism and determines the bioavailability of ethanol in humans. The reduced capacity for alcohol in women is in part due to their low gastric mucosal isoenzyme content.

Clinical effects

Ethanol stimulates or inhibits gastric acid production depending upon the chronicity of the usage, and the concentration and type of alcohol. Gastritis, peptic ulceration and duodenitis are important complications of ethanol ingestion, sometimes leading to haematemesis. The efficacy of H_2 blockers, e.g. cimetidine, and the healing of peptic ulceration are impaired by prolonged alcohol consumption.

Liver

Metabolic effects

Alcohol can cause an enlarged liver as a result of fatty infiltration. Lipids accumulating in the liver may originate from dietary lipids, from adipose tissue as free fatty acids, or be synthesized in the liver. Serum concentrations of γ-glutamyltransferase are uniformly increased in the fatty liver, possibly because of microsomal induction. Fat is found under light microscopy in two distinct patterns which may coexist and are reversible. There may be an accumulation of large intracytoplasmic fat vacuoles with displacement of the nucleus. or the cytoplasm may show minute fat droplets, **microvascular steatosis**. One theory is that this hepatic fat deposition represents a malnutrition state similar to Kwashiorkor, though these lesions can occur in the presence of an apparent sufficiency of nutrients.

Hepatic microtubules are decreased in alcoholic liver disease. The acetaldehyde may bind to the sulphydryl groups of the cysteine residues in the microtubules. A function of microtubules is the promotion of the intracellular transport of proteins and their secretion. Long-term alcohol feeding delays the secretion of proteins with a corresponding retention in the liver which includes albumin and transferrin and fatty acid-binding proteins. The latter may also influence the accumulation of fat in the liver.

Protein accumulation may be associated with an increase in liver-bound water, which, with the increased fat, will enlarge the liver cells. This swelling has a distinct centrilobular distribution and may alter key cellular functions by physically separating the reaction sites.

Acetaldehyde binds to cysteine and glutathione with a resultant reduction in the amount and turnover of hepatic glutathione, a compound which is important in the scavenging of free radicals and protection against reactive oxygen species. A severe reduction favours peroxidation and may be facilitated by iron from ferritin.

Alcohol and the mitochondria

There are striking morphological and physiological alterations in the liver mitochondria in alcoholics due directly to the effects of ethanol rather than to malnutrition. There is swelling, abnormal cristae, decreased respiratory capacity, reduced cytochrome a and b content and altered oxidative phosphorylation. There may also be altered function in the mitochondria as a result of changes in cellular membranes, plasma membrane glycoproteins and plasma membrane structure.

Hepatotoxicity Many of the toxic effects of ethanol are linked to its ability to shift the redox equilibrium towards a reduced state. Possible mechanisms include increased triglyceride synthesis, decreased fatty acid oxidation, increased mobilization of fatty acids from fatty stores, and decreased release of lipids. When ethanol is oxidized in the liver there is a depression of the citric acid cycle activity. A major site of inhibition is 2-ketoglutarate dehydrogenase activity. The end result is decreased fatty acid oxidation. There may also be alterations in mitochondrial structure which decrease fatty acid oxidative processes. Lipogenesis is increased, possibly by the elongation pathway or transhydrogenation by nicotinamide adenine dinucleotide phosphate (NADPH). This pathway is limited in its load-carrying potential and will not invariably be effective.

The partition of fatty acids between oxidation and esterification is determined by the outer membrane of the mitochondria by the release of carnitine acyltransferase and glycerophosphate acyltransferase. An increase in the concentration of sn-glycerol-3-phosphate would favour esterification, and thence ethanol-induced fatty liver.

During alcoholic liver decompensation acute hepatitis-like changes can be found in central zones of the hepatic acinus. Chronic alcohol ingestion results in an increased consumption of oxygen, due largely to increased mitochondrial reoxidation of NADH. In the liver this is associated with a steeper oxygen gradient along the entire sinusoid length so that necrosis is found in Zone 3 (also called perivenular or centrilobular). Thus, the necrosis may be hypoxic in origin and is more likely to be due to defective oxygen uptake than a lack of blood oxygen.

Ethanol metabolism may also exaggerate the redox shift prevailing in the centrilobular acinar zones. The zonal distribution of some enzymes may influence the selective centrilobular toxicity. Following chronic ethanol excess there is proliferation of the smooth endoplasmic reticulum particularly in the centrilobular zone and associated enzyme inductions. There may be more ADH enzyme activity in the centrilobular zone, though this may be a result of ethanol ingestion.

The alteration in the NADH/NAD ratio depresses citric acid oxidation of 2-carbon fractions and hence suppresses fatty acid oxidation (the major source of 2-carbon fragments) and favours the accumulation of triglycerides. There is also an increase in hepatic α-glycerophosphate which favours the accumulation of hepatic triglycerides by trapping fatty acids. Ethanol increases the lactate/pyruvate ratio. Hypoxia increases NADH, which in turn inhibits the activity of NAD^+-dependent xanthine dehydrogenase and favours the xanthine oxidase pathway. Increased acetate from ethanol metabolism results in the accumulation of purine metabolites, e.g. uric acid. Oxygen radicals are produced which are toxic to liver cells. Acetaldehyde, also derived from alcohol, may be a substrate for xanthine oxidase.

Pathological changes Liver cells exposed to continuous dosage of ethanol show **lytic necrosis**. In the cytoplasm are to be found clumps of refractile dense eosinophilic materials, the so-called alcoholic hyaline lines of Mallory. These are complex glycoproteins with antigenic properties. Polymorphs surround the necrosing liver cells, cholestasis and jaundice develop. The portal zones show stellate fibrosis which in the severely malnourished can obliterate the hepatic venous radicals, heralding fibrosis. Unless ethanol consumption is stopped this may progress to cirrhosis.

In the final stages the collagen in the liver forms a network, dividing the residual liver into small regular nodules. It is possible that alcohol itself may have a direct effect on collagen metabolism in the liver. Alcohol causes proliferation of myelofibroblasts It is possible that ethanol metabolism may result in an increased lactate production, consequent increased peptidylproline hydroxylase activity, inhibition of proline oxidase and increased collagen synthesis. With progression of fibrosis the estimated regional hepatic tissue haemoglobin concentration may decrease, and this decreased oxygen supply to the liver may have an important role in the progression of the pathological process. The zonal nature of ethanol induced liver disease is important in relation to fibrosis.

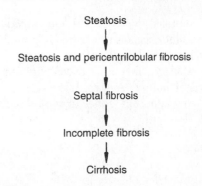

FIGURE 6.3 The progression of alcohol-induced hepatic fibrosis.

suggests that the usual explanation that regenerative nodules press on the efferent venous flow may be only part of the explanation. Hepatocyte swelling may be a prime event in portal hypertension and fibrosis may be secondary.

Clinical effects

Alcoholic liver disease can lead to jaundice, fever, anorexia and right hypchondrial pain. There is a distinct mortality among such individuals with complications which include acute liver failure. A poor prognosis is reflected in ascites and extended prothrombin time.

The basis of treatment is abstinence from alcohol. Even in advanced cases the inexorable advance of the condition can be slowed by desisting from alcohol.

Cirrhosis A further complication of prolonged alcohol exposure is **cirrhosis**. If a 70-kg man drinks 210 g of alcohol per day then he has a 50% chance of developing cirrhosis. Initially the liver may be uniformly nodular, but it may assume an irregular appearance with long-standing cirrhosis. The common signs of uncomplicated cirrhosis are weight loss, weakness and anorexia. Other signs include splenomegaly, ascites, testicular atrophy, spider naevi, gastrointestinal haemorrhage, gynaecomastica, palmar erythema and Dupuytren's contracture. Other complications include portal hypertension, hypoalbuminaemia and hypoglobulinaemia.

Alcohol and liver collagen

The mechanism of the biosynthesis and degradation of collagen is not yet clear. Interstitial collagens are synthesized intracellularly as procollagens that contain extension propeptides at the carboxy ends of their three polypeptide chains. After secretion, there is conversion of procollagen and assembly into fibrils. Aminopropeptides of Type I procollagen are released at an early stage of fibrillogenesis, those of type III are retained in collagen fibrils in the liver.

Alcohol may promote fibrogenesis directly, which results in pericellular, perisinusoidal and percentrilobular fibrosis which is coupled to an increased collagen mRNA. This process is shown diagrammatically in Figure 6.3.

The development of fibrotic distortion of the liver structure can lead to re-routing of the portal vasculature and the formation of varices at the lower end of the oesophagus and fundus of the stomach and an increase in the portal vein pressure. The variability of this further complication

The complications of cirrhotic portal hypertension include ascites which may spontaneously become infected, bleeding from oesophageal and gastric varices, hepatic encephalopathy and hepatorenal failure. A transudate type of ascites with a low albumin concentration (< 250 g/1) may develop within the peritoneal cavity.

Pancreas

Pathological effects

Heavy alcohol intake is associated with chronic relapsing pancreatitis. A possible mechanism for pancreatitis is oedema or spasm of the papilla. Alternatively, the pathogenesis involves the precipitation of protein in peripheral pancreatic plugs. These plugs may calcify and thereafter block larger ducts with resultant atrophy and fibrosis. A further extension of the problem is pancreatic pseudocyst formation and common bile duct obstruction. Pancreatic secretions become insufficient for nutrient digestion.

Clinical effects

Pancreatitis may be extremely painful but progressive loss of pancreatic function may also be painless. Weight loss and steatorrhoea develop with a reduction in fat-soluble nutrient absorption and an extension of the problem to a more general deficiency, e.g. vitamin D deficiency and bone problems. A particularly brittle form of diabetes mellitus may ensue in association with the alcoholic aetiology.

Intestine

Alcohol may alter intestinal function. The effect may be directly on transport, motility, metabolism, circulation and cellular structure in the small intestine. The problem is compounded by pancreatic and hepatic insufficiency and dietary inadequacies, leading to deficiencies which include folic acid, pyridoxine, thiamin, iron, zinc and fat-soluble vitamin deficiency. The activation and inactivation of vitamins is also affected. Secondary malabsorption can develop and must be recognized and treated. Malnutrition is a major problem in these subjects but may very readily, if temporarily, be corrected by abstention from alcohol, good diet and vitamin supplements.

Endocrine abnormalities

Metabolic effects

Alcohol has effects on the endocrine system in a number of target organs including the pituitary and hypothalamus. The peripheral metabolism of a number of hormones may be affected by alterations in hepatic blood flow, protein binding enzymes, cofactors or receptors and by malnutrition.

Clinical effects

Rarely alcohol can cause a pseudo-Cushing's type of syndrome. Alcoholic hypoglycaemia can occur after prolonged fasting or malnutrition, which may on first sight be indistinguishable from a drunken state.

6.4.2 Effects on body systems

Cardiovascular system

Metabolic effects

Any protective influence of alcohol against coronary heart disease may be secondary to an increased high-density lipoprotein concentration. However, the increase is in HDL_3, not the cardioprotective HDL_2. The apparent protective influence of light daily ethanol drinking is probably due to a consequence of multiple advantageous characteristics, low blood pressure, lower mean body mass index and low cigarette smoking.

Clinical effects

Prolonged excessive ethanol intake may be manifested by breathlessness, easy fatigue, palpitations, anorexia and oedema. This results from a cardiomyopathy, cardiomegaly, arrhythmias, intraventricular conduction abnormalities, pathological Q waves and decreased voltages. These changes may result from nutritional or direct toxic effects. Transmural cardiac infarction can occur in the absence of significant coronary artery disease. Alcoholic binges can result in fatal or incapacitating arrhythmias, usually supraventricular in nature. The consumption of alcohol is associated with the development of **hypertension,** in part due to a pressor effect

Blood system

Alcoholic patients often have a red cell macroytosis (MCV > 98 fl) due to a direct toxic effect on bone marrow. Vitamin B_{12} and folate deficiency may be contributory factors, resulting from folate malabsorption, a folate-poor diet, the blockage of storage, methylfolate utilization and increased urinary loss. Quantitative and qualitative changes can occur in white cells with effects on resistance to, for example, respiratory infections. Alcoholic thrombocytopenia is common, leading occasionally to disseminated intravascular coagulopathy. The reduced platelet count and function is a direct effect of alcohol toxicity. The mechanism may be mediated through shortened platelet survival, ineffective platelet function and decreased thromboxane A_2 release. It is also possible that the altered platelet function may in part be responsible for the reduction in coronary thrombosis reported in moderate drinkers.

Musculoskeletal system

The incidence of osteoporosis is increased in alcoholics. The aetiology is complicated but may result from malnutrition, lack of exercise (thought by some alcoholics to be a waste of a drinking man's time), alterations in endocrine status and absorption. Such osteoporosis may contribute to the 5- to 10-fold increase in bone fracture rate, a rate not unrelated to the increased incidence of falling while drunk. Other complications include aseptic necrosis in the femoral head.

Gout occurs in alcoholics in whom there is an accumulation of purine metabolites, e.g. uric acid. It may also be related to the frequent obesity, hypertriglyceridaemia, and activation of the xanthine oxidase pathways which occur in alcoholics.

Skeletal muscle myopathy

A proximal metabolic myopathy is reported in alcoholism. It selectively involves Type II fast twitch glycolytic fibres and may result in a loss of 25% of muscle mass. This may relate to α-tocopherol availability.

Skin

Changes include facial oedema, rosacea and rhinophyma, all contributing to the rubicund face of the committed drinker. Other cutaneous changes include spider naevi and porphyria cutanea tarda.

Respiratory system

Clinical effects

Clinically, fractured ribs and pneumonia from inhaled vomit are common problems.

Central nervous system

Intoxication is an important consequence of alcohol drinking and follows the subjectively pleasant effects of the accumulation of alcohol in the blood and, more importantly, in the brain and nervous system. The brain adapts to continued exposure to ethanol, with larger amounts being required to obtain an equivalent effect. The adaptation leads to tolerance and thence to dependence.

The central tenet of the adaptation/tolerance/dependence hypothesis is that the central nervous system sets up an adaptive state which opposes the acute effect of the drug. Central neurones become more excitable and this state becomes established and only represents instability when alcohol is withdrawn. Withdrawal from alcohol results in central nervous system hyperexcitability. Other complications of withdrawal include tremulousness, withdrawal seizures some 17–48 hours after alcohol withdrawal, and delirium tremens after 3 to 5 days. Such problems require careful evaluation, including the exclusion of cerebral trauma consequent upon a fall, hypoglycaemia and fluid and electrolyte problems.

The molecular basis of intoxication

The molecular basis of ethanol intoxication is not known. There is some evidence that the action is through the same pathway as the barbiturates and benzodiazepines whose major sites of action are the receptor protein for γ-aminobutyric acid (GABA), the major inhibitory neurotransmitter in the mammalian central nervous system.

Ethanol enters neuronal membranes and disrupts the packing of lipid membrane molecules. The anaesthetic effect of ethanol depresses neuronal activity by acting on receptors, e.g. GABA$_A$ and ion channels. Both processes involve depolarization of the nerve cell membrane in which Ca^{2+} is very important. Prolonged exposure to ethanol can cause neuronal death and hence brain damage. The mechanism appears to be damage by lipid peroxidation or activating intracellular enzymes which break down membrane lipids and proteins.

Behavioural changes of a violent nature can occur in bout drinking though these are probably associated with increased irritability rather than central nervous system pathology. Peripheral neuropathies and brain atrophy occur in prolonged drinking in part due to nutritional and particularly vitamin deficiencies. These include Wernicke–Korsokoff syndrome which consists of nystagmus, ataxia and impaired memory and is due to thiamin deficiency and resolves with replenishment of the vitamin. Alcoholic cerebellar degeneration responds in its earlier stages to thiamin. Nutritional amblyopia and pellagra which present with depression and poor memory response to niacin. Pyridoxine deficiency may result in cerebral symptoms and signs.

There is a relationship between ethanol abuse and the incidence of haemorrhagic strokes independent of other causes of stroke. The mechanism of this is obscure. In female drinkers there is an increased risk of subarachnoid haemorrhage.

Reproductive system

In men, ethanol leads to loss of sexual hair, initially increased and later decreased libido, reduced potency, testicular and penile shrinkage, reduced or absent sperm production and hence infertility. In women, chronic excessive ethanol intake leads to menstrual irregularities and shrinkage of breasts and external genitalia.

Immune system

There are disturbances in the reticuloendothelial system of the liver through interference with the mobilization and activation of macrophages and their phagocytic activity. Cell-mediated immunity is also inhibited. These are direct toxic effects, in contrast to undernutrition which may also be associated with effects on immunoprotein production. Heavy intake of alcohol can also alter the production and turnover of B and T lymphocytes in the thymus and spleen. Alcoholic subjects may also have a leucopenia, reversible on stopping ethanol intake.

6.5 Alcohol in the young

6.5.1 Alcohol and the foetus

It has been calculated that the ingestion of more than 90 ml of absolute ethanol (six drinks) per day by the pregnant mother poses identifiable risks to the foetus. It has been suggested that there is no safe threshold, and that ethanol and pregnancy are not compatible. The peak alcohol concentration is more important than daily intake.

6.5.1 Foetal development

The placenta is a critical organ for foetal development. The placenta is a developed multi-functioning organ by 12 weeks of gestation. Ethanol and extraneous and metabolically created acetaldehyde affect placental growth, transport and metabolism. The delivery of oxygen to the foetus may be diminished by ethanol-induced reductions in prostaglandin metabolism, alterations in vascular tone in the placental blood flow, and foetal hypoxia.

Ethanol is well established as a **teratogen**. There are also indirect effects, on the supply of essential nutrients, of alterations in foetal immunity and possible mutagenic effects of paternal ethanol exposure.

Clinically, the foetal central nervous system is a major target organ for the deleterious effects of maternal ethanol ingestion. The mechanism is possibly through impaired protein synthesis, central nervous system cell division and reduced cell number. Another postulate for poor brain growth is diminished tissue zinc. The **foetal alcohol syndrome** has characteristic features, facial structure,

intrauterine growth deficiency and learning disabilities. These may also result from alcohol-related foetal injury. It has been calculated that in the USA this problem is responsible for some 11% of the total cost of the care for learning disabilities.

6.5.2 Alcohol and children

Drinking ethanol can lead to acute intoxication in children. It is characterized by hypoglycaemia, hypothermia and depressed respiration. Prolonged exposure may have effects on the child's physical development. There are significant effects on the upbringing of a child through excessive alcohol intake by the parents.

Summary

1. Ethanol is a nutrient whose principal popularity depends upon its effects on mood and an induced sense of well-being. Alcohol drinking is, in general, a social activity.
2. Alcoholic drinks are produced in many different forms. It is the constituents other than alcohol that give the beverage its particular taste.
3. Alcohol intake is measured by the unit; 1 unit is equal to 8–10 g of ethanol.
4. The existence of a U- or J-shaped relationship between alcohol intake and morbidity is discussed.
5. Alcoholism occurs more frequently in certain families. Cultural and genetic factors have been implicated.
6. Ethanol is metabolized initially by alcohol dehydrogenase to acetaldehyde, which is converted to acetate by the enzyme aldehyde dehydrogenase. The type and amount of isoenzymes of this enzyme affect the susceptibility to alcohol of women and some Mongol races.
7. Ethanol has destructive effects on all organs when drunk in excess, with consequences which affect the individual's health.

Further reading

Goodwin, D.W. (1985) Alcoholism and genetics – the sins of the fathers. *Archives of General Psychiatry*, 28–1, 238–43.

Lucas, E.G. (1987) Alcohol in industry. *British Medical Journal*, 294, 460–1.

Murray, R.M., Clifford, C.A., Gurling, H.M.D., Topham, A., Clow, A. and Bernadt, M. (1983) Current genetic and biological approaches to alcoholism. *Psychiatric Developments*, 2, 171–92.

Palmer, T.N. (1991) *The Molecular Pathology of Alcoholism*, Oxford University Press, Oxford.

Pokorny, A.D., Miller, B.A. and Kaplan, H.B. (1972) The brief MAST: a shortened version of the Michigan alcoholism screening test. *American Journal of Psychiatry*, 129, 342–8.

Shaper, A.G., Wannamethee, G. and Walker, M. (1988) Alcohol and mortality in British men: explaining the U-shaped curve. *Lancet*, 2, 1267–72.

Chapter 7

Vitamins

7.1 Introduction

- Vitamins are essential dietary organic substances which the body cannot synthesize and are required in small amounts for metabolic processes
- Vitamins may be either water-soluble or fat-soluble
- Vitamins are absorbed in the upper gastrointestinal tract with the exception of vitamin B_{12} which is absorbed in the ileum
- Vitamins which are ingested in a form which is not absorbable in the intestine require chemical modification
- Some vitamins, e.g. the B vitamins, may act as cofactors for enzymes
- Some vitamins, e.g. vitamins A and D, alter metabolic activity through receptors on membranes
- A deficiency of a vitamin has defined and possibly profound effects on body metabolism

Vitamins are organic substances which the body requires in small amounts for metabolism, but is incapable of synthesizing, at least not in sufficient quantity for the overall needs of the body. Vitamins are not related chemically and differ in their physiological actions.

7.1.1 Nomenclature and classification of vitamins

As the vitamins were discovered each was identified with a letter. Once each vitamin had been isolated and its chemical structure identified, a specific chemical name became possible. Many of the vitamins, e.g. A, D, K, E and B_{12}, each consist of several closely related compounds of similar physiological properties.

The vitamins may be subdivided into water- and lipid-soluble groups. The water-soluble vitamins are vitamin C (ascorbic acid), vitamin B_1 (thiamin), nicotinic acid (niacin) and nicotinamide, riboflavin, vitamin B_6 (pyridoxine), pantothenic acid, biotin, folic acid and vitamin B_{12}

(cyanocobalamin). The fat-soluble vitamins are vitamin A (retinol), vitamin D (cholecalciferol), vitamin K and vitamin E (tocopherols). Vitamins may be single chemical entities, as is ascorbic acid, or consist of a family of closely related compounds, as are A, D, K, E and B_{12}.

The vitamin B complex is a useful concept as the vitamins, while being unrelated chemically, often occur in the same foodstuff. The B vitamins are involved in intermediary metabolism, being coenzymes in the glycolytic, tricarboxylic acid and pentose pathways.

The classification of vitamins becomes more difficult as an understanding of their action increases. Vitamins may have actions which have blood-forming importance (folic acid and vitamin B_{12}), anti-oxidant (ascorbic acid and vitamin E), energy metabolism (thiamin, riboflavin and pyridoxine), bone formation (vitamin D) and protein metabolism (vitamins K and A). Some vitamins act specifically in enzymes as coenzymes.

Vitamins as coenzymes

Vitamin	Coenzyme form	Function
Thiamin	Thiamin pyrophosphate	C–C and C–X bond cleavage
Riboflavin	Flavin mononucleotide	One or two electron transfer reactions
	Flavin adenine nucleotide	
Pyridoxine	Pyridoxal phosphate	Various reactions with α-amino acids
Nicotinic acid	Nicotinamide adenine dinucleotide	Hydride transfers
	Nicotinamide adenine dinucleotide phosphate	
Pantothenic acid	Coenzyme A	Acyl group transfer
Biotin	Biocytin	Carboxylation reactions
Folic acid	Tetrahydrofolic acid	One-carbon transfers
Vitamin B$_{12}$	Deoxyadenosyl cobalamin	Rearrangements on adjacent carbon atoms

---

The lipid-soluble vitamins have functions which are less easily defined. Vitamin D can be synthesized in the skin and has many similarities to a hormone, having influences which extend further than that of bone formation. Vitamin A as a pigment–protein complex acts as an absorber of light in the eye. Vitamins D and A act on a variety of receptors which are now beginning to be understood. Vitamin K, though best seen as important in clotting reactions, is also involved in the formation of γ-carboxyglutamate in a number of processes which involve calcium and calcium-regulated metabolic processes.

The classification of vitamins may therefore be as water- and fat-soluble, or according to function as blood forming, anti-oxidant, involved in energy or protein metabolism, or bone-forming. These classifications are important from the historical perspective and belong to a brilliant period of discovery in nutrition. The deficiency conditions or diseases which brought vitamins to attention were regarded as central to the action of the vitamins. The identification of the role of vitamins as coenzymes and with hormone-like actions makes all of these classifications no longer useful.

Therefore, the vitamins are described individually, with the exception of the metabolically interconnected folic acid, vitamin B$_{12}$, choline and methionine system.

7.1.2 Factors influencing utilization of vitamins

Bioavailability

Not all vitamins may be ingested in an absorbable form in the intestine, e.g. nicotinic acid which is derived from cereals is bound in such a way that it is not absorbed. Fat-soluble vitamins may be malabsorbed if the digestion of fat is in any way impaired.

Water-solubility

The intestinal absorption of vitamins is by specific pathways many of which are sodium-dependent. In some instances, absorption includes chemical change, e.g. phosphorylation, which is a feature of the absorption process of riboflavin and pyridoxine.

Anti-vitamins

These are present in natural food. Several synthetic analogues of vitamins are highly poisonous, e.g. aminopterin, tesoxypyridoxine. These substances inhibit the activity of true vitamins and enzyme systems.

Pro-vitamins

These, while not themselves being vitamins, can be converted to vitamins in the body. Carotenes are pro-vitamins of vitamin A and the amino acid tryptophan can be converted to nicotinic acid. Vitamin D is synthesized in the skin by the action of sunlight on a derivative of cholesterol. Because vitamins have traditionally been regarded as dietary constituents, it is anomalous that vitamin D is synthesized in the skin in response to sunlight. In some respects vitamin D could better be regarded as a hormone.

Biosynthesis in the gut

The normal bacterial flora of the gut can synthesize some vitamins, e.g. vitamin K, nicotinic acid, riboflavin, vitamin B_{12} and folic acid. Because these are synthesized in the colon it may be that these are not nutritionally relevant, as they may not be absorbed.

Interaction with nutrients

Examples of this phenomenon are where a diet rich in carbohydrates and alcohol requires additional thiamin for the body's metabolic requirements. Similarly, when there is a high intake of polyunsaturated fats, vitamin D requirements are increased.

Summary

1. Vitamins are organic substances which the body requires in small amounts for metabolism, and is incapable of synthesizing, or does not synthesize in sufficient quantity for its overall needs. They are not related chemically and differ in their physiological actions.
2. Vitamins may be classified as water- or lipid-soluble. The water-soluble vitamins are vitamin C (ascorbic acid), vitamin B (thiamin), nicotinic acid (niacin) and nicotinamide, riboflavin, vitamin B_6 (pyridoxine), pantothenic acid, biotin, folic acid and vitamin B_{12} (cyanocobalamin). The fat-soluble are vitamin A (retinol), vitamin D (cholecalciferol), vitamin K and vitamin E (tocopherols).
3. Vitamins may be single chemical entities, as is ascorbic acid, or comprise a family of closely related compounds, as are A, D, K, E and B_{12}.
4. Vitamins may have actions which have blood-forming importance (folic acid and vitamin B_{12}), anti-oxidant (ascorbic acid and vitamin E), energy metabolism (thiamin, riboflavin and pyridoxine), bone formation (vitamin D) and protein metabolism (vitamin K and A). Vitamins may act in a general systemic manner as anti-oxidants, e.g. ascorbic acid and vitamin E.
5. Vitamins may also act as enzymatic cofactors, have hormone-like actions on receptors or act as anti-oxidants.
6. Many of the classifications of vitamins are of historical interest only, as more is understood of their biological activity.

Further reading

Bates, C.J., Thurnham, D.I. *et al.* (1991) *Design Concepts in Nutritional Epidemiology* (eds D.M. Margetts and M. Nelson), Oxford Medical Publications, Oxford.

Davidson and Passmore (1986) *Human Nutrition and Dietetics*, Churchill Livingstone, Edinburgh.
Dietary Reference Values for Food, Energy and Nutrients for the United Kingdom (1991) Report on Health and Social Subjects, no. 41. HMSO, London.

7.2 Vitamin A

7.2.1 Introduction

Vitamin A consists of a group of biologically active compounds closely related to the plant pigment carotene. The carotenoid family consists of approximately 100 naturally occurring pigments, which provide the yellow-red colour of vegetables and some fruits. β-carotene is a pro-vitamin A. β-carotene has widespread distribution in plants and is associated with chlorophyll. It is singular in that it is the only carotenoid in which both halves of the molecule are identical to retinol. Other carotenoids, e.g. xanthophyll, which is another yellow pigment associated with chlorophyll and lycopene, the red pigment of tomatoes, do not have pro-vitamin A activity.

The term 'retinol' means vitamin A alcohol, while vitamin A includes all compounds of vitamin A activity. One international unit of vitamin A is equivalent to 0.3 μg of retinol. In terms of biological activity, 6 μg of β-carotene are equivalent to 1 μg of retinol. Originally, vitamin A activity was described as international units, but now that crystalline retinol is available, international units which are functional are no longer necessary.

Retinol structure

The retinol molecule consists of a hydrocarbon chain with a β-ionone ring at one end and an alcohol group at the other. The usual form is the all-*trans* stereoisomer. An isomer with the *cis* configuration at the 11 or 13 position exists, but is less potent biologically. Vitamin A_2 is 3-dehydroretinol, which has half the biological activity of retinol. This is extremely uncommon, occurring only in the liver of some Indian fish.

The terminal alcohol group of retinol can be oxidized to an aldehyde (retinal) or carboxylic acid group (retinoic acid). In foodstuffs the alcohol is usually esterified with fatty acid (retinyl esters).

Retinol and carotene are soluble in fat but not in water; they are also stable to heat at ordinary cooking temperatures but liable to oxidation and destruction if the fats turn rancid. Vitamin E can protect such oxidation. Retinol is also chemically changed by exposure to sunlight. Drying of fruit and food in the sun results in loss of vitamin A. Fish liver oils in clear glass lose their potency on exposure to light. Canned vegetables may retain their carotene over many years.

Retinol is present in milk, butter, cheese, egg yolk, liver and fatty fish. The liver oils of fish are the richest natural source of vitamin A. Carotenes are found predominantly in green vegetables associated with chlorophyll. The green outer leaves of vegetables are a good source of carotenes, whereas white inner leaves contain little. Yellow and red fruits and vegetables, particularly carrots, are good sources. Vegetable oils with the exception of red palm oil which is found in west Africa and Malaysia, do not contain vitamin A. Retinol is present in breast milk.

In a typical Western diet, about 15% of beta-carotene and about 8% of other dietary carotenoids intake are converted to vitamin A in the intestinal mucosa. This varies between individuals and different food types. Carotenoids in fruit juices or oily solution are better absorbed than native carotene in carrots.

7.2.2 Action of vitamin A

Vitamin A is essential for growth and normal function of the retina and development of epithelial surfaces in the retina. Recent discoveries have shown that most actions of vitamin A in development, differentiation and metabolism are made possible by nuclear receptor proteins that bind retinoic acid, the active form of vitamin A.

The biological functions of the different molecular types of vitamin A and the retinoids, all-*trans*-retinal, 11-*trans*-retinal, 11-*cis*-retinal and all-*trans*-retinoic acid (T-RA) and N-*cis*-retinoic acid (C-RA) (Figure 7.1) act through specific interaction with a nuclear receptor protein. The nuclear retinoid receptors belong to the steroid receptor family of proteins which act as ligand-dependent regulators of gene transcription. These all have a highly conserved sequence specific DNA-binding domain and a ligand-binding domain. These form dimers and bind to DNA. Other members of this family include the receptors for progesterone, oestradiol, glucocorticoids and 1,25-dihydroxy vitamin D_3.

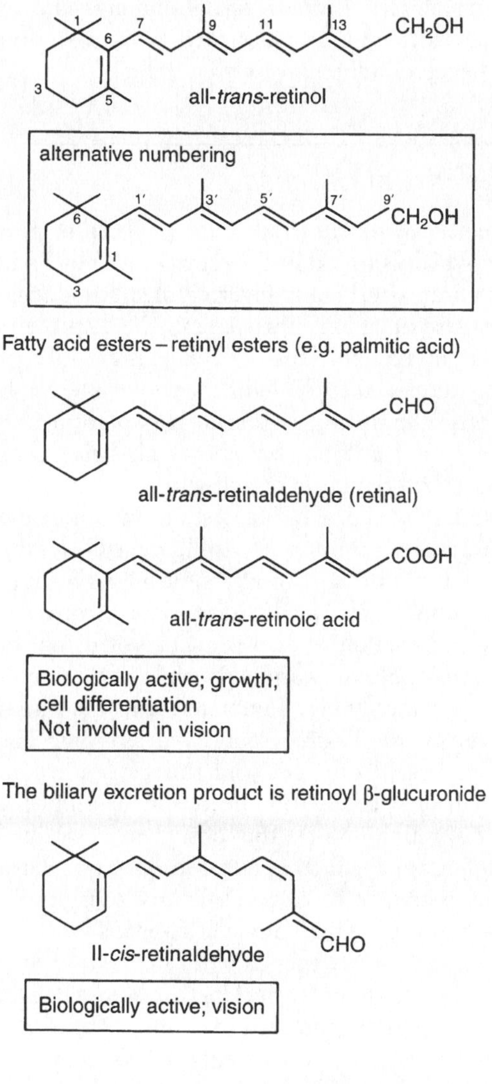

FIGURE 7.1 Structure of substances with vitamin A activity; all-*trans* retinol, retinyl esters, retinal, retinyl β-glucuronide and 11-*cis* retinaldehyde.

Nuclear retinoid receptors

These receptor proteins are divided into functional domains. From the amino-terminal to the carboxy-terminal of the receptor the A/B domain is either an immunogenic or a transcriptional domain. Region C is the DNA-binding domain which binds to a specific DNA sequence adjacent to the gene under regulation which is stimulated to be transcribed; region D is a hinge or linker domain and region E is the ligand-binding domain conferring specificity upon the protein.

The high-affinity receptor proteins for retinoic acid (RAR and RAR with α, β, γ isoforms and two separate activating ligands C-RA and T-RA) each have relative molecular masses of approximately 50 000 Da. The receptors differ in their transcription activation domains. They are coded for by three different genes which are expressed at different times and places during development and differentiation. Another group of retinoic acid receptors are the RXR. 9-*cis*-retinoic acid which is formed by cells from all *trans*-retinoic acid binds to RXR.

The RXR proteins are involved in the activation of genes which are involved in development and differentiation. It is possible that RXR proteins act in metabolic regulation.

The retinoid receptors regulate the expression of genes. They act through cellular retinol binding protein type II. RAR also increases the binding of other receptors to and activation of their response elements. It is quite possible that RXR and the binding with N-*cis*-retinoic acid is central to intermediate metabolism with activation of enzymes, for example the medium-chain acyl-CoA dehydrogenase. It is also possible that 9-*cis*-retinoic acid and RXR α are necessary for the activation of the apoliprotein A_1 synthesis in the liver.

7.2.3 Availability of vitamin A

The first step in vitamin A metabolism is the absorption or uptake of retinol from the intestine.

Retinyl esters are hydrolysed in the small intestine by pancreatic hydrolases or an intestinal brush border hydrolase. The retinol is made more soluble by inclusion in a micellar system. Absorption requires a saturable passive carrier-mediated system.

β-carotene is split by an enzyme in the small intestinal mucosa, β-carotene 15, 15'-oxygenase yielding two molecules of retinol. Within the enteric cell pro-vitamin A carotenoids are oxidatively cleaved to produce retinal and apocarotenoids. Retinal is reduced to retinol.

After absorption, retinol is esterified with long-chain fatty acids. This reaction is catalysed by two microsomal enzymes:

- lecithin : retinol acyltransferase (LRAT) which uses the sn-1 fatty acid of phosphatidylcholine as the fatty acid donor
- acyl-CoA : retinol acyltransferase (ARAT) which uses acylated free fatty acids

Retinol is carried from the intestine as retinyl palmitin in chylomicrons to be taken up by the liver.

Vitamin A is stored as retinyl esters with long-chain fatty acids in animal tissues, especially the liver. Release from the liver is in the form of retinol, which circulates bound to a specific transport protein, retinol binding protein, which forms a complex with plasma prealbumin. These can be measured by immunoassay. Concentrations are low in malnourished children. After ingestion, 8% of retinol is absorbed; 30–50% is stored in the liver; and 20–60% is conjugated and excreted in bile as a glucuronide. Stores of retinol are substantial – of the order of 400 mg – and last for many months, even years.

Almost certainly there is an enterohepatic circulation of retinoids, since retinoyl β-glucuronides and other retinoid metabolites are found in bile.

7.2.4 Vitamin A deficiency

The clinical effects of vitamin A deficiency are usually seen only where the diet has been deficient in dairy produce and vegetables over a prolonged period. Malabsorption is another important contributor in vitamin A deficiency.

Vitamin A deficiency results in a reduction in the rhodopsin content of the rods of the retina and this leads to night blindness. Epithelial surfaces undergo squamous metaplasia in vitamin A deficiency, that is, the cells become flattened and heaped one upon the other with the surface being keratinized. This is particularly marked in the conjunctiva covering the sclera and cornea of the eye, which is known as **xerophthalmia**. This can lead to softening and destruction of the cornea and hence blindness.

The visual process

The photopigment rhodopsin is a receptor protein which is found in the retinal rod cells of all vertebrates and many invertebrates. Rhodopsin consists of a membrane-embedded protein, opsin, and a light-sensitive pigment group, retinal. Retinal absorbs light in the visible range (400–600 nm) and is found in two forms, an 11-*cis* form and a lower-energy all-*trans* form. The 11-*cis* retinal is attached through a Schiff base linkage to a lysine residue in opsin and can absorb a photon of visible light (495 nm). The photon of light converts the 11-*cis* form to the *trans* form which dissociates from the opsin. Following this, a sequence of biochemical, biophysical and physiological events occurs which when summated into the visual cortex of the brain is perceived as sight. The 11-*cis* retinal unbends and this causes a conformational change in the opsin. Possibly the 11-*cis* retinal holds the opsin in a strained configuration which is released once the Schiff base linkage is broken. The message is amplified by cyclic cGMP and the G-proteins.

Colour vision depends upon pigments which absorb at different wavelengths; blue 420 nm, green 530 nm and red 560 nm. These are found in outer cone segments. It would appear that a single ancestral opsin duplicated twice to give a rod opsin and red and blue cone opsins. The different absorption spectra are due to three different opsins which bind retinal.

Xerophthalmia

This is one of the most important deficiency diseases in the world. It is widespread in south-east Asia, the Middle East and Africa. It occurs in the first year of life among artificially fed infants but is rare among breast-fed infants. If, however, the mother is deficient in vitamin A then the milk is also deficient, and hence the child has an inadequate intake of vitamin A. Dietary protein calorie malnutrition also compounds the problem. Xerophthalmia is common in young children, with a significant proportion becoming permanently blind. Vitamin A deficiency may well worsen keratoconjunctivitis of other aetiology, as with the measles infection. Initially, xerophthalmia occurs in the conjunctiva but with more pronounced deficiency the cornea is affected, with a danger of corneal ulceration and permanent visual damage. There may be softening of the cornea, **keratomalacia** which predisposes to blindness.

In vitamin A deficiency the epithelial cells of the cornea develop squamous metaplasia. Its clinical forms are:

- conjunctival xerosis
- corneal xerosis
- keratomalacia – this leads almost certainly to blindness
- night blindness – this is an early symptom of vitamin A deficiency and may occur without any evidence of xerophthalmia
- xerophthalmia fundus
- corneal scars

Vitamin A is also involved in the maintenance of epithelial surfaces and a deficiency leads to epithelial metaplasia in the respiratory tract, mucous membranes (especially the eyes), gastrointestinal tract and genitourinary tract. The mucosa is replaced by inappropriately keratinized stratified squamous epithelium. In the skin, vitamin A deficiency results in keratinization which blocks the sebaceous gland with plugs, producing a condition known as follicular keratosis.

Treatment

Prophylaxis is necessary by teaching local populations to eat dark green vegetables which are rich in vitamin A. This is particularly important for pregnant women, the weaning child, the growing infant and the adult.

Where xerophthalmia is endemic then there may be merit in giving vitamin A prophylactically in capsule form. It is said that a population should be regarded as being at risk of keratomalacia if more than 2% of the children have conjunctival xerosis or if 5% have a plasma retinol level of < 100 µg/litre. Another approach is fortification of foodstuffs with vitamin A in a water-soluble form, that is, added to table sugar or monosodium glutamate.

Vitamin A (30 mg retinol daily, as oral halibut oil or intramuscular retinol palmitate) can be given for 3 days. Thereafter, retinol in the form of fish liver oil will prevent recurrence. The diet should be monitored, if possible to ensure continued adequate vitamin A intake.

Vitamin A deficiency has also been suggested to be a possible risk factor for childhood illness and death. In the mountainous region of Nepal, Jumla, malnutrition is prevalent. A single oral dose of vitamin A (50 000–200 000 units; 15–60 mg; according to age) was given to children under 5 years old. There was a 26% reduction in death rate in the treated group which averted death resulting from diarrhoea, pneumonia and measles.

7.2.5 Vitamin A excess

β-carotene is not toxic but high intakes lead to a yellow appearance. An excess intake of food rich in carotenoids can result in a distinct orange yellow colour of the skin, called **hypercarotenaemia**. The eyes do not become yellow. The persistence of this change in colour is dependent upon continued intake of carotenoids.

Animal livers contain on average 1300–40 000 µg/100 g. The liver of the polar bear is rich in retinol (600 mg/100 g). Eating the liver of the polar bear can cause drowsiness, headache, vomiting and excess peeling of the skin. Husky dog livers contain half this amount. Excess administration of retinol to young children can also lead to anorexia, irritability, dry itching skin, coarse sparse hair and swelling over the long bones. Children are more sensitive than adults to a high retinol intake and great care should be taken not to overdose.

There is also evidence that retinol is **teratogenic**. Consequently, it has been suggested that pregnant women or those who are trying to become pregnant, should not eat liver, liver products or vitamin supplements.

7.2.6 Recommended requirements

Adults

For the average adult the EAR of 500 μg/day for a 74-kg male and 400 μg/day for a 60-kg female is reasonable. The LRNIs are 300 μg/day for men and 250 μg/day for women; RNIs are 700 μg for men and 600 μg for women.

Infancy

The RDAs for infants are usually based on the vitamin A provided by breast milk. A daily intake of 350 μg retinol equivalents meets a young child's requirements, allowing for growth and maintaining liver stores. This means the EAR is 250 μg/day and the LRNI 150 μg/day.

Children

Children are growing as well as requiring vitamin A for the loss of body stores. The recommended intakes are of the same order as for adults.

Pregnancy

In pregnancy extra vitamin A is required for the growth and maintenance of the foetus, providing reserves and for maternal tissue growth. This is particularly important during the third trimester. An increment of 100 μg/day during the pregnancy, increasing the maternal RNI to 700 μg/day should meet all requirements. A word of caution – there are dangers of large intakes of vitamin A (see below).

Lactation

The diet should contain an increment of 300 μg/day for milk production.

7.2.7 Body store measurements

As vitamin A is stored in the liver in an esterified form, the liver content of the vitamin is the best measure of retinol status, though this is not a readily accessible measurement. Plasma retinol is an insensitive indicator of vitamin A status. Only during the latter stages of extreme deficiency does the plasma retinol fall below 0.7 μmol/l, while levels below 0.35 μmol/l indicate deficiency. Borderline values of 0.35–0.70 μmol/l may also be affected by inadequate protein intake, parasitic infestation, liver disease and other conditions.

The major markers of vitamin A status are plasma retinol and the 'relative dose–response' (RDR) test. The control of retinol transport by the retinol binding protein is used as a functional test for retinol stores.

Retinol dose–response test

In vitamin A deficiency there is a continuous synthesis of retinol binding protein which remains in the liver as apo-RBP. A blood sample is taken and a loading dose of retinol given, followed by a second sample of blood taken after 5 hours. The effects differ from the conventional loading test in that the plasma response is greater in deficient subjects.

Apo-retinol binding protein (apo-RBP)

The deficient liver protein readily combines with the incoming retinol and is excreted into the plasma. If the pre-loading concentration is increased by more than 20% this indicates a deficient liver retinol below 20 μg/g. Little increase in plasma retinol is associated with a normal vitamin A status. The only biochemical marker for carotenoids is a plasma concentration which reflects short- to medium-term intakes.

The time lag in improvement tends to overestimate requirements.

An alternative approach to the dose–response test is based on body size pool. Adequate vitamin A status is defined in terms of an adequate body pool, based on the amount of vitamin A in the liver, which contains the majority of vitamin A in the body. A liver retinol concentration of 20 µg/g tissue is the basis of the FAO/WHO recommendations.

Summary

1. Vitamin A consists of a group of biologically active compounds closely related to the plant pigment carotene. The carotenoid family consists of approximately 100 naturally occurring pigments, which provide the yellow-red colour of vegetables and some fruits.
2. The retinol molecule consists of a hydrocarbon chain with a beta-ionone ring at one end and an alcohol group at the other. The usual form is the all-*trans* stereoisomer.
3. Vitamin A is essential for growth and normal function of the retina and development of epithelial surfaces in the retina. The photopigment rhodopsin is a receptor protein which is found in the retinal rod cells. Rhodopsin consists of a membrane-embedded protein, opsin, and a light-sensitive pigment group, retinal.
4. Vitamin A and the retinoids act through nuclear receptor proteins which regulate gene transcription.
5. Vitamin A deficiency is an important cause of eye malfunction, night blindness and xerophthalmia – a lesion of the conjunctiva and cornea. Other epithelial tissues, e.g. skin, are affected.

Further reading

De Luca, L.M. (1991) Retinoids and their receptors in differentiation, embryogenesis and neoplasia. *FASEB Journal*, 5(14), 2924–33.

Dietary Reference Values for Food, Energy and Nutrients for the United Kingdom (1991) Report on Health and Social Subjects, no. 41. HMSO, London.

Levin, A.A., Sturzendecker, L.J., Kazler, S. *et al.* (1992) N-cis-retinoic acid stereoisomer binds and activates the nuclear receptor RXR. *Nature*, 355, 359–61.

McClaren, D.S. (1980) *Nutritional Ophthalmology*, Academic Press, London.

Mangelsdorf, D.J., Onges, E.S., Dyck, J.A. and Evans, R.M. (1990) Nuclear receptor that identifies a novel retinoic acid response pathway. *Nature*, 345, 224–9.

Smith, C.U.M. (1989) *Elements of Molecular Neurobiology*. John Wiley & Sons, Chichester.

World Health Organization/Unicef International Vitamin A Consultative Group Task Force (1988) *Vitamin A Supplementation. Guide to their Use in the Treatment and Prevention of Vitamin A Deficiency and Xerophthalmia*, WHO, Geneva.

7.3 Vitamin C (ascorbic acid)

7.3.1 Introduction

Vitamin C (ascorbic acid) is a simple sugar with molecular weight 176 Da. The important sources of the vitamin are fresh fruit and fruit juices, with blackcurrant and guavas particularly rich sources, as are green leafy vegetables. Liver and fresh milk are other sources. Vitamin C is lost by oxidation during cooking, particularly when boiled in water or cooked in deep fat. This oxidative process may be accelerated by traces of copper in an alkaline medium.

7.3.2 Synthesis of vitamin C

Many plant and animal tissues can synthesize vitamin C from glucose through an intermediary L-gulonic acid and L-gulonolactone. This hepatic metabolism does not take place in man and the primates, the guinea pig, an Indian fruit-eating bat, the red-vented bulbul and some birds. Consequently these creatures, including humans, are susceptible to **scurvy**, through a deficiency of vitamin C. It is possible that there is modest synthesis of vitamin C in some humans, as not every sailor on long sea voyages developed scurvy.

Synthetic pathway of vitamin C

α-D-glucose →→ UDP-D-glucuronate → D-gluconate → L-gulonate → L-gulono-γ-lactone → 2-keto-L-gulonolactone → L-ascorbic acid

Human beings, other primates, guinea pigs and several other creatures do not synthesize ascorbic acid as they do not possess the last enzyme required by the pathway L-gulono-γ-lactone oxidase.

Ascorbic acid is converted to L-dehydroascorbate through the donation of two electrons (Figure 7.3). This requires:

• dehydroascorbate reductase which uses reduced glutathione as a co-substrate resulting in ascorbate
• oxidized glutathione reductase using NADPH for reduction of oxidized glutathione may convert dehydroascorbate to ascorbate

The conversion of the lactone ring of L-dehydroascorbate to 2,3-diketo-L-gulonate requires an enzyme lactonase. This enzyme is not present in humans. It may well be that other reducing agents such as glutathione, cysteine, tetrahydrofolate, tetrahydrobiopterin, dithiothreitol and 2-mercaptoethanol may be used in place of ascorbate.

7.3.3 Action of vitamin C

Vitamin C (ascorbic acid) is a powerful reducing agent and electron donor and as such has a central role in the relative states of oxidation/reduction of

other metabolically important water-soluble substances. It is a source of electrons for the reduction of oxygen and is a reducing agent which maintains elements in the reduced state. This is important in the intestinal luminal contents wherein dietary iron in the ferric state is converted

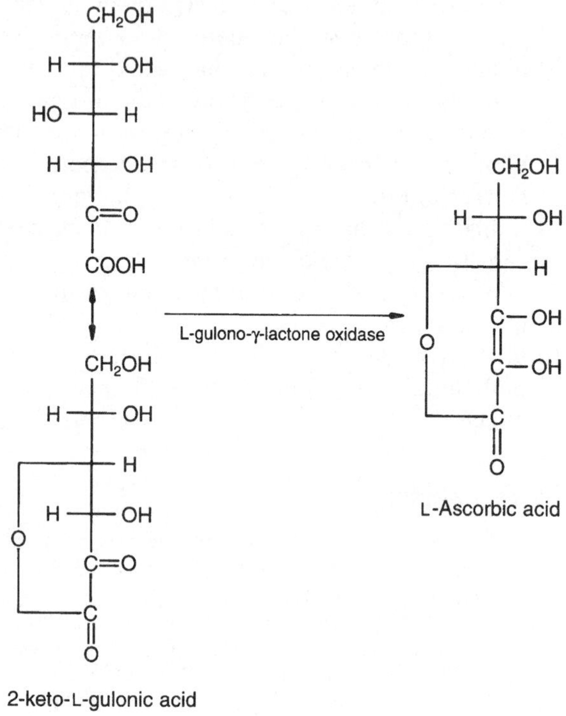

FIGURE 7.2 The synthesis of 2-keto-L-gulonic acid to L-ascorbic acid is not possible in humans because of an absence of the enzyme L-gulono-γ-lactone oxidase.

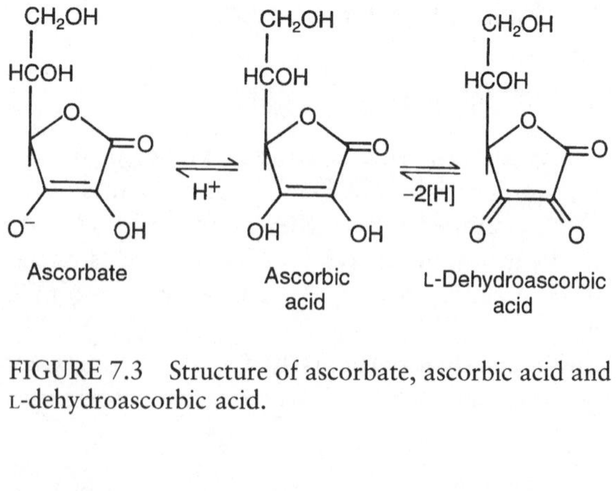

FIGURE 7.3 Structure of ascorbate, ascorbic acid and L-dehydroascorbic acid.

to the ferrous state. Ascorbic acid readily gives up an electron to convert Fe^{3+} into Fe^{2+}, which facilitates iron absorption.

Ascorbic acid is present in all tissues, especially the aqueous phase, and is important in synthetic processes and energy exchanges. It serves as a co-substrate in a number of oxidoreduction reactions, acting synergistically with vitamin E. Urate, another water-soluble antioxidant protects vitamin C. Interactions between iron and vitamin C are essential to iron metabolism in the body. Vitamin C is also involved in copper absorption and subsequent caeruloplasmin copper transport activity.

Ascorbic acid has an important role in a number of biological systems, synthesis of hormones, neurotransmitters, collagen, carnitine and in the detoxification system through cytochrome P_{450} activity.

Enzyme reactions which require ascorbic acid include hydroxylations utilizing oxygen with Fe^{2+} or Cu^{2+} as cofactors. The interaction between ascorbic acid and iron can increase the oxidative potential of iron, and iron acts as a pro-oxidant in the presence of ascorbic acid. The enzymes involved include the synthesis of hydroxyproline and hydroxylysine in pro-collagen. Ascorbic acid may be relevant in the hepatic microsomal mono-oxygenase system, which is important for steroid hormones and xenobiotics.

Other enzyme reactions which require ascorbic acid include:

- mono-oxygenases which require copper, molecular oxygen and a reducing agent such as ascorbate. Ascorbate is active at the level of the metal to activate the oxygen as in dopamine β-hydroxylase and peptidylglycine α-amidating mono-oxygenase.
- a di-oxygenase reaction in which both atoms of a dioxygen molecule are entrapped into homogentisate utilizing 4-hydroxyphenylpyruvate dioxygenase.

All the dioxygenases have a requirement for ferrous iron which is held in the ferrous state by a reducing agent of which ascorbic acid is the most important. Other dioxygenases use α-ketoglutarate as a substrate and incorporate one atom of oxygen into succinate and one into a product of oxidation of the specific substrate and require iron in the ferrous state, thus utilizing ascorbate.

Dopamine β-hydroxylase is a final and rate-determined reaction in the conversion of tyrosine to noradrenaline. This is found in catecholamine storage vesicles in nervous tissue and in granules of the chromaffin cells of the adrenal medulla. The enzyme is formed from four identical subunits arranged as dimers joined by disulphide bonds and contains 2–12 atoms of copper as Cu^{2+} per tetramer. Ascorbate is considered to be a reductant in the reaction. The adrenal medulla contains a very high concentration of ascorbate.

Many peptides, active as hormones, hormone-releasing factors and neurotransmitters, have a carboxy-terminal residue which is amidated. The amidation is catalysed by a copper-requiring enzyme that oxidatively cleaves the carboxy-terminal residue using molecular oxygen. The peptides which are amidated by this enzyme include bombesin, calcitonin, cholecystokinin, ACTH, gastrin, growth hormone-releasing factor, α- and γ-melanotropin, neuropeptide, oxytocin, vasoactive intestinal peptide and vasopressin. The enzymes involved in the amidation are activated by the presence of ascorbate. Ascorbic acid appears to be involved in the conversion of 4-hydroxyphenylpyruvate to homogentisate in the oxidation of tyrosine to carbon dioxide and water.

Prolyl and lysyl hydroxylases are important in collagen formation. The known hydroxylases for collagen metabolism, prolyl 4-hydroxylase, prolyl 3-hydroxylase and lysyl hydroxylase, are all α-ketoglutarate-dependent dioxygenases that require ferrous iron. All require a reductant, the most effective being ascorbic acid. However, the main function of ascorbic acid may not be in the hydroxylation of prolyl and lysyl residues but rather in protein biosynthesis.

The hydroxylation of other proline residues in elastin in the aorta and other tissues requires ascorbic acid. Ascorbic acid deficiency does not decrease elastin synthesis but results in under-hydroxylation. It may also be involved in carnitine biosynthesis. Carnitine is required in fatty acid metabolism for formation of acylcarnitines that can cross mitochondrial and peroxisomal membranes. Acylcarnitines are required for transport of fatty acids into mitochondria for oxidation. Two hydroxylation reactions are involved in the conversion of lysine to carnitine. Both enzymes require a reducing agent to keep the iron reduced.

The wide requirements of ascorbic acid account for the many structural, regulatory, metabolic and immune disorders associated with scurvy. Their participation is probably indirect. Ascorbic acid function may be to bring the iron constituent of enzymes, when oxidized, back to the active ferrous form.

7.3.4 Availability of ascorbic acid

Ascorbic acid is readily and rapidly absorbed in the small intestine, is distributed in the blood and taken up by tissues to sustain metabolic functions. The plasma concentration is related to dietary intake and is in the order of 68 μmol/litre on a daily intake of 100 mg. Binding to proteins does not appear to be a central feature of vitamin C metabolism, though an association with albumin may be important. It is probable that the transport of ascorbic acid requires a membrane carrier and active transport which is probably specific to ascorbic acid.

Metabolism

A major pathway of ascorbic acid metabolism is to 2,3-dioxo-L-gulonate, which is further metabolized to oxalate and L-threonate. Ascorbic acid is excreted in the urine either as the free ascorbic acid or as oxalate. During excessive vitamin C intake there is always the possibility of the development of oxalate stones.

Tissue levels of vitamin C

High concentrations of vitamin C are found in the adrenal glands of the order of between 1–3 mmol/kg. There are high concentrations in all tissues at birth which steadily reduce with increasing age. Tissue concentrations in the infant vary between 0.4 mmol/kg in the heart to 3 mmol/kg in the adrenal glands. This is in contrast to the middle-aged and elderly adult where the concentrations range from 0.1 mmol/kg in the heart and 1.0 mmol/kg in the adrenals. There is good conservation of vitamin C by the kidneys and urinary excretion only occurs when the plasma concentration exceeds 70 μmol/litre. The rate of utilization of vitamin C appears to be determined by the size of the vitamin C pool.

Other examples of vitamin C levels in various tissues, cells and fluids are:

Cervicovaginal tissue 16   mmol/kg
Adrenal medulla ....10.5 mmol/kg
Monocytes ....8.0 mmol/kg
Brain ....1.3 mmol/kg
Aqueous humour ....1.0 mmol/kg
Tears ....0.77 mmol/kg

7.3.5 Ascorbic acid deficiency and excess

Scurvy

Scurvy is a nutritional disease which results from prolonged subsistence on diets which do not include fresh fruit and vegetables. Lack of ascorbic acid results in disturbance of the structure of connective tissue leading to swollen, bleeding gums and haemorrhage into the skin. Scurvy tends to occur at the extremes of ages, in the young or in elderly people living in a socially isolated situation.

Pathological effects
There is a failure of the body's supporting tissues to produce and maintain intercellular substances. Capilliary haemorrhage results from a defect in the capilliary basement and the intracellular link-

ages between the endothelial cells. Wound healing and cartilage, bone and dentine growth are adversely affected. There is also a defect in the extracellular matrix where chondroblasts, osteoblasts and ondontoblasts lay down calcium. The matrix of collagen is important in bone and cartilage structure. The defect in scurvy is in the formation of collagen, which is an omnipresent support protein in the body. Collagen is assembled outside cells from pro-collagen, which is a coil with repeating units of –glycine–hydroxyproline–proline–.

Hydroxyproline metabolism requires the enzyme proline hydroxylase, which is present in fibroblasts. This enzyme requires ascorbic acid for activation. Another observation in animals is that scorbutic guinea pigs have hypertrophy of the adrenal glands.

Aetiology
A lack of ascorbic acid is responsible for the characteristic disease. In addition to a dietary deficiency of vitamin C the vitamin may have been destroyed by cooking.

Clinical signs
Haemorrhages, either large or microscopic, may occur anywhere in the body, including the gums, subcutaneous tissues, synovia of joints and beneath the periosteum of bones. Haemorrhages may also occur into the brain or heart muscle, with probably fatal consequences.

There is failure of wound healing, and old wounds which have healed break down. Lind, a Scots naval surgeon who conducted the first clinical trial on the treatment of scurvy with fruit and lemons, said that the pathognomonic sign of the disease was the appearance of the gums, with the characteristic gingivitis. The gums, particularly in the region of the papillae between the teeth, are swollen and the scurvy buds may protrude beyond the biting surface of the teeth. The spongy gums are livid in colour and bleed readily. Superinfection is also a problem. Perifollicular bleeding around the orifice of a hair follicle is often found in the lower thighs and below the knees but may appear on the buttocks, abdomen, legs and arms,

followed by petechial haemorrhages which are not confined to the hair follicles. This must be differentiated from the follicular keratosis associated with vitamin A deficiency. In vitamin A deficiency there is a horny plug of keratin projecting from the orifice of the hair follicle whereas in scurvy, there is a heaping up of keratin-like material on the surface around the mouth of the follicle, through which a deformed corkscrew hair projects. Other signs that have been described are ocular haemorrhages particularly in the bulbar conjunctiva, Sjögren's syndrome (a loss of secretion of salivary and lacrimal glands), femoral neuropathy, oedema of the lower limbs, oliguria and psychological disturbances, hypochondria and depression. Anaemia which may be normoblastic or megaloblastic is a common finding. **Osteoporosis** may also occur in scurvy.

Vitamin C plasma levels in scurvy

On a deficient diet the plasma concentration falls from 110 μmol/l to 70 μmol/l after 4 weeks, a concentration which is compatible with the appearance of scurvy. The white cell content of vitamin C is a good indication of ascorbic acid status with an adequate dietary intake being suggested by a white cell level in excess of 0.85 mmol/l. Concentrations below 400 μmol/l suggest that there is a risk of scurvy.

Scurvy in infants
Until the teeth have developed, gingivitis is not apparent but scurvy buds do occur. Bleeding is usually seen as a large subperiosteal haemorrhage over the long bones, for example the femur.

7.3.6 Recommended requirements

In the normal adult approximately 3% of the body pool of ascorbic acid, independent of size, is degraded each day. Leucocyte or buffy coat vitamin C concentrations reflect tissue concentrations and there is a lower satisfactory limit of

0.09 μmol/10^8 cells, concentrations below which indicate deficiency. Plasma vitamin C concentrations reflect recent intake and values less than 2 mg/l (11 μmol/l) indicate biochemical deficiency. While the necessary dietary intake of vitamin C has not been identified, dietary intakes of less than 10 mg/day are probably too small.

Vitamin C – recommended intake

There is a sigmoidal relationship between vitamin C intake and plasma ascorbate concentrations. The RNI is probably 40 mg/day. An LRNI of 10 mg/day for adults will probably prevent the development of scurvy. The EAR is 25 mg/day. These are the British recommendations.

The recommendations for intake vary from expert committee to expert committee. The variation in opinion is not unrelated to the definition of a normal vitamin status, and fulfilment and analytical measurement of that defined normality:

- Pregnancy and lactation: the RNI should increase by 10 mg/day during the third trimester. During lactation an intake of 70 mg/day is probably satisfactory.
- Children: clinical scurvy has not been observed in fully breast-fed infants. The vitamin C content of breast milk varies from 170 to 450 μmol/l, which provides 25 mg/day. The LRNI for infants is 6 mg/day.
- Elderly: there is no need routinely to increase vitamin C intake in the elderly, though as the diet of the elderly is not always ideal, vitamin C deficiency should always be considered.
- Smokers: smokers have an increased turnover of vitamin C and their intake should be increased to over 80 mg/day.

7.3.7 Body stores measurements

Ascorbic acid can be estimated in blood plasma or whole blood. Possibly the best index of vitamin C status is the white cell concentration of ascorbic acid.

There is a quantitative relationship between vitamin C intake and its levels in plasma (characteristically an S-shaped curve) and leucocytes. Urinary vitamin C is a good mark of a high intake, although chemical instability is a problem. Vitamin C measurements are difficult because of the vitamin's instability.

The body can be saturated with approximately 5 g ascorbic acid so that 250 mg given orally four times a day should rapidly restore ascorbic acid concentrations. Alternatively, fresh fruit and vegetables, sprouting peas or extracts of pine needles are all that is required.

Summary

1. Ascorbic acid is a simple sugar with a molecular weight of 176 Da.
2. Ascorbic acid is a powerful reducing agent and an electron donor. As such it has a central role in the relative states of oxidation/reduction of other metabolically important water-soluble substances.
3. Enzyme reactions requiring ascorbic acid are hydroxylations utilizing oxygen with Fe^{2+} or Cu^{2+} as cofactors. The interaction between ascorbic acid and iron can increase the oxidative potential of iron, and iron acts as a pro-oxidant in the presence of ascorbic acid. The enzymes involved include the synthesis of hydroxyproline and hydroxylysine in pro-collagen, carnitine from lysine, and the hepatic microsomal mono-oxygenase system which is important for steroid hormones and xenobiotics.
4. A dietary deficiency of ascorbic acid results in scurvy.
5. Recommended requirements for vitamin C are defined for various ages and states.

Further reading

Block, G., Henson, D.E. and Levine, M. (eds) (1990) Ascorbic acid. Proceedings of a conference at NIH, Bethesda. *American Journal of Clinical Nutrition*, 54(6), Supplement vii (Foreword).

Englard, S. and Seifter, S. (1986) The biochemical function of ascorbic acid. *Annual Review of Nutrition*, 6, 365–406.

Linder, M.C. (1985) *Nutritional Biochemistry and Metabolism with Clinical Applications*, Elsevier, New York.

Schorah, C.J. (1992) The transport of vitamin C and effects of disease. *Proceedings of the Nutrition Society*, 51, 189–98.

7.4 Biotin

7.4.1 Introduction

Biotin contains a ureido group in a five-membered ring fused with a tetrahydrothiophene ring with a five-carbon side chain terminating in a carboxyl group. The molecular weight is 244 Da.

Biotin is found in yeast, bacteria, liver, kidney, yeast extracts, pulses, nuts, chocolates and some vegetables. Most meats, dairy products and cereals are relatively poor sources. Liver, egg yolks and cooked cereals are rich sources of biotin containing between 20–100 µg/100 g. Human breast milk contains 0.03 µmol/l of biotin, and cows' milk 0.10 µmol/l. This is less if the mother is deficient in biotin. Another source of biotin is endogenous colonic bacterial synthesis. It is synthesized biochemically, by microorganisms, from pimelic acid, a seven-carbon dicarboxylic acid, L-alanine and L-cysteine.

- acetyl-CoA carboxylase catalyses the formation of malonyl-CoA from acetyl-CoA, bicarbonate and ATP. Malonyl-CoA is then used in fatty acid synthesis and fatty acid chain elongation
- pyruvate carboxylase converts pyruvate to oxaloacetate leading into the tricarboxylic acid cycle

This feeds the tricarboxylic acid cycle as well as providing a source of the carbon skeleton from the cycle necessary for the synthesis of the amino acids aspartate and glutamate.

In the tissues, biotin is important in gluconeogenesis, especially in the liver and kidney, where oxaloacetic acid is utilized for the synthesis of glucose.

7.4.2 Action of biotin

Biotin is a cofactor for the acetyl-CoA, propionyl-CoA and pyruvate carboxylase systems. These enzymes are involved in the synthesis of fatty acids and involved in gluconeogenesis and carboxylation in this metabolic pathway.

The types of enzymatic reactions involved include:

- all carboxylases involved in the fixation of carbon dioxide and requiring adenosine triphosphate (ATP)

Biotin binding

As a covalently bound cofactor in enzymes, biotin is a site for formation of a carboxylated intermediate, being bound to the enzyme by an amide linkage between ε-amino groups of enzyme lysine and carboxyl groups of biotin's valeric acid side chain. The coenzyme function of biotin allows the carboxylation reactions to proceed by receiving the ATP-activated carboxyl group and transferring to the carboxyl acceptor substrate (Figure 7.4).

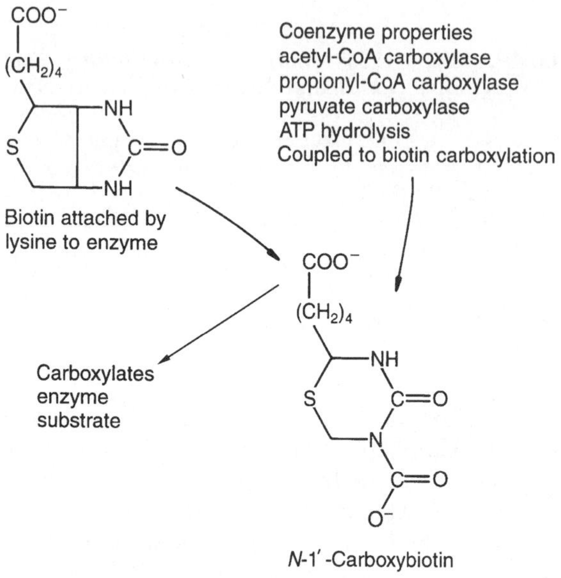

FIGURE 7.4 The structure of biotin and its activity on the dependent coenzymes which are attached by lysine to the enzyme. ATP hydrolysis is coupled to biotin carboxylation.

Biotin in fatty acid synthesis

Propionyl-CoA carboxylase catalyses the carboxylation of propionyl-CoA to methylmalonyl-CoA which is then converted to succinyl-CoA and enters the tricarboxylic acid cycle. Biotin is essential for the catabolism of propionic acid which is derived from intestinal flora, for the catabolism of isoleucine, valine, methionine, and threonine, the side chain of cholesterol and the oxidation of odd-numbered fatty acids. 3-Methylcrotonyl-CoA carboxylase forms 3-methylglutacrotonyl-CoA from 3-methylcrotonyl-CoA in the catabolic path of leucine.

Acetyl-CoA carboxylase is found in the cytosol of the liver, where it catalyses the first step in the biosynthesis of fatty acids. The activity of this enzyme is important in the control of fatty acid synthesis. The carboxylases are synthesized as inactive apocarboxylases lacking biotin. Biotin binds to the apocarboxylases requiring the enzyme holocarboxylase.

7.4.3 Availability

Little is known about the enteric absorption of biotin, but it is probably absorbed in the upper gastrointestinal tract. Raw egg white contains the protein **avidin** (molecular weight 68 000 Da), which binds biotin with a high affinity. Ingestion of high amounts of avidin leads to the formation of biotin–avidin complexes and prevents the absorption of biotin. Denaturation of the avidin releases biotin.

In plasma, most of the biotin is bound to albumin and α- and β-globulin, but some is free. Biotin is stored in the liver and excreted as free biotin in the urine.

7.4.4 Biotin deficiency

Biotin deficiency may include fatigue, depression, sleepiness, nausea and loss of appetite, muscle pain, hyperaesthesiae and paraesthesiae without reflex changes or other signs of neuropathy. These symptoms and signs are similar to those of thiamin deficiency. The tongue becomes smooth, with loss of papillae, the skin is dry with fine scaly desquamation, and anaemia and hypercholesterolaemia develop.

There are no indications that **excess** biotin can be harmful.

7.4.5 Recommended requirements

The dietary requirement of biotin is not known with certainty. The average intake in the British male is 39 (range 15–70) μg/day and in women 26 (range 10–58) μg/day. It is estimated that biotin intakes of between 10 and 200 μg/day are safe and adequate. The human neonate is born with higher biotin levels in the blood than those of adults, the biotin concentration of cord blood being 35–50% greater than that of maternal blood.

Summary

1. Biotin contains a ureido group in a five-membered ring fused with a tetrahydrothiophene ring with a five-carbon side chain terminating in a carboxyl group.
2. Biotin is a cofactor for the acetyl-CoA, propionyl-CoA and pyruvate carboxylase systems. These enzymes are involved in the synthesis of fatty acids and involved in gluconeogenesis and carboxylation in this metabolic pathway.
3. Biotin is a covalently bound cofactor in enzymes, as a site for formation of a carboxylated intermediate, being bound to the enzyme by amide linkage between ε-amino groups of enzyme lysine and the carboxyl group of biotin's valeric acid side chain.
4. The coenzyme function of biotin allows the carboxylation reactions to proceed by receiving the ATP-activated carboxyl group and transferring to the carboxyl acceptor substrate.
5. A dietary deficiency of biotin results in fatigue, depression, sleepiness, nausea and loss of appetite, muscle pain, hyperaesthesiae and paraesthesiae without reflex changes or other signs of neuropathy. The tongue becomes smooth with loss of papillae, the skin is dry with fine scaly desquamation, and anaemia and hypercholesterolaemia develop.

Further reading

Sweetman, L. and Myhan, W.L. (1986) Inheritable biotin-treatable disorders and associated phenomena. *Annual Review of Nutrition*, 6, 317–43.

7.5 Nicotinic acid

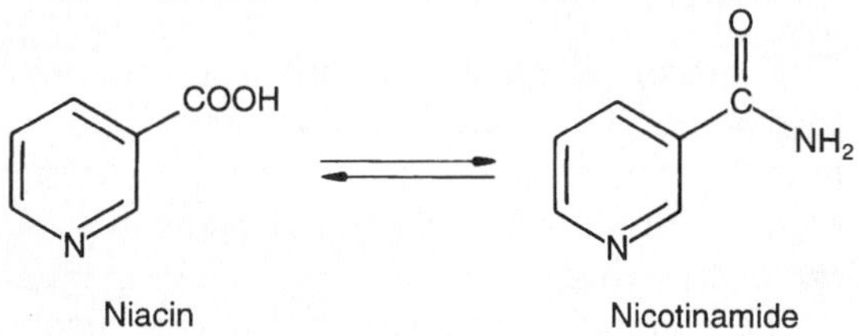

FIGURE 7.5 The structure of niacin and nicotinamide.

7.5.1 Introduction

Nicotinic acid (also known as niacin or vitamin B_3), molecular weight 123 Da, occurs as a pyridine derivative and is found in the body as an amide, nicotinamide (Figure 7.5).

Nicotinic acid is found in plants and animal foods in small amounts, with the exception of meat, fish, wholemeal cereals and pulses. However in many cereals – particularly in maize and perhaps potatoes – the nicotinic acid is held in a bound, unabsorbable form. Nicotinic acid can be liberated from the bound form, niacytin, in an alkaline medium. In central America maize is an important foodstuff and often eaten as tortillas. The tortillas from maize are treated with lime water. Cooking results in little destruction of nicotinic acid, though it does pass out into the cooking water which must then be consumed if the nicotinic acid is not to be lost.

Humans are not entirely dependent upon dietary nicotinic acid as this vitamin can be synthesized from tryptophan.

7.5.2 Synthesis

The rate of conversion of tryptophan to niacin varies from individual-to-individual and from day-to-day. Increased dietary intake of tryptophan results in a greater efficiency of conversion to niacin. The two enzymes involved in tryptophan metabolism, kynureninase and kynurenine hydroxylase are vitamin B_6- and riboflavin-dependent (Figure 7.6). Consequently a deficiency of either of these vitamins may result in deficiency of niacin despite adequate intakes of tryptophan.

Approximately 60 mg of tryptophan in the diet are required to produce 1 mg of dietary nicotinic acid. The nicotinic acid equivalent, that is the nicotinic acid content plus 1/60th of the tryptophan content of the food or diet, is a good measure of the dietary intake, rather than the nicotinic acid content alone.

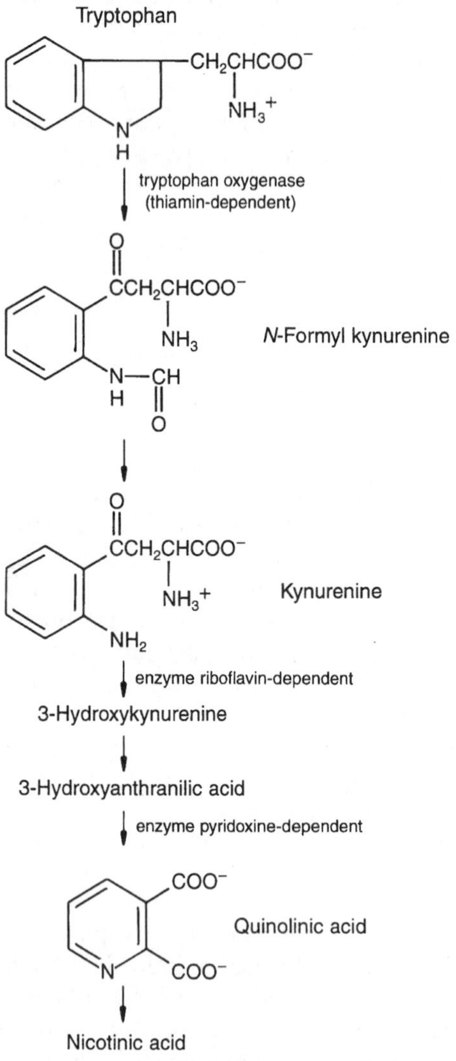

FIGURE 7.6 The synthesis of nicotinic acid from tryptophan. The enzymes are thiamin-, riboflavin- and pyridoxine-dependent.

7.5.3 Action of niacin

Nicotinamide is a component of the coenzymes nicotinamide adenine dinucleotide (NAD^+) and nicotinamide adenine dinucleotide phosphate ($NADP^+$). NADP is also known as triphosphopyridine nucleotide (TPN^+). These coenzymes are electron carriers. In the reaction $NAD^+ \rightarrow NADPH$, two electrons and a proton are accepted from the substrate which is being oxidized. NADH is then reoxidized, accepting O_2 with the reversal to NAD^+ and the formation of ATP. NAD^+ is a cellular electron carrier (Figure 7.7).

Typical reactions which require NAD^+ include alcohol dehydrogenase, glutamine dehydrogenase, and glyceraldehyde dehydrogenase. NAD^+ acts as a true bound coenzyme in such reactions as epimerizations and aldolizations. Enzymes involved are UDP galactose-4-epimerase and S-adenosylhomocysteinase.

7.5.4 Availability

Free niacin is readily absorbed in the upper gastrointestinal tract. Nicotinic acid held in a bound form, niacytin, is not absorbed from the gastrointestinal tract. There is no apparent storage of this vitamin.

The vitamin is excreted as 5'-methylnicotinamide in the urine.

7.5.5 Niacin deficiency

Pellagra

A deficiency of niacin results in **pellagra**, a disease found among poor populations living largely on maize, and possibly accompanied by protein energy malnutrition, anaemia and deficiencies of thiamin and other vitamins.

Pellagra is a chronic and relapsing problem with a seasonal incidence. The main features are loss of weight, increasing debility, erythematous dermatitis (affecting skin exposed to sunlight),

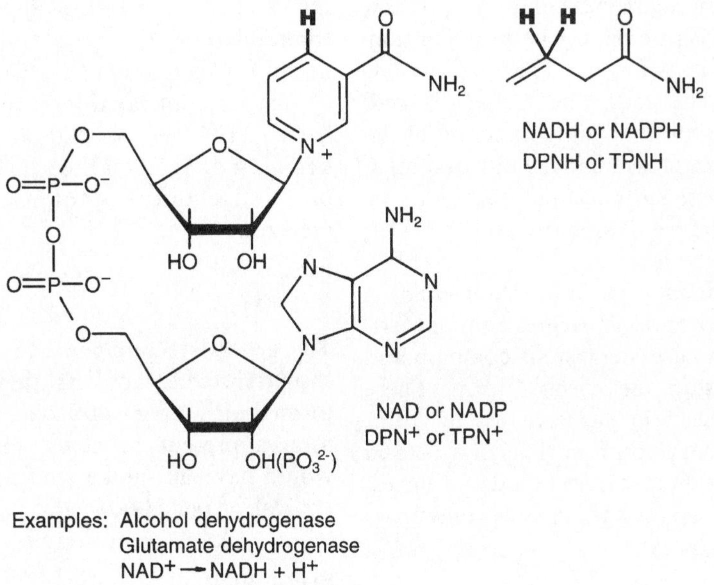

FIGURE 7.7   Nicotinamide as a coenzyme, indicating the hydrogens involved in the coenzyme's activity.

gastrointestinal disturbances (especially diarrhoea and glossitis), and mental changes. Pellagra is called the disease of the 3Ds: dermatitis, diarrhoea and dementia. The latter two are usually found in more advanced cases and rather than dementia, depression may be the problem. Pellagra became a common problem following the cultivation of maize, which was introduced into many countries as an easily grown crop in dry areas.

Aetiology
Pellagra is due to a deficiency of nicotinic acid. Tryptophan, from which nicotinamide is synthesized in the body, is present in ample amounts in most dietary proteins, but not in zein, the chief protein in maize. If pellagra is not to result then the maize must be eaten in conjunction with, for example, milk.

Pellagra was not found in central America because the maize is carefully cooked with a thin paste of slaked lime, heated for 18 hours, and then eaten as tortillas. The prolonged heating liberates the bound nicotinic acid. Pellagra can occur in alcoholics and in patients who have been starved as in malabsorption syndrome.

There is a rare inbuilt inborn error of metabolism which resembles pellagra and is characterized by skin lesions, cerebellar ataxia and biochemical abnormalities, notably amino aciduria. This is called Hartnup's disease. The problem is the transport of tryptophan which affects the absorption of the amino acid in the small intestine as well as in the renal tubules.

Clinical features
The patient is underweight and there is increasing debility:

- **Skin**: an erythema develops with the appearance of sunburn with a symmetrical distribution over the parts of the body exposed to sunlight, backs of the hands, wrists, fore-arms, face and neck. Areas exposed to mechanical irritation or trauma are particularly affected. The skin is initially red and slightly swollen, itchy and hot. In acute cases the skin lesions may progress to vesiculation, cracking, with exudates and ulceration. There may be secondary infection. In chronic cases the dermatitis is a roughening and thickening of the skin with dry scaling and brown pigmentation.

- **Digestive system**: diarrhoea may be present, also nausea, epigastric pain. The problem may be aggravated by parasites. There is angular stomatitis and cheilosis. The tongue is red, swollen and painful. Secondary infection of the mouth is common. There may be mucosal atrophy affecting the gastrointestinal tract.
- **Reproductive organs**: vaginitis and amenorrhoea may occur.
- **Nervous system**: in mild cases there will be weakness, tremor, anxiety, depression and irritability. In severe cases delirium is common and dementia occurs in the chronic form. There may be paraesthesia in the feet, loss of vibration sense and proprioception. This may lead to ataxia, spasticity and increased tendon reflexes. It is difficult to separate these symptoms from vitamin B_{12} deficiency.

Treatment

Mild cases improve with treatment with nicotinamide or nicotinic acid or by an improved diet. Mental symptoms, especially dementia, may not respond to treatment.

Nicotinamide, in contrast to nicotinic acid, does not cause unpleasant flushing and a burning sensation; the oral dose is 100 mg 4-hourly. The response is rapid, within 24 hours the erythema is less, the tongue is restored to normal and diarrhoea ceases. Complementary vitamin B complexes, folic acid and B_{12} should also be given. It is important to restore the individual to a diet containing an array of good-quality protein from the sources compatible with that individual's culture.

7.5.6 Niacin excess

A very high dosage of niacin, 3–6 g/day, can affect liver structure and function with hepatotoxic consequences.

7.5.7 Recommended requirements

Recommendations for niacin are associated with tryptophan ingestion. The median intake of protein in Britain is 84 g/day for men and 62 g/day for women, containing approximately 12.6 mg tryptophan per gram, equivalent to 17 mg/day of niacin for men and 13 mg/day for women.

The recommendations are for niacin equivalents. The RNI is 6.6 mg/1000 kcal and the LRNI 4.4 mg/1000 kcal. These estimates are based on requirements of niacin to prevent or cure pellagra.

Infants

The general recommendation is that infant milk should provide not less than 3.3–3.85 mg preformed niacin per 1000 kcal. Because of the tryptophan present in cows' milk protein, infants would have an intake similar to adults of niacin equivalent per 1000 kcal.

Pregnancy

Because of hormonal changes in tryptophan metabolism in late pregnancy, 30 mg of tryptophan is equivalent to 1 mg of dietary niacin. There should be simultaneous increased metabolism of tryptophan, so it is unlikely that there is a need for an increased dietary intake of niacin.

Lactation

Mature human milk provides pre-formed niacin (2.7 mg/litre). It is assumed that maternal tryptophan metabolism would compensate for this.

7.5.8 Body store measurements

There are no absolute laboratory assessments of niacin status. The most sensitive measurement of niacin, nicotinamide and related compounds in serum, urine and food, is by microbiological methods. In such methods an organism which requires niacin for growth is cultured in the presence of the test material, and growth calibrated against standard solutions of niacin. An alternative is to measure the main urinary metabolites of nicotinamide, namely N-methyl nicotinamide over a defined time or as a ratio of creatinine in urine. The measurement of urinary N-methyl-2-pyridone-5-carboxamide (2-pyridone) excretion is a more sensitive method used in marginal deficiencies of nicotinamide, and is expressed as mmol/creatinine. N-Methyl nicotinamide is an excretory product of nicotinic acid and is present in the urine in reduced amounts in pellagra, e.g. below 0.2 mg per 6 hours.

Summary

1. Nicotinic acid occurs as a pyridine derivative and is found in the body as an amide, nicotinamide.
2. Nicotinamide is a component of the coenzymes nicotinamide adenine dinucleotide (NAD^+) and nicotinamide adenine dinucleotide phosphate ($NADP^+$). These coenzymes are electron carriers. In the reaction $NAD^+ \rightarrow NADPH$, two electrons and a proton are accepted from the substrate being oxidized.
3. Dietary nicotinic acid deficiency leads to pellagra, characterized by dementia, diarrhoea and dermatitis.

7.6 Pantothenic acid

7.6.1 Introduction

Pantothenic acid, molecular weight 219 Da, is the dimethyl derivative of butyric acid joined by a peptide linkage to the amino acid β-alanine. The biochemically active form of the vitamin is 4'-phosphopantetheine which is present in all tissues (Figure 7.8). Pantothenic acid is widely available in natural foods. There is destruction of the molecule in heat in alkaline or acid conditions.

7.6.2 Action of pantothenic acid

4'-Phosphopantetheine is a constituent of coenzymes:

- as coenzyme A, bound covalently to an adenyl group
- to the functional part of a protein, e.g. serine hydroxyl group

The sulphydryl group of the 4'-phosphopantetheine coenzyme (β-mercaptoethylene) is the functional part of the coenzyme's activity in enzymatic reactions. Coenzyme A has a central role:

- in intermediary metabolism (transfer of acyl groups to oxaloacetic acid), in energy metabolism

- as an acyl carrier in the synthesis of lipids, fatty acids, β-oxidation of fatty acids and acylation
- in the synthesis of glycerol, glycerides, cholesterol, ketone bodies and sphingosine

The creation of coenzyme A esters of carboxylic acids is important in enolization and acyl group transfer reactions.

7.6.3 Availability

Pantothenic acid is absorbed and then excreted as the free acid in the urine. There is no storage of the vitamin.

7.6.4 Pantothenic acid deficiency and excess

No specific deficiency syndrome has been identified. In malnutrition there will be a general deficiency of pantothenic acid but this will be difficult to separate from the overall problems associated with malnutrition.

No toxicity as a result of pantothenic acid excess has been identified at the dosages used.

7.6.5 Recommended requirements

Most human diets provide 3–10 mg derived from a variety of natural foods. Estimates in the British diet give values of the order of 5–6 mg/day. There is no evidence of pantothenic acid deficiency at intakes of between 3–7 mg; therefore this must be an adequate intake, even during pregnancy and lactation.

The intake for children is of the order of 3 mg/1000 kcal.

7.6.6 Body store measurements

No biochemical method identifies pantothenic acid status in humans. Blood levels and urinary excretion have been measured but these are difficult to interpret in terms of dietary need.

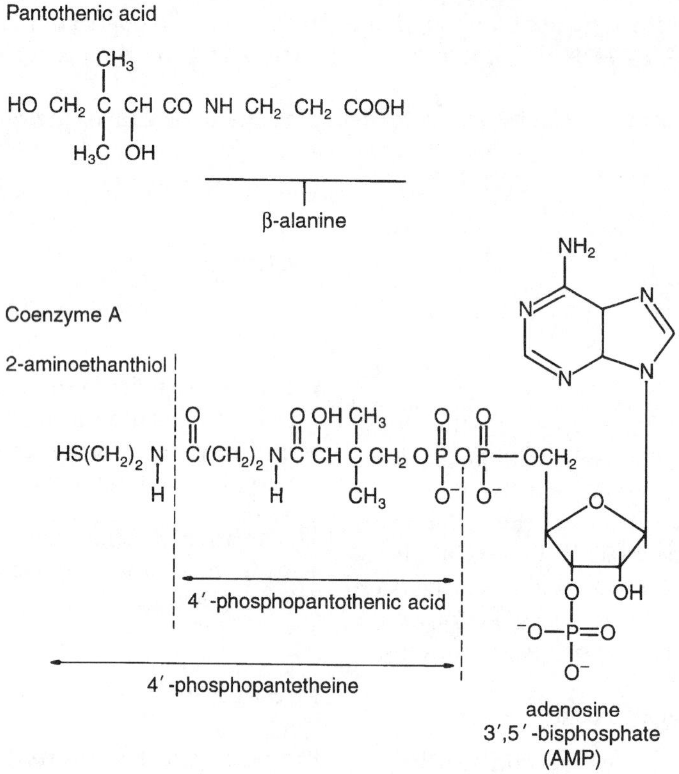

FIGURE 7.8 Structure of pantothenic acid and the active principle of coenzyme A.

Summary

1. Pantothenic acid is the dimethyl derivative of butyric acid joined by a peptide linkage to the amino acid β-alanine. The biochemically active form of the vitamin is 4'-phosphopantetheine.
2. Pantothenic acid is the precursor of coenzyme A.
3. No deficiency condition has been described in humans.

7.7 Riboflavin

7.7.1 Introduction

Riboflavin, molecular weight 219 Da, is a substituted alloxazine ring linked to ribotol, an alcohol derived from the pentose sugar ribose. Riboflavin is stable in boiling acid but not in boiling alkaline solution or after exposure to light.

Dietary sources are liver, milk, cheese, eggs, some green vegetables and beer. Other sources are yeast extracts, e.g. Marmite and meat extracts, e.g. Bovril.

7.7.2 Action of riboflavin

In plant and animal tissues riboflavin links with phosphoric acid as flavin mononucleotide or riboflavin-5'-phosphate (FMN) which with adenosine monophosphate forms flavin adenine dinucleotide (FAD) (Figure 7.9). FMN and FAD are similar in their coenzyme activity. These are the prosthetic groups of the flavoprotein enzymes. The functional part of the coenzyme is the isoalloxane ring. These flavoproteins are stronger oxidizing agents than NAD^+ and are versatile redox coenzymes. They are involved in one- or two-electron reactions with free radicals or metal ions. In the reduced form they react with O_2 in hydroxylation reactions.

Flavoproteins are involved in metabolism with the oxidation of glucose and fatty acids and the production of ATP, which is significant in anabolic processes. Important metabolic processes involving FAD/FMN include:

- acting as a coenzyme in oxidation/reduction reactions
- electron transport; oxidative phosphorylation, e.g. succinic dehydrogenases
- fatty acid synthesis and oxidation
- amino acid oxidases
- xanthine oxidase
- glutathione reductase

7.7.3 Availability

Riboflavin is absorbed in the upper gastrointestinal tract, does not appear to be stored, and is excreted unchanged in the urine. Chronic infection can affect urinary riboflavin excretion.

7.7.4 Riboflavin deficiency and excess

Riboflavin deficiency causes minimal morbidity but is associated with cheilosis, angular stomatitis and superficial interstitial keratosis of the cornea. Nasolabial seborrhoea can occur.

No toxic effects have been shown for riboflavin.

7.7.5 Recommended requirements

Adults

The RNI is 1.3 mg/day for men and 1.1 mg/day for women. The LRNI in adults, male and female, is 0.8 mg/day; the EAR is 1.0 mg/day for men and 0.9 mg/day for women.

Infants

The average riboflavin content of breast milk in Britain is approximately 1.4 µmol/l. The range varies enormously, depending on the mother's intake. The probable requirement of RNI is 0.4 mg/day. The RNIs for children range between 0.4 mg/day for infants aged up to 3 months and 1.0 mg/day for those aged 7–10 months.

Pregnancy and lactation

Because of the extra demands for riboflavin the requirement has been increased by 0.3 mg/day during pregnancy, and 0.5 mg/day during lactation.

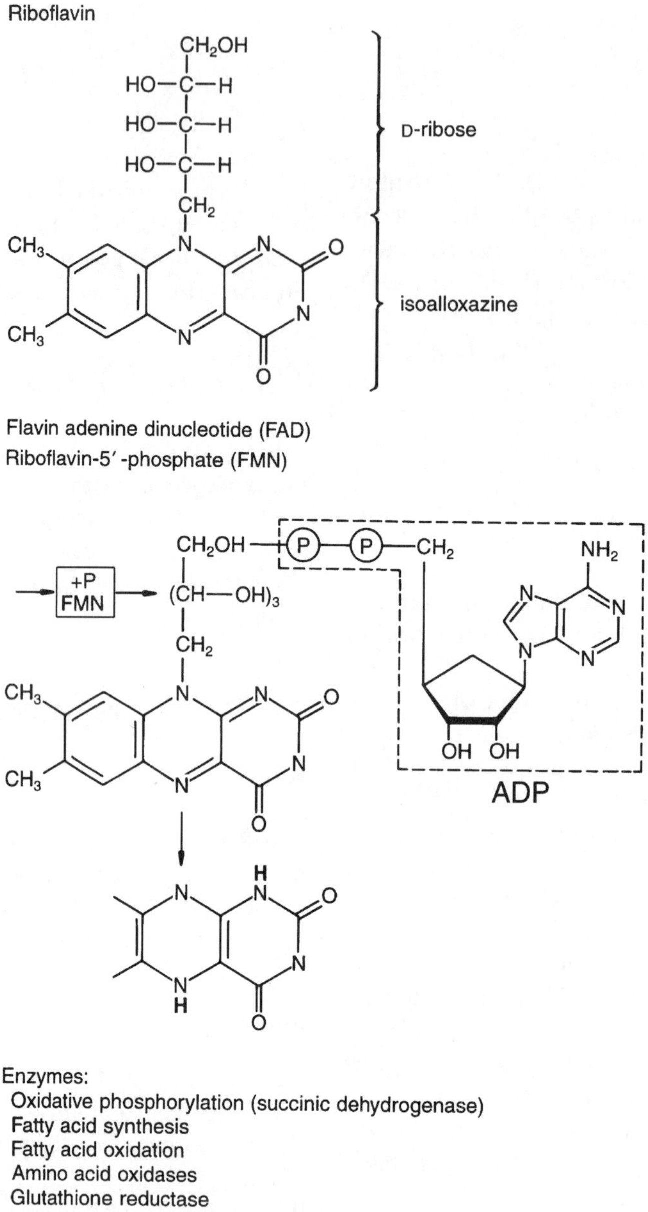

Riboflavin

Flavin adenine dinucleotide (FAD)
Riboflavin-5′-phosphate (FMN)

Enzymes:
Oxidative phosphorylation (succinic dehydrogenase)
Fatty acid synthesis
Fatty acid oxidation
Amino acid oxidases
Glutathione reductase

FIGURE 7.9   Structure of riboflavin and role in flavin adenine dinucleotide and riboflavin-5'-phosphate.

Elderly

While the resting metabolism and riboflavin intake decreases with age, the RNI for elderly individuals should be the same as for young people.

7.7.6 Body store measurements

The brilliant greenish-yellow fluorescence in ultraviolet light provides a mechanism for detecting riboflavin. Riboflavin status can be estimated by measuring:

- urinary riboflavin
- red cell riboflavin
- erythrocyte glutathione reductase activation coefficient (EGRAC)

The EGRAC test is a measure of tissue saturation and long-term riboflavin status. EGRAC values below 1.3 imply complete saturation of the tissue with riboflavin.

Methods to measure tissue saturation depend on protein intake and positive nitrogen balance. A negative nitrogen balance complicates the measurement of riboflavin requirements, because riboflavin and protein are involved in the formation and storage of flavoproteins in lean tissue. A negative niacin balance and tissue breakdown increases urinary excretion of riboflavin and the saturation of red cell glutathione reductase.

Urinary riboflavin measurement was the initial method used to assess riboflavin status. There is a linear relationship between intake and urinary excretion when there is a riboflavin intake in excess of dietary requirements. However, when riboflavin intake is less than tissue requirements there are adaptive changes in the utilization of riboflavin coenzymes, reducing riboflavin excretion.

Glutathione reductase activity depends on riboflavin. This is used in the erythrocyte glutathione reductase stimulation test, which measures tissue saturation and long-term riboflavin status. A finger-prick blood sample is used and is not affected by the age and sex of the subject. It is sensitive only at low levels of riboflavin intake and is affected by circulating riboflavin when nutritional status is poor. An EGRAC of between 1.0 and 2.5 indicates adequate intake.

Summary

1. Riboflavin is a substituted alloxazine ring linked to ribotol, an alcohol derived from the pentose sugar ribose.
2. Riboflavin links with phosphoric acid as flavin mononucleotide or riboflavin-5'-phosphate (FMN) which with adenosine monophosphate forms flavin adenine dinucleotide (FAD). These are the prosthetic groups of the flavoprotein enzymes. The functional part of the coenzyme is the isoalloxane ring.
3. These flavoproteins are strong oxidizing agents and are versatile redox coenzymes.
4. Riboflavin deficiency is associated with cheilosis, angular stomatitis and superficial interstitial keratosis of the cornea.

7.8 Thiamin

7.8.1 Introduction

Thiamin has a molecular weight of 337 Da. Thiamin hydrochloride consists of a substituted pyrimidine ring linked by a methylene group to a sulphur-containing thiazole ring (Figure 7.10). Oxidation results in the inactive product thiochrome which is strongly fluorescent in ultraviolet light; this effect can be used as a chemical estimation of the vitamin.

All animal and plant tissues contain thiamin. The important sources are plant seeds and the germ of cereals, nuts, peas, beans, pulses and yeasts. Green vegetables, roots, fruit, meat and dairy products contain modest amounts of the vitamin. Loss occurs during cooking when rice and vegetables are boiled. Alkaline conditions result in the destruction of the vitamin.

7.8.2 Action of thiamin

Thiamin pyrophosphate is the coenzyme in α-ketoacid decarboxylation reactions (Figure 7.11). Thiamin pyrophosphate is the coenzyme of the carboxylase which is involved in the oxidative

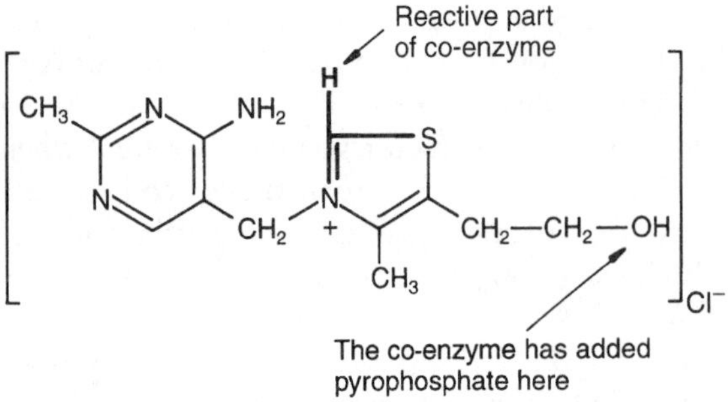

FIGURE 7.10 Structure of thiamin and structure as coenzyme. The reactive part of the coenzyme is shown in bold type.

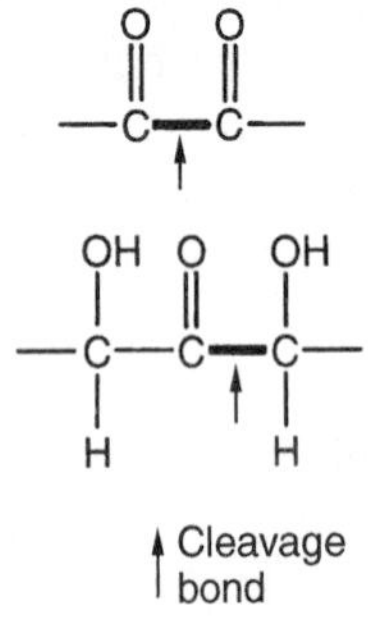

FIGURE 7.11 Thiamin pyrophosphate is involved in the cleavage of ketoacids (ketoacid decarboxylation).

decarboxylation of pyruvic acid to acetyl-CoA. Thiamin pyrophosphate is also required for the decarboxylation of α-ketoglutarate in the Krebs citric acid cycle and also in the transketolase reaction in the hexose monophosphate shunt.

7.8.3 Availability

Absorption is by passive diffusion and active absorption from the upper gastrointestinal tract. Thiamin is not stored in the body and the only reserve is the thiamin bound functionally to enzymes. Metabolism is extensive and multiple end-products are excreted in the urine.

7.8.4 Thiamin deficiency

Thiamin deficiency results in **beri-beri**, which was first recognized as a widespread problem among the rice-eating people of the East. The reason for

the widespread occurrence of the condition was a result of eating highly polished rice, the bran being sold for cattle food. The disease occurs in three forms:

- wet beri-beri (oedema and high output cardiac failure)
- dry beri-beri (polyneuropathy)
- infantile form

Thiamin deficiencies are also found in chronic alcoholics:

- alcoholic neuropathy, which is similar to that of dry beri-beri
- thiamin-responsive cardiomyopathy
- encephalopathy (the Wernicke–Korsakoff syndrome)

In these thiamin deficiency conditions the metabolism of carbohydrates is impaired because of the role of thiamin pyrophosphate as an essential coenzyme in the decarboxylation of pyruvate to acetyl CoA. Pyruvic acid and lactic acid accumulate in tissues and fluids, because of the lack of thiamin pyrophosphate, the coenzyme for transketolase in the hexose monophosphate pathway and for decarboxylation of 2-ketoglutarate to succinate in the citric acid cycle. Such accumulation results in peripheral dilatation and oedema. This may be compensated for by an increased cardiac output. When the heart is unable to sustain this increased cardiac output then cardiac failure results. This is a high output failure.

Clinical features

Initially the symptoms are anorexia, malaise, a feeling of heaviness and weakness of the legs, and problems with walking. There may be a little oedema of the legs and face and precordial pain and palpitations. There may be calf tenderness, pins and needles, numbness in the legs and reduced tendon jerks. Anaesthesia of the skin over the tibiae is a feature which may persist over prolonged periods. The consequence of this mild affliction is a reduction in stamina.

Wet beri-beri
Oedema affects the legs, face, trunk, lungs and peritoneal cavity. Palpitations, breathlessness,

anorexia and dyspepsia are common, with exercise-related leg pain a feature. There is evidence of congestive cardiac failure with increased jugular venous pressure, cardiomegaly with displacement of the apex beat and hypotension. The extremities are warm, the pulse as in aortic regurgitation, is a water-hammer pulse. With cardiac failure evidence of decompensation, coldness and cyanosis become apparent. The individual with cardiac failure may rapidly deteriorate and die suddenly.

Dry beri-beri

In the later stages the muscles become progressively wasted and weak, resulting in great difficulty in walking. The patient becomes emaciated and there may be evidence of Wernicke encephalopathy.

Infantile beri-beri

This occurs in breast-fed infants, usually between the second and fifth month. Infantile beri-beri may be acute or chronic; in the acute form cardiac failure develops rapidly, the child becomes restless, cries frequently, is oliguric, oedematous, cyanotic, dyspnoeic and develops a tachycardia. They may develop convulsions and become comatose.

In the chronic form, symptoms of gastrointestinal upset, constipation and vomiting appear, and the child is fretful and sleeps poorly. The muscles may be soft and toneless, but not markedly wasted. The skin may show pallor with cyanosis about the mouth. Cardiac failure and sudden death are common.

Alcoholic neuropathy

This is a neuropathy affecting both sensory and motor nerves. Sensory nerve dysfunction may manifest itself by paraesthesiae and severe nerve pain, the loss of sensation, numbness of the extremities and loss of proprioception. Motor nerve lesions include foot drop, muscle wasting and impaired knee and ankle jerks. In addition to this there may be cardiomyopathy.

Wernicke–Korsakoff syndrome

This consists of a weakness of eye muscles so that the patient cannot look upwards or sideways. The patient may be disorientated and apathetic. There may be nystagmus, and ataxic confabulation is a not infrequent finding. The anatomical lesions appear symmetrically in the brainstem, diencephalon and cerebellum. In severe cases there may be tissue necrosis with destruction of myelin. If large doses of thiamin are given sufficiently early there may be substantial, but not complete, improvement.

Treatment

Treatment of wet beri-beri is by intramuscular thiamin, 25 mg twice daily for 3 days; a dose of 10 mg two or three times daily should be sufficient thereafter. The reversal of the clinical signs is rapid, with improvement of heart size, respiration and physical performance.

Improvement following the treatment of dry beri-beri is slow. The neurological abnormalities in particular reverse slowly.

In infantile beri-beri, treatment of the mother benefits the child if there is breast feeding. The mother should receive 10 mg thiamin twice daily and the child may be given thiamin intramuscularly 10–20 mg once a day for 3 days and thereafter 5–10 mg twice daily.

Prevention

Beri-beri can be prevented by the use of unmilled rice, by the fortification of rice with thiamin, by the use of pulses and other foods containing thiamin, or by medicinal preparations of thiamin.

7.8.5 Thiamin excess

Long-term intakes in excess of 50 mg/kg body weight, or more than 3 g/day are toxic, leading to headaches, irritability, insomnia, rapid pulse, weakness, contact dermatitis, pruritis and even death. At 'normal' dosages, toxicity is unknown.

7.8.6 Recommended requirements

Adults

Thiamin requirements are related to energy metabolism. The average requirement is 0.3 mg/

1000 kcal. The RNI is 0.4 mg/1000 kcal. However, intakes should not be less than 0.4 mg/day. The requirements for men and women are probably the same.

Infants

The thiamin concentration in human milk is of the order of 0.5 μmol/l, which is equivalent to 0.3 mg/1000 kcal. The RNI for infants is 0.3 mg/1000 kcal.

Children

The RNI for children is the same as for adults.

Pregnancy and lactation

There is no evidence of any increased need for thiamin during normal pregnancy. The loss of 0.14 mg thiamin per day in milk should be met by a recommended increase in energy intake. The RNI is 0.4 mg/1000 kcal during pregnancy or lactation.

Therapeutic use

Thiamin is used in the treatment of cardiovascular and infantile beri-beri, in Wernicke encephalopathy and some cardiomyopathies.

7.8.7 Body store measurements

It is possible to estimate dietary status by measurement of urinary thiamin, though this is laborious and difficult to achieve. The thiamin : creatinine ratio has been used for random samples. Loading doses of 1–5 mg of thiamin are given orally or intramuscularly and the urinary thiamin measured over 4–24 hours. If the proportion of urinary thiamin is below 20% of the administered dose over the subsequent 24 hours then thiamin deficiency is suggested.

An alternative method is the reactivation of the cofactor-depleted red cell enzyme transketolase. However, the biochemical response to a given intake varies enormously between individuals

Summary

1. Thiamin hydrochloride is a substituted pyrimidine ring linked by a methylene group to a sulphur-containing thiazole ring.
2. Thiamin is the precursor of the coenzyme thiamin pyrophosphate. This is the essential coenzyme involved in the action of enzymes that catalyse cleavages of C–C and C–X bonds, e.g. in many α-keto acid decarboxylations.
3. Thiamin deficiency causes beri-beri.

7.9 Vitamin B$_6$

7.9.1 Introduction

Vitamin B$_6$, molecular weight 168 Da, comprises a family of free and phosphorylated compounds which includes pyridoxine, pyridoxal and pyridoxamine (Figure 7.12). Vitamin B$_6$ is found in cereals, meat (particularly liver), fruits, leafy and other vegetables. The free form is most commonly found in plants; the phosphorylated form, pyridoxamine phosphate, is more commonly found in animal tissues.

7.9.2 Action of pyridoxine

Pyridoxal-5'-phosphate is a coenzyme which has a major role in the intermediary metabolism of amino acids, in α-decarboxylation, aldolization, and transamination reactions (Figure 7.13). The coenzyme is also involved in the β-carboxylation of aspartic acid. The aldehyde group of pyridoxal-5'-phosphate forms a Schiff base with the α-amino acid group of amino acids. Pyridoxal-5'-phosphate catalyses several different types of bond cleavage

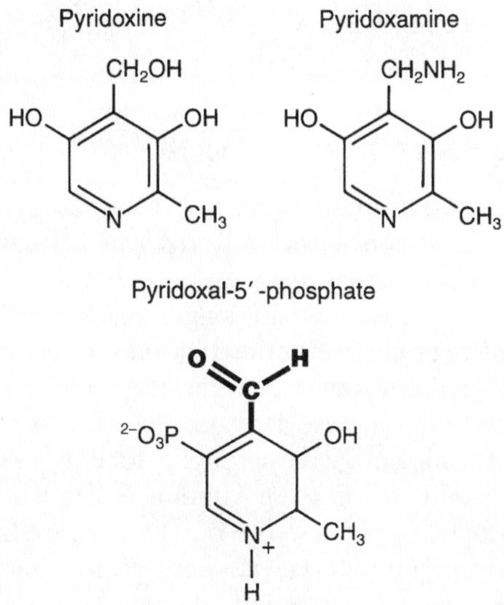

Pyridoxine

CH_2OH

Pyridoxamine

CH_2NH_2

Pyridoxal-5'-phosphate

FIGURE 7.12 Structure of pyridoxine, pyridoxamine and pyridoxal-5'-phosphate. The reactive part of the molecule is shown in bold type.

by stabilizing the electron pairs at the α- or β-carbon atoms of α-amino acids.

Metabolic roles of pyridoxine include:

1. Transamination enzymatic reactions: the vitamin forms pyridoxamine phosphate, which accepts the α-amino group of the amino acid, and then is released with the transfer of the amino group to a keto acid. Some of the vitamin B$_6$-dependent enzymes are rate-limiting in metabolic processes, e.g. the degradation

S——R (side chain)

O——R (side chain)

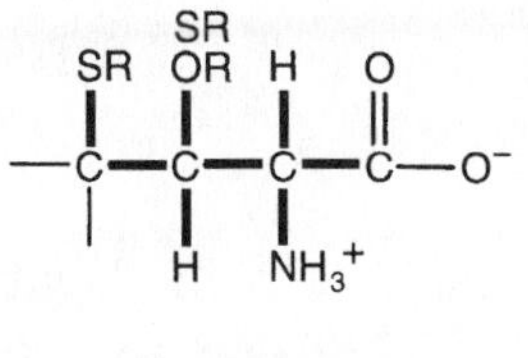

FIGURE 7.13 The bonds (bold type) cleaved by enzymes requiring pyridoxal-5'-phosphate including α-amino acids in transamination, α-decarboxylation and α,β-eliminations.

of tyrosine (tyrosine aminotransferase). The amount of these enzymes is modulated by corticosteroids, glucagon substrate concentration.

2. Decarboxylation in the synthetic pathway of neuroactive amines serotonin, tyramine, histamine and γ-aminobutyric acid. Other decarboxylation reactions include δ-aminolevulinic acid and intermediates in the synthesis of lecithin and taurine. A further important decarboxylation pathway and removal of the sulphur occurs with cysteine.

3. Oxidative removal of the amino group from serine and threonine as ammonia to form α-keto acids.

4. Formation of glycine and formate from serine, removal of alanine from kynurenine and 3-hydroxy-kynurenine and the splitting of cystathione in the degradation of methionine.

7.9.3 Availability

The vitamin is absorbed in the free form and subsequently phosphorylated for use in enzymes. It is excreted in the urine largely as 4-pyridoxic acid.

7.9.4 Pyridoxine deficiency and excess

Primary dietary deficiency has not been reported in adults, largely because of the widespread distribution of the vitamin in foods. Nevertheless, a deficiency may occur in alcoholics and in patients taking certain drugs, e.g. *p*-aminosalicylic acid. This deficiency condition may result in increased urinary excretion of urea, xanthurenic acid, kynurenine and hydroxyurenine and oxalic acid. The hyperoxaluria may lead to urinary stone formation. In infants there may be hyperirritability, convulsions and anaemia.

The anti-tubercular drug isoniazid (hydrazide) inactivates pyridoxal phosphate by forming a hydrazone and results in a peripheral neuropathy. Pyridoxine is given with isoniazid to prevent such problems.

Few toxic effects have been observed, except in women taking pyridoxine supplements for premenstrual tension.

7.9.5 Recommended requirements

Cereals, meat, fruits, leafy and other vegetables contain vitamin B_6 in the range of 0.1–0.3 mg/100 g, whereas liver contains 0.5 mg/100 g.

The total body pool of vitamin B_6 is of the order of 40–250 mg, with a half-life of 33 days. This would imply a daily intake of 0.6–3.78 mg. Deficiency develops more rapidly on high protein intakes the order of 80–160 g/day, than on protein intakes of 30–50 g/day.

For adults, male and female, the RNI is 15 μg/g dietary protein. The LRNI and EAR are 11 and 13 μg/g protein, respectively. EARs for infants vary from 6 μg/g protein at less than 3 months up to 13 μg/g protein at 7–10 years. There is no evidence of an increased requirement during pregnancy and in the elderly.

7.9.6 Body store requirements

Biochemical markers include plasma pyridoxal phosphate concentrations, red cell transaminase activation and urinary excretion of B_6 degradation products. There is no single marker which is sensitive at all levels of dietary intake. The plasma pyridoxal phosphate concentration and the erythrocyte transaminase activation coefficient all have shortcomings. Two metabolic loading tests are being used to measure vitamin B_6 status. In the tryptophan load test the potentially rate-limiting enzyme kynureninase is sensitive to vitamin B_6 depletion. The excretion of kynurenic and xanthurenic acid is measured before and after the loading dose of tryptophan. Another test is the methionine load test.

Summary

1. Vitamin B_6 is found in several forms; pyridoxal, pyridoxamine and pyridoxine, as the free form in plants and the phosphorylated form, pyridoxamine phosphate, in animal tissues.
2. Pyridoxal-5'-phosphate is a coenzyme with a major role in the intermediary metabolism of amino acids, in α-decarboxylation, aldolization, transamination reactions, and the β-carboxylation of aspartic acid.
3. The aldehyde group of pyridoxal-5'-phosphate forms a Schiff base with the α-amino acid group of amino acids. Pyridoxal-5'-phosphate catalyses several different types of bond cleavage by stabilizing the electron pairs at the α- or β-carbon atoms of α-amino acids.
4. Dietary pyridoxine deficiency has not been demonstrated in humans.

7.10 Carnitine

7.10.1 Introduction

L-Carnitine [(β-hydroxy-)-γ-N-trimethylaminobutyrate], molecular weight 161 Da, (Figure 7.14) is a natural constituent of the flesh of higher animals.

Acylcarnitine is found in a variety of food sources, but primarily those of animal origin. Red meat and dairy products are particularly rich sources.

Human milk contains 28–95 nmol/ml of carnitine. Milk-based formulas also contain carnitine, though artificial formula milks manufactured

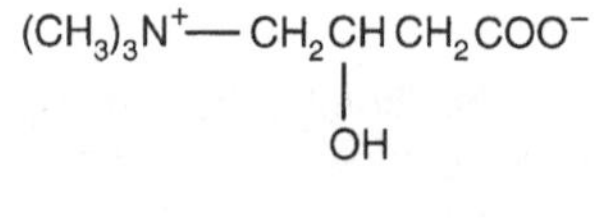

$$(CH_3)_3N^+ \!\!-\!\! CH_2CHCH_2COO^-$$
$$|$$
$$OH$$

FIGURE 7.14 Structure of carnitine.

from soya bean protein or casein contain little or no carnitine, with consequently reduced plasma carnitine concentration.

7.10.2 Availability

Carnitine is not an essential nutrient in the diet of adult humans, as their tissues are able to synthesize the amino acid. It is not known whether humans can synthesize sufficient however to meet requirements. Individuals consuming cereal-based diets, low in carnitine, can maintain similar plasma carnitine concentrations to those of populations where dietary carnitine is readily available.

Carnitine is derived from lysine which is a limiting amino acid in diets. The precursors of carnitine are lysine and methionine. *S*-Adenosylmethionine provides the methyl groups for enzymatic trimethylation of peptide-linked lysine (Figure 7.15). Proteins which contain ϵ-*N*-trimethyllysine residues include histones, cytochrome C, myosin and calmodulin. An important enzyme in the synthesis of lysine to carnitine is γ-butyrobetaine hydroxylase. The activity of this enzyme is dependent on age. It is only by the age of 15 years that the hepatic enzyme activity is within the range of adult values. This enzyme is present in human liver, kidney and brain, but not in skeletal muscle or heart.

Skeletal muscle contains over 90% of total body carnitine in humans. Normal carnitine muscle levels range from 11–52 nmol/mg noncollagen protein, the concentration being approximately 70-fold greater than in plasma. The turnover time for carnitine in skeletal muscle and heart is approximately 8 days, and for other tissues, primarily liver and kidney, 12 hours, in extracellular fluid 1 hour, and for the whole body 66 days. Carnitine is conserved in the human by renal reabsorption.

Glucose is a major metabolic fuel for the foetus. At birth the infant adapts to lipid as a major source of calories when there is a rapid increase in blood free fatty acids and β-hydroxybutyrate concentrations due to release of free fatty acids from adipose tissue. Fatty acids become the preferred

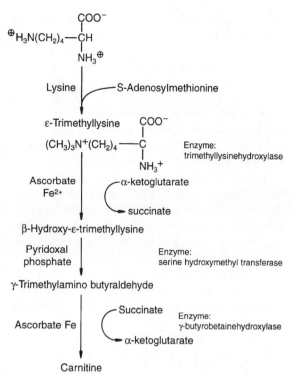

FIGURE 7.15 Carnitine synthesis from lysine. The reactions involved require ascorbate, iron, pyridoxal phosphate and *S*-adenosylmethionine.

fuel for heart and skeletal muscle and so carnitine becomes an important cofactor for energy production in the neonate.

Relatively reduced hepatic γ-butyrobetaine hydroxylase activity may limit the rate of carnitine biosynthesis in infants. It has been speculated whether low plasma and tissue concentrations of carnitine influence lipid utilization by the infant, for example during parenteral nutrition and the use of infused triglycerides. Experimental studies do not show a relationship between carnitine levels and the ability to handle triglycerides by the infant.

7.10.3 Action of carnitine

Carnitine plays a role in the transport of long-chain fatty acids into the mitochondrial matrix.

Fatty acid acyl-CoAs are unable to cross the inner membranes of mitochondria, but after they are converted into the acylcarnitine derivative the membrane can be crossed (Figure 7.16). The enzymes involved are carnitine acyltransferases, each with a specificity according to fatty acid chain length. Within the mitochondria the reaction is reversed by carnitine acyltransferase II with the restoration of the fatty acid CoA. There are a number of mitochondrial reactions resulting in CoA-esters of short- and medium-chain organic acids. These esters may be further metabolized to regenerate free coenzyme A. During stress, when there is an excess production of these esters, the organic acid may be trans-esterified to carnitine, releasing reduced coenzyme A for involvement in other mitochondrial pathways, e.g. the tricarboxylic acid cycle. Carnitine is a mitochondrial buffer for excess organic acids, though this role is probably minor. During abnormal conditions this role may become important.

Renal excretion of carnitine esters may be a mechanism for removing excess short- or medium-chain organic acids. Hyperthyroidism markedly increases urinary carnitine excretion, whereas hypothyroidism decreases urinary loss of carnitine. Prolonged fasting in normal subjects decreases renal excretion of free carnitine, but there is increased excretion of acylcarnitine esters.

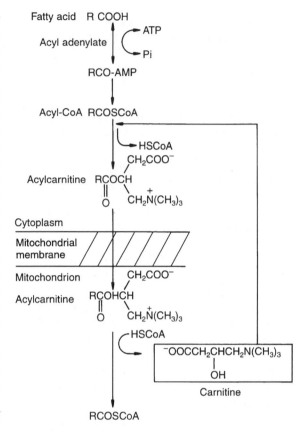

FIGURE 7.16 Coenzyme A and its derivatives are unable to pass across the mitochondrial inner membrane. Acyl groups bonded to coenzyme A cross the inner membrane combined with carnitine. Acyl carnitine is exchanged at either side for acetyl-CoA.

7.10.4 Carnitine deficiency

Carnitine deficiency syndromes in humans result from inborn errors of metabolism. Clinical symptoms of carnitine deficiency in infants are rarely seen.

Primary genetic carnitine deficiency

There are two clinical types: systemic and myopathic. Primary muscle carnitine deficiency is characterized by mild to severe muscle weakness and various excesses of lipids in skeletal muscle fibres due to defective transport of carnitine into muscle. Other symptoms are: metabolic encephalopathy, hypoglycaemia, hypoprothrombinaemia, hyperammonaemia and lipid excess in liver cells. This resembles Reye's syndrome.

Carnitine deficiency can also be associated with organic aciduria, for example in long- and medium-chain acyl-CoA dehydrogenase deficiency, isovaleric acidaemia, glutaric aciduria, propionic and methylmalonic acidaemia and short-chain acyl-CoA dehydrogenase deficiency.

Carnitine deficiency has been reported in ageing, diabetes and chronic heart failure.

Summary

1. L-Carnitine [(β-hydroxy-)-γ-N-trimethylaminobutyrate] is a natural constituent of higher animals.
2. Carnitine is derived from lysine which is a limiting amino acid in diets. The precursors of carnitine are lysine and methionine. S-Adenosylmethionine provides the methyl groups for enzymatic trimethylation of peptide-linked lysine.
3. Carnitine plays a role in the transport of long-chain fatty acids into the mitochondrial matrix. Fatty acid acyl-CoAs only cross the inner membranes of mitochondria, after they are converted into the acyl carnitine derivative. The enzymes involved are carnitine acyltransferases.
4. Carnitine deficiency syndromes in humans result from inborn errors of metabolism. Clinical symptoms of dietary carnitine deficiency in infants are rarely seen.

Further reading

Rebouch, C.J. and Paulson, D.J. (1986) Carnitine metabolism and function in humans. *Annual Review of Nutrition,* 6, 41–66.

7.11 Inositol

7.11.1 Introduction

Myo-inositol (Figure 7.17), molecular weight 180 Da, is found in the diet in the free form, as inositol containing phospholipid and as phytic acid (inositol hexaphosphate). The cyclitols include the inositols, of which there are nine possible isomers of hexahydroxycyclohexane.

Inositol (myo-inositol) is found throughout the animal and plant kingdom. In a mixed North American diet an adult eats approximately 1 g of inositol each day from plant foodstuffs in the form of phytic acid. Inositol hexaphosphate found in cereals is an important contributor to the total dietary phosphorus intake. Inositol is also present in animal products such as fish, poultry, meat and

dairy products in the form of free and inositol-containing phospholipids (phosphatidylinositol) (Figure 7.18). In cows' milk the myo-inositol concentration is 170–440 μmol/l.

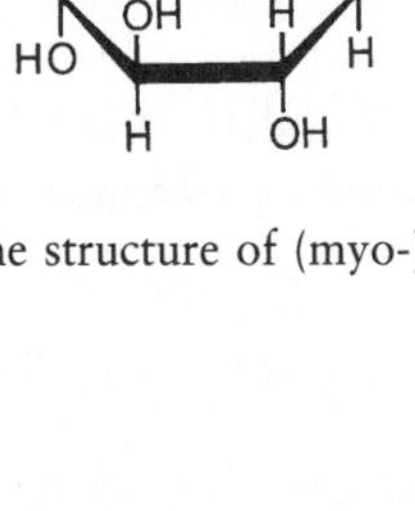

7.11.2 Action of inositol

Glucose 6-phosphate can be converted into inositol 1-phosphate (enzyme inositol 1-phosphate synthase) followed by dephosphorylation (enzyme inositol 1-phosphatase). Inositol can be incorporated into phosphoinositides. The *de novo* synthesis involves the reaction of inositol with the liponucleotide, CDP-diacylglycerol (enzyme CDP-diacylglycerol : inositol phosphatidyltransferase), a microsomal enzyme. Free inositol can react with endogenous phospholipid to become phosphoinositide by a microsomal manganese-stimulated exchange reaction.

The fatty acid composition and distribution in inositols is unusual in being rich in stearic and arachidonic acids. Stearate is found in the sn-1-position and arachidonate in the sn-2-position of the sn-glycerol backbone of the phosphoinosi-

FIGURE 7.17 The structure of (myo-) inositol.

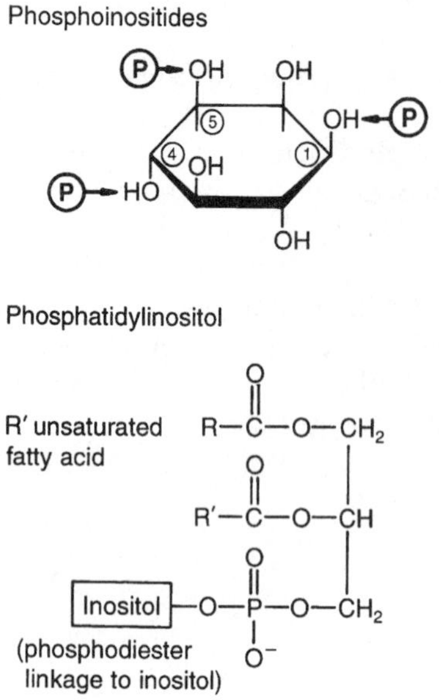

Phosphoinositides

Phosphatidylinositol

FIGURE 7.18   The structure of phosphoinositides and 1,4,5-phosphatidylinositol.

tides. Phosphatidylinositol synthesis is found in the endoplasmic reticulum of cells. Phosphatidyl-inositol exchange proteins are found in the cyto-plasm of several tissues.

Myo-inositol is an essential growth factor for many cells and is important in promoting the growth of the young. This requirement depends upon the diet composition. Inositol acts as a lipotropic factor (see section 7.12). The type of dietary triglyceride may influence the function of dietary inositol as a lipotrope which is not simply related to the degree of saturation of the fat. Membrane phosphatidylinositol can regulate enzyme activity and transport processes, while providing a source of free arachidonic acid for the synthesis of eicosanoids.

Myo-inositol functions in the cell in membrane phosphoinositides by stimulating the release of the second messengers 1,2-diacylglycerol and ino-sitol triphosphate in stimulated cells.

Phytic acid (Figure 7.19) is found in plant foodstuffs and is hydrolysed in the gut by the enzyme phytase, an enteric enzyme. This releases free inositol, orthophosphate and intermediate products including the mono-, di-, tri-, tetra- and

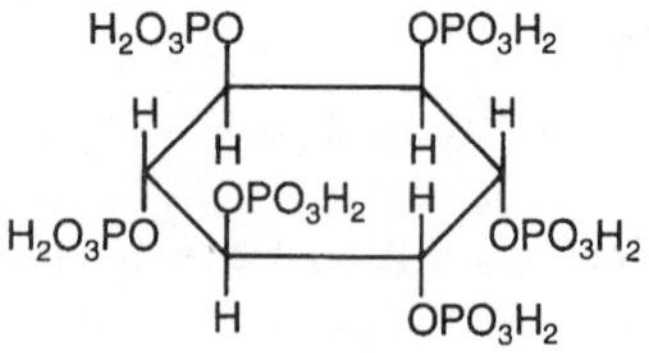

FIGURE 7.19   Phytic acid (inositol hexaphosphate).

pentaphosphate esters of inositol. Dietary phytic acid can reduce the bioavailability and utilization of both calcium and zinc. Phytic acid may bind these metals and prevent their enteric absorption. Calcium phytate is excreted in the faeces with the loss of both phytate and calcium. Inositol crosses the small intestinal brush border membrane by active transport against a concentration gradient which is Na^+- and energy-dependent and inde-pendent of the D-glucose pathway. It is possible that dietary polyphosphoinositides are hydrolysed by a pancreatic phospholipase A in the intestinal lumen. The product can be reacylated through acyltransferase activity upon entering the intesti-nal cell or further hydrolysed with the release of glycerylphosphorylinositol.

7.11.3 Availability

The plasma concentration of free inositol in nor-mal human subjects is approximately 30 μmol/litre. The human reproductive tract is particularly rich in free inositol. Unbound inositol concentra-tions in the brain, cerebrospinal fluid and choroid plexus are also higher than in plasma. The con-centration of inositol is about 0.6 mmol/litre at 3–7 months' lactation in human breast milk. There is also a disaccharide form of inositol 6-β-galactinol in milk, approximately 17% of the total non-lipid inositol at the 18th day of lactation.

In tissues and cells inositol occurs in the free form, bound covalently to phospholipid as phos-phatidylinositol. Lower concentrations of the polyphosphoinositides (phosphatidyl 4-phosphate and phosphatidyl 4,5-biphosphate) are to be found.

Summary

1. Myo-inositol is found in the diet in the free form, as inositol-containing phospholipid and as phytic acid (inositol hexaphosphate). The cyclitols include the inositols of which there are nine possible isomers of hexahydroxycyclohexane.
2. Myo-inositol functions in the cell in membrane phosphoinositides by stimulating the release of the second messengers 1,2-diacylglycerol and inositol triphosphate in stimulated cells.

Further reading

Holub, B.J. (1986) Metabolism and function of myo-inositol and inositol phospholipids. *Annual Review of Nutrition*, 6, 563–97.

7.12 Lipotropes

7.12.1 Introduction

The lipotropes include choline, methionine, vitamin B_{12} and folic acid. Choline and folic acid are found in both animal and plant foods. Animal products and microorganisms are the sole dietary sources of methionine and vitamin B_{12}.

7.12.2 Action of lipotropes

These are a group of biologically active compounds which have a major role in cellular metabolism. They are essential to the synthesis and methylation of DNA, the metabolism of lipids and the maintenance of tissue integrity. The lipotropes interact with each other and with other nutrients (Figures 7.20 and 7.21). The role of folate, B_{12} and methionine are important in the transfer of one-carbon units (Figures 7.22–7.24). Methyl groups are involved in:

- purine ring formation
- pyrimidine biosynthesis
- amino acid interconversions
- formate metabolism

Vitamin B_{12} and folate are essential for growth and proliferation of mammalian cells. Rapid availability of nucleotide precursors is important to the lymphatic system, which depends on proliferation and cell division in response to a foreign stimulus. Folate, and B_{12} deficiency, lead to a defect in thymidylate synthesis and hence DNA synthesis. Vitamin B_{12} is involved in the isomerization of methylmalonate to succinate which is a link between carbohydrate and lipid metabolism.

The methylation of homocysteine to methionine is the metabolic link between B_{12}, folate and methionine metabolism. The enzyme transmethylase, which requires vitamin B_{12}, is required for the conversion of homocysteine to methionine. By this process tetrahydrofolate (THF) is regenerated. Since methionine is available from other sources, i.e. the diet and amino acid pool, the rate-limiting step is a regeneration of THF. Folic coenzymes carrying single-carbon units in different states of reduction are involved in reactions in which methyl groups are transferred. Cellular DNA synthesis depends on the availability of the four nucleotide precursors. While thymidylate can be formed directly from thymidine most cells convert deoxyuridine monophosphate to thymidine monophosphate utilizing the enzyme thymidylate synthetase. This is the rate-limiting step in DNA synthesis and requires 5,10-methylene-THF (Figure 7.25) as a cofactor. The cofactor is reduced to dihydrofolate, which can be further reduced through dihydrofolate reductase to THF. On the other hand, THF may be regenerated through the B_{12}-dependent methyl transferase reaction in which the methyl group is transferred from 5-methyl-THF in the synthesis of methionine. The folate-derived methyl groups go through successive stages of reduction and in so doing are converted to 5,10-methylene-THF, a cofactor for thymidylate synthetase (Figure 7.26) or may be further irreversibly reduced to 5-methyl-THF. This folate coenzyme must be converted to THF for the methyl group to re-enter the methyl pool. In vitamin B_{12} deficiency this cannot take place and 5-methyl-THF accumulates. Patients with vitamin B_{12} deficiency show an increase in excretion of formiminoglutamic, for-

mate and 4(5)-amino-5(4)imidazole-carboxamide. These require folate cofactors for further conversion and are restored by dietary methionine. Methionine is an essential amino acid which also serves as a methyl donor.

There is a liver enzyme for the direct methylation of homocysteine through betaine, but most cells use the vitamin B_{12}, 5-methyl-THF-dependent methyl transferase reaction to synthesize methionine. This can then be further converted to S-adenosylmethionine which is an inhibitor of the 5-methyl-THF : homocysteine transmethylase reaction and of 5,10-methylene-THF reductase which reduces the amount of

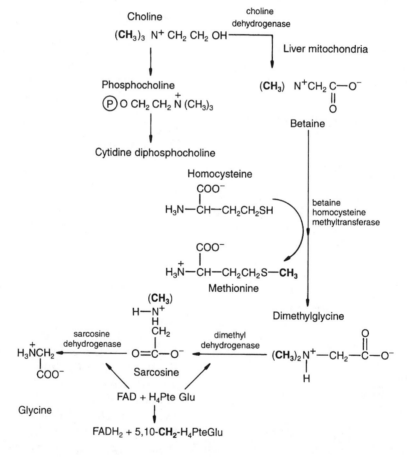

FIGURE 7.20   Metabolism of dietary choline. Synthesis to cytidine diphosphocholine or, alternatively, to betaine, dimethyl glycine, sarcosine and glycine. Choline acts as a methyl donor in the biosynthesis of methionine and in the generation of 5, 10-CH_2-H_4-Pte Glu.

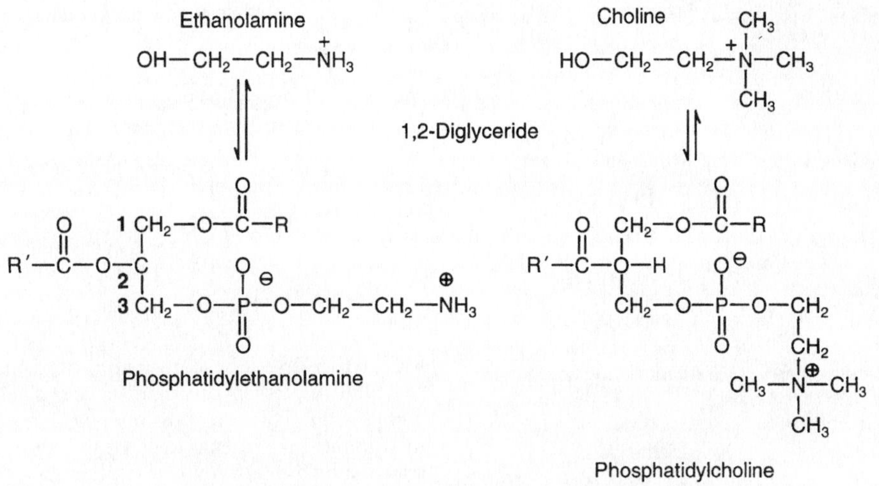

FIGURE 7.21   Ethanolamine and choline are synthesized into phospholipids, namely phosphatidylethanolamine and phosphatidylcholine. R = fatty acid; R' = usually unsaturated fatty acid.

5-methyl-THF formed and increases the availability of the folate cofactors. Methionine is also involved in the conversion of formate into carbon dioxide.

7.12.3 Availability

The lipotropic nutrients methionine, choline, folic acid and vitamin B_{12} are stored in the body and

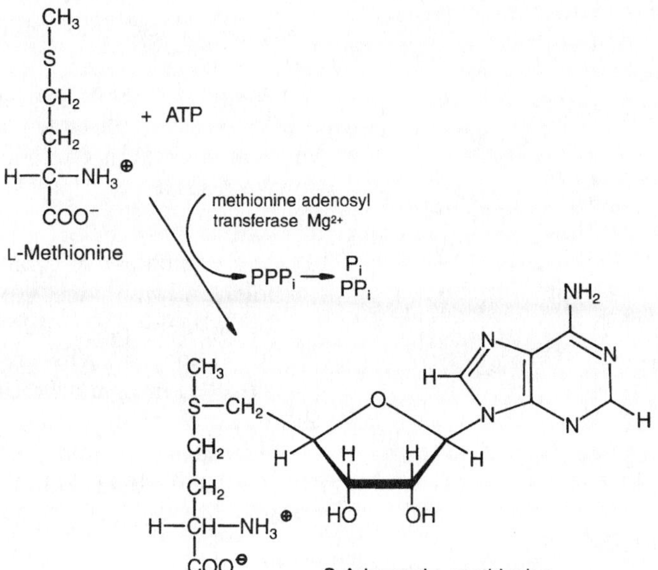

FIGURE 7.22   Methionine is a methyl donor (e.g. to amines or alcohols) in the form of S-adenosyl-L-methionine, a reaction catalysed by methionine adenosyl transferase. S-adenosyl-L-methionine is the form in which methionine is activated for methylation reactions.

213

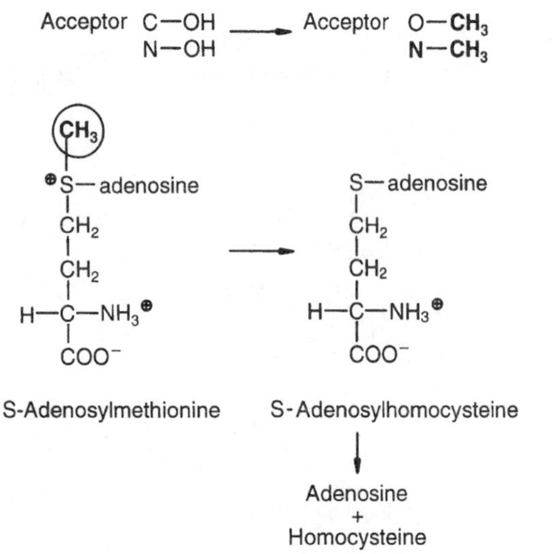

Acceptor C—OH $\longrightarrow$ Acceptor O—CH$_3$
         N—OH   N—CH$_3$

S-Adenosylmethionine $\longrightarrow$ S-Adenosylhomocysteine

$\downarrow$

Adenosine
+
Homocysteine

FIGURE 7.23 *S*-adenosylmethione is a contributor of CH$_3$ to acceptor groups containing N or O. The *S*-adenosylmethionine loses the CH$_3$ group to form *S*-adenosylhomocysteine and adenosine and homocysteine.

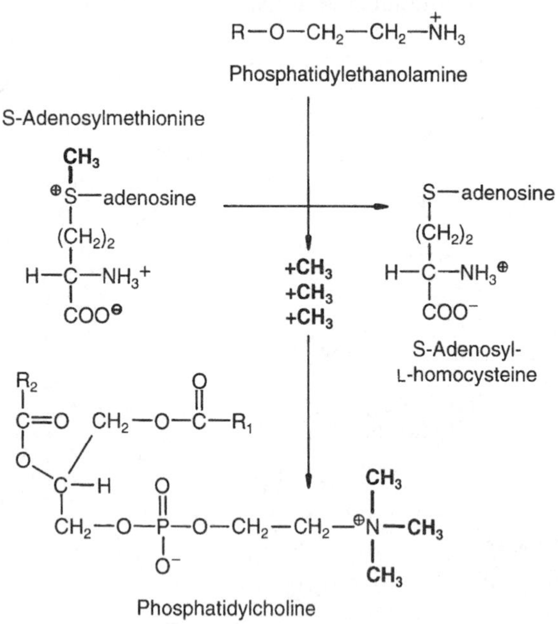

R—O—CH$_2$—CH$_2$—$\overset{+}{N}$H$_3$
Phosphatidylethanolamine

S-Adenosylmethionine

+CH$_3$
+CH$_3$
+CH$_3$

S-Adenosyl-
L-homocysteine

Phosphatidylcholine

FIGURE 7.24 Phosphatidylcholine is a ready source of methylating groups and is synthesized from phosphatidylethanolamine with three CH$_3$ groups (donated by *S*-adenosylmethionine) available for transfer.

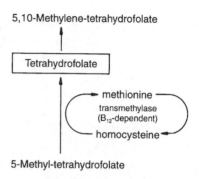

5,10-Methylene-tetrahydrofolate

Tetrahydrofolate

methionine
transmethylase
(B$_{12}$-dependent)
homocysteine

5-Methyl-tetrahydrofolate

FIGURE 7.25 The conversion of 5-methyl-tetrahydrofolate to tetrahydrofolate is a critical step in 1-C metabolism. This is a transmethylase reaction in which methionine is regenerated from homocysteine; the transmethylase enzyme is vitamin B$_{12}$-dependent. In this reaction methionine, folic acid, vitamin B$_{12}$ and (indirectly) choline inter-react.

turn over at different and specific rates. Vitamin B$_{12}$ has the largest store relative to its daily requirements. Methionine is stored in tissue proteins which are constantly turning over, and as a t-RNA derivative and in an active form, *S*-adenosylmethionine is essential to transmethylation. Folic acid and choline stores are quite small.

Choline is stored as choline phospholipids which is widely distributed in tissues. Phospholipids are important components of the membrane in two respects:

1. As an amphipathic compound which has both polar and non-polar groupings, which makes for a membrane with varied charges. The membranes of cells are where enzymes are active, as in lipid synthesis.
2. A source of important precursors of biologically active substance, e.g. fatty acids as with arachidonic acid and eicosanoids and choline and acetylcholine (Figure 7.27).

Phosphatidylcholine and phosphatidylethanolamine are important phospholipids. Their biosynthesis begins with dietary choline entering the cell. Choline is phosphorylated to phosphocholine by

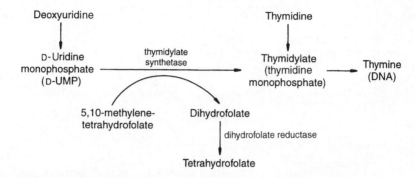

FIGURE 7.26   5,10-Methylene-tetrahydrofolate is important in the methylation of D-uridine monophosphate (D-UMP) to thymidine monophosphate, which is a rate-limiting step in DNA synthesis.

choline kinase. Phosphocholine and cytidine-5'-phosphate (CMP) form CDP-choline (enzyme CTP : phosphocholine cytidyltransferase) a rate-limiting enzyme. This enzyme is inactive in the cytosol and activated on the endoplasmic reticulum by phospholipids. The CDP-choline reacts with diacylglycerol, the enzyme involved being attached to the endoplasmic reticulum. Phosphotidylethanolamine is synthesized from ethanolamine in a similar series of reactions to phosphatidylcholine. In the lung the phosphatidylcholine is dipalmitoylphosphatidylcholine (palmitic acid in sn-1 and sn-2 positions) and acts as the lung surfactant.

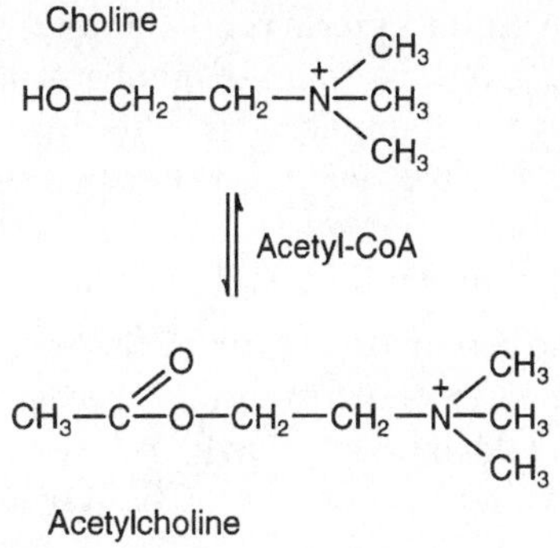

Choline

$$HO—CH_2—CH_2—\overset{+}{N}\Big\langle\begin{matrix}CH_3\\CH_3\\CH_3\end{matrix}$$

Acetyl-CoA

$$CH_3—\overset{O}{\overset{\|}{C}}—O—CH_2—CH_2—\overset{+}{N}\Big\langle\begin{matrix}CH_3\\CH_3\\CH_3\end{matrix}$$

Acetylcholine

FIGURE 7.27   Choline is a contributor to the synthesis of acetylcholine, the neurotransmitter.

Summary

1. The lipotropes include choline, methionine, vitamin B_{12} and folic acid.
2. These are a group of biologically active compounds with a major role in cellular metabolism. They are essential to the synthesis and methylation of DNA, the metabolism of lipids and the maintenance of tissue integrity. The lipotropes interact with each other and with other nutrients. The role of folate, vitamin B_{12} and methionine are important in the transfer of one-carbon units as methyl groups
3. The methylation of homocysteine to methionine is the metabolic link between vitamin B_{12}, folate and methionine metabolism. The enzyme transmethylase containing vitamin B_{12} is required for the conversion of homocysteine to methionine. By this process tetrahydrofolate (THF) is regenerated.

Further reading

Newberne, P.M. and Rogers, A.E. (1986) Labile methyl groups and promotion of cancer. *Annual Review of Nutrition,* **6,** 407–32.

7.13 Folic acid

7.13.1 Introduction

Folates, which are derivatives of folic acid, are found in many forms throughout nature. Folic acid, molecular weight 441 Da, (pteroylglutamic acid) consists of a pterin ring (2-amino, 4-hydroxy-pteridine) attached to a *p*-aminobenzoic acid conjugated to L-glutamic acid (PteGlu). The compound is stable in acid solutions, sparingly soluble in water but unstable in neutral or alkaline media. The variations on folic acid include (Figure 7.28):

- di- and tetrahydro forms of the pteridine ring
- a single-carbon substitution (methyl $-CH_3$, formyl $-CHOH$, methenyl $=CH-$, methylene $=CH_2$ or formimino $-CHNH$) at N-5 or N-10
- a chain of glutamates attached to the L-glutamate (in humans the number varies between four and six)

In human tissues and fluids, folic acid is found principally as monoglutamate derivatives, the majority as 5-methyltetrahydrofolate (methyl-THF) and some as 10-formyl tetrahydrofolate (10-THF).

Important sources of folate include liver, yeast extract and green leafy vegetables. Folic acid is found in diets which contain other B vitamins.

7.13.2 Action of folic acid

Folic acid undergoes a number of metabolic changes which involve the transfer of 1-carbon groups on the N-5 or N-10 position to other compounds (Figure 7.29). Of these, the most important folic acid (tetrahydrofolate, THF) 1-carbon derivatives are 5-methyl-THF, 5-10-methylene-THF, 5-10-methenyl-THF, 5-formyl-THF and 10-formyl-THF, and 5-formimino-THF.

The folic polyglutamates may be the active coenzyme within the cell. In some reactions concerned with purine and pyrimidine synthesis some folate is oxidized to the dihydro form. The enzyme dihydrofolate reductase reduces dihydrofolate to the tetrahydo form. There is, in the role of folic acid as a coenzyme, some irreversible splitting of the C-9–10 bond. This is increased during increased DNA turnover, when there are extra requirements for dietary folic acid. The reactions which folic acid is involved in include amino acid interconversions and DNA synthesis.

Amino acid interconversions

- Serine into glycine: this requires the transfer of a formaldehyde group from serine to the tetra-folate cofactor (enzyme serine hydroxymethyltransferase). This reaction is important particularly in the dividing cell.
- Homocysteine to methionine: (enzyme 5-methyl-THF methyltransferase); this reaction also involves vitamin B_{12}.

DNA synthesis

- Purine nucleotide synthesis (5,10-formyl-THF and 10-formyl-THF), e.g. glycinamide ribotide to formyl glycinamide ribotide (enzyme glycinamide ribotide transformylase) (Figure 7.31).
- Pyrimidine nucleotide synthesis (5,10-methylene-THF as coenzyme), e.g. deoxyuridine monophosphate into thymidine monophosphate (enzyme thymidylate synthetase).
- Methylation of transfer RNA.

The folates provide methyl groups for many methyltransferase reactions. This is achieved

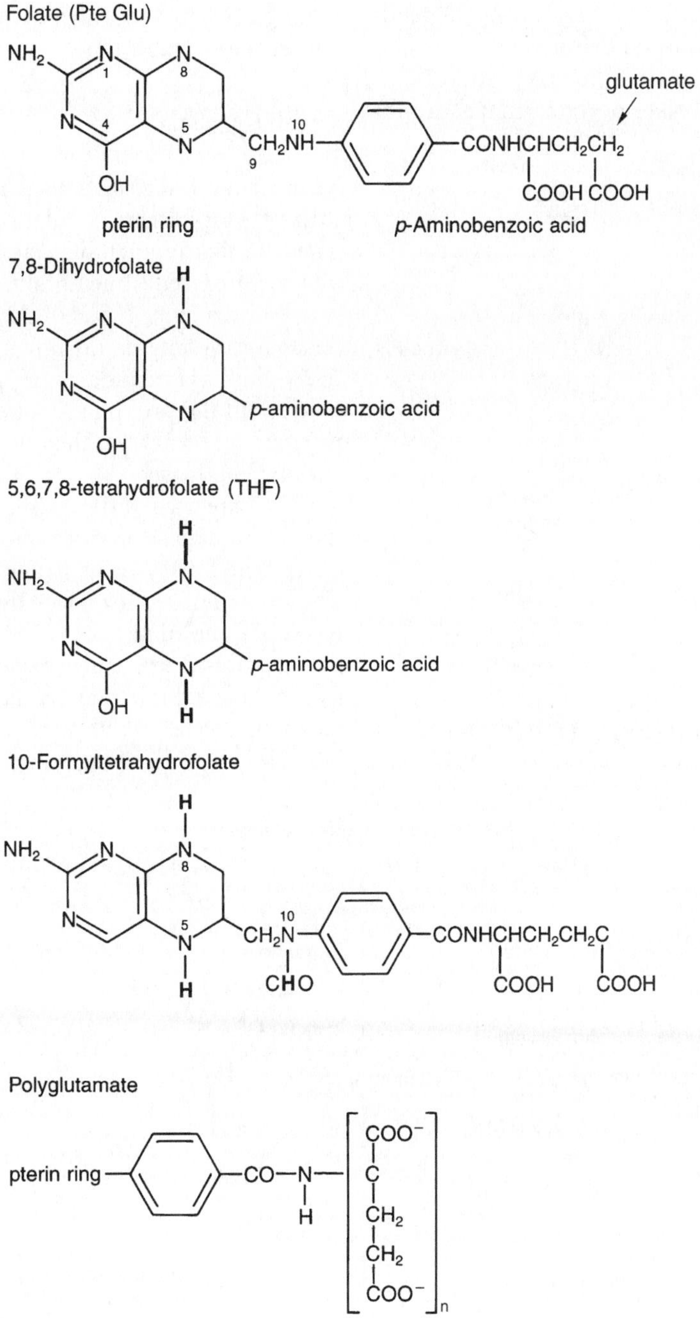

Folate (Pte Glu)

7,8-Dihydrofolate

5,6,7,8-tetrahydrofolate (THF)

10-Formyltetrahydrofolate

Polyglutamate

FIGURE 7.28   Structure of folate: 7,8-dihydrofolate; 5,6,7,8-tetrahydrofolate (THF); 10-formyltetrahydrofolate; polyglutamate.

by the conversion of 5,10-methylene-THF to 5-methyl-THF, which in turn methylates homocysteine to methionine by the vitamin B_{12}-dependent enzyme, methionine synthase.

Folate polyglutamates are the main intracellular forms of folic acid. Folate polyglutamates do not cross or are poorly transported across cell membranes. Therefore, the metabolism of pterolymonoglutamates to polyglutamate allows cells to concentrate folates.

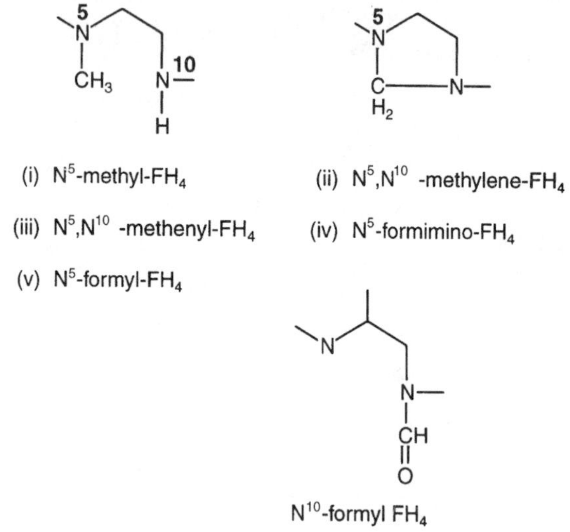

7.13.3 Availability

Most of the folate in food is in the polyglutamyl form. Dietary forms of folate are typically conjugated, so that luminal digestion to the monoglutamyl form precedes membrane uptake, which is a rate-limiting step (Figure 7.32). Gastric and duodenal juice contents do not appear to hydrolyse the γ-glutamyl peptide chain. Absorption is most efficient in the proximal duodenum. The enzymes in the intestinal epithelium are γ-L-glutamyl carboxypeptidases.

There are two distinct folate conjugase activities in humans, one in the brush border and the other which is lysosomal and is within the cell. The brush border enzyme – the most important – has a pH optimum of 6.5–7.0, is an exopeptidase and requires Zn^{2+}. This enzyme may be inhibited by alcohol and this may be important in the folic acid deficiencies which may accrue from alcohol intemperance. The drug salazopyrine, used in the

(i) N^5-methyl-FH$_4$

(ii) N^5,N^{10}-methylene-FH$_4$

(iii) N^5,N^{10}-methenyl-FH$_4$

(iv) N^5-formimino-FH$_4$

(v) N^5-formyl-FH$_4$

N^{10}-formyl FH$_4$

FIGURE 7.29 Tetrahydrofolate (THF; FH4) is an important source of 1-carbon units; N-5 or N-10.

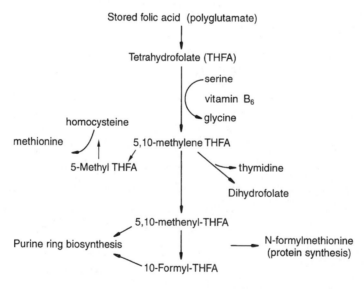

FIGURE 7.30 Stored folic acid is available as tetrahydrofolate in reactions including homocysteine, thymidine, purine ring biosynthesis and protein synthesis.

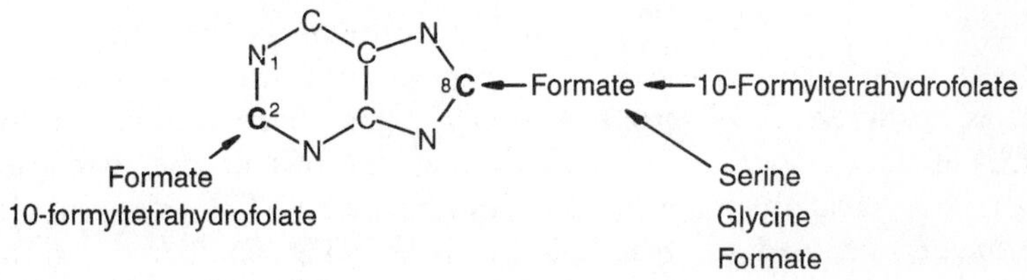

FIGURE 7.31 Tetrahydropteroyl glutamate (10-Formyl-H_4 Pte Glu) supplies the carbon-2 and carbon-8 in purine synthesis.

treatment of ulcerative colitis, may have a similar effect. The intracellular hydrolase has a pH optimum of 4.5, is an endopeptidase and has no metal requirements. Intestinal absorption of the polyglutamate is very efficient.

Monoglutamyl folate is primarily absorbed by a carrier-mediated system, though there may be an element of passive absorption. Folate uptake can occur by saturable and non-saturable mechanisms. Membrane vesicle studies have shown that pH affects both saturable and non-saturable components, probably by anion (folate/hydroxyl ion) exchange. A mildly acid environment in the intestinal lumen facilitates absorption. Subsequent intracellular disposition of absorbed folate includes reduction and methylation and formylation. This may influence the rate of release of absorbed folic acid. Folate bound to a high-affinity folate binding protein in milk is absorbed from the ileum. This may be important in the suckling animal.

Folic acid is stored in the liver at a concentration of approximately 5–15 µg/g.

7.13.4 Deficiency

Folic acid deficiency may arise for a number of reasons:

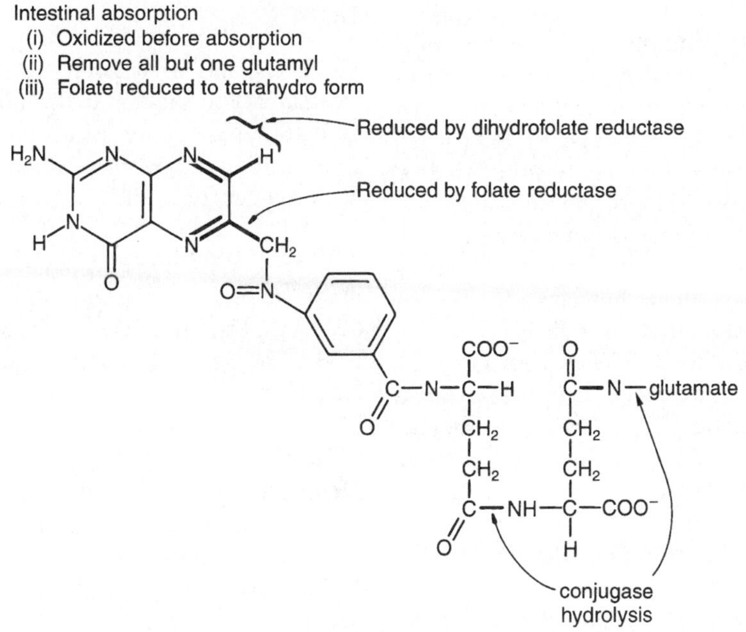

FIGURE 7.32 Dietary folate and the bonds susceptible to enzymatic cleavage

- as a dietary defect
- from malabsorption, as in coeliac disease
- excess demands, as in increased cell proliferation, e.g. as in leukaemia
- interference with folic acid metabolism by drugs, e.g. anticonvulsants used to treat epilepsy
- in inborn errors of folic acid metabolism (though these are extremely rare)

Folic acid deficiency is an important cause of megaloblastic anaemia. The functional relationship between folic acid and vitamin B_{12} is in the tetrahydropteroylglutamate methyltransferase reaction determining intracellular folate. A cellular deficiency develops and the synthesis of purines and pyrimidines is reduced. Red cell formation is also reduced in the bone marrow and a megaloblastic anaemia results (see Lipotropes, section 7.12).

Pregnancy

Neural tube defects (NTDs) is a collective term for congenital deformities of the spinal cord and brain. NTDs include spina bifida (50% of cases), anencephalus (40%), encephalocoele and iniencephaly. NTDs have a prevalence rate which varies for different populations. Folic acid has long been suggested to be a cause of NTDs. It is believed that a genetic predisposition may be influenced by environmental factors, the prevalence rising with increasing parity. The range of prevalence among populations ranges from 1 to 6 per 1000, and is increased 10-fold if there has already been a pregnancy and a baby affected with NTDs.

Epidemics of increased NTDs-affected births have been recorded. Inadequate nutrition, infectious diseases, climatic factors, seasonal usage of chemical fertilizers and pesticides are possible aetiological factors. Seasonal changes in dietary folic acid availability may alter the aetiological significance of folate in NTDs throughout the year. Excessive intake of vitamin A or zinc have frequently been suggested as possible causes of NTDs. Most studies have also considered vitamin B_{12}, vitamin C and zinc dietary status.

Closure of the neural tube occurs early in pregnancy, before the first antenatal visit. No difference has been noted in blood folic acid concentrations at the first antenatal visit from mothers whose babies are subsequently diagnosed to have NTDs compared with mothers of normal babies. Pregnancy, however, places a demand on folate and may deplete reserves, though this is not reflected in low blood levels early in pregnancy. However, red cell folate concentrations are more significant, those in affected mothers being reduced. A congenital disorder of folate metabolism has been suggested.

In intervention studies mothers with a previously affected baby were given an iron and multivitamin preparation with 360 μg of folic acid. The result was impressive, the NTDs rate being 0.6% compared with 5% in an unsupplemented control group. Other larger studies have demonstrated a protective effect of folic acid containing multivitamins during the first 6 weeks of pregnancy.

7.13.5 Recommended requirements

Adults

The LRNI has been estimated to be 100 μg/day. Median folate intakes in Britain are 300 (range 145–562) μg/day for men, and 209 (range 95–385) μg/day for women. A RNI of 200 μg/day has been set for adults.

Infants and children

Breast milk provides 40 μg/day; a formula milk 60–70 μg/litre (equivalent to 50–60 μg/day). A RNI of 50 μg/day has been suggested for formula-fed infants.

Pregnancy

The mean intake of additional folic acid required to maintain plasma and red cell folate concentrations above those of non-pregnant women is 100 μg/day. National authorities have recommended that women planning a pregnancy should increase their intake of folic acid from the usual 0.2 mg/day to 0.4 mg by capsule supplement. A dietary supplement would require eight glasses of orange juice, 10 servings of broccoli or three servings of Brussels sprouts.

Lactation

Total breast milk folic acid excretion averages 40 μg/day and in order to replace this from the diet an intake of 60 μg/day (DRV) has been suggested.

Elderly

There is little evidence that the elderly require enhanced intakes of folic acid.

7.13.6 Body store measurements

Until recently the most efficient way of measuring folic acid was by microassay methods, but now high-pressure liquid chromatography (HPLC) measurements are providing more consistent results.

The best assay of folate status is to measure folate concentration in serum and red cells. Red cell folate is a better estimate of long-term status, as it reflects body stores. If the concentration in red cells falls below 100 ng/ml then the individual may be considered to be deficient for requirements. Folic acid is susceptible to oxidation and therefore if long-term storage is required suitable antioxidants should be added to the sample. Perhaps to get a total picture of folate status, serum and red cell folate with serum vitamin B_{12} are important measurements. Radioassay kits have replaced microbiological assays in the measurement of folate.

The lower limit of normal serum folic acid is 3 ng/ml.

Summary

1. Folates are found in many forms throughout nature. Folic acid (pteroylglutamic acid) is a pterin ring (2-amino, 4-hydroxy pteridine) attached to p-aminobenzoic acid conjugated to L-glutamic acid (PteGlu). The variations on folic acid include di- and tetrahydro- forms of the pteridine ring; a single-carbon substitution (methyl $-CH_2$, formyl $-CHOH$, methylenyl $= CH-$, methylene $= CH_2$ or formimino $-CHNH$) at N-5 or N-10 and a chain of four to six glutamates attached to the L-glutamate.
2. The tetrahydrofolates function as co-substrates for enzymes involved in one-carbon (1-C) metabolism.
3. Dietary folic acid deficiency is a cause of megaloblastic anaemia. Supplementation of dietary folic acid before conception and during pregnancy reduces the incidence of neural tube defects in the foetus.

Further reading

Chanarin, I. (1990) *The Megaloblastic Anaemias*, 3rd edn, Blackwell Scientific Publications, Oxford, London.

Herbert, V. and Zalushy, R. (1962) Interrelationship of vitamin B_{12} and folic acid metabolism: folic acid clearance studies. *Journal of Clinical Investigation*, **41**, 1263–76.

Scott, J.M., Kirke, P.N. and Weir, D.G. (1990) The role of nutrition in neural tube defects. *Annual Review of Nutrition*, **10**, 277–95.

Shane, B. and Stokstadel, R. (1985) Vitamin B_{12}/folate inter-relationships. *Annual Review of Nutrition*, **5**, 115–41.

Wald, N.J. and Bower, C. (1995) Folic acid and the prevention of neural tube defects. *British Medical Journal*, **310**, 1019–20.

7.14 Vitamin B$_{12}$

7.14.1 Introduction

Vitamin B$_{12}$ contains cobalt and has a molecular weight of approximately 1350 Da. The structure of vitamin B$_{12}$ is complex (Figure 7.33), consisting of four linked pyrrole rings (a corrin) coordinating with a cobalt atom at the centre. The prefix cob implies the presence of cobalt. As the corrin forms the core of the molecule, such a molecule is a corrinoid.

The term cobalamin is used for those cobinamides which play a part in human metabolism. The naturally occurring forms of the vitamin B$_{12}$ are methylcobalamin and adenosylcobalamin. These carry a carbon–cobalt bond which is not found anywhere else in nature. The other form of cobalamin found in tissues is hydroxycobalamin, which can be converted to the methyl or adenosyl form, but only when the valency of the cobalt is reduced from three to one. This hydroxy form may be the stored form which is the precursor, readily converted to the coenzyme form. In plasma the predominant form is the methyl cobalamin, in red cell the adenosyl cobalamin.

Cyanocobalamin is an artefact of the extraction process from the sources and is converted into active forms. This form, which is given therapeutically, is soluble in water, stable in boiling water at neutral pH, but not in the presence of alkali.

7.14.2 Synthesis of vitamin B$_{12}$

Vitamin B$_{12}$ is not found in any plant, cobalamin synthesis being confined to microorganisms. Colonic bacteria produce cobalamin but it is not absorbed in the colon. Yeast is a source of cobalamin which is also found in several forms in animal

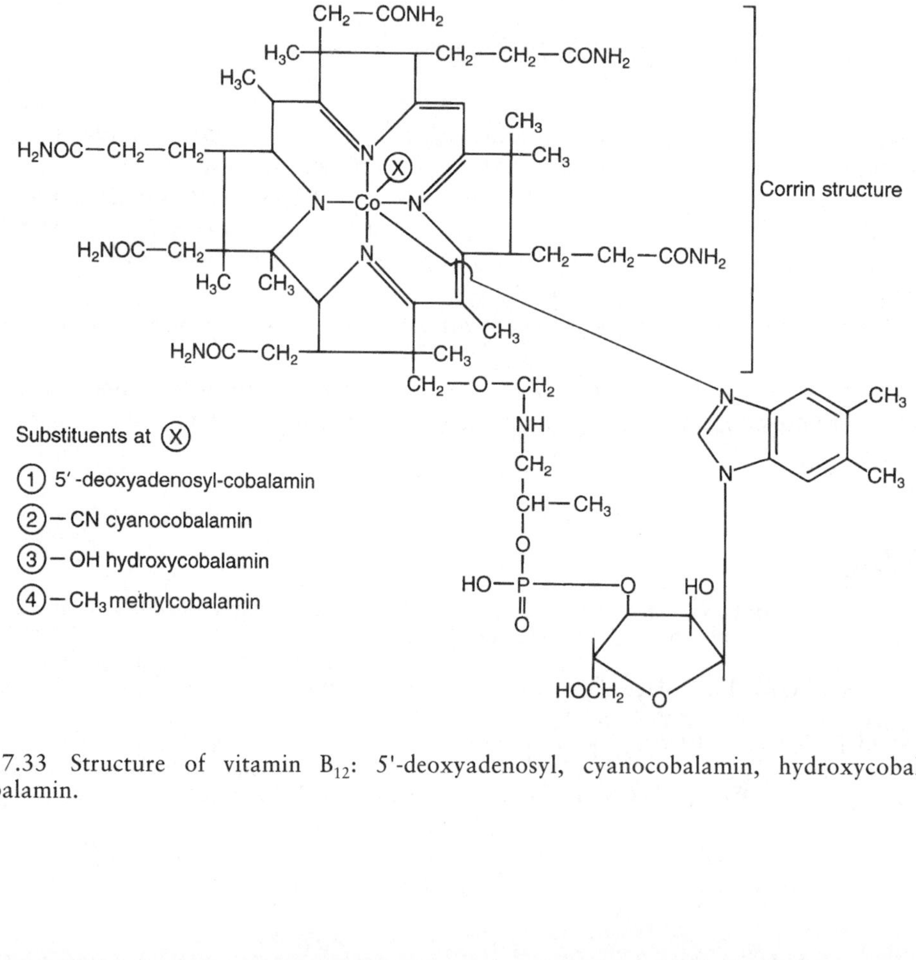

FIGURE 7.33 Structure of vitamin B$_{12}$: 5'-deoxyadenosyl, cyanocobalamin, hydroxycobalamin and methylcobalamin.

food, primarily as adenosyl- and hydroxycobalamin, of which one-third or one-half respectively are absorbed. Methyl cobalamin is found in egg yolk and cheese, and sulphitocobalamin in some foods. Little or no cyanocobalamin occurs in food, except for cows' milk which contains 3 μg/litre.

A vegetarian diet free of eggs, milk and other foods of animal origin leads to the risk of vitamin B$_{12}$ deficiency.

7.14.3 Action of vitamin B$_{12}$

Vitamin B$_{12}$ itself is not active as a coenzyme but has two enzymatically active derivatives, Co-5'-deoxyadenosylcorrinoid and aquacorrinoids (methyl B$_{12}$). Co-5'-deoxyadenosylcorrinoid accounts for 80% of tissue cell stores principally in the mitochondria. Only three reactions require vitamin B$_{12}$ (Figure 7.34):

1. Isomerization of methylmalonyl-CoA to succinyl-CoA (enzyme methylmalonyl-CoA mutase; coenzyme Co-5'-deoxyadenosylcorrinoid).
2. Isomerization of α-leucine to β-leucine.
3. Methyl transferase reactions, e.g. homocysteine to methionine.

The enzyme methyltetrahydrofolate homocysteine methyltransferase requires methyl cobalamin in the transfer of a methyl group from N-5-methyl-THF to homocysteine, which converts homocysteine to methionine. These reactions occur in the cytoplasmic and mitochondrial fractions of mammalian cells.

In nature, vitamin B$_{12}$ is combined with a protein. At the pH of the stomach, cobalamin is separated from the dietary protein–cobalamin complex by acid and pepsin. Vitamin B$_{12}$ forms complexes with haptocorrin at the pH of the stomach. Haptocorrin (formerly known as R-type binder) mediates cobalamin uptake through an asialoglycoprotein receptor. The cobalamin is released from the haptocorrin by pancreatic enzymes and binds to intrinsic factor. Intrinsic factor is a glycoprotein secreted by the parietal cells of the stomach. Secretion is stimulated by histamine, pentagastrin and cholinergic agents.

The production of intrinsic factor does not follow the same post prandial time-course as hydrochloric acid, intrinsic factor secretion being completed earlier. Vitamin B$_{12}$ is absorbed from the ileum in an intrinsic factor complex. There is binding to a mucosal receptor which is specific to the absorption of this complex. The receptor has a molecular weight of 75–80 kDa and requires Ca^{2+} ions. Binding of the complex is very species-dependent and the intrinsic factor must preferably be from the same species in order to be absorbed. As ileal receptors are limited in number, absorption rates are low and both dietary and biliary-excreted vitamin B$_{12}$ must be absorbed. Absorption from the receptor is probably as the intrinsic factor–cobalamin complex.

Vitamin B$_{12}$ has an important role in the maintenance of myelin in the nervous system. This may be due to a dependence of propionate catabolism on vitamin B$_{12}$. The normal sequence for propionyl-CoA reactions is through methylmalonyl-CoA to succinyl-CoA, which is metabolized in the citric

1. Intramolecular group transfers, e.g. methyl malonyl-CoA mutase, requires deoxyadenosyl cobalamin:

2. Methyl transferase reactions, e.g. tetrahydropteroyl glutamate methyltransferase requires methyl cobalamin:

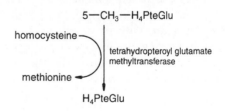

FIGURE 7.34 Two enzyme families require vitamin B$_{12}$: intramolecular group transfers and methyl transferase reactions.

223

acid cycle. Deoxyadenosyl vitamin B_{12} is essential for the last step.

7.14.4 Availability

Cobalamin is carried in the blood by three proteins, transcobalamins I, II and III. Of these, II is the most important and rapidly releases B_{12} to the tissues. The other two are more permanent binders of the vitamin.

Some 80% of the body storage (between 2–5 mg) is contained in the liver and the turnover is 0.05–0.2% of the body pool per day. The kidneys and pituitary also contain stored cobalamin. There is efficient conservation by the kidneys and the enterohepatic circulation.

It is possible that the release of protein-bound cobalamin may be affected by cooking, the nature of the food binding the cobalamin and the presence of gastric, pancreatic or ileal disease or advancing age. In pregnancy, plasma concentrations of cobalamin fall, but there are increased absorption rates.

The tape worm, *Diphyllobothrium latum*, can infect humans and may grow to a length of 15 metres. This worm competes with the host for the vitamin, with a resultant failure of vitamin B_{12} absorption. Similarly, bacteria can modify vitamin B_{12} absorption, as in enteric colonization which occurs spontaneously in the elderly or also in conjunction with a surgical blind loop, or small-bowel diverticuli. Some drugs, e.g. *p*-aminosalicylic acid, biguanides, slow-release potassium and colchicine, can also interfere with the absorption of vitamin B_{12}.

Most vitamin B_{12} is excreted in the urine, though some is excreted in bile; up to 2 µg daily is excreted in the faeces, though faecal vitamin B_{12} may be of bacterial synthetic origin.

Cobalamin-binding proteins, which are found in amniotic fluid, are synthesized by the human foetus as early as 16–19 weeks' gestation.

7.14.5 Vitamin B_{12} deficiency

A deficiency of vitamin B_{12} results in megaloblastic anaemia and neurological disorders, especially in the posterolateral columns of the spinal cord.

Pernicious anaemia is characterized by large red cells in insufficient numbers, a failure to produce acid in the stomach, gastritis, and antibodies in the blood to parietal cells and intrinsic factor. Intrinsic factor is not produced and secreted into the stomach, with resultant changes in release of cobalamin from the haptocorrin complex and ileal absorption. Secondary causes include the consequences of gastric surgery, pancreatic insufficiency, Zollinger–Ellison disease and ileal resection. The megaloblastic anaemia is believed to occur due to the methyl availability, and the amount and activity of methionine reductase (see Lipotropes, section 7.12).

In vitamin B_{12} deficiency odd-number carbon (15- and 17-) fatty acids and branched-chain fatty acids appear in the nervous system. Neurological problems ensue which are believed to result from a deficiency of mutase activity, and the accumulation of methylmalonyl-CoA which can be utilized in fatty acid synthesis instead of acetyl-CoA, with resulting fatty acid biosynthesis impairment. The abnormal odd-numbered fatty acids incorporated into myelin result in an unstable myelin sheath. S-Adenosylmethionine (SAM)-dependent methyltransferases are involved in the methylation of arginine residues of myelin basic proteins. Inability to regenerate methionine from homocysteine in nerve tissue might lead to an inadequate supply of methionine and SAM, and hence demyelination.

Deficiency in the young

Defects that affect adenosylcobalamin lead to metabolic ketoacidosis in the young and methylmalonic acidaemia and methylmalonic aciduria. Methylcobalamin deficiencies result in failure to thrive, neurological problems, homocysteinuria and hypomethionaemia. Treatment is with the hydroxylated cobalamin.

7.14.6 Vitamin B_{12} excess

Vitamin B_{12} has extremely low toxicity and may be taken at as much as 3 mg/day.

7.14.7 Recommended requirements

Adults

The average dietary intake requirements appear to be less than 1 μg/day. The LRNI is set at 1 μg/day. The RNI of 1.5 μg/day for adults would meet requirements and withstand a prolonged period, e.g. up to 5 years without any vitamin B$_{12}$ intake. There is no evidence that elderly subjects have an increased requirement, though release of protein-bound cobalamin may be affected by age.

Pregnancy and lactation

It is not known whether additional intake is required during pregnancy, but 1.5 μg/day is probably sufficient to cover the needs of a pregnant women. During lactation an increment of 0.5 μg/day should ensure an adequate supply for breast milk, which contains approximately 0.2–1.0 μg/litre vitamin B$_{12}$ when the lactating woman has an adequate intake of vitamin B$_{12}$.

Infants

In infants the LRNI is 0.1 μg/day.

7.14.8 Body store measurements

Vitamin B$_{12}$ status is estimated from serum concentrations. The relationship between dietary intake and serum concentrations is not linear because the body stores of B$_{12}$ are largely in the liver. Thus, the serum concentration only identifies individuals into broad nutritional categories.

The plasma concentration in a healthy person lies between 200 and 960 pg/ml (150–710 pmol/l). A value of under 80 pg/ml (60 pmol/l) indicates vitamin B$_{12}$ deficiency. The measurement of B$_{12}$ absorption is by the use of a radioactive cobalt B$_{12}$, with or without intrinsic factor (Schilling test). An alternative method of measurement is to use a whole-body scanner.

Summary

1. Vitamin B$_{12}$ is a complex consisting of four linked pyrrole rings (a corrin) coordinating with a cobalt atom at the centre.
2. Vitamin B$_{12}$ has two enzymatically active derivatives, Co-5'-deoxyadenosylcorrinoid and aqua-corrinoids (methyl B$_{12}$). Co-5'-deoxyadenosylcorrinoid accounts for 80% of tissue cell stores, principally in the mitochondria.
3. Only three reactions require vitamin B$_{12}$: (I) isomerization of methylmalonyl-CoA to succinyl-CoA (enzyme methylmalonyl-CoA mutase; coenzyme Co-5'-deoxyadenosylcorrinoid); (ii) isomerization of α-leucine to β-leucine; and (iii) methyl transferase reactions homocysteine to methionine. The enzyme methyltetrahydrofolate homocysteine methyltransferase requires methyl cobalamin in the transfer of a methyl group from N-5-methyl-THF to homocysteine, which converts homocysteine to methionine.
4. A deficiency of vitamin B$_{12}$ results in megaloblastic anaemia and neurological disorders, especially in the posterolateral columns of the spinal cord. Vitamin B$_{12}$ has a role in the synthesis of fatty acids in the myelin of nerve tissue.

Further reading

Newberne, P.M. and Rogers, A.E. (1986) Labile methyl groups and the promotion of cancer. *Annual Review of Nutrition*, 6, 407–32.

Scott, J.M. (1992) Folate–vitamin B$_{12}$ interrelationships in the central nervous system. *Proceedings of the Nutrition Society*, 51, 219–24.

7.15 Vitamin D

7.15.1 Introduction

Compounds with vitamin D activity occur in several forms, including vitamin D_3, molecular weight 384 Da, and vitamin D_2, molecular weight 397 Da.

Vitamin D_3 (cholecalciferol)

Most of the vitamin D_3 required by humans is produced in the skin by the ultra-violet irradiation of 7-dehydrocholesterol (pro-vitamin D) present in animal fats (Figure 7.35).

Cholecalciferol is also found in those fatty fish which eat the plankton that live near the surface of the sea. Oils from these fish, e.g. cod liver oil, are an important source; other sources include eggs and chicken liver.

Cholecalciferol may be regarded primarily as a hormone rather than a vitamin. The rate of production is dependent on skin exposure to sunlight. The extent of conversion is very dependent on the melanin content of the skin which reduces the irradiation of 7α-dehydrocholesterol.

Vitamin D_2 (ergocalciferol)

The major dietary source of vitamin D in humans is ergocalciferol. This results from the artificial exposure of the natural sterol ergosterol to ultra-violet light. Such a photosynthetic process results in a number of products, some of which are toxic, but only ergocalciferol has a beneficial effect on rickets. Ergocalciferol differs from cholecalciferol with an extra methyl group at C-24 and a double bond at C-22.

7.15.2 Synthesis of Vitamin D

Both vitamin D_2 and vitamin D_3 are biologically inactive, lipid-soluble and bound to α-globulin (transcalciferol) for transport in the blood to the liver. A proportion is converted into 25(OH) vitamin D (calcifediol) which has modest biological activity. In the liver, calcifediol is converted to 25(OH) vitamin D (calcitriol) or to 24,25(OH)$_2$ vitamin D. Because the hydroxylation process in the liver is a cytochrome P_{450}-dependent process there can be induction of the hydroxylating enzymes by drugs, e.g. phenobarbitone, with consequences for vitamin D status.

The active form of vitamin D is 1,25-dihydroxycholecalciferol (1,25(OH)$_2$ vitamin D) which is formed in the kidney by a specific mitochondrial hydroxylase acting on 25(OH) vitamin D. 1,25(OH)$_2$ vitamin D is the biologically active form of vitamin D and is 100 times more potent than 25(OH) vitamin D.

Vitamin D and its various products are stored in fat with a very long half-life; several months for vitamin D and 2–3 weeks for calcifediol. Both are held in the enterohepatic circulation. The half-life of 1,25(OH)$_2$ vitamin D is less than 24 hours, the vitamin being converted to more polar inert products within the liver which are excreted in bile and urine.

7.15.3 Actions of vitamin D

Calcium homeostasis

This requires an interaction of parathyroid hormone and 1,25(OH)$_2$ vitamin D. The action of 1,25(OH)$_2$ vitamin D is to regulate calcium and phosphate metabolism. A reduction in plasma calcium concentration stimulates the secretion of parathyroid hormone, which stimulates 1-hydroxylation of 25(OH) vitamin D in the renal tubule mitochondria.

1,25 Dihydroxycholecalciferol (1,25(OH)$_2$ vitamin D) is required for optimum intestinal absorption of calcium. It increases the transport of calcium across the brush border of the intestinal epithelial cell and export from the intestinal cell into the blood stream. A number of mechanisms may be involved:

1. Through production of a specific messenger RNA and synthesis of a specific calcium-binding protein (CaBPs); or integral membrane calcium-binding protein (IMCAL) or alkaline

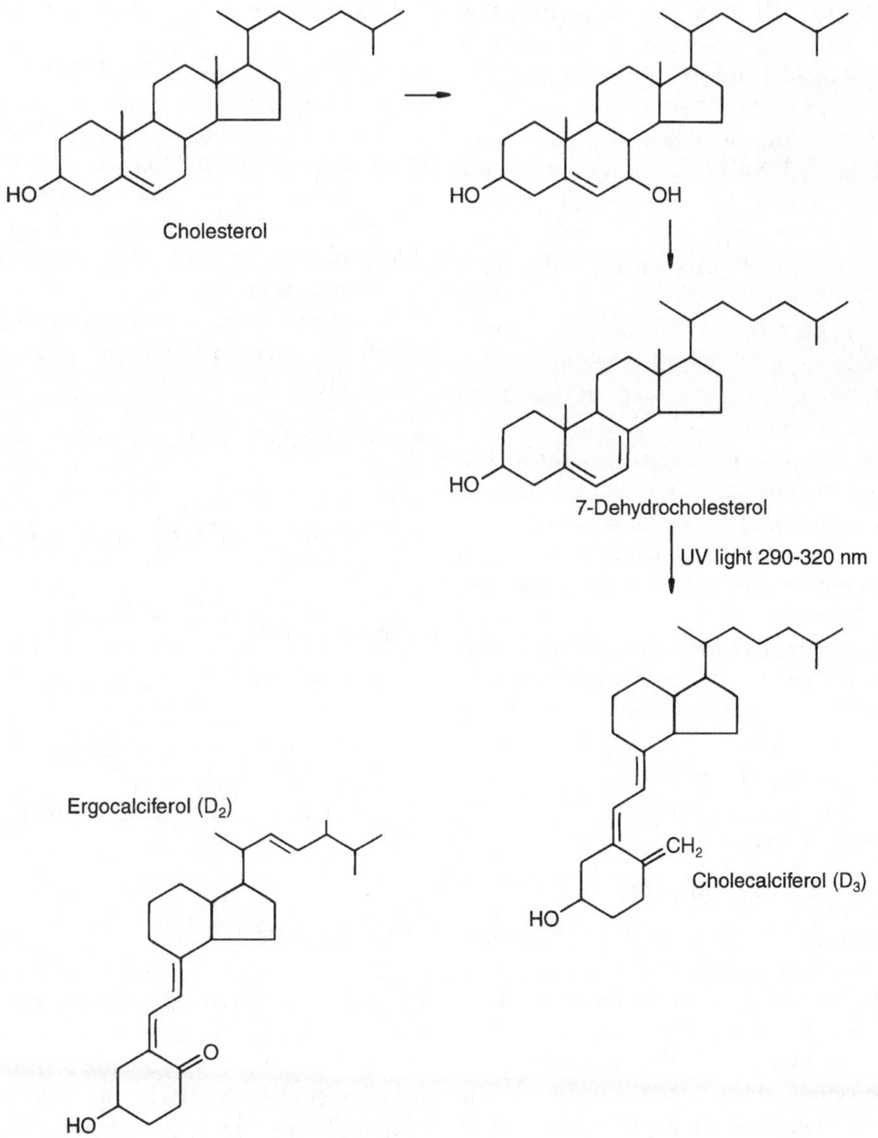

FIGURE 7.35 Synthesis of vitamin D$_3$. Structure of cholesterol; 7-dehydrocholesterol (produced from cholesterol) is converted by ultra-violet light to cholecalciferol (D$_3$). D$_2$ is ergocalciferol.

phosphatase; or changes in the lipid composition and fluidity of brush border membranes; increased endocytosis of calcium; and increased calmodulin binding to specific proteins in the brush border. The overall effect is possibly on the rate-limiting step in calcium absorption.

2. Calcium must then move across the cytoplasm of the enteric brush border cell. This is facilitated by the action of vitamin D, the intracellular diffusion of calcium or calcium binding to intracellular proteins.

3. Vitamin D may also have a role in the movement of calcium into the blood stream by either

227

stimulating the calcium pump, stimulating Na^+/Ca^{2+} exchange, or increasing exocytosis or open voltage-dependent Ca^{2+} channels.

4. In the presence of parathyroid hormone there is activation in the osteoblast of a number of functions related to bone formation. These include stimulation of the osteoblasts and osteoclasts, mobilizing calcium from bone. $1,25(OH)_2$ vitamin D acts on the osteoblast, stimulating non-matrix proteins. This allows parathyroid hormone-activated mobilization of calcium from bone and the activation of the giant osteoclast, an early step in the bone remodelling process.

5. Phosphate absorption is stimulated through a separate phosphate transport mechanism in intestinal epithelial cells. Vitamin D stimulates the sodium-dependent component of phosphate absorption and movement across the basolateral membrane.

6. $1,25(OH)_2$ vitamin D acts on the distal renal tubular cells. Renal reabsorption of calcium in the distal tubule involves both parathyroid hormones and $1,25(OH)_2$ vitamin D.

7. $1,25(OH)_2$ vitamin D acts directly on the parathyroid cells to suppress parathyroid hormone production and secretion.

Vitamin D as a steroid hormone

$1,25(OH)_2$ vitamin D functions through nuclear receptors (molecular weight 55 000 Da) which belong to the steroid thyroid hormone receptor family. The gene for this receptor is on chromosome 12 in humans. Vitamin D binds to a response element in the promotor region of the target gene, and requires an accessory protein. It is possible that this protein may be the retinoic acid X receptor. Such binding initiates a sequence of events, including phosphorylation, which stimulates transcription of the gene, production of mRNA and synthesis of the accessory protein for vitamin D. Vitamin D may have two modes of action, acting on gene function (genomic) and on receptors not controlling genes (non-genomic functions). The concentrations required for non-genomic actions are high and may not be physiological.

Vitamin as a more general hormone

Receptors for $1,25(OH)_2$ vitamin D are also found in the parathyroid glands, the islet cells of the pancreas, the keratinocytes of skin, mammary epithelium, endocrine cells of the stomach and some cells in the brain. Vitamin $1,25(OH)_2$ vitamin D may be a developmental hormone inhibiting proliferation and promoting differentiated function in cells

The many genes and hormones influenced by $1,25(OH)_2$ vitamin D fall into four main groups, which are:

- gene products associated with mineral metabolism
- regulators of vitamin D secosteroids
- differentiators of events in the skin and immune system
- regulators of DNA replication and cellular proliferation

Vitamin D probably has little influence on magnesium absorption, but increases the absorption of aluminium, lead and selenium. This may have consequences when excesses of aluminium or lead are present in the diet. Impairment of calcium absorption may result.

7.15.4 Availability

Dietary vitamin D is absorbed in the small intestine, by similar mechanisms to fat digestion and absorption. There is subsequent transport to the liver in chylomicrons. In the liver all forms of vitamin D are converted by hepatic microsomes to the 25-hydroxylated form (Figure 7.36), which is carried in the plasma on a specific transport globulin. Storage is primarily in adipose tissue. Vitamin D may be excreted in bile in the form of hydroxylates and conjugates (glucuronides).

7.15.5 Vitamin D deficiency

The most important consequence of vitamin D deficiency is **rickets**, in which there is a failure to

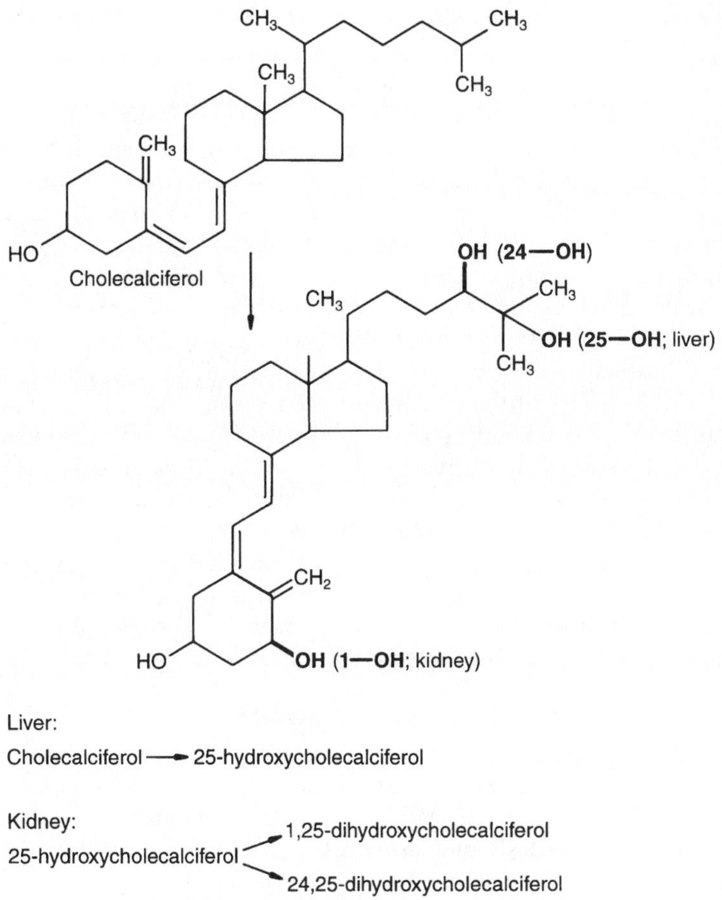

FIGURE 7.36 Metabolism of cholecalciferol to 25-hydroxycholecalciferol in the liver, 1,25-dihydroxycholecalciferol in the kidney or 24,25-dihydroxycholecalciferol.

mineralize the bony skeleton (see also Bone structure, section 12.3).

Rickets

Vitamin D deficiency affects the growth of bone. Bone growth occurs at the inner band of the epiphyseal cartilage, between the shaft and the epiphysis. The site of new cartilage formation is at the epiphyseal end of the bone. In contrast at the diaphyseal end, the cartilage is removed by capillaries and osteoblastic-forming osteoid tissue in which calcium salts are deposited. The two processes are in equilibrium, and hence the bone extends. Growth ceases when no further new cartilage is formed and the bone and the epiphysis meet and fuse.

In vitamin D deficiency the epiphyseal cartilage grows but the replacement of cartilage becomes defective so there is a widening of the zone between the shaft and the epiphysis. There is little calcification taking place so that this leads to an overgrowth of osteoid tissue below the periosteum.

Vitamin D deficiency is associated with a failure of mineralization of the bone's osteoid matrix, which results in the formation of a soft bone. The bony abnormalities which accrue depend on the age of onset and the way in which gravity stresses the bone. The appearance of the child does not necessarily indicate the development of rickets and it may well be that the child is well-built. However, they may be restless with somewhat flabby and hypotonic muscles. The limbs are able to twist into various positions, which is called 'acrobatic rickets'. There is excess sweating on the head and the abdomen is distended because of weak abdominal muscles. Diarrhoea, respiratory infection and a delayed development of teeth are not uncommon. The most common finding is enlargement of the lower end of the radius and the costochondrial junction of the ribs – 'ricketic rosary'. Later there is bossing of the frontal and parietal bones and delayed closure of the anterior fontanel, and 'pigeon chest' – an undue prominence of the sternum and a transverse depression passing outwards from the costal margins towards the axilla. All of these are due to pressure on the soft bones when the child is lying and standing and are associated with breathing. This pressure deformation continues when the child is upright, so that these somewhat softened bones bend, leading to kyphosis of the spine and enlargement of the lower ends of the femur, tibia and fibula. Consequently, the legs are bowed with, for example anterolateral bowing of the tibia at the junction of the middle and lower third. Kyphosis is often replaced by lordosis and also pelvic deformity develops. Pelvic deformity has long-term consequences for childbirth in the fertile adult female.

Tetanic spasm
If the concentration of ionized calcium in the plasma is reduced then infantile tetany can result. This is manifested by spasm of the hands and feet and vocal chord resulting in high-pitched cries and breathing problems.

Diagnosis
Other than the clinical picture described above, plasma measurements of raised plasma alkaline phosphatase and particularly plasma $1,25(OH)_2$ vitamin D are important.

Risk factors
- Inadequate exposure to sunlight: in northern communities there is less exposure to sunlight and hence less conversion of cholesterol to vitamin D_3. It is particularly marked with children and adolescents who require vitamin D for growth.
- Strict vegetarianism: there is a risk of excluding vitamin D from the diet with an increased risk of osteomalacia or rickets.
- Breast feeding: if an infant is fed for more than 3 months on milk, or from a mother deficient in vitamin D for more than 3 months, then the risk of infantile rickets increases.
- Skin pigmentation: this is a minor risk factor wherein there is screening of the metabolically active skin sites by melanin and reduced conversion of cholesterol to vitamin D_3. Pigmented races in areas of reduced sunlight are particularly vulnerable to vitamin D deficiency.
- Secondary osteomalacia and rickets occurs in conjunction with gastrointestinal disease, the malabsorption syndrome following peptic ulcer surgery where lipid absorption and hence vitamin D absorption are impaired.

Prevention
The risk of rickets can be removed by a supplement of 10 µg of vitamin D daily or regular exposure to sunlight. Protein–energy malnutrition may be associated with rickets.

Osteomalacia

This may present with pain and muscular weakness. Pain may be found in the ribs, sacrum, lower lumbar vertebrae, pelvis and legs. A waddling gait may be adopted, and even tetany which is characterized by carpopedal and facial twitching. Spontaneous fractures may also occur.

X-ray features include rarification of the bone and translucent bands (pseudofractures, Looser-s zones) which may be symmetrical and occur at points where there is compression stress. This may occur in the ribs, the scapula, the pubic rami and the cortex of the upper femur. Such zones are characteristic of osteomalacia.

Renal disease, osteomalacia and other bone disorders may occur in individuals with chronic renal failure, with impaired hydroxylation and hence impaired synthesis of $1,25(OH)_2$ vitamin D in the kidneys. This can occur in chronic renal failure and congenital conditions such as Fanconi's syndrome where activation of cholecalciferol by 1 : 25 hydroxylation is impaired. A failure of 25-hydroxylation of vitamin D can lead to osteomalacia of hepatic origin.

Other diseases related to vitamin D

- Renal osteodystrophy results from a renal inability to produce $1,25(OH)_2$ vitamin D. Patients on renal dialysis may have calcium replaced but these patients can develop severe secondary hyperparathyroidism. Treatment with $1,25(OH)_2$ vitamin D is very effective and suppresses parathyroid hormone production.
- Vitamin D-dependent rickets type 1 is an autosomal genetic disorder in which children have rickets despite normal intake of vitamin D. This defect is in the 1α-hydroxylation of $25(OH)$ vitamin D. The defect is treated by physiological amounts of $1,25(OH)_2$ vitamin D or large doses of vitamin D.
- Vitamin D-dependent rickets type 2 is an end-organ-resistant defect, with autosomal recessive inheritance. This is due to a mutation of the gene for the $1,25(OH)_2$ vitamin D receptor. Mutations include changes to the two zinc fingers of the receptor which binds vitamin D response elements from target cells. There is a mutated nucleotide resulting in a premature stop codon with a truncated receptor which does not function. This form of rickets is resistant to $1,25(OH)_2$ vitamin D.
- Osteoporosis has been suggested to be an oestrogen deficiency disorder. However, age-related osteoporosis is in part a defect in $25(OH)$-D 1-α-hydroxylase.
- In X-linked hypophosphataemic vitamin D-resistant rickets there is a renal phosphate leak and severe hypophosphataemia and rickets. This is treated by frequent administration of oral phosphate. Such treatment results in increases in ionized calcium and secondary hyperparathyroidism. All phosphate therapy requires additional hydroxylated vitamin D compound supplements to correct the secondary hyperparathyroidism. However, this condition is not due to defective vitamin D function.

7.15.6 Vitamin D excess

There are instances of individuals developing hypervitaminosis D. This occurs during infant supplementation and also with replacement therapy in people who previously had osteomalacia. The consequence is an increase in plasma calcium concentration, tetany, electrocardiogram changes, convulsions and occasionally, death.

If vitamin D is given in milligram amounts then the consequences are lethal, to the extent that vitamin D has been used as a rodent killer at 0.1% of the diet. The action may be through the blocking of receptors normally used by $1,25(OH)_2$ vitamin D by vitamin D and $25(OH)$ vitamin D

7.15.7 Recommended requirements

Adults

Plasma $25(OH)$ vitamin D concentrations range from 15–35 ng/ml in summer and 8–18 ng/ml in winter. No minimum dietary intake has been identified for individuals living predominantly exposed to the sunlight, but for those without this exposure an RNI of vitamin D of 10 μg/day is suggested.

Infants and children

Breast milk has a seasonal variation in vitamin D concentration; with winter milk containing very little vitamin D. Vitamin D intakes usually decline on weaning as most weaning foods are modestly fortified and are low in vitamin D, as is whole cows' milk. Plasma $25(OH)$ vitamin D is therefore lower in the second 6 months of life – a much more vulnerable period for the growing infant than the first 6 months. After this, plasma $25(OH)$ vitamin D concentrations are satisfactory. Pigmented races living in northern climes with a reduced exposure to sun, should receive dietary supplements of vitamin D.

Low vitamin D status is more common in individuals of Asian origin living in Northern Europe, especially children, adolescents and women, mainly due to vegetarian diets, low calcium intake and limited exposure to sun because of the mode of dress.

Pregnancy and lactation

It is suggested that pregnant and lactating women should receive supplementary vitamin D to ensure an intake of 10 μg/day, despite seasonal variations.

Elderly

The elderly may have reduced stores because of insufficient exposure of skin for an adequate time in the summer. A daily 30-minute exposure of the face and legs will raise 25(OH) vitamin D levels at 37° latitude, whereas 1–2 hours may be necessary in the north of Britain. To maintain winter 25(OH) vitamin D concentrations above 8 ng/ml would require summer plasma concentrations of 16 ng/ml, which is two to three times more than

values often recorded in the elderly. Such deficiency may result in osteomalacia developing in a small proportion of this age group. It is recommended that the elderly should consume 10 μg/day of vitamin D to achieve satisfactory 25(OH) vitamin D concentrations.

7.15.8 Body store measurements

The best measure of the vitamin status in humans is the plasma 25(OH) vitamin D concentration. This reflects the availability of vitamin D in the body, and 1,25(OH) vitamin D metabolic need. There is a reasonable prediction of dietary intake from measurements of plasma 25(OH) vitamin D concentrations in individuals minimally exposed to sunlight. The threshold of risk associated with osteomalacia is a concentration of 25(OH) vitamin D less than 12.5 nmol/l. In contrast, where there is exposure to sunlight, the lower limit of 25(OH) vitamin D should be 12–25 nmol/l.

Summary

1. Compounds with vitamin D activity occur in several forms, including vitamin D_2 (ergocalciferol) and vitamin D_3 (cholecalciferol)
2. Vitamin D_3 is produced in the skin from 7-dehydrocholesterol. Cholecalciferol is also found in fatty fish. Cholecalciferol may be regarded primarily as a hormone rather than a vitamin.
3. The major dietary source of vitamin D in humans is ergocalciferol. This differs from cholecalciferol in having an extra methyl group at C-24 and a double bond at C-22.
4. Both vitamin D_2 and vitamin D_3 are biologically inactive. The active form of vitamin D is 1,25-dihydroxycholecalciferol (1,25(OH)$_2$ vitamin D); this is formed in the kidney by a specific mitochondrial hydroxylase acting on 25(OH) vitamin D and is regulated by parathyroid hormone and plasma phosphate concentration.
5. The actions of the vitamin D family include calcium homeostasis (an interaction of parathyroid hormone and 1,25(OH)$_2$ vitamin D), steroid hormone action (1,25(OH)$_2$ vitamin D functions through nuclear receptors) and as a more general hormone.
6. 1,25(OH)$_2$ vitamin D may be a developmental hormone inhibiting proliferation and promoting differentiated function in cells
7. The most important consequence of vitamin D deficiency is rickets, in which there is a failure to mineralize the bony skeleton. The site of the consequent boney deformities depends upon weight-bearing on the softened bones.

Further reading

De Luca, H.D. (1993) Vitamin D. *Nutrition Today;* 28, 7–114.

Fraser, D.R. (1995) Vitamin D. *Lancet,* 345, 104–7.

Pike, J.W. (1988) Vitamin D$_3$ receptors. *Annual Review of Nutrition,* 11, 189–216.

Suda, T., Shink, T. and Takahadi, N. (1990) Role of vitamin D in bone and intestinal cell differentiation. *Annual Review of Nutrition,* 10, 195–211.

7.16 Vitamin K

7.16.1 Introduction

Vitamin K is a naphthoquinone which occurs in two forms, vitamin K_1 and K_2 (Figure 7.37). Vitamin K_1, molecular weight 450 Da, was isolated from the plant lucerne and is a phytylmenaquinone (an alternative name is phylloquinone), but in the pharmacopoeia is called phytomanadione. The vitamin consists of 2-methyl-1,4-naphthoquinone (menadione or menaquinone) attached to a 20-carbon phytyl side chain. This is a yellow oil and is the only form that occurs in plants. Vitamin K_2, molecular weight 649 Da, is one of the family of chemical homologues produced by bacteria with 4 to 13 isoprenyl units in the side chain. These are called menaquinone-4 up to menaquinone-13, dependent on the number of isoprenyl units.

Vitamin K_1 is present in fresh green vegetables, e.g. broccoli, lettuce, cabbage and spinach. Beef liver is a good source, but other animal tissues, cereals and fruit are poor sources. Vitamin K_2 as menaquinones is produced by intestinal bacteria.

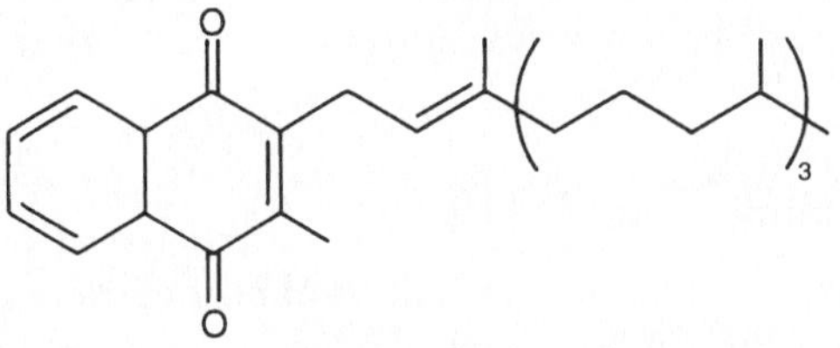

Vitamin K_1 (phylloquinone)

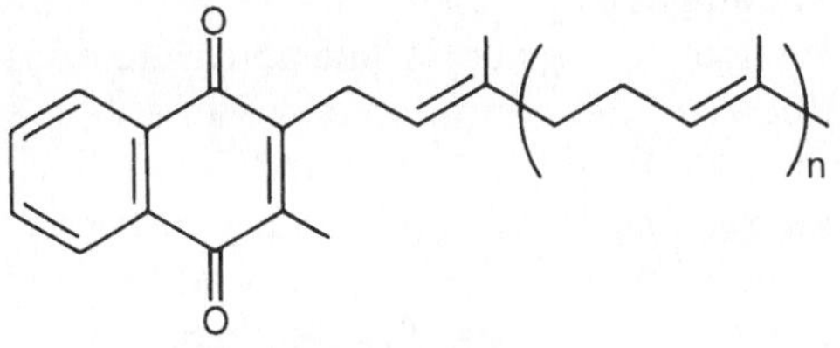

Vitamin K_2 (menaquinone series)

FIGURE 7.37 Structure of vitamin K_1 (phylloquinone) and K_2 (menaquinone).

7.16.2 Action of vitamin K

Vitamin K is involved in the synthesis of proteins central to blood coagulation, prothrombin and Factors VII, IX and X (Figure 7.38). Vitamin K is necessary for the post-translational carboxylation of glutamic acid in the coagulation proteins (Figure 7.39). γ-Carboxyglutamate allows the binding of calcium and phospholipids in the formation of thrombin. When warfarin, an anticoagulant, is given γ-carboxyglutamate addition does not occur to the proteins. The proteins are ineffective in clotting mechanisms.

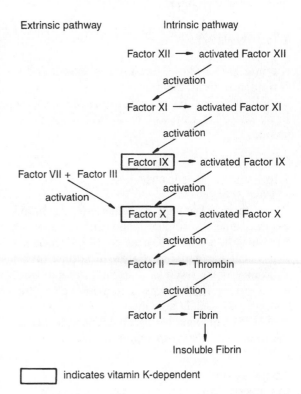

indicates vitamin K-dependent

FIGURE 7.38 Coagulation and fibrinolysis. There are two pathways of fibrin formation: extrinsic and intrinsic. The production of Factor IX and Factor X is vitamin K-dependent.

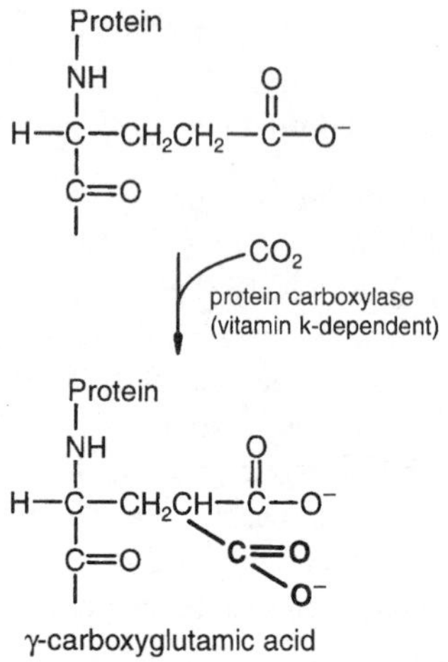

FIGURE 7.39 Vitamin K is a coenzyme in the carboxylation of glutamic acid residues in proteins and forms γ-carboxyglutamic acid. This reaction is important in the clotting cascade.

7.16.3 Availability

Vitamin K_1 is absorbed as any other lipid and is transported from the intestine in the blood in chylomicrons as β-lipoproteins. Vitamin K of bacterial origin can also be absorbed from the colon.

7.16.4 Vitamin K deficiency

When there is vitamin K deficiency, the blood clotting time is prolonged and the activities of Factors VII, IX and X are reduced. Deficiency may occur in infants but rarely in adults. Infant deficiency is the result of a sterile intestinal tract and a dietary deficiency as human and cows' milk contain only small amounts of vitamin K. The problem is compounded by the immature liver of the infant being slow to synthesize prothrombin.

Acquired deficiencies of vitamin K malabsorption

This includes any condition where fats are malabsorbed including biliary obstruction, malabsorption, bacterial colonization and liver disease.

7.16.5 Vitamin K excess

In general natural vitamin K preparations are free from toxic effects. Synthetic preparations of menadione have little biological activity and the high reactivity of the unsubstituted 3-position can lead to haemolysis and liver damage in the newborn.

Antagonists and anti-coagulants

Spoilt sweet clover in which a dicoumard is produced, prolongs the prothrombin time of the cow, causing a bleeding condition. From this observation was developed dicoumarol, an analogue of vitamin K, which has been used to prolong prothrombin time in clinical medicine. This has been replaced by warfarin and phenindione.

7.16.6 Recommended requirements

Adults

The accepted criteria for vitamin K needs is the maintenance of normal plasma concentration of the vitamin K-dependent coagulation factors, assessed by prothrombin times. Most estimates of the dietary requirements of phylloquinone give figures between 0.5 and 1.0 μg/kg body weight per day.

Infants

Vitamin K in human milk is almost entirely in the form of phylloquinone and may vary between 1 and 10 μg/litre. A reasonable DRV for an infant fed on breast milk would be approximately 8.5 μg phylloquinone.

Vitamin K is given in **haemorrhagic disease of the newborn**. In approximately 1 in 800 newborn babies, bleeding occurs into the tissues (including skin, peritoneal cavity, alimentary tract or central nervous system) between the second and fifth day. In some countries, e.g. Britain, vitamin K is given routinely at birth, though this not a universal practice. Water-soluble analogues of vitamin K are not used in the newborn, particularly the premature newborn, because this may cause hyperbilirubinaemia. The dose given is 1 mg of vitamin K_1 intramuscularly. Vitamin K preparations may be important supplements when there is lipid malabsorption in the intestine and in reversing the effects of anticoagulants.

7.16.7 Body store measurements

Vitamin K deficiency can be detected by the blood prothrombin time, which measures prolongation of clotting time.

Summary

1. Vitamin K is a naphthoquinone. There are two forms: vitamin K_1 is from the plant lucerne, a phytylmenaquinone (phylloquinone); vitamin K_2 is produced by bacteria and has 4 to 13 isoprenyl units in its side chain.
2. Vitamin K is involved in the synthesis of blood coagulation proteins, prothrombin and Factors VII, IX and X. Vitamin K is necessary for the post-translational carboxylation of glutamic acid in the coagulation proteins. γ-Carboxyglutamate allows the binding of calcium and phospholipids in the formation of thrombin.
3. When warfarin, an anticoagulant, is given γ-carboxyglutamate addition does not occur to the proteins which are then ineffective in clotting mechanisms.
4. When there is vitamin K deficiency, the blood clotting time is prolonged and the activities of Factors VII, IX and X are reduced.

Further reading

Dowd, P., Ham, S.W. and Hershline, R. (1995) The mechanism of action of vitamin K. *Annual Review of Nutrition*, 15, 419–440.

7.17 Vitamin E

7.17.1 Introduction

The term vitamin E is a term for any mixture of biologically active tocopherols. Eight tocopherols and tocotrienols with vitamin E activity are known. The tocopherols are the most potent of the vitamin E compounds, while the tocotrienols are less potent, the differences being in the number and positions of the methyl groups around the ring of the molecule (Figure 7.40). While all have the same physiological properties, α-tocopherol, molecular weight 430 Da, a synthetic product, is the most potent; β and γ-tocopherol and β-tocotrienol have an activity of 48% and 20% relative to β-tocopherol. Other analogues have little nutritional activity.

Abundant sources are vegetable oils, wheatgerm, sunflower seed, cottonseed, safflour, palm, rape seed and other oils. Vitamin E is found in all cell membranes where it inhibits the non-enzymatic oxidation of polyunsaturated fatty acids by molecular oxygen.

7.17.2 Action of vitamin E

The biological function of vitamin E is not specific, e.g. as a cofactor for an enzymatic reaction. Vitamin E acts as antioxidant or a free radical scavenger in chemical systems. Vitamin E is the only known lipid-soluble anti-oxygant in plasma and in red cell membranes.

Ascorbic acid may reduce tocopheroxyl radicals formed by the scavenging of free radicals during metabolism. This enables the single molecule of tocopherol to scavenge many radicals. Vitamin C is therefore protective in membranes against free radical damage. Vitamin C is very water-soluble

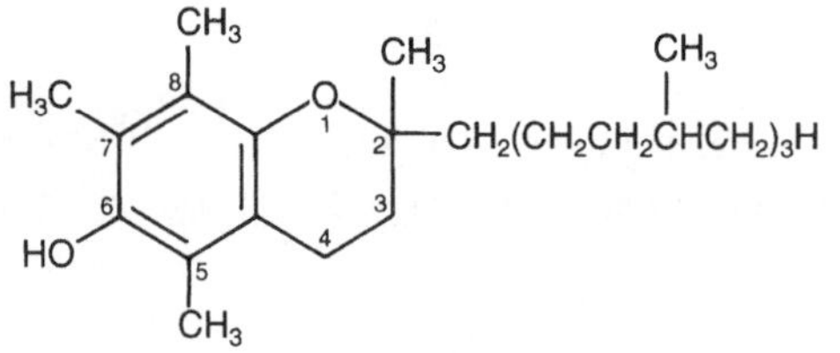

In β-tocopherol, the 7-methyl group is absent

In γ-tocopherol, the 5-methyl group is absent

In δ-tocopherol, the 5- and 7-methyl groups are absent

In tocotrienols, the side chain is $-CH_2(CH_2CH=CCH_2)_3H$ with CH_3

FIGURE 7.40 The structures of vitamin E (tocopherol), α-tocopherol, β-tocopherol, γ-tocopherol (no 5-methyl group), δ-tocopherol and tocotrienols.

but as vitamin E is buried within the membrane organelles in cells, the mechanisms wherein these substances may interact is not known. It is possible that glutathione may be the donor of electrons in most tissues.

Free-radical scavengers

The special radical-scavenging properties of vitamin E are found in the fused chroman ring system; the phytyl side chain does not affect the inhibitory properties of this chroman. The antioxidant properties are due to the positioning of the pair of small pi electrons of the ethereal oxygen in the ring. Vitamin E is associated with membranous organelles, largely because of its lipid solubility, miscible with lipids of the biological membrane. The phytyl chain allows the tocopherol molecule to enter the hydrophobic environment of the membrane.

7.17.3 Availability

The absorption characteristics are as all lipid-soluble nutrients. Normal lipoprotein concentrations of vitamin E as α-tocopherol range from 11–37 μmol/litre. α-Tocopherol forms 90% of vitamin E found in tissues. Because vitamin E is not soluble in water, transport through aqueous fluids requires transport in lipid transport systems.

Vitamin E is transported in plasma by all plasma lipoproteins; there does not appear to be a specific carrier protein. This system has the advantage that there is protection of polyunsaturated fatty acids which are also being transported by similar mechanisms. Tocopherol enters the systemic circulation in chylomicrons and intestinal very low-density lipoproteins (VLDL), which are the major carrier of vitamin E.

Efficient removal of vitamin E from the circulation depends upon lipoprotein lipase. The vitamin E is taken up by the liver or transferred to other lipoproteins. α-Tocopherol is secreted into the blood stream within nascent VLDL, whose concentration is important in maintaining the plasma concentration of α-tocopherol due to differential intracellular binding. α-Tocopherol is secreted preferentially to other stereoisomers of tocopherols.

7.17.4 Deficiency and excess of vitamin E

Deficiency

Deficiency may occur as a result of gastrointestinal malabsorption and in premature infants.

Excess

There appears to be no adverse effects from large doses of vitamin E, up to 3200 mg/day.

7.17.5 Recommended requirements

Adults

The average intake in Britain is 6 mg/day, of which 26% is derived from fats and oils and 9% from cereals. It would appear that the amount of vitamin E which is necessary is dependent upon the requirements of sites in membranes which are determined by the polyunsaturated fatty acid content of tissues; this in turn reflects the polyunsaturated fatty acid content of the diet. The relationship between polyunsaturated fatty acid intake and vitamin E requirements is neither simple nor linear; hence identifying vitamin E requirements is very difficult.

Intakes of 4–10 mg and 3–8 mg of α-tocopherol equivalents per day respectively for men and women have been recommended by various committees. An alternative provision would be 0.4 mg α-tocopherol equivalents/dietary polyunsaturated fatty acids (PUFA) per day. This formula might also be used for infant formulas. On a PUFA intake of 6% of dietary intake then the vitamin E requirement would be approximately 7 mg/day.

Infants

Human milk has a varied vitamin E content, up to concentrations of 1 mg α-tocopherol equivalents/100 ml in colostrum, which may reduce to 0.32 mg/100 ml at 12 days and remains constant thereafter. This would give the infant consuming 850 ml of breast milk a daily intake of 2.7 mg.

7.17.6 Body store measurements

Vitamin E status can be measured by the plasma tocopherol concentration. To some extent dietary vitamin E intake is reflected in blood tocopherol concentrations. However, the value of all the biochemical indices and dietary intake calculations are modest. The absorption of the vitamin is incomplete and varies from 20–80%. Furthermore, because of the variable biological activity of the different tocopherols, the value of these measurements is somewhat limited. Plasma concentrations of α-tocopherol of less than 11.5 μmol/l suggest vitamin E deficiency. Vitamin E is also correlated with the total lipids in the blood, particularly the cholesterol fraction. The molar cholesterol: vitamin E ratio is approximately 200 : 1. An alternative ratio is vitamin E : cholesterol. A plasma tocopherol concentration of 11 mmol/litre or a tocopherol : cholesterol ratio of 2.25 μmol/mmol is considered normal. The ratio less than 2.2 μmol α-tocopherol per mmol of cholesterol is suggestive of high risk. A functional test of vitamin E status is the hydrogen peroxide haemolysis test (erythrocyte stress test).

Summary

1. Vitamin E is a term for any mixture of biologically active tocopherols. Eight tocopherols and tocotrienols with vitamin E activity are known, the differences being in the number and positions of the methyl groups around the ring of the molecule.
2. Vitamin E acts as antioxidant or a free-radical scavenger in chemical systems and is associated with membranous organelles, miscible with lipids of the biological membrane.
3. Vitamin E is the only known lipid-soluble anti-oxygant in plasma and in red cell membranes.
4. Ascorbic acid may reduce tocopheroxyl radicals formed by the scavenging of free radicals during metabolism. This enables the single molecule of tocopherol to scavenge many radicals. Vitamin C is therefore protective in membranes against free-radical damage.

Further reading

Cohn, W. *et al.* (1992) Tocopherol transport and absorption. *Proceedings of the Nutrition Society,* 51, 179–88.

McCay, P.B. (1985) Vitamin E. Interaction with free radicals and ascorbate. *Annual Review of Nutrition,* 5, 323–40.

Chapter 8

Water, electrolytes, minerals and trace elements

- Water is essential to life.
- Water acts as a bulk, cellular and compartmental solvent in the body.
- The properties of chemicals and polymers in water are central to metabolism.

8.1 Water

8.1.1 Introduction

Water is the basic chemical of life, acting both as a bulk and localized solvent for the body. Water is an angular molecule with two vertical planes of symmetry and is an acceptor and donator of protons. Water freezes at 0°C to form a stable phase (ice) with a variety of structural formations. The chemical potential of ice is much lower than that of liquid water. As water is warmed the structure becomes more open and is at its maximum volume at 4°C. Water molecules are held apart by hydrogen bonds between structures.

Water readily dissolves a number of chemicals and such solubility is important in biological processes: (i) cell structure; (ii) blood; and (iii) excretory systems, e.g. urine and bile.

Equally important are water-insoluble lipid phases which form separate and distinct functional units: (i) cell membranes; and (ii) hydrophobic domains in enzymes.

The charges on the water molecule allow other atoms and molecules to be variably charged.

Solutions

A **solution** is a homogeneous mixture of two components. In a solution atoms of A, the solute are surrounded by atoms of B, the solvent and other atoms of A. A sample of the solution, however small, will be representative of the whole.

In solution, ions will be charged positively or negatively and often the water provides the complementary charge to the ion. The solution must be electrically neutral and the counter ions move over each other, anion over cation and cation over anion, to create a neutral ionic atmosphere.

Osmolality and osmolarity

Osmolality is a measure of the number of osmoles of solute/kg of solvent. **Osmolarity** is the number of osmoles/litre of fluid. Thus, 1 mmol of a non-polar solute, e.g. sucrose, gives a 1 mosmol solution; 1 mmol of a salt, e.g. NaCl, dissociates to give two ions and therefore a 2 mosmol solution. In the body the major contributors to osmolality are sodium, and its anions chloride, bicarbonate and sulphate and glucose and urea.

Water as a solvent

In any given situation the water molecule is surrounded by other molecules or atoms. In pure water, other water molecules surround each molecule at the corner of a tetrahedron, similar to the formation of ice. Other chemicals may dissolve in water to a varying extent, dependent upon ions, molecular charge and hydration – which is the ability to tolerate a shell of water molecules. Water is the **solvent**, the substance dissolved is the **solute**. Solutes will dissolve to varying degrees in water. Sucrose and salt are very soluble in water, triacylglycerides are very insoluble. A large protein molecule will have regions which associate with water (hydrophilic) and other regions which are hydrophobic and are not associated with water. Phase diagrams can be constructed which portray solubility of one or several solutes in a solvent at differing concentrations of each solute. Various thermodynamically stable boundaries will develop where the phases will exist in equilibrium, e.g. soluble, partially insoluble and totally insoluble.

Osmolality is measured in the blood and urine by depression of freezing point measurements; 1 g molecule in 1 kg water of any unionized sub stance is equal to 1 osmol. For monovalent ions 1 equivalent weight has an osmolality of 1 osmol; divalent ions 0.5 osmol. A mixture can be identified in terms of the total concentration of ions.

The body is a complex phase diagram with differing areas defined by boundaries osmolality and by solubilities. The movement of substances between these phases, e.g. intra- and extracellular, is important metabolically and physiologically.

8.1.2 Water balances in the body

Water forms 50–60% of body weight. One-third is extracellular fluid, two-thirds intracellular (e.g. in a 65-kg man, 15 and 30 litres, respectively). These compartments are separated by cell membranes, often freely permeable to water movement dictated by osmolality.

Water is absorbed throughout the gastrointestinal tract. Water by mouth is largely absorbed in the jejunum, by a process which is passive, though glucose- and sodium-dependent, but is also absorbed in the colon. Water absorption occurs largely through the paracellular pathways in the jejunum. There is also transcellular water flow by lipid-mediated osmosis. Over 2 litres of intestinal fluid enter the caecum and normally all but 100 ml are absorbed.

Thirst

Humans satisfy their requirements for water by drinking at regular intervals, usually dictated by social practices, or by drinking water after or during meals, stimulated by thirst and regulated by water-retention mechanisms through anti-diuretic hormone (vasopressin, ADH). During severe dehydration, craving for water becomes of paramount import. When there is thirst there is hypersecretion of ADH from the neurohypophysis. Water intake and ADH secretion are regulated by enteroreceptors. There is probably a lower stimulus for initiating ADH secretion than that governing the desire to drink. The physiological situations that arouse thirst and hypersecretion of ADH are:

- deficit of water without corresponding loss of sodium (hypovolaemic hypernatraemia)
- osmotic shift of water from the cells to the extracellular fluid produced by excess sodium intake (hypervolaemic hypernatraemia)

In both situations there are cellular dehydration, hyperosmolar body fluids and increased extracellular sodium concentration. Dehydration causes increased osmotic pressure in the blood and in severe instances in cells, the consequences are thirst and ADH secretion. Some 90% of the osmolality of the interstitial fluid and the blood plasma is provided by sodium and associated anions. Sodium is excreted continuously from cells in exchange for potassium by active enzymatic cation transport. The osmotic regulation of water intake is a protector of normal plasma sodium concentration. There is a close relationship between plasma ADH concentration and plasma osmolality.

Osmolality and thirst

The average normal plasma osmolality is 287 mosmol/kg. A 2% increase in total body water suppresses ADH secretion below detectable level and induces maximal urine excretion and dilution. The thirst threshold is reached at 2% deficit of body water, average plasma osmolality 294 mosmol/kg.

The thirst threshold provides an effective source of water in response to the increase in plasma ADH when the renal action of the hormone can no longer prevent an undue increase of plasma sodium and osmolality.

There is also volume regulation of water intake and ADH release which complements the osmotic regulation. These volume regulators include the effective circulating blood volume, the cardiovascular reflexes and the renal renin–angiotensin system. This volume regulation is secondary to osmotic regulation during moderate fluctuations of the extracellular fluid volume. More than 20% of the blood volume needs to be lost before the thirst mechanism becomes activated. About a 10% reduction of the blood volume is required before there is an increase in plasma ADH. Afferent impulses moderate the osmotic regulation of water intake and ADH secretion. This regulation is predominantly by cerebral sensors which are stimulated when the carotid blood osmolality is increased with sodium salts and other cell-dehydrating substances, e.g. fructose and sucrose. The sensors for thirst and ADH secretion appear to be located close to the cerebroventricular system,

particularly in the anterior wall of the third ventricle. Arterial baroreceptors and left arterial pressure receptors create a tonic inhibition of neurohypophyseal ADH release.

Thirst is not permanently satisfied until sufficient water has been absorbed to bring the activity of cerebral sensors to a level below that required to stimulate drinking. Drinking-induced temporary depression of the thirst drive is mainly activated by the oropharangeal region and also from the stomach, suggesting mechanical, thermal and chemical factors are involved. It is possible that there are water taste fibres in the oropharangeal region. If too much fluid is drunk then the consequence is a water diuresis.

A craving for water that persists in the absence of known osmotic and non-osmotic thirst is called **primary polydipsia**.

Water is lost from the body as urine, in faeces and by evaporation from the skin and lungs. The amount lost from the skin and lungs is very temperature dependent. The visible water loss is readily measured as urine and faecal water, though the faecal water weight is usually very small. The evaporation water loss is 'invisible weight loss' which is the weight of food and liquid consumed plus/minus any change in body weight minus the weight of urine and faeces.

The evaporation water loss equals the invisible water loss, minus weight of CO_2 expired plus the weight of CO_2 absorbed. The intake and the sensation of thirst depend upon an equilibrium between fluid intake and fluid loss. **Sweating** is an important cause of water loss with rates as high as 2500 ml/hour in hot climates. Water loss of 500 ml/hour is not unusual. Expired air is saturated with water vapour and the water loss is of the order of 300 ml/day. When the air is very dry or during hyperventilation, losses may be considerably increased.

Water is also lost through the gastrointestinal tract. In Britain the usual faeces weight is 100–200 g/day but in diarrhoea the water loss may become considerable, as in cholera. Other causes of profound loss are intestinal fistula or persistent vomiting.

Renal excretion of water in part is dependent upon fluid intake, which in turn determines urine output. Urine is more concentrated than blood and there are renally induced concentration mechanisms. The blood usually has an osmolality of just under 300 mosmol/kg and urine of the order of 1200 mosmol/kg. The main constituents of the urine are 80% nitrogen end-products and sodium chloride, though there are a host of other substances present in the urine. These account for less than 50% of the osmolality.

Diet and urine osmolality

In general, the effect of diet on urine osmolality is: 1 g of dietary nitrogen leads to approximately 2 g of urea. Therefore, when an individual eats 100 g/day of protein, this yields 30 g of urea and results in urine with an osmolality of 500 mosmol. Thus, 2 g of sodium chloride will give an osmolar load of 340 mosmol. The protein and salt in the diet will lead to a urine osmolality of 840 mosmol. In order to excrete this with an osmolality of 1200 mosmol, the individual must pass 830 ml of urinary water, which is the obligatory water required to excrete the chemicals excreted by the kidney. The additional water excreted which reflects excess fluid intake is known as **free water**.

Osmolality in the urine is also under the regulation of the posterior pituitary gland, antidiuretic hormone and osmoreceptors present in the hypothalamus.

8.1.3 Clinical signs of water depletion

Water depletion may result from lack of available water, inability to ingest water, increased losses from skin, lungs, alimentary tract and urine. These occur in association with a hot environment, excessive exercise, hyperventilation, high altitudes, prolonged vomiting and diarrhoea, osmotic diuresis (as in diabetes mellitus) and loss from fistula or nasogastric tube suction.

Evidence of loss of water includes: sunken features (particularly the eyes), the skin and tongue are dry, and the skin becomes loose and lacks elasticity. A useful symptom and sign is a reduced urine output; this indicates the need for increased water intake. The most important sign, however, is **haemoconcentration** in which there is an increase in the blood urea and possibly – but not always – increased plasma sodium and potassium.

In severe water loss, caused for example by diarrhoea in cholera and other enteric infections, oral water with glucose and sodium chloride is the basis of therapy and restoration of fluid volume.

8.1.4 Excessive water intake

Excess water intake may rapidly induce hyponatraemia and cause pulmonary oedema. Alternatively, cerebral damage may lead to essential hypernatraemia in which there is an effect on the osmotic regulation of water intake and ADH release.

Water on its own is drunk in excess only in illness (polydipsia). More frequently, fluid intake is dictated by social circumstances, and the vehicle for the water, e.g. beverage or alcoholic. The effect of excess then becomes entwined with the congener, e.g. excess beer or coffee. An immediate effect of increased fluid intake is increased urinary output which is dictated by plasma osmolality, plasma ADH concentration and urinary osmolality.

If an individual drinks water in excess symptoms do not occur until the plasma sodium concentration falls below 120 mmol/litre. Confusion and headache may be followed by coma and fits in extreme instances.

Oxytocin infusions can occasionally result in severe and dangerous water intoxication in pregnant women.

Abnormal fluid intakes are usually due to abnormalities in one or several of the control mechanisms.

Diabetes insipidus

- Primary: this condition results from an abnormality of the synthesis, storage or release of ADH in the anterior hypothalamus of the brain.
- Secondary: this usually follows damage to the hypothalamus either by trauma, birth injury, during some procedures, tumour growth, abscess formation, infection or vascular thrombosis.

Psychogenic polydipsia

Drinking bouts alternate with normal intake. This results in varying plasma osmolality; unlike the constant large urine output and consistently reduced plasma osmolality of diabetes insipidus. Fluid intake may exceed 40 litres/day with attendant low osmolalities of under 240 mosmol.

Psychoactive drugs

Drugs used in the treatment of depression and schizophrenia, e.g. thiothixene, amitriptyline and chlorpromazine, cause dryness of the mouth with resultant polydipsia.

Ecstasy (MDMA; 3,4-methlyenedioxymethamphetamine) has metabolic effects over and above the mood enhancing effects of other drugs used at dance events (raves), e.g. lysergic acid (LSD), herbal ecstasy, khat and Eve. Ecstasy has mild amphetamine stimulant actions which release brain messenger transmitters, especially serotonin (5-hydroxy tryptamine). This results in increased wakefulness, euphoria, sexual activity, adrenaline release with tachycardia, raised blood pressure and metabolic hyperactivity. Exercise increases these pharmacological effects which may result in malignant hyperthermia and severe dehydration especially if there is little access to fluids. Death results from hyponatraemia, convulsions and acute renal failure. Rarely another condition occurs, due to rapid release of ADH from the pituitary. This can lead to brain damage or death. In this variant condition, excess water is contra-indicated.

Renal resistance to ADH

This is found in inherited nephrogenic diabetes insipidus, potassium depletion, hypercalcaemia and may be induced by drugs, for example lithium carbonate (in 40% of patients), dimethylchlortetracycline, amphotericin B, methoxyfluorane anaesthesia, propoxyphene and gentamicin.

Kidney disease

Kidney disease of all types reflects an inability to concentrate urine.

Solute diuresis

This is one of the cardinal symptoms in diabetes mellitus (glucose) and uraemia (urea).

8.1.5 Recommended requirements

Adults

Water intake includes fluid drunk and the water in ingested food. There is also metabolic water produced by the oxidation of carbohydrates, protein and fat:

- 1 g starch produces 0.60 g water
- 1 g protein produces 0.41 g water
- 1 g fat produces 1.07 g water

Fluid intake is usually of the order of 2–2.5 litres/day for the average adult in a temperate climate. This intake will be affected by temperature, activity, diet and health. It is necessary for the kidneys to excrete a number of water-soluble compounds which if retained are toxic and lethal.

Babies

The renal system of newborn babies takes several days to adjust to extrauterine life. During the first 2 days, daily urine output is of the order of 20 ml. By 2 weeks, daily output is 200 ml, with a milk intake of 500 ml. At 3 months the daily milk intake is 800–900 ml and urine output 300 ml.

Loss by evaporation is high because of the high surface area-to-body weight ratio.

Elderly

Age may affect the thirst mechanisms. The secretory response to ADH with water deprivation is increased in the elderly; thirst is less of a feature in the elderly. Consequently it is important to monitor fluid intake in the elderly as fluid needs are not affected.

8.1.6 Body store measurements

Body stores are dependent upon fluid intake and urinary output. Methods to quantify water stores include:

- monitoring of body weight over a period of time
- water intake/output volume measurements
- blood urea levels
- blood osmolality
- tritiated water dilution studies

Summary

1. Water is both a bulk and localized solvent for the body. Water is an angular molecule with two vertical planes of symmetry and is an acceptor and donor of protons.
2. Water is the basic chemical of life which readily dissolves a number of chemicals. Such solubility is important in biological processes, cell structure, blood and excretory systems, e.g. urine, bile. Equally important are the lipid phases which allow a separate and distinct localization of function.
3. The body is a complex phase diagram with differing areas defined by boundaries of osmolality and by solubilities. The movement of substances between these phases is important metabolically and physiologically.
4. Water forms 50–60% of body weight. One-third is extracellular fluid, two-thirds are intracellular. These compartments are separated by cell membranes, often freely permeable to water movement. This movement is dictated by osmolality.
5. Humans satisfy their requirements for water by drinking habits which are not necessarily dependent on true thirst. The efficient thirst mechanism and regulation of the water retention through antidiuretic hormone (ADH, vasopressin) is important for water balance.
6. It is essential to well-being that the body contains a sufficiency of water. Insufficient or too much water or reduced or excessive loss result in well-characterized clinical conditions. Such imbalances can occur both from natural or pathological causes.

Further reading

Anderson, B, Leksell, L.G. and Rundgren, M. (1982) Regulation of water intake. *Annual Review of Nutrition*, **2**, 73–89.

Atkin, P.W. (1990) *Physical Chemistry*, 4th edn, Oxford University Press, Oxford.

Desjeux, J.F., Nath, S.K. and Taminiau, J. (1994) Organic substrate and electrolyte solutions for oral rehydration in diarrhoea. *Annual Review of Nutrition*, **14**, 321–42.

Weatherall, D.J., Ledingham, J.G.G. and Warrell, D.A. (eds) (1996) *Oxford Textbook of Medicine*, 3rd edn, Oxford Medical Publications, Oxford.

8.2 Minerals and electrolytes

- Minerals and electrolytes are essential dietary constituents.

- Minerals and electrolytes act as charge carriers, e.g. sodium and potassium or are important in the conformation of proteins or act as enzyme cofactors or redox catalysts.
- Minerals and electrolytes are important in body structure, the skeleton and proteins including haemoglobin.

8.2.1 Sodium

Introduction

The distribution of sodium in the body is quite distinct from that of potassium, despite their total body pools being similar.

- atomic weight, 23; valency 1
- natural isotope 23
- relative abundance on earth's crust, 2.64%

Availability

Common sources of sodium in the diet are the addition of salt to food at the table, to cooking, and in processed foods. Salt is an important food preservative which has been used since the beginning of time in the salting of meat and fish in order that populations could survive long winters and periods of hardship.

The sodium content of natural food varies between 0.1 to 3.3 mmol/100 g. In contrast, processed foods have a sodium content of 11–48 mmol/100 g, partly for taste, and in part because sodium nitrate is used as a preservative.

Sodium is a major cation and contributor to the osmolality of the extracellular fluid of the body – one-third of the body water in the case of adults.

Sodium transport

Sodium is important in the transport of chemicals across selectively permeable cell membranes. This may depend upon osmotic gradients, as with water. The transport of other chemical substances in and out of the cell requires transport systems which range in complexity from passive diffusion processes to selective active energy-requiring transport systems.

Glucose and amino acids cross the cell membrane by a process of facilitated diffusion in response to a concentration gradient. Such a system is dependent upon carrier molecules. This accelerated process often takes place against a concentration gradient, achieved by a sodium co-transport system, with work being needed for the process to continue.

The binding of glucose to the cell membrane carrier is increased in the presence of sodium. Sodium is present in high concentrations at the luminal border of the membrane, but levels at the inner cell surface are low, and kept low by the Na^+/K^+ exchange pump. When the carrier–sugar–sodium complex reaches the inside of the cell the glucose leaves the carrier, the affinity of the receptor for glucose falls, the sugar is released and the sodium is pumped back to the outside.

Sodium/potassium balance

There is an influx of two potassium ions as three sodium ions are pumped out. In this way the negative potential within the cell is maintained. The presence of Mg-ATP is essential for this transport system. Energy for the carrier is provided by ATP hydrolysis, and a sodium/potassium ATPase enzyme.

The sodium co-transport is a widely distributed cellular mechanism. The active transport of sodium out of the cell uses some 20–45% of the total energy derived from cellular metabolism.

Sodium and potassium movement across nerve membranes occur during **nerve conduction**. The changes in concentration of these ions are involved in the excitation of the neural membrane. There are changes in the membrane permeability during nerve conduction and sodium and potassium ions move in and out. Sodium penetrates faster than potassium, which increases the electrical charge.

Absorption of sodium

Intestinal sodium absorption is virtually complete in the small intestine and colon. Sodium is absorbed by a variety of processes. In the proximal intestine sodium is absorbed in part by a solute-dependent co-transport system and is involved in nutrient absorption. In the more distal intestine and colon sodium absorption is a sodium/hydrogen interchange; in the colon this process is coupled to chloride/bicarbonate exchange. In the distal intestine and colon the process is coupled and electroneutral. These processes involve protein carriers. In the distal colon active sodium transport occurs against an electrochemical gradient.

Water absorption is a passive process which requires active transport of sodium and chloride. The optimum absorption of water occurs when the concentration of glucose in the intestinal lumen is around 110 mmol/litre. This finding has been of great importance in the development of oral rehydration solutions (ORS) (see below, 'Treatment of sodium loss').

Thirst

The requirement for salt expressed as thirst sensation is controlled by the thirst – adrenocortical hormone–renal response to changes in plasma sodium concentration. Hyponatraemia reduces ADH secretion, which is followed by renal loss of water and correction. Hypernatraemia results in thirst and there is an increased water intake. Changes in sodium concentration result from changes in water intake rather than the converse.

Body sodium content is affected by renal regulation of urinary loss of sodium of over 1–500 mmol daily.

Sodium regulation

Most of the body's sodium pool is contained in the extracellular fluid compartment. Sodium is found in significant amounts in bone, but this pool is not readily available at times of rapid loss of sodium.

Sodium content of the body

A male adult weighing 65–70 kg has a total body sodium content of 4 mol (92 g):

- 500 mmol (11.5 g) in intercellular fluid (concentration 2 mmol/l)
- 1500 mmol (34.5 g) in bone
- 2000 mmol (46 g) in extracellular fluid (concentration 130–145 mmol/l)
- Daily dietary intake is 50–200 mmol (1.15–4.6 g)

The regulation of extracellular fluid sodium content relates very closely to extracellular fluid volume control. The control of the latter is a result of changes in pressure and distension in the cardiac atria and right ventricle, the pulmonary vasculature, the carotid arteries and the aortic arch. These activate centres in the medulla and hypothalamus of the brain. When extracellular fluid or blood volume falls, neural sympathetic activity increases, and the response is vasoconstriction and a redistribution of renal blood flow. There is reduced glomerular filtration and increased sodium and water retention. In addition, there is increase in renin production, circulating angiotensin II, noradrenaline, adrenaline and adrenocorticotrophin and antidiuretic hormone.

Sodium excretion

Sodium is filtered from the plasma in the kidneys, the reabsorption of sodium occurring as an osmotic phenomenon in the proximal tubule, Henle's loop and distal tubule. Distal tubular absorption is very important, and under the control of the atrial natriuretic factor. Renal sodium excretion is also controlled by angiotensin II, prostaglandins and the kallikrein–kinin system.

Sodium depletion

A reduced body sodium results in reduced extra-cellular fluid volume. Sodium is lost largely via the urine, with only minimal loss occurring via the faeces or skin – unless there are abnormal situations such as diarrhoea or excessive sweating.

Increased sodium loss in urine can occur in diabetes mellitus, in Addison's disease (adrenal cortical insufficiency), following excessive doses of diuretic drugs, and in cases of renal tubular damage, as in chronic renal failure.

Sweating

It is possible to lose significant amounts of water and electrolytes during sweating. Up to 3 litres/hour may be lost in continuous hard physical exercise. The sodium content of sweat is of the order of 20–80 mmol/litre. Consequently, exercising hard in a hot climate may result in the loss of a significant amount of water and sodium. In situations where there is increased loss of sweat and exercise, for example marathon running, daily losses of sodium in the sweat may be up to 350 mmol.

Alimentary tract losses

It is possible to lose substantial amounts of water, sodium and potassium in diarrhoea. These losses may be quite rapid during infections such as cholera.

Clinical signs of sodium depletion

The signs of sodium depletion are often non-specific psychological and behavioural changes. The rate of decrease in body sodium is perhaps closely associated with the onset of these symptoms. A slow change is often more tolerable than a rapid loss.

In true sodium depletion there is an avid retention of sodium in the plasma, even if the concentration is reduced. The conservation is reflected by a urinary sodium of less than 10 mmol per 24 hours.

A low blood sodium is not always due to sodium depletion. In inappropriate ADH excess there is a normal body sodium but a retention of water; this leads to an apparent decrease in plasma sodium. The phenomenon is recognized by the plasma osmolality being reduced or the urinary sodium being more than 50 mmol over 24 hours. The causes of inappropriate ADH syndrome include malignancy, e.g. carcinoma of the lung. Other causes include intakes of fluid in excess of urinary excretion (or use of the drug ecstasy, see page 246). The normal excretion of water is 10–20 ml/minute. This may be exceeded by intake in certain patients with emotional problems and by avid beer drinkers (beer drinker's potomania). Excessive intravenous administration of water with glucose may produce a similar hyponatraemia. Cardiac failure, cirrhosis or renal failure may also produce a dilutional hyponatraemia.

Excess sodium intake

An excess of sodium leads to an increase in body extracellular volume, unless the urine is cleared by the kidneys. The most common cause is, however, secondary to water loss, or iatrogenic sodium excess. Insufficient water intake relative to sodium is also an important cause.

Infants can develop this problem if milk powder is made up with insufficient water, as the infant's ability to excrete sodium is somewhat slow to develop. With a high plasma sodium there is even a risk of infantile convulsions.

In adults, especially the elderly, congestive cardiac failure may occur.

Recommended requirements

The dietary intake of sodium varies from population to population (100–200 mmol/day in Britain). Conventional dietary wisdom is to reduce the intake of sodium to assist blood pressure control. The recommended dietary intake is not universally agreed; suggested figures include an LRNI of sodium for adults of 25 mmol/day with an RNI of 70 mmol/day. The LRNI in infants up to 6 months, based on calculations for breast-fed infants, should be approximately 6 mmol/day and an RNI of 9–12 mmol/day.

Treatment of sodium loss

In the severe loss of water and sodium through infectious diarrhoea of a secretory nature, e.g.

Vibrio cholerae or enterotoxigenic coliforms, the treatment is initially using oral (water, sodium and glucose) replacement solution (ORS) to replace the water volume. Following the enterocyte absorption of the glucose and sodium, water is absorbed by a passive process. Sucrose or rice powder are also useful sources of carbohydrate. The additional benefit of rice powder, other than its availability in poor areas of the world is the

---

Oral rehydration solutions

A solution of 90 mmol/litre of sodium and sugar or rice powder of 30 g/litre for adults and 30 mmol/litre of sodium for infants and the very ill has been recommended by some authorities. Potassium at concentrations of 20 mmol/litre is important. Bicarbonate at 30 mmol/litre is helpful in correcting the metabolic acidosis in severe dehydration.

---

constituent amino acids. Glycine (at 30 mg/100 g) stimulates sodium and water absorption. There is still debate over the correct concentration of sodium in ORS for adults and the young. In

children there is concern of hypernatraemic dehydration.

Such ORS should be the basis of clinical practice for diarrhoea treatment. Millions of babies have been saved by this simple preparation, though the very ill may need admission to hospital where it is often necessary to give intravenous fluids which are isotonic with blood using either sodium chloride and glucose.

Measurement of sodium status

The 24 hour urinary excretion of sodium is generally a good indicator of dietary intake. Normal faecal excretion of sodium is of the order of 2–4 mmol/day, and of little physiological consequence, though in diarrhoeal states the faecal sodium becomes significant and must be measured. Twenty-four-hour urinary excretion of sodium accounts for 95–98% of dietary sodium intake. The 'within-person' variability in sodium excretion is 30%; 24-hour urine collections may be incomplete and hence there will be poor correlation between individual estimates of diet and individual estimates of urine sodium excretion. For good results at least 7 days of urine and diet collections are needed to get good correlations.

Summary

1. The distribution of sodium and potassium in the body is quite distinct while the total body pools are similar. Sodium is a major cation and contributor to the osmolality of the extracellular fluid of the body. Sodium is important in the transport of chemicals across cell membranes. Sodium and potassium movements across nerve membranes occur during nerve conduction.
2. The requirement for salt, or rather thirst sensation, is controlled by the adrenocortical hormone–renal response to changes in plasma sodium concentration. Hyponatraemia reduces ADH secretion, which is followed by renal loss of water and correction. Hypernatraemia results in thirst and there is an increased water intake. Changes in sodium concentration result from changes in water intake rather than the converse.
3. Sodium is lost in urine, sweat or faeces under a variety of physiological and pathological circumstances.
4. In severe secretory diarrhoea, oral replacement solutions (ORS) i.e. water : glucose, sucrose or rice powder (30 g/l) and sodium 90 mmol/l or 30 mmol/l in infants should be used to replenish sodium levels.

Further reading

Dietary Reference Values for Food Energy and Nutrients for the United Kingdom (1991) Report on Health and Social Subjects, no. 4l, HMSO, London.

Weatherall, D.J., Ledingham, J.G.G. and Warrell, D.A. (eds) (1996) *Oxford Textbook of Medicine*, 3rd edn, Oxford Medical Publications, Oxford.

8.2.2 Potassium

- atomic weight, 39; valency 1
- natural isotopes, 39,40
- abundance in earth's crust, 2.4%

Availability

In natural and processed foods the potassium content varies from 2.8 to 10 mmol/kg. Dietary potassium intake tends to be derived from fresh vegetables and meat.

Transport and absorption of potassium

The transport of potassium into cells is under the control of the Na/K ATPase enzyme, and allows transport of potassium against a concentration gradient. The ratio of extracellular to intracellular potassium concentration is important in the membrane potential difference in cells.

Over 90% of dietary potassium is absorbed in the proximal small intestine. In the small intestine potassium absorption is passive but in the colon, it is an active process. In the sigmoid colon absorption is mediated by a K/H^+ mechanism.

Body stores of potassium

Most of the potassium is intracellular; that is, in the cell fluid compartment. An adult male of approximately 70 kg, would contain 2800–3500 mmol (110–137 g), of which 95% is intracellular (150 mmol/litre). Cellular potassium concentrations are affected by pH, aldosterone, insulin and the adrenergic nervous system

The plasma concentration at 3.5–4.5 mmol/litre is dependent upon intake, excretion and the balance between extracellular and intracellular compartments. There is a direct reciprocal relationship between plasma potassium and aldosterone production. Control is mainly through urinary loss with some additional colonic loss. Insulin excretion is increased when the plasma potassium increases, possibly provoking cellular uptake of potassium.

Potassium homeostasis

The body homeostasis of potassium is controlled by renal glomerular filtration and tubular secretion.

Renal conservation of potassium

Proximal tubular reabsorption is partially an active process and is complete by the end of the proximal segment. There is potassium excretion in the pars recta and descending limb of Henle's loop with further control through absorption in the ascending limb.

Chronic increased dietary potassium intake increases potassium secretion via the kidneys. There is an associated degree of hyperaldosteronism. Increased sodium coming to the distal nephron results in an increased simultaneous urinary loss of potassium.

Excretion

Potassium is largely lost in the urine, though 10% of the daily loss occurs through the distal ileum and colon. Small amounts are lost in sweat and vomit.

Potassium deficiency and excess

Deficiency

This can occur as a result of vomiting and diarrhoea and chronic usage of purgatives. It may also occur as a result of urine loss in wasting disease and starvation, overdosage with drugs (for example, diuretics and corticosteroids), endocrine disturbances, aldosterone excess, in Cushing's syndrome, and in hypertension. Deficiency may also occur in hepatic cirrhosis and in renal disease.

During the breakdown of tissue, there is an

important loss of potassium as in underfeeding, diabetes and after injury. The loss of 1 kg muscle mass results in the loss of 105 mmol of potassium and 210 g of protein (equivalent to 34 g of nitrogen).

Potassium depletion results in muscular weakness and mental confusion, and is reflected in electrocardiographic changes and loss of smooth muscle motility, as in the intestine.

Rarely, potassium deficiency is due to familial periodic paralysis when there is episodic over-secretion of aldosterone by the adrenal cortex. Such a non-specific diagnosis may be sought when there is excess vomiting or at any time when the patient appears not to be in good health.

Excess

Plasma potassium concentrations in excess of the normal may occur in total parenteral nutrition, renal failure dialysis and in liver failure. If not corrected, cardiac arrhythmias and even cardiac arrest may result.

Recommended requirements

Habitual daily potassium intakes are maintained at suitable levels to ensure optimal metabolism of potassium. Reported potassium intakes by Western populations are in the range of 40–150 mmol/day. Potassium replacement, if performed, should be gradual; a 4-g tablet of potassium chloride provides 53 mmol potassium.

Measurement of potassium status

The urinary excretion of potassium is generally a good indicator of dietary intake. Faecal losses vary from 5–13 mmol/day, which is approximately 11–15% of the dietary intake. The 'within-person' variation in potassium excretion is 24% for each 24-hour collection. In populations where diarrhoea is endemic, potassium loss in the faeces may be 30% of the dietary intake.

Sources of inaccurate results in balance studies are loss through sweating and breast feeding.

Summary

1. The body distribution of potassium differs from that of sodium; body pools are, however, similar.
2. The extracellular : intracellular potassium ratio is important in the membrane potential difference in cells. Most of the potassium is in the cell fluid compartment.
3. The body homeostasis of potassium is controlled by renal glomerular filtration and tubular secretion. Proximal tubular reabsorption is partially an active process and is complete by the end of the proximal segment.
4. Potassium is lost in urine or faeces under a variety of physiological and pathological circumstances.

Further reading

Dietary Reference Values for Food Energy and Nutrients for the United Kingdom. Report on Health and Social Subjects, no. 4l. HMSO, London.

Weatherall, D.J., Ledingham, J.G.G. and Warrell, D.A. (eds) (1996) *Oxford Textbook of Medicine*, 3rd edn, Oxford Medical Publications, Oxford.

8.2.3 Calcium

- atomic weight, 40; valency, 2
- natural isotopes, 40, 42, 43, 44, 46
- abundance in earth's crust, 3.39%

Availability

The most important source of dietary calcium is **milk**, which may provide over half of the intake. The calcium content of milk is 35 mg/100 ml for human milk and 120 mg/100 ml for cows' milk. Other important sources of calcium include cheese (hard 400–1200 mg/100 g, soft 60–75 mg/100 g). Of less but certainly significant value are nuts (13–250 mg/100 g), herring, vegetables, eggs, cereals, fruit (20–70 mg/100 g). Meat and rice are very modest sources.

Function as a nutrient

Calcium is concentrated in the organelles and blood, and is also very important in the structure of the skeleton and in the maintenance of the extracellular fluid calcium concentration.

Skeletal calcium

A positive calcium balance is required before bone growth can occur. Calcium intake and skeletal modelling and turnover determine calcium balance during growth. Of the total calcium pool most is retained in the bony skeleton which contains some 1 kg of calcium. (see Bone structure, section 12.3).

Non-skeletal calcium function

Half of the plasma calcium (2.25–2.6 mmol/litre) is ionized, the remainder being bound to albumin (40%) and globulin (10%). This ionized fraction affects humoral controls, important in dictating intestinal absorption, renal loss and calcium bone metabolism. A small (approximately 25 mmol) but very important part of the total calcium pool is that of calcium in soft tissues and extracellular fluid.

Intracellular concentrations of calcium are 100 times less than those in the extracellular fluid. There is a substantial concentration of calcium in mitochondria. Hormonal and pharmacological activation of cells, membrane function and enzyme activity may all be affected by the local concentrations of calcium. Calcium is bound within the cell to enzyme proteins; this binding alters the protein configuration and hence enzyme activity. Calcium has a wide range of activity, being involved in muscle contraction, endocytosis, exocytosis, cell mobility, the movement of chromosomes and release of neurotransmitters. Calcium acts as a intracellular messenger.

Calcium-regulating hormones

The major regulating hormones for calcium are parathyroid hormone, calcitonin, vitamin D and to a lesser extent growth hormone, thyroid hormone, adrenal steroids, sex hormones and some gastrointestinal hormones.

Parathyroid hormone (PTH)

This is secreted as a single peptide chain of 84 amino acids (molecular weight 5500 Da), of which only the first 32–34 amino acids have biological activity. The extra amino acids form a pro-hormone. There is some cleavage of these extra amino acids in the liver producing biological activity. The separated polypeptide fragments are degraded in the liver, kidney and skeleton. A reduction in the ionized status of calcium activates PTH excretion. A number of hormones also influence PTH secretion. PTH directly affects body calcium through bone resorption and kidney proximal and distal tubular calcium reabsorption.

PTH stimulates the 1-α-hydroxylase enzyme which is involved in the hydroxylation of 25(OH) vitamin D_3 to 1,25(OH)$_2$ vitamin D_3. This conversion results in increased calcium absorption from the intestine.

Calcitonin

This consists of 32 amino acid residues with a disulphide bond between cystine residues in positions 1 and 7. Increased secretion results from increases in calcium concentration. The source is the parafollicular cells of the thyroid. Hormone secretion is increased by glucagon, gastrin and beta-adrenergic hormones. Bone resorption is inhibited, tubular resorption of calcium, sodium, phosphate, magnesium and potassium is decreased. Gastrointestinal hormone secretion is inhibited.

Calcitriol

Calcitriol [1,25(OH)$_2$ vitamin D] increases the availability of calcium and phosphorus by increasing bone resorption.

Calcium absorption and balance

Calcium absorption is largely from the jejunum but may also occur in the ileum and colon. The predominant absorptive process is by active transport and there is also some simple passive diffusion in the ileum. Calcium absorption is by paracellular and transcellular pathways. The transcellular pathway consists of:

- crossing the brush border
- transport though the cytoplasm complexed with specific calcium-binding protein; these calcium-binding proteins are synthesized following the binding of a vitamin D metabolite to a nuclear receptor in the enterocyte
- active extrusion of calcium from the basal lateral membrane into the blood stream

Calcium and phosphate may well be absorbed separately from each other. The average dietary intake of calcium is 25 mmol (1000 mg), of which 10 mmol is absorbed (Figure 8.1). How much is absorbed is very much dependent upon age, skeletal and metabolic requirements. The amount absorbed and retained will be regulated by $1,25(OH)_2$ vitamin D_3, hormone and PTH activity. During infancy and adolescence a greater proportion of the dietary calcium in the diet is absorbed than at other times of life. Intestinal absorption of calcium decreases with advancing age. Phosphorus has no effect on calcium absorption and retention.

Phytate binds calcium within the intestinal lumen, which affects its absorption. Phytic acid, a constituent of wholemeal flour, is the hexaphosphoric acid ester of inositol [(1,2,3,4,5)-(6-hexakis) (dihydrogen phosphate, myo-inositol)]. Calcium forms insoluble salts with phytate which are not absorbed from the intestine. Calcium absorption is compromised when there is heavy dependence on wholemeal flour.

Fatty acids form insoluble soaps with calcium, though the effect on absorption of calcium is unknown.

The renal filterable calcium is approximately 60% of the total plasma calcium. In health, 97%

of the calcium is reabsorbed from the kidney. The urine contains usually between 100–350 mg of calcium per day (Figure 8.1). There is effective conservation and reabsorption of calcium by the proximal and distal renal tubules. Several hormones are involved, including PTH. There is increased absorption of calcium and decreased tubular absorption of phosphate. This is achieved by cyclic AMP and activation of adenylate cyclase in the renal cortex. Bicarbonate proximal renal reabsorption is decreased. All of this is determined by dietary calcium, protein and sodium intake. With age there is a reduction in calcium glomerular filtration rate. This is dependent on the individual and may be increased during summer, presumably because of sunlight and vitamin D status. There is some increase in urinary calcium in women after the menopause. Urinary excretion of calcium is less when dietary protein is reduced and rises when it is increased.

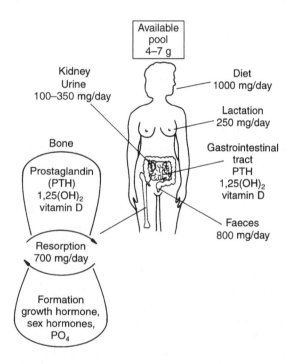

FIGURE 8.1 Calcium fluxes in the body through the gastrointestinal tract, lactation, bones and urine. PTH, parathyroid hormone.

Urinary excretion figures are imprecise, particularly if intestinal absorption is low: 70% of calcium at low intake to 30% at high intake. Calcium excretion is also dependent on salt and protein intake.

Intestinal biliary, pancreatic and intestinal secretions account for a daily enteric loss of 5 mmol calcium. Calcium is excreted in urine but there is also loss in faeces, sweat, skin, hair and nails. The loss to the gastrointestinal tract of approximately 400 mg of calcium per day is through bile, pancreatic secretions and desquamated cells from the mucosal lining.

There is some colonic absorption of calcium dependent on poorly understood factors.

Calcium deficiency and excess

Deficiency

Parathyroid hormone deficiency leads to hypocalcaemia. A reduction in plasma calcium results in tetany, which is due to hyperactivity of the motor muscles. This results in facial spasm, spasm in the wrist and metacarpophalangeal joints.

Excess

Hypercalcaemia may occur in infants who have been given an excess of vitamin D as with cod liver oil fortification of infant foods. The infant loses appetite, vomits, loses weight, is constipated and has a characteristic facial appearance. The calcium concentration in the plasma is increased with that of urea and cholesterol. The blood pressure may increase. Prolonged increased plasma concentrations result in calcification of the heart and kidneys and even cerebral damage and death.

In adults, hypercalcaemia can occur from hyperparathyroidism or excessive dosage of vitamin D. This can occur in individuals who have had peptic ulcer surgery and to whom calcium has been given on an empirical basis.

Recommended requirements

It is difficult to define calcium requirements because of its long-term accumulation; hence the study of calcium metabolism is extremely complex.

The skeleton of a newborn infant contains 25 g of calcium and, during the first years of life, the calcium content of the body increases quite quickly. Vitamin D and dietary calcium are important dictates of calcium balance.

Calcium accumulation in early **infancy** is approximately 150 mg/day and this declines to approximately 100 mg/day by the age of 3 years. Thereafter, there is a steady increase to a maximum at puberty, during the **adolescence** growth spurt, an effect which occurs to a greater extent – and later – in boys than in girls. Thereafter there is a fall in calcium uptake, with a peak in adults of between 250 and 400 mg/day. In children, approximately 20–30% of dietary calcium is absorbed, compared with the infant where the absorption rate is 40–60%. Calcium uptake increases in pre-adolescence and puberty, during which time 45% of the adult skeleton is formed. Bone mineral content increases at approximately 9% a year. There is also substantial urinary loss of calcium so that an enhanced absorption rate is necessary. The majority of individuals have reached the peak bone mass by the age of 30.

Between the ages of 18 and 30 years, bone density increases by a further 10% and 120 g of calcium (about 25 mg/day) is added to the skeleton during this period (see Bone structure, section 12.3).

Another period of enhanced calcium requirement is during **pregnancy**, which requires the absorption by the mother and utilization by the foetus of a total of up to 30 g of calcium.

Within Great Britain, the average daily calcium intake for adults is 940 mg (24 mmol) for men and 730 mg (18 mmol) for women. The average intake varies from country to country (range 350–1200 mg; 9–30 mmol), the range being influenced by milk intake and food fortification policies. A calculation of required calcium intake requires an identification of calcium needs for growth and maintenance.

The bioavailability of calcium for absorption in the intestine is important in establishing the DRV.

It has been estimated that the dietary intake necessary to achieve calcium absorption of 4 mmol (160 mg/day) is 20 mmol/day (800 mg) to achieve 20% absorption, and 5.8 mmol/day (230 mg) to achieve 70% absorption. It has proved exceedingly difficult to calculate EAR and figures are based on the retention of calcium. These vary considerably with age and sex.

Calcium intakes

Dietary calcium average intake in the UK is approximately 20 mmol (800 mg) per person per day, though this varies geographically. Typically, calcium is obtained:

- 12.8 mmol (512 mg) from milk and milk products
- 1.2 mmol (46 mg) from vegetables
- 4.5 mmol (181 mg) from fortified cereals
- 3.9 mmol (154 mg) from hard water

Infants
In early infancy calcium balance may be negative. The daily rate of calcium retention is approximately 4 mmol (160 mg). Breast-milk calcium is absorbed at about 66% efficiency, so that in the first year 6 mmol/day (240 mg) would be adequate. In contrast, absorption from infant formulae is about 40% so that the EAR is 10 mmol/day (400 mg). The RNI would be 13 mmol/day (520 mg).

Children
Calcium retention for skeletal growth increases from 1.8–3.8 mmol/day (70–150 mg) between the ages of 1 and 10 years. Absorption is approximately 35%; therefore this degree of calcium retention requires 7 mmol/day (280 mg) and 10.6 mmol/day (425 mg) at age 10 years. Corresponding RNIs are 8.8 and 13.8 mmol/day.

Adolescence
During adolescence 6.3 mmol/day (250 mg) for girls and 7.5 mmol (300 mg) for boys are retained. Absorption effectiveness is of the order of 40%. The EAR are 15.6 mmol/day (625 mg) and 18.8 mmol/day (750 mg).

Adults
The EAR has been recommended as 3.8 mmol/day (150 mg) plus an estimated 0.25 mmol/day (10 mg) for losses through skin, sweat, hair and nails. Absorption is estimated to be 30%. Therefore the EAR is 13.1 mmol/ (525 mg), RNI is 17.5 mmol/day (700 mg) and the LRNI is 10 mmol/day (400 mg).

Pregnancy
There is some mobilization of maternal calcium depots rather than a dietary increment during foetal growth. Bone density may diminish in the first 3 months of both pregnancy and lactation in order to provide an internal calcium reservoir which is replenished by 6 months. A deficiency of absorbed calcium arises during pregnancy compared with non-pregnant women. However, if pregnancy occurs in adolescence, then the growth requirements of the mother and foetus require a doubling of calcium provision.

Lactation
Food intake is increased during lactation with consequent benefits on calcium intake. A lactating mother will secrete some 150–300 mg (4–8 mmol). Approximately an additional 14 mmol/day (550 mg) of calcium may be required during lactation.

Postmenopause
Osteoporosis results from the loss of all components of bone, not just of calcium. Therefore the postmenopausal increase in bone loss – the main cause of osteoporosis – is not due to calcium loss. Oestrogen and other hormone deficiencies may be much more important. Oestrogen replacement can prevent and possibly restore bone loss.

Elderly

It is not clear that there is any need for an increase in calcium intake by individuals aged 60 or over.

Body store measurements

Plasma calcium, especially the ionized form, is the most consistent value in calcium metabolism. The

maintenance of a positive calcium balance is required to preserve the skeleton. However, study of the conservation of the skeleton is not easy in other than short-term studies. It has not proved easy to generate meaningful data in studies of calcium loss. Urinary calcium excretion reflects dietary intake and, if increased, may lead to the formation of kidney stones.

Summary

1. Calcium is concentrated in the body in organelles and blood. Calcium is very important in the structure of the skeleton and in the maintenance of the extracellular fluid calcium concentration. A positive calcium balance is required before growth can proceed. Calcium intake and skeletal modelling and turnover determine calcium balance during growth.
2. Hormonal and pharmacological activation of cells, membrane function and enzyme activity may all be affected by the local concentrations of calcium. Calcium is bound within the cell to enzyme proteins, this binding alters the protein configuration and hence enzyme activity. Calcium has many roles, including that of an intracellular messenger.
3. The major regulating hormones for calcium are parathyroid hormone, calcitonin, vitamin D and to a lesser extent growth hormone, thyroid hormone, adrenal steroids, sex hormones and some gastrointestinal hormones.
4. A reduction in plasma calcium results in tetany, which is due to hyperactivity of the motor muscles. This results in facial spasm, spasm in the wrist and metacarpophalangeal joints.
5. A sustained deficiency of dietary calcium can result in osteomalacia, a loss of calcium in the bone.

Further reading

Dietary Reference Values for Food, Energy and Nutrients for the United Kingdom. Department of Health Report on Health and Social Subjects, no. 41, HMSO, London.

Matkovic, V. (1991) Calcium metabolism and calcium requirements during skeletal modelling and consol-idation of bone mass. *American Journal of Clinical Nutrition,* 54, S245–60.

Weatherall, D.J., Ledingham, J.G.G. and Warrell, D.A. (1996) *Oxford Textbook of Medicine*, 3rd edn, Oxford Medical Publications, Oxford.

8.2.4 Iron

- atomic weight, 56
- iron is found in two forms, ferrous and ferric, which interchange: $Fe^{2+} \rightleftharpoons Fe^{3+} + e^-$
- natural isotopes, 54, 56, 57, 58
- abundance in earth's crust, 4.7%

There are many important sources of iron including meat, meat products, cereals, vegetables and fruits. A prime source is black pudding (20 mg/100 g), and most foods contain 1–10 mg/100 g. There are regional variations in iron content of plant foods dependent upon the soil in which the plants are grown. Milk is a poor source of iron.

In the normal adult there are 4–5 g of iron, 75% of which is in the form of haemoglobin (2.5 g), myoglobin (0.15 g), haem enzymes and non-haem enzymes. The remainder is stored as **ferritin** and **haemosiderin** in the hepatic reticulo-endothelial system, spleen and bone marrow and hepatic parenchymal cells. Women, because of menstruation, have a reduced iron content.

The iron must be readily available for the synthesis of essential iron proteins, e.g. haemoglobin and myoglobin and enzymes, e.g. catalase.

Iron is present in all cells of the body and plays a key role in many biochemical reactions.

Absorption and transport of iron

Absorption

Between 10 and 15% of dietary iron is absorbed to compensate for daily losses. Although 10–15 mg/day of iron is ingested, only 1–2 mg is absorbed in a steady state and is either taken up by bone marrow for haemoglobin or into the reticular endothelial tissue stores. There is increased iron absorption during growth, blood loss or pregnancy. Absorption takes place in the upper small intestine. Iron absorption is controlled by the mucosal cells and mediated by specific receptors on the intestinal mucosal surface. Dietary iron absorption is dependent upon overall iron stores.

Iron absorption from food depends on the form of the iron and other constituents in the diet. Iron is absorbed both as haem and non-haem iron. Most iron is present as haem iron in animal foods, in the Fe^{2+} form. Haem iron is absorbed more readily than inorganic iron from vegetable foods and is little affected by other constituents in the diet. Haem iron is absorbed from the intestine by a process which is different from non-haem iron. Within the intestine, haem is released from haemoglobin, myoglobin or cytochromes by proteolytic degradation of the protein fraction. Haem is then transported through the brush border of epithelial cells bound to a receptor. Once absorbed within the cell the iron is liberated enzymatically from haem by a haem oxygenase. Iron in vegetable food is present as non-haem complexes as Fe^{3+}, bound to protein, phytates, oxalates, phosphates and carbonates. Most of the non-haem iron is in a high-molecular weight form and is less well absorbed than solubilized iron. Binding to low-molecular weight chelators, sugars, amino acids, ascorbic acid and glycoproteins forms soluble iron complexes and increases absorption. Non-haem iron is taken from the gut lumen through border membranes at receptor sites which may be glycoproteins. The absorbed iron passes into the plasma and is bound to the transferrin protein. Inorganic iron uptake is facilitated by ascorbic acid, where the reducing ability of ascorbic acid expedites the conversion of Fe^{3+} to the more water soluble Fe^{2+}. Gastric hydrochloric acid facilitates the absorption of non-haem iron by converting ferric to ferrous iron; achlorhydria and iron-deficiency anaemia are closely associated. Prolonged iron deficiency may lead to gastric atrophy. Partial gastrectomy and malabsorption can cause iron-deficiency anaemia.

Phytic acid in cereals, phosphate, carbonates, oxalate, pancreatic bicarbonate bind iron and decrease iron absorption.

Enhancers of iron absorption

Physiological

- Iron deficiency, an anaemic state, fasting, pregnancy

Dietary

- Ascorbic acid, citric acid, lactic acid, malic acid, tartaric acid, fructose, sorbitol, alcohol, amino acids, e.g. cystine, lysine, histidine.

Inhibitors of iron absorption
Physiological

- Iron overload, achlorhydria, copper deficiency

Dietary

- Tannins, polyphenols, phosphates, phytate, wheat bran, lignin, proteins, egg albumin and yolk, legumes, protein, inorganic elements, calcium, manganese, copper, cadmium, cobalt

---

Iron absorption is greatly increased during active production of red cells. In iron-deficient individuals, the absorption of iron may double. The control of iron uptake is through the activity of the bone marrow and the amount of iron in stores, though these controls appear to be independent of each other. Iron stores fall progressively during pregnancy and this is a stimulus to increased iron absorption.

Transport

Iron is transported into the blood stream bound to a globular protein transferrin, which is synthesized in the liver. The transferrin concentration in the plasma is 2–2.5 g/litre. The concentration of iron in the plasma is 16–18 μmol/litre.

Transferrin

Transferrin (Tf) is a glycoprotein of molecular weight 80 000 Da. Two Fe atoms are carried on each molecule, on two high-affinity sites. Tf is the major supplier of iron for most cells. The diferric transferrin binds to a Tf receptor which is present on the cell membrane, and the diferric transferrin complex is taken into the cell by receptor-mediated endocytosis. The apoferritin remains bound to the receptor until released without Fe and then recycles. There are specific receptors for transferrin and the transferrin iron complex on reticulocytes.

There are wide diurnal variations (the range varying by 100% over 24 hours) in plasma iron concentrations. The iron is carried to the marrow (80% of the total) where new red cells are formed. The life of a red cell is in the order of 120 days and each day one-120th of body haemoglobin is degraded and resynthesized. Thus, 20 mg of iron each day passes from the spleen, liver and other lymphoreticular tissue, the site of red cell breakdown. The total amount of intracellular iron is approximately 500 mg.

TABLE 8.1 Distribution of iron in the body of an adult male

Organ	Physiology	Amount (g)
Liver, spleen, bone marrow stores	Ferritin	0.70
	Haemosiderin	1.00
Tissues	Myoglobin	0.30
	Haem enzymes	0.10
	Non-haem enzymes	0.40
Blood	Haemoglobin	2.50
Total		5.00

Storage of iron

The typical pool of iron increases from 300 mg in infants to 4000 mg in adults, an increase of 0.5 mg/day. Both ferric and ferrous iron form complexes with organic and inorganic ions. Ionic iron is very toxic so iron is held throughout the body in a bound form. The distribution of iron in the body is shown in Table 8.1.

Iron storage compounds

Haemoglobin The haemoglobin molecule is central to oxygen transport. It has a molecular weight of 64 500 Da and is formed from four haem groups which are linked to four polypeptide chains. Haemoglobin can bind four molecules of oxygen. Divalent iron (Fe^{2+}) in haem reversibly binds oxygen for transport to tissues, while oxidation of iron to the ferric (Fe^{3+}) in methaemoglobin causes haemoglobin to lose its capacity to carry oxygen.

Myoglobin This is important in the storage of oxygen in muscle. It has a molecular weight of 17 000 Da and is formed from one polypeptide chain and one haem molecule. Myoglobin is a reservoir of oxygen for muscle metabolism, having a higher affinity for oxygen than haemoglobin. Oxygen is released to cytochrome oxidase, which has a greater affinity for oxygen than myoglobin. Iron is an integral part of haem protein cytochromes, catalase, peroxidases or as a cofactor.

All of these are reversible acceptors or donors of electrons.

There is a controlled interaction with molecular oxygen by haem iron proteins, iron–sulphur proteins, and non-haem iron-containing oxygenases. Ribose is converted to DNA by the iron-containing ribonucleotide reductase.

Cytochromes

The cytochromes transport electrons to molecular oxygen through the reversible valency changes of iron atoms present in these molecules. There is a stepwise release of energy by the haem-containing proteins of the mitochondrial electron transport apparatus.

Iron complexes are all octahedral, and paramagnetic because of unpaired electrons in the 3d orbital. Haem coenzymes contain Fe^{2+} and are directly involved in catalysis. Many redox enzymes including electron transferring protein NADH dehydrogenase, contain iron–sulphur complexes which catalyse one-electron transfer reactions. The complex is usually two or four iron and sulphur ions bound as an equal complex to the cysteinyl-sulphur group of the protein. In the cytochrome P_{450} system there are haem, iron–sulphur complexes, flavin coenzymes and nicotinamide coenzymes in a multienzyme system (see the section on cytochrome P_{450}).

Ferritin Haemosiderin and ferritin contain approximately 1 g of iron, mostly as ferritin. Ferritin is present in all cells, but particularly those of the liver, spleen and bone marrow.

Normal iron stores amount from 0.75–1 g in men and 0.3–0.5 g in women, largely in the liver as ferritin. Iron is stored in the reticuloendothelial cells of the liver, spleen and bone marrow.

Ferritin

Ferritin is a soluble complex, with Fe^{3+} in the core surrounded by a coat of apoferritin proteins. Iron is deposited within the core as insoluble ferric hydroxyphosphate. This prevents excessive iron from damaging cells and allows up to 4500 ferric atoms to be stored. The removal of free iron from the cytoplasm protects against the peroxidation of cell lipids, DNA and some proteins. Ferritin consists of 24 protein subunits of two types (H) heavy and (L) light. These are formed as truncated molecular pyramids. The shell formation is formed of subunit proteins, four long and one short helical segments that result in a rigid structure. The rest of the subunit proteins consist of non-helical segments connecting the helices. The amounts of these proteins are controlled by the effectiveness of their mRNAs. There is probably only one expressed copy of the L-gene and one of the H-gene in human genomes. The expressed H- and L- genes of the human have three introns separating four exons. Gene translation from small mRNA to amino acids is modified to provide protection of cell molecules against peroxidation by iron accumulation. This is achieved by accumulating dormant ferritin mRNA activated by iron when present in excessive amounts. This occurs through an iron-dependent protein (IRE-BP) which binds to regions of the mRNA for each regulatory protein. Iron removes a specific protein that blocks the translation of ferritin mRNA. Iron administration causes the cytoplasmic pool of dormant ferritin mRNA to become active by association with polysomes. The L subunit of ferritin is transcribed preferentially in response to increased iron intake, whereas the H subunit of ferritin responds to cell differentiation of the factors. This affinity decreases as iron availability increases. This is a simple control mechanism. The mRNA for ferritin binds upstream of the protein-coding region and the passage of the translation complex is blocked when there is such absorption. Therefore, synthesis of ferritin is inhibited when the binding affinity increases. The shortage of iron tends to facilitate binding downstream of the coding region and stabilizes the transferrin receptor mRNA to increased levels of the receptor and hence iron uptake.

A balance between the synthesis of ferritin and the plasma membrane receptor for transferrin regulates iron availability. The transferrin receptor is required for the uptake of iron and ferritin in necessary for the storage of any iron which is in excess of requirements.

Ferritin has a molecular weight of 450 000 Da which increases to 900 000 Da when loaded with iron.

Iron in the infant

During pregnancy some 700 mg of iron passes from the mother, through the placenta to the foetus, this transfer being most marked in the second half of pregnancy. This poses problems for the premature infant, as the period of transfer is shortened and maternal milk is not rich in iron. Late clamping of the umbilical cord can increase the infant's iron stores, and can be of significance in determining transferred haemoglobin. The pool of iron is related to body weight. During the first 6 weeks of life there are major shifts of iron stores from haemoglobin to other iron protein complexes.

Iron loss, deficiency and excess

Iron loss

The amount of iron absorbed equals the amount of iron lost by dead cells desquamated from surfaces. Iron is also lost in bleeding and in female menstruation. Iron-deficiency anaemia in the male is uncommon, though when this happens in adults a pathological cause must be sought, since it usually results from a source of bleeding in the gastrointestinal tract. In parts of the world where there are hook-worm infestations (*Ancylostoma duodenale* and *Nector americanus*) there is loss of blood from the gastrointestinal tract. Urinary loss of iron is small, less than 100 µg/day. The loss of

iron in sweat is small and is in the order of 250–500 µg daily.

Daily losses of endogenous iron include: desquamated gastrointestinal cells (0.14 mg), haemoglobin (0.38 mg), bile (0.24 mg) and urine (0.1 mg).

Iron deficiency

Iron deficiency is associated with a low serum iron and usually an excess loss of blood either in menstruation, pregnancy, insufficient iron stores (as in the premature baby), or loss from lesions in the gastrointestinal tract or dietary insufficiency.

Iron deficiency results in anaemia. It has also been shown that iron deficiency results in impaired psychomotor development with reduced attention span and cognitive function in the young and during growth. Iron deficiency also reduces work performance, particularly endurance work. In pregnancy, iron deficiency is associated with low birth-weight babies and perinatal mortality. In the older population, anaemia resulting from iron deficiency can cause tiredness, and risk of complications associated with atherosclerosis, e.g. angina, myocardial infarction and strokes.

Iron excess

An excess of iron in the body is known as **siderosis**. The body store of iron is reflected by and measured by the plasma ferritin.

Haemochromatosis is a failure to control iron absorption from the small intestine, with a resulting progressive increase in total body iron stores. The increase in total body iron results in iron absorption inappropriate to the levels of body iron stores, either alone or in combination with parenteral iron loading. Excess deposition of iron may result in cellular damage and functional insufficiency of the organs. The causes of haemochromatosis are:

- **genetic**: primary idiopathic haemochromatosis
- **acquired**: secondary haemochromatosis. This can be secondary to: (i) anaemia and ineffective erythropoiesis, thalassaemia, and major sideroblastic anaemia; (ii) high oral iron intake, prolonged ingestion of medicinal iron, or intake

of iron with alcohol; or (iii) liver disease, including porphyria cutanea tarda, alcoholic sclerosis, or following portacaval anastomosis.

Haemochromatosis

There is a genetic basis for idiopathic haemochromatosis. The mode of inheritance is autosomal recessive. Heterozygous subjects may show some phenotypic expression of the disease by an increased serum transferrin saturation. The abnormal gene lies close to the histocompatibility locus antigen (HLA) complex on chromosome 6. The prevalence is 1 in 10 of European stock and is homozygous in 1 in 300 individuals. Excess body iron leads to hepatic fibrosis and cirrhosis, excess skin pigmentation, and diabetes mellitus due to pancreatic iron loading.

Pathology

Over a period of time, iron accumulates and there is an increase in plasma iron and a percentage saturation of transferrin. The tissues may contain over 20 g of iron. The excess iron is deposited in the lysosomes of the parenchymal cells of the liver, pancreas and heart. This may result in tissue injury from disruption of iron-laden lysosomes or mitochondrial lipid peroxidation.

In the early stages of haemochromatosis haemosiderin is deposited in the periportal hepatocytes, particularly in the pericanalicular cytoplasm within lysomes. This progresses to perilobular fibrosis and deposition of iron in bile duct epithelium, Kupffer cells and fibroceptor cells. An irregular macronodular cirrhosis develops later.

A nutritional iron overload occurs when the dietary iron intake exceeds 40 mg/day. This can occur from absorption from iron vessels used for cooking and making alcoholic drinks.

Recommended requirements

Infants

Breast-fed infants can absorb approximately 50% of the iron present in milk, whereas iron absorption from formulated milks may be only 10% of that in the milk. The LRNI for infants 0–3 months is 0.9 mg/day; the estimated average requirement is 1.3 mg/day and the RNI 1.7 mg/day. This should treble during the next 6 months of life to an EAR of 3.3 mg/day.

Menstruation

Median and mean losses of blood in menstruation are 30 and 44 ml, respectively. The calculated loss of iron is therefore approximately 20 mg, averaging 0.7 mg/day. In Britain the average iron intake among fertile women is in the order of 12 mg/day. To allow for the somewhat skew distribution of menstrual blood loss, an EAR of 11 mg/day, an LRNI of 8 mg/day and an RNI of 15 mg/day have been suggested. By the age of puberty the intake should be 1 mg/day.

Pregnancy

It has been estimated that the toll of pregnancy on the mother's iron stores is in the order of 700 mg. This should be met by pre-existing body stores. In part this will be met by the cessation of menstrual losses, increased intestinal absorption and the mobilization of maternal stores. However, this may be inadequate if the iron stores are insufficient at the beginning of pregnancy.

Lactation

Breast milk contains 7 μmol/litre at 6–8 weeks postpartum, which decreases thereafter to 5 μmol/litre during weeks 17 to 22. The total loss is of the order of 5–6 μmol/day. This secretion will be balanced by the amenorrhoea generally associated with lactation.

Blood donors

A pint of blood donated every 6 months is more than compensated for by increased iron absorption. However, more frequent donations – especially by women of child-bearing years – are particularly likely to upset iron status.

Iron intakes in Britain average 14 mg/day for men and 12 mg/day for women.

Body store measurements

Haemoglobin in part reflects iron status. Normal values for plasma iron range from 12–30 μmol/litre and the total plasma iron-binding capacity 45–70 μmol/litre. Plasma ferritin closely reflects body iron stores in the absence of inflammatory disease (normal range 12–250 μg/litre).

There are large amounts of ferritin in the storage tissues but little in serum. A plasma ferritin level of 1 μg/l corresponds to 8–10 mg storage iron in the average adult. Serum ferritin is relatively increased in the newborn and falls rapidly in the first few months as iron is required for the increasing red cell mass. The concentration increases slowly during childhood until late adolescence, males having values three times those of females. The female ferritin level is low during the child-bearing years and increases thereafter. Below a plasma ferritin of 12 μg/litre the measurement is not useful in assessing iron deficiency, especially in infants and during pregnancy.

Summary

1. Iron is found in two forms, ferrous and ferric, which interchange: $Fe^{2+} \quad Fe^{3+} + e^{-}$
2. The iron must be readily available for the synthesis of essential iron proteins, e.g. haemoglobin and myoglobin and enzymes, e.g. catalase.
3. Iron is central to oxygen metabolism in that 2.5 g circulate in haemoglobin and 0.3 g are present in myoglobin. Haemoglobin has a molecular weight of 64 500 Da and is formed from four haem groups which are linked to four polypeptide chains. Divalent iron (Fe^{2+}) in haem reversibly binds oxygen for transport to tissues, while oxidation of iron to the ferric (Fe^{3+}) state in methaemoglobin causes haemoglobin to lose its capacity to carry oxygen. Iron is present in all cells of the body and plays a key role in many biochemical reactions.
4. Iron deficiency is associated with an excess loss of blood in menstruation, pregnancy, insufficient iron stores (as in the premature baby), or loss from lesions in the gastro-intestinal tract. Iron deficiency results in anaemia and impaired psychomotor development and reduced cognitive function in the young and during growth. Iron deficiency also reduces work performance (particularly endurance work). In pregnancy, iron deficiency is associated with low birth-weight babies and perinatal mortality.

Further reading

Chesters, J.K. (1992) Trace element/gene interactions. *Nutrition Reviews*, 50, 217–23.

Cook, J.D., Baynes, R.D. and Skikne, B.S. (1992) Iron deficiency and the measurement of iron status. *Nutrition Research Reviews*, 5, 189–202.

Mascott, D.P., Rup, D. and Thach, R.E. (1995) Regulation of iron metabolism. *Annual Review of Nutrition*, 15, 239–62.

Munro, H. (1990) The ferritin genes – their response to iron status. *Nutrition Reviews*, 51, 65–73.

Powell, L.W. and Kerr, J.F. (1975) The pathology of the liver in hemochromatosis, in *Pathobiology Annual 1975* (ed. H.L. Joachim), Appleton-Century-Cross, New York, pp. 317–36.

8.3 Trace elements

- Trace elements have vital roles in the body, including the establishment of potential differences across membranes, acting as the active principle of enzymes (e.g. zinc), and maintaining protein and hormone structure.

8.6.1 Introduction

Trace elements are important in the body by virtue of:

- establishing potential differences across membranes; such differences have to be maintained by energy-requiring reactions, e.g. Na^+, K^+, Ca^{2+}
- allowing oxidation–reduction (redox) enzymatic reactions to take place, e.g. iron, copper
- acting as cofactors in enzymatic reactions, e.g. zinc
- maintaining the structure (conformation) of proteins, e.g magnesium
- maintaining the structure of hormones, e.g. iodine in thyroxine

---

Trace elements and protein structure

The trace elements adsorb to protein in specific configurations. Such ligands may act as cofactors or structural elements by virtue of high concentrations of positive charges, directed valences for interacting with two or more ligands, i.e. acting as bridges and able to exist in two or more valency states. Alkali metals, Na^+ and K^+ with single positive charges and no d-electronic orbital for sharing, rarely form complexes with proteins. Occasionally the divalent alkaline metals Mg^{2+} and Ca^{2+} complex with proteins. The other elements in biological systems are from the first transition series of the periodic table and with partially filled 3d orbitals. They exist in more than one oxidative state which is important in oxidation–reduction reactions. Zinc is unusual in having a full 3d orbital. Zinc chelates with nitrogen, in histidine side chains or sulphur-containing amino acid chains, e.g. cysteine and makes the structure of a protein rigid.

Trace elements are required in small amounts, though their concentrations in the diet will vary. In communities totally dependent upon locally grown food the trace element content of the local soil and drinking water becomes important. In communities buying from sources from all over the world such restrictions may not apply.

8.6.2 Dietary interactions of trace elements

The amount and chemistry of dietary inorganic constituents affect the efficiency of absorption of the essential elements. This is shown for example by the inhibition of absorption of calcium and trace metals, e.g. zinc with dietary phytate. The effectiveness of release of a range of metals from their complexes with phytate is in order: copper > cadmium > manganese > zinc > lead.

Zinc–phytate interactions

Dietary phytate : zinc ratios exceeding 25 : 1 can inhibit zinc utilization, provided that dietary calcium is maintained at or above the currently accepted RDA. There is a synergistic effect of dietary calcium upon the phytate : zinc relationship. The action of phytate is dependent upon the available phytate for binding, i.e. that which is not already bound to some other element, e.g. calcium. Phytate is lost by bacterial hydrolysis by phytases in the intestinal mucosa. Free amino acids (histidine, cysteine, methionine) can desorb zinc from an insoluble (ZnCu–phytate) complex. The yield of soluble copper or zinc from such complexes is directly proportional to the amino-nitrogen content of the intestinal soluble phase.

Copper absorption is effected by interactions affecting its intraluminal solubility and competitive interactions modifying its transport through the mucosa.

A mild degree of iron depletion increases the uptake of other metals. Dietary iron in such depletion states, not only increases the efficacy of iron absorption but also of lead, zinc, cadmium, cobalt and manganese. A clue possibly lies with apolactoferrin, which also has affinity for zinc and other metals, and this may influence absorption.

Serum zinc, copper, manganese, chromium, molybdenum and vanadium are not good indicators of nutritional status and dietary intake. Measurement of trace elements in hair or nail clippings is of little value in determining dietary intake or body status.

8.6.3 Individual trace elements

Aluminium

- atomic weight, 27; valency, 3
- natural isotope, 27
- natural abundance in earth's crust, 7.51%

Aluminium is found in all biological materials and the amount depends on the local soil and environmental availability. Aluminium hydroxide used to be used therapeutically as an antacid in the treatment of duodenal ulcers. Daily intake is approximately 20 mg/kg of food and 0.5 mg/1500 ml fluid. Fluoride and silicon reduce bioavailability. Aluminium is excreted in urine.

High concentrations of aluminium are found in tea and orange juice. The tea plant absorbs aluminium from acid soil, concentrating 500–1500 mg/kg in its leaves. Brewed black tea contains 2–6 mg/litre of aluminium. The bioavailability of aluminium in tea is low because it binds with organic compounds. Aluminium is also found in herbs, processed cheeses, baking powder and pickles.

The only biological role for aluminium could be as part of the succinic dehydrogenase–cytochrome C system.

Aluminium may be involved in the development of **Alzheimer's disease** and is present in increased amounts in the brains of sufferers. If renal dialysis fluids contain aluminium, dementia can occur in patients receiving such treatment. Such dialysis dementia is an acute condition which is often reversible with treatment. The pathology is quite different from that of Alzheimer's. There are no geographical or temporal variations in dementia.

Antimony

- atomic weight, 122; valencies, 3, 5
- natural isotopes, 121, 123
- abundance in earth's crust, $2.3 \times 10^{-5}\%$

Antimony does not appear to be an essential element. The intake in the North American diet is in the range of 2–10 μmol/day. The efficiency of absorption is about 15% and antimony is accumulated in the liver, kidneys, skin and adrenals.

Toxic effects include gastrointestinal symptoms and irregular respiration. Antimony poisoning may arise from the storage of soft drinks in enamel containers.

Arsenic

- atomic weight, 75; valencies, 3, 5
- natural isotope, 75
- abundance in earth's crust, $5.5 \times 10^{-4}\%$

Arsenic is present in many plant and animal foods but has yet to be established as an essential element in the diet. Shell-fish are particularly rich in arsenic. Arsenic has been given as a dietary contaminant to poultry and pigs and may be present in muscle and liver up to 6.5 and 26 μmol/kg. In seafood, the arsenic is present as arsenobetaine. It is rapidly absorbed and excreted in urine and bile. Inorganic forms of penta- and trivalent arsenic are rapidly methylated in the digestive tract. Intakes are said to be within 0.1–0.4 μmol/day.

Tolerable daily intake
Arsenic is extremely toxic and an upper limit of tolerable daily intake is 2 μg/kg body weight.

Boron

- atomic weight, 11; valency, 3
- natural isotopes, 10, 11
- relative abundance in earth's crust, 0.0014%

Boron may form complexes with organic compounds such as sugars, polysaccharides, adenosine-phosphate, pyridoxine, riboflavin, dehydroascorbic acid, pyridine nucleotides and steroid hormones. Boron is an essential nutrient in plants but there is no evidence that it has any place in animal metabolism. Boron is excreted in the urine.

Recommended requirements
Human diets provide 2 mg/day. Toxic effects are experienced with intakes 50 times this amount.

Bromine

- atomic weight, 80; valencies, 1,3
- natural isotopes, 80, 81
- abundance in earth's crust, $6 \times 10^{-4}\%$

Bromine does not have an evident physiological function but may be substituted for chlorine in some reactions. It has been reported that bromine may concentrate in the thyroid gland. Bromine is found widely in the environment and it is unlikely that deficiency will occur.

Excess
Bromide was formerly used in the belief that it reduced sexual drive and as a sedative.

Cadmium

- atomic weight, 112; valency, 2
- natural isotopes, 106, 108, 110, 111, 112, 113, 114, 116
- abundance in earth's crust, $1.1 \times 10^{-5}\%$

Cadmium is not an essential nutrient. Cadmium exposure results from industrial waste. Small amounts are absorbed from water and accumulate in tissues. There may be 200–300 μmol in the body, with the highest concentration in the kidneys. Body cadmium deposits have a biological half-life of 15 to 30 years. Cadmium may react with the intestinal mucosa and with zinc, copper, manganese and iron. It is possible to inhale cadmium by smoking cigarettes. During pregnancy this may be associated with increased accumulation of cadmium, with adverse consequences for the foetus.

Caesium

- atomic weight, 133; valency, 1
- natural isotope, 133
- abundance in earth's crust, $7 \times 10^{-5}\%$

Caesium is available predominantly as the radioisotope caesium 137. It is metabolized in the same way as potassium, being concentrated in the muscle. Following radioactive fallout, caesium may concentrate in cattle and sheep feeding on grass contaminated with this element.

Cobalt

- atomic weight, 59; valencies, 2,3
- natural isotope, 59
- abundance in earth's crust, 0.0018%

Wheat, especially wholemeal flour (0.5–0.7 μg/g), and seafoods (1.6 μg/g) are good sources of cobalt.

Cobalt is a constituent of vitamin B_{12}. This appears to be the only biological function of the element.

Absorption of cobalt is facilitated by incorporation into vitamin B_{12}. The intestinal absorption of the cobalt salt is high at 63–97% efficiency. The plasma concentration ranges from 0.007 to 6 μg/100 ml, bound to albumin.

The main organs/tissues of accumulation are liver and fat; cobalt is excreted in the urine.

Recommended requirements
Average intakes of cobalt are approximately 0.3 mg/day and the total body content about 1.5 mg.

Cobalt can have serious toxic effects, including goitre, hypothyroidism and heart failure. In Quebec, cobalt was added to beer to improve the head. This led to a cobalt concentration of 15 μmol/litre, which was toxic to men with heavy alcohol consumption. They developed severe cardiomyopathy from which some died.

Chromium

- atomic weight, 52; valencies, 2, 3
- natural isotopes, 50, 52, 53, 54
- abundance in earth's crust, 0.033%

Chromium is present in all organic matter. Wheat (1.8 μg/g) and wheat-germ (1.3 μg/g) are especially good sources of chromium, as is molasses (1.2 μg/g). There is little chromium in milk.

Dietary intakes range from 5 to 100 μg/day, but absorption is poor at 1% of available chromium in the diet. Plasma concentration is 0.15 μg/ml, bound to transferrin. There is accumulation of chromium in the spleen and heart. Chromium is excreted in urine.

The trivalent (CrIII) form is biologically active, but cannot be oxidized to the hexavalent (CrVI) form in tissues. The role of chromium in human nutrition is uncertain, but may act in an organic complex which influences and extends the action of insulin. There may also be a role in lipoprotein metabolism, in stabilizing the structure of nucleic acids and in gene expression.

Recommended requirements

The recommended requirements for chromium are debatable. It has been suggested that a reasonable intake for adults would be 0.5 μmol/day and between 2 and 19 nmol/kg/day for children and adolescents. The usual body content in adults is 100–200 μmol. A not unreasonable requirement is 0.4 μmol/day. The chromium content of breast milk is 0.06–1.56 ng/ml; this gives the infant 15–1300 ng/day intake.

Copper

- atomic weight, 64; valencies, 1,2
- natural isotopes, 63, 64
- abundance in earth's crust, 0.010%

Sources of copper include green vegetables, fish, oysters and liver. They provide copper at approximately 4 μmol/kcal. Other foods, milk, meats, bread, provide less than 0.5 μmol/kcal.

Action

Copper is an important component of enzymes including cytochrome oxidase and superoxide dismutase. Copper may also have a role in iron metabolism, wherein Fe^{2+} released from ferritin is oxidized to Fe^{3+} for binding to transferrin. Copper is excreted in bile and eliminated via the faeces.

Absorption

The absorption efficiency of copper appears to be 35–70%, which may well diminish with age. The bioavailability of copper from milk-based formulas is approximately 50%. Absorption is facilitated by a complicated specific transport system, which is energy-dependent. Factors involved in absorption include the intestinal luminal concentration of amino acids, chelating agents, e.g. gluconate, citrate or phosphate and the effect of other minerals, e.g. zinc. Cupric copper is more soluble than cuprous, so that reducing agents such as ascorbic acid inhibit absorption.

Availability

Copper is bound to a specific blood copper-binding protein, caeruloplasmin, and to a lesser extent to albumin. Caeruloplasmin is present in serum at a concentration of 25–45 mg/100 ml. Each molecule of the protein contains six atoms of copper. The copper is not exchangeable, so the protein is not a transporter protein. The protein has an *in vitro* oxidase activity to aromatic amines, cysteine, ferrous ions and ascorbic acid.

The total copper in an adult is approximately 50–80 mg, mainly concentrated in muscle and liver. In pregnancy, copper is transferred from the maternal to the foetal liver, where there is accumulation for early extrauterine life.

Deficiency

Copper deficiency has not been reported in humans because of the variety of sources in the diet. There is accumulation of copper (8 mg) in the infant's liver before birth. Premature birth may result in low copper stores. A rare congenital condition, Menke's syndrome, involves failure of copper absorption, the consequence being poor mental development, failure to keratinize hair, skeletal problems and degenerative changes in the aorta.

Copper excess

Copper can accumulate in the body in rare, genetically determined conditions: Wilson's disease is a hepatolenticular condition resulting from an accumulation of copper in the body due to a failure to excrete copper in bile. Efficient synthesis of caeruloplasmin leads to copper being transported in the plasma and into albumin. Excess copper is deposited in tissues, particularly the liver and the basal nuclei of the brain, leading to sclerosis, kidney, corneal and brain abnormalities.

This is now treated with D-penicillamine, a chelating agent which facilitates copper excretion.

Acute toxicity may present with haemolysis and brain and hepatic cellular damage. In chronic excess there is interference in the absorption of zinc and iron.

Recommended requirements

The normal adult diet provides 1.5 mg/day. The RNI for an infant is 0.3 mg/day. For children, the RNI is 0.7–1.0 mg/day and for adults an RNI of 1.2 mg/day is required.

Pregnancy Increasing requirements have to be met during the first, second and third trimesters at 0.033, 0.063 and 0.15 mg/day. This increasing need would be met by improved absorption through adaptation of enteric absorption rates.

Lactation Milk has a copper content of 3.5 μmol/litre and the absorption efficiency is said to be 50%. Therefore, the maternal increment required during lactation is of the order of 0.4 mg/day.

Body store measurements

Copper is associated with two main plasma proteins, albumin and caeruloplasmin. The major copper transport protein may be albumin. Concentrations of caeruloplasmin and a second major copper protein, copper–zinc superoxide dismutase, are unrelated to dietary copper.

Fluorine

- atomic weight, 19; valency, 1
- natural isotope, 19
- relative abundance in earth's crust, 0.027%

The function of fluoride appears to be in the crystalline structure of bones; fluoride forms calcium fluorapatite in teeth and bone.

Absorption and metabolism

Fluoride is absorbed passively from the stomach but protein-bound organic fluoride is less readily absorbed. Fluoride appears to be soluble, rapidly absorbed and is distributed throughout the extracellular fluid in a manner similar to chloride.

Concentrations in blood where fluorine is bound to albumin and tissues are small. Fluoride is rapidly taken up by bones but equally rapidly excreted in the urine.

Fluoride excess

High intake of fluoride, that is in excess of 1 mg/litre, results in mottling of the teeth: the enamel is no longer lustrous and becomes rough, an effect particularly marked on the upper incisors. In concentrations well in excess of 10 parts per million, fluoride poisoning can occur, with a loss of appetite and sclerosis of the bones of the spine, pelvis and limbs. There may be ossification of the tendon insertion of muscles.

Recommended requirements

There is no known requirement for fluoride and no RNI is recommended. There is an overall acceptance of a role for fluoride in the care of the teeth. When drinking water contains 1 mg/litre (1 ppm) there is a coincidental 50% reduction in tooth decay in children.

It has been suggested that adults should have a mean dietary intake of 95 μmol/day or 150 μmol from fluoridated water. Tea is an important source of fluoride; food provides only 25% of the required intake.

Body store measurements

A high proportion of the dietary intake of fluoride appears in the urine. Urinary output in general reflects the dietary intake.

Germanium

- atomic weight, 73; valencies, 2, 4
- natural isotopes, 70, 72, 73, 74, 76
- abundance in earth's crust, 1×10^{-4}%

This abundant element is present in the diet at trace levels and is consumed at the rate of 1 mg/day and rapidly excreted in urine.

Excess

Consumption of germanium of the order of 50–250 mg/day (0.7–3.4 mmol) over a prolonged period may result in morbidity and even death.

Iodine

- atomic weight, 127; valencies, 2,4
- natural isotope, 127
- natural abundance in the earth's crust, $6 \times 10^{-6}\%$

Thyroid hormones

Thyroxine is bound to thyroglobulin which may be stored in the vesicles of the thyroid gland. Thyroid hormone release is dependent upon thyrotropic hormone activity from the pituitary gland. Thyroxine is important in cellular metabolism by inducing an increase in the number, size and activity of mitochondria and constituent enzymes. Cellular protein synthesis is increased by thyroid hormones by the increased transcription of mRNA. Thyroid hormones increase the utilization of ATP through increased transmembrane transport. Fatty acids are released from adipose tissue under the action of thyroid hormones. Selenium is important in thyroid hormone production, particularly in the conversion of T_4 to the active T_3. It is possible that zinc is also involved. All T_4 is synthesized in the thyroid, whereas 80% of plasma T_3 is derived from 5'-monoiodination of T_4 in liver, kidney and possibly muscle. The enzyme involved in the peripheral conversion of T_4 to T_3 in liver and kidney is Type 1 iodothyronine deiodinase (idi), the activity of which is reduced in selenium deficiency. The enzyme is a selenoenzyme.

Most foodstuffs except seafood are poor sources of iodine. Fruit, vegetables, cereals, meat and meat products may contain up to 100 μg/kg depending on the soil content where grown. Milk is a major source of iodine with a content of the order of 0.2–23 μmol/kg. Water is not an important source (1–50 μg/litre).

Iodine-supplemented cattle feed, iodinated casein, is a lactation promoter in cows. The contaminations of milk from iodophors, the sterilizing agents, is an important source of iodine.

Water is not an important source of iodide and contains only small amounts (1–50 μg/litre), though the iodine content of the soil water in which plants and cereal are grown is important.

Action

Iodine is required for thyroid hormones, thyroxine 3,5,3',5'-tetraiodothyronine (T_4), 3,5,3'-triiodothyronine (T_3). Iodide is oxidized to iodine and bound to tyrosine forming mono- and diiodotyrosines. Subsequent metabolism produces T_3 (triiodothyronine) and T_4 (thyroxine).

Absorption

Dietary iodine from food and water is absorbed as inorganic iodide. Some of this is transported to the thyroid gland depending on the activity and requirements of the gland.

The body content of iodine is 20–50 mg (160–400 μmol). About 8 mg is to be found in the thyroid gland.

Iodine deficiency

Iodide is of great importance in the prevention of thyroid enlargement, namely goitre, in areas where iodine is deficient in the soil. Iodine deficiency disorders have consequences for normal growth and development.

Iodine-deficient food crops are found in areas with soils from which iodine has been leached by glaciers, high rain fall or flooding. This is found in the Himalayas, the Andes and the vast mountain ranges of China, the Ganges valley and Bangladesh. All the food grown in such soil is iodine-deficient and the iodine deficiency will persist in the local population until such times as there is dietary diversification and an increased provision of iodine in the diet. There are 800 million people in the world at risk from developing iodine deficiency, of whom 190 million may develop goitres and more than 3 million are cretinous. In some populations apathy has been found even among domestic animals as a result of iodine deficiency.

Goitre Iodine deficiency results in reduced thyroid iodine stores and production of T_4. This causes an increased production of pituitary thyroid stimulating hormone and ineffective hyperplasia of the thyroid. Iodine deficiency can be demonstrated by urinary iodine excretion measuring 24-hour samples (normal range 100–150 μg/day). Goitre may arise through eating a group of chemical known as goitrogens, e.g. thiocyanate. Staple foods from the third world, cassava, maize, bamboo shoots, sweet potatoes, lima beans and millet contain cyanogenic glucosides which produce cyanide that may in turn give rise to thiocyanate. In general these are not a problem; they are found in the inedible portions of the plant.

Iodine deficiency in the foetus results from iodine deficiency in the mother; reversed by thyroid hormone replacement. Maternal deficiency is associated with a high incidence of still-births, spontaneous abortions and congenital abnormalities. Perinatal death is also increased.

An effect of foetal iodine deficiency is endemic **cretinism**, found in mountainous regions of India, Indonesia and China. This occurs in populations where the iodine intake is less than 25 μg/day – the required being 80–150 μg/day. The thyroid hormone is essential for brain development. The human brain has reached only one-third of its full size at birth and continues to grow rapidly until the second year. Severe iodine deficiency affects neonatal thyroid function and hence brain development, mental deficiency, deaf mutism and spastic diplegia. This is the nervous or neurological type, in contrast to the myxoedematous type found with hypothyroidism and dwarfism. This community problem is reversible with replacement of iodine.

Iodine deficiency in children causes goitre. The incidence of goitre increases with age and is maximal in adolescence; it is particularly marked in girls. The result is poor school performance and reduced intelligence quotients. However, assessment of IQ is often difficult as the iodine-deficient areas are likely to be remote and to suffer from social deprivation, resulting in poorer school facilities, poverty and poorer general nutrition.

The problems of iodine deficiency may be reversed by the injection of iodized oil.

Iodine excess
A modest increase in the incidence of hyperthyroidism has been described following iodized salt programmes. This condition is found in individuals over 40 years of age who form a small proportion of the population of a developing country. The condition is readily controlled with anti-thyroid drugs.

Correction of iodine deficiency
People in deficient areas can receive iodine as an additive to food or water or by direct administration of iodized oil or potassium iodide or iodine in Lugol's solution.

Iodinated salt Salt is usually produced from the solar evaporation of seawater or brine, or from rock deposits. Crude granulous salt is collected from the salar, dried in the sun and packaged. The crude salt is refined by recrystallization or milling. Iodine can be added by dry mixing the iodine compound with salt or by dripping a solution of iodine onto the salt on a conveyor belt. Potassium iodide and iodate are the most common additives. The drip feed system is the simplest and cheapest, but a spray–mix method allows more uniform distribution of iodine with very fine salt. Iodide is cheaper but iodate is more stable in warm humid climates. Iodine is usually added at a ratio to salt of 1 : 10 000 and 1 : 50 000; this gives an iodine intake range of 50–150 μg with a salt intake of 3–15 g/day.

Iodized oil Iodine covalently bound to vegetable oils is given intramuscularly or orally. A single dose of iodized oil is adequate iodine for at least 3 years, e.g. iodinated poppy seed oil, lipiodol, contains 38% iodine by weight. Iodinated walnut or soya bean oil contains 25% iodine. A single oral administration is effective for at least a year, with the added advantage that injections may not always be clean and have the danger of disease spread, e.g. hepatitis and AIDS. Iodized oil is used as an emergency stop-gap measure to control iodine deficiency in severely affected areas until there is an effective programme for iodinated salt. High intakes do not appear to be deleterious.

Recommended requirements

Adults An intake of 70 μg/day. The RNI has been suggested as 140 μg.

Infants and children The LRNI is of the order of 40–50 μg/day and the RNI 50–100 μg/day.

Body store measurements
Thyroid hormone measurements reflect iodine status. A high proportion of the dietary intake of iodide appears in the urine, so that urinary output in general reflects the dietary intake.

Lead

- atomic weight, 207; valencies, 2, 4
- natural isotopes 204, 206, 207, 210, 211, 212, 214
- natural abundance in earth's crust, 0.002%

There is no evidence that lead is of any importance physiologically. In an industrialized society, 1–2 μmol are ingested in the food daily, and 90% of this is unabsorbed. Plasma concentrations are in the range of 15–40 μg/100 ml, protein bound. Lead is deposited in bone and excreted in bile.

Toxicity
Lead has been widely used in cooking utensils and in water pipes. Consequently, it has been an important, albeit toxic, element in the diet. Lead poisoning may lead to anaemia, peripheral neuropathy or encephalopathy. This has been a problem in lead workers, e.g. plumbers, in the past and also in those areas where the water supply passes through lead pipes. Water, especially when slightly acidic and which has been held overnight in the pipes, is particularly dangerous. Other lead hazards are pewter vessels, lead toys, paints and leaded petrol, from which may develop a substantial concentration of lead in the atmosphere. An undue exposure to lead may affect the IQ of children growing in this environment. Lead is used in the manufacture of crystal decanters. Lead can be leached from the glass by alcoholic beverages over a period of time.

A blood concentration of more than 1.4 μmol/ litre is not encouraged.

Lithium
atomic weight, 7; valency, 1

- natural isotopes, 6, 7
- relative abundance in earth's crust, 0.005%

There is no known place for lithium in normal physiology. The main interest in lithium has been that it has an important role in the prophylactic management of manic depressive psychosis. Lithium is absorbed efficiently by the intestine and excreted in the urine. The blood therapeutic concentration is 0.6–1.0 mmol/litre and side effects occur at concentrations above 2 mmol/litre. Toxicity is associated with abnormalities of glucose metabolism, teratogenicity, hypothyroidism and renal problems.

Magnesium
- atomic weight, 24; valency, 2
- natural isotopes 24,25,26
- natural abundance in earth's crust, 1.94%

Magnesium is present in most foods, particularly those of vegetable origin containing chlorophyll, e.g. green leaves and stalks. Typically, a diet contains 200–400 mg/day. Magnesium is a component of magnesium–porphyrin complexes in chlorophyll in some plant sources.

Availability and action
The whole body content of magnesium is about 1 mol (25 g). Almost two-thirds of body magnesium is to be found in bone in association with phosphate and bicarbonate. The remaining 30% is found intracellularly in soft tissues, bound to protein. This magnesium is complexed with ATP in ATP-dependent enzyme reactions, e.g. glycolysis and Krebs cycle, adenyl cyclase in cAMP formation, phosphatases, and in protein and nucleic acid synthesis.

Magnesium is an important cofactor for co-carboxylase and is involved in the replication of DNA and synthesis of RNA.

The homeostasis of magnesium is controlled by a system which involves parathyroid hormones. Excretion is via the kidneys and is related to dietary intake.

Absorption

Magnesium is absorbed from the distal intestine by processes which require active transport though there is also additional passive diffusion. The maintenance of magnesium homeostasis includes the re-absorption of endogenous magnesium in enteric secretions.

The involvement of vitamin D in the absorption process appears to be at the most, modest.

The plasma concentration is 0.6–1.0 mmol/litre; within the cell the concentration is about 10 mmol/litre.

Deficiency

Magnesium deficiency is manifested by progressive muscle weakness, failure to thrive, neuromuscular dysfunction, tachycardia, ventricular fibrillation, coma and death. Alcohol abuse and diuretics are important causes of a low serum magnesium.

Recommended requirements

Adults Magnesium balance is achieved in adults with an intake of 50 mg/day. The effectiveness of absorption increases from 25% on a high magnesium diet to 75% on a magnesium-restricted diet. The LRNI for magnesium for adult men is 190 mg/day and for women, 150 mg/day. The EAR for men is 250 mg and for women, 200 mg/day. The RNI is 300 mg for men and 270 mg/day for women.

Infants Human milk contains approximately 0.12 mmol per 100 ml, so that infant intake is approximately 1 mmol/day (25 mg). By 3 months the intake is 6 mg (0.25 mmol)/kg/day.

Pregnancy The foetus requires approximately 8.0 mg/day over 40 weeks, so that the maternal requirement is of the order of 16 mg/day. Increased effectiveness of absorption and drawing upon maternal stores ensure adequate provision for the foetus.

Lactation The magnesium content of breast milk is approximately 1.2 mmol/litre, producing about 25 mg/day. It is suggested that the lactational increment should be 50 mg/day.

Body store measurements

Urinary magnesium is the best method available of assessing dietary intake, but is not very accurate. Urinary output in general reflects dietary intake.

Manganese

- atomic weight, 55; valencies, 2, 3, 4
- natural isotope 55
- natural abundance in earth's crust, 0.085%

Tea is a source of dietary manganese, as well as unrefined vegetarian diets for example, cereals, legumes and leafy vegetables. Meat, milk and refined cereals are poor sources. Good sources include wheat: wholemeal, 50 μg/g; wheat-germ, 130 μg/g; and nuts, 17 μg/g.

Action

Manganese is important for enzyme activity, e.g. pyruvate carboxylase, mitochondrial superoxide dismutase and arginase. It may also activate other enzymes, e.g. glycosyl transferases, hydrolases, kinases, prolinase and phosphotransferases.

Absorption

Manganese absorption occurs along the entire intestine but overall is only 3–4%. The absorption efficiency in the small intestine is low. High concentrations of calcium, phosphorus, fibre and phytate reduce manganese absorption through interactions. Plasma concentrations are 1–2 μg/g, bound to transferrin.

Manganese accumulates in liver and bone. The body pool is largely intracellular and generally

0.2–0.4 mmol. The pancreas and liver have the highest concentration and there is about 25% in the skeleton. Manganese is excreted in bile, and also in intestinal secretions.

Manganese deficiency
Manganese deficiency has not been reported in humans.

Recommended requirements
The average intake in Britain is said to be in the order of 4.6 mg per person per day, half of which is derived from tea. The content of manganese in breast milk in the first 3 months postpartum is 1.9 μg/day and 1.6 μg/day thereafter. Healthy infants fed cows' milk will have an intake of 28–42 μg/kg/day.

Mercury

- atomic weight, 200; valencies, 1, 2
- natural isotopes, 196, 198, 199, 200, 201, 202, 204
- natural abundance in earth's crust 2.7 × 10^{-6}%

Mercury is present in trace amounts in food, possibly because it has important uses in industry and therefore is a widespread contaminant. Mercury is not an essential element.

Toxicity
The toxic forms of mercury are the alkyl derivatives, methyl mercury and ethyl mercury, which can produce an encephalopathy. This was a problem in the manufacture of top hats in which mercury was used; hence, the 'Mad Hatter' in *Alice in Wonderland*. Alkyl mercury derivatives may be ingested after eating carnivorous fish who have lived in polluted waters. Seed grain previously treated by mercurial fungicides is equally poisonous. Microorganisms may be contaminated with inorganic mercury from water. This is converted into methyl mercury which may then pass into the food chain through plant-eating small fish to carnivorous large fish, e.g. tuna, sword-fish and pike. These may then be eaten by humans and other predators. Lakes may contain up to 25 μmol of mercury per litre of water which then passes into the food chain. Most deep-sea fish contain small amounts of mercury, for example cod may contain 400 nmol/kg, while fish living in coastal waters near estuaries may contain 2.5 μmol/kg. This may not be a real hazard but it is noteworthy. Mercuric poisoning may occur with contamination of food in excess of 145 μmol/kg. Grain which has been sprayed with alkyl mercury compounds to prevent fungal disease has on occasions been eaten by unsuspecting peasant families. This may lead to ataxia and visual disturbances and even paralysis and death.

Molybdenum

- atomic weight, 96; valencies, 2, 3, 4
- natural isotopes, 92, 94, 95, 96, 97, 98, 100 natural abundance in earth's crust, 7 × 10^{-4}%

The amount of molybdenum in plants is dependent on where they are grown and upon the soil content. Vegetables grown in neutral and alkaline soils with a high content of organic matter have a higher content of molybdenum. Important dietary sources are wheat flour and germ (0.7 and 0.6 μg/g, respectively), legumes (1.7 μg/g) and meat (2 μg/g).

Action
Molybdenum is essential for the enzymes xanthine oxidase/dehydrogenase, aldehyde oxidase and sulphite oxidase which are important in the metabolism of DNA and sulphites.

Absorption
Intestinal absorption efficiency is high at 40–100%; there is a carrier-dependent active process in the stomach and proximal intestine.

Availability of molybdenum
Plasma concentration is 1 μg/100 ml, bound to protein. Storage is in the liver and excretion in urine.

Deficiency and excess
As yet, there have been no reports of molybdenum deficiency in humans.

It has been suggested that a high incidence of gout may be attributed to high intakes of molybdenum (10–15 mg/day). Molybdenum intake at this level may also be associated with altered metabolism of nucleotides and with impaired copper bioavailability.

Recommended requirements
These 50–400 µg/day for adults. For breast-fed infants a requirement of 0.5–1.5 µg/kg/day has been suggested.

Nickel

- atomic weight, 59; valencies, 2, 3.
- natural isotopes, 58, 60, 61, 62, 64
- natural abundance in earth's crust, 0.018%

Nickel may be essential in some animals and birds but deficiency in humans has never been proven.

Absorption
Absorption is meagre at 3–6% from the diet. Plasma concentrations are 2–4 µg/100 ml, some bound to albumin, the remainder in free solution. Nickel is excreted in urine.

Deficiency
Nickel deficiency might result in depressed growth and haemopoiesis, but there is uncertainty if nickel is essential.

Phosphorus

- atomic weight, 31; valencies, 3, 5
- natural isotope, 31
- natural abundance in earth's crust, 0.12%

Phosphorus is present in all natural foods. The usual diet in Britain provides 1.5 g of phosphorus daily. Approximately 10% of dietary phosphorus is present as food additives. An excessive intake of aluminium hydroxide antacids may bind dietary phosphate and result in secondary phosphate depletion. This results in muscle weakness and bone pains.

Action
Phosphorus is an important component of the crystalline structure of the bony skeleton with calcium (see Bone structure, section 12.3). Phosphorus is important in oxidative phosphorylation as part of adenosine triphosphate (ATP). Under normal physiological conditions, electron transport is tightly coupled with phosphorylation. ATP generation depends upon electron flow, which only occurs when ATP can be synthesized.

Other critical roles are in nucleic acids through the phosphorylation of sugars as a base–sugar–phosphate nucleotide, the phosphate group being a connector between nucleotides in the polynucleotide chain. Phosphorus is important in the control of activity of enzymes through phosphorylation.

The activity of an enzyme may be changed through alteration in its covalent structure. One way is by phosphorylation of hydroxyl groups in the enzyme, usually the side chain of serine, threonine or tyrosine residues. This is effected by a protein kinase and reversed by a phosphoprotein phosphatase. The enzyme can therefore be in an active or inactive conformation. The control becomes complex, as in addition to substrate availability there are two enzymes exerting control. Glycogen phosphorylase is controlled in this manner. The kinase enzyme acting as an enzyme activator allows amplification of a effector action by a hormone.

Absorption
Phosphorus absorption is as free inorganic phosphorus from the diet and of the order of 50–70%, reaching 90% at low dietary intakes. Phosphorus in the form of phytate may not be hydrolysed or absorbed. Other sources of phosphorus in the intestine include salivary and intestinal secretions of the element. Absorption is by active transport controlled by $1,25(OH)_2$ vitamin D, at both the brush border and basolateral membrane. The process is active, sodium-independent, and non-saturable. The jejunum absorbs better than sites further along the intestine. The plasma concentration is 0.8–1.4 mmol/litre and is dependent on excretion.

Availability

About 80% of the phosphorus in the human body (19–29 mol; 600–900 g) is in the form of the calcium salt in the skeleton. The remainder is as inorganic phosphate and forms part of essential metabolic components such as ATP.

Intracellular concentrations are higher than extracellular, the former being about 5–20 mmol/litre; therefore there is an action as an intracellular buffer.

Plasma phosphate concentrations are controlled by renal excretion, though there is some excretion in faeces. Urinary phosphate secretion undergoes diurnal variation, which can be reduced by exercising. This may be governed by adrenocortical hormones through phosphate re-absorption in the renal tubules. Vitamin D metabolites, glucocorticosteroids and growth hormones increase urinary phosphorus excretion. Oestrogens, thyroid and parathyroid hormones and increased plasma calcium concentrations increase renal reabsorption and hence conservation.

Phosphate metabolism may be disturbed in a number of conditions, particularly those affecting kidneys and bone.

Recommended requirements

In general, phosphorus requirements are estimated as equimolar to calcium.

Selenium

- atomic weight, 79; valencies, 2, 4
- natural isotopes, 74, 76, 77, 78, 80, 82
- abundance in earth's crust, $8 \times 10^{-5}\%$

Selenium is found in a number of forms in food, selenoamino acids, e.g. selenocysteine and selenomethionine, in selenoproteins and as selenide, selenite or selenate.

The main source of selenium is cereal, meat and fish. Milk vegetables and fruit contain little: meats, 2 µg/g; seafoods, 0.5 µg/g; nuts, 0.7 µg/g; and wheat flour, 0.3 µg/g.

Action

Selenium is an essential nutrient. Selenoenzymes, selenium-dependent enzymes, e.g. iodothyronine, deiodinase and glutathione peroxidases, protect the cell from peroxidative damage. Selenium is part of the active site of both enzymes as selocystine. Selenoprotein activity includes the hepatic microsomal deiodination of thyroxine.

Metallothioneins are small cysteine-rich proteins which bind divalent metal ions. They act as a store for metals such as zinc. Their concentration in tissues is dependent on metal ion availability. They also act as transcriptional enhancers. Proteins (metal-responsive transcription factors) interact with one of the inducing metal ions and facilitate the transcription of the metallothionein gene.

> Genetic control of the conversion of selenocysteine into selenoproteins involves the codon U(T)GA. This is normally a stop codon but marks a site of insertion of a selenocysteine residue in each selenoprotein. Insertion of selenocysteine is through an unusual tRNA which is charged with a serine residue which is converted to selenocysteine. The concentration of the mRNA responsible for glutathione peroxidase is linked with dietary selenium intake.

Absorption

Absorption is efficient at 35–85%. Selenium absorption may be through a wide variety of mechanisms, e.g. amino acid pathways as seleno amino acids.

Plasma concentration is 7–30 µg/100 ml, protein bound. Excretion is via the urine and possibly bile.

Deficiency

There is an overlap between selenium deficiency and vitamin E deficiency in a number of animals. This is because selenium is involved in glutathione peroxidase which destroys lipid hydroperoxides and is important in stabilizing lipid membranes by inhibiting oxidative damage. In New Zealand, Venezuela and China the soil is poor in selenium and deficiency problems are reported. In China, selenium deficiency has been complicated by a cardiomyopathy in children (Keshan's disease).

Selenium deficiency results in a decrease in glutathione peroxidase activity.

Recommended requirements

The adult RNI has been suggested at 1 μg/kg, which is approximately 70 μg/day in a British adult; 40 μg/day is a suitable LNRI.

Pregnancy and lactation In pregnancy, fertility is dependent upon an adequate selenium intake. There are adaptive changes in metabolism during pregnancy. In lactation, the concentration of selenium in colostrum varies from 50–80 ng/ml. This falls during the first month of lactation to 18–30 ng/ml. This requires an increase in dietary intake of about 15 μg/day.

Infants Breast-fed infants will receive approximately 5–13 μg/day. Intakes from formula feeds are generally lower, of the order of 2–4 μg/day. The non-breast-fed infant's RNI is of the order of 1.5 μg/kg/day at 4–6 months.

Children The RNI has been estimated to be in the order of the adult per kg body weight.

Body store measurements

Urine is the major route of excretion and is a reasonable marker of intake. Plasma concentrations reflect intake to a degree, though there is a wide range of variation. Red cell selenium or glutathione peroxidase activity are markers of medium-term status. Hair and toe-nail concentrations reflect long-term status, though these measurements may be made invalid by the use of shampoos.

Silicon

- atomic weight, 28; valency, 4
- natural isotopes, 28, 29, 30
- natural abundance in earth's crust, 25.8%

Silicon occurs as a silicate, which is very insoluble in water.

Cereal grains and other sources of dietary fibre are important sources of silicon. Other sources include drinking water (2–12 μg/ml).

Action

The role of silicon in human nutrition is unclear; silicon is an essential nutrient for the growing chick and rat. Silicon may be important in the proteoglycans of cartilage and of the ground substance of connective tissue. The human aorta, trachea, lungs and tendon are rich in silicon. The aortic silicon content may decline with age, particularly in the presence of atherosclerosis.

Absorption and storage

Silicic acid in foods and drink is absorbed quickly. The body storage pool is approximately 3 g in a 60-kg man. Maximal levels occur in skin and are found as the free monosilicic acid in plasma (500 μg/100 ml). Only trace amounts are found in tissues, especially skin and cartilage and tissues containing glycosoaminoglycans.

Recommended requirements

The dietary requirements of silicon are not known.

Silver

- atomic weight, 108; valency, 1
- natural isotopes, 107, 109
- natural abundance in earth's crust, $4 \times 10^{-6}\%$

Silver occurs in low concentration in soils, plants and animal tissues. It may interact with copper and selenium but has no known essential function in humans.

Strontium

- atomic weight, 88; valency, 2
- natural isotopes, 84, 86, 87, 88 89
- natural abundance in earth's crust, 0.017%

Strontium is widely distributed in the environment and in plants, particularly in wheat bran, rather than the endosperm of grains, and in the peel of root vegetables. The strontium content of drinking water varies from 0.02–0.06 mg/litre, though higher values have been recorded.

Strontium is present in foods which are rich in calcium, for example milk and fresh vegetables, and is stored in bone. The concentration is approximately 1000 times less than that of calcium. Cows' milk has a higher content of strontium than human milk. This means that bottle-fed babies have a greater intake of strontium than breast-fed infants. The amount of strontium retained in the body is dependent on urinary loss. Strontium is lost in urine, sweat, hair and other fluids.

Absorption and storage
The body store is of the order of 3.5–4 mmol, of which 99% is present in bones. The efficiency of absorption of strontium is of the order of 20%.

Dietary intake
The dietary intake is of the order of 1–3 mg/day. When strontium 90, resulting from atomic explosions is present, this may result in bony malignant lesions, e.g. sarcoma. Strontium has no known function or essentialness in humans.

Sulphur

- atomic weight, 32; valencies, 2, 4
- natural isotopes, 32, 33, 34, 36
- relative abundance in earth's crust, 0.048%

Action
Sulphur occurs in tissues as the sulphate, SO_4^{2-}. This is a component of the proteoglycans, which are important in extracellular matrices as dermatan sulphate, chondroitin sulphate and keratin sulphate. These are present in cartilage, vascular and reproductive systems. Disulphide cross-linkages are important in the specific three-dimensional folding of proteins.

Sulphate is involved in the hepatic enzymatic detoxification systems of phenols, alcohols and thiols, in part by increasing water solubility. Sulphate is involved in an active form as phosphoadenosine and phosphosulphate and is derived from cysteine and methionine using the molybdenum-dependent sulphite oxidase system. Sulphur forms part of the glutathione and coenzymes including coenzyme A.

There is an important reduction of sulphate to sulphide in the colon which influences the production of hydrogen sulphide or methane.

Sulphur is also important in sulphur-containing amino acids. L-Methionine is metabolized by transmethylation and trans-sulphuration.

S-Adenosylmethionine

In methionine metabolism there is a formation of the high-energy sulphonium, S-adenosyl-L-methionine. The latter is both a methyl donor for transmethylation reaction and the precursor of decarboxylated S-adenosyl-L-methionine, which is the aminopropyl donor for the synthesis of polyamines. S-Adenosyl-L-methionine is the methyl donor for all known biological methylation reactions with the possible exception of those involved in methylation of L-homocysteine. S-Adenosyl-L-homocysteine is hydrolysed to produce L-homocysteine which can be remethylated to methionine or condensed with serine to form cystathionine. S-Adenosyl-L-methionine may be decarboxylated, which is the donor of aminopropyl groups for synthesis of spermidine and spermine. The methyl group of adenosyl methionine is transferred to a nitrogen, oxygen or sulphur group of a wide range of compounds in reactions catalysed by numerous methyltransferases.

Absorption
Sulphur absorption is largely as amino acids conjugates which are subsequently desulphated. It is excreted as free sulphate or as organic and inorganic sulphates.

Recommended requirements
Dietary intake is of the order of 0.7 mg/day.

Body store measurements

Free sulphate in the diet is freely absorbed and freely excreted in the urine.

Tin

- atomic weight, 119; valencies, 2, 4
- natural isotopes, 112, 114, 115, 116, 117, 118, 119, 120, 122, 124
- natural abundance in earth's crust, $6 \times 10^{-4}\%$

Absorption

Human diets may contain 150–200 µg/day, though how much is from food and how much from tin cans is not clearly identified. The lacquering of the interior of cans reduces the amount of tin available for absorption. It is not known whether tin is an essential nutrient.

Ingested tin is poorly absorbed and mainly excreted in the faeces. Tin (SnII) is four times more easily absorbed than the (SnIV) form. Tin accumulates in the skeleton, liver, spleen and lung and is excreted in urine.

Deficiency and excess

No naturally occurring tin deficiency has been reported.

High intakes of inorganic tin lead to gastro-intestinal symptoms.

Recommended requirements

The upper limit permissible in a canned food is 2.1 mmol/kg. Average intakes in Britain have been estimated at 190 µg/day for an adult, but 99% of the tin is excreted in faeces.

Vanadium

- atomic weight, 51; valencies, 2, 3, 4
- natural isotopes, 50, 51
- natural abundance in earth's crust, 0.016%

Vanadium is present in most human foods, particularly shell-fish, mushrooms and peppers. The daily intake in the American diet for example may be 25 µg.

Action

The biological function and nutritional requirement of vanadium has yet to be identified.

Absorption

Absorption efficiency is meagre at 0.1–1.5%. This is in part because of the number of non-absorbable complexes which vanadium forms. Vanadium may occur in oxidation states from -1 to $+5$. Tetravalent and pentavalent ions are the most common form in foods. The reduced vanadyl ion (VO_2^{2+}) binds to ferritin and is transported in the blood in that form. Plasma concentrations are of the order 0.5–2 µg/100 ml and there is binding to transferrin. Excretion is via the urine.

Recommended requirements

These are of the order of 1–2 µg/day, but this is only a rough estimate.

Zinc

- atomic weight, 65; valency, 2
- natural isotopes, 64, 66, 67, 68, 70
- natural abundance in earth's crust, 0.02%

Zinc-finger 'proteins'

There are a number of generic classes of transcriptional regulators, including those based on helix–turn–helix, leucine-zipper and zinc-finger transcription motifs. The number of different zinc-finger proteins may exceed 100.

Most zinc-finger proteins relate to genes transcribed by RNA polymerase II and bind to a promoter lying outside the coding region. Zinc-finger proteins include at least two distinctly different groups of transcription factors, the C2–H2 and C2–C2 series and GAL4. Polypeptide loops are stabilized by being held together by a zinc ion tetrahedrally coordinated to a cysteine residue. The C2–C2 zinc-finger proteins are important as nuclear receptors for the steroid and thyroid hormones and for retinoids. All of the zinc-finger proteins have been shown to be transcriptional control factors. The zinc ions are essential for their function.

Dietary sources of zinc are meats (3–5 mg/ 100 g), whole grains and legumes (2–3 mg/100 g), and oysters (70 mg/100 g). White bread, fats and sugars are not very good sources.

Action

Zinc is required for many enzymatic functions, DNA synthesis, cell division and protein synthesis. Zinc is involved in enzyme activity, including carbonic anhydrase, alcohol dehydrogenase, alkaline phosphatase, lactate dehydrogenase, superoxides, dismutases and pancreatic carboxypeptidase. It has long been believed that zinc is important for wound healing.

Absorption

Approximately 20% of dietary zinc is absorbed from the intestine. During digestion free copper is complexed with ligands such as amino acids, phosphates and organic acids. Phytates and oxalates may form insoluble complexes which inhibit absorption. Phytate may bind zinc and affect zinc absorption. The coincidental presence of copper and iron may inhibit absorption.

Normal plasma concentration is 11–22 μmol/ litre, where the zinc is associated with albumin.

The adult body content of zinc is over 2 g (30 mmol). The prostate gland, choroid of the eye and semen have high concentrations of zinc, though the greatest is in bones (about 200 μg/g). Red cells contain 13 μg/ml, hair 120–250 μg/g, both in newborns and adults.

Urinary loss is of the order of 6–9 μmol/day. Most zinc is lost in the faeces.

Deficiency

Zinc deficiency is rare, as zinc is so widely available in foods, but where it has been described it has been associated with increased zinc losses from the body. The clinical syndrome of growth retardation, male hypogonadism, skin changes, mental lethargy, hepatosplenomegaly, iron-deficiency anaemia and geophagia has been reported from studies of male dwarfs in Iran and are said to be due to zinc deficiency. This secondary sexual retardation is in part due to reduced serum testosterone, dihydrotestosterone and androstenedione concentration. These deficiencies are treatable with zinc supplementation (15 mg, three times a day as zinc acetate) for 12 months or more. The mechanism by which zinc affects testosterone concentration in zinc-deficient subjects is not known, but a zinc-dependent enzyme may be important in sex hormone synthesis.

The **dermatological signs** of severe zinc deficiency include progressive bullous–pustular dermatitis at the extremities and the oral, anal and genital areas, combined with paronychia and generalized alopeciae. These respond to dietary supplementation of zinc sulphate.

Zinc deficiency has been reported in alcoholics in whom the serum zinc concentration is lower compared with controls. Similarly, hepatic cirrhosis is associated with low serum and hepatic zinc, and at the same time an increased urinary zinc excretion. Zinc deficiency has been reported in patients with steatorrhoea in whom presumably zinc has not been absorbed, or when there is loss of zinc protein complexes into the intestinal lumen to various degrees in patients with neoplastic conditions. Patients with widespread burns may also have a reduced plasma zinc concentration due to loss of zinc through the skin. In renal disease an excess of zinc may be lost in the urine, with a consequent reduction of plasma and tissue concentrations.

Iatrogenic causes for deficiency of zinc include the use of antimetabolites and diuretics.

Genetic disorders Acrodermatitis enteropathica is a lethal autosomal recessive trait which manifests itself in the early months of life after weaning. Acrodermatitis enteropathica consists of pustular and bullous dermatitis, alopecia and diarrhoea. Zinc supplementation is necessary. The condition is due to zinc malabsorption. Sickle-cell anaemia may be associated with zinc deficiency. Zinc supplementation for sickle-cell anaemia subjects increases weight gain, growth of pubic hair, serum testosterone concentrations, plasma zinc, tissue zinc and neutrophil alkaline phosphatase activity.

Zinc excess
Excessive zinc can lead to nausea, vomiting and fever.

Recommended requirements
The recommended intake varies between 7 and 15 mg/day, though only 20–30% of this is absorbed.

Adults Adult requirements of zinc are of the order of 2–3 mg/day (30–40 μmol). Minimal losses are in the order of 30 and 20 μmol/day in men and women, respectively. Since absorption is only 20–30% efficient, this suggests RNIs of 145 and 110 μmol/day for men and women, and LRNIs of 85 and 60 μmol, respectively.

Infants Human milk is not a rich source of zinc and the infant depends very much on the stores obtained during the last 3 months of intrauterine life.

Assuming a daily requirement of 1 mg/day and an absorption efficiency of 30%, an EAR of 3 mg/day has been suggested, and an RNI of 4 mg/day.

Children EAR are of the order of 3.8–5.4 mg/day with an RNI of 5–7 mg/day.

Pregnancy Extra zinc is required during pregnancy. Additional zinc is accumulated during the last 3 months of gestation of the order of 8 mg, which means that there is a requirement of an additional 2 mg, assuming an absorption efficiency of 20–30%. This means that the dietary requirement is 6–14 mg/day.

Lactation It is not known if there is an increased dietary need during lactation.

Supplementary zinc may be given as zinc sulphate.

Body store measurements
Zinc concentration in plasma which is neither haemolysed or contaminated is a measure of zinc status. An alternative is the plasma copper : zinc ratio. An increase in this ratio of greater than 2 is suggested to be related to zinc deficiency. Zinc in the red blood cells and hair gives a long-term assessment of zinc status, as the zinc turnover in red cells and hair is slow.

Neutrophil alkaline phosphatase activity has been used to assess zinc status. Urinary excretion of zinc is decreased during zinc deficiency.

Summary

1. Trace elements are present in the diet in less than milligram amounts.
2. They are important in that they establish potential differences across membranes; such differences have to be maintained by energy requiring reactions.
3. They also allow oxidation–reduction reactions to take place by enzymes; the active principle may be a trace element, e.g. copper, iron, magnesium, manganese or sulphur; in addition, phosphorus has a role in controlling the enzyme active sites.
4. They help to maintain the structure of proteins and nucleic acids; trace elements, e.g. zinc, may adsorb to proteins in specific configurations. Such trace elements act as cofactors or structural elements by virtue of high localized concentrations of positive charges, and can exist in two or more valency states, e.g. copper, molybdenum and selenium.
5. Trace elements are also important in the structure of hormones and vitamins, e.g. iodine in thyroxine and cobalt in vitamin B_{12}.
6. Trace elements are involved in skeletal structure, e.g. fluorine, magnesium and phosphorus in bone and teeth, and silicon in cartilage.
7. Na^+ and K^+ with single positive charges, rarely form complexes with proteins. Occasionally the divalent alkaline metals Mg^{2+} and Ca^{2+} complex with proteins. The other elements in biological

systems are from the first transition series and exist in more than one oxidative state which is important in oxidation–reduction reactions.

8. Some trace elements are toxic, e.g. lead, aluminium, mercury and arsenic. These may reduce intellect in the young and elderly. The function of some trace elements has still to be established, e.g. vanadium.

Further reading

Anonymous (1992) Aluminium, a dementing ion. *Lancet,* 339, 713–14.

Anonymous (1992) Essential trace elements and thyroid hormones. *Lancet,* 339, 1575–6.

Bates, C.J., Thurnham, D.I. et al. (1991) in *Design Concepts in Nutritional Epidemiology* (eds B.M. Margetts and M. Nelson), Oxford Medical Publications, Oxford.

Chesters, J.K. (1992) Trace elements and gene interactions. *Nutrition Reviews,* 50, 217–23.

Cousins, R.J. (1994) Metal elements and gene expression. *Annual Review of Nutrition,* 14, 449–69.

Dietary Reference Values for Food, Energy and Nutrients for the United Kingdom (1991) Report on Health and Social Subjects, no. 41, HMSO, London.

Documenta Geigy Scientific Tables, 6th edn (1962) (ed. K. Diem), Geigy Pharmaceutical Company, Manchester.

Hetzel, B.S. and Dunn, J.T. (1989) The iodine deficiency disorders. Their nature and prevention. *Annual Review of Nutrition,* 9, 21–38.

Johnson, L.R., Alpelpers, D.H., Christensen, J., Jacobson, E.D. and Walsh, J.H. (1994) *Physiology of the Gastrointestinal Tract,* Raven Press, New York.

Linder, M.C. (ed.) (1985) *Nutritional Biochemistry and Metabolism,* Elsevier, New York.

Mill, C.F. (1985) Dietary interactions involving the trace elements. *Annual Review of Nutrition,* 5, 173–93.

Prasad, A.S. (1985) Clinical manifestations of zinc deficiency. *Annual Review of Nutrition,* 5, 341–63.

Vulpe, C.D. and Packman, S. (1995) Cellular copper transport. *Annual Review of Nutrition,* 15, 293–322.

Chapter 9

Agricultural chemicals

- Agricultural chemicals are widely used in the control of infestations and plant diseases.
- Agricultural chemicals can enter and be concentrated in the food chain, accumulate and persist in the environment and storage organs of animals.
- Agricultural chemicals are metabolized differently by vulnerable organisms such as insects and weeds than by humans or by crops.

9.1 Agricultural chemicals in the food chain

9.1.1 Introduction

The agricultural industry is the source of our readily available food supply. The industry looks for a predictable harvest which requires good weather, adequate equipment, fertile soil and freedom from infestation and disease in the plants.

Food price competition places undue pressure on farmers to produce cheap food. Such pressure requires more intensive use of better land. Such efficiency is seen as being sustainable only with the use of fertilizers and the elimination of pests and weeds, often by chemical methods. Man, as a hunter–gatherer, had to wander over large areas to find randomly distributed sources of foods, as did those other forms of life which also lived off these particular crops and animal foods. Intensive farming spared man from hunter–gathering. Such concentration of food did not pass unnoticed by the predators who would also enjoy the concentration of food in one area. The use of chemicals is intended to redress the balance in man's favour. This has introduced other unexpected imbalances which appear to benefit no one.

The control of insects which are parasitic on plants used for food has a very long history. Sulphur was used to control pests by the Egyptians, the Ancient Greeks and the Romans. In the 19th century, the French developed the use of copper sulphate and lime as a fungicide. For generations farmers in the United States have combated the Colorado beetle using the arsenical poison, Paris green.

Large volumes of pesticide liquid are sprayed onto plants throughout the world, and have entered every food chain. In the USA in 1984, the value of pesticides sold by the top 16 USA manufacturers was approximately $6500 million. In 1989, British gardeners spent £23M on pesticides, £9M on herbicides, £7M on insecticides, £1M on fungicides, and £5M on herbicide fertilizer mixtures. Local authorities use pesticides including amitrole and atrazine for weed control on footpaths. Agents such as dichlophen are used to control grass moss in sports grounds. Metaldehyde is used for slugs and MCPA is used in road margins.

Intensive animal production is heavily dependent on disinfectants and other measures against infection such as formaldehyde, phenolics, iodine-containing sheep dips, vaccines and organophosphates, given in bolus form to ruminants.

Therapeutic antibiotics are permitted for prophylactic use when prescribed by veterinary practitioners. Nevertheless, the number of strains of bacteria resistant to these antibiotics increases. Feeds containing preservatives, propionic acid, minerals, vitamins and urea will supplement the requirements of the ruminal flora for nitrogen. Gut flora is manipulated by copper and zinc bacitracin to remove bacteria.

Lindane, carbaryl, permethrin, malathion and phenothrin are used for head lice control in schools.

Pesticides are used in the control of wood rot and mould. Pentachlorophenol is a hazardous wood preservative. Following the spraying of affected wood within a confined space, pentachlorophenol residue concentrations may be three times higher than recommended. The clothing of operatives who have been spraying chemicals may be contaminated by pesticides.

Large volumes of pesticides are transported around the world. There is always a concern that there might be leakage of chemicals into the environment during the clearance and cleaning of the tankers. The dumping of surplus pesticides is also a problem for both industry and individuals with ecological interests.

These chemicals are present in every aspect of life.

Some 40 million sheep in Britain were dipped each year until recently. The pesticide residue presents formidable disposal problems including concerns over seepage into deep waters. In 1985, when 1500 samples of sheep's kidney fat were analysed for pesticide residues, 71% contained residues of lindane. Pesticide residues have been detected in a wide range of foods, 10% of bran-based breakfast cereals, 55% of wheat germ, 93% of pure bran, 16% of processed oats, 24% of rice, one-third of sausages and nearly half of sampled burgers, cheeses and apples.

During spraying, particularly with aerial pesticide spraying, pesticides may contaminate the air and be carried over from the sprayed area into the local area. This is known as spray drift wherein tiny drops of pesticide float away from the intended crop. Only certain pesticides are approved for aerial spray and these include captan, benomyl, chlorpyrifos, 2,4-D, dichlorvos, malathion, metaldehyde and 2,4,5-T.

It is important to make a distinction between the risks associated with different stages in the usage of pesticides, which are associated with diverse industrial, application and residue concentrations.

Some of the more toxic chemicals, including dioxin and the polychlorinated biphenyls (PCBs), are appearing in the Arctic. The north-east Atlantic is the largest reservoir for PCBs. They appear to be evaporating from soils, waste dumps and polluted lakes and condensing on the snow and ice. The organochlorines are readily stored in fat and the animals in these frozen conditions rely upon an abundance of fat to survive. As the organochlorines are genotoxic the effect is to reduce the fertility of many animal breeds.

Farm workers are readily contaminated during crop spraying and it has been claimed that diminution in sexual potency can result from contact with some sprays.

9.1.2 Agrochemicals

Agrochemical pesticides can be divided into four main groups:

1. Traditionally compromising inorganics, e.g. arsenate, sulphur and copper.
2. Biologically selective agents – plant extracts such as nicotine and pyrethroids (see section 9.2.3). More recent developments include synthetic pyrethroids which are stable against exposure to light and water.
3. Organochlorine compounds – DDT, dieldrin and aldrin (see section 9.2.2).

4. Organophosphorus products (see section 9.2.1).

The frequency, amount and types of agrochemicals applied vary from year to year, but in 1987 the most commonly used in the United Kingdom were:

- **Herbicides**: isoproturon, metsulfuron, methyl triallate
- **Fungicides**: propiconazole, prochloraz, fenpropimorph
- **Insecticides**: dimethoate, deltamethrin, pirimicarb

Many structurally different fungicides are classified into non-penetrative protective types, e.g. dithiocarbamates and phthalimides, and penetrative types, e.g. benzimadazoles.

Herbicides have had a profound effect on crop yields, allowing direct drilling, early harvesting and proven product quality. They include phenoxy compounds, the carbamates and glyphosates.

There is a basic need in agricultural practice for nitrogen, potassium and phosphate. The crop yield becomes more abundant as the amount of nitrogen applied increases. Inorganic nitrogen is usually applied as ammonium nitrate. Organic nitrogen in the form of manure and blood, etc. is both smelly and bulky for the amount of nitrogen delivered.

The battle is not being won by the agrochemical industry. Insects and weeds are developing resistance to pesticides and herbicides, while opportunist weeds grow in the spaces left by the disappearance of other weeds.

9.1.3 Labelling and classification of pesticides

In the UK, pesticide labels must conform to the *Data Requirements for the Control of Pesticide Regulations*, which covers the classification, packaging and labelling of dangerous substances regulations 1984, and includes various EU directives. A label must include:

- the trade name

- the common name with concentration
- restrictions for use
- manufacturer's recommendations on the method of usage, e.g. rate of application

The law also requires further instructions to be followed, for example:

- precautions
- name and address of holder of approval
- registration number approval.

Dangerous pesticides are classified as those which are:

1. Very toxic, with an oral LD_{50} in rats of 5 mg/kg or less.
2. Toxic, i.e. LD_{50} of 5–50 mg/kg.
3. Harmful, i.e. LD_{50} of 50–500 mg/kg.

Other classifications are inflammable, flammable, oxidizing or corrosive. It is also important that the public are aware of the presence of residues in food, including the labelling of post-harvest application.

9.1.4 Pollution of water supplies

The agricultural industry can contaminate water both at point source and over a wide area. Point source contamination arises in heavily farmed areas, e.g. farm slurries. Diffuse pollution from organic fertilizers, nitrates and pesticides is a problem. The pollution of river water by fertilizers is highly seasonal, greatest in autumn from leaching of land run-off water following heavy rain with resulting pollution of water supplies, resulting in high ammonia concentrations. In summer, concentrations are relatively low and the applied nitrate is immediately taken up by the crop. Nitrates may contaminate underground water stores as a consequence of increased fertilizer use. Long-term storage of nitrate-containing water in reservoirs leads to a 50% decrease of nitrates to nitrogen gas by bacterial reduction. Nevertheless, nitrate pollution of rivers and aquifers which supply half of the water is a growing problem, more complicated in aquifers where some water takes up to 30 years to percolate from the surface soil to the water table. This means that the pollution of these underground water sources may take years to establish and even longer to reduce. The most certain health risk from high nitrate levels in water is methaemoglobinaemia. Maximum concentrations of nitrate in drinking water were established at 100 ppm in 1974 by the World Health Organization (WHO).

Water-soluble pesticides and herbicides such as atrazine and simazine contaminate drinking water. A survey of British tap water found that two-thirds of samples contain pesticide levels in excess of European and British Government guidelines. In the United States aldicarb (an insecticide), triazines (a herbicide) and EDB, DCP and DBCP (soil fumigants) have been detected in drinking water.

9.1.5 Monitoring pesticide contaminants of food

In the UK there are two organizations which test for pesticides and pesticide residues. The members of the Association of Public Analysts test products on behalf of local authorities. The Government's Working Party on pesticide residues (WPPR) publishes figures every year. Such testing shows that overall, 2% of foods, 6% of cereal and cereal products and 6% of fruit and vegetables contain more than the maximum residue levels. Some 99% of apples receive pesticide treatment, the residual pesticide declining with time. Peeling apples removes 90% of the pesticide residues. Meat is not analysed and there may be several residues present in any one food.

The potential size of the problem of applying chemicals to growing plants is illustrated by the use of the plant growth regulator daminozide, which 'plumps up' apples. The United States Environmental Protection Agency has calculated that if a million people were to eat apples sprayed with daminozide over a life time then 45 would develop cancer. This carcinogenic action is mediated through a metabolite, UDMH. The fruit and vegetable marketing system insists on standards for perfect appearance in the produce, e.g. no surface blemishes, perfect ripeness, size and colour. Neither taste nor nutritional value are included in for example the EU and US Government Federal Grading Systems.

Those farmers who use a minimum of synthetic chemicals or organic farmers cannot completely eradicate pests, so there is some cosmetic damage. If the produce is sent to the food processing industry then blemishes are less important. The potential biological and pathological hazards to humans of the infections that cause the blemishes do not appear to be a cause of concern. However, it is possible that many consumers would accept less attractive fruit and vegetables containing a lower pesticide residue content. Organic foods tend not to be marketed by supermarkets in the same way as the mass-produced food. The smaller volume results in higher costs, which with the overall poorer appearance, militates against sales.

In a survey in the US the majority of interviewees felt that the health benefits of fruit and vegetables outweighed the risks accruing from possible pesticide residues.

There is a general deep concern for the environment and for children. On the other hand the reappearance of ergotism in Germany from untreated rye indicates the need for a balance between pest and pesticide control and residue content of crops. Chemicals may cause either no toxic effect or an acute toxic effect after a single dose, or may have a chronic effect after repeated small and non-lethal doses. Acute toxicity is measured as LD_{50}, the dose at which 50% of the organisms die from a randomly chosen group of a batch of a species. The dosage is described as mg/kg.

Short-term effects and problems of under-reporting

The primary hazard with exposure to pesticides is acute toxic reactions from skin contact or inhalation over relatively brief periods. Such exposure can lead to acute eye and upper respiratory tract irritation and contact dermatitis, and even serious poisoning. Individuals who work regularly with pesticides are most at risk. Much of the confirmed poisoning accidents are the result of drift from nearby spraying operations. Domestic timber treatment is another potential problem.

Long-term effects

There are possible long-term effects of insecticides on reproduction, by for example the insecticide DBCP. Pesticides also have both chronic as well as acute neurotoxic effects. Immune response may also be modified by these substances. Such case control studies show possible associations between herbicidal exposure and disease, but cannot as yet be seen as absolute proof. Nevertheless, this is an area where considerable and continued thought and monitoring are necessary.

Cancer

The International Agency Research on Cancer (IARC) classifies chemicals by their carcinogenic potential. The IARC describes evidence for carcinogenesis under a series of categories: sufficient; inadequate; or limited.

It cannot be overestimated that a carcinogenic potential in animals does not necessarily apply to carcinogenesis in the human situation and vice versa. Risks to humans are categorized under:

- Group I, proven to be carcinogenic in humans
- Group IIA, probably carcinogenic
- Group IIB, possibly carcinogenic
- Group, III not classifiable
- Group IV, probably not carcinogenic in humans

Examples of Group I are arsenical pesticides and vinyl chloride. One probably carcinogenic pesticide is ethylene dibromide. There are 17 possible carcinogenic pesticides and 26 not classifiable.

Classification of pesticide concentration

All pesticides are tested on animals before application to crops eaten by humans. Such tests establish an Acceptable Daily Intake (ADI), which is calculated to be a safe intake over a lifetime. ADI is defined as the amount of chemical on a body weight basis, which can be consumed daily in the diet over a whole lifetime in the practical certainty, on the basis of all the known facts, that no harm will result.

Food

The observed no effect level (NOEL) is the maximum dose which is safe and specific treatment-related effects do not occur. The NOEL is used to calculate the ADI. The safety factor is set at a figure of 100-fold; this is based on a figure of 10 times to allow for variation between animals and humans, and 10 times for the possible variation among individuals. Different safety factors are given for different foods or chemicals.

The ADI is usually set by the WHO through an international committee, the Codex Alimentarius Commission of the United Nations, who also set Maximum Residue Levels (MRL). These are the maximum residue concentrations allowed for any particular pesticide in the food when leaving the farm. MRLs are calculated as the maximum quantity of the given product that anyone could eat to ensure that the ADI is not exceeded. The European Commission has made the MRL into a legal limit to be used for each pesticide. As yet, only a limited number of available pesticides have a defined MRL.

Water

The European Commission (1985) and the UK – Water Supply (Water Quality) Regulations 1989 have established Maximum Admissible Concentrations (MACS) for pesticides in drinking water at 0.1 parts per billion (ppb) for individual pesticides and 0.5 ppb for total pesticide content.

Occupational exposure

Occupational Exposure Limits (OEL) (based on the long-established USA Threshold Limit Value, TLV) are not Maximum Exposure Limits (MELs). The TLV was established by the American Conference of Government Industrialist Hygienists (ACGIH) and refers to airborne concentrations under which workers may be repeatedly exposed day after day without adverse effects. This does not allow for the wide range of individual susceptibility or for the effects of metabolites over a prolonged period. The ACGIH committee now recommends the use of a Short-Term Exposure Limit while awaiting definitive information which would enable sensible comment on long-term effects. In the USA there is also another set of values called Integrated Risk Information System (IRIS) which is used to calculate long-term risks. IRIS uses maximum time-weighted air concentrations (WAC), which should cause no adverse effects in humans over a 40-year exposure. Carcinogens are considered not to be safe at any concentration. The risk factor is the amount by which the OEL exceeds the level obtained for IRIS.

Risk to human health

The estimated intake of pesticide residues is measured in micrograms per day, whereas acute toxic doses are many orders of magnitude greater. There have been reports of poisoning from carbamate and organophosphate residues in the USA and Israel. Significant pesticide residues of DDT have been measured in the blood and placentas of Indian women who have had spontaneous abortions. This was not found in women with full-term deliveries. In the USA a survey has shown a reduction in sperm density which is related to pesticide concentrations in semen.

Pesticides may undergo chemical changes during food processing and hence qualitative changes in toxicity potential. The measurement of these by-products at various stages of cooking and processing is a difficult logistic and analytical problem. Residues may be concentrated in particular tissues, e.g. organochlorines in oils, meat, milk and fatty tissues.

Among fungicides, benomyl breaks down spontaneously to carbendazin, which also has fungicidal activity. It is claimed that some crops may safely be harvested the same day as spraying, but concentrations double that of the MRL (UK) i.e. 5 g/kg (UK Consumers Association, July 1991) have been found on the same day as spraying occurred. The residue concentrations should be less than the legal Maximum Residue Level permitted within UK or within the limit set by the Codex.

9.1.6 Control of pesticides

International trade in pesticides is covered by two United National Conventions. The United Nations Food and Agriculture Organization (FAO) publishes the International Code of Conduct on the use of and distribution of pesticides: 'the Code'. The UN Environmental Programme has produced guidelines for the exchange of information on chemicals in international trade, known as the London Guideline.

The FAO code covers the management and use of pesticides, their availability, distribution and trade, and recommendations for labelling, packaging, storage, disposal and advertising. The Code is intended for the guidance of Governments and industry and other interested organizations.

The European Commission has increasing control over pesticide use in Europe. Following the passing of the single European Act, Article 100A of the Act gives a strong commitment to establish high standards of environmental consumer and public health protection. EU directives identify residue limits for a number of pesticides and prohibit some products.

Under the UK 1986 Control of Pesticide Regulations, when a chemical is submitted for approval it must:

- be effective against the designated pests
- be safe against plants and wildlife in general
- have no undesirable effects on the environment
- be safe to humans

Pesticide residues

Government agencies are now publishing details of pesticide used, pesticide residue analysis and the results during the production of foods. Within the United Kingdom the UK Pesticide Guide 1990 is published by CAB International and ACP (Advisory Committee on Pesticides) or MAFF (Ministry of Agriculture, Fisheries and Food) Usage Reports. These cover various aspects of the use of agricultural pesticides for England and Wales.

9.1.7 Natural ways of controlling pests

Fungi

There are no natural organic fungicides but some naturally occurring mineral fungicides, e.g. copper salts and sulphur are available. It is possible to minimize the spread of some fungal diseases by growing resistant, less vulnerable varieties, growing young plants in good light and well-ventilated sites and removing infected plants. Good growing conditions are important; however, the breeding of new fungus-resistant plants may also result in unexpected, unwanted modifications of the overall characteristics or specific chemistry of the plant.

Pesticides

Pesticides which are acceptable to the organic organizations, such as the Soil Association, include pyrethrum, derris, soft soap and aluminium sulphate. Derris and pyrethrum are natural products extracted from plants. These products may also contain piperonyl butoxide which adds to the effectiveness of the pyrethrum. Both of these compounds are short-lived but kill a wide range of creatures, including ladybirds, bees, butterflies, fish, toads and tortoises. Soft soap kills soft-bodied insects by removing the protective wax covering of their skins.

9.1.8 Principles of pesticide metabolism

Metabolism has an important role in dictating the selectivity of action of pesticides and protects some species, e.g. animals and humans. Most pesticides are poorly soluble in water. Oxidation or hydrolysis are primary metabolic reactions with the insertion or the uncovering of polar groups. Sometimes the products of primary actions undergo further secondary changes, e.g.

conjugation before biliary and urinary excretion. The secondary reactions are very species-dependent. There is therefore diverse metabolism of foreign compounds in organisms as remote as mould and man.

A toxic substance has to accumulate in a living organism until a concentration is achieved which allows for toxicity. Enzymes that metabolize foreign compounds catalyse reactions which:

- alter molecular structure, to a less toxic product
- increase the polarity and water-solubility and facilitate urinary and biliary excretion

The majority of the enzymes responsible for the primary metabolism of foreign compounds are hydrolases and oxygenases (Figure 9.1).

The degree of metabolism dictates the persistence of pesticides in soil, plants and animals. Resistance to the pesticides results from developing or possessing metabolic processes which reduce the biological activity or increase the rate of urinary or biliary excretion.

Hydrolases

Hydrolytic enzymes are widely found in plants, animals and in various parts of individual cells including vertebrate cytoplasm, microsomal membranes and the endoplasmic reticulum of liver cells.

Hydrolytic enzymes are found as membrane-bound hydrolases in microsomes. The microsomal polysubstrate mono-oxygenases responsible for the oxidation of drugs and pesticides vary considerably between species in both the plant and the animal kingdom, but the primary metabolites are remarkably constant from phylum to phylum.

Carbamates have a high affinity for esterases. Hydrolysis plays an important part in the metabolism of the oxime subgroup of carbamates after an initial oxidative step.

Many pesticides contain $-O-CH_3$ or $-O-C_2H_5$ groupings. These alkyl groups are removed by a glutathione transferase system (Figure 9.1).

Amide groups are removed by an amidase, e.g. dimethoate.

Propanil is a selective herbicide used to control weeds in rice fields. Rice plants are unaffected as they reduce the herbicidal activity of propanil by an aryl carboxylamidase enzyme. Some carboxyamidases act as carboxylesterases and are able to select amides as substrates. This is important in cross-resistance over a wide range of effects of different pesticides, e.g. the ester malathion and amide dimethoate. Resistance in insects to some insecticides results from possessing or developing high concentrations of carboxylesterase.

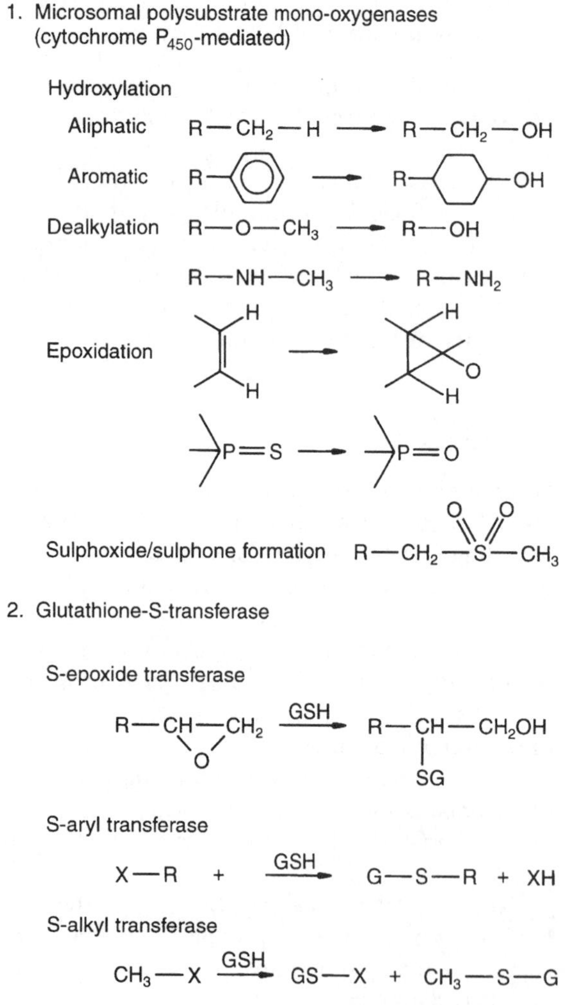

Hydrolase actions

Many pesticides have ester, amide or phosphate linkages which are split by hydrolases:

- organophosphorus, pyrethroid, carbamate – insecticide
- dithiocarbamate, dinitrophenol – fungicide
- urea, carbamate – herbicide

There are a number of substrate specificities:

- R–O–P linkage – phosphatase
- R–COOR' linkage – carboxylesterase, carboxyesterase
- R–CONHR' linkage – carboxyamide

The classification is difficult and ill-defined. Esterases are of two types:

- A-esterase – hydrolytic reaction increased by organophosphates, e.g. paraoxan
- B-esterase – inhibited by organophosphates

The difference may rest in the rate of dephosphorylation after the reaction.

Malathion is detoxified by a carboxylesterase:

$$R–COO–C_2H_5 \rightarrow RCOOH + C_2H_5OH$$

whereas hydrolysis of esters of 2,4-D within the plant leads to the development of an active metabolite within the plant cell.

---

1. Microsomal polysubstrate mono-oxygenases (cytochrome P_{450}-mediated)

2. Glutathione-S-transferase

FIGURE 9.1 Metabolism of pesticides. Microsomal polysubstrate mono-oxygenases and glutathione-S-transferases. GSH, glutathione.

Epoxides are transitory intermediates in the metabolism of unsaturated compounds. The epoxide of the organochlorine aldrin is stable, persists in fat tissues and is more toxic than the parent. Epoxide hydrolases are important in biodegradation; they are hepatic enzymes which modify epoxides, highly active substances that bind to DNA. Cyclic compounds with one or two double bonds readily form labile epoxides. Organic pesticides form epoxides, e.g. dieldrin and endrin. These are poor substrates for epoxide hydrolase.

Oxidation reactions

The oxidation of drugs and pesticides requires microsomal polysubstrate mono-oxygenases involving cytochrome P_{450}. Such enzyme systems are found in plants, vertebrates and invertebrates. The oxidation reactions involve mono-oxygenase reactions:

1. (i) Aliphatic hydroxylation
   (ii) Aromatic hydroxylation
2. (i) O-dealkylation
   (ii) N-dealkylation
3. Epoxidation
4. Substitution of oxygen for sulphur
5. (i) Formation of sulphoxides and sulphones
   (ii) Formation of amino oxides

Microsomal polysubstrate oxygenases are found in the smooth endoplasmic reticulum of liver cells:

$$R–H + O_2 \rightarrow R–OH.$$

These are mixed-function oxidases utilizing cytochrome P_{450}, and yielding O^-, which is hydrolysed to water.

Flavine adenine dinucleotide oxygenases are of limited relevance to pesticide metabolism, and require NADH. There are three components, a cytochrome P_{450} fraction, a flavoprotein fraction, NADPH-cytochrome P_{450} oxidoreductase and a phospholipid fraction. These enzymes form a group of isoenzymes, varying from organ to organ, are found in all species and phyla, and appear to be under hormonal control.

Cytochrome P_{450} mono-oxygenases are inhibited by methylene dioxyphenols, e.g. piperonyl butoxide. These delay the metabolism of pesticides, e.g. carbamates which then accumulate.

Aliphatic hydroxylation is a common metabolic route for aromatic or heterocyclic pesticides; for example, a methyl group in the acid moiety of pyrethrin can be converted to a hydroxymethyl group.

Oxidative desulphuration is important in the metabolism of thiophosphate insecticides, which are NADPH-dependent. The S atom is replaced to yield a more potent choline esterase inhibitor than the original. This phenomenon is known as lethal metabolism, e.g. parathion to the more lethal paraoxon.

Sulphoxidation is the oxidation of sulphide sulphur to sulphoxide. This involves a FADH-containing mono-oxygenase system, e.g.

demeton–S–methyl $\rightarrow$ sulphoxide
aldicarb $\rightarrow$ aldicarb sulphoxide $\rightarrow$ aldicarb sulphone

O-Dealkylation is a common metabolic pathway for pesticides involving at least three mechanisms.

N-Dealkylation is common in pesticide metabolism as many pesticides are substituted amines or amides, e.g. carbamate, carbufuran or diuron, as well as the herbicide atrazine. Resistance to atrazine occurs when this metabolic pathway is well developed in the plant.

Glutathione is important in pesticide degradation in glutathione-S-transferase activity:

1. Glutathione-S-epoxide transferase
2. Glutathione-S-aryl transferase
3. Glutathione-S-alkyl transferase

Reaction 1 opens the epoxide ring and inserts glutathione as an additive reaction; this is important in the metabolism of potentially carcinogenic oxides, e.g. dieldrin, endrin and heptachlor epoxide.

Reaction 2 is important in animal liver but may also occur in plants. The activity varies between species. The reaction involves the transfer of active thiol groups, with elimination of a hydrogen halide, e.g. dichloronitrobenzene, atrazine and other chlorinated triazine herbicides. Maize, which is resistant to atrazine, contains an abundance of this enzyme.

Reaction 3 is with alkyl halides and involves the removal of methyl groups from organophosphorus insecticides containing CH_3–O–P.

Conjugation

These require membrane-bound enzymes requiring cofactors in the cytosol. The conjugation reactions involve glucuronic acid, glucose, arginine, glutamic acid, glycine, sulphate and acetylation.

Glucuronide formation is important and occurs in terrestrial vertebrates possessing a microsomal glucuronyl transferase. The fungicide ferbam is converted to the *S*-glucuronide in some animals.

Glucose can be conjugated to a alcoholic thiol amino acid. The herbicide propanil is so metabolized in rice plants.

Sulphate esters: a sulphotransferase is found in mammalian livers and kidneys and may be found in insects but rarely in plants.

9.2 Categories of agrochemicals

9.2.1 Organophosphorus insecticides

These compounds have a wide range of physicochemical and biological properties and variable uses in agriculture and plant hygiene. They have a common basic structure called the leaving group (Figure 9.2).

Organophosphorus compounds have now replaced organochlorine compounds as the most common insecticides for the control of aphids and other soft-bodied insects. These compounds are less persistent but need to be sprayed at more frequent intervals, e.g. demeton-*S*-methyl, malathion, parathion, phorate. Organophosphates inhibit acetylcholine esterases by phosphorylating the enzyme, whereas carbamates compete with acetylcholine for the enzyme surface.

The toxicity of the organophosphorus compounds varies from the instantly fatal to negligible. In addition to acute and prolonged toxic anticholinergic effects, the organophosphorus compounds can be teratogenic in some avian species. Chromosomal abnormalities may be found more frequently than expected in humans poisoned by malathion. The problem of safety is made more complicated by contaminating impurities, e.g. isomalathion which potentiates the toxicity of malathion. The impurity strongly inhibits the β-esterases which detoxify malathion, and hence potentiate its action. On the other hand, resistance in insects results from the development of high carboxylesterase activity, complicated by the substrate specific activity of the range of various carboxyesterases.

The insect's diet may alter its hepatic P_{450} activity and therefore the rate of metabolism of the pesticide.

Enzymatic hydrolysis is often rapid and depends upon two types of esterase and amidase enzymes (Figure 9.3). Esterases attack the bond on the side of the carbonyl group attached to the oxygen atom, whereas amidases attack the bond on the side attached to the nitrogen atom. In most animals the primary attack on most carbamates

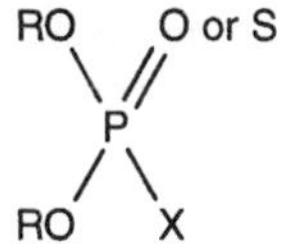

R is usually a methyl or ethyl group
X is the leaving group
These are joined by ester or thioester
They are usually complex, either aliphatic, homocyclic or heterocyclic in nature

Examples

Orthophosphate   e.g. dichlorvos

Thion phosphate   e.g. bromophos

Dithiophosphate   e.g. malathion

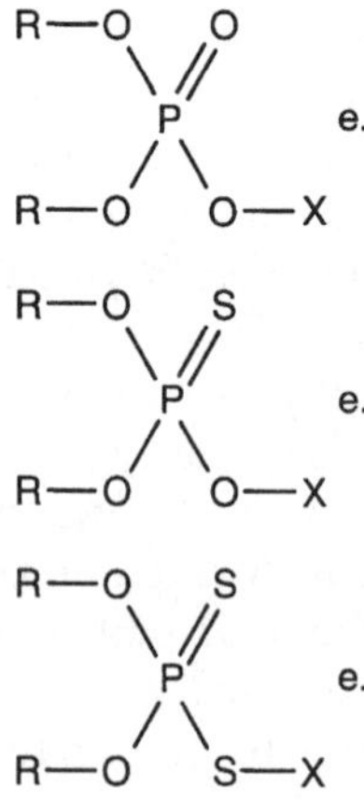

FIGURE 9.2   Organophosphorus insecticides – general structure.

290

involves oxidative N-demethylation, ring hydroxylation and epoxide formation. This weakens the molecular structure and allows more rapid enzymatic hydrolytic change. In the insect the initial reaction may involve a mono-oxygenase reaction, so the rate of detoxification may be inhibited by a methylene dioxyphenol synergist. The carbamate insecticide **carbaryl** (Figure 9.4) may inhibit cellulase activity by soil bacteria. Carbaryl is leached from soil to deeper water layers following chemical or bacterial chemical decomposition. Soil microbiological metabolism of the carbamates may increase over the years with prolonged exposure. In soils and animals some carbamates are degraded to sulphoxides. In anaerobic soils containing ferrous ions degradation is rapid. Soil moisture content may also affect degradation rate. In anaerobic soil the degradation products are a nitrate and an aldehyde.

The parathion family of organophosphorus insecticides are toxic to mammals. These compounds are readily oxidized by mono-oxygenases in animals, insects and plants and are converted to derivatives containing the $P = O$ group which are even more powerful cholinesterase inhibitors than the parent compound. Degradation of parathion is by an oxidative NADPH-dependent oxidase reaction. Quite different mechanisms are involved in methyl parathion and fenitrothion metabolism wherein there is a rupture of a $P–O–CH_3$. In general the mammalian liver performs this reaction faster than insects.

Oxidation of the insecticide may precede conjugation. If the oxidation product is more toxic than the parent compound, e.g. carbofuran, retention in the enterohepatic circulation is not ideal. The N-methyl carbamate derivatives of oximes, e.g. aldicarb, methomyl and oxamyl, are toxic to higher animals. Excretion in humans may be facilitated by conjugation to endogenous compounds. In animal liver the principal primary metabolite is sulphoxide. Some carbamates are excreted in bile to the intestine and subsequently are retained in

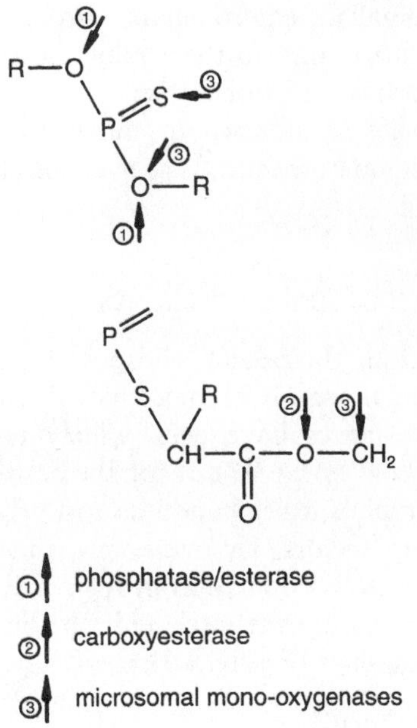

① phosphatase/esterase

② carboxyesterase

③ microsomal mono-oxygenases

FIGURE 9.3 Some enzymes cleaving organophosphorus compounds, phosphatase/esterase, carboxyesterase and microsomal mono-oxygenases.

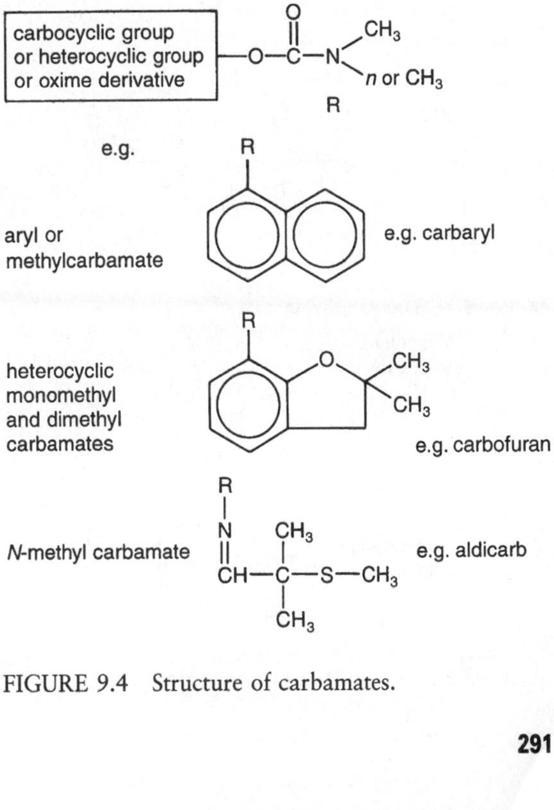

FIGURE 9.4 Structure of carbamates.

291

the mammalian enterohepatic circulation. The toxic **aldicarb** has further value in controlling phytophagous nematodes, hence its wide usage. The half-life of aldicarb in soil is variable and depends upon how much is applied and the nature of the soil.

9.2.2 Organochlorine insecticides

These include the potent, cheap and widely used DDT and gamma-HCH (hexachlorocyclohexane). These substances have been widely used, especially during World War II for the eradication of malaria, typhus, river blindness and yellow fever-harbouring vectors. Unfortunately, there was an element of indiscrimination in their use.

The organochlorine insecticides belong to three major groupings (Figure 9.5):

1. DDT-related.
2. Gamma-HCH.
3. Compounds related to aldrin.

All of the organochlorine compounds are neurotoxic substances which interfere with ion transfer in the nerves.

The chemical stability of these compounds is

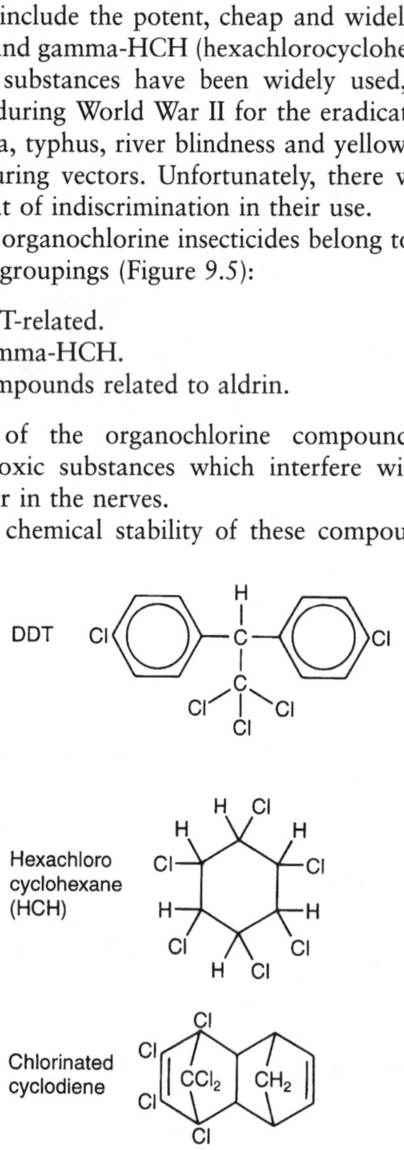

FIGURE 9.5   Structure of organochlorine insecticides.

due to the C–C, C–H and C–Cl bonds, which makes for slow dissimilation and accumulation in the environment. In 1970, it was calculated that rain water contained 2×10^{-4} ppm, air 4×10^{-6} ppm and seawater 1×10^{-6} ppm of organochlorines (see also Biomagnification, pages 26 and 293).

By 1972, concentrations of DDT and their derivatives were reported as 3×10^{-4} ppm in plankton, 1×10^{-3} ppm in aqueous invertebrates, and 5×10^{-1} ppm in marine fish. Vegetables contained 2×10^{-2} ppm, meat 2×10^{-1} ppm and human adipose tissue 6 ppm. Many components of the food chain are involved and contaminated.

DDT

DDT is effective against a wide variety of pests but is relatively ineffective against aphids and spider mites. The initial effect of DDT, which is upon the peripheral nervous system, is temperature-dependent; hence efficacy is reduced in regions of warmer climate.

DDT metabolism

The major pathway for the degradation of DDT is dehydrochlorination to DDE. Removal of one chlorine from the trichloromethyl group results in p,p'-DDD (TDE) (tetrachlorodiphenyl-ethane) a metabolic breakdown product of DDT. DDE – which is less toxic to insects – is the major DDT residue found in animal tissue. The reaction is dependent upon a glutathione-*S*-transferase reaction. A second route is reductive dechlorination to DDD and a third is oxidative and leads to dicofol. In many species a further metabolite is a water soluble DDA in which –CCl_3 in DDT is replaced by –COOH. This can be excreted as a free compound or conjugate in bile and faeces.

DDT is very insoluble in water. In higher animals the DDT accumulates in the central nervous system. The most damaging effects of DDT appear to be close to postsynaptic membranes of neuron–neuron or neuron–muscle contacts, affecting the sodium ion channel. An alternative role for DDT may be on the mitochondrial membranes responsible for transmembrane conduction. Such a lipophilic substance may well distort the activities of photosynthesis, oxidative phosphorylation, active transport and nuclear division.

From acute experiments the LD_{50} for DDT given in one dose to a 70-kg man would be 14 g. These substances are being phased out, but their use is still allowed in less environmentally aware countries.

Hexachlorocyclohexane (gamma-HCH, formerly BHC)

The γ-isomer (lindane) is more toxic to insets than the α- or δ- form. γ-HCH is odourless, whereas the other isomers are smelly and taint food. The mechanism whereby this one isomer is so potent compared with the others is not known. The toxic action on the insect is on the central nervous system, and may well be involved in γ-aminobutyric acid (GABA), dopamine and N-acetyldopamine brain receptors. The complicated metabolism of γ-HCH requires mono-oxygenases and glutathione-S-transferases and is species-dependent. In humans, γ-HCH acts on the central nervous system.

Chlorinated cyclodiene family

These include the stereochemically related dieldrin and aldrin. **Aldrin** is a soil fumigant which kills wireworms and larvae of root flies, whereas dieldrin is for root dip and seed dressing. Aldrin is metabolized to dieldrin by an NADPH-dependent reaction in hepatic microsomes. Slow reactions are important in retaining a lethal concentration.

The chlorinated cyclodiene insecticides are the most toxic and persistent of all pesticides. Many similarities exist between the central nervous system neurotoxic properties of cyclodiene and lindane.

Aldrin and dieldrin metabolism

The olefinic cyclodienes aldrin, isodrin and heptachlor are oxidized by microsomal oxidases to the corresponding stable epoxide. Other reactions include secondary alcohol substitution or mono-oxygenase reactions in mammalian liver. Hydrolysis of the oxirane ring by epoxide hydrolase leads to formation of a *trans*-diol. These substances induce liver enzymes and hence hepatic excretion of the organochlorines. Another metabolite is 2-keto-dieldrin. Dieldrin can be degraded in mud and soils by anaerobic organisms. These epoxide cyclodienes are toxic and persistent and are found in all animal fats and are very toxic to birds.

Biomagnification

Organochlorines persist in clay and manure rich soils for months, even years. There is little effect on soil bacteria. However the phenomenon of **biomagnification** means that there is a concentration effect in certain plant species. In plants which transpire rapidly there is substantial uptake through the roots. Organochlorine residues accumulate in soil and run off into the sea, rivers, reservoirs and lakes. Water contamination is universal. Despite the reduced usage of these chemicals there is persistence, e.g. in human fat tissue (12 ppm in 1951 in the USA).

Meat, eggs and milk can be contaminated. If cows are exposed to DDT, the DDT appears in the milk at a steady state within 2 weeks. Concentrations in human milk are even higher than in cows' milk, with consequences to the food chain.

9.2.3 Natural and synthetic pyrethroids

These originate from naturally occurring compounds found in the dried inflorescences of *Chrysanthemum cinerariaefolium* (Figure 9.6). These are 'knock down' substances which stun insects in their track, without toxic effects to warm-blooded animals. A second lethal compound is required to give the coup de gras.

Pyrethroids

The four active principles in pyrethrium flowers are pyrethrin I and II and cinerin I and II, and small amounts of jasmolins I and II. All four are esters with an acid containing a three-carbon ring joined to an alcohol containing a five-carbon ring. Compounds designated I contain chrysanthemic acid and II pyrethric acid. The naturally occurring acids are in the *trans* form and its esters are more toxic than the synthetic stereoisomers. The natural alcohols are in the *cis* form. The esters are unstable;

storage results in 20% loss in activity but anti-oxidants can protect in the dry and dark. Water and insect tissue activity leads to hydrolysis. The esters are insoluble in water and are soluble in lipid solvents. Some methylene dioxyphenyl compounds are powerful synergens of pyrethroids. This suggests that the metabolism of the pyrethroids is by liver microsome activity. Metabolism occurs not by hydrolysis of the ester linkage but rather by oxidation of the methyl group in the isobutenyl side chain of the acid moiety to a hydroxymethyl group and then to a carboxyl group in a NADPH-dependent reaction.

Pyrethroids are attractive insecticides as they are minimally toxic to humans and mammals, and are readily destroyed by cooking and digestive

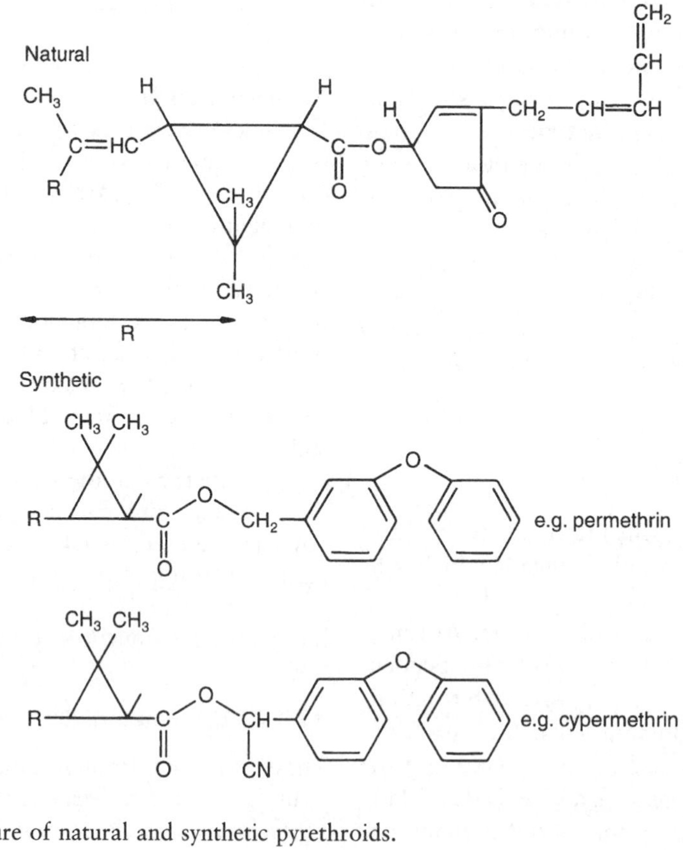

FIGURE 9.6   Structure of natural and synthetic pyrethroids.

juices. They are, however, expensive, readily decomposed by light, and are toxic to fish.

Synthetic pyrethroids are more stable in the light with varying knock-down propensity and relative toxicity to flies and vertebrates and are more expensive, e.g. allethrin, bioresmethrin, permethrin, cypermethrin and deltamethrin. It may be that different types of pyrethroids act at different sites in the nervous system.

Permethrin

This is a non-systemic, moderately persistent insecticide, useful in domestic and veterinary medicine and effective against a wide range of phytophagous insects and for the control of mosquitos. It consists of a mixture of isomers, related *cis* and *trans* with respect to the spatial arrangement of the ester linkage and the dichlorovinyl linkage. The isomeric mixture gives rise to a large number of metabolites. Most of the initial metabolic changes involve hydrolysis or oxidation.

> *Trans*-permethrin is usually metabolized more rapidly than is *cis*-permethrin. The initial reaction is specific, but it is difficult to follow the oxidative and hydrolytic reactions in sequence. Many metabolites are conjugates formed by secondary metabolism. There are differences in the hepatic microsomal metabolism of the *trans*-permethrin compared with the *cis*-permethrin which may be due to the *trans*-isomer being hydrolysed by esterase whereas the *cis*-isomer is cleaved by an oxidase. The metabolism of permethrin by pseudoplusia is inhibited by low concentrations of organophosphorus compounds, particularly relatively non-polar compounds. Such synergism is important in pesticide formulations.

Cypermethrin

A similar compound to permethrin but with an α-cyano group which improves photostability and leads to a powerful and rapid debilitating effect

on insects but also increases the number of possible metabolic products. The degradation of cypermethrin is initiated by bond cleavage or by hydroxylation. Many birds are tolerant to cypermethrin, partially as the compound passes rapidly through the intestine and is rapidly metabolized if absorbed. The avian nervous system is relatively insensitive to this compound. The compound has a wide biological spectrum and is used on numerous crops against biting and sucking insects. However, when tolerance develops in certain strains of flies, effectiveness can be restored by giving organophosphorus compounds which inhibit these esterases.

The insecticidal effect of both DDT and pyrethroids usually increases as the temperature falls. While the action of pyrethroids is understood at the physiological level, less is known of the chemistry which underlies the physiological action.

9.2.4 Fungicides

Fungi may attack plants through the soil or be transmitted through the seeds. Fungal mycelia have an almost unlimited ability to regenerate from a few surviving hyphal strands. This makes an established fungus very difficult to eradicate. The fungicide that kills on contact can check the growth of mycelium and limit the production of reproductive structure. This delays spread from infected plants to healthy ones. Non-systemic fungicides are usually insoluble in water.

The site-specific fungicidal agents affect single receptors or enzymes and are systemic in action. Fungicides with multiple sites of action have low systemic and eradicant properties, affect non-specific targets, and may react with thiol groups and disorganize lipoprotein membranes (Table 9.1).

Copper fungicides

Copper fungicides are used against a wide variety of fungi but are quite toxic to many organisms. Bordeaux mixture is a concentrated solution of copper sulphate added to a slight excess of lime suspended in water. Bordeaux mixture is insoluble in water, is gelatinous, and the copper binds to the leaf surface. The fungicidal efficiency of Bordeaux

TABLE 9.1 Groups of fungicides. (Based on MAFF, 1984, with additions.)

Possible multiple sites of action	Site-specific
Copper and tin compounds	
Mercurials	Dinitrocompounds (e.g. dinocap)*
Sulphur	Benzimidazoles (e.g. benomyl)
Dithiocarbamates (e.g. thiram, maneb)	Oxathins (e.g. carboxin)
Phthalimides (e.g. captan)	Steroid synthesis blockers
Phthalonitriles (e.g. chlorothalonil)	(a) Morpholines (e.g.tridemorph)*
	(b) C-14 demethylation inhibitors (e.g. prochloraz, triforine,
Site(s) of action uncertain	triarimol)
Dicarboximides (e.g. vinclozolin)	Hydroxyaminopyrimidines (e.g. ethirimol)
	Antibiotics (e.g. kasugamycin)
	Phenylamides (e.g. metalaxyl)
	Organophosphorus compounds (e.g. pyrazophos, fosetyl)
	Others (e.g. guazatine, cymoxanil, prothiocarb)

*Systemic activity may be low or absent.

precipitate decreases on storage, probably due to the changes in the extent and type of crystalline aggregation. The activity of copper may be through the chelation of amino acids, e.g. glycine and keto acids which have exuded through the leaf surface. However, the observed leakage of amino and keto acids may also be a consequence of the toxic effect of the copper.

Mercuric compounds

Inorganic mercury compounds are still used in some countries for defined purposes but in decreasing amounts. Mercury is potentially toxic to higher animals and the accumulation of mercury in soil and animals represents a serious environmental hazard. Microorganisms under aerobic and even anaerobic conditions can interconvert different forms of mercury, organic to inorganic, inorganic to organic, and inorganic to elemental mercury. Various forms of mercury are used for sealing pruning cuts and for seed treatments used to protect cereal seeds from a variety of diseases. All treated grain must of necessity be sown, it must never be employed to prepare food for humans or livestock. To reduce the risk of treated grain being consumed by accident it is conventional to colour the seed dressing, the colour acting both as a warning and a code to indicate the nature of the fungicide or insecticide in the dressing. Sometimes, however, starving communities wash the treated grain, assuming that the removal of colour indicates that the poison had been removed. Treated grain can be a major hazard to grain-eating birds, especially when treated with alkyl mercury but not aryl and alkoxyalkyl compounds.

Sulphur and lime sulphur

Sulphur is widely used by farmers in the United States in considerable amounts as a fungicide, though this usage is declining. The quantity used is in part because of the low potency of the preparation, of the order of kilograms per hectare, whereas most organic fungicides are applied at a dosage of less than a hundredth of this amount. It is not known how sulphur works; it may be that sulphur oxides are the active principle, that the sulphur is hydrolysed by water to active polysulphides, or that hydrogen sulphide is the agent.

Non-systemic organic fungicides

Protection by fungicides can be divided into two types:

1. Those that are focused at the point of delivery.
2. Systemics which enter the plant through the roots or leaves and are transported within the plant in xylem or phloem.

The correct and efficacious use of fungicides is much more difficult than that of insecticides or herbicides, especially in the developing world. Most organic fungicides, both non-systemic and systemic, tend to be selective. Consequently, success or failure depends on accurate identification of the fungal infection which is not a universal farming skill. The effectiveness of a fungicidal application may also depend on timing, and an understanding of the life-cycle of the particular fungus.

Dithiocarbamates

There are two fungicide dithiocarbamates, dimethyldithiocarbamate and bisdithiocarbamate (Figure 9.7). Thiram, a disulphide oxide, is used as a fungicide; its diethyl analogue is used medically to treat alcoholism. The dithiocarbamates enter fungi either as unionized molecules, as a weak acid or disulphide derivative, or as a covalent complex. Mixtures of diethyl carbamates with other fungicides having a different chemical action are often more effective than the individual components used alone. The dimethyldithiocarbamates may form toxic complexes with copper or alternatively may sequester essential trace elements.

The dimethyldithiocarbamates have a bimodal action when applied to moulds. An initial effect which declines is followed by a second phase of

activity, due to the action of varied products of dithiocarbamate generated during metabolism.

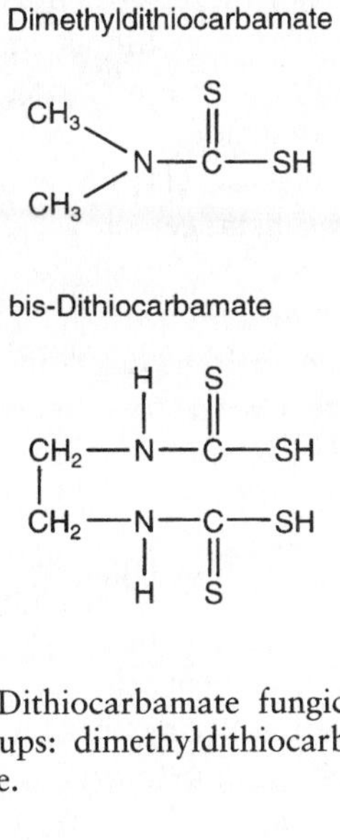

Bisdithiocarbamate action is on the thiol group of an enzyme's coenzyme or biological carriers with resulting breakdown to isothiocyanates and thiourea derivatives, and probably inhibits fungal respiration and growth. The active species is not known but may be an isothiocyanate disulphide or an ethylenethiuram disulphide.

All dithiocarbamate fungicides are unstable and can be broken down chemically and photochemically as well as by enzymes in plants and fungi. There is formation of significant amounts of ethylene thiourea during the cooking of food treated by the dithiocarbamate spray, maneb.

Phthalimide group

This group, which includes captan (Figure 9.8), folpet, captafol and dichlofluanid, reacts with thiol groups. The breakdown of phthalimides is accelerated by the endogenous thiol and glutathione. Fungicides of the captan group are toxic to fish but oral toxicity to mammals is very low. Some of these compounds give indications of mutagenicity with the Ames' test. Captan and related compounds have an unusual residue problem, by accelerating the corrosion of tin cans. This problem for the food industry is prevented if the captan is destroyed by heat processing.

Dinitrophenol derivatives

Dinocap (Figure 9.8) and binapacryl have been used as insecticides, fungicides, herbicides and for mothproofing since 1892, as a herbicide since 1932, and as a fungicide since 1949. Simple dinitrophenates uncouple mitochondrial oxidation from phosphorylation. It is possible that this is an example of lethal metabolism and these fungicides are hydrolysed by fungal enzymes to liberate free dinitrophenols which are then fungitoxic.

Chlorine-substituted aromatic fungicides

Cationic detergents have both bactericidal properties and defined but useful antifungal properties (Figure 9.8). They are safe and are chemically related to domestic detergents. They attack lipoprotein membranes and disrupt vital membrane-

Dimethyldithiocarbamate

bis-Dithiocarbamate

FIGURE 9.7 Dithiocarbamate fungicides. There are two major groups: dimethyldithiocarbamate and bis-dithiocarbamate.

dependent processes such as selective permeability and oxidative phosphorylation.

Imazalil and prochloraz
These imadazoles inhibit steroid synthesis and have a weakly systemic action, especially against seed and air-borne pathogens. When prochloraz is fed in high doses to rats it induces the cytochrome P_{450} monooxygenase system and glutathione-*S*-transferases in the endoplasmic reticulum. This leads to an accelerated metabolism of foreign compounds, including pesticides.

Systemic fungicides

A systemic fungicide attacks internal mycelium and penetrating haustria and thereby prevents the regeneration of even small pieces of surviving mycelium.

Most of the main plant fungal pathogens are attacked by members of one or other of the systemic fungicide families which attack very specific processes. Such site-specific activities are usually influenced by single gene selection amplification or modification and include interference with nucleotide base synthesis with polynucleotide or protein formation and the synthesis of steroids and components of lipoprotein membranes. Such selectivity means that resistance readily develops and persists.

Benzimidazole (Figure 9.9) is the parent substance of a family of systemic fungicides including benamyl, thiophanate-methyl and thiabendazole. There are three types, carbamates, non-carbamates and the thiophanate family which are converted to benzimidazoles after application. Some of these compounds break down quite readily in soil. The half-life in soil of detectable benzimidazole derivatives is 6 months or more and 30% of the original benomyl could be accounted for as breakdown products 2 years after application.

Adsorption to soil constituents increases with rising organic matter content and is affected by pH. The acute oral toxicity of most benzimidazoles to higher animals is low. In the mammalian gastrointestinal tract, breakdown of benomyl occurs rapidly following oral ingestion. Within 24 hours of feeding, 40% appears in the urine, partially conjugated to glucuronic acid, cysteine and acetylcysteine. The major mode of action of the benzimidazoles is to interfere with the division of cell nuclei by disrupting the assembly of tubulin into microtubules. The underlying mechanism is binding of the fungicide to proteins. Resistance to the benzimidazoles develops very readily, particularly in single substance sprays.

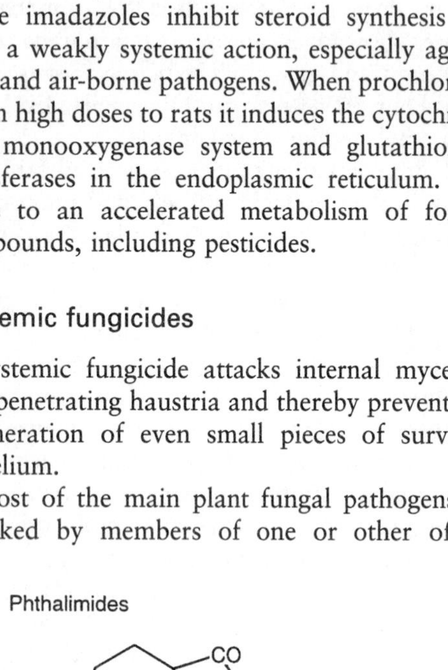

Oxathiins or carboxamides
Carboxin (Figure 9.9) is used in Britain exclusively as a cereal seed treatment against rusts, smuts and bunts. Carboxin is fungitoxic because it inhibits the respiration of sensitive fungi, possibly acting on succinate dehydrogenase and possibly causing mitochondrial damage. The *cis*-crotonanilide grouping is believed to be the active group. In some organisms the metabolism of carboxin consists of *p*-hydroxylation of the phenyl moiety, particularly in peanuts. In barley, there is

FIGURE 9.8 Non-systemic organic fungicides. Structure of phthalimides, dinitrophenols and chlorine-substituted aromatic fungicides, e.g. captan, dinocap-6 and pentachloronitrobenzene (PCNB).

oxidation of the sulphide sulphur to sulphoxide, which has slow fungitoxic activity. Adding a further atom of oxygen gives the sulphone, which reacts with lignin to produce insoluble complexes and aniline derivatives. In animals, oral doses of carboxin are excreted largely unchanged. In soil, carboxin loses its activity over 3 weeks.

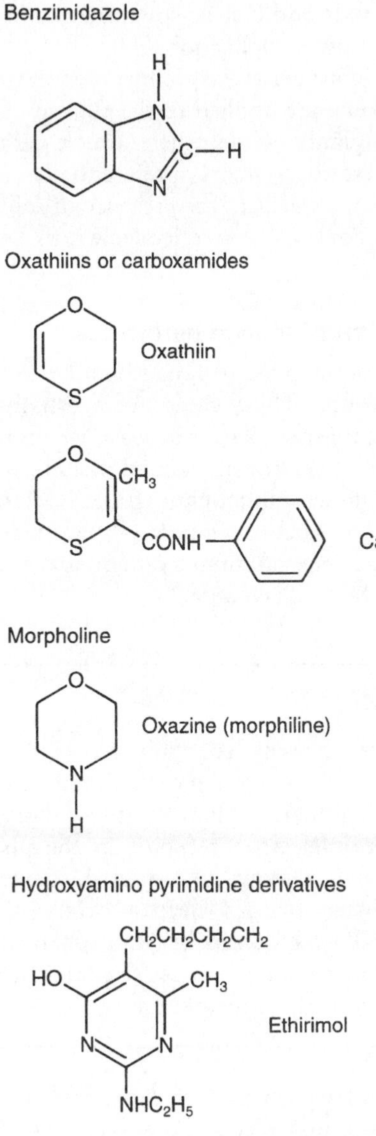

Benzimidazole

Oxathiins or carboxamides

Oxathiin

Carboxin

Morpholine

Oxazine (morphiline)

Hydroxyamino pyrimidine derivatives

Ethirimol

FIGURE 9.9 Systemic fungicides: benzimidazole, oxathiins (carboxamides); morpholine and hydroxyaminopyrimidine derivatives.

Morpholine inhibitors of sterol synthesis

The most common such nitrogen-containing heterocyclic fungicides are compounds based on the triazole ring, pyrimidine, pyridine, piperazine or imidazole (Figure 9.9). These control powdery mildew infections but some are also effective against rusts. The mode of action of morpholines is to inhibit the normal growth of fungal hyphae. This results from inhibition of one or both of two steps in the complex process of biosynthesis of sterols by the inhibition of C-4-demethylation of the sterol lanosterol by the cytochrome P_{450} system. This leads to an accumulation of unwanted intermediate sterols.

Hydroxyaminopyrimidine derivative

These compounds, e.g. ethirimol (Figure 9.9) are effective only against powdery mildews. Their use is restricted due to high levels of resistance. The primary metabolism of ethirimol is similar in plants and animals, with N-de-ethylation to a primary amine of low biological activity. The butyl group can be hydroxylated, although this route is less important in plants than in animals. Conjugation (with glucuronic acid in animals, glucoside in plants) can occur on the hydroxyl C-4 without any preliminary change. Ethirimol has a half-life of less than 4 weeks in barley plants when uptake is through the roots. Ethirimol interferes with the enzyme adenosine deaminase which catalyses the conversion of the amino groups of adenine in the form of adenosine to the amino groups of the hypoxanthine in the form of inosine.

9.2.5 Herbicides

The global weed control programme is somewhat different from the control of insects and pathogens. Weeds only seriously reduce crop yield or quality when they compete with a crop for available moisture, nutrient and light. Soil quality does not in itself limit food production. In general, weed control may be by hand and hoe by farmers. In more extensive farming chemicals may be used. Crop rotation is another method for weed and pest control.

Herbicides are used in agriculture to remove weeds which would otherwise compete with a crop. This can be achieved in a number of ways, not all of which require the weed killer to possess an intrinsic selectivity between the weed species and crop plants.

Biochemical differences between crops and common weeds are exploited. The absence of β-oxidase in some legumes allows a tolerance of 2,4-DB. Maize is tolerant of atrazine. The selective action of herbicides depends upon differences in life-style and plant morphology so that the crop is exposed to a lower effective dose than the weed receives.

Herbicide uptake

The physical and chemical properties of herbicides applied to foliage are very different to those of compounds normally applied through the soil. A toxic substance which is lipid-soluble when applied to leaves, will penetrate the waxy cuticle.

The factors influencing herbicide uptake are many – as are the interactions in the total metabolic process – and in consequence their action is not fully understood. Some herbicides may persist in the soil, particularly those with low solubility in water. Volatile materials such as diclobenil, certain thiolcarbamates and nitronilines, have relatively short half-lives in many soils. A problem inherent in the use of soil acting herbicides is an accumulation over years.

Following continuous use of a herbicide, the herbicide-degrading enzymes in the soil microorganisms increase with an attendant increase in degradation of herbicides. Most herbicides are readily degraded by microorganism oxidative reductive and hydrolytic enzymes. The type of soil also influences the persistence and effectiveness of any one herbicide. Some substances are strongly adsorbed on to soil organic matter, e.g. urea derivatives, triazines, carbamates and nitrophenyl ethers while others are strongly adsorbed onto certain types of soil particles, e.g. diquat. Thus, the type of soil affects the toxicity to weeds of a given dose of the herbicide and the retention of the herbicide within the soil. The effectiveness of many herbicides may be dependent on the level of

rainfall within a few weeks following soil application.

It is therefore necessary carefully to select a herbicide for a particular purpose, especially when using soil-acting compounds. The rate of application, rainfall, soil type, temperature and microbial population all affect persistence in the soil. Some crops are relatively tolerant of some of the compounds and can be sown or planted in a short period after application.

There is considerable specificity of herbicidal function. Some are applied to the foliage of well-established plants, while others attack germinating weed seeds. Others preferentially attack dicotyledonous weeds in the presence of foliage of monocotyledons. A few selectively kill seeding cotyledons.

Classification of foliage herbicides

The herbicides applied to foliage can be classified into five groups. These compounds usually control well-established plants, at least 10 cm high.

Herbicides that kill all foliage include quaternary ammonium compounds (Figure 9.10); in addition, two such members – **paraquat** and **diquat** – are bipyridinium compounds and kill exclusively by foliage contact.

Paraquat

Paraquat inhibits the reduction of $NADP^+$ and thereby prevents reduction of carbon dioxide in the photosynthetic carbon cycle by replacing ferredoxin as electron receptor in the photosystem. Light energy is re-routed, paraquat being reduced by a 1-electron transfer and automatically reoxidized by atmospheric oxygen to form superoxides. These attack unsaturated fatty acids in membrane lipids.

Small amounts of paraquat can irritate the skin or if inhaled may cause nose bleeding. When larger amounts are ingested a non-cancerous multiplication of lung cells occurs, accompanied by proliferation of mitochondria which continues long after all traces of the herbicide have disappeared, and leads to respiratory failure and death.

Paraquat and diquat undergo little metabolism in plants or animals. These quaternary ammonium compounds accumulate in water plants and hence are a danger to aquatic animals. Mammals may die after running through treated areas and then licking their contaminated fur.

Glyphosphate
Glyphosphate (Figure 9.10) is an acidic substance containing phosphonate and derived from glycine. These quaternary ammonium compounds are inactivated by soil.

Glyphosphate has little selectivity of action and kills all foliage on contact. Plants that have been exposed to glyphosphate usually develop chlorosis as they die. The main action appears to be on the biosynthetic pathway whereby aromatic amino acids are synthesized in plants, the shikimic acid route of synthesis (phenylalanine, tyrosine and tryptophan). Decomposition of glyphosphate in plants appears to occur rather slowly. The main product is aminomethylphosphonic acid. Chemical decomposition in soil is also slow but there is rapid microorganism degradation with the release of carbon dioxide.

Aminotriazoles are foliage herbicides and rapidly disappear after killing the weeds, but are not inactivated by soil. The major use is to control weeds in uncultivated land, which is intended for planting in the near future. Aminotriazole blocks carotenoid synthesis in which dehydrogenase enzymes remove hydrogen atoms to form double bonds.

In the plant, detoxification involves conjugation, producing higher molecular weight products which are poorly excreted.

Selective herbicides for broad-leaf weeds

Phenoxyacetic acid derivatives These include 2,4-D and MCPA and are used to control broad-leaved weeds in cereal crops from grasslands (Figure 9.10). The metabolism of phenoxyacetic acid involves side chain degradation with a substituted

phenol as a final product. Ring hydroxylation is usually the main metabolic route. Aromatic compounds chlorinated on C-4 are often hydroxylated in this position and thereafter the chlorine group may migrate to the C-3 or C-5 position. Conjugation with glucose can occur; such glucosides form a reservoir for active phenoxyacetic acids.

Phenoxyacetic acid derivatives readily pass through cuticular lipids at a rate dictated by their molecular polarity. Phenoxy acid herbicides, 2,4-D and MCPA, have persistent auxin-like characteristics. They react with the plasma membrane

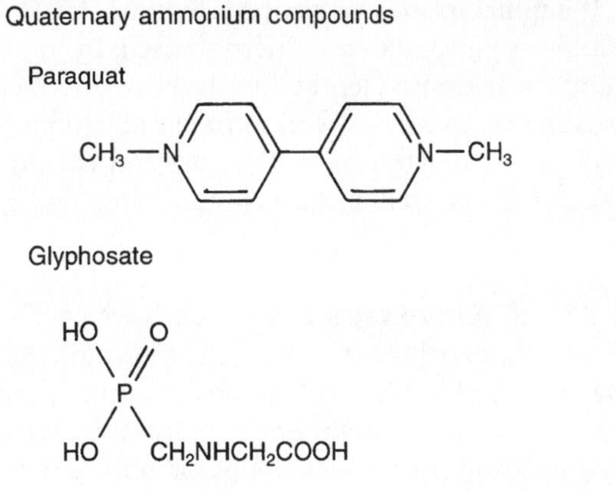

FIGURE 9.10 Herbicides. Quaternary ammonium, e.g. paraquat; glyphosate and phenoxyacetic acid derivatives.

and then by their continued presence cause transcription to be unnaturally prolonged. Whatever receptor or enzyme for auxin-like activity is the target of phenoxyacetic acid herbicides, activity is very structurally specific.

The primary effect of these phenoxyacetic acid herbicides is to cause aberrant growth of young, rapidly growing tissues near the meristem. There is evidence that auxins stimulate cell wall development and phenoxyalkanoic acid herbicides may interfere with the metabolism of pectin, methyl esters or some other components of young cell walls.

If applied to soils, salts of MCPA and 2,4-D are readily washed away or decomposed by microorganisms. Consequently, low levels of herbicide usually disappear 1–4 weeks after application, but may persist much longer when the soil is cold or dry. Enhanced degradation occurs with multiple application.

Risks of phenoxyacetic acid derivatives These herbicides may persist and can, for example, alter the quality of fodder. At high dosage, animals lose the ability to maintain their body temperature when moved into either a hot or cold environment, though the reason for this loss of homeostasis is not known.

In humans, only mild side effects have been claimed; a mild neural disorder, peripheral neuropathy has been reported.

The most important hazards of these compounds lie with their analogues containing three chlorine carbons. This is largely because of the highly toxic contaminating dioxin (Figure 9.11) which is formed as a side reaction in the chemical preparation of 2,4,5-trichlorophenol (2,4,5-T). There have been fears for individuals working in or near factories making such agrochemicals or their precursors. It has been suggested that soft tissue carcinomas are more common in workers who spray 2,4,5-T than in uncontaminated groups.

Auxin-like herbicides derived from benzoic acid
The most important of these are dicamba and 2,3,6-TBA (Figure 9.11). These are growth regulators and elicit growth responses in broad-leaved weeds. They are applied to the soil, and persist over long periods. Their main function is to add range and versatility to the herbicidal spectrum of phenoxyalkanoic compounds. The main metabolism of dicamba is hydroxylation at C-5 in the benzene ring and conjugation with glucose.

Bromoxynil and ioxynil These herbicides (Figure 9.11) attack the seedlings of several of the broad-leaved weeds that are not readily controlled by phenoxyacetic acid derivative herbicides. The herbicidal action is mediated by interference in a wide range of biochemical processes in plant organelles. They destabilize lipoprotein membranes, and thus affect mitochondrial electron transport and inhibit protein synthesis. These herbicidal nitriles affect the light reaction of photosynthesis, possibly through the electron transport system. They prevent the incorporation of carbon dioxide into acetyl-CoA and the formation of the malonyl-CoA for fatty acid biosynthesis. They also inhibit the reduction of $NADP^+$, thereby interfering with the photosynthetic reduction of carbon dioxide. Microorganisms present in some soils appear to be able to degrade bromoxymil to carbon dioxide.

Herbicides to control grassy weeds
There are two chlorinated aliphatic acids, e.g. dalapon and trichloroacetic acid (TCA). Dalapon is internally translocated after appliance to the foliage; TCA is usually applied before planting in the soil to control couch and other grasses. Little is known about the mode of action of dalapon but it may be by non-specific binding and precipitation of proteins and also by inhibiting RNA synthesis.

Some herbicides selectively kill grasses, after the emergence of the crop, e.g. phenoxybutyric acid and derivatives such as MCPB and 2,4-DB. These are of low toxicity until converted into the acetic acid member of the species by a specific β-oxidase. The degradation involves an acyl-CoA derivative found in plant and animals during fatty acid oxidation.

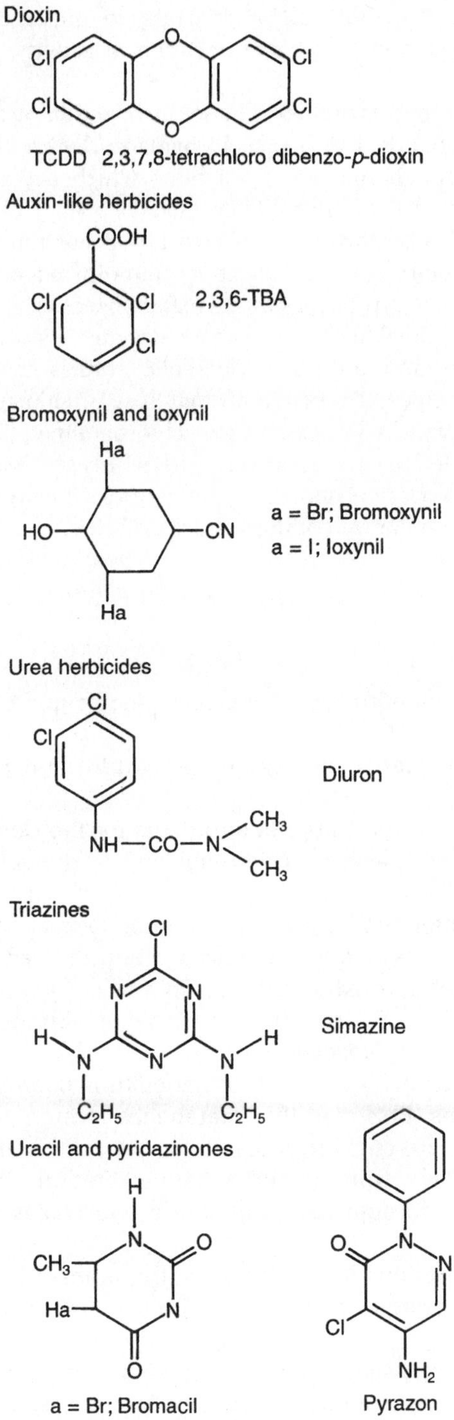

Dioxin

TCDD 2,3,7,8-tetrachloro dibenzo-*p*-dioxin

Auxin-like herbicides

COOH

2,3,6-TBA

Bromoxynil and ioxynil

a = Br; Bromoxynil
a = I; Ioxynil

Urea herbicides

Diuron

Triazines

Simazine

Uracil and pyridazinones

a = Br; Bromacil

Pyrazon

FIGURE 9.11 Herbicides. Dioxin; auxin-like herbicides; bromoxynil and ioxynil; urea herbicides; triazines; uracil and pyridazinones.

Dinitrophenol derivatives, e.g. Dinoseb, are extremely toxic and should not be used. This group uncouples oxidation from phosphorylation in the mitochondrial electron transport system. Grasses can flourish in the absence of competitors, therefore causing considerable crop losses. Post-emergence herbicides selectively control grass growth. Herbicides in these families have names that end in fop, e.g. fluazifop. A second family have names ending in -dim, of which sethoxydim is the best-known example. Diclofop is metabolized by microorganisms, and has a half-life of 10–30 days in a normal range of soil. Sethoxydim, in prairie soils has been shown to have a half-life similar to diclofop, but in air-dried soils 94% of an applied dose was recovered unchanged 28 days later. The primary mode of action seems to be on chloroplast fatty acid biosynthesis in susceptible plants, the membranes of which are strikingly different from those in plants and animals.

Soil-acting herbicides against seedlings
These herbicides are inhibitors of photosynthesis.

Urea herbicides This family of substituted ureas, e.g. diuron, fluometuron (Figure 9.11) is trisubstituted to form ureides. One of the amino groups carries either two methyl groups or one methyl group and one methoxy group. The other amino group is substituted with a benzene ring which may contain halogen atoms. The main action of ureides, triazines and ureciles is to disrupt the light reaction of photosynthesis. Ureides are adsorbed onto soil organic matter. Microbial action is largely limited to the proportion of the herbicide which is dissolved in soil water. Degradation rates are determined by the concentration of herbicide in water and therefore by adsorption on, or desorption from, soil particles. Two major routes of degradation occur: *N*-demethylation and ring hydroxylation. Differences in metabolism with or without differences in uptake and translocation by the plant are major factors contributing towards a selective response and sensitivity

303

between various crops and weeds for a particular ureide.

Triazines Triazine herbicides, e.g. atrazine, simazine and ametryne (Figure 9.11) persist in soils for months depending upon the soil type. Their mode of action resembles that of the ureides, in that they block photosynthetic processes. Metabolism of triazines is through one or more major routes. Chlorinated triedes undergo non-enzymatic but catalysed hydrolysis. Some plants contain one or more glutathione-S-transferases which allow glutathione to conjugate directly with herbicidal triazines, of the chlorotriazine family, to form inactive products. This enzyme system and reaction may be how the herbicidal triazine is innocuous to mammals. The secondary amine groups on C-4 or C-6 allow triazines to undergo N-dealkylation with some retention of activity.

Uracil pyridazinones The selective action of these compounds, e.g. bromacil, pyrazon (Figure 9.11) is through metabolic reactions which act at or before cell division. They include phenyl carbamate esters wherein a phenyl group has replaced an amino hydrogen group in aminoformic acid.

Uracil pyridazinones are metabolized by similar microsomal mono-oxygenase enzyme systems in plants and animals. Metabolic studies in rats show that 30% of the normal dose of chlorpropham was hydrolysed to give chloroaniline and its N-acetylated derivative. There is subsequent hydroxylation and then acetylation. Glucuronide and sulphation conjugates are excreted.

Summary

1. The use of agricultural chemicals has contributed to a plentiful and inexpensive food supply. The real cost may be environmental.
2. Large volumes of pesticide liquid are sprayed onto plants throughout the world, and have entered every food chain.
3. In the United Kingdom pesticide labels must conform to the Data Requirements for the Control of Pesticide Regulations which cover the classification, packaging and labelling of dangerous substances.
4. Agricultural chemicals are designed to act on the intended target and be innocuous to other creatures. This may change following metabolism or passage into the food chain. Side effects may be acute, cumulative and chronic. A toxic substance must accumulate in a living organism until a concentration is achieved which allows for toxicity. For example, the organochlorines are toxic, readily stored in fat and reduce the fertility of many species.
5. Metabolism has an important role in dictating the selectivity of action of agricultural chemicals and protects some species, e.g. animals and humans. Enzymes that metabolize foreign compounds catalyse reactions which: (i) alter molecular structure to a less toxic product; and (ii) increase the polarity and water-solubility to facilitate urinary and biliary excretion. Most enzymes responsible for the primary metabolism of foreign compounds are hydrolases and oxygenases.
6. Oxidation or hydrolysis are the primary metabolic reactions in pesticides, with the insertion or exposure of polar groups. Occasionally, products of primary actions undergo secondary changes which are species-dependent.
7. The degree of metabolism dictates the persistence of pesticides in soil, plants and animals. Resistance to pesticides results in the possession or development of metabolic processes that reduce biological activity or increase rates of urinary or biliary excretion.

Further reading

Anonymous (1982) Phenoxy herbicides, trichlorophe-nols and soft tissue sarcomas. *Lancet,* **319**, 1051–2.

Bingham, S.A. (1989) Agricultural chemicals in the food chain. *Journal of the Royal Society of Medicine,* **82**, 311–15.

Coggan, D. and Acheson, E.D. (1982) Do phenoxy herbicides cause cancer in man? *Lancet,* **319**, 1057–9.

Hassall, K.A. (1990) *Biochemistry and Uses of Pesticides,* 2nd edn, MacMillan Press, New York.

Lang, T. and Clutterbuck, C. (1991) *Pesticides,* Ebury Press, London.

9.3 Non-nutritional effects of plants

- Not all chemicals within a plant have nutritional value.
- Some plant chemicals have one function in plants, but quite different functions in animals and humans.
- There are various forms of nomenclature of herbs.
- Some plant chemicals are advantageous, as in herbal remedies, but others are toxic.
- Caffeine is commonly used, and though of no nutritional value, has a variety of physiological and psychological actions.

9.3.1 Introduction

Ethnobiology is the study of the traditional lore of plants and animals. This is the wisdom of the herbal healers, shamans and medicine men. Some 25 000 plants, 10% of all species, are used medicinally throughout the world. Plant chemicals have an effect in mammals and humans both to foster health and minimize disease, but the mechanisms are not fully understood. It is not known how a chemical which has one function in the plant is able to function in a quite different manner in animals and humans.

Herbs traditionally added to food may also have effects which are distinct from their flavour-enhancing properties. Many herbal substances have non-nutritional properties, physiological, pharmacological or even pathological.

9.3.2 Herb nomenclature

Herb nomenclature may be:

- the English common name
- a transliteration of the herb name
- the latinized pharmaceutical name
- the scientific name

For example, corresponding names for ginseng would be ginseng, ren-shen, radix ginseng and *Panax ginseng.* The naming is in fact imprecise, as the term ginseng is given to oriental American as well as Siberian ginseng.

9.3.3 Beneficial effects

There are many examples of traditional plant remedies which, following careful extraction and pharmacological studies, result in the identification of a drug which has a general and accepted use in medicine. The bark of the willow was chewed to relieve agues, fevers, pains and rheumatism. The active principle identified was salicylic acid (aspirin). The rosy periwinkle was originally thought to be useful for diabetes and eventually proved to be an effective antileukaemic agent.

Qinghaosu (*Artemisia annua* L.), also known as annual or sweet wormwood in the Western World, has a variety of medical uses including the treatment of fevers, haemorrhoids, and has antimalarial properties. The active principle is artemisinin, a sesquiterpine.

Herbal remedies contain active principles in varying amounts which are dependent upon how the plant is grown and stored and specific characteristics of that particular plant. Herbs must be precisely identified and given in the correct amount to minimize the risk of toxicity. This was apparent in the development of digitalis from the foxglove which, until proper biological assays and digoxin were developed, had very varying efficacy.

9.3.4 Adverse effects

The traditional herbal cures are not without their complications which may include hepatic and renal failure. Comfrey is a traditional cure for broken bones, bronchitis and ulcers, but contains chemicals which may be associated with liver cancer. The purple pennyroyal taken for indigestion, headaches and menstrual pain contains substances causing irritation of the skin and urinary tract and abortion. Many garlic lovers believe that garlic is protective against ischaemic heart disease but garlic may cause allergic skin reactions in some individuals. The onion has well-known effects on the tear glands, inducing crying in the cook. Table 9.2 details some recently appreciated hazards of traditional herbal remedies. (See also section 1.4.6, page 266)

The **betel-nut** or supari is widely used in India, chewed alone or as 'pan', which is a mixture of supari, lime and sometimes tobacco. There is a causal relationship between this habit and oral cancer.

Qat, commonly chewed in Somalia and other countries in the Horn of Africa, has addictive properties. The most potent form of qat, miraa, grows on small farms alongside coffee and tea plants. Reddish-green twigs are plucked from the trees daily and placed in bundles. The average user may chew bundles of 100 fresh twigs a day which must be used rapidly as they wither and lose potency.

9.3.5 Caffeine

Caffeine is currently the most used plant chemical which does not have a nutritional benefit. It has a variety of physiological and psychological actions which are said to be mediated by the blockade of adenosine receptors. The direct central nervous-stimulating effect of caffeine has not been established, yet such an action is widely believed to be the case. Increased alertness, improved mood and behaviour are well-recognized and valued responses to caffeine-containing drinks. The elimination of caffeine from the body is dependent upon acetylation, a process which is genetically determined to be either slow or fast. Slow acetylators will accumulate caffeine with the potential for intolerance, whereas fast acetylators may never achieve these concentrations. Vegetables in the Brassica (cabbage) family are known stimulators of the cytochrome P_{450} enzyme system in the liver.

TABLE 9.2 Traditional remedy and modern warnings

Plant	Traditional remedy	Modern warnings
Monkshood	Sedative	Poisoning
Camomile	Sedative	Skin rashes
Cuckoo pint	Aphrodisiac	Poisoning
Greater celandine	Liver complaints	Poisonous
Lily of the Valley	Heart complaints	Poisonous
Broom	Laxative	Poisonous
		Large doses can cause miscarriage
Larkspur	Skin parasites	Poisonous
Foxglove	Heart complaints	Poisonous
Alder, buckthorn	Constipation	Poisonous
Ivy	Hangovers	Poisonous
St John's wort	Sedative	Skin blisters
Juniper	Improved eyesight	Irritates the intestine
Colts foot	Respiratory problems	Linked to liver cancer
Mistletoe	Sedative	Gastroenteritis and/or liver damage

Several consecutive helpings of cabbage reduces the half-life of caffeine by approximately 20%, illustrating the interaction of dietary constituents and the metabolism of xenobiotic compounds.

Summary

1. Ethnobiology is the study of the traditional lore of plants and animals, the wisdom of the ancients. Some 10% of all plant species are used medicinally. Herbs, traditionally added to food, have effects which are distinct from their flavour-enhancing properties. Many herbal substances have non-nutritional properties, physiological, pharmacological and pathological.
2. Herb nomenclature includes the English common name, the transliteration of the herb name, the Latinized pharmaceutical name and the scientific name.
3. Traditional plant remedies include the bark of the willow which was chewed and whose active principles include salicylic acid (aspirin). Herbal remedies contain active principles in varying amounts which are dependent upon specific characteristics of the plant and how it is grown and stored.
4. Plants may have adverse effects on hepatic and renal function and may be carcinogenic.
5. Caffeine is the most ubiquitous plant chemical which has no nutritional benefit but a variety of physiological and psychological effects.

Further reading

D'Mello, J.P.F., Duffus, C.M. and Duffus, J.H. (eds) (1991) *Toxic Substances in Crop Plants*, Royal Society of Chemistry, Cambridge.

James, J.E. (1991) *Caffeine and Health*, Academic Press, London.

Pui-Hay But, P. (1993) Need for correct identification of herbs in herbal poisoning. *Lancet*, 341, 637.

Thurnham, D. (1996) Physiologically active substances in plant food. *Proceedings of the Nutrition Society*, 55, 371–446.

Chapter 10

Eating and digestion

- Body weight and composition is a balance between nutrient intake and energy utilization.
- Apoptosis is the removal of old, damaged or abnormal cells.
- Energy balance is between 'energy consumed' and 'energy expended + change in body energy stores'.
- Important energy stores are glycogen (short-term) and lipids (long-term).
- Lipid stores include triglycerides whose metabolism is controlled by hormones.
- Brown adipose fat is involved in thermogenesis and energy regulation, particularly in infants.

10.1 Regulation of nutrient balance

10.1.1 Introduction

The regulation of body weight is dependent upon a balance between nutrient intake and utilization, though there are other important factors (Figure 10.1).

Energy balance

Energy consumed = energy expended + change in body stores

Amino acid oxidation adjusts to amino acid intake

Carbohydrate oxidation adjusts to intake

Fat balance is not regulated and nutrient excess goes to fat stores

FIGURE 10.1 Factors involved in the regulation of body weight.

10.1.2 Apoptosis

Apoptosis is a coordinated cellular response to noxious stimuli that are not immediately lethal. A multicellular organism must sustain equal rates of cell generation and cell death to maintain a constant size. Senescent, damaged or abnormal cells that could interfere with organ function must be removed. Physiological cell death is not a random process. Apoptosis is a rapid process and is directed towards scattered individual cells rather than all the cells in a particular area. It occurs during embryogenesis in the process of tissue turnover and after withdrawal of a trophic hormone from its target tissue. An example of apoptosis in the adult is a shedding of intestinal cells.

During apoptosis there is no leakage of cellular contents or inflammatory response. Early in apoptosis there is an increase in cytosolic ionized calcium which activates latent enzymes important in cleaving DNA and cross-linking cytosolic proteins. The distinct morphological features include compaction of chromatin against the nuclear membrane, cell shrinkage and preservation of organelles, detachment from surrounding cells and nuclear and cytoplasmic budding to form membrane-bound fragments (apoptotic bodies). These are rapidly removed by adjacent parenchymal cells or macrophages. Apoptosis may be important in autoimmune disease, degenerative diseases and ageing. The decline in body cell mass with age is probably controlled genetically, as it is related to life span.

10.1.3 Energy balance

The general principle of energy balance is:

energy consumed = energy expended + change in body energy stores.

Amino acid oxidation adjusts to dietary protein intake. Glycogen reserves are of the order of an average day's carbohydrate intake. The reserves are maintained at a balanced amount which prevents hypoglycaemia but excess carbohydrates are channelled into lipid stores (Figure 10.2). Fat balance does not appear to be regulated in the same manner. Fat oxidation and metabolism is not dependent on the fat content of the meal, fatty meals lead to fat accumulation and obesity.

10.1.4 Fat tissue

The function of fat tissue is to receive, store and release lipids. Adipose tissue is the tissue both quantitatively and qualitatively most directly affected by diet. Storage of triacylglycerol-fatty acids reflects dietary intake. Fat-soluble xenobiotics accumulate in fatty tissue.

The accumulation of fat in white adipose tissue results from two processes:

1. The uptake of circulating triacylglycerol under the control of lipoprotein lipase.
2. *De novo* synthesis of fatty acids from glucose (lipogenesis).

There are two routes whereby fat is deposited in fat tissue:

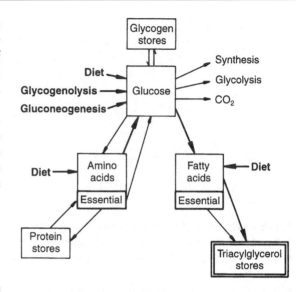

FIGURE 10.2 Macronutrient interactions between glucose, amino acids and fatty acids.

- uptake of pre-formed fatty acids
- uptake of other substrates which can be converted into triacylglycerol

Lipoprotein lipase

Fatty acids are delivered to fat tissue as chylomicron–triacylglycerol (TAG) or very low density lipoprotein (VLDL)-TAG. Circulating triacylglycerol is hydrolysed by lipoprotein lipase in the capillary lumen. The enzyme is synthesized within adipocytes, modified intracellularly and transferred to the capillary endothelium (Figure 10.3).

Chylomicrons are a preferred substrate to VLDL for lipase. The whole of this process is initiated by insulin over a period of hours. Some of the released fatty acids pass to the fatty tissue, are esterified and stored. Some of the fatty acids spill over postprandially into the plasma.

Lipoprotein lipase activity is low at birth and increases during the first 10 days of life. This is during the active phase of adipocyte proliferation and decreases to very low levels before further increasing after weaning. In humans there is little evidence for *de novo* lipogenesis during normal life as a mode of fat deposition.

Lipoprotein lipase

Lipoprotein lipase is released from capillary walls and released into the blood stream in the presence of heparin and sulphonated dextran and other highly charged compounds. Lipoprotein lipase is specific for 1- and 3- linkages in triacylglycerols. The fatty acid at 2-monoacylglycerol position isomerizes to 1(3)-monoacylglycerol before hydrolysis. Adipocytes contain a very active monoacylglycerol lipase. Removal of the first fatty acid of the triacylglycerol is a rate-limiting step. Lipase also removes the

second fatty acid, monoacylglycerol lipase removes the third. The result is non-esterified fatty acids and glycerol. The glycerol appears to be transported elsewhere, whereas the fatty acids are available for re-esterification.

The amount of fat in the fat cell is an equilibrium between the rate of lipolysis and the rate of lipid synthesis. Lipolysis is catalysed by lipase which is activated by phosphorylation by cyclic-

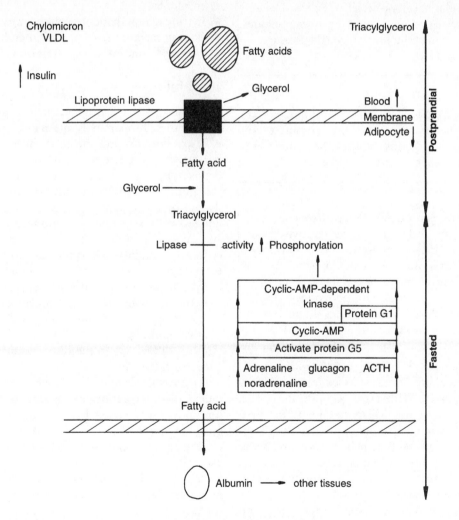

FIGURE 10.3 Fat deposition and release postprandially and during fasting.

AMP-dependent kinase under the control of hormones.

> Adrenaline, glucagon, ACTH and also nor-adrenaline stimulate adenylate cyclase-A-kinase signal transduction. Each of the peptide hormones has a receptor on the plasma membrane and noradrenaline reacts with the β-adrenergic receptor. All of which activates GTP-binding protein, G_S. This initiates an enzyme cascade whereby adenylate cyclase is stimulated to produce cAMP which itself activates A-kinase which phosphorylates and activates the hormone-sensitive lipase. The enzyme cascade is modulated by noradrenaline and adrenaline which act through prostaglandins (E_1 and E_2)

These are linked to a second GTP-binding protein, G_1. Catecholamines can both stimulate and inhibit lipolysis, dependent upon the number of β- and α_2-adrenergic receptors of the fat cell.

> Insulin influences lipolysis, for example controlling the phosphatase which dephosphorylates and inactivates hormone-sensitive lipase. A second kinase phosphorylates hormone-sensitive lipase on a serine next but one to that phosphorylated by A-kinase. Phosphorylation on one serine prevents phosphorylation on the other. Increased concentrations of palmitoyl-CoA activate the AMP-stimulated kinase.

An accumulation of unesterified fatty acids may inhibit lipolysis. Within the fat cell there is a filamentous structure which surrounds the lipid droplet. Access of the lipase to the droplet is controlled by proteins which in turn are under control of A-kinase. Hormone-sensitive lipase is a hydrophobic protein, difficult to solubilize, whose gene has been identified.

There is metabolic exchange across the membrane of the adipocyte: triacylglycerol, glucose, oxygen, acetoacetate and 3-hydroxybutyrate are extracted from plasma and non-esterified fatty acids, glycerol, lactate and carbon dioxide are released. The tissue also takes up glutamate and releases alanine and glutamine.

Adipose tissue is an important source of energy in the post-absorptive state. Net nutrient uptake into adipose tissue after a carbohydrate load requires suppression of fat release. Ethanol appears to act on adipose tissue metabolism by unknown but distinct mechanisms.

Diet can influence lipolysis by:

- altering the serum concentration of acute acting hormones
- altering sympathetic nervous activity
- altering membrane composition and fluidity
- increased concentrations of hormones

10.1.5 Fat stores

Fat is stored inside the abdomen in amphibians and reptiles. In mammals and many birds the fat is distributed around the body in discrete compartments in close contact with other tissues. The deposition, size and site of individual fat deposits is a continued topic for conjecture. Intermuscular depots of fat are the most metabolically active. Fat cell volume may be important in a particular site. Activity is peculiar to any one site and inferences from studies on that site may not necessarily be extrapolated to the entire fat component of the body. Fat distribution differs from species to species and from individual to individual. Appetite and exercise may be important in dictating the relative lipid masses. In humans there are important sex differences in the distribution and abundance of fat tissue.

It is reasonable to define obesity as a body mass index (BMI; weight/height^2) greater than 30. It is not known why people become fat. An important question is whether or not there is a inherited propensity to obesity. Biological inheritance accounts for only 5% of the chances of having a particular BMI index and amount of subcutaneous fat, but for 20% of the propensity for the manner in which the fat is distributed. Excessive fat is stored in the white adipose tissue, when energy intake persistently exceeds energy output. There is very little evidence to suggest a genetic basis for human obesity. The condition has a strong cultural and social background.

An inverse relationship between the total number of fat cells in the body and the age of onset of obesity was demonstrated in the early 1970s. The result was a belief that there is an inevitable progression to an adult body habitus. 'Cells for storing fat in the body develop primarily in the first year of life. They persist throughout life, so a fat baby becomes a fat adult who is unable to control weight.' Later work showed the need for caution in the interpretation of fat cell counts. It is possible that the number of such fat cells reflects the degree of obesity rather than the age of onset of obesity. The fat cell pool contains mature fat cells, mature fat cells which are reverting to precursor cells, fat cell precursors and replicating fat cells, an actively changing system. The cellular development that accompanies fat tissue growth involves both cellular hyperplasia and hypertrophy at all stages of life.

Brown adipose tissue

Brown adipose tissue (BAT) may be regarded as an arrested embryonic form of white adipose tissue. White adipose tissue is a storage organ and brown adipose fat a thermogenic organ. There is probably a continuum between the two types of cell. BAT is regarded as a factor in the development of obesity, and its defective functioning may lead to reduced thermoregulatory energy requirements. BAT thermogenesis is conducted through a specific membrane protein (molecular weight 32 000 Da), the uncoupling protein. The amount of this enzyme in the BAT is related to the thermogenic capacity of the tissue.

In animals born to a cold environment and who have no protective coat, an ability to alter metabolic rate in response to changes in ambient temperature is important. Brown fat is an important tissue in promoting such a protective flexibility in metabolism, in part a result of increased conversion of thyroxine to triiodothyronine. Increased ATP activity is the result of the activity of a

mitochondrial protein, uncoupling protein or thermogenin.

In the developing foetus the thermogenic capacity of BAT remains at a low level of activity and increases before delivery. Growth and development of BAT in the foetus has three phases:

- tissue growth as a result of hyperplasia and hypertrophy
- differentiation and ability to generate triiodothyronine
- expression of thermogenin during late pregnancy and birth

BAT multilocular cells appear at mid-gestation. During the last month *in utero* the BAT becomes innervated with sympathetic nerves, there is an increase in 5'-monodeiodinase activity which converts thyroxine to triiodothyronine and the expression of genes for thermogenin and lipoprotein lipase.

There appears to be a slow increase in BAT activity over the first few days of neonatal life. The diet and nourishment of the infant thereafter may determine the rate of loss of BAT lipid.

At birth, BAT provides 1–2% of birth weight and is to be found in the axillary and perirenal regions. Over subsequent months, it is replaced by white adipose tissue. Following this, shivering – rather than non-shivering – thermogenesis is the mechanism of body temperature control.

10.1.6 Weight, morbidity and life expectation

Little is known about the effect of weight change on longevity. There is a gradual increase in risk of death with malnutrition, particularly in children. However, it would appear that there is no increased death rate in the mildly or moderately malnourished child either in the short or long term. Anthropometric measures of nutritional status do not distinguish between children who will die or survive. Nutritional status is related to deaths from diarrhoea and anaemia but not to deaths from acute malaria or acute respiratory infection. A useful policy therefore would be to target nutritional intervention, for example to provide appropriate and extra food in order to protect children with mild muscle wasting or growth stunting from progressing to marasmus or kwashiorkor.

In the adult who changes weight there are often concomitant changes in smoking habits and exercise levels. In a long study of weight changes and longevity in Harvard graduates, the lowest all-cause mortality was in individuals who maintained a stable body weight between 1962–1977. If there was an increase or decrease in weight of over 5 kg during that period then there was an increase in mortality. This was primarily due to an increase in coronary heart disease rather than to deaths from cancer. These changes in mortality were not influenced by smoking habits or exercise levels.

A long-term study between 1922 and 1935 followed up adolescents aged 13–18 who were overweight (BMI greater than 75th percentile). Mortality rates were compared with lean contemporary subjects with a BMI between the 25th to 75th percentile. In middle age (analysed in 1968) the only health risk from adolescent obesity was diabetes mellitus. When the group became elderly, then there was an increased mortality among those males who had been fat in adolescence. There was an increase in death from all causes, including coronary heart disease, stroke and colorectal cancer. The morbidity associated with early obesity in women included an increased incidence of coronary heart disease, atherosclerosis and arthritis.

Summary

1. The regulation of body weight is in part dependent upon a balance between nutrient intake and utilization

2. A multicellular organism must balance cell generation and cell death to maintain a constant size. Senescent, damaged or abnormal cells are removed. Physiological cell death, apoptosis, is directed towards scattered individual cells rather than all the cells in a particular area.

3. The general principle of energy balance is that 'energy consumed' = 'energy expended + change in body energy stores'. Amino acid oxidation adjusts to dietary protein intake. Glycogen reserves are maintained at a balanced amount which prevents hypoglycaemia, excess carbohydrates are channelled into lipid stores. Fat balance, oxidation and metabolism are not dependent on the fat content of the meals; fatty meals lead to fat accumulation and obesity.

4. The accumulation of fat in white adipose tissue results from: (i) the uptake of circulating triacylglycerol under the control of lipoprotein lipase; and (ii) *de novo* synthesis of fatty acids from glucose (lipogenesis). The two routes whereby fat is deposited in fat tissue are: (i) uptake of preformed fatty acids; and (ii) uptake of other substrates which can be converted into triacylglycerol.

5. The amount of fat in the fat cell is an equilibrium between the rate of lipolysis and the rate of lipid synthesis.

6. Diet can influence lipolysis by altering the serum concentration of hormones, altering sympathetic nervous activity, and altering membrane composition and fluidity.

7. Obesity is a body mass index (weight/height^2) greater than 30. Appetite and exercise may be important in dictating the relative lipid masses. In humans there are important sex differences in the distribution and abundance of fat tissue.

8. Brown adipose tissue thermogenesis is conducted through a specific membrane protein, the uncoupling protein. The amount of this enzyme in the brown adipose tissue is related to the tissue's thermogenic capacity.

9. There is a gradual increase in risk of death in prolonged severe malnutrition, particularly in children. There is no increased death rate in the moderately or mildly malnourished child either in the short or long term. A sustained weight carries a better prognosis than a fluctuating weight.

10. Young fat males have an increased incidence of vascular disease in old age compared with those who were not obese in youth.

Further reading

Ashwell, M. (1992) Why do people get fat: is adipose tissue guilty? *Proceedings of the Nutrition Society,* **51**, 353–65.

Carson, D.A. and Ribeiro, J.N. (1993) Apoptosis and disease. *Lancet, 341,* 1251–4.

Cryer, A., Williams, S.E. and Cryer, J. (1992) Dietary and other factors involved in the proliferation, determination and differentiation of adipocyte precursor cells. *Proceedings of the Nutrition Society,* **51,** 379–85.

Flatt, J.P. (1992) Body weight, fat storage and alcohol metabolism. *Nutrition Reviews, 50,* 267–70.

Frayn, K.N., Coppack, S.W. and Potts, J. (1992) Effect of diet on human adipose tissue metabolism. *Proceedings of the Nutrition Society, 57,* 409–18.

Lee I-Min and Paffenbarger, R.S. (1992) Change in body weight and longevity. *Journal of the American Medical Association, 268,* 2045–9.

Must, A. *et al.* (1992) Long-term morbidity and mortality of overweight adolescents: a follow up of the Harvard Growth study of 1922 to 1935. *New England Journal of Medicine, 327,* 1350–5.

Pond, C.M. (1992) An evolutionary and functional view of mammalian adipose tissue. *Proceedings of the Nutrition Society, 51,* 367–77.

Symonds, M.E. and Lomax, M.A. (1992) Maternal and environmental influences on thermoregulation in the neonate. *Proceedings of the Nutrition Society, 51,* 165–72.

van den Broeck, J., Eeckels, R. and Vuylsteke, J. (1993) Influence of nutritional status on child mortality in rural Zaire. *Lancet, 341,* 1491–5.

Vernon, R.G. (1992) Effects of diet on lipolysis and its regulation. *Proceedings of the Nutrition Society, 57,* 397–408.

10.2 Smell and taste

- The sensations of taste and smell are age-dependent.
- Olfaction (smell) is experienced in the mouth, tongue and nose.
- Olfaction or the perception of smells is a very complex sensation which has not been totally defined chemically.
- Gustation (taste) distinguishes salt, sour, sweet, bitter and umami.
- Taste receptors recognize a wide range of chemical substances as having sweet and bitter properties.

10.2.1 Introduction

Gustation is the term for the concept of taste, olfaction for smell; both are subunits of taste. There are a large number of tastes and smells which add to the joy of eating and life in general. Good nutrition must include an enjoyment of food which is enhanced by good company and food of an appetising appearance and taste.

Taste qualities are divided into four or five sensations; salt, sour, sweet, bitter and umami. The latter is described as an amino acid type of taste, e.g. sodium glutamate. Each of these tastes is immediately recognized. Intensity only becomes relevant when a taste sensation is too strong or too weak.

Each of these tastes involves a single transducive nerve sequence and is associated with recognition of specific chemical structures. The appreciation of the flavour of food involves several sensory systems, mechanoreceptive, thermoreceptive and chemoreceptive. Many of the non-taste and non-olfactory sensations are carried by the trigeminal nerve (V). Taste is largely dependent upon multiple sensations, predominantly taste and olfaction (smell). Taste is appreciated by specialized receptor cells in the mouth and palate which recognize chemical stimuli. The recognition of a chemical involves altering the firing rate of the sensory nerve.

10.2.2 Olfaction

The olfactory system is capable of recognizing thousands of smells. The system is therefore very complex and it is not clear whether recognition takes place peripherally or centrally in the brain.

The olfactory system is a three-compartment sensory system specialized for the detection and processing of molecules called odorants. There are also three anatomical areas, the olfactory epithelium, the olfactory bulb and the periform epithelium. The olfactory epithelium in humans is found predominantly on the dorsal aspect of the nasal cavity, the septum and part of the superior turbinates. There are approximately 6 million ciliated olfactory receptor neurons in the 2 cm^2 area of the human olfactory tissue.

The are three layers of human olfactory mucosa:

1. The superficial acellular layer composed of the mucociliary.
2. The olfactory epithelium with three morphologically defined cell types, olfactory receptor neurons, sustentacular cells and basal cells.
3. The lamina propria consisting of olfactory and trigeminal nerves.

Smell is perceived by bipolar neurons of the olfactory cranial (I) nerve (Figure 10.4). The receptors are located on the ciliary processes that arise from the olfactory receptor neurons. The life span of these receptors is 30 days.

The detection of odours depends on solubility in the mucus. The axons of the receptor cells pass through the cribriform plate to innervate the secondary projection neurons found in the olfactory bulb. Information is carried to higher cortical regions including the periform cortex. Odour information is subsequently distributed to both cortical and subcortical structures throughout the nervous system.

Smell in prenatal infants

The ability to detect odour may develop prenatally. The olfactory bulb and receptors have an

adult pattern by the middle of the 11th week of gestation. Around 28 weeks the olfactory system is capable of detecting chemical stimuli. It would appear that infants can detect and discriminate among a variety of quite distinct odorants. The breast fed baby is able to distinguish the mother's odour from that of another lactating female. Bottle-fed infants do not make similar discriminations. Breast-feeding infants do not discriminate the axillary odours of their own father from another father. At 7 weeks, breast-fed infants will turn towards a perfume which has been worn on the mother's breast during feeding in preference to perfume worn by other mothers.

Odour classification

Modern odour classifications define the inherent characteristics of a stimulus or relationships among a set of qualitatively distinct stimuli rather than defining the sensory capability of the subject or patient. Attempts to classify odours into discrete qualitative characters have tantalized scien-

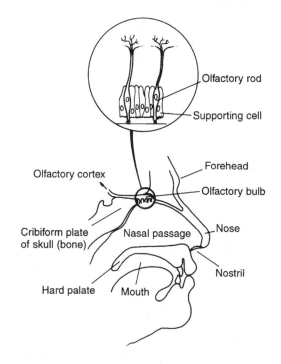

FIGURE 10.4 Olfaction. Anatomy of cranial nerve I and olfaction.

tists from the ancient Greeks to the modern olfactory chemists. Aristotle divided flavours and smells into pungent, succulent, acid and astringent. Amor classified – by physicochemical parameters – odour qualities on the basis of a set of primary odours, ethereal, camphoraceus, muscae, floral, antipungent and putrid.

Many factors influence the ability to smell, including prior exposure to the test stimulus or related compound, or the subject's age, sex, smoking habits, exposure to environmental toxins, drug usage and state of health. Exposure to an odorant, if recent and relatively continuous, can result in a temporary decrease in its perceived intensity. This is called **olfactory fatigue**. Little is known about the degree to which the ability to smell can be influenced by training. Repeated testing within a threshold odorant concentration results in decreased thresholds or enhancement of signal detection sensitivity. Repeated exposure to odorant makes unpleasant odours less unpleasant and pleasant odours less pleasant. Practising with feedback increases the ability to name odours. This has been used to a high degree in training food and beverage tasters and also in testers in the perfume industry.

The transport of volatile smells (odorants) to the chemoreceptive membrane of the mouth and palate occurs in two different physical phases. Initially, the air-borne odorants pass in the air stream to the nasal chambers and the olfactory mucosa. In the second phase, fluid-borne odorants are transported through the olfactory mucus to the receptor membranes. The response is dependent upon a partitioning between the air and mucus which is dependent on water solubility. The olfactory mucus can concentrate relatively water-insoluble odorants by a factor of 10-fold.

The physicochemical properties of olfactory mucus regulate the access of odorants to, and their clearance from, olfactory cells where the stimulus occurs. The viscosity of olfactory mucus varies enormously, being related primarily to the glycoprotein and water content. The olfactory mucus consists of a shallow 5 μm superficial watery layer and a deeper 30 μm more viscous mucoid layer. Viscosity microstructure and depth of olfactory mucus determines the diffusion times for odorants and therefore regulates the speed with which stimulants reach the olfactory cilia. The physical properties of mucus and its chemical composition of electrolytes, glycoconjugates, proteins and enzymes are regulated by a complex interplay of odorants, cellular and neural factors. Odorants, in general are organic molecules which have a molecular weight of less than 350 Da. They range in water solubility from infinitely soluble to poorly soluble. 'Musks', which are of higher molecular weight are nearly always water-insoluble, with a low partition coefficient and are classified as hydrophobic. Most organic odorants have molecular weights of less than 150 Da and are moderately water-soluble. There are two complementary hypotheses which describe how odorants cross the mucus layer. The first suggests a free diffusion and the other a facilitated diffusion of odorants by transport proteins.

Olfactory receptor neurons have four primary functions to:

- detect odorants as a chemical stimulus
- transmit the information about the molecular identity of the odorant as well as its concentration and duration of stimulation
- couple this information to the electrical properties of the neuron
- transmit this information to the brain

Odorant receptor molecules and odorant-regulated channels are located in the distal ciliary region where sensory transduction occurs. Three categories of membrane mechanisms have been proposed for olfactory transduction. The first includes the odorant receptor-specific stimulation of the adenylate cyclase cAMP second messenger and the phosphoinositide-derived messenger systems that activates ion-gated channels. It is also feasible that G proteins and the cAMP second messenger system are involved in human odour perception.

Nerve fibres sensitive to noxious chemical stimuli on exposed or semi-exposed mucosal membranes respond to pungent spices, lacrimators and compounds; the result is sneezing and skin irritation.

Another classification of chemical stimuli is based on a proposed mechanism of interaction with the nerve endings:

- those which have an affinity for reaction with –SH groups on proteins
- molecules such as sulphadioxide which do not react with –SH groups and may cleave S–S bonds in proteins.
- sensory irritants belonging to neither, e.g. ethanol

The olfactory receptor nerve fibres for pungent chemicals serve as affectors, relaying chemically induced stimuli to the trigeminal nuclei and to the brainstem. This sensory input triggers several sensory-mediated reflexes and peripheral action reflexes to protect the animal from further exposure to the irritant and eliminate the irritating stimulus from the nose, eye or mouth, e.g. sneezing or withdrawing or spitting out the chemical.

Pungent-tasting vegetables contain compounds, e.g. isothiocyanate in horseradish and 1-propanyl-sulphenic acid in onion and garlic. Spices such as mustard, cloves, chilli pepper, black pepper and ginger contain active irritant principles, allyl and 3-butenyl isothiocyanate, eugenol, capsaicin, piperine and 6-gingerol and 6-shogaol respectively. Piperine and 6-gingerol and 6-shogaol include an aromatic ring and an alkyl side chain with a carbonal function. Changes in the alkyl side group or in the amide function near the polar aromatic end abolish or reduce the pungent taste. Side chains of 9 carbon atoms' length for capsaicin and piperine and 10 carbon atoms for gingerol result in the most potent irritants.

Olfactory system measurements

The measurement of olfactory sensitivity causes problems because there is no simple, available physical measurement analogous to that of colour or sound pitch. Mixtures of odorants do not give predictable psychological and physiological effects in the same way as mixtures of light and sounds. The problem is not made easier when compounds with different chemical structure can smell similarly to humans. Repeated exposure to

smells can reduce the response sensitivity. Valid measures of olfactory sensitivity require reliable and reproducible procedures for presenting stimuli to the subjects. Such methods include the draw tube olfactometer of Zwardemaker, glass sniff bottles, plastic squeeze bottles, air dilution olfactometers, glass rods, wooden sticks or strips of blotting paper dipped in odorants. The results obtained include threshold measurements, the method of constant stimuli, methods of determining the limits of detection and a staircase or up/down method. A variety of techniques have been developed that use threshold stimuli to assess olfactory function. These can be divided into three general classes:

1. Those that require subjects to judge the relative amount of one or more attributes with a set of stimuli such as intensity or pleasantness (supra-threshold attributes scaling).
2. Those in which different thresholds are determined, i.e. the minimum increase in concentration required to make concentration A perceptibly more intense than concentration B (difference threshold measurement).
3. Those that specifically test the ability of subjects to detect, recognize and identify and clarify, or remember after supra-threshold stimuli.

10.2.3 Gustation

The chemoreceptor cells which experience taste are arranged into buds in the oral cavity. Most taste buds are found on the upper surface of the tongue, though there are some in the soft palate, larynx, pharynx and epiglottis. Taste buds are found in the lingual epithelium associated with connective tissue called papillae.

Fungiform papillae are found on the anterior two-thirds of the tongue and are concentrated near the anterior tip. Each fungiform papilla may contain up to 36 taste buds. A typical taste bud contains 50–150 individual cells. Taste cells are innervated by sensory nerve fibres from the facial glossopharyngal and vagal cranial nerves (Figure 10.5). More than half of the fungiform papillae contain no taste buds. Circumvallate papillae (large mushroom-shaped structures) are found on the posterior tongue and contain more taste buds than individual fungiform papillae. Gustatory stimulation of a single fungiform papilla is sufficient to identify correctly any of the four basic tastes, though this accuracy varies with the function and number of intact taste buds in that papilla. Taste sensitivity varies according to both regional taste bud density and the number of taste buds per papilla. There is enormous variation in the taste bud density between individuals. The anterior tongue is covered not only by the fungiform and folate papillae but also by many coarse conical filiform papillae which may respond to mechanical stimuli.

Taste receptors are innervated by the VII, IX and X cranial nerves. The anterior two-thirds of the tongue is innervated by the VIIth cranial nerve, the posterior third including vallate and foliate papillae by the IXth, and the soft palate, glottis and epiglottis by the Xth nerve (Figure 10.5). Taste cells are neuroepithelial in type and have a life-span of 9 days. They appear able to respond to more than one taste quality. Gustatory information is carried by the VII, IX and X cranial nerves, which project to the haustral half of the medulla. Axons from the gustatory region of the medulla enter the central tegmental tract and finish in the ventrobasal thalamus. There is a close linkage between the gustatory, somatosensory and trigeminal system. There is probably no single taste area in the cortex.

Development of taste

Taste cells are developed in the human foetus at 7–8 weeks of gestation and morphologically mature cells are to be found at 14 weeks. Foetal taste receptors may be stimulated by chemicals present in the amniotic fluids. The foetus begins to swallow from 12 weeks of gestation and may swallow 200–700 ml/day by delivery. Amniotic fluid contains glucose, fructose, lactic acid, pyruvic acid, citric acid, fatty acids, phospholipids, creatinine, uric acid, amino acids, polypeptides, proteins and salts. The chemical composition of the amniotic fluid varies over the course of gesta-

tion and may be abruptly altered by foetal urination. It is possible that foetuses show a preference for sweet tastes and reject bitter tastes. Swallowing amniotic fluid appears to be necessary for the proper development of the gut.

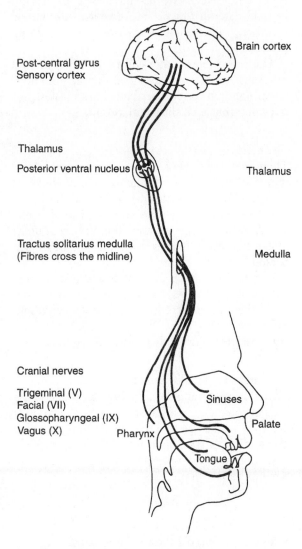

FIGURE 10.5 The sensory fibres from the taste buds transmit sensation to the trigeminal nerve (V), facial (VII), glossopharyngeal (IX) and vagus (X) cranial nerves. These innervate the tongue, palate and pharynx, pass to the tractus solitarius in the medulla, cross the midline to the thalamus in the posterior ventral nucleus and then to the post-central gyrus in the sensory cortex.

Newborn infants

Facial expression suggestive of contentment and liking or discomfort and rejection is used to evaluate the response of newborn babies to taste stimulation. Other response measurements which have been used to measure neonatal taste perception include lateral tongue movements, autonomic reactivity, differential ingestion and sucking patterns.

The newly born baby enjoys sweet, sour and bitter tastes and can discriminate sweet from non-sweet solutions and sour from bitter tastes, but does not appear to differentiate salt.

Older infants and young children

There have been few studies in infants aged 1–24 months. This is probably due to the unwillingness of such children to cooperate or to indulge scientists by eating strange and unwanted food items. Children of this age are unwilling to accept unfamiliar bottles or food items from an unfamiliar person.

Familiarity with specific foods and taste in specific contexts has been shown to play an important role in the preferences of pre-school children. Young children tend to show preferences for more concentrated sugar and salt than do adults.

Ageing and the taste system

There is a wide range of taste bud densities on the human tongue (varying from 3 to 514 taste buds per cm^2) which has no relationship to age or sex. The number of taste buds in the gustatory papillae does not decrease with age, and taste response remains robust in old age.

Saliva

Saliva is produced by three large paired salivary glands, the parotid, submaxillary and sublingual glands. Mechanical stimulation during chewing of both food and inert substances promotes a flow of saliva, with about 500–750 ml being produced each day; most is swallowed and absorbed from the gut.

Saliva performs a variety of functions:

- It is essential for normal taste function; it is difficult to taste food with a dry mouth. Saliva acts as a solvent for chemical stimuli in food and carries these stimuli to the taste receptors. At rest, gustatory receptors are covered with a layer of fluid that extends into the taste pores and bathes the receptors of the microvilli. Because gustatory stimulation alters salivary flow and composition, the fluid environment may alter during the activation of sensations of food.
- It has a role in the taste process; drugs injected intravenously can be tasted as they circulate as they are secreted into saliva onto the tongue. Such drugs include the intensely bitter sodium dehydrocholate and saccharine. It is not possible to taste drugs which are not secreted.
- It has a cleansing and antimicrobial action and also a buffering action which protects the teeth. Salivary glands secrete a large number of physiologically active substances, growth factor, vasoactive peptides and regulatory peptides.

The composition of saliva

Saliva is a dilute aqueous solution containing inorganic and organic constituents (Table 10.1). Its composition varies with the type of glands, species, time of day and degree and type of stimulation. Salivary flow during sleep is almost nil. With an increase in flow rate the composition of saliva changes and the concentrations of most salivary components change. Sodium concentrations increase with increasing flow rate while potassium decreases. Protein secretion by salivary glands is preceded by an initial uptake of amino acids and peptide synthesis in the gland followed by exocytosis.

High flow rates of saliva result from parasympathetic stimulation, and increased amylase secretion results from sympathetic nerve activation. The parasympathetic secretomotor neurons controlling the salivary glands arise from the salivatory nucleus of the vagus nerve. The sympathetic nerve supply to the salivary glands originates from the superior cervical ganglion.

There is an increase in salivary secretion even before, and certainly after, food is placed in the

TABLE 10.1 Composition of saliva

Electrolytes	Resting (mEq/l)	Stimulated (mEq/l)
Sodium	2.7	63
Potassium	46	18
Chloride	31	36
HCO_3^-	0.6	30
Magnesium	4.5*	0.4*
Calcium	42*	38*

Organic components
Proteins of acinar cell origin
Amylase, lipase, mucus, glycoproteins, proline-rich glycoproteins, basic glycoproteins, acidic glycoproteins, peroxidase

Proteins of non-acinar cell origin
Lysozyme, secreted IgA growth factors, regulatory peptides

*mg/l.

mouth. Reflex secretion results from stimulation of the oral mechanoreceptors, especially periodontal ligament mechanoreceptors as well as taste buds. Not all taste stimuli by food are equally effective in promoting salivary flow. Citric acid produces a copious flow of saliva. Ingestion of sucrose is followed by significant production of salivary amylase whereas salt stimulus produces secretion of saliva with a much higher protein content.

Taste papillae contents may control access and removal of stimuli and hence perception of taste. During feeding and drinking, muscles controlling the jaws, tongue and face move food and fluid around the mouth and expose the solubilized tastes to the whole population of taste receptors. This is particularly important for taste receptors situated in the clefts of the circumvallate and foliate papillae. Muscle movements determine the rate and direction of delivery of stimuli to the receptors. Since saliva contains various ions which are themselves gustatory stimuli, taste receptors are continuously stimulated by salivary components. For example, to detect dietary salt the concentration must exceed the concentration in saliva. Detection thresholds are reduced when the mouth is rinsed with distilled water and increases with higher concentrations of sodium chloride.

Of the many organic constituents of saliva,

proline-rich proteins increase the ability to taste bitter compounds, quinine, raffinose, and cyclohexamide.

Most tastes increase salivary flow in a concentration-dependent manner. Salivary proteins serve as carrier proteins for trophic substances. Zinc has long been recognized as having an important role in taste perception as well as taste bud maintenance.

Taste partitioning

Most tastes are water-soluble and non-volatile, weak acid (sour), salts (salty and bitter), sugars (sweet), amino acids (sweet, bitter and umami) and protein (sweet and bitter). These dissolve during chewing and pass through saliva and the mucus layer covering the taste pore to reach the microvilli processes of taste receptor cells. Some tastes are lipophilic molecules, such as alkaloids, caffeine and quinine. Most hydrophobic molecules are extremely bitter-tasting compounds and their taste detection thresholds tend to be lower than for their hydrophilic counterparts.

Taste sensation is initiated when chemicals interact with sites on the apical membranes of taste receptor cells. Some tastes bind receptors in ligand channels directly, while others stimulate signal transduction pathways for GTP-binding proteins and second messengers. Salty, sour and some bitter stimuli do not require specific membrane receptor proteins for stimulation. These tastes appear to interact directly with specific ion channels located on the atypical membrane. Specific membrane receptors are required for other tastes, however, including sweet stimuli, amino acids and at least one bitter compound.

When substances with different tastes are mixed there is often suppression of one or both tastes. Mixture interactions with food and beverages are the sum of interactions that occur at different sites in the taste system. Each of these interactions probably depends on more than just the qualities of the mixtures. The area of tongue stimulated, the temperature of the stimuli and the way in which they are tasted, e.g. the rate of flow over the tongue will determine the precise effect.

Some textbooks on taste show a picture of the

tongue showing sweet receptors concentrated at the tip, salt receptors at the front, sour at the rear edges and bitter at the back of the tongue. Such tongue 'maps' are incorrect. The original work by Hanig showed that the thresholds for each of the four basic tastes do not remain constant as the locus is altered. The threshold for sweet was slightly lower at the front of the tongue, the threshold for bitter was slightly lower at the rear of the tongue etc. The differences in sensitivity across each site were actually very small. All tastes are perceived well on all loci with taste receptors. Quantitative studies of taste show that there is a direct relationship between spacial features of the stimulated regions and intensive features of taste response. There are large variations in taste sensitivity between human subjects.

Taste perception

Salty taste

Salty taste is experienced with sodium chloride; other alkali halides have less marked salty tastes. Larger cations retain the salty taste but they tend to take on a bitter taste. That is why no salt substitute has been found that does not taste bitter. There is only one cation that is smaller than sodium, namely lithium, but while lithium has a nice salty taste it can be toxic. Salty taste transduction is mediated through an epithelial ion channel wherein the influx of Na into the cell depolarizes the cell membrane. This leads to alterations in voltage-sensitive ion channels, which modulates intracellular calcium activity leading to neurotransmitter secretion. Sodium ions are removed from these cells through a Na/K ATPase pump.

Sour taste

Sourness, while being a major entity, has been poorly studied. The primary event in sourness is a stimulation through an ion channel involving potassium ions and cellular depolarization.

Sweet taste

Sweet taste is seen as good and bitter as dangerous. This difference is because many beneficial foods are sweet and some poisonous compounds are very bitter.

Sweet and bitter tastes are typical of organic compounds. As the anion size is increased there is a change in the balance of bitter and sweet taste dependent upon structure.

Sweet tastes

Sweet-tasting compounds from natural sources include:

- low-molecular weight sugars (sucrose, fructose, glucose)
- amino acids (alanine, glycine)
- terpenoid glycosides (glycyrrhizin acid), osladin, sevioside, baiyunoside
- proteins, thaumatin, monellin, mabinilin, pentadin

Other sweet tasting compounds include:

- amino acyl sugars, methyl 2,3-di-O-(l-alanyl)-α-D-glucopyranoside
- alanine D-amino sugars
- peptides, aspartame
- chlorinated hydrocarbons, chloroform, halogenated sugars, sucralose
- N-sulphonyl amide, saccharine
- sulphamates (cyclamate)
- polyketides, neogasperidin, anilines and ureas

Sweet-tasting compounds have no characteristic chemical structure. Many attempts have been made to develop a general theory which relates chemical structure to sweetness, though such theories have not been generally successful.

There are bulk sweeteners which give body and viscosity to foods, e.g. glucose, sucrose, isomalt. The second group are intense sweeteners, used at low concentration, e.g. saccharine, aspartame, thaumatin.

Most sugars taste sweet at a concentration of 10^{-1} M, aspartame at 10^{-3} M, aspartic derivatives at 10^{-6} M and thaumatin at 10^{-7} M. The problem of identifying sweet taste receptors is that the chemical structural diversity of sweet compounds is so wide. Sugars stimulate adenylate cyclase in a concentration-dependent manner, indicating a role for G-proteins.

There is no perfect sweetener that is without additional tastes, e.g. bitterness, sourness or differences from surface tension and viscosity. A real problem is persistence which is a prolonged sweet taste.

Many of the very sweet, naturally occurring compounds are at least 50–100 times sweeter than sucrose and are mainly terpinoids, flavinoids and proteins. Many are constituents of green plants rather than microorganisms, insects, marine plants and animals. Plants from quite different taxonomic groups often synthesise similar classes of biochemical sweet compounds. However, if there are sweet-tasting constituents in one plant it does not follow that such sweetness will occur in other species of the same genus. As with attempts to define sweetness chemically there is no logic which enables the prediction of finding sweet compounds using plant taxonomy. Field investigations and literature sources have been used to identify sweet compounds. For example, agranandulcin, a novel sesquiterpene sweetener, was rediscovered from a monograph entitled *Natural History of New Spain (Central America)* written between 1570 and 1576 by Francisco Hernandez.

Two proteins, monellin and thaumatin, found in African berries are very sweet (100 000 times sweeter than sucrose) at concentrations of 10^{-8} M. Thaumatin is a single chain protein of 207 residues and monellin two peptide chains of 45 and 50 amino acid residues each. Native conformation of the protein is essential for the sweet taste; although both proteins are intensely sweet they are somewhat different chemically. Antibodies raised against thaumatin cross-react with and compete with monellin and other sweet compounds but not for chemically modified non-sweet monellin. This suggests a common chemical domain or portion of structure which confers sweetness. These proteins have little calorific content, are safe, natural and neither introduce non-natural metabolites into the body to distort the

balance of the amino acid pool. The disadvantage is that the taste profiles of these proteins differ from sugars and therefore have limited applicability. Monellin loses its sweet taste on heating as the two peptide chains separate. Thaumin has eight disulphide bonds and is more heat-stable than monellin, but once denatured does not regain potency.

Interactions of molecules and taste

Shallemberger and Acre proposed that the sensation of sweetness was appreciated by a sweet receptor. This was called the AH,B system (Figure 10.6). A and B are electronegative atoms (usually oxygen) and AH is a hydrogen bond. which functions through intermolecular hydrogen bonds. The hydrogen donor AH and hydrogen bond acceptor are separated by about 0.3 nm. The bipolar system is a electrophilic–nucleophilic (e–n) system. There is an additional important role in the sweet receptor site for the hydrophobic region of the molecule at another portion of the sweet receptor. The quality and intensity of a given molecule's taste depends strongly on the e–n system, the size and shape of the hydrophobic moiety and its position relative to the e–n system.

Kier proposed a third complementary binding site involving dispersion forces. Sweetness was induced in cooperation with the AH,B receptor site. It is now thought that an ideal sweetener contains up to eight optional and cooperative binding sites.

Structurally different sweeteners, amino acids, sweet peptides, aspartame or nitroaniline derivatives, have a common binding at the sweet taste receptor. The sweet taste of sugars, amino acids, saccharine, chloroform, olefin alcohols and meta-nitroanilans are attributed to their ability to form two hydrogen bonds with a complementary B,AH entity at the receptor. A structure–activity relationship is based upon a three-dimensional molecular theory rather than one-dimensional structural approaches. The problem is the peculiar relationship between sweet- and bitter-tasting molecules. Simple amino acids and peptides can change their taste from sweet to bitter when the chirality of the carbon adjacent to the amino and carboxyl groups is changed. This seems to point to a two-mirror-image receptor active site. A large number of synthetic sweeteners are flat rigid molecules. It might be possible to account for a change in taste of amino acids with chirality simply by inverting the AH,B entities of the twin receptors for sweet and bitter tastes. Saccharine substituted in position 6 of the aromatic ring remains sweet when the hydrogen is substituted with a methyl group, an amino group, a fluorine or a chlorine but loses its sweet taste when hydrogen is substituted with an iodine or methoxy group.

Amino acids and peptides Amino acids may have tastes, some of which are very sweet. The short-chain neutral amino acids (both L- and D- configuration) are sweet, e.g. glycine. Aspartyl dipeptide is very sweet. Such sweetness does not depend on the L- or D- configuration of the second amino acid ester but rather on the size and shape of the amino acid ester and side chain substitutions. When R1 and R2 amino acid side chains are sufficiently dissimilar in size the sweeteners' potency is very high. If space remains in the dipeptide ester receptor binding site at the C- or N-terminus of the sweet aspartyl dipeptide then peptides extended at the C- or N-terminus of the sweet peptides may taste sweet. A dipeptide N-terminus is required for high sweetness potency. Both sterically small and large hydrophobic moieties are required on the second amino acid for the specific sweet spatial orientation. It would appear that amine branching is important for a successful sweet taste. The increased steric bulk either from trimethyl or cyclopropyl substitution of the α-carbon is required for high sweetness potency in the aliphatic series. L-aspartyl-L-phenylalanine, methyl ester (aspartame) is 200 times sweeter than sucrose. The sweet region lies in the phenylalanine.

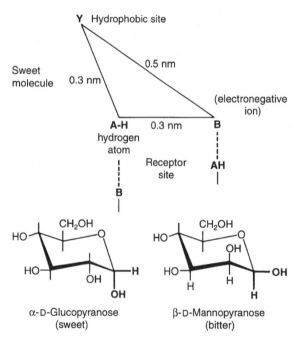

FIGURE 10.6 Sweetness perception – Shallenberger's saporous unit. For a molecule to taste sweet there must be A and B electronegative atoms; AH acts as an acid while B acts as a base. The AH,B unit forms a double hydrogen bonded complex with a similar AH,B system on the taste receptor. The distance between AH and B is critical at 0.3 nm for sweetness. The entire molecular geometry as well as the AH,B system affects the hydrogen-bonded complex, the quality and intensity of the sweet response. Bitterness may well have a similar receptor shape of different dimensions. This may account for bitter–sweet overlap in molecules which have similar dimensions fitting roughly into both sites.

Sugar A simple conformational change in an asymmetrical carbon can alter a sweet to a bitter taste. β-D-glucopyranose is sweet and β-D-mannopyranose is bitter. The sweet taste of sucrose can be increased by selective halogenation. The halogenation of hydroxyl groups results in an increase of sweetness of up to several thousand times. This is very stereo-specific, with many compounds being entirely tasteless. Other chlorinated carbohydrates, including derivatives of maltose and lactose, are extremely bitter.

Substitution at the carbon 6 position results in compounds 400 times sweeter than sucrose. The 6-*O*-methyl derivative is 500 times sweeter, whereas the 6-*O*-isopropyl derivative has no sweetness. Chlorination of the C-2 position produced an extremely bitter compound, suggesting that the presence of a hydroxyl group at the 2 position was essential for sweetness. Increasing the size of substitution at the 4' position has a positive effect on sweetness. Sucralose, the 4,1',6-trichloro derivative of sucrose is 650 times sweeter than sucrose, is very stable and tastes sweet.

It is possible that the intense sweetness of some of these sugar derivatives is a consequence of the molecules that occupy multiple binding sites.

Bitter tastes

Bitter taste receptors appear to involve a G-protein which activates the production of inositol triphosphate which releases calcium from internal stores.

A large number of chemicals taste bitter. Some are chemically related, quaternary amines, acetylated sugars, alkaloids, amino sugars, L-isomers, and some inorganic acids. There are anomalies in some people in, for example, the genetic dimorphism on the ability to taste thioureas and some other chemicals.

β-D-mannose and gentiobiose have a bitter taste whereas their α-isomers, e.g. α-D-mannose and isomaltose are sweet-tasting. The lipophilicity of the aglycone glycosides is associated with bitterness. The methyl glycosides of glucose, galactose, mannose, fructose, arabinose and benzyl glycosides are all bitter. The bitterness increases with chain length, with the methyl glucoside being sweet, ethyl glucoside bitter-sweet, and propyl glucoside bitter. The esters of sugars are bitter. The bitter-sweet interrelationship for amino acids is related to substitutions in the hydroxyl groups of alanine or serine; or bitter-sweet, e.g. threonine or valine, while most L-isomers are bitter (L-proline and L-glutamine). Increasing the steric bulk of the amino acid with lipophilic moieties in the side chain increases the bitter taste. Substitution of the α-carbon abolishes taste. Aromatic side chains increase the bitterness, e.g. phenylalanine and tryptophan are very bitter. Bitterness decreases with polar substitutions in the side chain, hydroxy, amino and carboxylic.

Peptides may taste bitter in both L- and D-isomers. The taste, however, of a peptide does not appear to be related to the constituent amino acids. The sequence in which the amino acids are paired in dipeptides, in some unknown manner, dictates the taste potential. This is regardless of the taste of the constituent amino acids.

Phenylthiocarbamide (PTC) tastes bitter to some but is tasteless to others. Non-tasting is a simple Mendelian trait. Substances with genetically based taste thresholds similar to PTC contain an N–C = S group, e.g. 6-*N*-propylthiouracil. In the USA and Europe, non-tasters account for one-third of the population, whereas among other ethnic groups the taster threshold frequency is higher.

Umami

This modality of taste is derived from the Japanese word for delicious or savoury. An example is monosodium glutamate.

Taste interactions

A variety of unusual taste interactions have been reported. The berries produced by *S. dulcificum* plant (miracle fruit) have the effect of making sour substances taste sweet. A glycoprotein in the berry adds a sweet taste to acidic substances. This may occur through a conformational change in the glycoprotein in miracle fruit. Acids change the conformation of the glycoprotein so that the sugars bind to it and stimulate the sweet receptor sides.

The leaves of the plant *G. sylvestre* contain substances that temporarily reduce the ability to taste sweet substances. Artichokes for some, but not all, individuals cause other substances to taste sweet for up to 15 minutes.

Summary

1. Gustation is the term for taste, olfaction for smell. In addition to the basic tastes of sweet, sour, bitter and salty there are a large number of tastes and smells.
2. The recognition of the flavour of food involves several sensory systems, mechanoreceptive, thermoreceptive and chemoreceptive.
3. The olfactory system is capable of recognizing thousands of smells. The system is very complex and it is not clear whether recognition takes place peripherally or centrally in the brain.
4. The olfactory system is a three-compartment sensory system specialized for the detection and processing of odorants.
5. The olfactory epithelium in humans is found predominantly on the dorsal aspect of the nasal cavity, the septum and part of the superior turbinates.
6. The olfactory bulb and receptors have an adult pattern by the middle of the 11th week of gestation. At about 28 weeks the olfactory system is capable of detecting chemical stimuli.
7. Attempts to classify odours into discrete qualitative characters have yet to be successful. The physical and chemical properties of olfactory mucus regulate the access of odorants to, and their clearance from, olfactory cells where the stimulus occurs.
8. The chemoreceptor cells which experience taste are arranged into buds in the oral cavity. Most taste buds are found on the upper surface of the tongue, though there are some in the soft palate, larynx, pharynx and epiglottis.
9. Taste cells are developed in the human foetus at 7–8 weeks' gestation and morphologically mature cells are to be found at 14 weeks. The newborn baby enjoys sweet, sour and bitter tastes, and can discriminate tastes. The number of taste buds in the gustatory papillae does not decrease with age, and taste response remains in old age.
10. Saliva is produced by three large, paired salivary glands: the parotid, submaxillary and sublingual glands. It is a dilute aqueous solution that contains inorganic and organic constituents and has a cleansing, antimicrobial and buffering action that protects the teeth. The composition of saliva varies with time of day and degree and type of stimulation.
11. The sensation of sweetness is said to be appreciated by a sweet receptor called the AH,B system. There is a role in the sweet receptor site for the hydrophobic region of the molecule elsewhere in the sweet receptor.

Further reading

Beets, M.G.J. (1978) *Structure Activity Relationships in Human Chemoreception*, Applied Science Publishers Ltd, London.

Coulante, T.P. (1984) *Food and the Chemistry of its Components*, Royal Society of Chemistry, London.

Dobbing, J. (ed.) (1987) *Sweetness*, ILSI Human Nutrition Reviews, Springer-Verlag, London.

Getchell, T.V., Batoshuk, L.M., Doty, R.L. and Snow, J.B. (1991) *Smell and Taste in Health and Disease*, Raven Press, New York.

Patterson, R.L.S., Charlwood, B.V., MacLeod, G. and Williams, A.A. (1992) *Bioformation of Tastes*, Royal Society of Chemistry, Cambridge.

Shallemberger, R.S. and Acre, T.E. (1967) Molecular theory of sweet tastes. *Nature*, **216**, 480–2.

Walters, D.E., Orthoefer, F.T. and DuBois, G.E (eds) (1991) *Sweeteners, Molecular Design and Chemoreception*, ACS Symposium Series 450, American Chemical Society, Washington DC.

10.3 Intake and satiety

- Satiation is important in the control of nutrient intake.
- Choice of food and taste, pleasure, volume and weight of food, energy density, osmolarity and proportions of nutrients are important dictates of satiety.
- Behavioural and cultural practices can override biological constraints.

10.3.1 Introduction

The amount of food that is eaten should ideally meet the needs of the individual. Nutritional intake is important in supplying metabolic processes which will vary with the individual in different phases and energy expenditures of life. There is also a strong sociocultural influence on food intake (Figure 10.7).

It is possible to fast, to eat sufficient or in excess of required or felt to be appropriate or desirable. If there is insufficient money or food available it may be imperative for a mother to reduce her energy intake in order to feed her children. If there is an excess of food available then this may lead to obesity and accumulation of body fat. These are not necessarily of paramount importance in controlling appetite or dietary intake. Eating behaviour may be determined by an interaction between the choice of food, satiation signals and the biological responses to food ingestion. Satiety is the inhibition of the sensation of hunger and as a consequence the limiting of food ingestion.

Satiety and nutrition selection

There are two approaches to eating behaviour studies:

1. A behavioural approach which measures eating patterns and food intake. Such qualitative aspects include eating, food choice, preference and the sensory aspects of food and hunger, fullness and hedonic sensations that accompany eating. The behavioural profile is the relationship between the pattern of intake of meals and snacks, and the taste and other sensory attributes of the foods.
2. A quantitative approach measures dietary nutrient and energy intake. Quantitative aspects include the consumption and energy value of food, their macronutrient value, the composition of food, impact and energy balance. A quantitative profile is the total food consumed over a 24-hour period, total energy intake and the proportions of macronutrients ingested and the overall fuel balance.

10.3.2 Behavioural and qualitative aspects of food intake

The sensory control of eating

Eating is organized into meals which are taken at variable time intervals. A meal is composed of a series of phases, each controlled by different mechanisms and associated with various emotional states. At least three distinct phases of the meal can be distinguished – meal initiation, which is associated largely with hunger, meal maintenance, which involves the act of eating, and meal termination, associated with satiety.

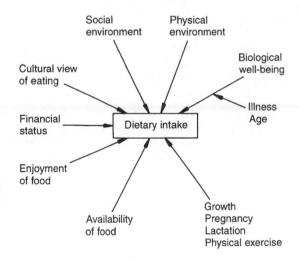

FIGURE 10.7 Dietary intake determinants are social, physical and biological.

327

Meal initiation

The transition from the non-eating to the eating state includes **anticipation** and the mechanisms involved in wanting to eat and anticipatory to eating food. **Perception** is an important factor in specific appetites. An animal ensures adequate nutrition by consuming a widely varied range of foods. Experiments in newly weaned infants and in adults show that when more than one food is available there is a natural tendency to switch between foods rather than to consume only the favourite meal. A nutritious but monotonous diet, given to military or even starving refugees may eventually be refused. The aversion may persist for several months. It is possible that the decline in acceptability of a particular food, consumed in excess is due to some innate automatic mechanism directing food selection, a sensory-specific satiety.

Food craving is one of the strongest sensations involved in eating. Among a normal population more than two-thirds of men and almost all women report intense desires for specific food items. Food cravings during pregnancy can be so profound that there are reports of pregnant women stealing food. Over 70% of bulimic women attribute binge episodes to previous carbohydrate cravings. Craving for sweet foods, especially chocolate, is frequent in premenstrual obese women. There are several theories about craving. One is that cravings indicate the need for a substance necessary to correct homeostatic imbalance, e.g. salt. Another theory is cravings may reflect the desire for specific sensory and not pharmacological or physiological stimulation.

Food-associated sensations such as smell and sight of food promote the desire for food and eating. Environmental stimuli associated with food or eating are strong controllers of meal initiation. Sensation arising from food itself, such as the sight or smell of food, enhance eating and induce craving, as do location, time of day and social environment. Another factor is what might be called energy depletion and loss of vital short term energy sources, e.g. glucose.

There are endogenous determinants of energy intake such as gender wherein a male, growing to a greater height and weight, will eat more than his contemporary female. As people grow older food intake declines, in part because of reduced physical activity.

Surveys of attitudes towards dietary fat show that fat intake is related more to the pleasantness of the taste rather than any views on health. Favourite foods are often fatty foods, desired because of their smells, flavours and palatable textures. Low-fat diets are seen as being monotonous and bland. It has been shown that body weight is related to a liking for fatty foods. As body weight increases so does the liking for fatty foods. High-fat, energy-dense foods are associated with a greater reduction in hunger than low-energy-containing foods, at the same weight or volume intake. The weight of food eaten tends to remain constant, the variable is the energy content of the intake.

The amount of food eaten is generally believed to be determined by its flavour and particularly by the palatability of its flavour. It is questionable, however, whether food taste and palatability influence food intake over a longer more nutritionally significant interval. The chemical senses are crucial for the recognition and selection of food. Food may be selected or rejected on the basis of inborn preferences or aversions. Innate preferences for a sweet or salty taste may help in the identification of food that contains sugars which are a source of calcium or sodium and other minerals which are essential nutrients. The aversion to bitter taste which is often characteristic of toxic substances may help reduce the risk of food poisoning. It has been suggested that there are alterations in taste in various conditions such as salt intake and hypertension and a number of nutrients and cancer. Similar changes may be noted in eating disorders, and in gastrointestinal, liver and kidney disease. Altered chemosensory response has yet to be implicated as a contributory factor in the changes in food intake and preferences associated with any of these.

Heavy sedation decreases food intake. Alcohol may decrease food intake because of gastritis, though alcohol itself is a food. Corticosteroids may increase the appetite and food intake.

Endogenous and exogenous determinants of food selection Society, religion and culture influence choice of food. The delicacies of peoples of one part of the world are elsewhere regarded with abhorrence. Finance may determine the availability of nutrients. Confusion over choice of food may occur after a period of deprivation, when the financial status improves and a wider range of food becomes available and factors in food choice, other than nutrition, e.g. ignorance and inappropriate models become important (Figure 10.7).

Meal maintenance
This is the period during which eating continues. The eating process passes through a series of stages. The cephalic phase response includes secretory and motor reflexes related to digestion, e.g. insulin or gastric acid secretions and motility changes which are activated by stimulation of receptors in the brain or oropharynx. Cephalic responses produced by brain receptors are often stimulated before eating by the sight and smell of food. Such responses develop as a result of learning. In contrast, cephalic responses produced by oropharyngeal receptors arise from contact with the food within the mouth. There is a good correlation between the palatability and the magnitude of the cephalic response produced.

Once eating has started, two types of sensory events are evoked, one excitatory and the other inhibitory. These are the main determinants of how much an individual eats at a meal. Obviously this assumes that there is an excess of food available. The palatability of food is a major determinant of the time over which the meal is eaten and consequently the amount eaten at the meal. Palatability includes:

- a sensory property of the food itself
- the physiological state of the person who is eating the food
- stimuli arising from previous associations with that particular food

Other factors include particular dietary likes and dislikes, e.g. sweet or savoury food. The more pleasant the sensory properties of the food, the more is eaten.

Meal termination
A meal is usually completed with a feeling of satiation. The excitatory stimuli which make for the eating of the meal have to be overcome. This occurs as a result of a suppression of the pleasant feeling for the food. This is called **alliaesthesia**. Signals are sent by the brain to suppress eating. The entire upper gastrointestinal tract produces sensory stimuli that signal meal termination. The stomach produces a whole range of sensations such as stomach distension transduced by gastric stretch receptors to be relayed to the brain.

Food intake controls

Controls of food intake include:

- hunger, cravings and hedonic sensation
- energy and macronutrient intake
- peripheral physiology and metabolic events, concentration of neurotransmitter and metabolic interactions in the brain.

The cephalic phase responses are a result of events in many parts of the gastrointestinal tract. The effect of ingested food in the mouth results in a positive feedback for eating, whereas from the stomach and small intestine there is negative feedback. The brain is informed about the amount of food ingested and its nutrient input through specialized chemo- and mechanoreceptors that control physiological activity. This is followed by a post-absorptive phase after the nutrients have been digested and enter the circulation. These nutrients may be metabolized peripherally or even pass into the brain. It used to be thought that there were opposing hunger and satiety centres in the hypothalamus. Now it is thought that a number of neurotransmitters and neuromodulators, pathways and receptors are involved in the central neurological process. It is thought that foods of varying nutritional composition act differently with the mediating processes and have different effects on satiation and satiety.

Factors involved in the modulation of appetite

- Taste (intensity and hedonics)
- Volume and weight of food
- Energy density
- Osmolarity
- The presence of different proportions of macronutrients

Biological responses include:

- oral stimulation
- gastric distension
- rate of gastric emptying
- release of hormones
- triggering of digestive enzyme secretions
- plasma concentrations of absorbed nutrients

10.3.3 Biological and quantitative basis of hunger and satiety

Food needs

The requirement to eat to provide energy may be regarded as continuous or intermittent. The basic vegetative activities of the cardiorespiratory, renal, endocrine, liver, nervous and brainstem systems ensure the 'housekeeping' of the body. In order to satisfy these needs, energy must be continuously available. Intermittent activities and their satisfaction with energy provision range from trivial movements to activity at maximal capacity where brain and the voluntary muscular systems are utilized to their limit.

Food intake may be continuous, as in the ever-eating sparrow or fieldmouse, or intermittent as in the eagle or the lion. These creatures appear quiescent between bursts of energy during which the prey is secured before the resumption of sloth.

In a freely eating society with readily available food, the intake of protein is tightly controlled between 11% and 14% of daily food intake. Fat and carbohydrate intake are less readily controlled. However, carbohydrate intake suppresses subsequent intake by an amount roughly equivalent to its energy content. The time-course of this suppressive action may vary according to the rate at which carbohydrates are metabolized. There is a relationship between energy intake and glucose, which is mediated through arteriovenous glucose differences. The rate of hepatic glucose utilization or the activation of glucoreceptive neurons in the brain are important controls. Cholecystokinin (CCK) is a hormone that mediates satiation and early-phase satiety. Protein or fat stimulates the release of CCK activating CCK-A receptors in the pyloric region of the stomach. From there, signals to the vagal afferents are passed to the nutrient nucleus of the tractus solitarius and to the medial zones of the hypothalamus. This is the mechanism by which dietary fat may trigger neurochemical responses which mediate satiety. Dietary fat may also control satiety through enzyme systems responsible for the digestion of fat, e.g. pancreatic procolipase. In such an activation there is release of an activation peptide, enterostatin. This may further decrease food intake. Oxidative metabolism of glucose and free fatty acids in the liver provides a significant source of information for the control of appetite. Alternatively, oxidation of fat may be an important signal in food intake.

Satiety can also be controlled through learned reflexes, though no centre controlling satiety has been identified in the brain. In humans the needs for metabolic fuels are substantial and continuous, yet eating is episodic. It is unlikely that there is a central receptor monitoring caloric flux and therefore moderating food intake. There may be important, non-central controls, e.g. the liver which monitors gastric emptying and the distribution of insulin and energy sources.

The passage of energy-rich food from the stomach to the intestine is regulated. Concentrated solutions empty slowly whereas dilute solutions empty rapidly, producing the same net delivery of calories to the intestine. Satiety may result from a feeling of gastric distension and rate of gastric emptying. After a meal, metabolism is stimulated by nutrients entering the circulation from the gastrointestinal tract, with excess energy being

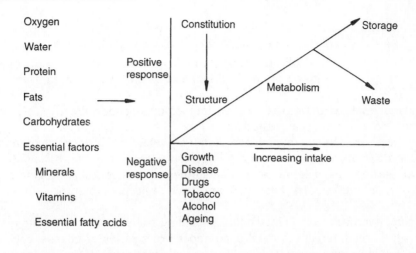

FIGURE 10.8 Primary food sources. Increasing dietary intake meets the needs of metabolism. Excess dietary intake is stored or excreted. Metabolism and conversion of nutrients into structure is dependent upon constitution. Possible negative effects on growth are shown

stored as glycogen or triglyceride. Intestinal absorption provokes insulin secretion and shifts hepatic metabolism from mobilization to storage, thus providing a second signal of satiety. When both satiety signals have disappeared, liver and adipose tissue energy stores are mobilized and hunger begins. Hunger may not reflect a biological need for food, as does thirst for water.

Following the ingestion of food, the stomach distends, nutrients are absorbed, there is hormonal release, and removal of nutrients from the blood; all of which may influence the feeling of satiation. Dietary fat passes through each of these stages at a slower rate than carbohydrates and proteins. There may be individual differences in response to different macronutrient loads which determine rate of metabolism and possibly satiation.

10.3.4 The role of the gut in controlling food intake

The mucosa of the gastrointestinal tract from the mouth to the terminal ileum is sensitive to chemical and mechanical stimuli of ingested food. A consequence is the release of gut hormones with negative feedback mechanisms which dictate the completion of feeding and the onset of postprandial satiation.

Increased contractions in the stomach cause the sensation of hunger. Insulin and hypoglycaemia can provoke hunger contractions. However, neither hypoglycaemia nor decreased glucose utilization occurs between meals. When the gastric motility response to insulin induced hypoglycaemia is abolished by dividing the vagus below the diaphragm, the urge to eat still occurs. Patients who have had a total gastrectomy experience hunger. As yet, however, no alternative theory to indicate the physiological mechanisms for hunger has been evolved.

There is a close association between epigastric fullness, observable abdominal protuberance, stomach distension and the cessation of eating. There are receptors in the wall of the stomach which are activated by distension.

Distending the stomach with balloons or food stops eating. Removing recently ingested food restarts the ability to eat and leads to over-eating. Gastric emptying is slowed by neural and hormonal duodenal mechanisms. It has been suggested that gut peptides are important in the physiology of postprandial satiety. Such gut hormones include CCK-8, bombesin, glucagon and somatostatin.

Summary

1. The amount of food eaten should meet the needs of the individual. Nutritional intake is important in supplying metabolic processes which will vary with the individual in different phases and energy expenditures of life. There is also a strong sociocultural influence on food intake.
2. Eating behaviour may be determined by an interaction between the choice of food, and the satiation signals of the biological responses to food ingestion. Choice of food or taste, intensity and pleasure given (hedonistic response), volume or weight of food, energy density, osmolarity and the proportions of macro nutrients are important factors in eating behaviour.
3. There are two approaches to eating behaviour studies: a behavioural approach measures eating patterns and food intake; a quantitative approach measures dietary nutrient and energy intake.
4. The behavioural approach to food intake looks at eating as being organized into meals which are taken at variable time intervals. A meal is composed of a series of phases, each controlled by different mechanisms and associated with various emotional states. At least three distinct phases of the meal can be distinguished: meal initiation, meal maintenance, and meal termination. Hunger is associated largely with meal initiation and satiety with meal termination.
5. The controls on food intake include: hunger, cravings and hedonic sensation; energy and macronutrient intake; peripheral physiology and metabolic events, and concentration of neurotransmitter and metabolic interactions in the brain.
6. The quantitative approach examines the requirement to eat as continuous or intermittent. The basic vegetative activities which ensure the housekeeping of the body require a continuous provision of energy.
7. No centre controlling satiety has been identified in the brain.
8. The passage of energy-rich food from the stomach to the intestine is regulated. Concentrated solutions empty slowly, whereas dilute solutions empty rapidly, producing the same net delivery of calories to the intestine. Satiety may result from a feeling of gastric distension and high rate of gastric emptying. Following the absorption of food, satiation may follow the storage of the absorbed nutrients.

Further reading

Blundell, J.E. and Halford, J.C.G. (1994) Regulation of Nutrient Supply – The Brain and Appetite Control. *Proceedings of the Nutrition Society,* 53, 407–18.

Kissileff, H.R. and Van Italhe, T.B. (1982) Physiology of the control of food intake. *Annual Review of Nutrition,* 2, 371–418.

Rolls, B.J. and Shide, D.J. (1992) The influence of dietary fat on food intake and body weight. *Nutrition Reviews,* 50, 283–90.

10.4 Food availability

- The absorption of food is dependent upon its chemical structure, the rate of its transit along the gastrointestinal tract, the intestinal absorption surface and receptors and the interaction between the nutrient components of the meal.
- Cooking may modify the rate of nutrient absorption.

10.4.1 Introduction

The uptake of food is dependent upon the total intake and bioavailability, i.e. absorption. No food is of value to the individual until it has been absorbed.

An important function of the intestine is to regulate the intake of minerals which are essential though poisonous in excess. Some essential nutrients are very labile, and consequently there is the potential for nutritional deficiency.

The availability of foods for absorption from the lumen of the intestine (Figure 10.9) is dependent upon a number of contributing factors, mucosal absorption, pathological factors altering intestinal absorption and alterations in luminal availability, e.g. interaction between accompanying nutrients within a meal. Other factors which may affect absorption include deficiency of necessary intestinal secretions and enteric bacteria, medicinal drugs and surgical procedures. Partial gastrectomies or gastroenterostomies will decrease gastric emptying time and accelerate absorption, endocrine response and cause hypoglycaemia, i.e. dumping.

The uptake by the mucosa of various nutrients may depend on the individual's nutritional needs whereby intestinal absorption may be either partially or totally controlled.

10.4.2 Stomach

Gastric mucosa

The stomach is a reservoir which stores food. The well-being of the mucosa and the rate of gastric emptying will influence the rate of release of food from the stomach to the small intestine for digestion and absorption. Alcohol can cause gastritis and gastric mucosal alcohol metabolism and can hence modify food intake

The mucoprotein **intrinsic factor** is essential for vitamin B_{12} absorption; its deficiency leads to a failure of vitamin B_{12} uptake, with consequent development of pernicious anaemia.

Gastric emptying time modifies the rate of intestinal absorption. A dilute glucose solution will empty from the stomach rapidly and hence be absorbed from the jejunum much more rapidly than a more concentrated glucose solution. Fat has a slower gastric emptying than carbohydrate.

Nutrient absorption may be slowed with a resulting modulation of nutrient uptake, resulting in a reduction of plasma concentrations followed by a modified endocrine response. A further consequence is that the renal excretion of nutrients may be reduced.

10.4.3 Small intestine and nutrient availability

Cooking may modify nutrient absorption; raw starch is poorly digested by amylase but during cooking starch granules are disrupted and become readily digested. Retrograde starch (see section 5.6.3), present in bread, potatoes and puddings, when stored in the cold after cooking is not readily hydrolysed by pancreatic amylase in the small intestine. This retrogradation is partially

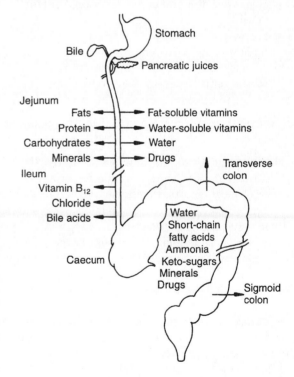

FIGURE 10.9 Absorption along the small intestine, stomach, jejunum, ileum and colon.

reversed by heating. Protein may have a reduced absorption efficiency after cooking through the formation of Maillard reaction products.

Enterocytes

The small intestine has been described as an 'active interface' between the external environment and the blood and lymph which distributes the absorbed materials for the metabolic needs of the body. The epithelial cells of the small intestine have three major functions: digestion, absorption and secretion. The **enterocytes**, that is the surface cells, are constantly being lost into the lumen from the tips of the villi and replaced.

Enterocytes

The life of an enterocyte is approximately 48 hours. Various constituents of food may alter enterocytes by:

- changes in the number of functional enterocytes
- changes in the carrier, enzymatic or metabolic process of the enterocytes
- changes in the rate of maturation or function in the enterocytes as they move from the crypts to the extrusion zone at the villus tip

Transfer across the enterocyte from the lumen involves:

- movement from the bulk phase across the 'unstirred layer' to the enterocyte surface.
- movement across the brush border membrane
- movement across the cytoplasm
- removal to the blood stream through the basolateral membrane

---

During starvation there is a progressive decrease in small intestinal enterocyte population, mucosal mass, enterocyte column number and villus height, particularly in the jejunum. Luminal factors, particularly nutrients and secretion, are necessary to maintain normal small intestinal

enterocyte numbers. The absorptive capacity of the small bowel is in part dependent on the number of enterocytes present. The maximum rate of absorption (J_{max} per unit length) is affected by the number of enterocytes.

The development of the intestinal mucosa and enzyme systems has been better studied in animals than in humans. The description given here is derived from such animal work in the expectation that the principles described will not be remote from what happens in humans.

Enterocytes produced in intestinal crypts express a genetically determined programme as

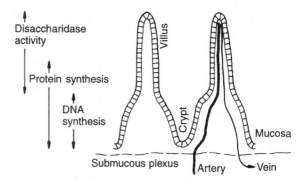

(a)

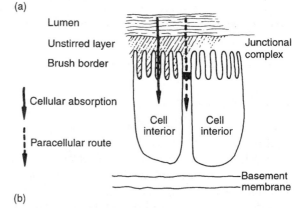

(b)

FIGURE 10.10 The intestinal mucosa and enterocytes. (a) The villus and crypt of the small intestinal mucosa showing crypt, villus, blood supply, submucous plexus and enzymatic activity along the villus from DNA to disaccharidase activity. (b) Enterocyte and absorption showing the unstirred layer, brush border and basement membrane. There are two forms of absorption; cellular and paracellular through the junctional complex.

they progress along the crypt. This can be affected by diet. Increasing the amount of food eaten or changing the protein content of the diet may alter enterocyte cell surface structure. Sucrase activity in the weanling and amino acid absorption in the adult rat are affected by diet. Increasing amounts of sugar in the diet increases the enterocyte content of sucrase and maltase. Lactase activity is not affected by diet. At the same time crypt proliferation is affected. Enterocytes change structure and ability to digest and absorb nutrients during maturation.

Slowing enterocyte migration rates to less than 6 μm/hour increases maximal microvillus length.

Enzymatic activity increases rapidly as the enterocytes migrate over the lower parts of villi, activity declining as the tip is approached. Amino acid transport occurs only in the upper regions of villi, suggesting that dietary constituents affect the expression of single or multiple genes at different stages of enterocyte development.

Enteric absorption

The lipid layer of a cell membrane acts as a container which enables the cell to separate its internal environment from the surrounding fluids. This requires that the cell surfaces are able to transport molecules in and out of the cells. The cells of the enteric border are particularly important in controlling the intake of molecules from the intestinal lumen and moving these molecules, either intact or modified, into the blood stream or lymphatic system. Water-soluble nutrients of low molecular weight may pass through the intestinal epithelium by two major routes, paracellular and transcellular. Paracellular transport is mostly if not entirely through the highly permeable tight junctions joining the brush border membranes of adjacent absorptive cells. This mechanism is by simple diffusion through pores or aqueous canals, the rate of transport being a function of concentration gradient. Paracellular transport is mediated by solvent drag; solutes dissolved in the bulk flow of solvent are carried through the transepithelial channels. This system transports at a rate proportional to the concentration of the solution and to the rate of solvent flow, but becomes less effective as the radius of the solute molecules approaches that of the channel.

The transcellular route involves entry across the brush border membrane of the absorptive cells. In the case of amino acids entry is by translocation through the cytosol of the cell interior, by diffusion and exit through basolateral membranes by

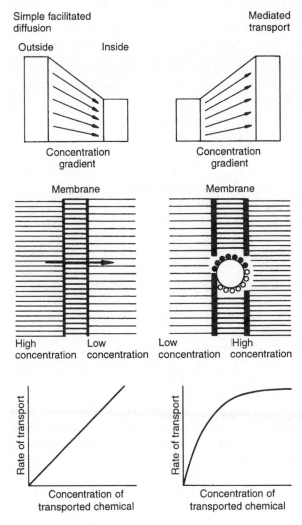

FIGURE 10.11 Simple facilitated diffusion and mediated transport. Simple facilitated diffusion passes down a concentration gradient and is concentration-dependent. Mediated transport passes up a concentration gradient and is dependent on carrier systems which can be saturated.

facilitated diffusion. There are two main methods of transport across the cell membrane:

1. Simple diffusion through minute aqueous canals in a predominantly lipid and protein cell membrane, though amino acids may diffuse through the lipid of a membrane. If the molecule is uncharged then the direction of flow is concentration-dependent. If charged then an additional influence is the electrochemical gradient. Facilitated diffusion is enhanced diffusion, not requiring metabolic energy, and is dependent upon a concentration gradient; the system is saturable and affected by competitive inhibition. Simple and facilitated diffusion are **passive transport systems.**
2. Mediated transport specific for each structurally related substance, e.g. amino acids across the membrane. Such a transport system involves a **specific carrier.**

Active transport is capable of transferring a substrate across a membrane against an electrochemical gradient driven by metabolic energy. Primary transport systems are directly dependent upon metabolic energy, whereas secondary active transport is coupled to the re-entry of sodium across the brush border membrane. The extrusion of Na^+ across the membrane is energy-dependent. When the transport of a substrate is linked to a metabolic energy supply, by a chain which involves transport of two ions, this is called **tertiary active transport.**

Facilitated and simple diffusion progress in one direction down a concentration differential, whereas active transport occurs in both directions.

Counter transport is a phenomenon wherein a substance being transported in one direction accelerates the transport of another in the opposite direction.

Many active transport systems including amino acids, monosaccharides and many other substances passing across the brush border membrane are coupled to the passive re-entry of Na^+ into the absorptive cell down an electrochemical gradient. Active transport requires transport proteins which belong to two categories. Channel proteins form hydrophilic pores which allow a suitable solute,

usually an ion, to pass along the channel. The active transport process also requires carrier proteins which bind the substance being transported and undergo conformational change in order to cross the membrane. The activity of the carrier protein is geared to an energy production process, e.g. ATP hydrolysis or ion gradient. Carrier proteins have substrate specific binding sites and follow Michaelis–Menten kinetics.

V_{max} and J_{max}

Using Michaelis–Menten saturation kinetics it is possible to measure 'apparent' K_m and V_{max}:

$$v = \frac{V[S]}{K_M[S]}$$

where v is the velocity of transport, the maximal velocity of transport is V_{max}, $[S]$ is the substrate concentration and K_m the substrate concentration at which the velocity of transport is half-maximal. K_m is an inverse measure of apparent affinity for transport, so a high value means a low affinity for transport and vice versa. This relationship does not mean that an enzyme system is involved, only that a receptor system is involved which can be saturated.

Sometimes in the intestine the term J_{max} is used instead of V_{max}. Apparent K_m is an indicator of the affinity of the binding mechanism between the nutrient and its receptor carrier mechanism. The J_{max} is the transport equivalent of the maximum velocity of the enzymes. This allows comparison of the different transport mechanisms.

'Apparent K_m' = 'real K_{max}' + $J_{max} d/D$, d being the thickness of the unstirred layer and D the diffusion coefficient of the solution being transferred.

The unstirred layer, that is the structured water layers close to the mucosal surface, and the glycocalyx or fuzzy coat of the microvilli influence the saturation kinetics and passive permeability coefficients. The unstirred layer is a rate-limiting step for the diffusion of solutes from the luminal bulk to the surface of enterocyte through which solutes must diffuse in order to reach the surface. This retards the rate of diffusion towards and away from the surface. In the intestine the unstirred layer has an important role in dictating the absorption rate. There is no definite thickness of this boundary, but it varies from 100–300 μm to 400–650 μm depending on conditions and measurement. The effect of the unstirred layer is to reduce the concentration of a solute at the membrane surface relative to the luminal concentration, thus reducing the absorption rate. The rate of movement across this unstirred layer and reaching the intestinal mucosa influences the rate of absorption.

The carrier-mediated systems consist of:

1. **Uniport**. simple transport from one side of the membrane to the other.
2. **Coupled transporters**. The transport of one solute is dependent upon the transport of a second solute. **Symport** is transport of both solutes in same direction; **antiport** is transport in opposite directions. An example of an antiport is the membrane Na^+-K^+ pump system, sodium out and potassium into the cell.

Effect of nutrient interactions on bioavailability

Nutrients may interact both chemically and physically to increase or decrease bioavailability (Figure 10.12).

The underlying mechanism between these seemingly unrelated enhancements and inhibitions of absorption is water solubility. Insoluble calcium salts, e.g. calcium phytate, and fatty acids are not absorbed. If a trace mineral, e.g. zinc or magnesium, is bound to phytic acid then calcium can displace the trace element which can then be absorbed, particularly if the latter is in the form of a soluble, readily absorbable amino acid salt.

After absorption some substances are metabolized on first contact with tissues, e.g. liver or enteric mucosal cells. This is called the **first pass metabolism**. Such metabolism of substances during absorption is important; an example is the immunosuppresant drug cyclosporin, the efficacy of which is marred by poor and unpredictable bioavailability. This drug, and possibly erythromycin, lignocaine and oestrogen, may be metabolized by the enzyme system P450IIIA, which is found in the liver and also in enterocytes.

Competition for nutrients

Small-intestinal bacterial colonization may develop in a surgically formed stagnant loop or in jejunal diverticulosis. There is nutrient competition between host and bacteria for vitamins, iron, proteins, fats and carbohydrates.

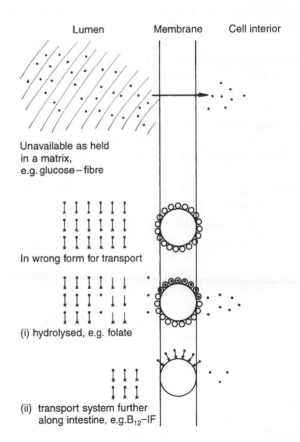

FIGURE 10.12 Interaction between nutrients and absorption across the intestinal cell membrane.

The most dangerous inhibitor of absorption is *Cholera vibrio*, the enterotoxin of which stimulates a cyclic AMP-mediated secretory process with profound water loss.

Drugs may cause malabsorption and diarrhoea, e.g. neomycin precipitates bile acids, so that insufficient bile acids are available to reach a critical micelle concentration and steatorrhoea (fat malabsorption) results.

Specific amylase blockers in beans slow the hydrolysis of amylose and retard the absorption of starch breakdown products.

Factors affecting bioavailability

The complexing of vitamin B_{12} with intrinsic factor allows the vitamin to be absorbed from the ileum. In contrast, niacin is held in cereals in a bound form which is not absorbed unless released by dilute alkali or roasting, a factor in the development of pellagra, vitamin B_6 deficiency. Biotin is bound to avidin, a protein in raw eggs which renders biotin unavailable for absorption. The binding of calcium to oxalate and phytate reduces calcium absorption, because the human intestinal tract mucosa contains no phytase. Yeast cells contain phytase which hydrolyse phytic acid during the leavening of dough. Calcium absorption is reduced in populations who habitually eat unleavened bread. Dietary fibre, a cation exchange binder may reduce mineral absorption.

Intestinal zinc absorption is reduced by cadmium, copper, calcium, ferrous iron, phytate and proteins which have undergone Maillard reactions. Zinc absorption is however increased by methionine, histidine, cysteine, citrate and picolinic acid.

Nutrient interactions may affect nutrient absorption. Alcohol may reduce thiamin absorption, particularly if there is coincidental folate deficiency. Intestinal calcium absorption is increased by lactose, unsaturated fatty acids, lysine, arginine and glucose polymers. Iron in the form of haem is readily taken up by the mucosal cells. Haem itself is poorly soluble in water but when haemoglobin is digested the resultant peptides render the haem more soluble. Non-haem iron is absorbed by a separate transport system in an ionic form. Absorption is increased by a coincidental presence of meat, fish and vitamin C and is inhibited by phytate and tannin. Selenium is better absorbed as the selenomethionine form than as sodium selenite.

10.4.4 The colon

The colon is a conserving organ, containing a large bacterial mass. The bacteria modify the enterohepatic circulation of chemicals secreted in bile, e.g. bile acids, hormones and drugs. The caecal bacterial breakdown of proteins is followed by the absorption of amino acid derivatives. Hydrogen, methane and other gases are also produced from fibre. Absorption from the colon is somewhat specific. Colonic bacteria are able to synthesis vitamin K_2 (multiprenylmenaquionones) and vitamin B_{12}, which are not absorbed. The faeces from subjects with pernicious anaemia contain vitamin B_{12} which is more than sufficient for their needs but is unavailable for absorption. The relatively modest large-bowel absorption of polar molecules, e.g. hormones, toxic bacterial products and drugs, is because of small colonic pores. Such pore size probably protects against toxins of bacterial origin. Changes in colonic pore permeability can be induced by drugs and disease, e.g. colitis. Fatty acids and dihydroxy bile acids increase permeability and pore size allowing for example oxalic acid to be absorbed.

10.4.5 Digestion and gastrointestinal hormone release

Endocrine cells in the gut are distributed along the length of the tract.

The endocrine cells of the gut are distinct and have similar staining properties to the adrenaline-producing chromaffin cells of the adrenal glands and therefore are called enterochromaffin cells. They have granules which are classified (G-, D-, I-, etc.) dependent on their shape, size, staining properties and number of granules. Endocrine cells are classified into open cells in which part of the cell surface is in direct contact with the gut lumen, and closed cells which are completely surrounded by other cells.

Some of the gut hormones are common to other tissues, for example the peripheral and central nervous system, e.g. CCK. These neuroendocrines and neuromodulators influence neurotransmission. They may also act as paracrine agents with effects limited to neighbouring cells. These are often spoken of as regulatory peptides, rather than gastrointestinal hormones, and can be considered in groups:

- The gastrin group (gastrin and cholecystokinin, CCK): gastrin increases acid secretion in the fundus of the stomach, and is released by luminal dietary peptides and amino acids in the antrum and duodenum. Gastrin is released after a meal, probably as a result of direct contact of food with the open gastrin cells of the antrum. The extent of the response is dependent on the type of food, particularly how much protein is present. Dietary fat and carbohydrate have little effect on the gastrin release. **Cholecystokinin** is released by the digestion products of fat and by protein and stimulates gallbladder contractions and pancreatic enzyme secretion. The best nutrient releasers of CCK are fatty acids of more than nine carbon atoms length. Glucose and amino acids may also be important in CCK release.
- Secretin group, consisting of secretin, glucagon, glucagon-like peptides (GLPS), gastric inhibitory polypeptides (GIP) and vasoactive intestinal polypeptide (VIP). Secretin is released by acid and results in the flow of water and bicar-

bonate from the pancreas. Somatostatin is released postprandially and has an inhibitory effect on the secretion of growth hormone, insulin and glucagon. Neurotensin is released by fats and is important in the control of motility. GIP is released by sugars, some amino acids and stimulates insulin secretion in the pancreas.

Other gut regulatory peptides

- Pancreatic polypeptide (PP), peptide YY and neuropeptide Y. Glucagon and PP are exclusively found in the pancreas. VIP, the enkephalins and neuropeptide Y are also found in neural tissue
- Gastrin (G cells) are predominantly found in the antrum and duodenal mucosa
- CCK and secretin cells (respectively I-cells and S-cells) in the duodenum and jejunum
- Somatostatin cells (D-cells) in all parts of the gut
- GIP-secreting K-cells are predominantly in the duodenum and jejunum, but are also present in the antrum and ileum
- M-cells that secrete motilin are found in the antrum but also in the upper duodenum
- Glucagon-like peptide-secreting cells (L-cells) and neurotensin (N-cells) are found in the ileum

Secretin is released by acid contact with the duodenal mucosa and causes bicarbonate secretion by the pancreas. Secretin also releases insulin, but glucose within the duodenum does not have any effect on secretin levels.

Somatostatin is present in the cell walls of neurones of the central and peripheral nervous system and in endocrine cells of the pancreas and gut. Its effect is largely inhibitory and it has an extremely short half-life. The somatostatin-containing endocrine cells are closely associated with the target cells with effects on gastric acid secretion, pancreatic endocrine and exocrine secretion, and possibly gut motility. Somatostatin may inhibit amino acid and glucose absorption.

Neurotensin delays gastric emptying time. Gastric inhibitory polypeptide (GIP) has an inhibitory effect on acid secretion and also increases insulin secretion under hyperglycaemic conditions.

Motilin is released during fasting, and is related to specific intestinal contraction patterns which are called the migrating motor complex. Motilin is excreted during the fasting period but it may also be involved in increasing gastric emptying.

Glucagon-like peptides (GLPs) or pancreatic glucagon are present in the intestine. The original gene product gives three sequences and the enzy-matic processes produce three different molecular forms.

Pancreatic glucagon

Pancreatic glucagon consists of 29 amino acids and is formed in the pancreas. In the small intestine the three sequences initiate glucagon-like peptides.

Enteroglucagon, which is homologous to pancreatic glucagon, has an extra 32 amino acids.

Glicentin, which contains the enteroglucagon sequence, has a further 32-amino acid extension at the N-terminal end.

Summary

1. The availability of foods for absorption from the intestinal lumen depends on a number of contributing factors, mucosal absorption, pathological factors altering intestinal absorption and alterations in luminal availability, e.g. interaction between accompanying nutrients within a meal. Deficiency of necessary intestinal secretions and enteric bacteria, medicinal drugs and surgical procedures may also affect absorption.
2. In the gastric mucosa and small intestine the epithelial cells of the small intestine have three major functions: digestion, absorption and secretion.
3. The enterocytes are constantly being lost into the lumen from the tips of the villi and are replaced every 48 hours.
4. Absorption from the lumen involves movement from the bulk phase across the 'unstirred layer' to the enterocyte surface, movement across the brush border membrane, movement across the cytoplasm, and removal to the blood stream through the basolateral membrane.
5. Water-soluble nutrients of low molecular weight pass through the intestinal epithelium by two major routes, paracellular and transcellular. Paracellular transport is through highly permeable tight junctions joining the brush border membranes, the rate of transport depends upon concentration gradient and solvent drag. The transcellular route is across the brush border membrane of the absorptive cells, followed by crossing the cytosol of the cell interior by diffusion and exit through basolateral membranes by facilitated diffusion.
6. There are two main methods of transport across the cell membrane: (i) simple and facilitated diffusion which are passive transport systems; and (ii) mediated transport, which is specific for each structurally related substance, e.g. amino acid across the membrane.
7. The colon is a conserving organ filled by a large bacterial mass.
8. Movement along and absorption from the gastrointestinal tract is dictated in part by local hormones.

Further reading

Dawson, D. (ed.) (1995) Intestinal mucins. *Annual Review of Physiology*, 57, 547–636.

Johnson, L.R., Alpers, D.H., Christensen, J., Jacobson, E.D. and Walsh, J.H. (1994) *Physiology of the Gastrointestinal Tract*, Raven Press, New York.

Seal, C.J. and Puigserver, A. (1996) Regulation of gastrointestinal secretion. *Proceedings of the Nutrition Society*, 55, 247–317.

Traber, P.G. and Silberg, D.G. (1996) Intestine-specific gene transcription. *Annual Review of Physiology*, 58, 275–98.

10.5 Drugs and nutrition

- Drugs may influence the rate of absorption and metabolism of nutrients.

10.5.1 Introduction

The long-term use of some drugs may influence the absorption and metabolism of some nutrients. Drugs may alter appetite, dietary intake, biosynthesis, absorption, transport, storage, metabolism or excretion of nutrients. Susceptibility to such drug-induced changes may be enhanced during growth, pregnancy and lactation. The concentration of drugs and their interaction with other drugs and the genetic heterogeneity of drug metabolic activity may affect the metabolism of drugs and hence their effect on nutrients.

10.5.2 Effects of drugs on food intake

Digoxin and non-steroidal anti-inflammatory agents may reduce appetite, either systemically through high blood concentrations or locally through gastritis. High plasma concentrations of sulphonamides may also reduce appetite. Sialadenitis (inflammation of salivary glands) has been reported in patients taking phenylbutazone.

10.5.3 Drug effects on hepatic function

Drugs are known to affect hepatic function, either by altering enzyme activity or through more direct effects on liver function. Many drugs are known to induce hepatic hydroxylating enzymes. The liver microsomal hydroxylating system is concentrated in the cytochrome P_{450} system. The P_{450}-dependent system includes enzymes involved in oxidation, dealkylation, deamination and sulphoxidation. Alcohol can induce increases in the metabolic turnover rates of drugs, for example blood concentrations of the anti-convulsant drug phenytoin. A constituent of cigarette smoke 3,4-benzpyrene influences the subsequent hydroxylation of itself to non-carcinogenic products. Vitamin D metabolism is affected by P_{450} activity in the presence of long-term anti-convulsant therapy. Thiazide diuretics can induce a diabetes-like state. Hyperglycaemia and glycosuria can develop in patients taking such drugs.

10.5.4 Drug interactions in the gastrointestinal tract

Many drugs interact within the lumen of the gastrointestinal tract. Some drugs are known to affect intestinal absorption by:

- a direct toxic effect causing morphological changes in the mucosa of the small intestine
- inhibition of mucosal enzymes with or without morphological evidence of mucosal change
- binding and precipitation of micellar components, e.g. bile acids and phospholipids
- altering the physicochemical state of other drugs

Neomycin may cause rapid and extensive microscopic damage to intestinal mucosa. A treatment for gout, colchicine, which is now less commonly used, can induce malabsorption causing small intestinal villous atrophy. Biguanides may reduce or slow the absorption of glucose in part due to the loss of matrix granules from the mitochondria of epithelial cells. This may be a sign of reduced energy metabolism. The anti-tuberculous drug p-aminosalicyclic acid can cause profound diarrhoea, steatorrhoea and malabsorption of xylose, folic acid and vitamin B_{12}. Methyldopa may cause partial villous atrophy in the small intestine with consequent malabsorption of both xylose and vitamin B_{12}.

Drugs which alter gastrointestinal motility alter the rate at which other chemicals, nutrients or drugs are absorbed. This is particularly important when tablet dissolution is a rate-limiting step, for example metoclopramide on digoxin, pethidine or diamorphine absorption. Malabsorption secondary to the ingestion of aluminium and magnesium-based antacids is well known. The polyvalent cations, e.g. Al^{3+} and Mg^{2+} form non-absorbable chelates with certain organic groupings, for example aluminium hydroxide may reduce the absorption of tetracyclines. Iron and tetracycline inhibit the absorption of each other. Liquid paraffin and magnesium sulphate can alter lipid absorption, but the effects are only apparent after long usage.

The anion-binding resin cholestyramine has been shown to influence fat-soluble vitamin absorption, e.g. vitamins A and E. Neomycin can produce a sprue-like malabsorption syndrome producing a malabsorption of fat, cholesterol, carotene, iron, vitamin B_{12}, xylose, glucose and nitrogen. The cationic amino groups in neomycin bind to the anions of detergents which are necessary for formation of micelles in the intestinal lumen, e.g. bile acids. This precipitation results in lipids coming out of solution and being malabsorbed.

Folic acid metabolism may be disturbed by anticonvulsants. The mechanism is altered hepatic microsomal enzyme activity with enhanced degradation of folic acid.

Pellagra is a rare complication of the anti-tuberculosis drug isoniazid which may be due to impaired niacin synthesis secondary to pyridoxine. Pyridoxine deficiency is common with isoniazid therapy leading to peripheral neuropathy. Cycloserine is another pyridoxine antagonist. The mechanism of this pyridoxine deficiency is increased urinary excretion of pyridoxine complexed with the drug or competitive inhibition of pyridoxal phosphate.

Mefenamic acid, a non-steroidal anti-inflammatory drug, is an inhibitor of prostaglandin synthetase and may also cause diarrhoea and malabsorption. The consequence of this may be profound malnutrition, an effect which may occur even after years of trouble free exposure.

The effects of opiates on the intestine are well known. The use of opium and morphine in the treatment of diarrhoea and dysentery is a long-standing therapy. In the 1970s several endogenous peptides were discovered whose action is similar to that of opiates. These are known as **endorphins** and include:

- the pentapeptides methionine enkephalin and leucine enkephalin
- The 31-amino acid peptide β-endorphins
- the dynorphins and β-neo-endorphins.

These are synthesized on three separate multi-component protein precursors, the products of three separate genes. They can be found throughout the central and peripheral nervous systems, in the pituitary and adrenal glands and in the enteric nervous system. The enkephalins are found in the intestine in extremely high concentrations, particularly in neurones of the myenteric plexus. The enkephalins act as neurotransmitters and possibly as circulating hormones. Similar receptors for these opiates exist in the intestine for acetylcholine, catecholamines and histamine. It has been suggested that there are five sub-types of opiate receptor. Drugs such as morphine act preferentially at μ receptors. The enkephalins act preferentially at δ receptors, β-endorphin at ε-receptors and dynorphin at κ-receptors. Opiates affect many sections of the gastrointestinal tract, including the enteric neurones, smooth muscle and epithelium. In all species studied morphine has a constipating effect with a marked increase in the segmenting or non-propulsive contraction of the small and large intestines. Part of the action of morphine in the intestine may be mediated through the central nervous system.

Opiates such as morphine have profound effects on intestinal fluid and electrolyte transport. Morphine increases absorption of fluids as well as sodium and chloride ions. Opiates decrease intestinal secretions by secretagogues, including cholinergic agonists, e.g. carbachol, prostaglandins, vasoactive intestinal peptide and enterotoxins, e.g. cholera toxin and *E. coli* heat-stable toxin. It is possible that the anti-secretory actions of the opioids act through the enkephalin-selective δ opiate receptors in the mucosa. In contrast, receptors in the longitudinal muscle–myenteric plexus appear to be μ or κ receptors.

10.5.5 Drugs which affect metabolism

Dihydrofolate reductase is inhibited by methotrexate, pyrimethamine, trimethamine, trimethoprim, pentamidine and triamterene. Nitrous oxide, the anaesthetic, may affect vitamin B_{12} metabolism by affecting methylcobalamin synthesis.

10.5.6 Drugs associated with increased loss of nutrients

A number of drugs may result in increased loss of nutrients. Oral diuretics including furosemide and ethacrynic acid can cause hypercalcuria, magnesium or potassium deficiency. The psychotropic drug chlorpromazine may influence riboflavin requirements. These chemicals have a similar structure. Chlorpromazine may inhibit the incorporation of riboflavin into flavin adenine dinucleotide through an inhibition of hepatic flavokinase.

Summary

1. Drugs may alter: (i) appetite and dietary intake, for example digoxin, which is plasma concentration-dependent; (ii) biosynthesis and absorption, by affecting liver metabolism, for example of cytochrome P_{450} activity or of phenytoin and folic acid function, intestinal absorption, e.g. of neomycin, and transport, e.g. of biguanides; (iii) intestinal motility, e.g. of codeine phosphate; and (iv) faecal and urinary excretion of nutrients, e.g. of cholestyramine and ethacrynic acid.
2. Susceptibility to such drug-induced changes may be enhanced during growth, pregnancy and lactation.

Further reading

Dollery, C. (1991) *Drugs*, Churchill-Livingstone, Edinburgh.

10.6 Carbohydrate digestion and absorption

- Dietary carbohydrate may be ingested as monosaccharides, disaccharides and complex polymers.
- Digestion requires the hydrolysis of complex polymers to monosaccharides and disaccharides by salivary and pancreatic enzymes.
- Monosaccharides and disaccharides are absorbed through specific intestinal mucosal transport systems.
- Carbohydrates may be metabolized within the enterocytes.
- Specific transport systems, different from those of the luminal side of the intestine, transport disaccharides and monosaccharides across the cell basal membrane to the body.

10.6.1 Introduction

Dietary carbohydrate is a major nutrient accounting for approximately half of the average western diet's energy intake. Some 60% of the carbohydrate is in the form of starch and glycogen; sucrose and lactose may contribute 30% and 10%, respectively. There may also be glucose and fructose in certain foods. Raffinose and stachyose are present in small amounts in beans and are not absorbed in the upper intestine but are fermented in the colon. The polysaccharides are digested by salivary and pancreatic amylase in the lumen of the intestine. Starch digestion is intraluminal. There is further hydrolysis of the glucosyl oligosaccharides by the digestive, absorptive brush border enterocytes.

Intestinal digestive and absorptive function declines with advancing age, so there may be reduced absorption of carbohydrates in about one-third of healthy individuals over the age of 65 years.

10.6.2 Factors affecting gastrointestinal digestion

The osmolarity of a sugar solution influences gastric emptying and intestinal transit time. The higher the osmolarity the slower the gastric emptying time. Other determinants of gastric emptying include duodenal pH, fat and caloric intake, viscosity, the solid content of the meal and whether the carbohydrate is eaten as a mono- or a disaccharide. Unabsorbed sugar results in accumulation of fluid within the intestine and hence a shortened transit time.

10.6.3 Dietary carbohydrate and the intestinal absorption of sugars

Oligosaccharides (glucose, galactose and fructose) are in general absorbed and metabolized after hydrolysis into the basic monosaccharides at the cell epithelial surface. Small amounts of larger molecular weight sugars may pass through the epithelial barrier. The enzymatic breakdown, with the exception of the hydrolysis of lactose, is extremely efficient. Sugar digestion and absorption is influenced considerably by the chemistry of the ingested sugar. Fructose as a constituent of

sucrose is very readily absorbed, whereas fructose as the monosaccharide is not. Glucose may enhance fructose absorption by influencing the intestinal transport system or by stimulating water absorption.

Monosaccharide absorption

Monosaccharides pass across the epithelial lining of the small intestine by three processes: simple diffusion, facilitated diffusion and active transport.

Monosaccharides move along a concentration gradient by simple diffusion to the surface of the enterocyte, passing through the unstirred layer (Figure 10.13). The rate of diffusion and therefore rate of arrival of the nutrient to the brush border may affect absorption kinetics and confine the products of hydrolysis to the cell surface.

Na⁺ active transport hexose

A co-transporter with two binding sites, one for the hexose and the other for the Na^+ ion. The Na^+ ion bound to the co-transporter increases the affinity of the hexose site for glucose and galactose. On the inside of the enterocyte the Na^+ and the hexose diffuse into the cell. The co-transporter then undertakes another transfer cycle. When the glucose co-transporter transfers Na^+ ions across the brush border an electrical potential difference is created across the membrane, the enterocyte and the intestinal wall. This is electrogenic, i.e. potential-producing or **rheogenic**, that is current-producing. The production of electrical activity allows for the measurement of absorption across the brush border and intestinal wall. These are called transfer potentials which follow Michaelis–Menten kinetics and can be characterized by apparent K_m. Electrical measurements are a measure of hexose kinetics. There may be more than one carrier in a membrane. The basolateral membrane is a barrier to free movement of hexoses in and out of the enterocyte.

The water-soluble monosaccharides cross the lipid brush border membrane slowly and inefficiently. The Na^+ active transfer of hexoses has a specific membrane carrier system.

Hexoses move out of the cell by diffusion and a facilitated transfer process which is Na^+-independent.

The transport systems for the absorption of sugars are found throughout biology, being similar in fish and humans. These are found in both the intestine and kidneys. Three major systems are known (Figure 10.13)

- brush border Na^+/glucose co-transporter (SGLT1)
- brush border fructose transporter (GLUT5)
- basolateral facilitated sugar transporter (GLUT2)

These are controlled by two gene families. SGLT1 belongs to a small family whereas the other two belong to a large family of genes. These govern a transporter system which is found in yeasts, plants and mammals. This transporter family includes 12 transmembrane transporters for a variety of sugars and organic acids. The activity of SGLT1 is the basis of the oral rehydration therapy for cholera. In the diarrhoea of cholera there are a series of enterotoxins released by the *Vibrio cholerae* organism which initiate a loss of water from the intestinal cell causing a secretory diarrhoea. An enterotoxin B subunit adheres to specific receptors on the cell surface and an A subunit enters the cell and permanently activates adenylate cyclase resulting in chloride, sodium and then water loss in large amounts. This continues until the cell is sloughed off in the usual turnover of cells. This very serious and life-threatening condition can be treated by oral rehydration therapy wherein sugar increases salt absorption and hence water absorption across the intestinal brush border.

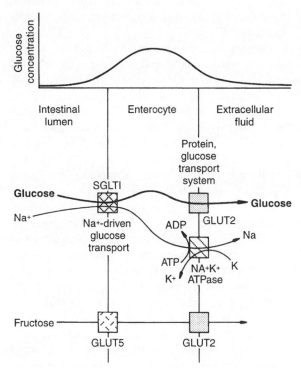

FIGURE 10.13 Glucose absorption across the enterocyte. Glucose passes from the intestinal lumen into the interior of the enterocyte by a Na^+-driven glucose transport system (SGLTI). Glucose passes across the enterocyte and into the interior of the body across the basement membrane along a concentration gradient of facilitated diffusion and a protein glucose carrier system (GLUT2). Sodium passes into the interior of the body through a Na^+ gradient generated by Na^+-K^+-ATPase in the basal membrane of the enterocyte. Fructose is transported across a cell down a concentration gradient across the brush border (GLUT5 brush border fructose transporter). GLUT2 transports fructose across the basement membrane. Galactose is transported by the same transport system as glucose.

Disaccharide absorption

The brush border disaccharidases hydrolyse maltose, sucrose and lactose to their constituent monosaccharides glucose, galactose and fructose. The rate of absorption of these monosaccharide products is the same as if they were monosaccharide solutions. However, there is a kinetic advantage of hexoses being liberated from disaccharides, leading to enhanced absorption rates over that for free hexoses from the lumen. The transport mechanism for glucose produced by sucrose hydrolysis is Na^+-independent. The amount of substrate increases the activity of the disaccharidase in the brush border and even its specific activity. A high dietary intake of sucrose or fructose, but not glucose, increases sucrase and maltase but not lactase activity. Lactase activity does not appear to be regulated by lactose. The monosaccharides released by disaccharide hydrolysis act as inhibitors of the disaccharidases. The greater the amount of monosaccharide formed, the greater will be the inhibition of the carbohydrase and reduction of activity, that is a local negative feedback system.

The brush border enzymes are specific for specific glycoside linkages. Brush border enzymes are specific for α-1–4, α-1–6 and β-1–4 linkages. The exceptions are sucrase and isomaltase which hydrolyse both sucrose and maltose.

Disaccharidases

The disaccharidases are all large protein heterodimers or single subunits, anchored with transmembrane domains, with most of the protein protruding into the intestinal lumen. These oligosaccharidases are synthesized as larger glycosylated precursor enzyme proteins. They pass to the apical membranes and are cleaved into active enzymes by the pancreatic enzymes. Large oligosaccharidases are removed from the enterocyte apical surface by pancreatic enzymes or by membrane shedding.

Lactose

Lactose is split by the brush border enzyme **lactase**. This is not an inducible enzyme. Both galactose and glucose are absorbed by active transport. Surface hydrolysis appears to be rate-limiting.

Lactase biosynthesis differs from other brush border hydrolases by initially being formed as a large precursor which is cleaved intracellularly. Lactase is processed intracellularly and rapidly to a mature 160-kDa enzyme. Complex glycosylation of the large precursor occurs before proteolytic cleavage. The lactase is anchored to the brush border membrane by a hydrophobic segment of its protein structure. The regulation of

lactase activity varies from species to species and race to race in humans.

Foetal intestinal lactase develops late in gestation and only achieves maximal activity at birth. Premature infants at 29–38 weeks have reduced lactase activities whereas other intestinal carbohydrases achieve adult concentrations by the time of birth. Lactase levels remain at a constant amount during the suckling period in mammals, but decline after weaning between the ages of 2 years and puberty. The jejunal brush border enzymatic activity decreases with age, whereas ileal activity increases.

Primary lactase deficiency is the norm throughout the human world and is of the order of 87% in China, 75% in Greece, 100% in Japan, 100% in Thailand, 100% in black Africans, 90% in South American Indians and 73% in black Americans. This is in contrast to the Caucasian races with 3% in Denmark, 5% in England, 16% in Finland, 20% in Northern France, 40% in Southern France, 50% in Northern Italy, 70% in Southern Italy and 3% in Sweden. The difference between white and aboriginal Australian primary lactase deficiency is 5% in the European stock and 70% in the Aboriginal stock. The persistence of lactose absorption is an autosomal dominant characteristic. Lactase deficiency is inherited as a single autosomal recessive gene which in the homozygous state suppresses the synthesis of intestinal lactase.

Similarly, congenital sucrase–isomaltase deficiency is inherited as an autosomal recessive defect. Heterozygotes have enzyme capability intermediate between total inactivity and individuals with full activity. They may also have sucrase intolerance.

Sucrose

Sucrose is absorbed by the brush border of the small intestine and is hydrolysed to glucose and fructose by sucrase. This is an inducible enzyme dependent upon the intake of sucrose. Fructose is absorbed by a facilitated diffusion mechanism and cannot be absorbed against a concentration gradient. It is absorbed at an appreciably slower rate than glucose.

Eskimos have the highest frequency of congenital sucrase–isomaltase deficiency at 10% of the population.

Sugar alcohols

Maltitol, isomaltol and lactitol, which are disaccharide sugar alcohols are only partially hydrolysed in the small intestine. The absorption of monosaccharide sugar alcohols xylitol, sorbitol and mannitol is by a passive diffusion process and they are absorbed less efficiently than glucose.

Oligosaccharides

The oligosaccharides raffinose and stacchyose are not hydrolysed and are absorbed in the large intestine after fermentation.

Starch

The plant of origin and the chemistry of the starch as digested affects the digestion of starches.

Ungelatinized starch

There is very variable pancreatic digestibility of ungelatinized starch granules. Cereal starches respond to pancreatic α-amylase more readily than legume and tuber starches. The digestibility of a starch is inversely related to the amylose content. Cooking removes this difference.

Gelatinized starch

Starch granules heated to 60–70°C in water swell, resulting in a weakening of intermolecular links. This allows amylose and amylopectin to form colloidal dispersions. When the starch cools there is reassociation to regions of varying polymer density. Cooking enhances the susceptibility of starches to amylase hydrolysis. The degree of gelatinization is affected by both cooking and the amount of water available during the process, and also varies within a given product, being greater in the soft parts of bread than in the crust. Within potato chips the degree of gelatinization on the outside is different to the inside. The coincidental presence of sucrose and fat may also reduce the extent of starch gelatinization. High amylose corn (amylomaize) is poorly digestible after cooking while waxy cereal starches are very digestible. Vegetables contain 30–40% of amylose starch. Starches from legume seeds (lentils and beans) are digested more slowly than starch from bread. This may be related to the entrapment of starch in the legume cell structure, preventing a complete swelling during cooking. The degree of milling is important in the digestibility of a cereal starch. Finely milled wheat flour is digested faster than the coarsely milled equivalent. While the presence of a fibrous cellular structure surrounding the starch itself is not sufficient to restrict carbohydrate digestion, the fibre may restrict access by the hydrolytic enzyme to the starch. Food processing with high temperatures and shear stress affects starch digestion so that commercially canned beans hydrolyse more quickly than home-cooked preparations. Extrusion cooking and explosion puffing are food manufacturing processes which alter the digestibility of starch in corn, rice or potato. Some starchy foods may produce a viscous microenvironment in the intestinal lumen. This affects the access of amylase to starch and the diffusion of hydrolytic products towards the intestinal mucosa.

Starch consists of two polysaccharides: (i) the linear 1–4-linked α-D-glucose (amylose); and (ii) the highly branched amylopectin with α-1–4 and α-1–6 glycosidic links. Both are found in the plant as insoluble semicrystalline granules. Salivary and pancreatic α-amylases act on the endo-α-1–4 links within starch, but do not digest the exo-glucose–glucose linkages (Figure 10.14). Salivary amylase cleaves α-1–4 glycosidic bonds. The protein is 94% identical with pancreatic amylase and is found in glycosylated and non-glycosylated forms. The free salivary amylase is inactivated at low pHs found in the stomach. However, starch and its end-products prevent this inactivation, partly by the bulk of the food which prevents the gastric acid from permeating throughout the food complex in the stomach. There is a stabilization of the pH above pH 4 in the stomach. Consequently some salivary amylase passes through the stomach without inactivation. Pancreatic α-amylase does not hydrolyse the α1–6 branching unit and the α-1–1 links adjacent to these branching points. Consequently, large oligosaccharides (α-limit dextrins) with five or more glucose units containing one or more α-1–6 branching links are produced by amylase action (Figure 10.14). The products are maltose, maltotriose, and α-limit dextrose which have 5–10 glucose units, but not glucose. The amylase is unable to split at links near α-1–6 branch points. Consequently α-1–4-linked disaccharides, maltose and trisaccharides (maltotriose) are final linear breakdown products. These small end-products are cleaved by brush border enzymes. There are active carbohydrases in the columnar epithelial cells of the duodenum with reducing activity in the ileal villus. The glucoamylase removes single glucoses from the non-reducing end of α-linear α-1–4 glucosyl oligosaccharide. α-Limit dextrins are split by the combined action of glucoamylase and sucrase–isomaltase. The released glucose is absorbed by the intestinal mucosal cells by a carrier-mediated process.

Starch may be classified nutritionally by the ability of enzymes to hydrolyse the material (Table 10.2):

- **Rapidly digestible starch.** Amorphous and dispersed starch which occur in freshly cooked starchy food.
- **Slowly digestible starch.** This form, while being completely digested in the small intestine, is digested slowly, over an hour or more. This includes most raw cereals and physically inaccessible amorphous starch.
- **Resistant starch** which is not hydrolysed by enzymes and includes (i) physically inaccessible starch, e.g. partly milled grain and seeds; (ii) resistant starch granules, e.g. raw potato and banana; and (iii) retrograded starch, e.g. cooled cooked potato, bread and cornflakes.

Resistant starch
Starch may retrograde to a form which is highly resistant to hydrolysis by α-amylase. In this way, hydrogen bonds are reformed and gelatinized starch partially recrystallizes. The proportion of starch in the enzymatically resistant form depends upon the food source and cooking. Resistant starch passes to the colon as it is variably hydrol-

ysed in the duodenum. In the colon is a large bacterial mass, the composition of which appears to be relatively stable and individual to any particular person. The microflora produce a wide range of enzymes to which most carbohydrates are susceptible. Under the conditions of the colon the major digestion products are: carbon dioxide, hydrogen and methane, short-chain fatty acids, acetate, propionate, butyrate, and organic acids such as lactic acid, pyruvic acid and possibly other breakdown products.

10.6.4 Complexes and interactions with other food components

Inclusion complexes may be formed between the helices of amylose and polar lipids. These may occur during processing or naturally as in cereal

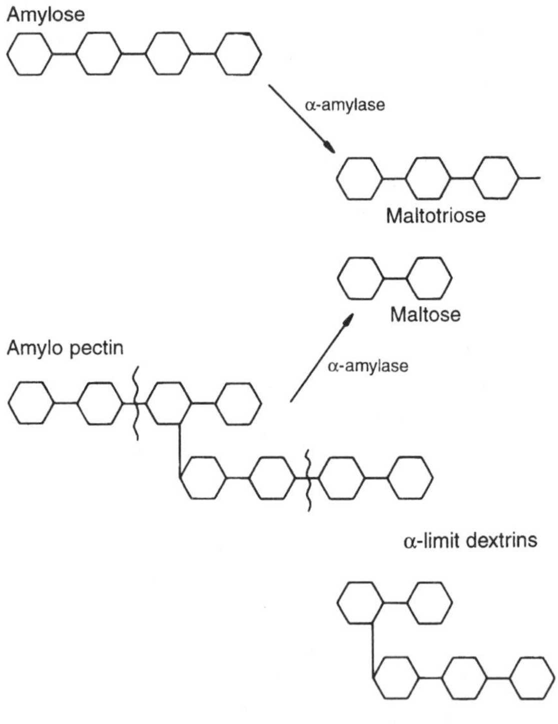

FIGURE 10.14 Digestion of starch. Amylose and amylopectin are digested by α-amylase. Amylose yields maltotriose and maltose, amylopectin maltotriose, maltose and α-limit dextrins, the limitation being the α-(1–6) branch points.

starches. Such amylose–lipid complexes may be formed during extrusion cooking resulting in a relative resistance to amylase *in vitro*. It has been suggested that in wheat flour there is a complex physical or chemical interaction between starch and gluten which affects digestibility. The gluten matrix which surrounds the gelatinized starch granules may limit access to amylase and reduce starch availability. Spaghetti has a high gluten content and is less readily digested than other wheat products with less gluten such as bread. The digestibility of starch is decreased by Maillard reactions.

Food processing, toasting, and excess heating storage all alter starch format and ideally enhance its digestibility by enteric enzymes. Other factors which affect starch digestibility include amylase inhibitors, lectins, phytic acid and tannins.

Maillard reaction

This is the non-enzymatic browning through a reaction between the amino acids of the protein, usually lysine or methionine and the carbonyl group of a sugar, the final reaction of which includes a polymerization reaction. The degree of browning is affected by process times and temperature, moisture and the presence of proteins and sugars as reducing agents.

10.6.5 Assessment of dietary carbohydrate digestibility

Tolerance tests

Carbohydrate loads of approximately 50 g are ingested, serial blood sugar estimations are there-

---

TABLE 10.2 Englyst has classified starch in foods by their ease of hydrolysis by amylase. The physical form of starch influences the resistance to hydrolysis by amylase and subsequently as a substrate for colonic microflora. Starches resistant to amylase pass to the colon and become a substrate for large intestinal bacteria. (From Englyst and Kingman, 1990).

Physical form of starch in food	*Examples of food*	*Susceptibility to hydrolysis by amylase*	*Available as substrate for microflora in large intestine*
Raw			
Granules: A structure	Most uncooked cereals	Readily hydrolysed	+
Granules: B and C	Potato and banana	Partially resistant to hydrolysis	++
Cooked			
Physically inaccessible	Granules within intact cell walls: whole or broken grains, many legumes	Resistant to hydrolysis	+++
Dispersed amorphous	Freshly cooked foods	Readily hydrolysed	+
Retrograded amylopectin	Cooled cooked potato	Partially resistant to hydrolysis	++
Retrograded amylose	White bread and many processed cereals	Resistant to hydrolysis	++++
Gelatinized starch dried at high temperature	Some cooked cereal foods	Readily hydrolysed	+++++

Granule structure is characterized by X-ray diffraction. The A type is associated with cereals and manioc (cassava) starches. The B types are found in tubers and legumes; amylomaize is also a B type. C is a mixture of A and B.
Rating of importance as a colonic bacterial substrate:
+, small amount reaches colon; +++++, majority reaches colon.

after obtained and a blood glucose curve is produced (Figure 10.15). Changes in blood glucose brought about may be estimated from carbohydrates of different food sources. In this way a **glycaemic index** is calculated from the area under the blood glucose curve after ingestion of 50 g of test carbohydrate compared with the area under the blood glucose curve following a 50 g standard carbohydrate (glucose or white bread). There is, however, a large scatter in results.

The glycaemic response to plant foods will depend on the source of the plant food, physical form, and the mode in which it is cooked. There are modifying effects on the absorption of starch by non-starch polysaccharides, fat and protein, through quite different mechanisms. Fats and proteins reduce the glycaemic response to carbohydrate by increasing insulin secretion in response to hyperglycaemia, and also by effects on gastric emptying. The hyperglycaemic effect of sweets and chocolates is reduced by the coincidental presence of fat and protein. Similarly, the glycaemic response to bread with butter is modified compared with bread on its own. These differences are due in part to changes in gastric emptying time and also to alterations in insulin response.

Dietary fibres, when they do act on absorption, may provide a mechanical barrier to the digestive juices acting on the starch granules, or interfering with the absorption of glucose through physicochemical mechanisms. This may involve interaction between the solid and soluble fibre and the glucoreceptor or the transporters on the surface of the enterocytes. In terms of increase in the blood glucose, glucose is the most potent. Sucrose, lactose, galactose and fructose have an effect on blood glucose in descending order. Pre-formed maltose is absorbed from the intestine even more quickly than glucose, yet when maltose is formed as an intermediate in the intraluminal hydrolysis of starch, this difference in absorption of maltose is not discernible.

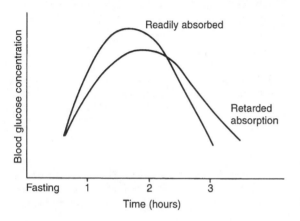

FIGURE 10.15 Glucose tolerance test. (a) Readily absorbable; (b) retarded absorption preparation of glucose. The readily absorbable glucose is absorbed in its free form, while the retarded absorption glucose is transported as a complex carbohydrate or in its natural form in food.

Summary

1. In many diets, the carbohydrate is 60% in the form of starch and glycogen, 30% as sucrose and 10% as lactose. Raffinose and stacchyose are present in small amounts in beans and are not absorbed in the upper intestine but are fermented in the colon.
2. The osmolarity of a sugar solution influences gastric emptying time and intestinal transit time. Other determinants of gastric emptying include duodenal pH, fat and caloric intake, viscosity, the solid content of the meal and whether the carbohydrate is a mono- or a disaccharide.
3. Oligosaccharides in general are efficiently absorbed and metabolized after hydrolysis into the basic monosaccharides. The hydrolysis of lactose is relatively slow.
4. Fructose is not absorbed readily as the monosaccharide, but as a constituent of sucrose is very easily absorbed.
5. Monosaccharides cross the intestinal epithelium by one of three processes: simple diffusion, facilitated diffusion and active transport. The Na$^+$ active transfer of hexoses has a specific

membrane carrier system, a co-transporter with two binding sites: one for the hexose and the other for the Na$^+$ ion.

6. Three major transfer systems are known: (i) a brush border Na$^+$/glucose co-transporter (SGLT1); (ii) a brush border fructose transporter (GLUT5); and (iii) a basolateral facilitated sugar transporter (GLUT2)

7. The brush border disaccharidases hydrolyse maltose, sucrose and lactose to the monosaccharides glucose, galactose and fructose. There is a kinetic advantage of hexoses being liberated from disaccharides, namely enhanced absorption rates from the lumen over those for free hexoses.

8. The brush border enzymes function on specific glycoside linkages. The disaccharidases are all large protein heterodimers or single subunits, anchored with transmembrane domains, most of the protein protruding into the intestinal lumen.

9. Lactase is an unusual enteric enzyme which is present at birth and persists through life for Caucasian races but not in most other races.

10. Starch consists of the linear α1–4-linked α-D-glucose (amylose) and the highly branched amylopectin with α1–4 and α1–6 glycosidic links. Salivary and pancreatic α-amylases act on the endo-α1–4 links within starch, but do not digest the exo-α-glucose–glucose linkages.

11. Starch may be classified nutritionally on the basis of its enzymatic hydrolysis: (i) rapidly digestible starch; (ii) slowly digestible starch; and (iii) resistant starch (which is not hydrolysed by pancreatic enzymes). Starch which is not digested in the intestine passes to the colon for fermentation by colonic bacteria.

12. Tolerance tests of carbohydrate loads allow changes in blood glucose to be estimated for carbohydrates of different food sources and compared against a standard carbohydrate (glucose or white bread). This allows a glycaemic index to be calculated.

Further reading

British Nutrition Foundation's Task Force Report (1990) *Complex Carbohydrates in Foods*, Chapman & Hall, London.

Crane, R.K. (l968) Digestive-absorptive circuits of the small bowel mucosa. *Annual Review of Medicine*, 19, 57–68.

Englyst, H.N. and Kingman, S.M. (1990) Dietary fiber and resistant starch. A nutritional classification of plant polysaccharides, in *Dietary Fiber* (eds D. Kritchevsky, C. Bonfield and J.W. Anderson), Plenum Publishing, New York.

Feibusch, J.N. and Holt, P.R. (1982) Impaired absorptive capacity for carbohydrate in the ageing human. *Digestive Diseases Science*, 21, 1095–100.

Jenkins, D.G.A., Wolever, T.M.S. and Taylor, R.H. *et al.* (1981) Glycemic index of foods: a physiological basis for carbohydrate exchange. *American Journal of Clinical Nutrition*, 34, 362–6.

Johnson, L.R., Alpers, D.H., Christensen, J., Jacobson, E.D. and Walsh, J.H. (1994) *Physiology of the Gastrointestinal Tract*. Raven Press, New York.

Levin, R.J. (1989) Dietary starches and sugars in man: a comparison, in *Human Nutrition Reviews; ILSI Europe* (ed. J. Dobbing), Springer-Verlag, London.

Parsons, D.S. (1976) Closing summary, Appendix 2, Unstirred layer, in *Intestinal Ion Transport* (ed. J.W.L. Robinson), MTP Press, Lancaster, pp. 407–30.

10.7 Protein absorption

- Amino acids are ingested as proteins and peptides.
- These proteins and peptides are hydrolysed in the stomach and duodenum by gastric and pancreatic enzymes.
- Dipeptides and amino acids are absorbed through specific absorptive intestinal mucosal transport systems.

- There is metabolism of dipeptides and amino acids within the enterocyte.
- Specific transport systems, different from those of the luminal side of the intestine, transport amino acids and peptides across the cell basal membrane to the body.
- Minute amounts of intact protein may be absorbed particularly in the infant.

10.7.1 Introduction

The dietary protein intake is approximately 70–100 g/day, with approximately 50–60% possibly of animal origin. In addition, 20–30 g of endogenous proteins, 30 g of desquamated cells and 1–2 g plasma proteins (1–2 g as albumin), enzymes and mucoproteins are secreted into the intestine. Protein of endogenous origin is in general digested and absorbed more slowly than exogenous protein. The faecal excretion of protein-derived nitrogen is about 10 g/day or less, demonstrating an effective absorption of protein, which is of the order of 95% in the small intestine. This depends upon the specific protein, for example cooked haricot bean protein absorption is poor. Faecal protein is largely bacterial in origin, whereas faecal nitrogen is of endogenous origin.

Protein absorption involves the breakdown of protein to tripeptides, dipeptides and amino acids. The site of maximal peptide or amino acid absorption may differ along the intestine and is species-dependent. The electrical gradient across the brush border is steeper in the jejunum than in the ileum. While most ingested protein is absorbed in the jejunum, there is evidence of protein being absorbed in the ileum and some, albeit a small amount, passing to the colon. Following a protein-containing meal there appear to be more intraluminal amino acids in the ileum than in the jejunum, suggesting that peptidases enter the ileal lumen. Absorption of amino acids from the ileum in humans may be more significant than peptide absorption. The colonic mucosa appears to be an effective absorber of amino acids. Nitrogen absorption from the colon consists of the fermentation products of bacterial metabolism. It is only possible to estimate how much protein entering the colon is of exogenous origin and how much is endogenous. It is possible

that 3–24 g of protein passes into the colon each day and of this, 40–60% is endogenous.

10.7.2 Digestion and absorption of proteins

There are six main phases in the digestion and absorption of proteins:

1. Whole protein absorption
2. Intraluminal digestion of protein and its breakdown products polypeptides, resulting from the sequential actions of the proteolytic enzymes of the stomach and pancreas
3. cellular uptake of protein hydrolysis products: (i) amino acids; (ii) peptides
4. Brush border digestion of small peptides
5. intracellular metabolism
6. The transfer of dipeptides and amino acids from the intestinal brush border to the blood stream.

Whole protein absorption

It is possible that 2% of a dietary protein intake is absorbed as whole protein. The absorption of whole proteins involves **pinocytosis**. This type of protein absorption is not thought to be important in adults, but may be more so neonates. Pinocytosis is the binding of a substance to a membrane and the membrane–substance complex is then brought into the cell interior as a whole vesicle which fuses with lysosomes. The pinocytotic vesicles formed from the brush border membrane fuse with lysosomes to form phagolysosomes in which there is some protein hydrolysis. Proteins which have escaped hydrolysis enter the intercellular spaces by exocytosis and reach the blood stream by the lymphatics. A mechanism of immunological significance is the pinocytotic uptake of antigens by membrane epithelial cells (M-cells) which are found in the columnar cells overlying the Peyer's patches. The M-cells which are related to lymphocytes (B- and T-cells), enter the blood stream through the lymphatics and travel to the intestinal mucosa and other sites, where they mature into the IgA plasma cells, important in local and secretory defence mechanisms. There is a significant utilization of nitrogen by the daily secretion of IgA (some 2–3 g/day in the adult male).

Intraluminal digestion

Protein digestion begins in the stomach with the enzyme pepsin which belongs to a class of enzymes aspartic proteinases, an enzyme group present in many forms of life (Figure 10.16). Another aspartic proteinase is renin, an example of which – chymosin (renin) – is found in neonates. The precursor pepsinogen is found as multiple isoenzymes. Proteolytic activity requires a pH less than 4. Pepsin is activated by the acid conditions of the stomach and inactivated in alkaline conditions, e.g. duodenum or the stomach in which acid production is suppressed. The pepsin produced activates the precursor pepsinogen by cleavage of an N-terminal peptide (44 amino acids) from the pepsinogen. The result is a mixture of very large polypeptides with molecular weight of many thousands (Figure 10.16). There is no gastric protein absorption.

The physical form of the protein solid or liquid phase may alter the rate of gastric emptying. The rate of intraluminal digestion is a rate-limiting factor in the absorption of proteins. Individual proteins are absorbed at very different rates. Casein disappears from the rat intestine three times faster than gliadin, which may be slowly attacked by trypsin. Gastric emptying time dictates the rate of release of protein into the small intestine and hence rate of digestion and absorption. The size of meal, chemical composition, presence of other nutrients and osmolality all affect the ease of digestion.

The next phase is the activation of proteolytic enzymes, secreted in an inactive form by the pancreas, a process initiated by enterokinase (Figure 10.16). Enterokinase converts the precursor trypsinogen to trypsin by the cleavage of a small terminal peptide. There is a positive feedback loop wherein the resultant trypsin activates more trypsin. The optimal pH is 7.5. Protein digestion involves hydrolysis of the peptide (CO–NH) bonds which link amino acids.

Pancreatic enzymes

The proteolytic enzymes of the pancreas may be classified according to the bonds hydrolysed:

- **Endopeptidases** split peptide bonds within the chain. Trypsin yields peptides with the basic amino acids (arginine, lysine) at the C-terminal end, while chymotrypsin A,B,C and elastin produce peptides with neutral amino acids at the C-terminal end.
- **Exopeptidases** hydrolyse C-terminal bonds at the end of protein and peptide chains. The enzymes act in concert, producing oligopeptides, and a proportion of free amino acids. The rate of release of individual amino acids varies widely. The release of arginine and lysine and many neutral amino acids is rapid; that of glycine, proline and the acidic amino acids glutamate and aspartate is slow or very slow.
- **Carboxypeptidases** release amino acids and oligopeptides 2–6 amino acids long, yielding 40% amino acids and 60% peptides.

Brush border enzymatic activity is stimulated by trypsin and removed from the apical membrane assisted by the action of bile acids. This influences brush border enzyme activity.

Intraluminal hydrolysis is far from complete. Some proteins are much less readily digested than others. Some are resistant to hydrolysis with the result that their subsequent absorption is incomplete. In the proximal jejunum in humans the rate of release of amino acids from milk powder proteins varies with lysine being released 140 times faster than glycine and 14 times faster than glutamic acid. More distally the variations are less. There are very small differences between the patterns of free amino acids in the lumen and the amino acid composition of the protein.

Vegetable proteins are less rapidly digested than those of animal origin. Overheating of proteins during storage and preparation, Maillard reactions and other complex chemical changes result in digestive products of small peptides which are not only resistant to hydrolysis but are also poorly absorbed. This may occur with dried milk which has been roller-dried. Gelatin is also resistant to later stages of hydrolysis. Many proteins which

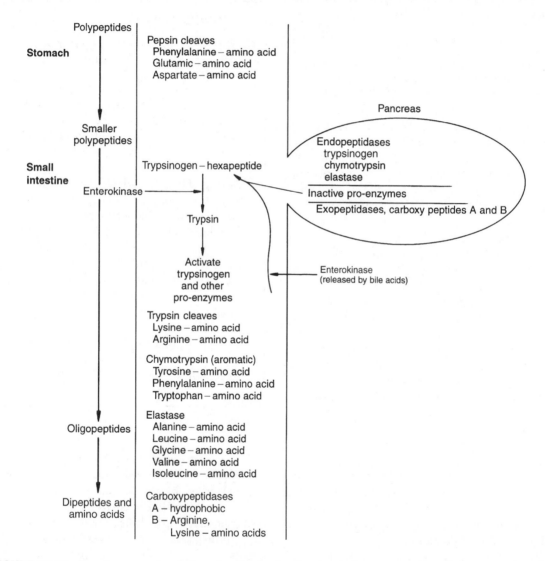

FIGURE 10.16   Proteins are sequentially reduced in molecular size during passage through the stomach and the upper gastrointestinal tract. Pepsin initiates the enzymatic process. The pancreas excretes inactive pro-enzymes which are activated by enterokinase, a mucosal enzyme. This activates trypsinogen which in turn activates chymotrypsin, elastase and carboxypeptidases A and B. Enzyme activity with an amino acid (specified) – amino acid peptide linkage.

have been used in physiological studies are unusual in their digestive behaviour. Casein digestion liberates phosphopeptidases which are initially absorbed but which resist further cellular digestion. Raw egg white and soy and other varieties of beans contain a trypsin inhibitor. Cooking may well increase the digestibility of proteins, and also inactivate trypsin inhibitors.

There is an enhanced peptide and amino acid absorption after **starvation**. There are changes in the thickness of the unstirred layer and possibly increases in the number of carrier proteins during starvation. There are diurnal and even seasonal changes in the intestinal ability to absorb nutrients. All of these changes may complicate experimental results.

Cellular uptake of protein hydrolysis products

Proteins are absorbed from the intestinal lumen largely in the forms of the predominantly small peptides and the hydrolysis products amino acids resulting from intraluminal digestion. Small quantities of whole proteins are also absorbed and enter the circulation in trace amounts. The absorptions of peptides and amino acids are complementary processes. Further hydrolysis of peptides occurs through the action of peptidases at the absorptive surface, the products being tripeptides, dipeptides and amino acids (Figure 10.17).

Ingesting a mixture of free amino acids is less efficient in maintaining N balance than an equivalent supply of amino acids in the form of peptides. This may in part be due to the amino acids being absorbed and delivered to and thus utilized by tissues at differing rates. This discordant delivery and utilization pattern is not optimum for protein synthesis.

Proteins of animal origin are more readily digested than vegetable proteins and yield a larger proportion of small peptides and amino acids. Egg protein is digested and absorbed with an efficiency of 85–90%. Casein, like other phosphoproteins, is relatively resistant to hydrolysis. Proteins with a high proline content, e.g. gluten or casein, are relatively resistant to pancreatic enzyme action. The nutritive value of a protein is

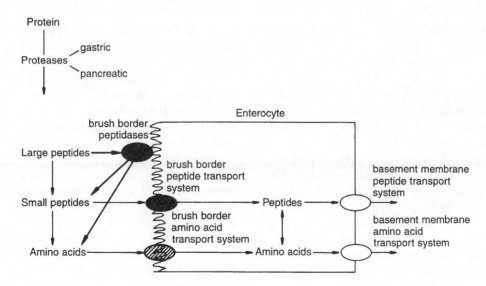

FIGURE 10.17 Digestion and absorption of protein. Proteins are cleaved by proteases. Large peptides are split by brush border peptidases to produce small peptides and amino acids. Small peptides are absorbed by brush border peptide transport systems. Amino acids are absorbed by specific brush border amino acid transport systems. Peptides and amino acids are transported into the body by basement membrane peptide transport systems and amino acid transport systems.

dependent upon its amino acid composition and digestibility. Peptide bonds resistant to hydrolysis may be important in ultimately determining the nutritive quality of a protein.

In general the rate of mucosal absorption from a partial hydrolysate of protein will be twice as fast as a mixture of equivalent amino acid content. It is possible that the absorption of peptides shows less intense competition than that of amino acids. The kinetics of brush border hydrolysis, rather than the kinetics of peptide or amino acid transport, govern the rate of absorption of the longer-chain peptides.

The relative rates of absorption of amino acids from a peptide and from an equivalent mixture of amino acids are strongly influenced by concentration. The relative rates of absorption of peptides and amino acids are dependent upon:

- the kinetics of mediated transport of peptides and amino acids
- the substantial uptake of amino acids released at the brush border; two independent transport systems become involved which accelerates total absorption as the two systems are not readily saturated.

Glucose usually stimulates the intestinal transport of amino acids, possibly due to an increase in water movement associated with glucose transport.

Amino acids have been reported to appear in the portal blood in approximate proportions to the amounts ingested. However, intraluminal digestion of protein releases amino acids in the free form at rates which differ very widely, not at rates proportional to their content in the ingested protein. There is also the possibility of substantial dilution or even complete swamping of exogenous by endogenous protein. Also free amino acid absorption – a minor mode of absorption of protein hydrolysis products – progresses at a slower rate than that of peptides.

Amino acids

Free amino acids are transported into the absorptive cells by a number of well-defined mechanisms which are mainly active and Na-linked. Several absorption mechanisms have defined specificity for certain groups of amino acids with common structural features. Amino acids, whether taken up as such or released in the absorptive cells following hydrolysis of peptides, pass from the cell to the blood by mechanisms in which facultative diffusion predominates. The amino acid transport system in the apical membrane is not the same as that of the basolateral membrane. The two transport systems are in contact with fluids and conditions of quite different composition. Some amino acids are extensively metabolized within the absorptive cell and all are utilized to some extent for protein synthesis which is very active. If the amino acid is neutral then electrical potentials are not relevant.

> Basic amino acids lysine and arginine have a positive charge in the physiological pH range. The interior of the cell is electronegative relative to the exterior, so they concentrate in the cell as a result of differences in electrical potential.

Charge is not important in dictating rate of mediated transport. Only amino acids are transported by the amino acid transport system. Multiple active transport systems are involved in the absorption of amino acids. There may well be five systems involved in the transport of neutral amino acids, imino acids and basic acids in the brush border. The system is complicated and incompletely understood.

The traditional 'main transport system' for neutral amino acids may possibly incorporate more than one system. The main transport system for most neutral amino acids requires a substrate with an uncharged side chain and a primary amino acid in the α-position. Proline and hydroxyproline are exceptions to this rule. The system has a strong preference for L-amino acids but does not transport basic or acidic amino acids, nor β-alanine.

The larger and bulkier the side chain of the amino acid and the stronger the lipophilic properties, the greater the apparent affinity for transport by the main system. Glycine and other amino acids with short aliphatic side chains may utilise the imino system and some may use the basic system.

In general the brush border transport of amino acids and peptides appear to be independent processes with little or no interaction between them.

> The imino system carries proline, hydroxyproline and N-methylated forms of glycine and to some extent glycine, β-alanine, γ-aminobutyric acid (GABA) and taurine. In humans a defect in this system leads to iminoglycinuria with impaired absorption of glycine, proline and hydroxyproline. There is an active intestinal transport system for the basic lysine, arginine and ornithine and for cystine. The investigation of the systems transporting acidic amino acids was hindered by the extensive transamination of glutamate and aspartate in the mucosa. They are transported by the same Na-coupled system.
>
> D-isomers appear to utilize the same transport systems as their comparable L-form, but with reduced affinity characteristics.

All animal cell membranes contain a large number of transport systems for amino acids differing in chemical and net charge. If one system is defective, another is there to retrieve the situation. There is a small role of non-mediated transport in the absorption of amino acids.

While much non-mediated transport takes place through water-filled channels, amino acids have a degree of lipid solubility which may allow some permeability through the lipids of the cell membrane. This is not likely to be a significant proportion. Diffusion through water-filled channels is likely to be greater with lower molecular weight amino acids. In diffusion through lipid membranes then the degree of lipid solubility of the amino acid becomes important. With amino acids this will be one with a large molecular volume and a bulky side chain and relatively lipophilic properties, e.g. phenylalanine > methionine > alanine > glycine. The transport of basic and acidic amino acids will be complicated by the effects of charge, e.g. the interior of the cell is electronegative relative to the exterior. Small lipid-insoluble molecules of sphere radius less than 0.4 nm and also a range of larger hydrophilic molecules cross the epithelium more slowly by non-mediated transport, e.g. hydrophilic cyanocobalamin (molecular weight 1357 Da).

> The paracellular pathways are believed to behave as water-filled channels with an equivalent pore radius of 0.4–0.8 nm. The pore size is said to be larger in the jejunum than in the ileum or colon (0.3 nm). Therefore the permeability of the intestine to water and water-soluble substances is higher proximally than distally. There are, in both the mucosal and serosal sides, a small number of electronegative pores with a larger size (6.5 nm) occupying 1% of the mucosal surface. Also there are cationic-selective pores (0.7 nm) with a negative charge, and also electroneutral pores filled with water which are 0.4 nm in diameter.

The absorption of certain amino acids is through highly specific mechanisms indicating a mediated absorption. The rate of absorption of amino acids from mixtures of free amino acids is related to their concentration and to the overall kinetics of their intestinal concentrations which is influenced not only by the kinetics of transport of each amino acid but also by mutual inhibitory and stimulatory effects. Methionine, leucine and isoleucine are absorbed quickly; glycine, threonine, glutamate and aspartate are absorbed relatively slowly regardless of the mixture. The rates are related to relative affinity of transport K_t.

Some of the amino acids with a strong transport affinity can be inhibitory to amino acids with a lower affinity. Neither essential amino acids nor peptides containing essential amino acids are necessarily those which are absorbed the most quickly.

Transport defects of amino acids, e.g. Hartnup's disease (dipolar amino acid) and cystinuria (cationic amino acids and cystine) are overcome

The amino acid transport systems (Table 10.3) can be classified as:

- **System B.** This is the major transport system for the transport of dipolar amino acids with the amino group in the α-position. Imino acids, β-dipolar, basic and acidic amino acids are not transported by this system. There is a transmembrane Na^+ gradient which uses an electrical potential. This system is found very early in foetal intestinal development.
- **System $B^{o,+}$.** This system transports dipolar amino acids, basic amino acids and cystine. The system is dependent upon a transmembrane Na^+ gradient and the transport system is electrogenic.
- **System $b^{o,+}$.** This is a high-affinity, Na^+-independent system for dipolar and basic amino acids. This system is found throughout the small intestine but not the colon.
- **System y^+.** Transports basic amino acids by a Na^+-independent system.
- **Imino system.** This system is exclusively for imino acids, e.g. proline, hydoxyproline and pipecolic acid. It is found throughout the small intestine and is Na^+- and Cl^--dependent.
- **β-system.** This system has a high affinity for taurine and other β-amino acids. There is no affinity for α-amino acids. The requirement is for Na^+ and Cl^- and it is very sensitive to the presence of calcium.
- **System X^-_{AG}.** This system is exclusive for the acidic amino acids aspartate and glutamate. The system is dependent upon Na^+.

when the amino acids are incorporated into peptides. Amino acids do not competitively interfere with absorption of peptides of similar amino acid composition. The features which make peptides resistant to hydrolysis may be associated with poor transport, e.g. D-amino acid residues.

Dipeptides and tripeptides

The transport of peptides and amino acids appears to be independent. Peptide absorption is probably the result of peptide transport followed by hydrolysis, supplemented by hydrolysis at the brush border membrane and is followed by uptake of amino acids from free solution.

There is much evidence that active transport is a major mechanism of transmembrane transport of peptides, and is probably limited to di- and tripeptides. Chain length is the limiting factor rather than molecular volume. There is evidence, however that biologically active peptides of very long chain length may be absorbed possibly by non-mediated transport or by mechanisms related to the uptake of whole proteins.

Peptide transport is stereospecific, favouring peptides containing only L-amino acids or glycine. Peptides with D-amino acids are poorly transported and hydrolysed. Methylation, acetylation or other substitutions of the amino-terminal group reduce the affinity for transport. Substitution of the carboxy-terminal group may also reduce affinity for the transport.

Free amino acid affinity for absorption transport systems is increased with increasing side chain bulk. Dipeptides of those neutral amino

TABLE 10.3 Brush border enterocyte membrane absorption systems for amino acids

Substrate						Gradient	
Acidic amino acids	Dipolar α-amino acids	Basic amino acids	Cystine amino acids	Imino amino acids	ß amino acids	Na⁺	Other
	ß					Yes	–
	$ß^{o,+}$	$ß^{o,+}$	$ß^{o,+}$			Yes	–
	$b^{o,+}$	$b^{o,+}$	$b^{o,+}$			No	–
				Imino		Yes	Cl⁻
					ß	Yes	Cl⁻
X^-_{AG}						Yes	K⁺

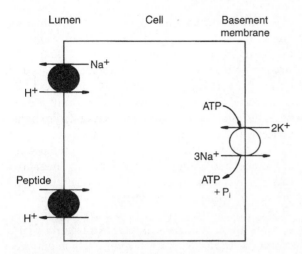

FIGURE 10.18 Peptide absorption across the entero-cyte brush border luminal surface using a Na^+-H^+ exchanger; peptide transporter. Basement membrane transport system Na^+, K^+-ATPase.

acids do not follow this relationship. Lipophilic properties are not important in determining the rate of transport of peptides.

Dipeptides, and to a lesser extent tripeptides, are transported across the brush border membrane by active proton-linked mechanisms (Figure 10.18). These products are hydrolysed to amino acids at a site, beyond the peptide transport mechanisms and probably in the cytosol of the absorptive cell. There may still be indirect linkage with Na transport.

Tertiary transport

The concentration of Na within the absorptive cells is kept low by the action of Na/K-ATPase which pumps Na out of the cells across the basolateral membrane. The re-entry of Na down an electrochemical gradient through the Na/H system in the brush border membrane causes the extrusion of protons into the lumen. Protons re-enter the cells across the brush border membrane down an electrochemical gradient by a mechanism coupling their entry to the inward transport of peptides. This is a tertiary transport system.

The effect of the unstirred layer on peptide absorption is the same as for amino acids; that is, to retard and hence to reduce the rate of absorption. The concentration of peptides at the absorptive surface will be greater than in the lumen.

The absorption of poorly hydrolysable peptides has been studied experimentally. The di- and tri-peptides share a common transport system or systems, and compete for brush border transport among themselves but not with amino acids, as they use different mechanisms. The di- and tripeptide brush border system is relatively indifferent to the net charge on the side chains, transporting neutral, basic and acidic peptides. There are possibly multiple peptide transport systems, but a classification is not, as yet, possible

Brush border digestion of small peptides

The majority of the peptidases of the intestinal mucosa are aminopeptidases, hydrolysing peptides sequentially from the amino-terminal end of the molecule. There are both brush border membrane and cytosol peptidases.

Peptides in which glycine is the amino-terminal residue undergo little brush border hydrolysis and are absorbed intact. The rate of hydrolysis is enhanced when the substrate consists of amino acid residues containing amino acids with large lipophilic side chains at the amino-terminus. Tri- and tetrapeptides are absorbed at faster rates than dipeptides. Peptidase activity against dipeptides containing proline is nil. Less than 10% of the total dipeptidase activity is to be found in the brush border, tripeptidases 60% and higher peptidases 100%. Individual di- and oligopeptidases are integral membrane glycoproteins with the carbohydrate-rich portion of the molecule projecting from the luminal surface of the brush border membrane.

It is also possible that there are some proteases associated with the enteric mucosa. There is a very active dipeptidase with a very broad specificity. The cytosol of the absorptive cells contains

90% of the total activity of the mucosa against dipeptides and less than 50% of that against tripeptides. It may well be that only di- and tripeptides are transported into the absorptive cells.

There are a large number of peptidases; some are singular to the enteric mucosa, e.g. enterokinase, and others are shared by other cell membranes. These peptidases are generally **metalloenzymes** and are found in the intracellular apical membrane. There are four families of peptidases:

- **endopeptidases**: the most important of these, neutral endopeptidase cleaves dipeptides at non-amino-terminal hydrophobic amino acid residues
- **aminopeptidases** with a variety of specificities for neutral, acidic, proline, tryptophan amino-terminal residues yielding amino acids
- **carboxypeptidases**, which hydrolyse peptides at the carboxy-terminal end
- **dipeptidases**, which hydrolyse over a wide range of amino acid residues.

Some peptidases in the brush border are glycosylated endopeptidases anchored at the amino-terminal end. Some amino acids, e.g. histidine, methionine, leucine, alanine and hydrophobic amino acids inhibit aminopeptidase activity. High protein intake increases and starvation decreases aminopeptidase activity.

Intracellular metabolism
After absorption free amino acids from whatever source enter a number of metabolic pathways, degradation and conversion into other amino acids, or proteins (Figure 10.19).

It is probable that at least 10% of amino acids taken up from the intestinal lumen, in free or peptide form are synthesized into protein and another 10–20% (especially those rich in glutamate) undergo metabolic change. The synthetic activity of the small intestinal mucosa is intense with rapid cell renewal, enzyme production, secretion of mucus and synthesis of apolipoproteins essential to the constituents of chylomicrons.

Glutamine, glutamate and aspartate are important amino acids in intestinal metabolism. There is extensive transamination of glutamic acid (+ pyruvic → alanine + α-ketoglutarate) and aspartic acid (+ α-ketoglutarate → glutamic acid + oxaloacetic acid). This metabolism may be a detoxification mechanism, protecting from high concentrations of glutamic acid which is then slowly metabolized by the liver. In some species glutamine and arginine are extensively metabolized by the small intestine which, together with the colon, are important sources of ammonia.

The unpleasant reaction to the savoury monosodium glutamate, the 'Chinese Restaurant Syndrome', is an idiosyncratic response which may be a result of a low level of intestinal alanine aminotransferase and hence an increased absorption of glutamate. When glutamic acid is given with glucose the reduced increment in plasma glutamic acid which occurs may be a result of increased metabolic activity in the liver and intestine.

Transfer of dipeptides and amino acids from brush border to blood stream
There is active transport of amino acids through the basolateral membrane by transport systems which are Na-independent and probably mediated by facilitated diffusion. A massive load of amino acids is imposed upon the exit mechanisms, the influx of peptides, and subsequent hydrolysis in the mucosa in addition to the dietary amino acids. Amino acid concentrations in absorptive cells postprandially do not increase very much so there is little impediment to flow from the cell to the blood.

Peptides, some of which are biologically active, may pass from the intestinal lumen to the blood; 10–30% of the absorbed partial hydrolysis products of protein have been estimated to cross into blood and red cells, the extent varying with the type of protein eaten. Of the 80% of protein digestion products which leave the lumen as peptides, approximately 50% might be transported across the brush border in peptide form. The proportion will vary with the protein, e.g. animal- and vegetable-derived proteins.

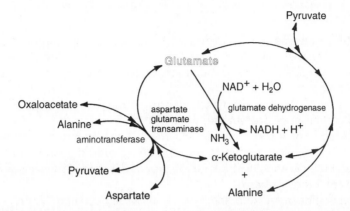

FIGURE 10.19   Intracellular metabolism in enterocytes. Relationship between glutamate, α-ketoglutarate, oxalo-acetate, pyruvate, alanine, aspartate and ammonia production.

Transport systems of amino acid transport

Na⁺-dependent systems
System A transports all α-amino acids including imino acids. This allows transport of all amino acids from the bloodstream into the cells. System ASC is also Na⁺-dependent and favours the transport of three- and four-carbon amino acids from the blood into the cell.

Na⁺-independent systems
A system, asc, is similar to ASC but is Na⁺-independent. System L is the most important of these basolateral transport systems and favours dipolar amino acids, glutamine and cysteine but not imino acids. System Y⁺ transports basic amino acids, lysine, arginine, ornithine and histidine.

Peptides and pharmacological agents, e.g. β-lactam antibiotics, captopril, bestatin and renin inhibitors, are absorbed through a common pathway.

Amino acid transport from cell to blood stream
There are five amino acid transport systems across the basolateral membrane. Two are Na⁺-dependent and the other three Na⁺-independent. The Na⁺-independent systems transport amino acids from the cell into the blood; the other Na⁺-dependent systems supply the cells with amino acids between meals.

10.7.3 Development of protein absorption

In newborn animals whole proteins may be absorbed, which is of major immunological importance. Proteolytic enzymes rapidly develop in the newborn. The ability to transport amino acids develops at different rates in foetal and neonatal life. In the rabbit the development of the active transport system for valine, methionine and lysine precedes that of proline and glycine. The absorption of peptides in the developing animal is very active. The foetus swallows large amounts of amniotic fluid (3 litres/day in late pregnancy) which contains amino acids and glucose in similar concentrations to maternal blood.

Summary

1. Protein absorption involves the breakdown of protein to tripeptides, dipeptides and amino acids. The site of maximal peptide or amino acid absorption may differ along the intestine.
2. Most ingested protein is absorbed in the jejunum, some is absorbed in the ileum and a small amount in the colon.
3. There are six phases in the digestion and absorption of proteins: (i) whole protein absorption; (ii) intraluminal digestion of protein and its breakdown products polypeptides, resulting from the sequential actions of the proteolytic enzymes of the stomach and pancreas; (iii) cellular uptake of amino acids and peptides; (iv) brush border digestion of small peptides; (v) intracellular metabolism; and (vi) transfer of dipeptides and amino acids from the intestinal brush border to the blood stream.
4. Protein digestion begins in the stomach with the enzyme pepsin in the presence of hydrochloric acid. The next phase is the activation of proteolytic enzymes, a process initiated by enterokinase which converts the precursor trypsinogen to trypsin by the cleavage of a small terminal peptide. Some proteins are resistant to hydrolysis with the result that their subsequent absorption is incomplete.
5. Proteins are absorbed from the intestinal lumen largely in the form of small peptides and amino acids. Small quantities of whole proteins are also absorbed and enter the circulation in trace amounts. Intraluminal digestion of protein produces a mixture of small peptides and amino acids in which peptides predominate. The absorption of peptides and amino acids are complementary processes.
6. Free amino acids are transported into the absorptive cells by a number of well-defined mechanisms which are mainly active and Na-linked. Several absorption mechanisms have defined specificity for certain groups of amino acids with structural features in common.
7. Active transport is a major mechanism of transmembrane transport of peptides. Active transport may well be limited to di- and tripeptides. Peptide transport is stereospecific, preferring peptides containing only L-amino acids or glycine. D-isomers appear to utilize the same transport systems as their comparable L-form but with reduced affinity characteristics.
8. Transport defects of amino acids include Hartnup's disease (dipolar amino acid) and cystinuria (cationic amino acids and cystine).
9. The majority of the peptidases of the intestinal mucosa are aminopeptidases, hydrolysing peptides sequentially from the amino-terminal end of the molecule.
10. The synthetic activity of the small intestinal mucosa is intense, requiring absorbed amino acids for rapid cell renewal, enzyme production, secretion of mucus and synthesis of apolipoproteins essential for chylomicrons.
11. There is active transport of amino acids through the basolateral membrane by transport systems which are Na-independent and probably mediated by facilitated diffusion.
12. There are five amino acid transport systems across the basolateral membrane. Two are Na^+-dependent and the other three Na^+-independent.
13. Whole protein absorption may be significant in newborn animals. This is of immunological importance. Proteolytic enzymes develop rapidly in the newborn.

Further reading

Johnson, L.R., Alpers, D.H., Christensen, J., Jacobson, E.D. and Walsh, J.H. (1994) *Physiology of the Gastrointestinal Tract*, Raven Press, New York.

Matthews, D.M. (1991) *Protein Absorption, Development and Present State of the Subject*, Wiley–Liss, New York.

10.8 Lipid absorption

- Dietary lipids, triglycerides, phospholipids, sterols and sterol esters are absorbed from the upper gastrointestinal tract following hydrolysis by pancreatic enzymes.
- Digestion requires the presence of bile acids and phospholipids of biliary origin.
- The lipid hydrolysis products, monoglycerides, fatty acids and cholesterol are incorporated into lipoproteins (chylomicrons) for transport to the body.

10.8.1 Introduction

The process of fat digestion takes place in the small intestine and involves pancreatic esterase and lipase enzymes which degrade the lipid into a form capable of being absorbed.

10.8.2 Phases of digestion and absorption of lipids

Gastric phase

The stomach creates a coarse oil and water emulsion stabilized by phospholipids. Proteolytic digestion in the stomach releases lipids from food lipoprotein complexes (Figure 10.20).

Jejunal phase

On leaving the stomach the fat emulsion is modified by mixing with bile and pancreatic juice. The bile contains bile acids as glycine and taurine conjugates of the trihydroxylated cholic acid and the dihydroxylated chenodeoxycholic acid and phospholipids. Biliary secretion is proportional to the amount of fat in the diet.

The pancreas secretes enzymes which release fatty acids from triacylglycerols, phospholipids and cholesterol esters. Pancreatic lipase catalyses the hydrolysis of fatty acids from positions 1 and 3 of triacylglycerols with little hydrolysis of the fatty acid in position 2. There is some modest isomerization to the 1-monoacylglycerols. The lipase hydrolyses triacylglycerol molecules at the surface of the large emulsion particles. Long-chain fatty acids ($20:5$ and $22:6$) in the 1 and 3 positions of the triacylglycerols are less readily hydrolysed by pancreatic enzymes than other shorter-chain fatty acids in those positions.

The pancreatic enzymes are modified to allow enzyme : lipid emulsion interaction to take place.

Bile acid molecules accumulate on the surface of the lipid droplet, displacing other surface-acting constituents. The bile acids give a negative charge to the oil droplets. This attracts **colipase**, a protein of 10 kDa which binds the pancreatic lipase to the surface of the lipid droplets, creating a ternary complex containing calcium, bile salts, colipase and pancreatic lipase.

As digestion proceeds the large emulsified particles are converted into mixed micelles (Figures 10.21 and 10.22). The micelles are stable small particles which may be insoluble or soluble. They contain monoacylglycerols, lysophospholipids and fatty acids. Fatty acids are in the form of soluble amphiphiles, since the pH in the proximal part of the small intestine is around 5.8–6.5. These soluble amphiphiles incorporate insoluble non-polar molecules such as cholesterol and vitamins into the micelles and consequently are important in absorption (Table 10.4).

Phospholipase A_2 hydrolyses the fatty acid in position 2 of phospholipids, particularly phosphatidylcholine. The enzyme is an inactive pro-enzyme in pancreatic juice and is activated by the tryptic hydrolysis of a heptapeptide from the N-terminus. Lysophospholipids accumulate in intestinal contents. Cholesteryl esters are hydrolysed by a pancreatic cholesteryl ester hydrolase.

Lipid absorption in humans occurs largely in the jejunum by passing through the brush border membrane of the enterocytes as monoacylglycerol and free-chain fatty acids (Figure 10.23). The rate-limiting step in the uptake of lipids is the unstirred water at the surface of the microvillus. Bile acids pass down to the ileum for absorption or into the caecum to be deconjugated and 7-α-dehydroxylated before reabsorption from the colon or excretion in the faeces.

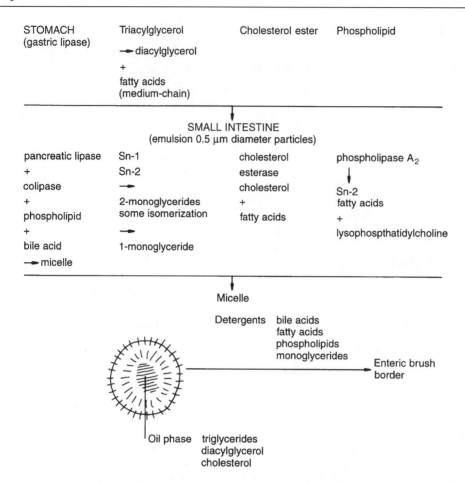

FIGURE 10.20   Lipid digestion takes place in the stomach and small intestine. Gastric lipase initiates the digestion of triacylglycerol. Fats pass from the stomach into the small intestine as small emulsions. Micelles are created when pancreatic lipase, cholesterol esterase, and phospholipase A_2 digest lipids to more polar substrates. Stable micelles are formed which pass through the unstirred layer to the enteric brush border where individual lipids are absorbed by specific pathways.

Absorption into the enterocyte

The process of absorption into the enterocyte involves an inward diffusion gradient of lipid products. Fatty acids enter the cell and bind to a fatty acid binding protein (or z protein) of molecular weight 12 kDa. The protein binds long-chain and unsaturated fatty acids in preference to saturated acids. The next phase is an energy-dependent re-esterification of the absorbed fatty acids into triacylglycerols and phospholipids. The free fatty acids are converted into acyl-CoA thioesters. Long-chain fatty acids are the preferred substrate for esterification to 2-monoacylglycerols, the major forms of absorbed lipids through the monoacylglycerol pathway.

The major absorbed products of phospholipid digestion are monoacyl phosphatidylcholines. Fatty acids are re-esterified to position 1 to form phosphatidylcholine by an acyl transferase in the tips of the intestinal brush border. This phospholipid stabilizes the triacylglycerol particles or chylomicrons.

Cholesterol absorption is slower and less complete than that of other lipids. There is also a loss

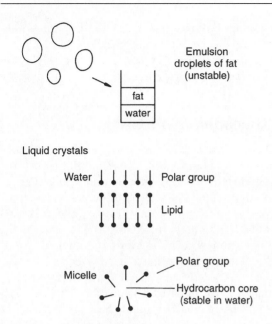

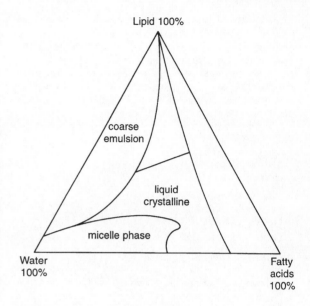

FIGURE 10.21 Lipid interactions with water: emulsions, liquid crystals and micelles. When the concentration of amphiphiles (detergents) reaches a certain level then a stable small structure called a **micelle** is formed. This concentration is called a critical micelle concentration. The hydrocarbon groups protrude into the core of the micelle. The polar groups interact with water to provide solubility. Micelles are very stable. Amphiphiles are lipids which sit at water–lipid interphases and facilitate the solubility of one phase into the other.

of absorbed sterol through desquamation of cells. Most of the absorbed cholesterol is esterified by cholesteryl esterase or acyl-CoA : cholesterol acyl transferase (Figure 10.24).

Transport of lipids into the blood

Lipids are transported in blood as apolipoproteins which stabilize the lipid particles with a coat of amphiphilic compounds of phospholipids and proteins. These vary in physical form but their function is to transport lipids from one tissue to another. Lipoproteins contain a core of neutral lipids, cholesterol esters and triglycerides and a surface coat of more polar lipids, unesterified cholesterol and phospholipids and apoproteins. The surface coat, which in many ways resembles cell plasma membrane is an interface between the plasma and the non-polar lipid core. Lipoproteins

FIGURE 10.22 Micelle formation. If an insoluble lipid and a fatty acid are mixed in varying concentrations then a phase diagram can be drawn in which various mixtures are identified with lipid (100%) at the apex and water (100%) and fatty acid (100%) at each base. Varying concentrations of each of these would be present at any mixture point shown in the interior of the diagram. When there is insufficient detergent and excess of lipid then a coarse emulsion is formed. If the amount of fatty acid increases a liquid crystalline formation results. If the ratios of detergent, water and lipid are correct then a micelle is created. In the intestine fatty acid monoacyl glycerols, phospholipids and bile acids are involved in micelle formation.

differ through the ratio of lipid to protein as well as having different proportions of lipids, triacylglycerols, free and esterified cholesterol and phospholipids. The biological function varies with the size of the lipoprotein molecule, from lowest to highest densities these are shown in Table 10.5. None of these groups is a single entity and contain a wide variety of particle sizes and chemical composition.

Fat dropules are found postprandially in the smooth endoplasmic reticulum where the enzymes of the monoacylglycerol pathways are found. Here, there is synthesis of phospholipids and apolipoproteins which coat the lipid dropules. Long-chain fatty acids are transported in chylomicrons. Fatty acids with a chain length of less than 12

TABLE 10.4 Classification of lipids in terms of interaction with water. Insoluble non-swelling amphiphile lipids have little solubility with water in the bulk phase. They form a thin layer of lipid when added to water. Insoluble swelling amphiphiles in water form laminated lipid water structures called liquid crystals. The non-polar groups of the lipid molecules face each other with water sandwiched between.

Non-polar
  cholesteryl ester
  hydrocarbons
  carotene

Insoluble non-swelling amphiphiles
  Triacylglycerols
  Diacylglycerols
  Fat-soluble vitamins

Insoluble swelling amphiphiles
  Monoacylglycerols
  Ionized fatty acids
  Phospholipids

carbon atoms are absorbed in the free form, passing into the portal vein and are metabolized directly by β-oxidation in the liver. This is because of their more ready hydrolysis from triacylglycerols and their water solubility.

10.8.3 Apolipoproteins

The protein moieties of lipoproteins solubilize lipid particles and ensure continued structural integrity. They are important in determining the lipoprotein type and how it is metabolized. They are cofactors in the activation of enzymes involved in the modification of lipoproteins. They interact with specific cell surface receptors which remove lipoproteins from the plasma. The lipoproteins are classified either by their density or by their electrophoretic mobility. The complete amino sequences of all of these are known. ApoBs are the largest apolipoprotein.

ApoAs are synthesized by both liver and intestine and exist in both chylomicrons and HDL. Types A-I, A-II and A-IV are recognized.

Apoprotein C is synthesized in the liver and forms part of the chylomicrons, VLDL, IDL and HDL. There are three types C-I, C-II and C-III; they are of low molecular weight 5800–8750 Da.

The apoproteins E, E-2, E-3 and E-4 are synthesized in the liver and peripheral tissues, have a molecular weight of 35 kDa and differ only in the amino acid occupying position 112 and 158.

Apolipoprotein B series

ApoB exists as two variants of different molecular mass (approximately 100 and 48 kDa) designated at apoB$_H$ (heavy) and apoB$_L$ (light) or apoB-100 and apoB-48 respectively. Human VLDL contains the heavy variant, whereas rat VLDL contains both. ApoB-48 is of intestinal origin and it forms part of the chylomicron. ApoB-100 is synthesized in the liver and forms part of the surface coat of the triacylglycerol-rich VLDL, IDL and cholesterol rich-LDL. ApoB has a considerable β-sheet protein structure in addition to regions of random coil and α-helix. ApoB is a glycoprotein linked to glucosamine through asparagine residues.

The isoforms of apoE are genetically transmitted, an isoform from each parent. Six genotypes exist: E-3/E-3 is found most frequently (two-thirds of the US population); E-3/E-4, 22%; E-3/E-2, 12%; and E-4/E-4, E-4/E-2 and E-2/E-2, less than 2%. These apoproteins are involved in chylomicrons, VLDL, IDL and HDL.

10.8.4 Lipoproteins

Chylomicrons

Chylomicrons (Figure 10.25(a)) are the largest and least dense of the lipoproteins, (diameter 80–500 nm); their function is to transport lipids of exogenous or dietary origin (Figure 10.26). Chylomicrons can be measured in the plasma after a fatty meal, their size depending on the rate of lipid absorption, and the type of dietary fatty acids which predominates. Larger chylomicrons are produced after the consumption of large

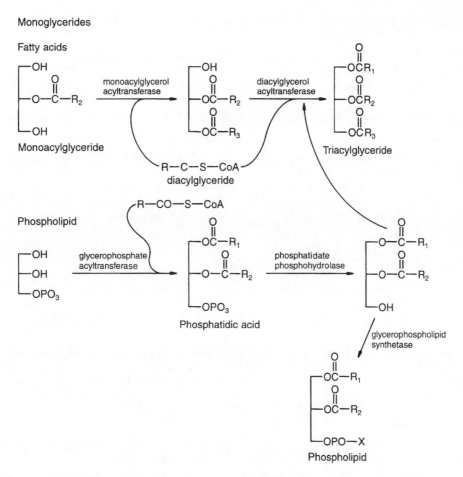

FIGURE 10.23 Intracellular metabolism of monoglycerides, fatty acids and phospholipids. Postprandially, the monoglyceride fatty acid pathway is the most important in the synthesis of triacylglycerol. During fasting the phospholipid pathway (α-glycerophosphate) becomes a major pathway for the formation of triacylglycerol. 2- Monoglyceride is inhibitory of the α-glycerophosphate pathway.

amounts of fat at the peak of absorption or when apolipoprotein synthesis is limiting. They contain triacylglycerols with small amounts of phospholipids and sufficient protein to cover the surface. The core lipid contains some cholesteryl esters and fat-soluble substances, fat-soluble vitamins, etc.

Several apoproteins are present in the chylomicron surface layer: apoB-48, apoA$_1$, apoA$_4$, the apoC group, and apoE. The apoB-48 and apoA series apolipoproteins are synthesized in the endoplasmic reticulum of the intestinal epithelial cells, whereas apoCs and apoE are acquired from other lipoproteins once the chylomicrons have entered the blood. The apoE and apoCs are important in the catabolism of the chylomicron.

The chylomicrons pass initially into the lymph and hence to the thoracic duct and into the blood stream. In the circulation, lipoprotein lipase, an enzyme on the capillary surface hydrolyses the triglyceride fatty acids. The free fatty acids, apoA and apoC are released. The chylomicron remnant returns to the liver and is removed from circulation. The released fatty acids are held in solution in the plasma bound to albumin. They are subsequently metabolized in two main ways:

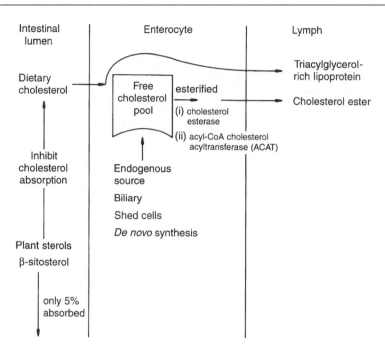

FIGURE 10.24 Absorption of cholesterol and plant sterols. Dietary cholesterol and cholesterol of endogenous sources, biliary shed cells and cholesterol from *de novo* synthesis enter the enterocyte. Dietary cholesterol passes into lymph as triacylglycerol-rich lipoprotein. Free cholesterol is esterified by cholesterol esterase or acyl-CoA cholesterol acyltransferase, the cholesterol ester being transported into the lymph. Plant sterols, e.g. β-sitosterol, are minimally absorbed and inhibit cholesterol absorption in the small intestine.

- they may be transported to adipose tissue for incorporation into triglycerides; when required these can be released through the action of adipose tissue lipase.
- they may be used by the liver as a fuel source or stored as triglyceride.

The triacylglycerols release the fatty acids; $apoC_2$ plays a key role in activating lipoprotein lipase. The peptide facilitates an interaction of the enzyme with the lipoprotein interphase catalysing hydrolysis of long-chain rather than short-chain fatty acids. This is a rapid hydrolysis, taking only 2–3 minutes; the apolipoprotein and the remaining apoC are transferred to HDL with the phospholipids. The remnant chylomicron particle, although returning to the same basic structure, contains less triacylglycerols and is rich in cholesterol esters. These are not able to compete for lipoprotein lipase and circulate in the plasma to be taken up by liver cells by a receptor-mediated endocytosis. The receptors in the liver bind to the apoE component of the remnant. There is then a complete hydrolysis of the lipid and protein component.

The regulation of lipoprotein lipase is crucial to the control of lipoprotein metabolism in different tissues in the body. The enzyme is synthesized in the parenchymal cells of the tissues and secreted into capillary endothelium, bound to the cell surface by sulphated glycosaminoglycan, activities of which are regulated by diet and hormones, particularly insulin. The secretion of insulin results in increased adipose tissue lipoprotein lipase activity. The equivalent muscle enzyme is suppressed. In fasting, the adipose tissue lipoprotein lipase activity is suppressed and the hormone-sensitive lipase activated, allowing the mobilization of free fatty acids. Muscle lipoprotein lipase activity is raised and fatty acids from circulating lipoproteins can be used as fuel.

TABLE 10.5 Lipoprotein molecules

Lipoprotein	Source
Chylomicrons	Intestine
Very low density lipoproteins (VLDL)	Liver
Intermediate density lipoprotein (IDL)	VLDL catabolism
Low density lipoproteins (LDL) and High density lipoproteins (HDL)	IDL catabolism Liver, intestine, other

During lactation, lipoprotein lipase is regulated by prolactin, which promotes the utilization of chylomicron triacylglycerol fatty acids for milk synthesis.

Very low density lipoproteins (VLDL)

These consist predominantly of triacylglycerols and transport triacylglycerols of endogenous origin derived from the liver and intestine (Figure 10.25(b)). They are spherical particles (30–80 nm), smaller than chylomicrons. There is a core of predominantly triacylglycerols and some cholesteryl esters with cholesterol, phospholipids and proteins on the surface. The composition and the classification depends on the method of isolation, the species, nutrition and physiological state of the animal. The major apoprotein is apoB-100, the amount of apoB per VLDL is dependent on the particle mass and is similar to the composition of LDL. Other lipoproteins include apoCI, CII and CIII and apoE.

As the apoB-100 passes towards the smooth endoplasmic reticulum, there is admixing with the triglyceride and cholesteryl esters at the junction of the smooth and rough endoplasmic reticulum. This creates nascent VLDL particles; these pass from the smooth endoplasmic reticulum to the Golgi apparatus. Secretory vesicles bud off and migrate to the surface with VLDL for release into the plasma.

Glucose is converted into the lipid precursor glycerol-3(sn) phosphate through the glycolytic pathway and long-chain fatty acids through the malonyl-CoA pathway. Some fatty acids arise from circulating free fatty acids bound to albumin. The triglycerides and small amounts of cholesteryl ester are synthesized by membrane-bound enzymes in the smooth endoplasmic reticulum.

The major site of synthesis of VLDL is in the liver, though some is produced in the enterocyte. VLDLs receive their full complement of apoproteins by synthesis in the liver or enterocyte rough endoplasmic reticulum (RER).

The VLDL circulating in the blood stream is converted into mature VLDL through incorporation of cholesterol esters and apoC-II and apoC-III and possibly apoE. These are transferred from HDL.

At the same time there is release of phospholipids and most apoCs and some apoEs which move onto HDL. Remnant VLDL is now rich in cholesterol esters which includes some transferred from HDL. The remnant VLDL may be removed by the liver through receptors on liver cells, including those for chylomicron remnants and LDL. The latter recognizes apoB-100 or apoE. Some VLDL remnant triglycerides are hydrolysed by hepatic triglyceride lipase, producing a cholesterol-rich LDL. This involves release of fatty acids and apoE and apoC.

Low density lipoproteins (LDL)

The function of LDLs is to transport cholesterol to tissues. They have a diameter of the order of 20–25 nm and a lipid core which is essentially cholesterol. The surface of the particle is unesterified cholesterol, phospholipids and apoB-100. The cholesterol is available for incorporation into membrane structure or conversion into various metabolites, e.g. steroid hormones. Each lipoprotein particle contains the same mass of apoB-100, but differs in the amount of bound lipid.

LDL is derived from VLDL by a series of steps which remove triacylglycerol, through the VLDL remnant which originates from the VLDL particle. This results in particles with progressively smaller proportions of triacylglycerol and increasing amounts of cholesterol and phospholipids. Such reactions take place initially in adipose tissue

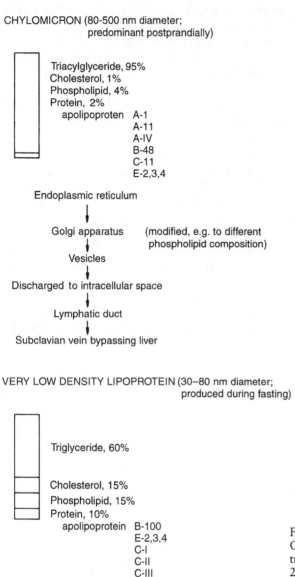

CHYLOMICRON (80-500 nm diameter;
              predominant postprandially)

Triacylglyceride, 95%
Cholesterol, 1%
Phospholipid, 4%
Protein, 2%
    apolipoproten   A-1
                    A-11
                    A-IV
                    B-48
                    C-11
                    E-2,3,4

Endoplasmic reticulum
          ↓
Golgi apparatus   (modified, e.g. to different
          ↓   phospholipid composition)
     Vesicles
          ↓
Discharged to intracellular space
          ↓
     Lymphatic duct
          ↓
Subclavian vein bypassing liver

VERY LOW DENSITY LIPOPROTEIN (30–80 nm diameter;
                             produced during fasting)

Triglyceride, 60%

Cholesterol, 15%
Phospholipid, 15%
Protein, 10%
    apolipoprotein   B-100
                     E-2,3,4
                     C-I
                     C-II
                     C-III

Endoplasmic reticulum
          ↓   [modified]
Golgi apparatus
          ↓
     Vesicles
          ↓
Discharged to intracellular space

FIGURE 10.25 (a) Intestinal lipoprotein assembly. Chylomicrons are produced postprandially: 95% triacylglycerol, 1% cholesterol, 4% phospholipid and 2% apolipoprotein. Apolipoproteins are assembled as pre-chylomicrons in the endoplasmic reticulum, and pass to the Golgi apparatus to be modified. The phospholipid composition is altered before discharge into the intracellular spaces by vesicles. (b) Very low density lipoproteins (VLDL) are produced predominantly during fasting. VLDL are formed as 60% triglycerol, 15% cholesterol, 15% phospholipid and 10% protein. The assembly of VLDL passes through different pathways to chylomicrons. VLDL pass through the endoplasmic reticulum and Golgi apparatus and are excreted via vesicles.

capillaries and subsequently in the liver. The apoB-100 remains with the LDL particle and apoC and apoE are progressively lost. The synthesis of LDL is dependent upon the amount of VLDL produced by the liver and the proportion of VLDL removed by the liver. This latter reaction is dependent upon the number of LDL receptors which remove VLDL remnants. These same receptors dictate the concentration of LDL and their removal from the circulation.

ApoB-100 is important in that it interacts with specific cell surface receptors before the LDL particle is taken up and metabolized by the cells. Macrophage receptors recognize modified LDL and are responsible for the degradation of LDL particles that cannot be recognized by normal cell surface LDL receptors. ApoE plays a role in receptor binding while the C group apolipoproteins are involved in reactions by which the particles are sequentially degraded by lipases.

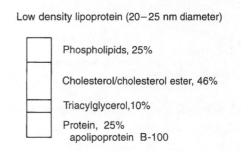

Low density lipoprotein (20–25 nm diameter)

Phospholipids, 25%

Cholesterol/cholesterol ester, 46%

Triacylglycerol, 10%

Protein, 25%
apolipoprotein B-100

FIGURE 10.27 Low density lipoproteins (LDL) transport cholesterol to the tissues. The coat, which is hydrophobic, consists of cholesterol and phospholipids and the hydrophobic core consists of cholesterol esters and triglycerides. Surface apoprotein B is important in LDL receptor recognition. The LDL are produced from very low density lipoproteins (VLDL). Lipoprotein lipase removes triglyceride from the core of the VLDL to form intermediate density lipoprotein (IDL) which may be further metabolized by lipoprotein lipase and hepatic lipase to LDL. Triglycerides are removed from LDL to VLDL and cholesterol esters pass from VLDL to LDL. The LDL lipoproteins belong to several classes: the smaller particles have a higher protein/lipid ratio; the larger particles have more triglyceride.

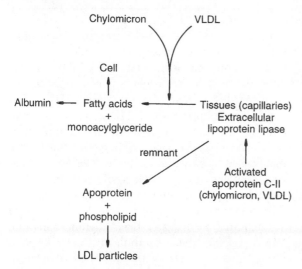

FIGURE 10.26 Chylomicrons carry triacylglycerol and cholesterol ester from the intestine to other tissues in the body. Very low density lipoproteins (VLDL) similarly carry lipids from the liver and the intestine. In tissue capillaries, extracellular lipoprotein lipase is activated by apoprotein C-II and hydrolyses triacylglycerol to produce fatty acids and monoacyl glycerides which are absorbed into the cell. Some fatty acids may bind to albumin to be transported to other tissues. The chylomicron and VLDL remnants, apoproteins and phospholipids become LDL particles.

LDLs are principally removed by specific extrahepatic LDL receptors. LDLs are less efficiently removed by hepatic receptors, which take the receptor–ligand complex into the cell lysosomes (Figure 10.28). The distribution of LDL to various tissues depends on the rate of trans capillary transport as well as the number of LDL receptors on the cell surface. Adipose tissue and muscles have few LDL receptors and take up LDL only slowly. The adrenal gland is important in the synthesis of steroid hormones and takes up LDL avidly. The LDL receptor has a recognition site for both apoE and apoB.

About 80% of the LDL receptors are concentrated in the clathrin-coated pits which, nevertheless, represent only 2% of the cell surface. Negatively charged residues bind electrostatically with the positively charged region of apoE. This lipoprotein binds at several receptor sites, apoB having a single site. Once bound to the receptor, the LDL receptor complex is taken into the cell and the LDL degraded by lysosomal enzymes. The receptor is recycled within minutes and re-utilized many times before eventual catabolism. The num-

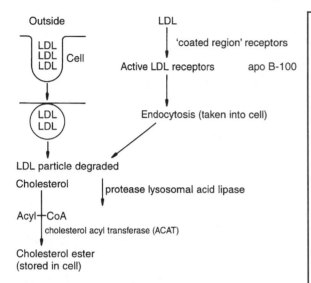

FIGURE 10.28 LDL catabolism. LDL is taken up by receptors in the liver and extrahepatic tissues. The LDL particles bind to coated regions which contain active LDL receptors. This is followed by the LDL particles being taken up into the cell by endocytosis. The LDL particle is degraded by protease and lysosomal acid lipase. Cellular cholesterol inhibits HMG-CoA reductase and stimulates cholesterol acyltransferase. The cholesterol ester produced is stored in the cell. The cholesterol ester inhibits the production of LDL receptors and limits the uptake of LDL.

ber of LDL receptors is regulated, by regulatory proteins and genes according to the amount of cholesterol in the cell.

In the cell, cholesterol esters are hydrolysed by cholesteryl ester hydrolase. The incorporation of cholesterol into the endoplasmic reticulum membrane inhibits β-hydroxy-β-methyl glutaryl-CoA reductase (HMG-CoA reductase), the rate-limiting enzyme in cholesterol biosynthesis. The cholesterol may be stored within the cell as the cholesterol ester, incorporated into membranes or exported to the plasma. In liver cells there is conversion to bile acids for excretion in bile.

High density lipoproteins (HDL)

The function of HDL (Figure 10.29) is to remove unesterified cholesterol from peripheral tissues and transport it to the liver to be degraded and excreted, possibly in the form of bile acids.

LDL receptors

of a number of domains. The domain which binds LDL consists of 292 amino acids and comprises a number of 40-amino acid repeat sequences. It is orientated to the luminal side of the cell membrane and is able to bind both apoB and apoE simultaneously. The second domain is approximately 400 amino acids and is similar in sequence structure to epidermal growth factor. The third, which is 58 amino acids, is where the carbohydrate residues are linked. The fourth domain (22 amino acids in size) crosses the plasma membrane, i.e. is transmembrane. The fifth (50 amino acids) projects into the cytoplasm and may have a role in the clustering of the receptors into coated pits. These migrate in the plane of the membrane until it reaches a pit coated with the protein **clathrin**.

HDL particle size is 5–15 nm; they are usually grouped into two classes HDL_2 and HDL_3. The smallest spherical form of HDL is HDL_3. HDL_2 appears to have a stronger inverse relationship with cardiovascular disease than HDL_3. The major apolipoproteins of HDL are $apoA_1$ and $apoA_2$. The former is a monomer of molecular weight 28 kDa; the latter is a dimer of this monomer, linked by a disulphide bond. The surface coat also contains some apoC, apoE and apoD, the latter being found only in HDL. Additional surface components, phospholipids and cholesterol are acquired by transfer from chylomicrons and VLDL during their catabolism by lipoprotein lipase. The first step is the synthesis of nascent HDL in both liver and intestine. Nascent HDL accepts unesterified cholesterol from cells or other lipoproteins. ApoA which is synthesized in the intestine and some apoC are transferred during the breakdown of chylomicrons. Further apoC is transferred from VLDL breakdown products.

As cholesterol accumulates in the plasma, subfractions of HDL containing $apoA_1$ and apoD become associated specifically with the enzyme lecithin: cholesterol acyltransferase (LCAT). This enzyme

High density lipoprotein (5–15 nm diameter)

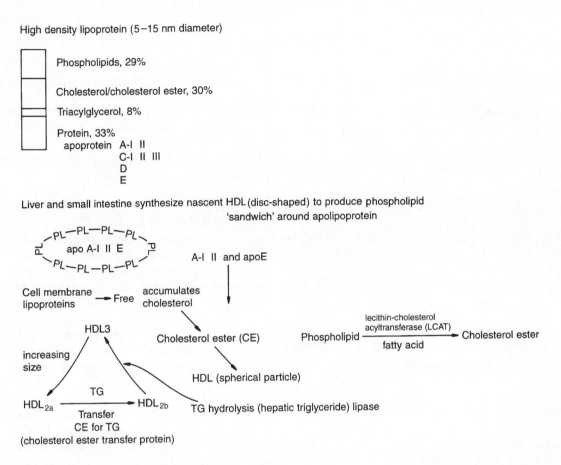

Phospholipids, 29%

Cholesterol/cholesterol ester, 30%

Triacylglycerol, 8%

Protein, 33%
 apoprotein A-I II
  C-I II III
  D
  E

Liver and small intestine synthesize nascent HDL (disc-shaped) to produce phospholipid 'sandwich' around apolipoprotein

FIGURE 10.29 Formation of high density lipoprotein (HDL). HDL are spheres consisting of phospholipids (29%), cholesterol and cholesterol ester (CE; 30%), triacylglycerol (8%) and protein (33%). They are formed in the liver and small intestine from nascent HDL and accumulate cholesterol, which is converted to cholesterol esters by lecithin–cholesterol acyltransferase (LCAT), the fatty acid being obtained from phospholipid. The smallest spherical form of HDL is HDL_3. As cholesterol accumulates, HDL_{2a} is formed. HDL_{2a} is converted into HDL_{2b} by the addition of triglyceride (TG) from triglyceride-rich lipoproteins, catalysed by cholesterol ester transfer protein. HDL_{2b} is converted to HDL_3 by triglyceride hydrolysis through hepatic triglyceride lipase.

catalyses the transfer of fatty acids from phosphatidylcholine to cholesterol to form a cholesterol ester. The phospholipid substrate is transferred from chylomicron remnants or IDL during the degradation of chylomicrons or VLDL. LCAT, by consuming cholesterol, enables its transfer from non-hepatic cells from plasma and other lipoproteins to a site of esterification. Molecules of cholesterol ester are transferred to lipoproteins containing apoB-100 or apoE, and are taken up by the liver. This transfers fatty acid from phosphatidylcholine to cholesterol on the surface of the disc-shaped HDL to form cholesterol esters which then accumulate in the core of the HDL particle.

Lysophosphatidylcholine is transferred to plasma albumin from which it is rapidly removed from blood and reacylated with fatty acids. During this distribution of lipid the particle changes from discoid to spherical in shape.

As HDL_3 increases in size, HDL_2 develops, initially as HDL_{2a} which in turn is converted into HDL_{2b} wherein cholesterol ester is exchanged for

373

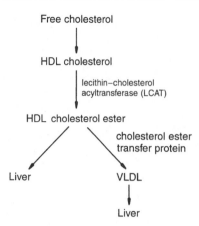

FIGURE 10.30   HDL carries cholesterol to the liver by reverse cholesterol transport. Free cholesterol attaches to the HDL sphere. The cholesterol is esterified by lecithin–cholesterol acyltransferase (LCAT). Some cholesterol ester is transferred to the VDL and LDL to be carried to the liver by cholesterol ester transfer protein. HDL also carries cholesterol to the liver.

triglycerides. These triglycerides are obtained from VLDL lipoproteins which in turn are replenished with cholesterol ester from the HDL. This exchange is facilitated by cholesterol ester transfer protein. HDL_{2b} is converted back to HDL_3 through the loss of triglycerides by hepatic triglyceride lipase. This whole process is called **reverse cholesterol transport** (Figure 10.30).

Other lipoproteins

Serum albumin binds many types of molecules including free fatty acids. As such it is the main transporter of free fatty acids released by lipases from adipose tissue into the blood. There are three classes of binding sites which bind 2, 5 and 20 molecules of fatty acids, respectively. Other fatty acid-binding proteins such as **z protein** transport fatty acids and acyl-CoAs within cells rather than in the blood. Phospholipids may be transported in cells by binding to phospholipid exchange proteins which introduce lipids into plasma membranes and organelle membranes remote from their sites of synthesis.

10.8.5 Lipid absorption in early life

The newborn animal adapts rapidly to a diet of breast milk with a relatively high fat content. The pancreatic secretion of lipase is rather low and the immature liver secretes insufficient bile salts for lipid digestion. This is particularly marked in the premature infant. However, the infant is able to digest fat as a result of the activity of a lipase secreted by the tongue, This is active in the stomach at a pH of 4.5–5.5, and does not require bile salts. Secretion of oral lipase is stimulated by sucking and the presence of fat in the mouth. The products are mainly 2-monoacylglycerols, the free fatty acids being predominantly medium-chain fatty acids. Milk fat in most mammals is relatively rich in medium-chain fatty acids. There may also be a lipase in human milk which facilitates the digestion of milk lipids.

On **weaning**, the site of fat digestion changes from the stomach to the duodenum.

Summary

1. Fat digestion takes place in the small intestine and involves pancreatic esterase and lipase enzymes to degrade lipid before being absorbed.
2. The fat emulsion, on entering the stomach, is modified by mixing with bile and pancreatic juice. The pancreas secretes enzymes which release fatty acids from triacylglycerols (lipase), phospholipids (phospholipase A_2) and cholesterol esters (cholesterol esterase).
3. Lipid absorption in humans occurs largely in the jejunum by passing through the brush border membrane of the enterocytes as monoacylglycerol and free-chain fatty acids. Fatty acids enter the cell and bind to a fatty acid binding protein (or z protein); there is next an energy-dependent re-esterification of the absorbed fatty acids into triacylglycerols and phospholipids.
4. The major absorbed products of phospholipid digestion are mono-acylphosphatidylcholines which are re-esterified to form phosphatidylcholine.

5. Cholesterol absorption is slower and less complete than that of other lipids.

6. Lipids are transported in the blood within apolipoprotein complexes, classified by their densities as chylomicrons, very low density lipoproteins (VLDL), low density lipoproteins (LDL), and high density lipoproteins (HDL).

7. Fatty acids with a chain length of less than 12 carbon atoms are absorbed in the free form and pass to the liver, where they are metabolized by β-oxidation.

8. The protein moieties of lipoproteins solubilize lipid particles, are cofactors in the activation of enzymes involved in the modification of lipoproteins, and interact with specific cell surface receptors which remove lipoproteins from the plasma.

9. Chylomicrons are the largest and least dense lipoproteins; they transport lipids of exogenous or dietary origin to adipose tissue for incorporation into triglycerides. These can be released through the action of adipose tissue lipase; alternatively, fatty acids may be used by the liver as a fuel source or stored as triglyceride.

10. VLDL consist predominantly of triacylglycerols and transport triacylglycerols of endogenous origin derived from the liver and intestine. The major site of synthesis of VLDL is the liver or enterocytes.

11. LDL transports cholesterol to tissues; it is derived from VLDL, initially in adipose tissue capillaries and subsequently in the liver. LDLs are removed by specific extrahepatic LDL receptors and less efficiently by hepatic receptors. The LDL receptor complex is taken into the cell and the LDL degraded by lysosomal enzymes.

12. In the cell, cholesterol esters are hydrolysed by cholesterol ester hydrolase. Cellular cholesterol inhibits β-hydroxy-β-methyl glutaryl-CoA reductase (HMG-CoA reductase), the rate-limiting enzyme in cholesterol biosynthesis.

13. HDL remove unesterified cholesterol from peripheral tissues for transport to the liver. HDL is synthesized in both liver and intestine and accepts unesterified cholesterol from cells or other lipoproteins. Lecithin : cholesterol acyltransferase (LCAT) catalyses the transfer of fatty acids from phosphatidylcholine to cholesterol to form a cholesterol ester.

14. The newborn adapts rapidly to a diet of breast milk with a relatively high fat content. The pancreatic secretion of lipase is rather low and the immature liver secretes insufficient bile salts for lipid digestion; this is particularly marked in premature infants.

Further reading

Johnson, L.R., Alpers, D.H., Christensen, J., Jacobson, E.D. and Walsh, J.H. (1994) *Physiology of the Gastrointestinal Tract*, Raven Press, New York.

Lairon, D. and Frayn, K.N. (1996) Lipid absorption and metabolism. *Proceedings of the Nutrition Society*, 55, 1–154.

10.9 Foetal and placental nutrition

- The foetus is totally dependent upon the mother for its nutrition.
- The placenta is the protector and mode of nourishment for the developing foetus.
- The transport of different nutrients across the placental barrier is specific to that nutrient.

10.9.1 Introduction

The foetus is separated from the mother by the placenta, an important organ which serves both as a protector of, and source of nourishment for, the developing foetus. During the first days following implantation of the fertilized human ovum, nutrition is from local sources. Thereafter the nutrition of the foetus is from the mother through the placenta. For the first 4–5 months, placental growth is greater than that of the foetus, but from thereon foetal growth exceeds that of the placenta. By full term at 40 weeks, the weight of foetus, placenta and liquor amni is 5 kg, 3.5 kg of which is the baby and 0.5 kg the placenta.

The placenta is a defined, transporting organ by the fourth week of pregnancy, which not only transports nutrients, but also is a source of oxygen to the foetus. The placenta is the means whereby carbon dioxide and substances excreted by the bowel and kidneys of the foetus are eliminated into the maternal circulation.

10.9.2 Placental circulation

There are three features to the circulation of the placenta:

- the relationship between the maternal and foetal circulation within the placenta
- the uterine circulation, which supplies maternal blood to the placenta
- the umbilical circulation, which supplies blood to the foetus

At the onset of pregnancy, the maternal blood vessels under the developing embryo become dilated. Small projections from the blastocyst, called the **chorionic villi**, grow into these blood vessels. There is erosion of the maternal tissue with the development of a blood-filled intervillous space. The foetal mesoderm invades the chorionic villi with a core of capillaries and connective tissue. Thus, the chorionic villi dip into the maternal blood, separated by three thin layers: the foetal vascular epithelium; the connective tissue of the villous; and the **trophoblast**. The trophoblast is tissue which attaches the ovum to the uterine wall. The maternal blood supply enters the intervillous space under pressure, spreads to the chorion plate and passes laterally and downwards past the capillary bed of the foetal villi to the basal plate. The blood then drains into the uterine veins.

10.9.3 Specific components of foetal nutrition

Monosaccharides

Glucose is an important foetal and placental fuel, accounting for substantial proportions of foetal oxygen consumption. The transfer system is stereospecific, can be saturated and is a mediated process, i.e. not a sodium-dependent concentrating glucose transporter. Thus, the placenta cannot concentrate glucose against a concentration gradient.

Monocarboxylates and dicarboxylates
The metabolism of glucose within the placenta produces lactate at a high rate which is delivered into the foetal and maternal circulation. The lactate carrier of the brush border membrane is sodium-independent. The system is specific for monocarboxylates e.g. lactate, pyruvate and β-hydroxybutyrate. Dicarboxylic acids do not interact with this transport system. There is a separate high-affinity transport system for dicarboxylic acids. This requires a transmembrane electrical sodium gradient.

Amino acids

Amino acid transport across the placental membrane involves mediated transport through both the microvilli and basal membrane.

Mono-amino, mono-carboxy amino acids
The microvillous membrane of the placenta uses transport systems common to many cell types. This is quite different from intestinal brush border systems which utilize specialized amino acid systems. Placental systems include:

- system A: a sodium-dependent transporter for alanine, serine, methyl amino isobutyric acid and proline

- system N: a sodium-dependent system for histidine and glutamine
- various sodium-independent systems including leucine, tryptophan, tyrosine and phenylalanine

There is a second sodium-independent system for alanine and serine and branched-chain and aromatic amino acids. Cellular regulation of amino acid transport involves regulation of transport systems with either of the maternal or foetal surface membranes.

Anionic amino acids

These are not concentrated in the foetal circulation and are not transferred between maternal and foetal circulations. Anionic amino acids are taken up from either or both circulations. Millimolar concentrations of aspartate and glutamate are present within the placenta, whereas concentrations in the maternal and foetal blood are in micromolar concentrations. There does not appear to be transport between mother and foetus.

Basic amino acids

The cationic amino acids, lysine and arginine, are concentrated in the foetal circulation and within the placenta. This concentration effect appears to be due to two sodium-independent transport systems in the basal membrane of the placenta.

β-amino acids

The amino acid with the highest concentration in the placenta is the β-amino acid taurine, of which the foetus requires an exogenous supply. The active transport of taurine by the placenta results in foetal concentrations that are greater than maternal concentrations.

Lipids and related compounds

Lipoprotein lipase on the maternal surface of the placental membrane hydrolyses triacylglycerol carried by maternal VLDL. The free fatty acids are taken up by the trophoblasts and may be used by the trophoblast or transferred to the foetus. The rate of fatty acid synthesis in the placenta is very high. LDL cholesterol is taken up by receptor-mediated endocytosis and released in lysosomes.

The cells of the growing foetus incorporate lipids into developing membranes. The foetus is dependent on the placental transfer of substrates from the mother which can then be metabolized into lipids through:

- biosynthesis from glucose in foetal tissue
- incorporation of fatty acids transferred from maternal to foetal circulation
- incorporation of fatty acids from circulating maternal lipoproteins after release by a placental lipoprotein lipase
- biosynthesis of lipids in the placenta itself which are then transferred to the foetal circulation.

In the foetus, glucose is a substrate for conversion into fatty acids and the glycerol moiety of glycerides. The placenta of most mammals is permeable to non-esterified fatty acids. There is an increased concentration of arachidonic acid in the placental circulation than on the maternal side, a form of **biomagnification**.

The development of human fat cells begins in the last third of the gestation period. At birth, a baby weighing 3.5 kg contains between 500 and 600 g of adipose tissue. As an important source of foetal fat reserves is circulating maternal lipids, the adipose tissue composition of the foetus and newborn infant will therefore reflect the fatty acid composition of the maternal diet. Concentrations of all lipoprotein classes increase in the maternal circulation during pregnancy mediated by the sex hormones.

Many of the essential fatty acids required during the perinatal period are for brain growth. Of these, 50% are long chain-polyunsaturated fats (PUFA), e.g. arachidonic acid 20 : 4 (n-6), adrenic acid 22 : 4 (n-6), docosapentaenoic acid 22 : 5 (n-6) and docosahexaenoic acid 22 : 6 (n-3) (DHA). The maximum rate of brain development in humans is in late gestation and in the early postnatal period. Long-chain derivatives of linoleic acid increase in the brain from mid-gestation to term. Little linoleic acid accumulates until after delivery, when the concentration increases three-fold. It is possible that linoleic acid is metabolized in the brain and neural tissue to long-chain polyunsaturated fatty acids throughout the perinatal period. The foetus obtains these fatty acids by placental transfer.

> The chemistry of the brain phospholipids is similar in most species. The concentration of essential fatty acids is low at 18 : 2 (n-6), 0.1–1.5%; and 18 : 3 (n-3), 0.1–1%, while arachidonic acid 20 : 4 (n-6) and docosahexaenoic 20 : 6 (n-3) acids predominate, 18–27% and 13–29%, respectively. This contrasts with the liver, where there is much variety. The precursor essential fatty acids are present in greater concentrations in the brains of carnivores and omnivores, with C22 : 6 predominating.

Phospholipids make up a quarter of the solid matter of the brain and are important for brain functioning. There appears to be no blood–brain barrier to fatty acid transfer in the foetus and infant. Variations in composition of dietary PUFAs may lead to different concentrations of long-term PUFAs in brain tissue. There appears to be powerful genetic control on the incorporation of these fatty acids into the cerebral cortex which overrides the effects of the wide variations in saturated and unsaturated fatty acid content of milk feeds fed to infants. It is not unreasonable that human breast milk with higher DHA and dietary n-3 fatty acids concentrations will be reflected in higher concentrations of long-chain PUFAs in brain cortical tissue. This may be beneficial to infant neurodevelopment. Excessive intake of linoleic acid should be avoided as there may be inhibition of α-linolenic metabolism

It is possible that deficiencies of fatty acids which should have been provided *in utero* will have consequences for brain and neurological development which are never retrieved in later development. This means that maternal nutrition, especially in the early months of development, are

all-important. In premature infants appropriate brain lipid development may not have been achieved. The diet after birth may be poorer in terms of long-chain fatty acids than that provided across the placenta.

At birth, the sole source of nutrition for the newborn baby is milk; some 50% of the available energy of milk is derived from fat. The baby's enzymes for fatty acid synthesis are suppressed and fat becomes the main source of energy and structure. Human fat contains a high proportion of long-chain PUFAs and this may be a response to the needs of the still-developing brain and nervous tissue. There are, however, wide differences in milk fat composition, depending on the mother's diet.

Vitamins

Water-soluble vitamins

Dehydroascorbate is taken up more rapidly by the syncytiotrophoblast membrane than is ascorbate. This is sodium-independent. After absorption the dehydroascorbic acid is metabolized to ascorbic acid and released into the foetal circulation.

Folic acid is probably transported by a specialist protein system. A low-capacity riboflavin transport system, which is saturable, results in a concentration in the placenta and partial metabolism to FAD and FMN. Thiamin transport has many of the features of riboflavin transport. Concentrations of vitamin B_{12}, biotin, vitamin B_6, pantothenic acid and nicotinic acid are higher in umbilical cord blood at delivery than in maternal blood. Water-soluble vitamins are actively transported across the placenta.

Lipid-soluble vitamins

Vitamin A is normally transported in the blood as a complex of retinol and retinol-binding protein. It is probable that maternal retinol-binding protein crosses the placenta. Foetal retinol-binding protein production occurs late in gestation. A considerable fraction of retinol is esterified and stored by the human placenta and subsequently hydrolysed and released into the foetal circulation unesterified. Concentrations of both 25(OH) vitamin D_3 and 1,25$(OH)_2$ vitamin D_3 are similar between maternal and cord blood. This suggests a diffusion of vitamin D across the placenta. Concentrations of lipid-soluble vitamins E and K are higher in maternal plasma than in umbilical cord plasma.

Inorganic nutrients, macrominerals and iron

Increasingly large quantities of **calcium** are required to support the development of the growing foetal skeleton. Concentrations of total and ionic calcium in cord blood exceed those in maternal blood, which indicates that there is an active placental transfer system. The transport system is similar to that in the intestine and kidney with partial saturation at physiological concentrations of maternal blood.

Little is currently known about the transfer of **magnesium** across the placenta.

Foetal serum concentrations of **phosphate** are higher than maternal concentrations. The substantial requirements of phosphate needed by the foetus in the last 3 months of pregnancy are transported against the concentration gradient.

Sodium transport across the brush border membrane of the placenta occurs by three mechanisms: (i) Na^+-H^+ exchange; (ii) co-transport with inorganic anions and organic solutes; and (iii) Na^+ conductants. The Na^+-H^+ exchange produces an electroneutral coupling of sodium influx from the maternal circulation into the placental membrane cells with H^+ reflux in the opposite direction. The Na^+-H^+ exchange at the placental brush border

may participate in functions that are essential to the normal growth and development of the placenta. The removal of H^+ from the cell by the exchanger is a significant factor in maintaining the intracellular concentration of H^+ and the regulation of intracellular pH.

The movement of hydrogen across the placenta is important for the maintenance of acid–base balance in the foetus and the placenta. Four potential mechanisms have been identified:

1. The Na^+-H^+ exchange at the placental brush border membrane transports H^+ along the Na^+ gradient across the brush border membrane.
2. A proton pump.
3. The coupled transport of H^+ and organic iron, e.g. lactate.
4. A protein-mediated H^+ transfer.

The foetus obtains its supply of **iron** from the maternal circulation in a receptor-mediated endocytosis of ferric transferrin.

There are two distinct mechanisms in the transfer of Cl^- across the placental brush border:

- An anion exchanger that accepts Cl^- as a substrate. The exchange ion appears to be HCO_3^-.
- There is a functional coupling between this transport system and the Na^+-H^+.

Sulphate, selenium, chromium, molybdenum and trace elements appear to be transported by active transport.

Summary

1. The foetus is separated from the mother by the placenta, an important organ which serves as both a protector of and nourishment source of the developing foetus. For the first 4–5 months, placental growth is greater than that of the foetus; thereafter foetal growth exceeds that of the placenta.
2. The placenta is a defined, transporting organ by the fourth week of pregnancy. It not only transports nutrients and is the source of oxygen, but also is the means of excretion via the bowel and kidneys and route of elimination of carbon dioxide.
3. Glucose is an important foetal and placental fuel. The transfer system is stereospecific, can be saturated, and is a mediated process.

4. Amino acid transport across the placental membrane involves mediated transport mechanisms through both the microvilli and the basal membrane.
5. There is metabolism of glucose within the placenta, producing lactate at a high rate.
6. Lipoprotein lipase on the maternal surface of the placental membrane hydrolyses triacylglycerol carried by maternal VLDL. The free fatty acids are taken up and may be used by the trophoblast or transferred to the foetus. The foetus is dependent on the placental transfer of substrates from the mother; these can then be used for lipid synthesis.
7. Much of the essential fatty acids required during the perinatal period are for brain growth. Of these, 50% are long-chain polyunsaturated fats (PUFA). The maximum rate of brain development in humans is in late gestation and in the early postnatal period.
8. Phospholipids comprise 25% of the brain's solid matter and are important for brain function. There appears to be no blood–brain barrier to fatty acid transfer in the infant or foetus. Variations in composition of dietary PUFAs may lead to different concentrations of long-term PUFAs in brain tissue.
9. Water- and lipid-soluble vitamins and inorganic nutrients are transported across the placenta by specific transport systems

Further reading

Battaglia, F.C. and Meschia, G. (1988) Foetal nutrition. *Annual Review of Nutrition*, 8, 43–61.

Farquharson, J., Cockburn, F., Patrick, W.A., Jamieson, E.C. and Logan, R.W. (1992) Infant cerebral cor- tex phospholipid fatty-acid composition and diet. *British Medical Journal*, 340, 810–13.

Zaret, K.S. (1996) Molecular genetics of early liver development. *Annual Review of Physiology*, 58, 231–52.

Chapter 11

The metabolism of nutrients

- The liver is an important organ in the metabolism and storage of nutrients.
- The liver is the first organ to which nutrients are exposed after absorption from the intestine.
- The liver consists of a cellular structure the acinus, composed of cells between the portal tract, which supplies the organ, and the hepatic vein, which drains it.
- There is a gradient of oxygen availability across the acinus; this influences the metabolism of nutrients in the liver.
- The liver excretes fat-soluble compounds such as cholesterol, bile acids, bilirubin glucuronide, hormones and drugs through the bile canaliculi.
- Plasma proteins and other important proteins are synthesized in the liver.

11.1 The liver

11.1.1 Introduction

The liver is the largest organ in the body, weighing 1200–1500 g – about 2% of the total weight of an adult. In the infant it is relatively larger and contributes to the characteristic rotund abdomen. The liver is important in the metabolism and storage of nutrients, particularly in that it is the first organ to which nutrients are exposed after absorption from the intestine.

11.1.2 Anatomy and blood supply

The liver has two blood supplies: the hepatic artery supplies the liver with arterial blood and the portal vein carries venous blood from the intestines and spleen. These vessels enter the liver through the porta hepatis on the lower surface of the right lobe of the liver. The hepatic artery and portal vein divide and perfuse the right and left lobes, which are joined by the right and left hepatic bile ducts, forming the common hepatic duct. The venous drainage from the liver is into the right and left hepatic vein and then into the inferior vena cava. Lymphatic vessels drain into glands around the coeliac axis.

The liver cells (hepatocytes) form 60% of the liver. They are polygonal in shape and are approximately 30 μm in diameter. The hepatocyte has three surfaces, one facing the sinusoid and the space of Disse, the second facing the canaliculus, and the third facing neighbouring hepatocytes. The space of Disse is a tissue space between the hepatocytes and sinusoidal lining cells.

Originally the liver was described in terms of hepatic lobules by Kiernan in 1833. Pyramidal lobules consist of a central hepatic vein and at the periphery a portal triad containing bile duct, portal vein radical and hepatic artery branch (Figure

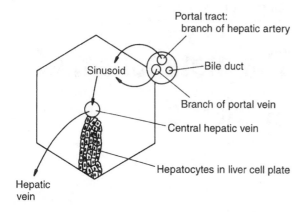

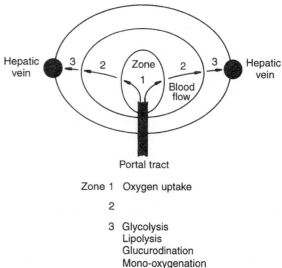

Zone 1   Oxygen uptake

2

3   Glycolysis
     Lipolysis
     Glucurodination
     Mono-oxygenation

FIGURE 11.1 Diagrammatic representation of a hepatic lobule with a central vein and portal tracts on its periphery, separated by radially arranged hepatocyte plates and sinusoids.

FIGURE 11.2 The hepatic acinus divided into zones, with the portal tract at the centre and the hepatic vein peripherally. The predominant metabolic processes in periportal and perivenous regions of the liver acinus are shown.

11.1). Columns of hepatocytes and blood vessels containing sinusoids divide these two systems. Hepatocytes radiate from a central vein and are interlaced by sinusoids. Liver cellular tissue is permeated by portal tracts and hepatic central canals. These run perpendicularly, are never in contact and are separated by about 0.5 mm. The sinusoids are irregularly placed, normally perpendicular to the plane of the central veins. The terminal branches of the portal vein carry blood into the sinusoids and the direction of flow is a function of the higher pressure in the portal vein compared with the hepatic vein. The central hepatic canals contain branches of the hepatic veins and are surrounded by a limiting plate of liver cells which is the interface between the portal tract and hepatic parenchyma. This is perforated by blood and lymphatic vessels and biliary radicles.

The **acinus** is the basis of a functional description of the liver, distinct from the concept of the central hepatic vein and surrounding liver cells. Rappaport first described the concept of a functional acinus. The liver acinus is a mass of hepatic parenchyma, dependent upon blood from hepatic arterial and portal venous branches which flows into sinusoids and drains to several veins supplying the intervening hepatic parenchyma. There is an oxygen gradient across the zone. Each acinus is

centred on the portal triad (Figure 11.2). The zone next to this is zone 1, which is well oxygenated. The hepatic venules are peripheral. The periphery of the acinus adjacent to terminal hepatic veins (zone 3) has a reduced oxygen supply and as such, is more vulnerable to injury, whether viral, toxic or anoxic.

Each efferent vein drains blood from several acini. Cells in zone 1 are nearest the portal tract and those in zone 3 are furthest from the tract and are at particular risk when blood flow is deficient. Complex acini are formed from three or more simple acini, supplied by a common large portal tract, the vessels of which divide to enter the terminal portal tracts of each simple acinus.

11.1.3 Metabolism in the liver

In the liver the periportal zone mitochondrial enzymes predominantly catalyse glucose production, oxidative energy metabolism, amino acid utilization, urea formation, bile acid and haem metabolism. The pericentral zone is important in glucose uptake and glutamine metabolism. This metabolic difference produces decreasing or increasing periportal to pericentral concentration gradients of glucose, amino acids, fatty acids, glycerol, ketone bodies and other nutrients. The magnitude and direction of gradients vary with the diurnal rhythm and during absorption and post-absorptive phases. The rate of enzyme synthesis, oxygen tension, and the effect of insulin and glucagon will also affect mitochondrial function. The nature of the diet, high carbohydrate, protein or fat can alter these gradients.

The liver contains a multitude of enzymes, sited within the liver structure in a manner that reflects metabolic needs. An example of this is glutaminase and glutamine synthetase activity. Glutamine synthetase activity is restricted to the small hepatocyte population surrounding the hepatic venules. Glutaminase activity is found in the periportal zone that produces ammonia which, in addition to the ammonia from the portal-drained visceral-derived blood, affects hepatic urea synthesis. Any ammonia escaping detoxification in the urea cycle is trapped in the perivenous glutamine synthetase-containing cells. The urea cycle is a low-affinity, high-capacity detoxification pathway. Urea which comes into contact with the perivenous glutamine synthetase system is detoxified by glutamine synthesis by a high-affinity, low-capacity detoxification pathway. The liver synthesizes and degrades glutamine.

The distribution of enzyme activity within the liver suggests that the periportal cells are important in oxidative respiration with a high activity of cytochrome oxidase. The hepatocytes in the periportal area may be more active in gluconeogenesis while those in the perivenous area may be more active in glycogenolysis. Perivenous hepatocytes contain more of the smooth endoplasmic reticulum and cytochrome P_{450} system which is important in drug metabolism. Under physiological conditions hepatocytes in the periportal region are involved in bile acid transport from the sinusoids to the biliary canaliculi.

Perivenous hepatocytes may be more involved in the synthesis of bile acids. In general the metabolic potential of many hepatocytes is not fully activated. Their activity will be dependent on blood flow, oxygen supply, hormone and nutrient substrates in the blood.

11.1.4 Biliary system

The excretory system of the liver is the **bile canaliculi**. These radicles converge to form increasing-sized bile ducts. The bile ducts amalgamate to form the common bile duct, which drains into the duodenum through the sphincter of Oddi. Bile is stored in the gallbladder so that a concentrate of bile can be mixed with the ingested meal to facilitate lipid absorption. The contraction of the gallbladder is subject to the control of the vagus nerve and hormones of which cholecystokinin is probably the most important. Through this system are eliminated chemicals with a molecular weight in excess of 300–400 Da, dependent on the species. Bile is also an important contributor to the digestion of fat in the duodenum. The primary or canalicular bile is modified by a mixing of ductular bile and the reabsorption of water and electrolytes in passage through the biliary tree. Canalicular flow can be measured using inert markers, e.g. erythritol or mannitol which pass unchanged through the biliary tree. In humans, canalicular flow is of the order of 450 ml per 24 hours. Canalicular secretion is dependent upon concentration gradients derived from osmotic forces. There is active secretion and a local osmotic pressure gradient. There are two secretion systems, each contributing half of the secretory drive:

- bile acid-dependent
- bile acid-independent, but sodium-dependent

There is a linear relationship between biliary secretion of bile acids and bile flow, called the **choleretic effect**. In humans the two prime bile acids, directly derived from cholesterol, are chenodeoxycholic acid 3,7-di-hydroxy-α-cholanoic acid and cholic acid 3,7,12-tri-hydroxycholanoic acid, which are conjugated to either taurine or glycine. Whether conjugation is principally to taurine or glycine depends on whether the diet is predominantly vegetarian (glycine) or meat (taurine) containing. The bile acids in the newborn child fed on milk are predominantly taurine-conjugated. The bile acids are excreted in the bile and are involved in lipid absorption through a series of steps:

- adherence to the fat emulsion
- activation of the lipase enzyme
- solubilization of the lipids through the formation of micelles

Following absorption of the lipid hydrolysate, the bile acids pass to the ileum to be absorbed. This is an **enterohepatic circulation**. Some bile acids, however, are not absorbed and pass into the colon where they lose their amino acid conjugate and the 7-α-hydroxyl grouping, producing lithocholic acid (3-α-hydroxycholanoic acid) and deoxycholic acid (3,12-α-dihydroxy acid). These are in general excreted in the faeces, though there is some colonic absorption.

Bilirubin, the final degradation of haemoglobin catabolism, is excreted in bile as the bilirubin glucuronide, the conjugation being catalysed by the enzyme, bilirubin glucuronyltransferase. The activity of this enzyme is modest at birth, a cause of jaundice. Bilirubin and other organic anions are extracted from the sinusoidal blood by binding to a protein ligand. This binding is affected by thyroxine, ethinyloestradiol and corticosteroids.

The biliary system also acts as an organ of excretion. Cholesterol is removed from the body, solubilized by bile acids. Other lipid-soluble compounds with a molecular weight of over 300 Da are excreted by this route. These include hormones such as the steroid hormones, thyroxine, xenobiotics such as antibiotics, and other drugs and dietary contaminants, e.g. DDT. These compounds may initially be oxidized, reduced or hydrolysed. They may then be made more soluble in bile through the addition of hydroxyl, carboxyl or amino groups, followed by conjugation with sulphate, glucuronic acid, or acetylation or methylation. UDP-glucuronyltransferase (UDP-GT) catalyses the conjugation of glucuronic acid with phenols, alcohols, carboxylic acids, thiols and amines; these are collectively called **aglycones**. UDP-GT enzymes are microsomal membrane-bound, this siting being important in the aglycone glucuronidation. Sulphation of endoge-

nous substances (including steroids and catecholamines) and foreign substances (including alcohols and hydroxylated compounds) is by cytosolic sulphotransferase. There are many substrate-specific forms of this enzyme. A substance may undergo both glucuronidation and sulphation, so the process is cooperative. Most hepatic glutathione-S-transferases are cytosolic. These catalyse the first step in mercapturic acid formation. The electrophilic metabolites of lipophilic compounds, e.g. epoxides of polycyclic hydrocarbons, products of lipid peroxidation and alkyl and aryl halides, are excreted into bile. Acetylation, which may be genotypically determined, is of amines and hydrazines and methylation of catecholamines, amines and thiols.

These compounds are excreted and pass along the small intestine to the colon without being absorbed. In the colon they are exposed to the colonic bacteria. A wide range of metabolic processes take place. Some of the metabolites are absorbed from the colon and reappear in bile. This constitutes the enterohepatic circulation. DDT, some drugs, e.g. digoxin, and hormones may remain in the circulation for a prolonged period. The non-absorbed fraction of the chemical is excreted in the faeces.

Kupffer cells and endothelial cells are in contact with the lumen of the sinusoids. These are mobile macrophages which take up particles such as old cells, tumour cells, bacteria, yeasts, viruses and parasites by endocytosis or phagocytosis.

11.1.5 Protein synthesis

The liver is the main site of synthesis of plasma proteins including albumin, fibrinogen, prothrombin, other clotting factors and caeruloplasmin. The normal liver produces approximately 10 g of albumin per day, the half-life of albumin being approximately 22 days. Prothrombin has a much shorter half-life.

In healthy humans, only 15–30% of hepatocytes in the liver contain albumin, though all hepatocytes are capable of its synthesis.

Summary

1. The liver is the largest organ in the body, is important in the metabolism and storage of nutrients, and is the first organ after the intestine to which absorbed nutrients are exposed.
2. The liver has two blood supplies: the portal vein carrying venous blood from the intestines and the spleen; and the hepatic artery which supplies the liver with arterial blood. Venous drainage from the liver is into the right and left hepatic veins which drain into the inferior vena cava. Lymphatic vessels drain into glands around the coeliac axis.
3. The liver cells (hepatocytes) constitute 60% of the organ. The space of Disse is a tissue space between the hepatocytes and sinusoidal lining cells.
4. The acinus is the basis of a functional description of the liver structure; this description differs from that of the hepatic lobule with a central hepatic vein and peripheral portal tracts.
5. The acinus depends upon blood from the hepatic arterial and portal venous branches. There is an oxygen gradient from zone 1 next to the portal triad to the periphery of the acinus adjacent to terminal hepatic veins (zone 3); this has a reduced oxygen supply and consequently is more vulnerable to injury, whether viral, toxic or anoxic.
6. The liver contains a multitude of enzymes, sited within the liver structure in a manner that reflects metabolic needs. The periportal zone mitochondrial enzymes predominantly catalyse glucose production, oxidative energy metabolism, amino acid utilization, urea formation, bile acid and haem metabolism. The pericentral zone is important in glucose uptake and glutamine metabolism.
7. The excretory system of the liver for chemicals, e.g. bile acids, cholesterol, bilirubin glucuronide, hormones and drugs, is the bile canaliculi. Bile is also an important contributor to the digestion of fat in the duodenum.
8. The liver is the main site of synthesis of plasma proteins including albumin, fibrinogen, prothrombin, other clotting factors and caeruloplasmin.

Further reading

Farrell, G.C., Murray, M., Mackay, I. and Hall, P. (1994) *Drug-Induced Liver Disease*, Churchill Livingstone, Edinburgh.

Kaplowitz, N. (1996) *Liver and Biliary Disease* 2nd edition, Williams and Wilkins, Baltimore, USA.

McGee, J. O'D. (1992) Normal liver: structure and function, in *Oxford Textbook of Pathology* (eds J.O'D. McGee, P.G. Issaacson and N.A. Wright), Oxford University Press; Oxford.

Rappaport, A.N. (1976) The microcirculatory acinar concept of normal and pathological hepatic structure. *Beitrage zur Pathologische*, **157**, 215–25.

Shearman, D.J.C., Finlayson, N.D.C. and Carter, D.C. (1989) *Diseases of the Gastrointestinal Tract and Liver*, Churchill Livingstone, Edinburgh.

Sherlock, S. (1989) *Diseases of the Liver and Biliary System*, Blackwell Scientific Publications, Oxford.

Smith, R.L. (1973) *The Excretory Function of Bile*, Chapman & Hall, London.

11.2 Thermodynamics and metabolism

- Thermodynamics describes the difference in energy between reactants and products.
- Thermodynamics does not describe how the reaction takes place.
- An understanding of thermodynamics is important in understanding the creation of energy stores and the utilization of nutrients.

11.2.1 Introduction

An important role of ingested nutrients separate from roles in growth and other functions is to provide **energy** to the body. The provision of energy has many similarities to other energy using systems such as were originally studied in machines. Consequently energy provision and utilization in the body has been described in the concepts used in thermodynamics. This section gives a simple non-mathematical introduction to thermodynamics, the laws of which are obeyed by all biochemical processes. Through such systems are determined: (i) whether or not a reaction will proceed spontaneously; (ii) how the reaction is designed; and (iii) how complex structures are folded.

Organisms can utilize carbohydrates, fats and proteins to produce energy by oxidation. Metabolic oxidation during respiration consumes oxygen and produces energy. Energy is stored work or the capacity to do work. Thermodynamics describes the difference in energy between the reactants and products, but not the mechanism whereby the reaction takes place. The latter part of this chapter incorporates the concepts of thermodynamics into the processes of intermediary metabolism.

The internal energy (E) of a molecule may be translational, rotational and vibrational. There are also electronic energies which involve electron–electron interactions, electron–nucleus interactions, and nucleus–nucleus interactions.

11.2.2 The first and second laws of thermodynamics

The **first law** of thermodynamics states that **the total amount of energy is constant**. The energy, however, may transform from one form to another; while the overall energy remains constant, energy may flow from the system to the surroundings, or from the surroundings to the system. Chemical energy may be translated into thermal, electrical or mechanical energy. Heat and work are equivalent and are a mechanism whereby the system may gain or lose energy to its surroundings. The change in energy is equivalent to the difference between the heat absorbed by the system and the work performed by it.

Definitions in energy and units of energy

$\delta E = q - w$.

E is the change in the energy of the system, q is the heat flow, and w is the work done.

Definitions

Joule: $1 J = 1 kg\ m^2\ s^{-1}$

$1 kg\ m^2 = 1 N\ m$ (Newton metre)

$= 1 W\ s$ (Watt second)

$= 1 CE$ (Coulomb volt)

Thermochemical calorie:

$1 Cal = 1 kcal = 4.184 kJ$

1 kilocalorie (kcal) is the amount of heat required to raise the temperature of 1 kg of water from 14.5 to 15.5°C

1 kilojoule (kJ) is the amount of energy needed to move 1 Newton of force over a distance of 1 km

---

The **second law** of thermodynamics has been expressed in a number of ways, but the gist is of an inevitable progression throughout the universe from a more ordered to a disordered state. This is the phenomenon of **entropy**, and is an index of the number of different ways in which a system can be arranged without changing its energy state. An irreversible process is accompanied by an increase in entropy. A reversible process is a finely balanced change, always in equilibrium with the surroundings. There is no loss of energy in a chaotic manner, nor increase in entropy. A system requires a certain amount of energy to be at any particular state, i.e. perpetual motion is impossible.

Enthalpy

The enthalpy of a **compound** is its internal energy. The enthalpy of a **reaction** is measured as the heat absorbed by the system at constant pressure. Chemical and biochemical reactions take place usually at constant pressure rather than at constant volumes. In most biochemical reactions there is little change in either pressure or volume. Where a reaction absorbs heat, the surroundings will cool, and the change in the enthalpy of the system is positive. If, however, heat is created by the system then the enthalpy (δH) of the reaction is negative, energy being lost from the system to the surroundings.

The change in the chemical internal energy is equal to the heat produced or consumed plus the work done on or by the reaction. The change in enthalpy in the reaction is:

$$\text{heat absorbed} - \text{work done}.$$

This is independent of the reaction mechanism, concentration of the substrate or product. The number of moles transferred in the reaction is important.

Entropy

The first law of thermodynamics cannot predict if a reaction will take place. Some reactions which are spontaneous may absorb heat. Natural systems allow only a proportion of their total potential energy to be available for work.

Entropy (S) is the randomness or disorder in a system. As S increases then the disorder increases. Entropy measures the extent to which the total energy of the system is unavailable for the performance of useful work.

Structural changes that make molecules more rigid reduce rotational, and vibrational energy and reduces entropy, e.g. double-bond formation or ring formation. This becomes important in the formation of comparatively rigid macromolecules from flexible polypeptide and polynucleotide structures. When a molecule is converted to a dimer, then the entropy is reduced, and this is a function of the molecular weight.

The physical state is a very important factor in the entropy of a compound. A gas has much more translational and rotational freedom than a liquid, which in turn has more freedom than a solid. In a reversible reaction there is a slow progression through intermediate states wherein the system is always in equilibrium. Entropy is increased on evaporation and melting; this can be measured from the heat of vaporization and from the heat of fusion. There is a greater increase in translational and rotational freedom in progressing from a liquid to a gas than in going from a solid to a liquid.

The entropy of solutions is affected by mixing two solvents, by placing solutes into solution, and by hydrogen bonding and other associations within the solvent or between solute and solvent. Each compound resulting from mixing will make a positive contribution to the entropy. An ideal solution is one in which there is no interaction between the molecules. Any intermolecular interaction causes a decrease in the entropy of mixing as the translational and rotational freedom of the individual molecules will be reduced. Solvation is the interaction between solute and solvent molecules and this is an important factor in the negative entropy of a compound in solution.

When a solute is added to a solvent, entropy decreases. This is due to the restricted movement of the solute and the restrictive movement of the solvent in the vicinity of the solute. The consequence is that small molecules become more highly hydrated than larger ones with the same charge. Anions are more readily hydrated than cations. The entropy of hydration becomes a larger negative number with increasing charge or decreasing radius.

Molecules and ions which have bipoles or hydrogen bond donor or acceptor groups interact strongly with water. Consequently this restricts the mobility of the water. An apolar molecule in water causes a decrease in entropy. The water is held on the surface of the molecule, forming a rigid structure of hydrogen bonds. Such an orientation of water is a feature of the apolar areas of proteins and other biological macromolecules.

These entropy changes are caused by binding or absorption of molecules on the surface of a macromolecule. This is important in the function of enzymes, substrates and inhibitors. When a substrate is bound onto an enzyme, water molecules are displaced and this results in a positive entropy change. Another entropy effect between enzyme and substrate is called the **chelation effect**. This occurs where if a molecule binds at several points to a protein the binding is stronger than through one point of attachment. This is important in enzyme kinetics. Any system tends to move towards the lowest enthalpy and the highest entropy. Entropy increases spontaneously.

Gibbs showed that for reactions occurring in equilibrium and at a constant temperature, the change in entropy is numerically equal to the change in enthalpy divided by the absolute temperature. Gibbs described the phenomenon of free energy. There is free or available energy and total energy in a system.

$$\Delta \text{ Free energy } =$$
$$\Delta\text{Enthalpy} - (\text{absolute temperature} \times \Delta\text{entropy})$$

where Δ = change. Free energy consists of enthalpy and entropy and determines whether or not a reaction can occur. This requires, however, that a chemical pathway is also possible and that the free energy change is negative and hence favourable.

The standard free energy of formation of a compound is the free energy difference between a compound and the state of the elements of which the compound is composed. If the system produces energy then it is negative or **exergonic** and will proceed without an external energy source, e.g. heat. If the system requires free energy to proceed (i.e. positive) or **endergonic** then the reaction can only take place with the addition of external energy; it is an **endergonic** reaction. **A reaction is only spontaneously possible if there is negative free energy.**

The molecules must, however, be in a reactive state. This reactivity is created by enzymatic activity in biological systems. Enzymes accelerate a reaction which is possible on energetic grounds. Proteins are highly ordered and have a defined constrained conformation. Amino acids are in a state of increased entropy and are more reactive than native proteins.

Free energies

Free energies are expressed in units of kilocalories per mole or kilojoules per mole. By subtracting the sum of the free energies of formation of the reaction from the total free energy of formation of the products, it is possible to calculate the free energy change of any reaction for which the free energies of formation of all reactants and products are known. There are tables available for free energies of formation of many compounds.

The standard free energy for a reaction can be calculated by adding or subtracting the standard free energies of two other reactions which will combine to give the desired reaction. When the concentration of reactants and products are at their equilibrium values, there is no change in free energy for the reactants to go in either direction. In biological systems, concentrations of compounds are maintained at values below their equilibrium values. This is achieved by removal of the end-products, either physically or by conversion to other compounds, so considerable free energy is generated by their reactions.

The free energy change of a system gives a measure of the maximum amount of useful work that can be obtained from a reaction. The larger the amount of work obtainable from a reaction, the further the reaction is from equilibrium. An important concept is the maximum amount of useful work that can be obtained from a reaction.

11.2.3 Work

The amount of work that is actually obtained depends on the pathway that the process takes. Work that can be derived from biological systems is of three types:

An enzyme may catalyse two reactions sequentially

An example is the phosphorylation of glucose by hexokinase/glucokinase in which ATP gives a phosphate to the glucose to yield ADP and glucose 6-phosphate. The reaction of glucose to phosphoric acid to form glucose 6-phosphate has an unfavourable equilibrium constant. The hydrolysis of ATP has a quite favourable equilibrium. When these reactions are coupled, then the overall reaction is favourable.

$$\text{Glucose} + \text{ATP} =$$
$$\text{glucose 6-phosphate} + \text{ADP}$$

This is the main source of free energy in energy-producing and energy-consuming systems. ATP can be hydrolysed by two quite different reactions. The α,β linkage can be hydrolysed to form adenosine monophosphate (AMP) and pyrophosphate ion. Or the β,γ linkage can by hydrolysed to form adenosine diphosphate (ADP) and phosphate ion. AMP can subsequently be hydrolysed in an irreversible reaction. This occurs in nucleic acid synthesis. ADP is involved in irreversible reactions. Less free energy is required to rephosphorylate the ADP product than that required for AMP. Considerable free energy is released on the hydrolysis of ATP or ADP, but not from AMP. The α,β and the β,γ phosphate links are high-energy linkages. Any compound can be phosphorylated by compounds which have a greater phosphate group transfer potential. So that ADP can be phosphorylated to ATP from glycerate 1,3-disphosphate (producing -11.8 kcal/mol) but not from glucose 6-phosphate (-2.2 kcal/mol). ATP has a standard free energy of hydrolysis of -7.5 kcal/mol. The ATP–ADP system is an acceptor and donor of phosphate groups. ADP accepts phosphate from high-energy compounds, e.g. phosphoenolpyruvate (-14.8 kcal/mol), glycerate 1,3-disphosphate (-11.8 kcal/mol) and phosphocreatine (-10.3 kcal/mol) and the lower energy compounds which ATP donates to phosphates, e.g. glucose 1-phosphate (-5.0 kcal/mol); glucose 6-phosphate (-3.3 kcal/mol) and glycerol phosphate (-2.2 kcal/mol). Regeneration of ATP is a requirement for all cells as part of the energy system of the cell.

1. **Mechanical work**: this is work that involves movement, e.g. muscle contractions, movement of chromosomes towards the opposite pole of a mitotic spindle.
2. **Osmotic and electrical work** (changes in concentration): this involves the movement of chemical compounds or ions against a concentration gradient. These processes include active absorption, and movement of chemicals from the blood into the cell.
3. **Synthetic work** (changes in chemical bonds): synthetic work is involved in the formation of the chemical bonds for complex organic molecules. Such large molecules are more complex and have a higher energy content than the molecules from which they are derived.

Some reactions are only possible by using high negative free energy of one reaction (exergonic reaction) to drive another reaction which has a low negative free energy (endergonic reaction). This means that there are many reactions happening at the same time, which are acting synergistically within a biological system. Reactions can occur simultaneously or rapidly in sequence without build-up of intermediate substrate.

The shortcoming of thermodynamic theory is that this applies to a closed system. In biology there is what is called an 'open system' in so far that there is removal and loss of compounds from a series of biologically and anatomically different areas.

In cells or organisms it is relevant that there is a high degree of collaboration between reactions. This results in diverse reactions taking place rapidly and with great specificity. Heat is generated and the reactions are chemically specific. The reactions require energy and use starting materials which are in general provided by nutrition.

11.2.4 Intermediary metabolism

This supplies the energy needed for:

- the synthesis of molecules
- the transfer of molecules across membranes
- the movement of molecules in tissues

Homeostasis or the development of a steady-state requires a constant and reliable supply of nutrients for the cell. The Japanese industrial system of 'just-in-time' is a man-made equivalent system for the cellular manufacturing system. That is, little is stored and just sufficient components are provided alongside the working station. The rate of flow of compounds along metabolic sequences is regulated to meet functional needs. The system, however, has to be economic in order not to spend excessive time or energy in wasteful processes. A series of reactions provide intermediates for initiating other reactions which branch from the overall reaction. The overall reaction passes along a pathway with a defined end-product. The conversion of glucose with 6 carbons to the 3-carbon pyruvate provides energy in the form of ATP in ten discrete enzymatically catalysed steps (Figure 11.3). In the first six reactions of this metabolic pathway, ATP is consumed, and in the last four steps ATP is regenerated.

Anabolic pathways, synthesis of proteins and polysaccharides always require energy (endergonic) and involve a decrease in entropy. **Catabolic pathways** are energy liberating (exergonic), and increase entropy. Acetyl-CoA is a chemical unit where many of the intermediate products of proteins, polysaccharides and lipids converge or emerge in new pathways. At the same time acetyl-CoA is a source of raw material for the synthesis of amino acids, sugars or fats. In general, catabolic pathways follow a converging pattern that results in common intermediates. Anabolic pathways start from common intermediates and end with quite different end products, i.e. a diverging pattern.

Just as reactions are organized into reaction sequences, enzymes that function in the different steps of this reaction are clustered together in the cell. Three groupings can be found:

1. The simplest in which all the enzymes for a particular pathway are different independent soluble proteins in the same cellular compartment. The reaction intermediates pass from one enzyme to another by diffusion through the cytoplasm or by transfer after contact with the sequential enzymes.
2. Where all the enzymes involved in a sequence of reactions form a **complex**. The intermedi-

ates are held in this complex until synthesis is complete.
3. Where functionally related enzymes are bound to a membrane. This applies to enzymes in the electron transport processes associated with oxidative phosphorylation.

An important aspect of biochemistry is that the direction of reactions in living cells is dictated in response to metabolic needs, rather than to

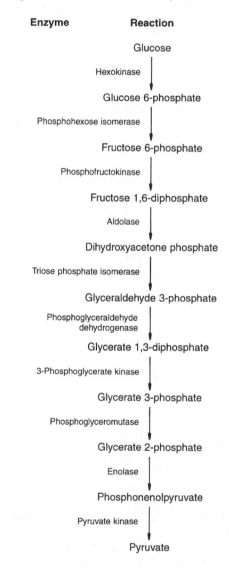

FIGURE 11.3 Metabolic pathway: the conversion of glucose to pyruvate.

thermodynamic factors. The direction of such changes must be independent of changes in concentrations of metabolic intermediates. The direction of the conversions, as well as their rates, is regulated by metabolic signals, for example the concentration of special local metabolites, available ATP or controlling hormones. Any reaction may be made thermodynamically possible by being coupled to a sufficient number of ATP to ADP conversions.

A number of different systems regulate metabolism. The **amount of enzyme** in a cell or compartment is regulated by protein synthesis and degradation. Enzymes are produced in their greatest amount when they are needed. Some are produced at moderate levels all the time and are constitutively expressed. Such enzymes include the enzymes involved in coenzyme synthesis.

Enzyme activity may be regulated by **non-covalent interactions** with small molecular regulatory factors, or by reversible covalent reactions, e.g. phosphorylation or adenylation of an amino acid side chain. Enzymes which are directly regulated occupy key positions in metabolic pathways. Often the first enzyme in a pathway is used to regulate the subsequent pathway. In a branched

chain pathway, end-product inhibition results in inhibition of the first enzyme after the point of branching in this reaction system.

$$\text{Energy charge} = \frac{\text{ATP} + 0.5\,\text{ADP}}{\text{ATP} + \text{ADP} + \text{AMP}}$$

The 0.5 in the numerator allows for ADP being half as effective as ATP in carrying chemical energy. The values for energy charge vary from 0 to 1. Some reactions in anabolic and catabolic pathways respond to variations in the value of the energy charge. The enzymes in catabolic pathways respond oppositely to enzymes in anabolic pathways.

If the ATP–ADP and occasionally ATP–AMP systems couple energy into biosynthetic sequences, this system is involved in controls which reflect the energy status of the cell. The energy status of the cell is called the energy charge and is the effective mole fraction of ATP in the ATP/ADP/AMP pool.

Summary

1. An important role of ingested nutrients is to provide energy to the body. Energy provision and utilization in the body has been described in the concepts used in thermodynamics, which describe the difference in energy between the reactants and products, but not the mechanism whereby the reaction takes place.
2. The total amount of energy is constant and this is described in the first law of thermodynamics. While the overall energy remains constant, energy may flow from the system to the surroundings, or from the surroundings to the system.
3. The enthalpy of a compound is its internal energy.
4. The second law of thermodynamics describes a transition from an ordered to a disordered state (change in entropy).
5. Only a proportion of total potential energy is available for work. There is free or available energy and total energy in a system: **Free energy** = **Enthalpy** – absolute temperature $\times$ **Entropy**. Entropy measures the extent to which the total energy of the system is unavailable for the performance of useful work.
6. A reaction is only spontaneously possible if there is negative free energy. The molecules must, however, be in a reactive state. This reactivity is created by enzymatic activity in biological systems. Enzymes accelerate a reaction which is possible on energetic grounds.
7. When a substrate is bound to an enzyme, water molecules are displaced and this results in a

positive entropy change. Another entropy effect between enzyme and substrate is called the chelation effect; this is important in enzyme kinetics.

8. Work that can be derived from biological systems: (i) mechanical work, which involves movements; (ii) osmotic and electrical work changes in concentration, or movement of chemical compounds or ions against a concentration gradient; and (iii) synthetic work with changes in chemical bonds.

9. Some biological reactions are only possible by using high negative free energy of one reaction (exergonic reaction) to drive another reaction which has a low negative free energy (endergonic reaction).

10. The direction of reactions in living cells is dictated in response to metabolic needs, rather than to thermodynamic factors. The direction of such changes must be independent of changes in concentrations of metabolic intermediates. The amount of enzyme in a cell or compartment is regulated by protein synthesis and degradation.

11. Enzyme activity is regulated by non-covalent interactions with small molecular regulatory factors, or by reversible covalent reactions, e.g. phosphorylation or adenylation of an amino acid side chain. Enzymes which are directly regulated occupy key positions in metabolic pathways.

Further reading

Atkins, P.W. (1990) *Physical Chemistry*, 4th edn, Oxford University Press, Oxford.

Harold, F.M. (1986) *The Vital Force. A Study of Bio-energetics*, Freeman & Co., New York.

Liang, Y. and Matschinsky, F.M. (1994) Mechanisms of action of non glucose insulin secretagogues. *Annual Review of Nutrition*, **14**, 59–81.

Newsholme, E.A. and Leech, A.R.. (1983) *Biochemistry for Medical Sciences*, John Wiley, Chichester.

Zubay, G. (1992) *Biochemistry*, W.C. Brown, USA.

11.3 Mitochondria

- Mitochondria are distinct cell structures central to metabolism.
- Mitochondria contain their own DNA derived entirely from the mother.
- Mitochondrial metabolism is important in oxygenation of all nutrients.
- Mitochondria are involved in the respiratory chain which creates a high energy difference between the two sides of the mitochondrial membrane. This enables ATP synthesis, an energy-requiring activity, to take place.

11.3.1 Introduction

Mitochondria are elongated cylinders, 0.5–1.0 m in diameter (Figure 11.4). They are important in metabolism, carry out most cellular oxidation reactions and produce most of the animal cell's ATP. The energy available in the form of NADH, NADH$_2$ and FAD is harnessed by electron transport chains to the synthesis of ATP. Energy is provided by the oxidation of nutrients.

The mitochondria consist of internal mem-branes, a smooth outer and a folded inner membrane. The infoldings of the inner membrane are called cristae and enhance the surface area of the inner membrane. The space within a crista connects to the intermembrane space through five to six tubular channels through the edges. The inner membrane separates the mitochondria into two distinct spaces:

- the internal or matrix space
- the intermembrane space

The outer membrane has few enzymatic activities, but is permeable to molecules with a molecular weight of up to 5000 Da. The inner membrane is impermeable to ions and polar molecules and limits the movement of energy-rich compounds between the mitochondrial matrix and the intermembrane space. The inner membrane is rich in enzymes.

The number of mitochondria in a tissue reflects the tissue's requirement for ATP; the more mitochondria, the more ATP produced. The mitochondrial enzymes and enzyme complexes are

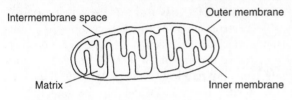

Intermembrane space

Outer membrane

Matrix

Inner membrane

Matrix: mix of enzymes including: citric acid cycle
DNA genes
ribosomes
tRNA

Inner membrane: folded into cristae to increase
surface area. Functions include
(i) Oxidative reaction of the
respiratory chain
(ii) ATP synthase
(iii) Transport protein for metabolites
(iv) Impermeable to small ions

Outer membrane: (i) Permeable to substances of molecular
weight <5000 Da
(ii) Lipid synthesis

Intermembrane space: enzymes using ATP

FIGURE 11.4 Mitochondria consist of an outer membrane, inner membrane and matrix. Two-thirds of the mitochondrial proteins are in the matrix, 20% in the inner membrane, 6% in the outer membrane. The matrix contains enzymes involved in the citric acid cycle and DNA genes, ribosomes, tRNA. The inner membrane is folded into cristae and is involved in oxidative reactions in the respiratory chain; contains ATP synthase, transport proteins for the transport of metabolites and is impermeable to small ions. The outer membrane is permeable for substances with a molecular weight < 5000 Da and contains enzymes involved in lipid synthesis. The intermembrane space contains enzymes using ATP.

organized in defined positions in the different compartments of the organelle.

11.3.2 Mitochondrial genes

The mitochondrion is unusual in that it contains its own DNA. Mitochondrial genes differ in that they are derived entirely from the mother through the ovum's contribution to the gamete. This means that the inheritance of the genes is non-Mendelian. The organelle DNA may evolve at its own rate. Organelle DNA is replicated by a different DNA polymerase to the nuclear DNA polymerase. Most organelle genomes take the form of a single molecule of DNA (mtDNA). Since there are several mitochondria in a cell there are a number of organelle genomes. Mitochondrial genomes in the human are of the order of 16 kilobases (kb). There may be 750 organelles in a cell, but the amount of DNA relative to the nuclear cell DNA is $< 0.1\%$. The mitochondria do not synthesize much protein; the enzymes encoded for include cytochrome b, cytochrome oxidase, NADH dehydrogenase and some units of ATPase. Most proteins are encoded by nuclear genes in free cytosolic ribosomes and enter the mitochondria by vectorial processing. These proteins undergo covalent modification, interact with coenzymes and are thereby activated. As both mitochondrial and nuclear gene products are required for effective mitochondrial activity, the expression of gene products must be coordinated. The failure to provide a necessary nutrient or mineral may upset this coordination.

11.3.3 Mitochondrial distribution and functions

Functions

These include the synthesis of ATP, terminal oxidation of pyruvate from carbohydrate and amino acid catabolism, β-oxidation of fatty acids and oxidation of acetate (acetyl-CoA) from ethanol and fatty acid, protein and carbohydrate oxidation. Mitochondria are also involved in oxidation of branched-chain amino acids, maintenance of nitrogen homeostasis and urea formation, oxidation of sulphite, activation of vitamin D_3 and synthesis of many important compounds, e.g. bile acids.

New mitochondria are formed by growth and division. Energy from oxidative reactions in the mitochondria is totally available for ATP synthesis. The control system includes rate limitation of the adenine nucleotide transporter system (delivery of ADP and removal of ATP from the matrix), supply of NADH from dehydrogenases, transfer of electrons through cytochrome oxidase (O_2).

Distribution

The distribution of mitochondria is in distinct patterns which are characteristic for different cell types. This is to provide ATP as required. In transport epithelia, e.g. renal proximal tubule cells and gastric parietal cells, mitochondria are found next to the pump transport systems in the membrane. In muscle and sperm cells, the mitochondria are found next to the contractile elements; with gap junctions in cardiac myocytes and synaptic terminals in neurones. Mitochondria are found between hepatocytes in the periportal region of the liver sinusoid and centrilobular region. Cells with different positions relative to blood supply are exposed to different nutrient provision. The nutrient supplies are a factor in determining the enzyme content and metabolic characteristics of the cells. In smooth muscle, glycolytic activity is separate so that the overall supply of nutrients to the mitochondria is independent of the supply of nutrients for glycolysis.

Mitochondrial electron transport and ATP

The outer membrane of the mitochondria has little enzymatic activity and is involved in the transmembrane transport of substances with a molecular weight of up to 5000 Da. The inner membrane is somewhat impermeable and allows protons or cofactors to move freely between the mitochondrial matrix and the intermembrane space. Contained in the membrane are proteins for oxygen consumption and the formation of ATP (Figure 11.5). The phospholipids of the inner membrane contain a high content of unsaturated

fatty acid, Cholesterol is found in the outer membrane but not in the inner.

The respiratory chain

The processes of glycolysis, β-oxidation and the tricarboxylic acid cycle generate NADH and $NADH_2$. These undergo oxidation–reduction reactions which provide a mechanism whereby electrons move from reduced coenzymes (NADH and $FADH_2$) to oxygen (Figure 11.6). Electrons are not passed directly from coenzymes to oxygen but progress through reversible oxidizable electron acceptors. Prosthetic groups that possess electrons or H^+ pass from one reversible oxidizable electron acceptor to another along the inner mitochondrial membrane. These reactions are coupled to the synthesis of ATP. At each step the electrons fall to a lower energy level until they are transferred to oxygen, which has the highest affinity of all of the carriers of electrons. Electrons

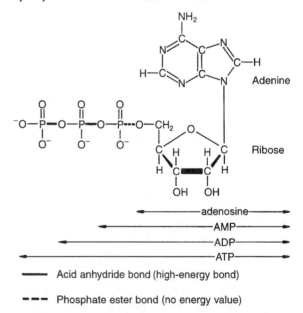

FIGURE 11.5 ATP (adenosine 5'-triphosphate). Adenosine monophosphate has no energy store but adenosine 5'-diphosphate, and particularly adenosine triphosphate, contain high-energy acid anhydride bonds. The conversion of ATP to ADP releases at least 8.4 kcal/mol.

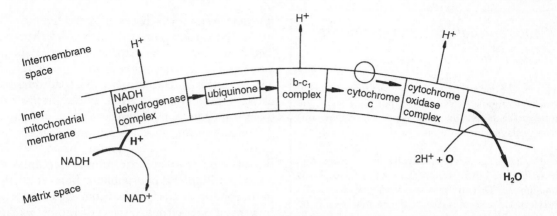

FIGURE 11.6 Respiratory chain across the inner mitochondrial membrane. NADH transfers an H^+ into the respiratory chain. The respiratory chain carries electrons from reduced coenzymes NADH and $FADH_2$ to oxygen. These reactions create a high energy difference between both sides of the membrane which enables ATP synthase and other energy-requiring reactions to take place.

bound to oxygen are in their lowest energy state. The free energy converted into an electrochemical gradient is harnessed for ATP synthesis.

The electron carriers include flavins, quinone, iron–sulphur complexes and copper atoms.

Flavins The dehydrogenases which catalyse the loss of an electron from succinate or NADH dehydrogenase require flavin adenine dinucleotide. There are many other flavoprotein dehydrogenases within the mitochondrial matrix. These transfer electrons to other membrane-bound flavoproteins. Some dehydrogenases (NADH dehydrogenase, succinate dehydrogenase and flavoprotein ubiquinone oxidoreductase) contain iron atoms bound to the sulphur atoms at the cysteine of the protein. These are non-haem iron proteins.

Ubiquinone This is a benzoquinone with a long side chain. The hydrocarbon tail makes the molecule strongly hydrophobic. Ubiquinone is reduced by the loss of two electrons to form dihydroquinone. Ubiquinone accepts electrons from the enzymatic activity of dehydrogenases and these electrons move on to the cytochrome system. Most of the major electron carriers, with the

exception of cytochrome c and ubiquinone, are to be found as large complexes. These contain polypeptide subunits, some of which have no electron-carrying groups. These enzymes also acquire some of the free energy released during the reaction.

The electron carriers include the cytochromes a, b and c, complexes of iron and porphyrin. Cytochrome c is found in the inner membrane of the mitochondria. The b- and a-type cytochromes are attached to large complexes on the membrane. The iron is attached to four pyrrole nitrogens in the porphyrin ring. The b cytochromes contain Fe-protoporphyrin IX in the same manner as haemoglobin and myoglobin. The a cytochromes include a modified prosthetic group haem A. The c cytochromes contain haem C wherein the vinyl groups of protoporphyrin IX are bound covalently to the protein by thioether links to cysteine residues. Cytochrome c is a globular protein with a molecular weight of 12 500 Da and consists of a planar haem group in the centre of the molecule which is surrounded by hydrophobic amino acids, including a strongly basic cluster of eight lysine residues. This positively charged area may be important in the activity of the cytochrome. The iron of the haem is bound to the sulphur atom of the methionine residue and the other to a histidyl nitrogen. The methionine is replaced by a second

Mitochondrial electron transport

Complex I – the NADH dehydrogenase complex

This is the largest complex in the mitochondrial inner membrane (26 different polypeptides). The total molecular weight is 10^6 Da. The complex consists of flavoprotein subunits with the iron–sulphur clustered at the centre, surrounded by a shell of hydrophobic proteins. Electrons move from one iron–sulphur centre to another and eventually to ubiquinone. This complex is placed very specifically between the two surfaces of the membrane of the mitochondria. The binding site for NADH faces towards the mitochondrial matrix space. This allows the oxidation of NADH which is produced in the matrix by the enzymes of the tricarboxylic acid cycle. This forms an enzymatic shuttle system. The ubiquinone then passes to the second complex, complex III, the cytochrome b c_1 complex in the central hydrophobic area of the membrane.

Complex II – the succinate dehydrogenase complex

Succinate dehydrogenase is embedded in the mitochondrial inner membrane, and consists of two iron–sulphur proteins, of molecular weight 70 000 and 27 000 Da. Oxidation takes place on the larger of the subunits and this is on the matrix side of the membrane. It possesses a 4-Fe centre and a molecule of FAD bound to a histidine residue of the protein.

Complex III

This consists of two different b-cytochromes and has a molecular weight of 450 000 Da. The subunits of complex III consist of stretches of hydrophobic amino acids which are transmembrane α-helices extending from one side of the inner membrane to the other. The oxidation of ubiquinone by the cytochrome b c_1 complex is related to the uptake of proteins from the matrix side of the mitochondrial inner membrane, and to the release of protons on the cytoplasmic side. This is called the **Q cycle**.

Complex IV – cytochrome oxidase

This contains atoms of copper in addition to the haems of cytochromes a and a_3. The copper in cytochrome oxidase may be important in preventing the production of superoxide in this reaction. Cytochrome oxidase consists of 6–13 subunits with a molecular weight range of 5000 to 50 000 kDa. Copper atoms are found in two of the large subunits. The three largest subunits are synthesized in mitochondria, the smaller ones in the cytoplasm.

histidine residue which leaves space for oxygen, water and carbon monoxide to bind to the iron. The iron atoms of the cytochromes are oxidized and reduced, cycling between the ferrous (Fe^{2+}) and ferric (Fe^{3+}). During anaerobic conditions the cytochromes become reduced. In the presence of oxygen they become oxidized.

The movement of electrons from NADH to the cytochrome b c_1 complex appears to progress by the diffusion of ubiquinone from one complex to another within the phospholipid bi-layer of the mitochondrial inner layer. From here there is movement of electrons to the cytochrome oxidase by the diffusion of reduced cytochrome c along the surface of the bi-layer. Cytochrome c is a water-soluble protein attached to the membrane surface by electrostatic weak interactions. NADH dehydrogenase can only oxidize NADH within the mitochondrial matrix. Protons are bound and released on either side of the inner membrane of the mitochondria when electrons progress down the respiratory chain. This is known as **chemiosmosis**.

Two electrons are used for every atom of oxygen that is reduced from O^{2-} to H_2O, The flow of two electrons through the complex of I,III,IV supports phosphorylation of ADP to ATP. This does not happen in the succinate dehydrogenase complex II.

There is a close relationship between respiration and phosphorylation. Respiratory control is the regulation of the rate of electron transport by ADP. Phosphorylation coupled to electron transfer in the respiratory chain is quite different from the mechanism of energy coupling in soluble enzymes. It has not been possible to identify 'high-energy' intermediate forms of electron carriers.

The **chemiosmotic coupling** system was first described by Peter Mitchell. Chemical reactions drive or are driven by the movement of molecules or ions between osmotically distinct spaces separated by membranes. Protons are pumped out of the mitochondria during respiration. The flow of electrons from reducing substrates to oxygen causes protons to move out of the matrix space to create a small pH gradient and an electrical potential difference across the mitochondrial inner

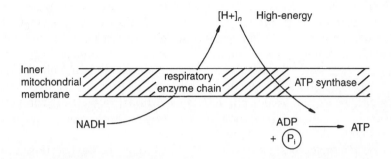

FIGURE 11.7 Reversible coupling. NADH releases hydrogen through the respiratory enzyme chain and enables a high energy state to be created whereby ATP synthase can generate ATP.

membrane (Figure 11.7). The electrochemical potential gradient for protons across the mitochondrial inner membrane is the mechanism by which electron transfer is coupled to phosphorylation; 10–12 protons are pumped out of the mitochondria for each pair of electrons that move down the respiratory chain from NADH to oxygen, while 6–8 protons are pumped for a pair of electrons from succinate. This creates an electrochemical potential gradient for protons across the membrane. This means that enzymatic reactions in such an organized situation as the membrane have what is called vectorial or a directional character. Protons move back into the matrix by ATP synthase activity and the formation of ATP. When mitochondria are oxidizing and form ATP at a constant rate, protons move inward at a rate that balances the rate at which protons are pumped out by the electron transport reaction. Three molecules of ATP are formed for each pair of electrons that pass along the whole system to oxygen. The oxidation of succinate involves the extrusion of eight protons per pair of electrons and the P/O ratio for this substrate is 2.0.

Proton ATP synthase

This multi-protein complex, F_1 is embedded on the mitochondrial inner membrane and includes five different polypeptide units with a total molecular weight of 360 000 kDa. A second complex, F_0, has five hydrophobic polypeptides and is an integral component of the inner membrane and holds F_1 to the membrane. This F_1–F_0 ATPase complex is to be found on the side of the mitochondrial inner membrane that faces the matrix. Such an orientation is very important for the function of this complex. Phosphorylation and electron transport are mediated by a proton motive force which is produced at one point of the membrane. The protons pass through the ATP synthase with the formation of ATP. It may well be that the movement of protons changes the enzyme conformation and is involved in the formation of ATP on the enzyme, the binding and release of ADP, phosphate and ATP.

ATP synthesis

ATP synthesized in mitochondria is transported from the mitochondria in exchange for ADP by an electrogenic process. The enzyme ATP/ADP exchange protein is the most abundant protein in the mitochondrial inner membrane. In contrast, the mitochondrial inner membrane does not transport NAD^+ or NADH. NADH is produced in the mitochondrial matrix by the tricarboxylic acid cycle or the oxidation of fatty acids. Electrons are transferred on the cytoplasmic NADH to the respiratory chain by a chain reaction which involves the conversion of dihydroxyacetone phosphate to glycerol 3-phosphate in the cytoplasm and the re-oxidation of the glycerol 3-phosphate to dihydroxyacetone phosphate by the mitochondrial glycerol 3-phosphate dehydrogenase. Another shuttle involves oxaloacetate and malate, malate being carried across the inner membrane by a specific transport system. The reverse reactions are important in gluconeogenesis. Oxaloacetate within the mitochondria can undergo transamination with the amino acid glutamate to form aspartate and α-ketoglutarate. The end-products can be transported from the mitochondria. Overall, the complete oxidation of glucose yields 36–38 molecules of ATP. Two molecules of ATP are formed in glycolysis; two more in the tricarboxylic acid cycle. Two molecules of NADH are produced in the cytoplasm by glycolysis. Eight molecules of NADH are generated in the mitochondrial matrix by the pyruvate dehydrogenase complex and the tricarboxylic acid cycle.

Succinate reduces two molecules of FADH to FADH$_2$. Re-oxidation of the eight mitochondrial NADHs and two FADH$_2$ generates 28 molecules of ATP and two of FADH$_2$. The glycerol 3-phosphate shuttle provides two electrons (FADH$_2$) allowing the formation of four molecules of ATP. The malate shuttle leads to the formation of six molecules of ATP. This gives a total of 36 or 38 molecules of ATP. However, the system is not 100% efficient, perhaps due to non-specific leakage of protons and other ions through the membrane. NADH from the cytoplasm may also be involved in other reductive biosynthetic reactions, e.g. formation of fatty acids.

11.3.4 Mitochondrial response to external factors

Mitochondria respond to metabolic and environmental conditions in ageing, training and pathological processes.

Giant mitochondria at least 10 μm in diameter develop during nutrient deficiency, when there are also degenerative changes in cristae structure and decreased oxidation of many oxidizable substances.

Riboflavin is an essential component of the prosthetic group of mitochondrial flavoproteins in the citric acid cycle, the respiratory chain and β-oxidation of fatty acids. In riboflavin deficiency, the mitochondria increase in size and volume. The structure of the cristae alters, increasing in number and size. Similar changes take place with protein-poor diets.

Variations in the protein content of the diet affect mitochondrial volume and number. The amount of hepatic urea cycle enzymes is directly proportional to the daily consumption of protein. Starvation also results in a net increase in the urea cycle enzymes.

Not only is the expression of the enzymes involved important but so also is the rate of degradation. Variations in carbohydrate diet can affect enzymes which relate to triglyceride synthesis.

The biochemical and morphological characteristics of skeletal muscle mitochondria change in response to submaximal endurance exercise training, increasing in number and size in those muscles involved in training. This means that this change is local and not the result of circulating hormones or metabolites. Pyruvate oxidation, citric acid cycle enzymes, electron transport, cytochrome content and fatty acid oxidation all increase in activity. Substrate utilization moves from carbohydrate to fat, thereby sparing glucose and allowing greater use of energy-utilizing processes with less glucose depletion and improved efficiency of energy expenditure. Efficiency of coupling of the mitochondrial proton motive force from biological oxidation with ADP phosphorylation is variable and under physiological control. Metabolism is inherently less efficient as ATP concentration increases relative to the concentration of ADP and activated phosphorus (P$_i$). ATP production increases before an increase in demand for ATP, thereby anticipating increased energy need, e.g. secretion, phagocytosis, or cell division. This may mean the release of the inhibition of ATP synthase by an endogenous inhibitor and stimulation of the activity of critical NAD$^+$-linked dehydrogenases. Exercise or inactivity affects energy efficiency. High ATP concentration means greater and immediately available utilizable energy, but at a reduced energy efficiency. More oxidizable substrate is needed to do the same productive work. Mitochondrial function is also affected by thyroid hormones and probably by the carbohydrate relative to protein content of the diet.

Thermogenesis in brown fat

Brown fat cells use electron transport which is uncoupled from phosphorylation for the generation of heat rather than for the formation of ATP. The heat produced by brown fat mitochondria is important for maintaining body temperature in the newborn. Brown fat produces heat at approximately 450 W/kg, which contrasts with 1 W/kg of other resting mammalian tissues. The inner membrane of brown fat mitochondria has a protein **thermogenin**; this is a channel for anions (OH^-, Cl^-). Thermogenin acts as an uncoupler, in that free energy is released in the electron transfer reaction in an electrochemical potential gradient and is then dissipated as heat. This requires ATP, ADP and GDP which bind to the protein and inhibit anion transport. This appears to be affected by fatty acid concentrations.

Summary

1. Mitochondria are elongated cylinders, important in metabolism and cellular oxidation reactions and produce ATP. The energy provided by nutrient oxygenation in the form of NADH, $NADH_2$ and FAD is harnessed by electron transport chains to the synthesis of ATP.
2. Mitochondria contain internal membranes, a smooth outer and a folded inner membrane. The inner membrane separates the mitochondria into two distinct spaces, the internal or matrix space and the intermembrane space. The outer membrane has few enzymatic activities, but is permeable to molecules with a molecular weight of up to 5000 Da. The inner membrane is impermeable to ions and polar molecules and limits the movement of energy-rich compounds and is rich in enzymes.
3. The mitochondrion is unusual in that it contains its own DNA and the genes are derived entirely from the mother.
4. The number of mitochondria in a tissue reflects that tissue's requirement for ATP.
5. Mitochondrial functions include the synthesis of ATP, terminal oxidation of pyruvate from carbohydrate and amino acid catabolism, β-oxidation of fatty acids and oxidation of acetate, fatty acid, protein and carbohydrate. Mitochondria are also involved in oxidation of branched-chain amino acids, maintenance of nitrogen homeostasis and urea formation, oxidation of sulphite, activation of vitamin D_3 and synthesis of many important compounds, e.g. bile acids.
6. The processes of glycolysis, β-oxidation and the tricarboxylic acid cycle generate NADH and $NADH_2$. These undergo oxidation–reduction reactions which provide a mechanism whereby electrons move from reduced coenzymes (NADH and $FADH_2$) to oxygen.
7. NADH dehydrogenase can only oxidize NADH within the mitochondrial matrix. Protons are bound and released on either side of the inner membrane of the mitochondria when electrons progress down the respiratory chain. This is known as chemiosmosis.
8. Mitochondria respond to metabolic and environmental conditions in ageing, training and pathological processes.

Further reading

Alberts, B., Bray, D., Lewis, J., Raff, M., Roberts, K. and Watson, J.D. (1994) *Molecular Biology of the Cell*, 3rd edn, Garland Publishers, New York.

Aw, T.Y. and Jones, D.P. (1989) Nutrient supply and mitochondrial function. *Annual Review of Nutrition*, 9, 932–51.

Harold, F.M. (1986) *The Vital Force; A Study of Bioenergetics*, W.H. Freeman & Co., New York.

Lewin, B. (1990) *Genes IV*, Oxford University Press, Oxford.

Zubay, G. (1992) *Biochemistry*, W.C. Brown, USA.

11.4 Cytochrome P$_{450}$

- Cytochrome P$_{450}$ is a group of enzymes which possesses mixed function oxidase activity.
- Cytochrome P$_{450}$ catalyse the oxidation of lipophilic chemicals.
- These allow the metabolism of a wide range of endogenous and exogenous compounds before excretion from the body in urine or bile or further metabolism.

11.4.1 Introduction

Cytochrome P$_{450}$ is the collective term for a family of haemoproteins found predominantly in the hepatic endoplasmic reticulum and which possess mixed function oxidase activity. Cytochrome P$_{450}$-dependent mono-oxygenases are a super gene family of enzymes which catalyse the oxidation of lipophilic chemicals through the insertion of one atom of molecular oxygen into the substrate. The enzyme systems include epoxide hydrolase, UDP-glucuronyltransferases, sulpho-transferases and glutathione-S-transferases. They are largely responsible for the metabolism of a wide range of endogenous and exogenous compounds before excretion from the body in urine or bile or further metabolism.

11.4.2 Expression and activity

While abundant in the liver, P$_{450}$ complex enzymes are also expressed at lower concentrations in almost all tissues in the body, with the exception of striated muscle and erythrocytes. Within a given species, different tissues may vary in their overall pattern of expression and activity of different cytochrome P$_{450}$ enzymes. All have characteristic ferrous carbon monoxide complexes and molecular weights of approximately 50 kDa. Their spectroscopic appearances show complex peaks near 450 nm. The P$_{450}$ molecule is formed from an apoprotein and a haem prosthetic group (iron protoporphyrin IX), linked by a cysteine residue. To date, 10 families of P$_{450}$ enzymes (some with subfamilies) have been found in mammals and there are probably 20 or more different mammalian P$_{450}$ proteins, all with distinct gene origins.

The activities of individual P$_{450}$ enzymes are altered by administration of, or exposure to, a wide variety of chemicals including carcinogens.

The P$_{450}$ enzymes show catalytic specificity determined by structural features in the substrate binding site. The specificity rests in the apoprotein region which is singular to each P$_{450}$ type. Different P$_{450}$ types vary in their rate of catalysis of a single reaction involving a particular substrate. Several can metabolize a single substrate but selectively catalyse reactions at different side chains of the molecule. Furthermore, enantiomeric pairs of a single substrate may be transformed at different rates by each P$_{450}$. When presented with a pro-chiral substrate a P$_{450}$ enzyme complex can selectively use one group, e.g. one hydrogen as opposed to another. Catalytic specificity has important consequences. Oxidation of a single compound may render the product more electrophilic and hence capable of reacting with macromolecules. Other oxidation reactions may result in products which are less biologically active and facilitate their elimination from the body.

P$_{450}$ oxidation is often sufficient for the elimination of a chemical though sometimes the generation of a more hydrophilic metabolite is required for elimination, e.g. glucuronide or sulphate. The cytochrome P$_{450}$ family contains microsomal enzymes which convert environmental organic compounds (xenobiotics) to either stable metabolites or to intermediate compounds that undergo additional metabolism by other enzyme systems. Some compounds are converted to highly reactive intermediates that covalently bind to cellular macromolecules.

11.4.3 Classification

Individual P$_{450}$ enzymes are classified on the basis of amino acid sequence similarities, those displaying less than 40% sequence homology being assigned to different families. In this way, 10 P$_{450}$

families can be isolated from mammalian liver, some of which contain more than one subfamily e.g. CYP II has eight subfamilies.

Of these 10 families, four are involved in steroidogenesis, two metabolize cholesterol in bile acid synthesis and one family is concerned with fatty acid and prostaglandin metabolism. The remaining three families (CYP I, CYP II, CYP III) are responsible for the metabolism of a wide variety of foreign compounds and xenobiotics including drugs and environmental carcinogens.

11.4.4 Metabolism by P$_{450}$

The cytochromes P$_{450}$ IA are responsible for the metabolism of compounds which include phenacetin (O-de-ethylation), caffeine (3-demethylation) and N-oxidation of carcinogenic arylamines.

Cytochromes P$_{450}$ IAI (CYP IAI) and IA2 (CYP IA2) are both members of the CYP IA gene family. This is an inducible enzyme system and induction is by many of its substrates as well as xenobiotics found in cruciferous vegetables, e.g. flavones and indoles. Aryl-hydrocarbon hydroxylase activity – an enzymatic reaction characteristic of CYP IA enzymes – is one such inducible reaction. Such an enzyme system also exists in the small intestine. It has been shown that the proton pump enzyme inhibitor omeprazole may induce CYP I genes in tissues such as human duodenum.

Polymorphism in human P$_{450}$ activity exists, most notably in the metabolism of debrisoquine (by CYP IIB6) which may vary in activity by several thousand-fold between individuals. Variant RNAs from defective IID genes are responsible for the poor metabolism of debrisoquine seen in some subjects. The clinical importance of this is unclear but rapid metabolizers of debrisoquine may be at increased risk of smoking induced lung cancer.

There are at least three P$_{450}$ IIIA proteins and these appear to be important in the metabolism of drugs such as nifedipine, cyclosporine A and many others. Omeprazole has been shown to be an inducer of human cytochrome P$_{450}$ IA and affects the clearance of diazepam and phenytoin.

Cigarette smoking can also induce members of this family. Substrates for cytochrome P$_{450}$ enzymes often bind to and are metabolized by the same proteins that they induce in the course of prolonged exposure. Another mechanism is 'down-regulation' of cytochrome P$_{450}$ II proteins, i.e. reduction in synthesis or activity which can occur with agents that induce members of the CYP IA family.

11.4.5 P$_{450}$ Gene control

In the genes controlling the synthesis of the P$_{450}$ enzyme system, there is significant genetic polymorphism. The diversity of P$_{450}$ enzymes appears to be controlled at the level of single genes, coding for single proteins and not by gene re-arrangements.

11.4.6 P$_{450}$ enzyme induction and inhibition

These P$_{450}$ enzyme systems may be induced and inhibited in many ways. There is competition between chemicals for processing by these enzymes. The imidazoles inhibit P$_{450}$ activity through a ligand interaction with the P$_{450}$ haem iron, thereby impeding the rate of P$_{450}$ reduction reactions. Cimetidine, used to reduce gastric hydrochloric acid production, is an imidazole derivative which inhibits P$_{450}$ activity and this effect resides in the hydrophobic nature of its side chain. In contrast, exposure to lipophilic substances may lead to an adaptive response by the P$_{450}$ system which in turn enhances the elimination of these substances. During induction of P$_{450}$ 1A1, for example, there is an interaction of the hydrocarbon with a hydrophobic cytosolic receptor (the 'Ah', i.e. aromatic hydrocarbon receptor). The substrate–ligand complex translocates to the nucleus and there is increased specific P$_{450}$ enzyme complex synthesis. Other inducers include glucocorticoids, e.g. dexamethasone and macrolide antibiotics also induce P$_{450}$ IIIAI activity. Phenobarbitone induces P$_{450}$ 2B1, 2C6 and 2A1 while ethanol is a potent inducer of P$_{450}$ 2E1 activity.

11.4.7 Cytochrome P$_{450}$ and cancer

The majority of known pro-carcinogens are converted by P$_{450}$-dependent reactions into electrophiles that bind covalently to DNA.

The cytochrome P$_{450}$ IA family appears to be particularly important in the biotransformation of pro-carcinogens and carcinogens. They activate or inactivate a broad range of pro-carcinogens which includes polyaromatic hydrocarbons, e.g. benzopyrene, arylamines and aflatoxin B$_1$. Enhanced activity of these enzyme systems may be important in the risk of cancer formation through bioactivation of pro-carcinogens. On the other hand, they may reduce this risk by the detoxification of carcinogenic parent molecules. Such bioactivation reactions include epoxidation and hydroxylation of the pro-carcinogen benzo[a]pyrene and related polyaromatic hydrocarbons. In contrast, protective pathways include 4-hydroxylation of the carcinogen aflatoxin to an inactive metabolite. The degree of induction of cytochrome P$_{450}$ IA depends on many factors, including dietary components and endogenous hormones. The extent of environmental exposure to particular chemicals as well is important as well as the variable extent of induction in individual subjects.

Summary

1. Cytochrome P$_{450}$ is the collective term for a family of haemoproteins found predominantly in the hepatic endoplasmic reticulum and which possess mixed function oxidase activity. This super gene family of enzymes catalyses the oxidation of lipophilic chemicals through the insertion of one atom of molecular oxygen into the substrate. The enzyme systems include epoxide hydrolase, UDP-glucuronyltransferases, sulphotransferases, and glutathione-S-transferases.
2. P$_{450}$ oxidation may be sufficient for the elimination of a chemical, but sometimes the generation of a more hydrophilic metabolite, e.g. glucuronide or sulphate is required.
3. In the genes controlling the synthesis of the P$_{450}$ enzyme system, there is significant genetic polymorphism. These P$_{450}$ enzyme systems may be induced and inhibited in many ways.
4. The cytochrome P$_{450}$ 1A family are important in the biotransformation of pro-carcinogens and carcinogens.

Further reading

Farrell, G.C., Murray, M., Mackay, I. and Hall, P. (1994) *Drug-Induced Liver Disease*, Churchill Livingstone, Edinburgh.

Guengerich, F.P. (1988) Roles of cytochrome P$_{450}$-enzyme in chemical carcinogenesis in cancer chemotherapy. *Cancer Research*, 48, 2946–54.

McDonnell, W.M., Scheiman, J.M. and Traber, P.G. (1992) Induction of cytochrome P$_{450}$ IA genes (CYP IA) by omeprazole in the human alimentary tract. *Gastroenterology*, 103, 1509–16.

Miles, J.S. and Wolff, C.R. (1991) Developments and perspectives on the role of cytochrome P$_{450}$ in chemical carcinogenesis. *Carcinogenesis*, 12, 2195–9.

Smith, R.L. (1973) *The Excretory Function of Bile*, Chapman & Hall, London.

11.5 Free radicals

- Metabolism is a balance of oxidation and reduction.
- Oxidation produces a number of free radicals.
- The anti-oxidant system controls the production of free radicals, which is an important protection to cells.

11.5.1 Introduction

Metabolism is a balance of oxidation and reduction. Oxidative processes have to be contained. Such a limitation requires defence mechanisms which are developed through an anti-oxidant system which nullifies excess free radicals engendered by oxidative processes. The anti-oxidant system includes mineral-dependent enzymes and small molecules, usually vitamins, which act as scavengers of reactive oxygen species. The enzymes include the selenium-dependent free radical scavenger, glutathione peroxidase. The small molecular weight molecules include the water-soluble ascorbic acid, glutathione and uric acid and the lipid-soluble carotenoids and vitamin E.

If there is a compound where A and B are two atoms covalently bonded. $_x^x$ represent the electron pair. Homolytic fission can be written as:

$A\,_x^x B \rightarrow A^x + B_x$

A^x is an A-radical, often written as A· and B_x is a B-radical (B·). Homolytic fission of one covalent bond in a water molecule will produce a hydrogen radical (H·) and a hydroxyl radical (OH·). A contrast to homolytic fission is heterolytic fission when one atom receives both electrons when a covalent bond breaks, i.e.

$A\,_x^x B \rightarrow A_x^{x-} + B^+$

This extra electron gives A a negative charge and B is left with a positive charge. Heterolytic fission of water gives a hydrogen ion H^+ and a hydroxyl ion OH^-.

Radicals are groups of atoms which behave as a unit. A free radical is any species capable of independent existence that contains one or more unpaired electrons. An unpaired electron is one that occupies an atomic or molecular orbital by itself. Radicals can easily be formed by homolytic fission when a covalent bond is split, when one electron from each of the pair shared remains with each atom.

11.5.2 Oxygen and its derivatives

- Oxygen is a good oxidizing agent, being a radical with two unpaired electrons, each located in a different π^* antibonding orbital. These two electrons have the same spin quantum number or parallel spins. This is the stable or **ground-state** of oxygen.
- **Oxidation** is the loss of electrons by an atom or molecule, e.g. the conversion of a sodium atom to the ion Na^+.
- **Reduction** is the gain of electrons by an atom or molecule, e.g. the conversion of a chlorine atom to the ion Cl^-.
- An **oxidizing agent** absorbs electrons from the molecule it oxidizes whereas a **reducing agent** is an electron donor.

$O_2 \xrightarrow[\text{reduction}]{\text{one-electron}} O_2^-$

$O_2 \xrightarrow[\substack{\text{reduction} \\ \text{(plus } 2H^+)}]{\text{two-electron}} H_2O_2$ (protonated form of O_2^{2-})

$O_2 \xrightarrow[\substack{\text{reduction} \\ \text{(plus } 4H^+)}]{\text{four-electron}} 2H_2O$ (protonated form of O^{2-})

O–O is relatively weak and hydrogen peroxide readily decomposes with a homolytic fission.

$H_2O_2 \longrightarrow 2OH^+$

When oxygen accepts electrons these electrons must be of antiparallel spin so as to fit in the vacant spaces in the π^* orbitals. Electrons in an atomic or molecular orbital would not meet this criterion as they would have opposing spins. This restricts electron transfer which tends to make oxygen accept electrons one at a time, and results in oxygen reacting slowly with many non-radicals. More reactive forms of oxygen known as **singlet oxygens** can be generated in pairs. There are thus no unpaired electrons; therefore this is not a radical. In both pairs of singlet oxygens the spin restriction is removed and the oxidizing ability is increased. If a single electron is added to the ground-state oxygen molecule it enters one of the π^* antibonding orbitals, producing a superoxide radical O_2^-. With only one unpaired electron superoxide is actually less of a radical than oxygen itself. Adding one more electron creates O_2^{2-}, the peroxide ion. As the extra electrons O_2^- and O_2^{2-} enter antibonding orbitals the strength of the oxygen bond decreases. Adding two extra electrons to the O_2^{2-} would remove the bond as they pass into the δ^*2p orbitals, so giving $2O^{2-}$ species.

Ozone

Ozone (O_3) is produced by the photodissociation of molecular oxygen into oxygen atoms.

$$O_2 \xrightarrow{\text{solar energy}} 2O$$
$$O_2 + O \longrightarrow O_3$$

Photodissociation of fluorinated hydrocarbons, e.g. CF_2Cl_2 and $CFCl_3$ produces chlorine atoms in the atmosphere which cause the breakdown of ozone. Nitric oxide (NO) and nitrogen dioxide (NO_2) can also affect ozone. These also involve free radicals.

11.5.3 Transition metals

All the metals in the first row of the d-block in the Periodic Table contain unpaired electrons and are radicals, with the exception of zinc. Copper is not exactly a transition element as the 3d-orbitals are full, but form the Cu^{2+} ion by loss of two electrons: one from the 4s- and one from the 3d-orbital, leaving an unpaired electron. The transition elements have a variable valency which allows changes in oxidation state involving one electron. Examples are iron, iron(II) ferrous ion and iron(III) ferric; copper(I) ion cuprous and copper(II) ion, which is cupric. Iron(III) is stable whereas iron(II) salts are weak reducing agents. Manganese can exist as Mn(III), Mn(IV) and Mn(VII). Zinc has only one valency, Zn^{2+}, and does not promote radical reaction.

Reactions of a free radical with a non-radical species may produce a different free radical which is more or less reactive than the original radical.

Radicals of OH^+ may be due to hydrogen removal, addition and electron transfer. Radicals produced by reactions with OH^+ are usually of reduced reactivity. An example of hydrogen removal is

$$CH_3OH \cdot \longrightarrow \cdot CH_2OH + H_2O$$
$$\cdot CH_2OH + O_2 \longrightarrow \cdot O_2CH_2OH$$

This may lead to two radicals joining to form a non-radical product joined by a covalent bond. $OH\cdot$ may react with aromatic ring structures and with the purine and pyrimidine bases in DNA and RNA. A thymidine radical may undergo a series of reactions.

11.5.4 Hydroxyl radicals in living systems

Ionizing radiation

The major constituent of living cells is water. Exposure of such water to ionizing radiation results in hydroxyl radical production and possibly damage to cellular DNA and to membranes.

Detection of hydroxyl radicals

Electron spin resonance (ESR) detects the presence of unpaired electrons. An unpaired electron has a spin of either $+\frac{1}{2}$ or $-\frac{1}{2}$ and behaves like a small magnet. If exposed to an external magnetic field an unpaired electron can align itself in a direction either parallel or antiparallel to that field and thus can have two possible energy levels. Electromagnetic radiation of the correct energy is absorbed and moves the electron from the lower energy level to the upper one. An absorption spectrum is obtained.

ESR spectrometers show not the absorption but the rate of change of absorption. ESR is very sensitive and can detect radicals at concentrations as low as 10^{-10} mol/l, provided that there is some stability. For very unstable radicals a number of other techniques are available. Flow systems are used whereby the radicals are continuously generated in the spectrometer so as to maintain a steady-state concentration. An alternative procedure is spin-trapping. A highly reactive radical is allowed to react with a compound to produce a long-lived radical, e.g. nitroso compounds (R·NO) in reaction with radicals may produce nitroxide radicals that have a prolonged lifetime.

Spin-trapping methods are often used to detect the presence of superoxide and hydroxyl radicals in biological systems, e.g. the formation of organic radicals during lipid peroxidation. Trapping molecules include *tert*-nitrosobutane, α-phenyl-*tert*-butylnitrone, 5,5-dimethylpyrroline-*N*-oxide, *tert*-butylnitrosobenzene and α-(4-pyridyl-1-oxide)-*N*-*tert*-butylnitrone. The ideal trap reacts rapidly and specifically with the radical under study to produce a stable product with a highly characteristic ESR. It has to be noted, however, that reactions with such radicals may inhibit the process and dynamics.

Aromatic hydroxylation

Aromatic compounds react rapidly with hydroxyl radicals to produce hydroxycyclohexadienyl radicals. Such a radical may undergo dimerization to give a product that may decompose to give biphenyl or result in a mixture of phenol and benzene. This is called a **disproportionation reaction**. Disproportionation is a reaction in which one molecule is reduced and an identical molecule oxidized. One radical molecule is reduced to benzene and another oxidized to phenol. The attack of OH· upon phenol itself produces a mixture of hydroxylated products.

Production of singlet oxygen

Two singlet states of oxygen exist when the spin restrictions are removed resulting in increased reactivity. The singlet state of oxygen is energetic. The singlet oxygen produced on illumination with light of the correct wavelength can react with other molecules or may attack the photosensitizer molecule itself. These are known as **photodynamic effects**. Photosensitization reactions involving singlet oxygen may be important in biological systems, e.g. chloroplasts of higher plants and the retina of the eye.

Singlet oxygen can interact with other molecules, to combine chemically or transfer its excitation energy by returning to the ground state while the molecule enters an excited state. This is known as **quenching**. The most important of these reactions of singlet oxygen involve compounds that contain carbon–carbon double covalent bonds. Such bonds are present in carotenes, chlorophyll and the fatty acid side chains in membrane lipids. Such double bonds separated by a single bond (conjugated double bonds) often react to give endoperoxides. If one double bond is present, a reaction can occur in that the singlet oxygen adds on and the double bond shifts to a different position. Damage to proteins by singlet oxygen is often due to oxidation of methionine, tryptophan, histidine or cysteine residues.

Reactions of the superoxide radical

Superoxide dismutase specifically catalyses removal of the superoxide O_2^-. The concept of the free radical suggests that O_2^- formation is a major factor in oxygen toxicity. Superoxide dismutase enzyme activity is an essential defence. Superoxide chemistry depends on whether the reaction is in aqueous solution or in organic solvents.

Superoxide in aqueous solution acts as a base, i.e. acceptor of protons (H^+ ions). When O_2^- accepts a proton, it forms the hydroperoxyl radical ($HO_2^·$). $HO_2^·$ can dissociate to release H^+ ions again. There is a supply of protons, i.e. it is an acid. When O_2^- and H^+ ions are mixed, an equilibrium is established

$$HO_2 \cdot \longleftrightarrow H^+ + O_2^-$$

Superoxide in aqueous solution is also a very weak oxidizing agent, i.e. an electron acceptor and oxidizes ascorbic acid. On the other hand, there is no significant oxidation of NADH or NADPH. It will, however, interact with NADH bound to the active site of the enzyme lactate dehydrogenase but not other dehydrogenases.

When superoxide is dissolved in organic solvents the ability to act as a base and as a reducing agent is increased and may reduce dissolved sulphur dioxide in organic solvents (which it does not do when in aqueous solution). O_2^- has a longer life and is able to act as a nucleophile, and is attracted to centres of positive charges in a molecule. Superoxide can displace chloride ion from chlorinated hydrocarbons such as chloroform.

The disappearance of O_2^- in aqueous solution is a so-called **dismutation reaction**, i.e.

$$O_2^- + O_2^- + 2H^+ \rightarrow H_2O_2 + O_2$$

This dismutation reaction is most rapid at acidic pH values needed to protonate O_2^- and will become slower as the pH rises and becomes more alkaline.

As well as acting as a weak base, O_2^- in aqueous solution is a reducing agent, i.e. a donor of electrons, and may reduce cytochrome c. Iron is reduced from the Fe^{3+} to the Fe^{2+} state.

Any biological system generating O_2^- may produce hydrogen peroxide by the dismutation reaction. On the other hand, the O_2^- may be intercepted by some other molecule, e.g. cytochrome c. Hydrogen peroxide, through O_2^-, may be formed in mitochondria and microsomes. Enzymes that produce hydrogen peroxide without the free O_2^- radical include glycollate oxidase, D-amino acid oxidase and urate oxidase.

11.5.5 Protection against oxygen radicals in biological systems

This is referred to as the superoxide theory of oxygen toxicity.

Protection by enzymes

It is important for cells to control the amount of **hydrogen peroxide** in the cell. This may be achieved enzymatically:

catalase: $2H_2O_2 \longrightarrow 2H_2O + O_2$
peroxidases: $SH_2 + H_2O_2 \longrightarrow S + 2H_2O$

where SH_2 is the substrate being oxidized. Catalase is present in all major organs, particularly the liver and red cells, and in modest amounts in the brain, heart and skeletal muscle.

Catalases

Catalases consist of four protein subunits containing a haem (Fe(III) protoporphyrin) group bound to the active site.

The catalase reaction is:

Fe(III) + $H_2O_2 \rightarrow$ compound I
compound I + $H_2O_2 \rightarrow$
catalase – Fe(III) + $2H_2O + O_2$

Glutathione peroxidase is found in animal tissues but not in higher plants or bacteria. The substrate is the thiol compound, glutathione (GSH). Most GSH exists as the free compound, but approximately one-third of the total cellular GSH may be present as mixed disulphides with other compounds that contain –SH groups, e.g. cysteine, coenzyme A and the –SH of the cysteine residues of proteins. Glutathione peroxidase enzymatically assists the oxidation of GSH to GSSG at the expense of hydrogen peroxide:

$$H_2O_2 + 2GSH \rightarrow GSSG + H_2O$$

This enzyme is found particularly in the liver and less so in the heart, lung and brain and hardly at all in muscle.

Glutathione oxidase consists of four protein subunits containing selenium at the active site.

Selenium belongs to group VI of the Periodic Table and has properties intermediate between a metal and a non-metal. Selenium may well be at the active site as selenocysteine wherein the sulphur atom is replaced by a selenium atom.

The GSH reduces the selenium and the reduced form of the enzyme reacts with hydrogen peroxide. The reduction of GSSG to GSH is by glutathione reductase:

$$GSSG + NADPH + H^+ \rightarrow 2GSH + NADP^+$$

The NADPH is derived from the oxidative pentose phosphate pathway. The pentose phosphate pathway is controlled by the supply of $NADP^+$ to glucose 6-phosphate dehydrogenase.

Glutathione reductase reduces the NADPH/$NADP^+$ ratio. The enzyme contains two protein subunits with FAD at the active site. NADPH reduces the FAD and the electron moves to a disulphide bridge (–S–S–) between the two cysteine residues on the protein. The two –SH groups interact with GSSG, which results in the reduction to 2GSH.

In mammalian red cells the pentose phosphate pathway is important in providing NADPH for glutathione reduction. However, red cell glucose 6-phosphate dehydrogenase deficiency results in haemolysis and is a common occurrence in populations in tropical and Mediterranean areas. Curiously this enzymatic deficiency is protective against infestation with the malarial parasite.

Superoxide dismutase

The superoxide radical is believed to be a major factor in oxygen toxicity and the superoxide dismutase enzymes are an essential defence mechanism. In general, oxidative metabolism results in water and carbon dioxide. A few enzymes, e.g. glycollate oxidase, produce hydrogen peroxide. A number of enzymes reduce oxygen to O_2^-.

The copper–zinc enzymes

The copper–zinc-containing superoxide dismutases are specific for the superoxide radical. These enzymes are found in all animal cells and have a molecular weight in the order of 32 000 Da. They contain two protein subunits, each bearing an active site containing one copper ion and one zinc ion. The reaction catalysed is:

$$O_2^- + O_2^- + 2H^+ \rightarrow H_2O_2 + O_2$$

The copper ions undergo alternate oxidation and reduction. The zinc does not appear to function in the catalytic cycle but stabilizes the enzyme.

The amino acid structure of the copper–zinc superoxidase dismutase from yeast, human red cell, horse liver and bovine red cell appears to be similar and consists of eight antiparallel strands of β-pleated sheet structure forming a flattened cylinder and three external loops.

The copper ion is held at the active site by interactions with the nitrogens in the imidazole ring structure of four histidine residues. Zinc is connected by a bridge to the copper by interaction with the imidazole of histidine. The surface of each protein is largely negatively charged repelling O_2^-, though there is a positively charged channel leading into the active site.

Manganese enzymes

The manganese superoxide dismutase is a tetramer and contains 0.5 or 1 ions of manganese per subunit which is essential for catalytic activity. The amino acid sequence of all manganese superoxidase dismutases is similar in animals, plants or bacteria.

Protection by small molecules

Ascorbic acid

Ascorbic acid is a cofactor for the enzymes proline hydroxylase and lysine hydroxylase involved in the biosynthesis of collagen. These enzymes contain iron at their active sites. Ascorbic acid is necessary for the action of dopamine-β-hydroxylase which is involved in the conversion of dopamine to noradrenaline. Ascorbic acts as a reducing agent, e.g. the reduction of Fe(III) to Fe(II). The loss of one electron by ascorbic acid results in the semidehydroascorbate radical which then oxidizes to dehydroascorbate. Dehydroascorbate is unstable spontaneously converts into oxalic and L-threonic acid. Ascorbic acid scavenges singlet oxygen.

Glutathione
Glutathione (GSH) is a substrate for glutathione peroxidase and for dehydroascorbate reductase. Glutathione is also a scavenger of hydroxyl radicals and singlet oxygen. Glutathione is a cofactor for such enzymes as glyoxylase, malonylacetoacetate isomerase, prostaglandin endoperoxide isomerase and may be involved in the synthesis of thyroid hormones.

Uric acid
Uric acid is present in plasma and is a scavenger of singlet oxygen and hydroxyl radicals and may, for example, inhibit lipid peroxidation.

11.5.6 Oxidative toxicity

It has been suggested that in tissue deprived of oxygen there is a rapid conversion of xanthine dehydrogenase to oxidase activity. During ischaemia there is degradation of ATP in the ischaemic cells and accumulation of hypoxanthine. However, when oxygen is restored there is reperfusion damage. The accumulated hypoxanthine is oxidized by xanthine oxidase and the excessive O_2^-

causes further damage. Other molecules that oxidise in the presence of oxygen to yield O_2^- include glyceraldehyde, the reduced forms of riboflavin, adrenaline and thiol compounds, e.g. cysteine.

Lipid peroxidation

Lipid peroxidation is the oxidative deterioration of polyunsaturated lipids. Superoxidation of a polyunsaturated fatty acid involves the removal of a hydrogen ion from a methylene ($-CH_2-$) group. This leaves an unpaired electron on the carbon $-CH-$ and the production of a conjugated diene to give a peroxy radical $R-OO^+$. The propagation stage of lipid peroxidation is the addition of a hydrogen atom from another lipid molecule so that a chain reaction develops to yield a lipid hydroperoxide, $R-OOH$. An alternative is the production of cyclic peroxides.

Vitamin E

Vitamin E concentrates in the interior of membranes and in tissues where a lipid antioxidant activity is important. Vitamin E quenches and reacts with singlet oxygen and reacts with the superoxide radical, though this reaction is slow. Vitamin E can react with lipid peroxy radicals to form vitamin E radicals, thereby interrupting the chain reaction of lipid peroxidation, i.e. a chain terminator. It is possible that the vitamin E radical may be reduced back to vitamin E by vitamin C.

Glutathione peroxidase

This enzyme disposes of hydrogen peroxide. It has an important role in protection against lipid peroxidation in lipoproteins.

Summary

1. Metabolism is a balance of oxidation and reduction. Oxidative processes have to be contained through defence mechanisms utilizing an antioxidant system which nullifies excess free radicals engendered by oxidative processes. The antioxidant system includes mineral-dependent enzymes and small molecules (usually vitamins) which act as scavengers of reactive oxygen species.

2. A free radical is any species capable of independent existence that contains one or more unpaired electrons. An unpaired electron is one that occupies an atomic or molecular orbital by itself.

3. Oxidation is the loss of electrons by an atom or molecule, e.g. the conversion of a sodium atom to the ion Na^+. Reduction is the gain of electrons by an atom or molecule, e.g. the conversion of a chlorine atom to the ion Cl^-. An oxidizing agent absorbs electrons from the molecule it oxidizes whereas a reducing agent is an electron donor.

4. The major constituent of living cells is water. Exposure of such water to ionizing radiation results in hydroxyl radical production and possibly damage to cellular DNA and to membranes.

5. It is important for cells to control the amount of hydrogen peroxide in the cell. This may be achieved enzymatically, e.g. catalase, peroxidases, superoxide dismutase and copper–zinc and manganese enzymes.

6. There is protection against oxygen radicals by small molecules, e.g. ascorbic acid, glutathione and uric acid. Lipid peroxidation is the oxidative deterioration of polyunsaturated lipids. Superoxidation of a polyunsaturated fatty acid involves the removal of a hydrogen ion from a methylene ($-CH_2-$) group. Protection is given by vitamin E and glutathione peroxidase.

7. It has been suggested that in tissue deprived of oxygen there is a rapid conversion of xanthine dehydrogenase to oxidase activity. During ischaemia there is degradation of ATP in the ischaemic cells and accumulation of hypoxanthine. However, when oxygen is restored there is reperfusion damage. The accumulated hypoxanthine is oxidized by xanthine oxidase and the excessive O_2^- causes further damage.

Further reading

Fridovich, I. (1995) Superoxide radical and superoxide dismutases. *Annual Review of Biochemistry*, **64**, 97–112.

Halliwell, B. and Gutteridge, J.M.C. (1985) *Free Radicals in Biology and Medicine*, Scientific Publications, Clarendon Press, Oxford.

Riemersma, R.A. (1994) Epidemiology in the role of antioxigents in preventing coronary heart disease: a brief overview. *Proceedings of the Nutrition Society*, **53**, 59–65.

11.6 Carbohydrate metabolism

• Glucose is a major source of energy production and it is from glucose that the synthesis of a range of chemicals occurs within the body.

• The rate of absorption of glucose may dictate blood glucose concentrations.

• Glucose is derived from dietary sources, or glycogen stores by glycogenolysis or from other nutrients through gluconeogenesis.

• Blood glucose may be disposed of as fat, glycogen stores, used in synthetic pathways or in glycolysis or excreted after metabolism as carbon dioxide.

• Glycolysis is the pathway by which glucose is metabolized into two 3-carbon molecules which are further metabolized into other compounds or pass into the tricarboxylic acid cycle with the generation of ATP.

• Excess glucose may be stored as glycogen or converted to fatty acids to be used as a fat store or as a fuel.

• Glycogen synthesis is dictated by glycogen synthase, which is controlled by a number of hormonal and other controls.

- Gluconeogenesis is the synthesis of glucose from non-sugar sources or the release of glucose by the liver into the blood.
- Ketone bodies are acetoacetate, β-hydroxybutyrate and acetone; they are produced during starvation or metabolic dysfunction such as uncontrolled diabetes mellitus.

11.6.1 Introduction

Despite the central importance of carbohydrates for metabolic processes, carbohydrates are not in the strict sense essential nutrients. Nevertheless, glucose is an important constituent in the provision of energy to the body and in that respect it is essential. If the carbohydrate content of the diet is reduced or low then more expensive and less immediately utilizable energy sources such as fat and protein have to be used. This is expensive nutritionally and metabolically.

Dietary carbohydrates, whether these be eaten as starches, glycogen or glucose are metabolized as their constituent monosaccharides. The presence of the monosaccharides, glucose, galactose and fructose in tissues is essential for normal nutritional development.

11.6.2 Individual sugars

Glucose

Dietary carbohydrates are 70–90% glucose. Starch that is absorbed in the jejunum is metabolized largely as glucose. The metabolic differences between different physical types of starch may be a result of digestion and absorption in different parts of the lumen of the gut.

Glucose in the blood is derived from food, the breakdown of preformed glycogen in the liver or by glucose molecules formed in the liver from circulating intermediate metabolites. Dietary glucose is absorbed, enters the body through the portal vein, and crosses the liver, where some of it is removed and converted by phosphorylation into glucose 6-phosphate and glycogen. A reliable blood glucose concentration is important for tissue metabolism. There are controls on carbohydrate metabolism which depend on whether there is an abundance or shortage of carbohydrate. Glucose may be taken up by tissues, particularly skeletal muscles. Insulin facilitates the active transport of glucose into the muscle and fat cells wherein glucose is converted into glucose 6-phosphate by hexokinase. The glucose membrane cell transport systems of the brain, the haematopoietic system, bone marrow and red cells, liver and endocrine pancreas do not require insulin.

Glucose is the prime fuel source for the red cell and kidney medulla and for the central nervous system. The liver and skeletal and cardiac muscle can use alternative fuels such as ketone bodies and fatty acids. Adipose tissue converts glucose to α-glycerol phosphate for triglyceride formation.

Postprandially in humans there is massive glycogen synthesis. Following a high protein meal much of the glucose incorporated into glycogen originates from pyruvate and oxaloacetate. The amino acids absorbed by the intestinal mucosa pass into the blood and are taken up by a variety of tissues including the liver. The carbon skeleton from the amino acids may be stored as glycogen. In contrast, during fasting glycogen may be converted to glucose 1-phosphate and thence to pyruvate. The liver is central to this process, exporting glucose to the blood from the glycogen which is stored postprandially. When the blood glucose concentration begins to fall, the liver supplies glucose to the blood following the breakdown of glycogen. When the blood glucose level is further reduced the hexose phosphate pool is supplemented from glycogen and from pyruvate and oxaloacetate. Glucose 6-phosphate is hydrolysed and glucose supplied to the blood.

Galactose

Galactose is absorbed through the same active transport system in the small intestine as glucose. Galactose enters the portal venous blood but is almost entirely removed by the liver at first pass so that little enters the peripheral circulation. The blood concentration of galactose is rarely greater than 1.0 mmol/l. Alcohol inhibits galactose uptake and metabolism by the liver and galactosaemia can occur. Galactose in the cells is converted by the specific enzyme galactokinase to galactose 1-phosphate and then to glucose 1-phosphate, catalysed by an epimerase. This is in contrast to fructose, which can enter the glycolytic pathway directly without being converted into glucose. While there are enzymes throughout the body which metabolize galactose, most galactose metabolism takes place in the liver through the utilization of dietary galactose. Any galactose which is required for structural reasons in body cells and connective tissue, e.g. galactosamine, a constituent of glycoprotein and mucopolysaccharides, is produced by the reversal of the normal galactose–glucose interconversions.

Galactose, when present in the blood, is taken up by specific tissues, particularly the lens of the eye. The galactose is converted by aldehyde reductase into galactitol. This may accumulate in tissues and is a factor in the development of **cataracts**. Cataracts are a complication of diabetes mellitus and may also occur where there are inborn errors of galactose metabolism caused by a deficiency of galactose 1-phosphate uridyl transferase and galactokinase deficiency. Galactokinase deficiency is not associated with any clinical problems other than galactosaemia, galactosuria, and cataracts.

Galactose 1-phosphate uridyltransferase deficiency usually has a fatal outcome in the first few days of life or results in mental retardation. Treatment depends upon instant recognition and the exclusion of galactose in the form of lactose in human or animal milk. In these instances it is necessary to give soya milk supplemented by sucrose, glucose or fructose. This deficiency condition is very rare. The heterozygotes have a reduced galactose 1-phosphate uridyltransferase activity and make up 1% of the population.

Fructose

In humans, fructose enters the portal venous system unchanged and is taken up by the liver; little passes to the systemic circulation. Within the liver fructose is phosphorylated to fructose 1-phosphate by a hepatic fructokinase. Fructokinase requires ATP and inorganic phosphate which is regenerated when fructose 1-phosphate is split into glyceraldehyde and dihydroxyacetone phosphate. The splitting of fructose 1-phosphate is catalysed by hepatic aldolase (aldolase B). The hepatic aldolase differs from the muscle aldolase in having an equal affinity for fructose 1-phosphate and fructose 1,6-diphosphate. Fructose 1,6-diphosphate is an important intermediary in the glycolytic and gluconeogenic pathways of the liver and kidney. Dihydroxyacetone phosphate is an intermediate in metabolism in both the glycolytic and gluconeogenic pathways.

Glyceraldehyde is a substrate for a number of enzymes before glycolytic intermediary metabolism. Much of the glyceraldehyde from fructose is converted to glyceraldehyde 3-phosphate, which can then be converted to glycogen. Alternatively, glyceraldehyde can be converted to glycerol 3-phosphate to be esterified as fatty acids to triglycerides. The activity of these reactions is dictated by the nutritional and hormonal state of the individual.

Following the ingestion of fructose or sucrose in substantial amounts there is an increase in the peripheral blood pyruvate and lactate as a result of intermediates entering the glycolytic rather than the gluconeogenic pathways of the liver. Fructose can increase the rate of alcohol metabolism by 10% when given intravenously with alcohol. This occurs through the inhibition of gluconeogenesis from pyruvate and lactate. Fructose, unlike glucose, does not require insulin to be transported into muscle and fat cells. Consequently, glucose and fructose have a different role in fat metabolism and insulin release.

In contrast to glucose, fructose can enter cells regardless of the insulin concentration and can be used as a substrate for hexokinase and conversion into fructose 6-phosphate and glycerol 3-phosphate. This is available for re-esterification of free fatty acids produced by intra-adipocyte lipolysis. The consequence of the production of the fructose 1-phosphate is an increase in ATP turnover and intrahepatic depletion of inorganic phosphate.

This may occasionally lead to a rise in body urate production, which may occur in chronic alcohol abuse with a predisposition to the development of gout. However, it is possible that there is increased urate production in hereditary fructose intolerance in the heterozygous state. This has led to the occasional treatment of gout by a fructose-restricted diet.

Polyols

Sorbitol is a sweetener which is absorbed more slowly from the gut than any of the monosaccharides and is removed from the portal blood by the liver. Sorbitol is converted into fructose by sorbitol dehydrogenase and then into fructose 1-phosphate by fructokinase. Sorbitol can be formed from glucose within the body by the action of aldose reductase.

Xylitol is an intermediate in the glucuronic acid pathway and is involved in vitamin C synthesis in some animals. In the benign inborn error pentosuria, L-xylulose is excreted into urine. L-xylulose reductase deficiency does not allow the conversion of L-xylulose into xylitol, xylitol being further metabolized to glucose 5- or 6-phosphate. Xylitol is sweeter than sorbitol.

Carbohydrates and vitamins

Most of the enzymes involved in carbohydrate metabolism require vitamin B metabolites as essential coenzymes. This requirement increases the dietary requirement for these vitamins. The coenzymes from vitamin B are cofactors in the glycolytic pathway, pentose phosphate shunt and tricarboxylic acid cycle.

Non-enzymatic glycosylation

Sugars may react with free amino groups on proteins to produce:

- a readily chemically reversible glycosylated product (a Schiff base), which may then undergo internal rearrangements to produce
- a more stable and possibly reversible Amadori-type glycosylation product, which can be converted into

- an irreversible compound which is an advanced glycosylation end product which may interact with other proteins.

Both aldose and ketose sugars can react with proteins to form Schiff bases when in the chair form. Glucose has only 0.002% in the chair form, whereas 0.7% of fructose is in the chair form. Fructose is thus more significant in the development of non-enzymatic glycosylated proteins.

11.6.3 Sugars and diabetes mellitus

The products of some metabolic pathways of glucose and galactose may result in many of the diabetic complications. Muscle tissue, fat cells and connective tissue require insulin for the transmembrane passage of glucose. Neurones as well as epithelial and endothelial cells do not require insulin for glucose entry and depend on the concentration of glucose in the plasma. At high glucose concentration glucose and other aldoses are reduced to the corresponding sugar alcohols, glucose forming glucitol and galactose to galactitol.

Diabetics have an increased concentration of an abnormal haemoglobin. This form of haemoglobin HbA_1 results from the attachment of a molecule of glucose to the N-terminus of the β-ring. A non-enzymatic reaction whereby there is the formation of a Schiff base leads to a ketoamine linkage by an Amadori rearrangement.

HbA_1

The HbA_1 concentration is dependent upon the mean blood glucose concentration. In the normal non-diabetic population on average 4–5% of the haemoglobin is of the A_1 variant and with diabetics this may increase to 8–15%, dependent upon the previous blood glucose concentrations. Measurement of HbA_1 enables a history to be given of the recent glucose control during several weeks preceding the assay.

Many proteins will condense with sugars even at normal concentrations of glucose. Collagen, the crystallins of the lens, serum proteins, nerve myelin, all membrane proteins, transferrin and fibronectin, all form glucose adducts. All these incorporations are increased in diabetics.

11.6.4 Glycaemic responses to sugars and starches

A carbohydrate meal, other than fructose, produces an increase in blood glucose. The capillary blood glucose concentration increases in response to starch and glucose.

Glycaemic response

Factors affecting the measured glycaemic responses to meals include:

- the method of measuring the blood sugar concentration
- whether arterial or venous blood is sampled
- whether long-term or acute response is taken into account
- the chemical nature of the carbohydrate under study, and especially the constituent monosaccharides
- whether the carbohydrate is taken alone or with other dietary constituents
- physical form and texture
- the amount of carbohydrate eaten at one sitting
- whether eaten raw, cooked or processed
- the energy content and composition of the normal diet
- the health status of the individual being studied

In the fasting state arterial and venous bloods contain equivalent amounts of glucose. Postprandially, however, the arterial blood concentrations of glucose rise rapidly and then fall to fasting levels. Venous blood rises more slowly and a peak is achieved within 2 hours of ingestion. This arterial–venous (A/V) blood glucose difference is a measure of the uptake of glucose by the tissues under the effect of insulin secreted in response to the diet. It is a useful indicator of peripheral utilization of the glucose.

Results of A/V glucose differences in diabetics can be quite specific. The rise in blood glucose concentration, particularly of venous blood, bears little relationship to the size of the glucose load. When large amounts of more than 50 g are taken there is a saturation of the system in the normal subject. In diabetics, the blood glucose concentration continues to rise with the glucose load until there is overflow and renal excretion.

Blood glucose measurements are usually made for 2–3 hours after a test meal. This enables the calculation of a glycaemic response or a graph of glucose concentration over time to be made. The results indicate the difference in rate at which glucose enters the body glucose pool and the rate at which it is removed. The rate of gastric emptying is another important determinant of carbohydrate absorption (see p. 333).

Capillary blood obtained from the hand is equivalent to arterial blood throughout the postabsorptive phase. This gives a more appropriate indication of the true glycaemic response to food than that of mixed forearm venous blood. Insulin is secreted in response to a meal and the more insulin-sensitive the tissue the greater the A/V glucose difference. In healthy young subjects with good glucose tolerance the A/V glucose difference can be 3.0 mmol/l or more.

It is possible in diabetic subjects to obtain a measure of fluctuations of blood glucose over the preceding few months. by the measurement of HbA_{1c} and glycosylated albumin.

There is often a hypoglycaemic period after the fall in blood glucose wherein the blood glucose falls to below fasting concentrations. This is particularly pronounced if venous rather than arterial blood is measured, but does not appear to occur following ingestion of mixed meals. The site of sugar absorption may be in the duodenum, jejunum or ileum. Therefore the effects of glucose may be seen as a hormone releaser or as a simple provision of metabolic fuel at each intestinal site.

11.6.5 Dietary carbohydrates and lipid metabolism

The long-chain fatty acids in the body, stored as triglycerides, are derived from the diet. However, it is possible that they can be synthesized from metabolites from the glycolytic pathway of carbohydrates and 'reducing equivalents' from the pentose phosphate shunt. These routes become relevant when dietary carbohydrates are eaten in excess of daily requirements. Fatty acids are either metabolized to carbon dioxide and water or stored as triglycerides. Glycerol is required to be phosphorylated in order to be brought back into the metabolic pool. Glycerol 3-phosphate, an intermediate in the glycolytic pathway, is used to re-esterify fatty acids. Glucose does not enter adipocytes from the blood unless insulin is present in the extracellular fluid at a concentration above a critical concentration. This concentration depends on the number and affinity of insulin receptors on the surface of the adipocytes. Below this insulin concentration, glucose does not enter the cells, glycolysis decreases and insufficient glycerol 3-phosphate is available for re-esterification of the free fatty acids produced by lipolysis. Free fatty acids, therefore are released into the blood stream where they can be used by heart and striated muscle as fuel or passed to the liver for conversion into ketone bodies (acetoacetate and β-hydroxybutyrate) or back to triglycerides in the form of very low density lipoprotein (VLDL).

Prolonged exposure to sucrose diets raises the fasting serum triglyceride concentration, being dependent on endogenous production of triglycerides. The triglyceride concentration after a 12- to 14-hour fast is a good indicator of triglyceride status. A substantial dietary sucrose intake results in an increase in the concentration of triglycerides compared with starch. This effect of sucrose is independent of the previous triglyceride concentrations. Plasma triglyceride concentrations are reduced when the unsaturated fat content of the diet is increased.

> Glucose given hourly to men reduces serum triglycerol concentrations with less marked reductions in phospholipid and cholesterol concentrations. Glucose given intravenously has the same effect. This is a *direct* effect – not an effect mediated by the absorption process. If insulin is added to the glucose infusion, the postprandial lipaemia is further decreased. Fructose on the other hand increases the postprandial hypertriglyceridaemia. All this may be due to increased removal of plasma triglycerides as a result of increased lipoprotein lipase activity, stimulated by insulin. However, this effect is not uniform. Variables include the age and sex of the individual – pre-menopausal women react to sucrose differently from men and post-menopausal women. It is possible that females clear plasma triglycerides more rapidly than males.

It is possible that there is a synergistic influence of sucrose but not starch and animal fat on plasma triglycerides. There is also a very individual response to these dietary changes, which presumably has a genetic basis, for example individuals with Type IV hyperlipoproteinaemia have a more pronounced increase in plasma triglycerides following sucrose ingestion than with starch compared to non-Type IV individuals.

Adipose tissue

Dietary carbohydrate can be converted to fat, an energy-consuming process which utilizes some 10–15% of the ingested glucose. The adipocyte converts glucose to fatty acids, a process which is less energy-consuming during fasting and following a high fat meal but the energy consumption is increased by a high carbohydrate diet. The process is mediated through insulin.

Fructose may also be taken up by the adipocyte but the transport across the cell membrane is slow as a result of the carrier system for fructose having a relatively high K_m.

Glucose is not converted to glycogen or fatty acid directly but through a C_3 molecule formed in tissues other than the liver. This C_3 unit is then recycled to the liver. Over two-thirds of glucose absorbed escapes uptake by the liver and is transported to the peripheral tissues.

Diets rich in dietary fibre produce a reduced stimulus to insulin secretion compared with diets free of fibre. This, depending upon the type of fibre, may decrease cholesterol and triglyceride concentrations.

11.6.6 Carbohydrates and the respiratory quotient

The ratio of carbon dioxide in the breath to oxygen, the respiratory quotient (R_Q), gives an indication of the type of substrate being metabolized and a hint as to which metabolic route is being followed. The R_Q should be 1; after the ingestion of fructose or galactose it is greater than after glucose. It is possible that in the release of energy after fructose digestion more of the carbohydrate is being broken down than with other sugars. The more rapid metabolism of fructose may be due to fructose metabolism following

---

The mean percentage increase in metabolic rate following glucose over a 3-hour period was 10%. The response in women is less than that of men. The metabolic response to the ingestion of a simple sugar may vary with the previous diet of the individual. The thermogenesis is greater after a high carbohydrate diet than after a high fat diet. Much of the carbohydrate load is deposited in glycogen rather than in fat. The metabolic cost of dietary carbohydrate may vary between 7% and 23% depending on whether the carbohydrate is stored as glycogen or fat.

---

first-order reactions and the utilization being proportional to concentration. There is an increase in

plasma uric acid, lactate and pyruvate concentrations after fructose ingestion.

The metabolism of dietary protein requires a greater energy expenditure than dietary carbohydrate or fat. Dietary fat and carbohydrate undergo less metabolism before being stored in adipose tissue. The resting metabolic rate measured for 3 hours after the ingestion of various di- and monosaccharides and mixtures shows an increase in metabolic rate with sucrose. This increase is greater than with the constituent glucose : fructose (0.90 compared with glucose 0.87 and another disaccharide lactose, 0.86).

11.6.7 Glucose oxidation and metabolism

In aerobic metabolism, organic carbon can be fully oxidized to CO_2 using oxygen as an electron acceptor. The ATP yield for each glucose metabolized under oxidative conditions for example, is almost 20 times greater than that under anaerobic conditions. The amount of energy that can be liberated from nutrients by anaerobic oxidation is limited. Every such anaerobic oxidation reaction involves an electron being removed and taken up by another organic compound. The energy difference between such coupled oxidation is very small.

Aerobic respiratory metabolism involves:

- oxygen being the ultimate electron acceptor
- complete oxidation of organic chemicals to CO_2 and water.
- the free energy being conserved as ATP

At the same time, there is a continuous re-oxidation of reduced coenzyme molecules, e.g. NADH and $FADH_2$ which link into the respiratory chain to produce ATP. These coenzyme molecules in their oxidized form are involved in the oxidation of organic intermediates from pyruvate. Therefore, under aerobic conditions a glycolytic pathway is the initial phase of glucose catabolism, following which are:

- the tricarboxylic acid cycle
- the oxidative phosphorylation of ADP to ATP

This system leads to the formation of 36–38 molecules of ATP for each molecule of glucose metabolized in this process.

11.6.8 Glycolysis

Glycolysis is the pathway in which glucose is metabolized before splitting into two interconvertible 3-carbon molecules (Figure 11.8). These reactions take place in the cell cytoplasm. The first three steps involve reactions which lead to a doubly phosphorylated fructose derivative

Glucose 1-phosphate, glucose 6-phosphate and fructose 6-phosphate readily interconvert and thereby form a single metabolic pool. Glucose 1-phosphate is the first product in the catabolism of storage polysaccharides, e.g. starch, glucose 6-phosphate is the first hexose phosphate generated when free glucose is metabolized and fructose 6-phosphate is the first hexose phosphate formed when carbohydrate is derived from non-carbohydrate precursors.

Glucose 6-phosphate (G6P) and fructose 6-phosphate (F6P) readily interchange through the action of the enzyme glucose phosphate isomerase. The amounts of these can be affected by the phosphorolysis of storage polysaccharides,

yielding glucose 1-phosphate, or by the phosphorylation of glucose yielding G6P or by gluconeogenesis of F6P. The subsequent catabolic steps are glycolysis with F6P as the starting point or the pentose phosphate pathway which uses G6P. Alternatively G6P and F6P can be converted into storage polysaccharides with G6P acting as the starting point. G6P can also be hydrolysed to yield free glucose to be transported in the blood to peripheral tissues.

The steps in glycolysis are (Figure 11.9):

1. Glucose is phosphorylated on the carbon 6 position by hexokinase or glucokinase. Both are Mg^{2+}-dependent enzymes. Hexokinase has a wide number of substrates, glucose, fructose, mannose and galactose. Hexokinase is inhibited by the concentration of the product G6P and ADP. Hexokinase is found as several isoenzymes with varying affinity for glucose. Glucokinase is solely a hepatic enzyme, is specific for D-glucose and is not affected by the product concentration. The K_m is 5–10 mM, rather than 0.1 mM for hexokinase.

2. G6P is isomerized to F6P. The generation of fructose provides a primary hydroxyl group on carbon-1 for subsequent phosphorylation. The carbonyl on carbon 2 is important for subsequent β-cleavage to produce two 3-carbon products. The enzyme is phosphogluco-isomerase.

3. The 1-carbon position is phosphorylated by phosphofructokinase (PFK), a Mg^{2+}-dependent enzyme and ATP. This reaction is rate-limiting. The product is fructose 1,6-diphosphate. ATP inhibits PFK, an effect counteracted by ADP and AMP. Citrate increases ATP inhibition, which is significant in this feeder system for the tricarboxylic acid cycle.

4. Fructose 1,6-diphosphate is reversibly split into two isomers, dihydroxyacetone phosphate (carbons 3, 2,1 of the fructose 1,6-diphosphate) and glyceraldehyde 3-phosphate (carbons 4,5,6 of the fructose 1,6-diphosphate). The enzyme involved is aldolase and

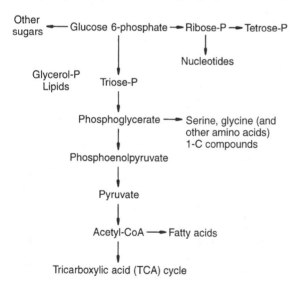

FIGURE 11.8 Glycolysis. The oxidative cleavage of glucose to pyruvate, acetyl-CoA and other metabolites.

the reaction is driven by the rapid removal of the glyceraldehyde 3-phosphate. There are isoenzymes of aldolase in muscle, liver and brain.

5. The two 3-carbon isomers are interconvertible by the action of the enzyme triosephosphate isomerase. The removal of glyceraldehyde 3-phosphate affects the equilibrium of this enzymatic reaction. This series of reactions is energy consuming in that 2 ATP have been used per glucose.

6. Glyceraldehyde 3-phosphate is converted to 1,3-diphosphoglycerate with the creation of a high-energy phosphoanhydride bond on carbon-1. The enzyme is glyceraldehyde 3-phosphate dehydrogenase. The reaction passes through a high-energy thioester stage, which accounts for the inhibitory effect of alkylating agents and heavy metals on the reaction. The phosphate is organic phosphate.

7. The carbon-1 phosphate of the 1,3-diphosphoglycerate is used to phosphorylate ADP to ATP, a substrate phosphorylation reaction. As two 3-carbon derivatives of glucose pass through this stage, then the two ATPs used in stages 1 and 2 are regenerated.

8. The phosphate on carbon-3 of 2-phosphoglycerate is moved to carbon-3. The enzyme is phosphoglyceromutase, a Mg^{2+}-requiring enzyme.

9. The removal of water by enolase produces a high-energy phosphate bond at carbon-3 next to a double bond, in phosphoenolpyruvate.

10. The high-energy phosphate is used to convert ADP to ATP with the production of pyruvate. This is an irreversible reaction requiring the enzyme pyruvate kinase. The enzyme is K^+- and Mg^{2+}- or Mn^{2+}-dependent and is inhibited by high ATP concentrations. The enzyme is activated by fructose 1,6-diphosphate and phosphoenolpyruvate.

The overall yield over the 10 stages of the process is 2 ATP per glucose oxidized. The final steps in the oxidation of glucose require the transfer of the pyruvate into the mitochondria. Pyruvate plays a major role as an intermediary metabolite and is central to the interconversion of glucose fatty acids and amino acids.

Pyruvate interconversion

The pathways that pyruvate may follow include conversion to:

- L-malate to oxaloacetate, then to the tricarboxylic acid (TCA) cycle
- acetyl-coenzyme A (acetyl-CoA), which is the substrate for (i) fatty acid synthesis; (ii) oxyacids; (iii) cholesterol; and (iv) TCA cycle.

In the TCA cycle the most important supply of acetyl-CoA is from pyruvate. Other suppliers include fatty acids and amino acids:

Pyruvate + CoA + NAD $\xrightarrow{\text{pyruvic dehydrogenase (PDH)}}$

Acetyl-CoA + NADH + CO_2

Pyruvic acid has to be converted to acetyl-CoA in the mitochondria, where the TCA cycle takes place. The first step is the transport of pyruvate into the mitochondria. The reaction is irreversible, very complicated, and is dependent upon three different enzyme activities and five coenzymes. The first step is the removal of a carbon dioxide by a decarboxylase which requires thiamin pyrophosphate. The next is a transfer to a lipoic acid cofactor and a dehydrogenation using a dehydrogenase. The acetyl group is then transferred by a transacetylase to form acetyl-CoA. In this reaction NADH is generated. This enzyme complex is very cofactor-dependent and sensitive to deficiencies of thiamin.

Pyruvate dehydrogenase (PDH) is inactivated by phosphorylation by PDH kinase and PDH phosphatase. In the presence of ATP concentrations the kinase phosphorylates the PDH enzyme. The process is inhibited by ADP and pyruvate, so that at times of nutrient shortage the enzymatic system is ready for the generation of acetyl-CoA. Inactivated PDH is activated by a phosphatase

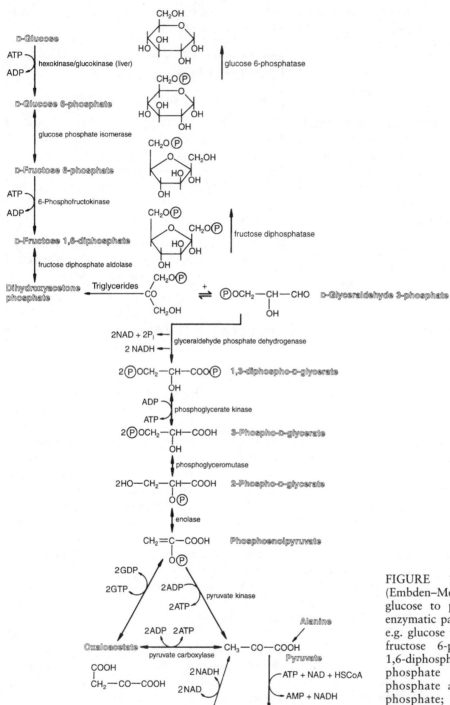

FIGURE 11.9 The glycolytic (Embden–Meyerhof) pathway from glucose to pyruvate. Some of the enzymatic pathways are irreversible, e.g. glucose to glucose 6-phosphate; fructose 6-phosphate to fructose 1,6-diphosphate; fructose 1,6-diphosphate to dihydroxyacetone phosphate and glyceraldehyde 3-phosphate; phosphoenolpyruvate to pyruvate, and pyruvate to acetyl-CoA.

which requires Mg^{2+} or Ca^{2+} as cofactors. These may be chelated by ATP or citrate which are also controlling systems. The nutritional and endocrine status dictates the activity of PDH. The availability of PDH is very tissue-dependent, being independent of nutrition in the brain and activated postprandially in heart and kidneys. Liver and fat PDH are largely present in the inactive form, regardless of nutritional status.

PDH is converted to the inactive form during starvation. The only other nutritional sources available for gluconeogenesis would be protein and glycerol.

11.6.9 Tricarboxylic acid (TCA) cycle

The TCA cycle is a mitochondrial system which starts with acetyl-CoA. This is derived either by oxidative decarboxylation of pyruvate which is derived from glycolysis, by oxidative cleavage of fatty acids, or from amino acids. Acetyl-CoA donates the acetyl group to oxaloacetate which is a 4-carbon acceptor. This produces citrate which has six carbons. Citrate subsequently passes through two successive decarboxylation processes and a number of oxidative processes, to yield oxaloacetate. While passing through the cycle two carbons are introduced from acetyl-CoA and two are released as CO_2.

The TCA acid cycle is the main oxidative process for the three major nutritional elements, carbohydrates, fats and proteins (Figure 11.10). The cycle is regulated to provide cell needs. This is important because the cycle's function is to:

- provide NADH and $FADH_2$ for the electron transport chain
- provide substrates for biosynthesis

The cycle depends on the availability of acetyl-CoA and also on the ratio of NADH/NAD$^+$ and the ATP/ADP ratio.

The decarboxylation of pyruvate requires an intermediate, stabilized by prior condensation of carbonyl with thiamin pyrophosphate. The PDH complex has a molecular weight of 9×10^6 Da. This enzyme system is regulated and is sensitive to the ATP/ADP ratio and the concentration of

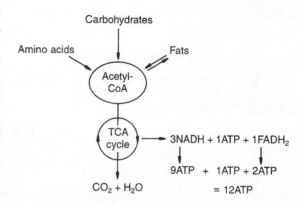

FIGURE 11.10 Overall scheme of the tricarboxylic acid (TCA) cycle. Amino acids, carbohydrates and fats in the form of acetyl-CoA enter the TCA cycle. Each rotation of an acetyl-CoA through the cycle yields 12 ATP. One molecule of glucose, after passing through the oxidative process of glycolysis and the TCA cycle, yields 38 ATP.

acetyl-CoA. The enzyme complex is also subject to regulation by enzyme modification in which a protein kinase catalyses the phosphorylation of specific serine –OH groups on the pyruvate carboxylase section of the enzyme complex. A phosphorylase removes these phosphoryl groups. The phosphorylated enzyme is relatively inactive. The action of the kinase and the decrease in the activity of the pyruvate complex is controlled by the ATP/ADP and NADH/NAD$^+$ ratios and by a high acetyl-CoA concentration (Figure 11.11).

$$\text{Acetyl-CoA + oxaloacetate} \xrightarrow{\text{citrate synthase}} \text{Citrate + CoA}$$

Citrate is isomerized to isocitrate by the action of aconitase. Isocitrate dehydrogenase is the enzyme for the conversion of isocitrate to α-ketoglutarate. NAD$^+$ is the electron acceptor in this oxidative step and Mg^{2+} and Mn^{2+} are required for the decarboxylation.

419

α-ketoglutarate $+ NAD^+ + CoA \longrightarrow$
Succinyl-CoA $+ NADH + H^+ + CO_2$

Succinyl-CoA is an activated intermediate.

Succinyl-CoA $+ GDP \longrightarrow$
Succinate $+ GTP$ Succinate thiokinase

The reaction includes an intermediate in which a phosphate group is attached to a histidine residue of the enzyme. The next stage is

Succinate $+$ FAD $\longrightarrow$
Fumarate $+ FADH_2$ succinate dehydrogenase

Pyruvate

NAD, CoA-SH, NADH, CO_2 — pyruvate dehydrogenase

Acetyl-CoA

citric: oxaloacetate lyase

Malate — L-malate: NAD oxidoreductase

Oxaloacetate

Citrate

citric hydro-lyase

L-malate hydrolyase

Fumarate

D=isocitrate

succinate: oxidoreductase

CO_2

isocitrate dehydrogenase

Succinate

succinate: CoA ligase

α-Keto glutarate

Succinyl CoA

2-ketoglutarate dehydrogenase

CO_2

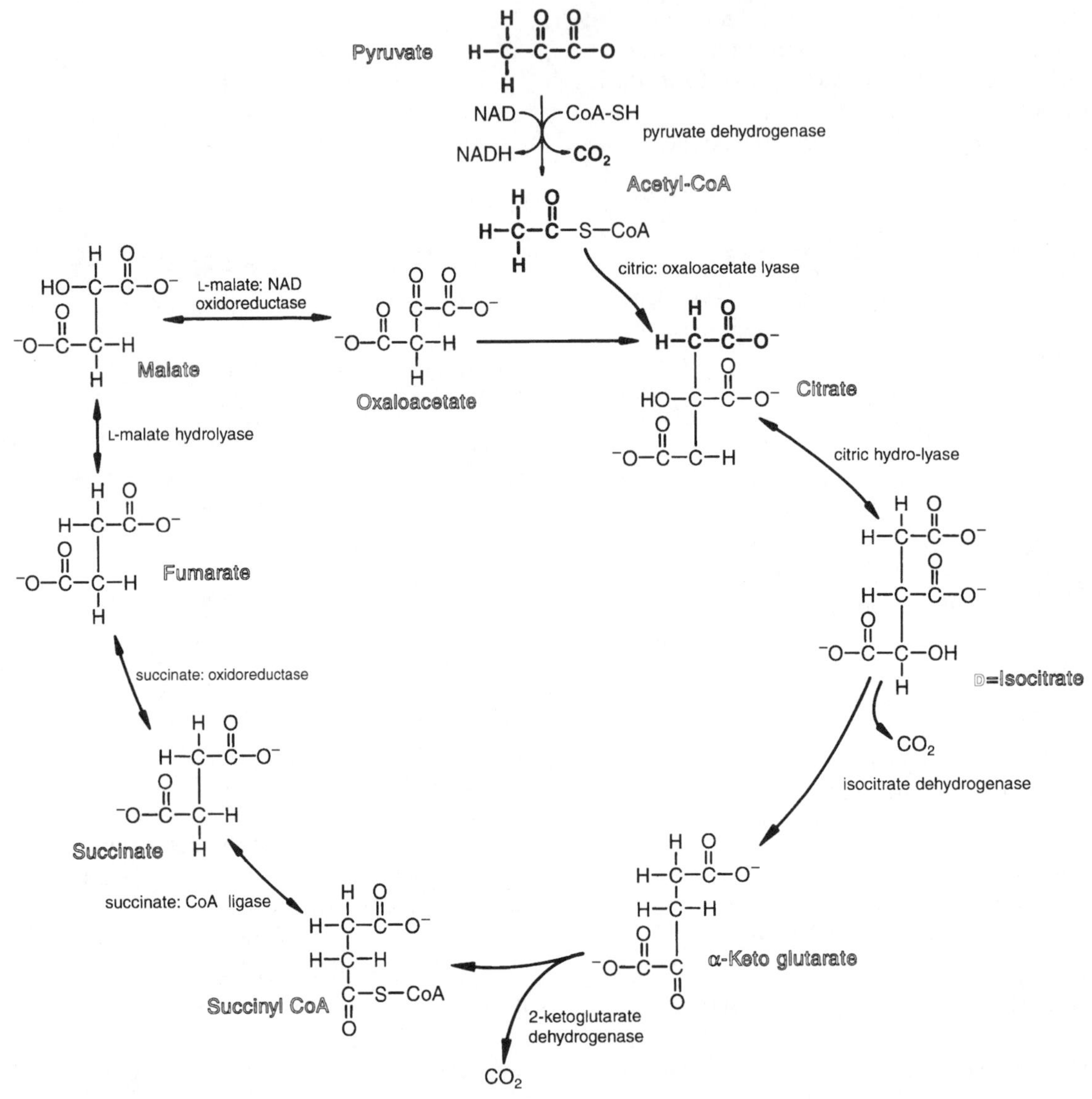

FIGURE 11.11   Tricarboxylic acid (TCA) cycle showing the detailed pathways. Most of the enzymes of the TCA cycle are found in the matrix of the mitochondria. Succinate dehydrogenase is a membrane protein on the inner mitochondrial membrane. The enzyme has a covalently linked flavin β-histidyl FAD. Succinate dehydrogenase is re-oxidized by the electron transport chain.

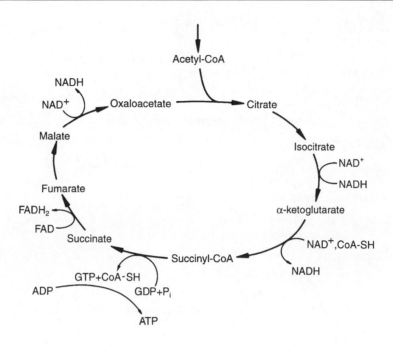

FIGURE 11.12 Tricarboxylic acid (TCA) cycle and electron carriers NAD^+ and FAD whose subsequent oxidation results in the synthesis of ATP.

This is a flavoprotein enzyme and is attached to the inner mitochondrial membrane:

$$\text{Fumarate} + H_2O \longrightarrow \text{L-malate}$$

This is a stereospecific addition of water to the double bond. The cycle is completed by

$$\text{L-malate} + NAD^+ \longrightarrow$$
$$\text{oxaloacetate} + NADH + H^+$$

There is considerable stereochemistry in the reaction between enzymes and substrates in the TCA cycle. This means that the enzymatic reactions are very stereospecific. The result of acetyl-CoA entering the TCA cycle is to yield 1 ATP molecule. However, there is storage of free energy as NADH and $FADH_2$ (Figure 11.12). Consequently, oxidation of 1 mole of acetyl-CoA leads to the overall production of 12 moles of ATP. In addition to this, pyruvate metabolism leads to 15 moles of ATP per mole of pyruvate, or 38 moles ATP per mole of glucose (Figure 11.10).

The metabolites in the TCA cycle are major starting materials for a number of biosynthetic pathways (Figure 11.13):

- Acetyl-CoA supplies the carbon atoms for fats and lipids and for the synthesis of some amino acids.
- Oxaloacetate is the initial chemical in the synthesis of aspartate, asparagine, threonine, isoleucine and methionine.
- α-Ketoglutarate is the initial chemical in the synthesis of glutamate, glutamine, proline and arginine.
- Succinyl-CoA is part of the synthetic path in the synthesis of haem.

This means that most of the starting materials required for the biosynthetic pathways of a cell can be derived from aerobic pathways fuelled by carbohydrates.

Most of these reactions are reversible, which means that the TCA cycle can be replenished at points other than acetyl-CoA. Acetyl-CoA is an important biosynthetic material, providing within the cytosol of the cell the carbon atoms of lipids, cholesterol and some amino acids. The acetyl-CoA is produced from fat degradation and pyruvate from carbohydrate degradation. In the

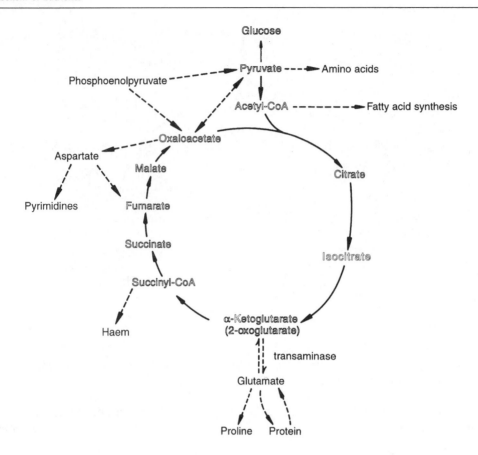

FIGURE 11.13   The tricarboxylic acid (TCA) cycle in relationship to other nutrients. Pyruvate can be synthesized to amino acids, acetyl-CoA to fatty acids, α-ketoglutarate transaminated to glutamate, and oxaloacetate synthesized into aspartate and phosphoenol pyruvate.

mitochondria the acetyl-CoA required for synthesis in the cytosol is derived from citrate which passes from the mitochondria to the cytosol. In the cytosol the citrate is cleaved to acetyl-CoA and oxaloacetate. Some of the oxaloacetate is converted to aspartate and is a precursor for the synthesis of asparagine. The excess is reduced to malate, which diffuses readily into the mitochondria and is convertible into pyruvate.

Citrate has a double role. It supplies the carbon atoms of fatty acids and also regulates fatty acid synthesis. The initiation of fatty acid synthesis is the carboxylation of acetyl-CoA to form malonyl-CoA. Citrate is an important regulator of this reaction. Citrate also indirectly inhibits PFK activity and hence the rate of glycolysis. PFK is inhib-

ited by ATP which is accentuated by the presence of citrate. Consequently increases in citrate concentration in the cytosol result in increased production of storage fats and reduce the rate of carbohydrate breakdown. Reduction in citrate concentration has the reverse effect.

The reduction of the carbonyl group of pyruvate to form lactate maintains supplies of NAD under anaerobic conditions. The enzyme is lactic dehydrogenase. This enzyme is found in five isoenzyme forms. These are various arrangements of A and B subunits, all with different K_m values. Lactic dehydrogenase isoenzymes with low K_m values are found in muscles where a high enzyme affinity for glycolysis is important for energy production. Cardiac muscle lactic dehydrogenase enzyme affinity is low. Lactate utilization is important during severe anaerobic skeletal exercise. Lactate may return to the liver and is converted back to pyruvate and then to glucose in the gluconeogenic pathway.

11.6.10 The pentose phosphate pathway

This is an alternative metabolic pathway to the glycolytic system (Figure 11.14). The initial reaction requires glucose 6-phosphate dehydrogenase as an enzyme and generates NADPH for other reductive reactions in fatty acid and cholesterol synthesis. The next product is D-ribulose 5-phosphate (enzyme, transaldolase), another NADPH-generating reaction. The NADPH production is important in the reducing conditions for some biosynthetic reactions, especially lipids.

Several ribose phosphates can be produced from ribulose 5 phosphate, e.g. xylulose 5-phosphate (enzyme, phosphopentose epimerase) and ribose 5-phosphate (enzyme, phosphopentose isomerase). These pentose phosphates are the starting point for formation of a whole array of three- to seven-sugar phosphates, from which are produced: triose phosphate, e.g. erythrose 4-phosphate which is converted to fructose 6-phosphate, ribose 5-phosphate for nucleic acid synthesis or generation of NADPH. NADPH is important in a variety of biosynthetic systems, e.g. oxidation of malate to pyruvate.

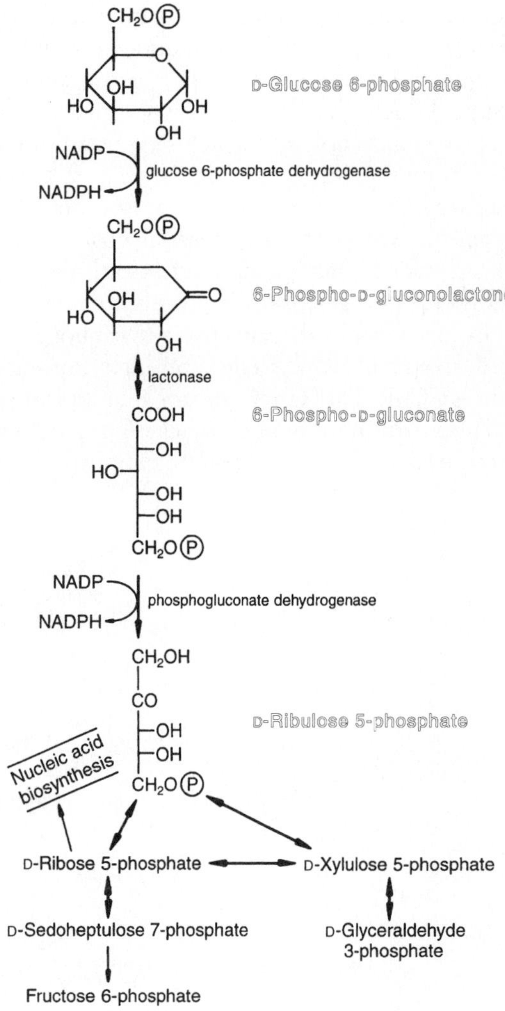

11.6.11 Metabolism in the presence of abundant glucose

In this situation the emphasis is on anabolic processes and storage (Figure 11.15). The liver converts some glucose to stored glycogen and some to fatty acids to be used as a fuel. Peripherally, glucose is metabolized to generate ATP. Excess is stored as glycogen and adipose tissue.

Glycogen synthesis

Glycogen is a convenient way to store sugar until required. Insulin stimulates the uptake of glucose

FIGURE 11.14 Pentose phosphate pathway. This pathway is a series of inter-related reactions providing a series of end products. It is also an alternative pathway for glucose oxidation with the formation of the reduced coenzyme, NADPH. The NADPH can be used to drive reductive anabolic pathways, e.g. synthesis of fatty acids and steroids. This pathway is a prime source of NADPH for biosynthetic reactions in most cells. The pathway also allows the interconversion of hexoses and pentoses. The end-products fructose 6-phosphate and glyceraldehyde 3-phosphate then pass into the glycolytic pathway. Ribose 5-phosphate is available for nucleic acid biosynthesis.

by various tissues and the synthesis of glycogen in the liver and reduces blood glucose concentrations. Glycogen is synthesized, using the enzyme glycogen synthase, by an addition reaction of UDP-glucose with an existing glucose polymer (Figure 11.15).

The resultant polymer structure is an $\alpha1$–4 with $\alpha1$–6 branch points (Figure 11.16). The branched chains of $\alpha1$–6 are synthesized by the branching enzyme amylo (1–4, 1–6) *trans*-glycosylase. The enzyme adds a chain of six to seven carbons to the C-6 group of a glucose in a glycogen chain.

Glycogen synthase is inactive when phosphorylated, is activated by insulin and high concentrations of G6P. This latter control is important in the liver but not in muscle where the enzyme hexokinase controls G6P concentrations.

$$\text{UDP-glucose} + \text{glycogen-1} \xrightarrow{\text{glycogen synthase}}$$
$$\text{Glycogen} + \text{UDP}$$

11.6.12 During glucose insufficiency or requirement

Mobilization of tissue glucose

Glucose stores in the form of glycogen and glucose are mobilized. Glucose synthesis from non-glucose sources (gluconeogenesis) is stimulated in liver and kidney.

Glycogen breakdown

Phosphorylase (Figure 11.17) releases glucose in the form of G6P:

$$\text{glycogen} + P_1 \xrightarrow{\text{phosphorylase a}} \text{Glycogen} + \text{G1P}$$

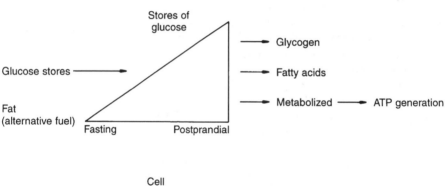

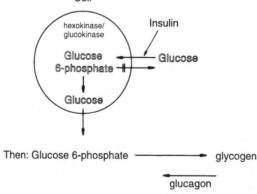

FIGURE 11.15 Carbohydrate metabolism. In times of excess dietary glucose then glucose is stored as glycogen or synthesized into fatty acids or metabolized to produce ATP. When dietary glucose is deficient then glucose stores are mobilized and fat is used as an alternative fuel. Glucose enters the cell by a transport system increased by insulin. In the cell, glucose is metabolized into glucose 6-phosphate which in turn can be synthesized into glycogen.

This process is active when phosphorylated and inhibited by insulin.

The phosphorylase cascade amplifies a regulatory signal with the resultant production of many molecules of cyclic AMP amplified at every step. Thus low concentrations of hormones have profound effects on metabolism.

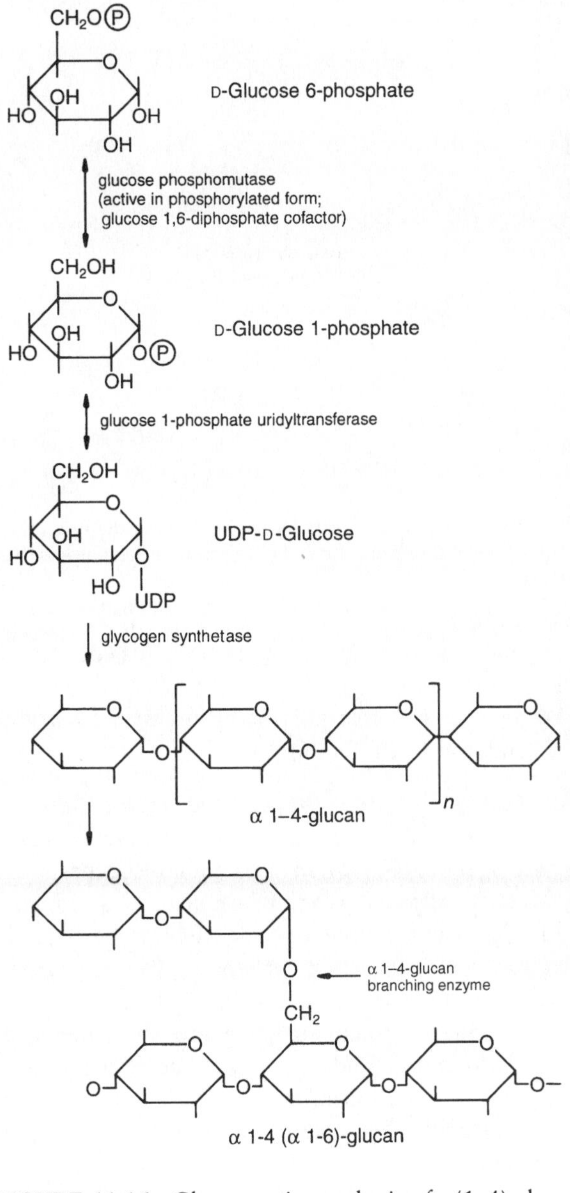

glucose phosphomutase
(active in phosphorylated form;
glucose 1,6-diphosphate cofactor)

glucose 1-phosphate uridyltransferase

glycogen synthetase

α 1–4 (α 1-6)-glucan

FIGURE 11.16 Glycogenesis: synthesis of α(1–4) glucan and α(1–4)-α(1–6) glucan.

The phosphorylase a and b 'cascade'

Skeletal muscle phosphorylase consists of two or possibly four identical proteins. A serine at position 14 in each constituent protein of the enzyme is phosphorylated by phosphorylase kinase b which requires ATP to yield the active phosphorylase a form. The inactive form phosphorylase b is produced by phosphorylase a phosphatase (Figure 11.18). The phosphorylase kinase is kept in an active and inactive form by protein kinase which is increased by noradrenaline and glucagon. The enzyme activity is affected by covalent modification of the enzymes. An increase in AMP is associated with a decrease in ATP, and activation of glycogen phosphorylase due to an increase of AMP. An increase in the amount of glucose 1-phosphate leads to an increase in the rate at which ATP is regenerated.

As the glucose concentration falls, glucagon and adrenaline increase the conversion of phosphorylase b to phosphorylase a. This is independent of AMP control. This overriding system is a membrane-bound enzyme, adenylate cyclase, which converts ATP to cyclic AMP. Activation of the cyclase requires GTP and a GTP binding protein called the G protein. Cyclic AMP activates protein kinase A. The inactive form of this enzyme consists of two catalytic (C) and two regulatory subunits (R). Cyclic AMP binds to the two regulatory subunits (R) causing a dissociation of the (R) and the (C) subunits. The C dimers dissociate to give two active monomers which catalyse a transfer of a phosphoryl group from ATP to specific sites of many proteins including phosphorylase kinase which results in activation of the phosphorylase kinase which then converts phosphorylase b to phosphorylase a by phosphorylation of a specific serine residue.

Noradrenaline converts phosphorylase b to phosphorylase a both in the liver and in muscle. In contrast, glucagon only affects the conversion of hepatic phosphorylase b to phosphorylase a, as

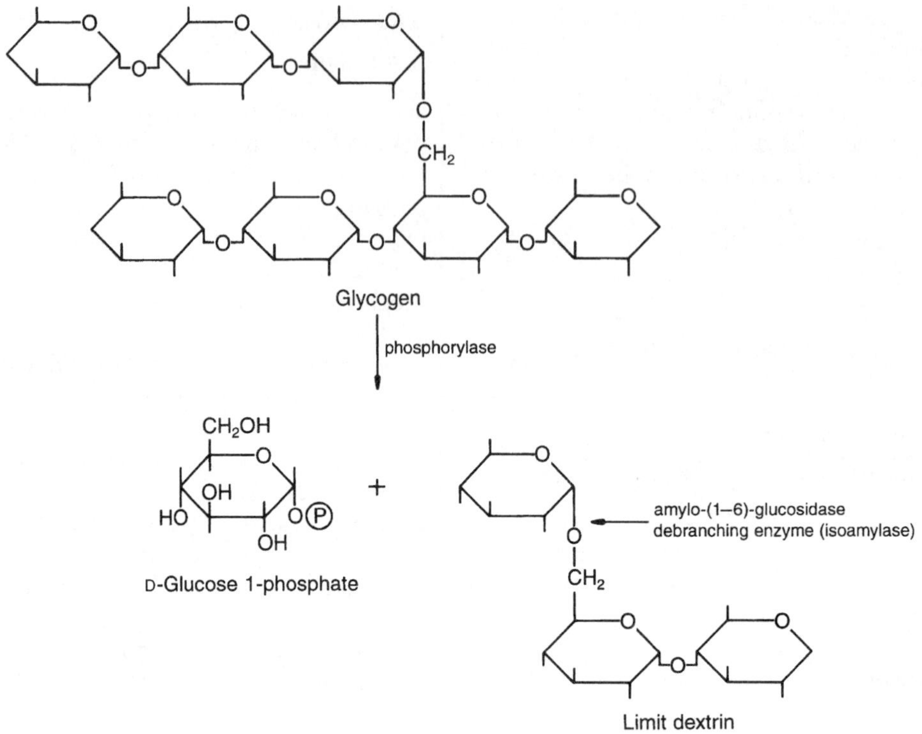

FIGURE 11.17 Glycogenolysis: the enzyme phosphorylase removes glucose units from the non-reducing ends as glucose 1-phosphate as far as 2–3 glucose units from an $\alpha(1-6)$ branch unit. The $\alpha(1-6)$ linkage-releasing enzyme is amylo-(1–6)-glucosidase, debranching enzyme or isoamylase.

glucagon is a gentle modulator of blood glucose concentrations.

11.6.13 Gluconeogenesis

Glucose synthesis

Glucose has a central role in the production and harnessing of energy and in the synthesis of other sugars. The ability to synthesize glucose is equally important as the brain and red cells are restricted to glucose as an energy source. While an individual is fasting the brain uses 80% of the glucose consumed. The liver only contains sufficient glucose to meet the requirements of the brain for 12 hours; the ability to synthesize glucose is therefore important.

Several systems are available:

- production of sugars from non-sugar sources, e.g. lactate, pyruvate, fatty acids or amino acids
- synthesis of free glucose by the liver for transmission in the blood

Gluconeogenesis is activated under different conditions and serves different functions in different species. Glycolysis and gluconeogenesis are directly opposed reaction sequences (see Figure 11.9). Most of the enzymes which function in glycolysis are also reversed in gluconeogenesis. This is an important property of carbohydrate metabolism, which can feed into all elements of carbohydrate, lipid and amino acid metabolism except for the essential fatty acids and essential amino acids. Fatty acids are not capable of gluconeogenesis.

The intermediates in the gluconeogenic and glycolysis pathways are the same, are in near equilib-

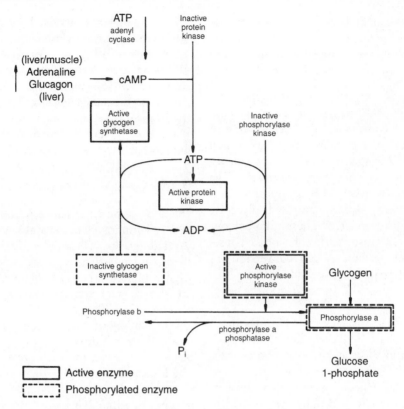

FIGURE 11.18   The synthesis of glycogen by glycogen synthase acts in the opposite direction to phosphorylase a, which cleaves glycogen to yield glucose 1-phosphate. This reciprocating enzyme activity is controlled by phosphorylation by protein kinase and phosphorylase kinase. Active phosphorylase a is phosphorylated. Active glycogen synthase is dephosphorylated.

rium concentrations, and can move in either direction dependent on small changes in concentration. All of these processes are coupled to the ATP–ADP system, and regulate the direction of subsequent metabolic pathways. A problem that has to be overcome within this process is that some steps in metabolic pathways are irreversible.

In glycolysis (see Figure 11.9) these are:

1. The phosphorylation of glucose and fructose 6-phosphate, which are ATP-dependent reactions. The irreversible nature of this reaction is due to the kinases involved. In the reverse reaction in gluconeogenesis this problem is bypassed by the enzymes being phosphatases which remove the phosphate in an inorganic form. These are cytosolic enzymes, remote from the mitochondrial enzymes of the TCA cycle.

2. Pyruvate to phosphoenolpyruvate: the bypassing of the enzymatic barrier imposed by the pyruvate kinase barrier is a two-step reaction utilizing ATP and GTP hydrolysis. Pyruvate is converted to oxaloacetate utilizing pyruvate carboxylase and ATP in the mitochondria. This enzyme requires Mg^{2+} and biotin. The CO_2 donor is a carboxylated derivative of the cofactor biotin. Oxaloacetate cannot gain access to the next enzymes which are cytosolic as it cannot pass across the mitochondrial membrane. Such passage is achieved by conversion to malate and conversion to oxaloacetate again in the cytosol. The merit of these stages is that

oxaloacetate and malate are therefore usable as substrates in gluconeogenesis. Oxaloacetate is a product of deamination of aspartate and both are constituents of the TCA cycle.

Oxaloacetate is converted to phosphoenolpyruvate from phosphoenolpyruvate carboxykinase and a GTP–GDP conversion. The regeneration of GTP is at the expense of ATP; consequently two molecules of ATP are used to reverse the conversion of pyruvate to phosphoenolpyruvate. The phosphorylated 3-carbon acids phosphoenolpyruvate, glycerate 2-phosphate and glycerate 3-phosphate are interrelated. These may be converted to the fructose 1,6-disphosphate/triosephosphate pool following the conversion of glycerate 3-phosphate to glyceraldehyde 3-phosphate. This is a reversible concentration-dependent reaction.

Fructose 1,6-disphosphate is hydrolysed to fructose 6-phosphate by the enzyme fructose disphosphate phosphatase at the phosphoryl ester bond at C-1. When fructose 6-phosphate is produced by gluconeogenesis an equivalent amount of glucose 1-phosphate is usually removed from the hexose monophosphate pool by conversion to glycogen, which requires a nucleoside triphosphate. Such activation is achieved by UDP-glucose, an important and perhaps critical hexose derivative in mammalian metabolism.

Of the enzymes involved in gluconeogenesis only pyruvate carboxylase is mitochondrial.

The energy difference between glycolysis and gluconeogenesis is 4 ATP molecules per hexose metabolized. This means that regulation of the glycolysis/gluconeogenesis systems is of prime importance to cellular metabolism. The direction of the flux is very sensitive to the needs of the animals.

11.6.14 Regulation of fructose phosphorylation

The conversion of fructose 6-phosphate to fructose 1,6-bisphosphate is catalysed by PFK and is affected by the ATP/ADP ratio in the opposite direction from the responses of typical kinases (Figure 11.19). When ATP binds at the regulatory site the enzyme activity decreases. Citrate also increases the influence of ATP by increasing the ease of binding of ATP to the regulatory site. This means there is a direct effect of concentrations of metabolites within the TCA cycle on PFK activity. The kinase that catalyses the production of fructose 2,6-disphosphate in the liver is inactivated by phosphorylation. This inactivation is mediated through the same cyclic AMP-dependent protein kinase which is responsible for the phosphorylation of phosphorylase kinase. The phosphorylated kinase is the enzyme hydrolysing the conversion of fructose 1,6-bisphosphate to fructose 6-phos-

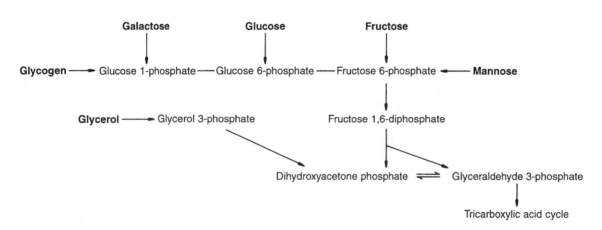

FIGURE 11.19 The interrelationship of carbohydrates glycogen, galactose, glucose, fructose, mannose and glycerol.

phate (Figure 11.19). As the blood glucose decreases, glucagon is secreted and increases the intracellular concentration of cyclic AMP. This in turn activates the cyclic AMP-dependent protein kinase that phosphorylates and inactivates the enzyme that generates fructose 1,6-bisphosphate. Glycolysis is inhibited and gluconeogenesis is stimulated, resulting in glucose being secreted into the blood.

During **starvation** or intake of high protein or fat diets, then the amount and activity of phosphoenolpyruvate carboxykinase increases and decreases when there is a restoration of carbohydrate to the diet.

Glucose is regenerated via the process of gluconeogenesis (Cori cycle), which is confined to liver and kidney cortex, with the consumption of 6 ATP. This is not an energy-creating system.

11.6.15 Ketone bodies

Ketone bodies are small molecules with a molecular weight less than 104 Da, water-soluble and weakly acidic. They are acetoacetate (AcAc), β-hydroxybutyrate (BHB) and acetone. AcAc and BHB are interconvertible, catalysed by α-hydroxybutyrate dehydrogenase within mitochondria and using the $NAD^+/NADH_2$ couple as a cofactor. The conversion of AcAc to acetone is a non-enzymatic reaction, and is followed by conversion to glucose (Figure 11.20).

Ketone bodies are synthesized almost entirely in the liver, but occasionally also in the muscle and kidneys. Free fatty acids, released from adipose tissue pass to the liver, enter the liver cells by hydrolysis and are transported to the mitochondria as carnitine ester. There is oxidative cleavage of long-chain fatty acid acyl-CoA to acetyl-CoA by β-oxidation. Two acetyl-CoA molecules condense to form acetoacetyl-CoA. Deacylation and reduction then results in 3-hydroxybutyrate. These diffuse from the liver cell into the plasma. Because the liver cell does not have the enzyme 3-ketoacid CoA-transferase the liver is unable to metabolize ketone bodies, which are either metabolized in peripheral tissues or excreted in urine.

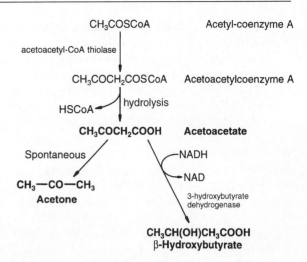

FIGURE 11.20 Ketone bodies include acetoacetate, acetone and β-hydroxybutyrate. They are produced when large amounts of acetyl-CoA are synthesized, as in uncontrolled diabetes and starvation with consequent fat mobilization.

Ketone bodies, as do free fatty acids and glucose, supply non-nitrogenous substrates to most tissues. Ketone bodies have an energy content of 4.2 kcal/g.

The brain, though primarily glucose-oxidizing and always needing some glucose, is able to metabolize ketone bodies. This valuable ability reduces the requirement for glucose when glucose is scarce. Red muscle is primarily free fatty acid-oxidizing. Liver, heart, lung and kidney can oxidize either glucose or free fatty acids. The metabolism of ketone bodies is dependent upon the concentration available.

Glucose is the principal endogenous substrate postprandially, while free fatty acids are the principal endogenous substrates during fasting conditions. Ketone bodies are intermediate in timing and substitute for amino acids.

As the meal is digested, glucose disappears from the portal system, there is a small drop in the systemic plasma glucose level and an increase in insulin concentrations. At this point ketone body synthesis begins and plasma concentrations of AcAc and BHB become detectable. Overnight fasts result in ketone bodies being the predominant fuel for muscles, with only the brain using glucose. At this stage one-third of glucose requirements are derived from gluconeogenesis.

If fasting continues then plasma concentrations of ketone bodies gradually increase. After 3 weeks of fasting, plasma glucose concentrations fall slightly and the concentrations of free fatty acids increase. The concentration of ketone bodies is four times that at the outset, with BHB concentrations twice those of the AcAc. At this stage hepatic ketone body production is maximum at 130 g/per day. The brain increases its oxidation of ketone bodies and the muscles utilize free fatty acids.

During a prolonged fast the kidney removes more BHB from the plasma than AcAc. There is a reduced loss of urinary ketone bodies and some reduced losses of ammonia/nitrogen, thereby conserving protein. This is called **starvation adaptation**, the main effects being on nitrogen metabolism.

During a prolonged fast ketone bodies replace carbohydrates during the onset and reduce carbohydrate requirement, even when the free fatty acids become the predominant metabolic fuel. Later, ketone bodies are used by the brain while muscles use free fatty acids. In this way there is a conservation of glucose of 60 g/day and muscle-sparing of 55 g/day. Prolonged starvation and diabetes result in very high concentrations of ketone bodies in blood, a condition known as **ketosis**. The high concentration of ketone bodies leads to serious related metabolic problems.

Summary

1. Glucose is an important constituent in the provision of energy to the body. Blood glucose is derived from food, liver glycogen or intermediary metabolites.

2. Glucose may be taken up by tissues, particularly skeletal muscles. Insulin facilitates the active transport of glucose into the muscle and fat cells. Postprandially there is massive glycogen synthesis. During fasting glycogen may be converted to glucose 1-phosphate and thence to pyruvate in the liver.

3. Galactose is absorbed through the same active transport system in the small intestine as glucose, but little enters the peripheral circulation. Galactose in the cells is converted to galactose 1-phosphate and then to glucose 1-phosphate. Galactose required for structure in body cells and connective tissue, e.g. galactosamine is produced from glucose.

4. Dietary fructose is taken up by the liver and enters the glycolytic pathway directly. Within the liver fructose is phosphorylated to fructose 1-phosphate which is split into glyceraldehyde and dihydroxyacetone phosphate; the latter is an intermediate in both the glycolytic and gluconeogenic pathways. Glyceraldehyde is converted to glyceraldehyde 3-phosphate then to glycogen, a source of glucose. Glyceraldehyde can be converted to glycerol 3-phosphate and esterified to fatty acids in triglycerides. These reactions are dictated by the nutrition and hormonal state. Fructose enters cells regardless of the insulin concentration.

5. Most of the enzymes involved in carbohydrate metabolism require vitamin B metabolites as essential cofactors in the glycolytic pathway, pentose phosphate shunt and tricarboxylic acid cycle.

6. Sugars may react with free amino groups on proteins to produce a chemically reversible glycosylated product (a Schiff base) and by internal rearrangement to produce a more stable Amadori-type glycosylation product, an advanced glycosylation end-product.

7. A carbohydrate meal produces an increase in blood glucose, unless the carbohydrate is fructose. The capillary blood glucose concentration increases as much from starch as following glucose. A number of factors affect the measured glycaemic response to meals.

8. Long-chain fatty acids esterified to triglycerides come from the diet or are synthesized from metabolites from the glycolytic pathway of carbohydrates and the pentose phosphate shunt.

Glucose requires insulin above a critical concentration to enter adipocytes from the blood. Below this insulin concentration, glucose does not enter the cells and glycolysis decreases. Prolonged exposure to high carbohydrate diets raises the fasting serum triglyceride concentration.

9. The ratio of carbon dioxide to oxygen in the breath (the respiratory quotient, R_Q), gives an indication of the type of substrate being metabolized and a hint as to which metabolic route is being followed. The R_Q after ingestion of fructose is greater than after glucose.

10. In aerobic metabolism, organic carbon is oxidized to CO_2 using oxygen as an electron acceptor. The ATP yield for each glucose molecule metabolized is 20-fold greater than under anaerobic conditions. The free energy is conserved as ATP and there is continuous re-oxidation of reduced coenzymes, e.g. NADH and $FADH_2$ which link into the respiratory chain to produce ATP. The oxidized coenzymes are involved in the oxidation of organic intermediates from pyruvate.

11. During aerobic conditions the glycolytic pathway is the initial phase of glucose catabolism with the tricarboxylic acid cycle and the oxidative phosphorylation of ADP to ATP, yielding 36–38 moles of ATP for each mole of glucose in this process.

12. Glycolysis is the pathway in which glucose is metabolized before splitting into two inter-convertible, 3-carbon molecules in the cell cytoplasm.

13. The catabolic steps are glycolysis with fructose 6-phosphate (F6P) as the starting point or the pentose phosphate pathway which uses glucose 6-phosphate (G6P). Alternatively, G6P and F6P can be converted into storage polysaccharides with G6P acting as the starting point. Fructose 1,6-diphosphate is reversibly split into dihydroxyacetone phosphate and glyceraldehyde 3-phosphate.

14. Glyceraldehyde 3-phosphate is converted to 1,3-diphosphoglycerate with a high-energy phos-phoanhydride bond on carbon-1. Two ATP are regenerated with the coincidental production of pyruvate which is transferred into the mitochondria. Pyruvic acid has to be converted to acetyl-CoA in the mitochondria, where the tricarboxylic acid (TCA) cycle takes place.

15. Pyruvate plays a major role as an intermediary metabolite and is central to the interconversion of glucose, fatty acids and amino acids.

16. In the TCA cycle the most important supply of acetyl-CoA is from pyruvate or fatty acids and amino acids.

17. The TCA cycle is regulated to provide for cell needs. The function of the cycle is to provide NADH and $FADH_2$ for the electron transport chain and to provide substrates for biosynthesis.

18. Acetyl-CoA entering the TCA cycle yields I ATP molecule and storage of free energy as NADH and $FADH_2$. Oxidation of 1 mole of acetyl-CoA leads to the overall production of 12 moles of ATP. In addition, pyruvate metabolism leads to 15 moles of ATP per mole of pyruvate or 30 moles per mole of glucose. The metabolites in the TCA cycle are major starting materials for a number of biosynthetic pathways.

19. The pentose phosphate pathway is a an alternative metabolic pathway. The initial reaction of glucose 6-phosphate dehydrogenation generates NADPH for fatty acid and cholesterol synthe-sis. The NADPH production throughout the pathway is important for some biosynthetic reactions, especially lipids.

20. In the presence of abundant glucose the emphasis is on anabolic processes and storage. The liver converts some glucose to stored glycogen and some to fatty acids to be used as a fuel. Peripherally glucose is metabolized to generate ATP and excess is stored as glycogen and adipose tissue.

21. Insulin stimulates the uptake of glucose by various tissues and the synthesis of glycogen in the liver.

22. Glycogen is an α1–4 polysaccharide with α1–6 branch points. Glycogen synthase is inactive when phosphorylated, and is activated by insulin and high concentrations of G6P.

23. Glucose stores in the form of glycogen are mobilized when needed. Glucose synthesis from non-glucose sources, lactate, pyruvate, fatty acids or amino acids (gluconeogenesis) is stimulated in liver and kidney.

24. Phosphorylase releases glucose from glycogen in the form of G6P. The enzyme phosphorylase a is active when phosphorylated. The inactive form phosphorylase b is produced by phosphorylase a phosphatase. The phosphorylase kinase is kept in an active and inactive form by protein kinase, the process being controlled hormonally.

25. Glucose has a central role in the production and harnessing of energy and in the synthesis of other sugars. While an individual is fasting the brain uses 80% of the glucose consumed. The liver only contains sufficient glucose (as glycogen stores) to meet the needs of the brain for 12 hours.

26. Glycolysis and gluconeogenesis are directly opposed reaction sequences. Most of the enzymes that function in glycolysis are reversed in gluconeogenesis. Fatty acids are not capable of undergoing gluconeogenesis.

27. The intermediates in the gluconeogenic and glycolysis pathways are the same and are coupled to the ATP–ADP system. Some steps in metabolic pathways are irreversible, hence alternative pathways are necessary.

28. The energy difference between glycolysis and gluconeogenesis is 4 ATP molecules per metabolized hexose. This means that regulation of the glycolysis and gluconeogenesis systems is of prime importance to cellular metabolism. There is a direct effect of concentrations of metabolites within the TCA cycle on phosphofructokinase activity.

29. During starvation or high protein or fat diets, the amount and activity of phosphoenolpyruvate carboxykinase increases and decreases when there is a restoration of carbohydrate to the diet.

30. Ketone bodies are acetoacetate (AcAc), β-hydroxybutyrate (BHB) and acetone. AcAc and BHB are interconvertible within mitochondria using the NAD^+/$NADH_2$ couple as a cofactor.

31. Ketone bodies are synthesized almost entirely in the liver, resulting in 3-hydroxybutyrate. There is diffusion from the liver cell into the plasma. The brain, but not the liver, is able to metabolize ketone bodies.

32. During starvation ketone body synthesis begins and plasma concentrations of AcAc and BHB become detectable. During prolonged starvation and diabetes mellitus high concentrations of ketones appear in the blood – the condition of ketosis.

Further reading

Coleman, L. (1980) Metabolic interrelationships between carbohydrates, lipids and proteins, in *Metabolic Control and Disease* (eds P.K. Bondy and L.E. Rosenberg), W.B. Saunders, Philadelphia, pp. 161–274.

Dodding, J. (ed.) (1989) *The Metabolism of Sugars and Starches; Dietary Starches and Sugars in Man: A Comparison*. Human Nutrition Reviews, ILSI, Springer-Verlag, London.

Harold, F.M. (1986) *The Vital Force; A Study in Bioenergetics*, W.H. Freeman & Co., New York.

Levine, R. (1986) Monosaccharides in health and disease. *Annual Review of Nutrition*, 6, 211–24.

Rich, A.J. (1990) Ketone bodies as substrates. *Proceedings of the Nutrition Society*, 49, 361–73.

Worth, H.G.J. and Curnow, D.H. (1980) *Metabolic Pathways in Medicine*, Edward Arnold, London.

Zubay, G. (1992) *Biochemistry*, W.C. Brown, Iowa, USA.

11.7 Fat metabolism

- Fatty acid synthesis involves the carboxylation of acetyl-CoA to malonate to which is sequentially added two carbon increments.
- Fatty acid synthesis is controlled by nutritional and hormonal status.
- Essential polyunsaturated fatty acids cannot be synthesized in the body.
- Fatty acids are stored as triacylglycerols in adipose tissue.
- Fatty acids are transported from organ to organ bound to albumin or as triacylglycerols in lipoproteins.
- Fatty acids are oxidized as an important source of metabolic energy in general by α, β or ω oxidation.
- Fatty acids and lipids are stored as triacylglycerols in lipid tissue.
- The synthesis and degradation of triacylglycerols are affected by the diet and by the fatty acid intake and organ involved.

11.7.1 Introduction

Most naturally occurring fatty acids have even numbers of carbon atoms. The entire chain of synthesized fatty acids derives from acetic acid. The major biosynthetic route to long-chain fatty acids is different from fatty acid degradation through β-oxidation. Liver and adipose tissue are the most important tissues for fatty acid biosynthesis.

11.7.2 Lipid synthesis

A series of enzymes are involved, which include acetyl-CoA carboxylase and fatty acid synthase. In humans it is probable that most fatty acids are synthesized in the liver and transported to the periphery for storage.

Acetyl-CoA carboxylase

The first step in fatty acid synthesis is the carboxylation of the 2-carbon fragment acetate, as acetyl-CoA, to malonate, catalysed by acetyl-CoA carboxylase, a biotin-containing enzyme which is an important control enzyme in fatty acid synthesis (Figure 11.21). Most of the animal cell acyl-CoA is derived from the oxidation of pyruvate in mitochondria. The feeder sources include lactate, glucose and amino acids. Pyruvate dehydrogenase is a mitochondrial enzyme, so that acetyl-CoA is formed within the mitochondria remote from fatty acid synthesis in the cytoplasm. The mitochondrial membrane is impermeable to acetyl-

CoA while the sites of fatty acid synthesis are mainly outside the mitochondria in the cytosol. The transfer may be mediated by conversion of the acetyl-CoA within the mitochondrion to citric acid as an intermediary.

acetate + ATP + HS–CoA $\longrightarrow$
acetyl-CoA + AMP + PPp$_i$
acetyl-CoA + oxaloacetate + ADP + P$_i$ $\longrightarrow$
citrate + ATP + HS–CoA

In the cytosol the reverse reaction occurs. The oxaloacetate is not readily transferred across membranes so there is a further conversion to malate and then pyruvate by the catalytic activity of malate dehydrogenase. This is the pyruvate–malate shunt (Figure 11.22). This series of reactions generates NADPH for fatty acid synthesis.

Citrate is also an activator of acetyl-CoA carboxylase, a rate-limiting enzyme in fatty acid synthesis.

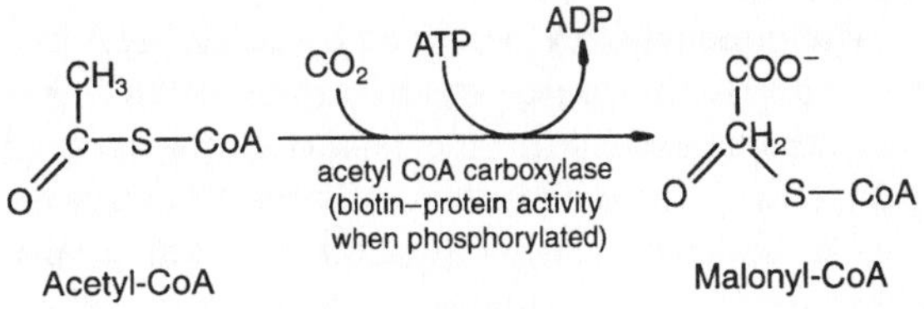

FIGURE 11.21 The first step in the synthesis of fatty acids is the synthesis of malonyl-CoA from acetyl-CoA (enzyme acetyl-CoA carboxylase, a biotin-requiring enzyme).

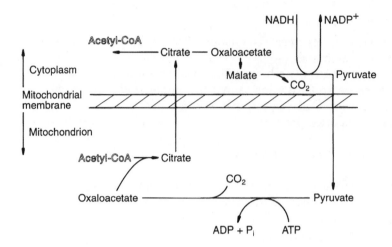

FIGURE 11.22 Pyruvate–malate shunt. Acetyl-CoA does not cross the mitochondrial membrane. This shunt system allows the effective transport of acetyl-CoA from the mitochondria to the cytoplasm.

$$\text{acetyl-CoA} + CO_2 + ATP \longrightarrow$$
$$\text{malonyl-CoA} + ADP + P_i$$

Citrate is converted to oxaloacetic acid by citrate lyase, and the oxaloacetate reduced by NADH to form malate:

$$\text{malate} + NADP \rightarrow \text{pyruvate} + CO_2 + NADPH$$

In the conversion of acetyl-CoA to fatty acids, NADPH is required, and the citrate/malate/pyruvate pump is a source.

Fatty acid synthase

Fatty acid synthesis from malonyl-CoA is a series of condensation additions and reductions from carbonyl to methylene groups. The reactions are catalysed by a multienzyme system, fatty acid synthase. There is a transfer of intermediates within the system from one active site to another during chain elongation.

The malonyl-CoA generated by acetyl-CoA carboxylase is the source of the atoms of the fatty acyl chain, except for the first two atoms from acetyl-CoA. In some instances butyryl-CoA is the initial molecule. Propionyl-CoA or branch primers permit the formation of odd chain length and branched-chain fatty acids respectively.

Acetyl-CoA synthase allows acetate to be transported from the liver and activated and oxidized in mitochondria of other tissues. Butyryl-CoA

synthetase is active for acids with chain length in the range 3–7 carbons. It is a mitochondrial enzyme in the heart but in the liver there is also a special propionyl-CoA synthetase.

Fatty acid synthase

Fatty acid synthase can be divided into Type I, II and III enzymes.

- Type I synthase are multifunctional proteins in which the proteins catalyse individual partial reactions.
- Type II and Type III the complexes of molecular weight 450–550 kDa. The multifunctional forms of fatty acid synthase have arisen by gene fusion. The genes for Type II fatty acid synthase found in lower bacteria and plants fuse to give two genes which would code for a yeast-type fatty acid synthase. The latter then fuse to give a single gene coding for the mammalian Type I fatty acid synthase.

The medium-chain acyl-CoA synthase is isolated from heart mitochondria and is active in acids with a chain length in the range 4–12 carbons. The catalysed reaction is the formation of

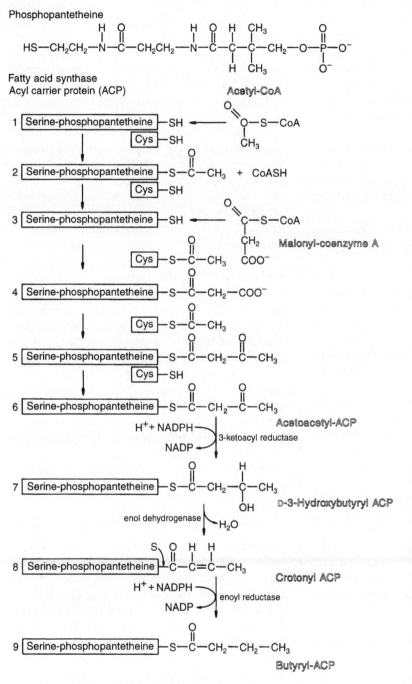

Phosphopantetheine

Fatty acid synthase
Acyl carrier protein (ACP)

1 | Serine-phosphopantetheine |—SH
Cys|—SH

Acetyl-CoA

2 | Serine-phosphopantetheine |—S—C—CH₃ + CoASH
Cys|—SH

3 | Serine-phosphopantetheine |—SH

Malonyl-coenzyme A

Cys|—S—C—CH₃

4 | Serine-phosphopantetheine |—S—C—CH₂—COO⁻
Cys|—S—C—CH₃

5 | Serine-phosphopantetheine |—S—C—CH₂—C—CH₃
Cys|—SH

6 | Serine-phosphopantetheine |—S—C—CH₂—C—CH₃

Acetoacetyl-ACP

H⁺ + NADPH → 3-ketoacyl reductase
NADP

7 | Serine-phosphopantetheine |—S—C—CH₂—C—CH₃
                                            OH
D-3-Hydroxybutyryl ACP

enol dehydrogenase → H₂O

8 | Serine-phosphopantetheine |—C—C=C—CH₃
Crotonyl ACP

H⁺ + NADPH → enoyl reductase
NADP

9 | Serine-phosphopantetheine |—S—C—CH₂—CH₂—CH₃
Butyryl-ACP

Then: cycle returns to stage 3 and adds malonyl-coenzyme A up to C-16 (palmitic acid).
Then: hydrolysis to produce the free acid.

FIGURE 11.23 Fatty acid synthase. Fatty acid synthesis begins with the binding of an acetyl group to fatty acid synthase. There follow incremental additions of malonyl-CoA. The active sites on the fatty acid synthase are phosphopantotheine and a cysteine grouping. There is an incremental addition of malonyl-CoA. The second active site (Cys-SH) binds the previous metabolite preparatory to the addition of further malonyl-CoA. The 3-ketoacyl product is reduced to the corresponding saturated acyl group, ready for the next addition of malonyl-CoA.

glycine conjugates of aromatic carboxylic acids such as benzoate and salicylate.

The fatty acid synthase (Figure 11.23) contains a functional group the 4'-phosphopantetheine group which is a flexible chain of 14 atoms that joins with the synthase protein through a serine residue. End products are transferred from one active site (4'-phosphopantetheine arm) to another (3-ketoacyl reductase). This has a terminal cysteine group on the acyl protein carrier protein (ACP) to which each successive condensation and reduction product is temporarily moved. A maximum of seven malony groups condense with the fatty acid chain.

Stages by which the malonyl groups condense with the fatty acid chain:

- acetyl transferase: the acetyl CoA is transferred first to the 4'-phosphopantetheine arm and then to a second site, leaving the 4'-phosphopantetheine arm free.
- malonyl transferase: each incoming malonyl group is transferred to the 4'-phosphopantetheine arm
- 3-ketoacyl synthase: the malonyl group on the 4'-phosphopantetheine arm is added to the acetyl group followed by decarboxylation to give 3-ketobutyryl derivatives
- D-hydroxybutyrate is produced by reduction through NADPH
- 3-hydroxyacyl dehydratase removes a water molecule
- this enoyl intermediate is reduced by NADPH, resulting in a fatty acid attached to the 4'-phosphopantetheine arm. This fatty acid is then transferred to the second active site 3-ketoacyl synthase leaving the 4'-phosphopantetheine arm free for the next malonyl-CoA
- thioester hydrolase releases palmitic acid from the 4'-phosphopantetheine arm

---

The end-product of animal fatty acid synthase enzymes is usually free palmitic acid. The cleavage of this acid from the enzyme complex is catalysed by a thioesterase, an integral part of the enzyme complex.

The rate of fatty acid synthesis is determined by the availability of malonyl-CoA, the concentration of palmitoyl-CoA, and possibly citrate.

Fatty acids have to be converted into metabolically active thiol esters before further anabolic or catabolic metabolism. The active form is usually the thiol ester of the fatty acid with the complex nucleotide coenzyme (CoA) or the small protein known as acyl carrier protein (ACP). This makes the acyl chains water-soluble. The formation of acyl-CoA is by acyl-CoA synthetases, dependent on chain length specificity and tissue distribution.

Fatty acid elongation

Elongation is catalysed by the Type III synthases or elongases (Figure 11.24). These enzymes are present in the endoplasmic reticulum. They catalyse the addition of malonyl-CoA to preformed acyl chains requiring NADPH as a reducing coenzyme. Two carbon units can be added up to 24 carbon atoms. Saturated fatty acids $< C-16$ are preferred as the initial substrate. The important function of elongation is in the transformation of dietary essential fatty acids to the higher polyunsaturated fatty acids. The starting point is linoleoyl-CoA which is first desaturated to a trienoic acid.

11.7.3 Lipid-controlling mechanisms

The activity of the enzymes involved in fat synthesis decreases during starvation or by a high dietary fat intake. The enzymes involved in fatty acid biosynthesis are most active during periods when a low fat, high carbohydrate diet is being eaten. Polyunsaturated fatty acids are effective in reducing fatty acid synthesis. Sucrose ingestion, probably as a result of the fructose content results in an increase in the enzymes associated with lipogenesis in the liver.

When fat is ingested – whether in the form of fatty acids or other lipids – blood insulin concentrations remain low. On an enriched fatty diet, the metabolic responses are those of a starved animal with increased concentrations of free fatty acids and low insulin, whereas increased glucose and insulin are found in the fed animal. Acetyl-CoA carboxylase, the rate-limiting step of fatty acid synthesis, is inhibited by fatty acyl-CoA and by free fatty acids.

Insulin is important in determining adipose tissue cell permeability. Following a high carbohydrate meal, blood glucose and insulin concentrations rise. Glucose enters the cell and may be converted to fatty acids and glycerol, and hence triglycerides. In contrast, on a high fat intake a limiting factor will be glucose for glycerol synthesis.

Within an adipose cell the hydrolysis of triglyceride yields diglyceride, monoglycerides, glycerol and free fatty acids. Triglycerides may be reformed or metabolized and the free fatty acids may enter the blood stream. This process is con-trolled by the concentration of glucose and cellular cyclic AMP (cAMP) and is therefore an important element in adipose tissue control. During starvation, cAMP concentrations increase and triglycerides are hydrolysed; the fatty acids then pass into the blood stream, to be bound to albumin. This complex passes to other tissues for utilization as a fuel. Muscle uses free fatty acids in preference to glucose. The permeability of glucose through cell membranes is inhibited; hence glucose metabolism is protected. Fatty acid metabolism creates raised concentrations of citrate and ATP which in turn inhibit glycolysis by inhibiting PFK activity.

Fatty acid synthesis is controlled by acetyl-CoA carboxylase, which requires a hydroxy tricarboxylic acid, e.g. citrate or isocitrate, as substrate. Citrate, being a precursor of acetyl-CoA, is a positive feed-forward activator. The main role of citrate in the catalysis is to keep the carboxylated form of the enzyme in its active conformation by shifting the equilibrium from the inactive to the active species of the enzyme. Long-chain acyl-CoA inhibits mammalian acetyl-CoA carboxylases.

Phosphorylation

Acetyl-CoA carboxylase, the dephosphorylated enzyme form is much more active than the phosphorylated.

Synthesis and degradation

Long-term regulation of acetyl-CoA carboxylase in animals can be due to changes in enzyme amounts. This depends on nutritional state, hormonal developmental and genetic factors. There is a decrease in enzyme content during fasting due to diminished synthesis or accelerated breakdown. The metabolite responsible for the control in the synthetic rate of the enzyme is unesterified fatty acid.

Metabolic controls of fatty acid synthase

The amount of liver fatty acid synthase can be decreased by starvation and glucagon and increased by feeding and by the hormones insulin, β-oestradiol, hydrocortisone and growth hor-

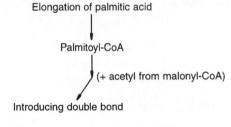

Reaction: occurs in endoplasmic reticulum and mitochondria

Double bonds: can only be added after C:16

FIGURE 11.24 Elongation of some mono-unsaturated fatty acids is possible by elongating the palmitoyl carbon chain and introducing double bonds, e.g. oleic acid 18 : 1. This reaction takes place either in the endoplasmic reticulum or mitochondria. Humans can introduce double bonds at C9, C6 and C3 but not towards the tail. Double bonds result from the catalytic activity of acyl-CoA desaturases in the endoplasmic reticulum, the enzymes being mixed-function oxidases.

mone. The increase in fatty acid synthase level between starved and refed is about 20-fold.

Dietary factors which affect fatty acid synthetase levels do not affect all tissues equally. The liver is highly influenced by such changes, yet the fatty acid synthase of brain is unaffected. The fatty acid turnover of the brain extends over a much longer time-scale than that of the liver. Fatty acid synthase increases in mammary gland tissue during mid- to late pregnancy and early in lactation. The synthase level in the brain is highest in the foetus and neonate and decreases with maturity. Differences in fatty acid synthase activity result from changes in enzyme amounts rather than changes in enzyme activity. These alterations result from changes in the balance of enzyme synthesis and degradation as a result of the amount of fatty acid synthase messenger RNA present in the cell. NADPH production is adjusted to cope with the altering amounts of fatty acid synthesis. The supply of malonyl-CoA by acetyl-CoA carboxylase activity is a major factor regulating overall fatty acid formation.

Control of fatty acid desaturases

Unsaturated fatty acids are present in all living cells. They are important:

- in regulating the physical properties of lipoproteins and membranes
- in regulation of metabolism in cells
- as precursors for physiologically active compounds, e.g. eicosanoids.

When diet causes an increased synthesis of fatty acid synthase there are similar increases in activity of the enzymes Δ^9 desaturase but not of Δ^4, Δ^5 or desaturase (the Δ numbering being the chemical numbering of the unsaturated bond). Polyunsaturated fatty acid desaturases are designated as Δ, e.g. 4,5 or 6 because they introduced double bonds between carbon atoms 4-5, 5-6 and 6-7 respectively.

Insulin influences desaturase activity. However, dietary fructose, glycerol or saturated fatty acids can also increase enzyme activity which means that the influence of insulin activity is modest or indirect.

11.7.4 Biosynthesis of unsaturated fatty acids

The most important pathway is an oxidative mechanism by which a double bond is introduced directly into a saturated long-chain fatty acid, using a reduced compound, e.g. NADH as cofactor. Most of the acids produced have a Δ^9 double bond, e.g. palmitic into palmitoleic acid. The double bond is introduced between carbon atoms 9 and 10 counting from the carboxyl end of the fatty acid chain (Figure 11.25).

In animals the usual elongation process is to introduce new double bonds between an existing double bond and the carboxyl group, whereas plants generally introduce the new double bond between the existing double bond and the terminal methyl group. The most abundant polyunsaturated acid produced by plants, linoleic acid (*cis,cis*-9,12 C-18 : 2) can not be synthesized by animals, yet this acid is necessary to maintain animals in a healthy state. Consequently, linoleic acid is an essential fatty acid for animals. In plants

n-9 series

Oleic acid 18 : 1 Δ^9
      → 18 : 2 $\Delta^{6,9}$
      → 20 : 4 $\Delta^{5,8,11}$

n-6 series (essential fatty acids)

Linoleic acid 18 : 2 $\Delta^{9,12}$
γ-linolenic acid 18 : 3 $\Delta^{6,9,12}$
Arachidonic acid 20 : 4 $\Delta^{5,8,11,14}$

n-3 series (essential fatty acids)

α-Linolenic acid 18 : 3 $\Delta^{9,12,15}$
      18 : 4 $\Delta^{6,9,12,15}$
      20 : 4 $\Delta^{8,11,14,17}$
Eicosapentaenoic acid 20 : 5 $\Delta^{5,8,11,14,17}$

n-system: count from CH_3 end
Δ-system: count from COOH end

FIGURE 11.25 Fatty acid synthesis. Elongation of polyunsaturated fats in n-9 (Δ^9), n-6 (Δ^6) and n-3 (Δ^3) series; the n-6 and n-3 series being essential fatty acids. From a dietary point of view linoleic and α-linolenic acids are essential. If these are present in the diet, the other longer fatty acids can be synthesized from these precursors.

and algae the precursor of polyunsaturated fatty acid formation is oleate.

Introduction of a double bond into a saturated fatty acid

Humans can introduce a double bond at the C-9 position (Δ^9 desaturase) and C-6 and C-3, but not beyond C-9. The double bonds are introduced with the catalytic action of mixed-function oxidases and acyl-CoA desaturase on the endoplasmic reticulum. Oxygen is activated through NADH or NADPH and a cytochrome b_5 or cytochrome P_{450}. To synthesize fatty acids with less than seven carbons beyond the double bond, then the reactant has to be a fatty acid of plant origin.

linoleate 18 : 2 (9,12) →
arachidonate 20 : 4 (5,8, 11,14)

But palmitate cannot be so transformed:

palmitoleoyl 16 : 1 (9) → 20 : 4 (4,7,10,13)

Polyunsaturated fatty acids of the same family can be recognized by subtracting the number of the last unsaturated bond from the number of carbons and these will be equal in a fatty acid family:

linoleate 18 : 2 (9,12) and
arachidonate 20 : 4 (5,8,11,14)
18–12 = 20–14 = n-6

Monounsaturated fatty acyl-CoA esters can be substrates for the Δ^6-desaturase enzyme present in animal cell membranes. Oleic acid is converted to a series of (n-9) fatty acids. Oleic acid 18 : 1 n-9 becomes 18 : 2 n-9. The enzyme is more usually involved in the reaction 18 : 2 n-6 to 18 : 3 n-6 and 18 : 3 n-3 to 18 : 4 n-3. When an 18-carbon chain is extended by two carbon atoms an extra double bond may be added between the carboxyl group and the first double bond. Linoleic acid is the precursor of a series of n-6 fatty acids. Δ^5-

Desaturase can produce arachidonic acid 20 : 4 n-6 and eicosapentaenoic acid 20 : 5 n-3. Further fatty acids for membranes and eicosanoid production result from Δ^4-desaturase which adds a double bond to produce docosapentaenoic acid 22 : 5 n-6 and docosahexaenoic acid 22 : 6 n-3.

The Δ^6-desaturase enzyme introduces a double bond at position 6 in the three families, n-3, n-6, n-9. There is competition for the enzymes, which have an affinity, that is greatest for α-linolenic acid, 18 : 3 n-3, less for linoleic acid 18 : 2 n-6 and least for oleic acid 18 : 1 n-9. The α-linoleic family can be converted to a series of n-3 fatty acids.

Introduction of a double bond and the subsequent polyunsaturated fatty acid families

In plants the double bonds are introduced at the 12,13 position to form linoleate followed by further desaturation of the 15,16 position to form α-linolenic acid. The inability of animals to desaturate oleic acid towards the methyl end of the chain gives rise to distinct families of polyunsaturated fatty acids that are not interconvertible. Polyunsaturation in animals is accompanied by three separate desaturases designated as Δ^4, Δ^5 and Δ^6 because they introduces double bonds between carbon atoms 4-5, 5-6 and 6-7. This involves cytochrome b_5 and NADH–cytochrome b_5 reductase, molecular oxygen and reduced nicotinamide nucleotide. Substrates for the first polydesaturation are oleic acid (n-9), linoleic acid (n-6) and α-linolenic acid (n-3).

Trans acids in the diet can be incorporated into the lipids of most tissues of the body, including liver, adipose tissue, brain and milk. The incorporation of *trans* fatty acids into tissues is proportional to the amounts present in the diet, particularly for *trans*-octadecenoates. However, this is also affected by the other constituents in the diet, for example the essential fatty acids reduce the accumulation of *trans* fatty acids in tissues.

Trans fatty acids are incorporated into all major classes of complex lipids, particularly triacylglycerols in adipose tissue. These are esterified to glycerol carbon positions of 1 and 3. In contrast, in heart, liver and brain *trans* fatty acids are incorporated into phospholipids, *trans*-octadecenoic acids behave like saturated fatty acids and are esterified preferentially into position C-1 of the phosphoglycerides. Oleic acid is randomly distributed among the carbons of the glycerol.

Trans fatty acids are catabolized as any other fatty acids and are readily removed from tissues and oxidized. In general, they behave as saturated fatty acids and are of no distinctive physiological or pathological significance.

Many of the biological effects of *trans* fatty acids are related primarily to the proportions of non-essential and essential fatty acids in the diet.

11.7.5 Cellular degradation of fatty acids

Fatty acids are transported from organ to organ as either:

- non-esterified fatty acids bound to albumin; or
- triacylglycerols associated with lipoproteins, particularly chylomicrons and very low density lipoproteins

Free fatty acids are transported across the cell membrane. Triacylglycerol is hydrolysed by lipoprotein lipase which is predominantly in adipose tissue. The fatty acids are transported bound to albumin and enter cells by active transport. Inside the cells, fatty acid acyl-CoAs are generated, by an ATP-dependent acyl-CoA synthetase. Free fatty acids and acyl-CoA fatty acids bind to distinct cytosolic fatty acid proteins. Such small molecular weight (14 kDa) proteins include the z-protein of liver. Acyl-CoA groups cannot transfer across the mitochondrial membrane. Transport occurs by transfer to carnitine (Figure 11.26; see also p. 208), which moves across the membrane to be reconstituted as acyl-CoA. The transport of fatty acids from the cytosol into the mitochondria

requires the catalytic activity of acylcarnitine translocase. Fatty acids of fewer than 10 carbons can be taken up by mitochondria.

Fatty acid breakdown involves oxidation at defined bonds in the acyl chain or oxidation at defined double bonds, in particular unsaturated fatty acids. The main forms of oxidation are termed α, β and ω, depending on the position of the carbon of the acyl chain which is oxidized (Figure 11.27).

β-Oxidation

β-Oxidation is quantitatively the most important mode of fatty acid oxidation (Figure 11.28). Long-chain fatty acids incorporated into triacylglycerols are the long-term storage form of energy in animal tissue. These fats are degraded as 2-carbon (acetyl-CoA) fragments by β-oxidation. The reaction takes place in mitochondria, peroxisomes and glyoxysomes. The latter two microbodies are particularly important in liver and kidney. Micro-

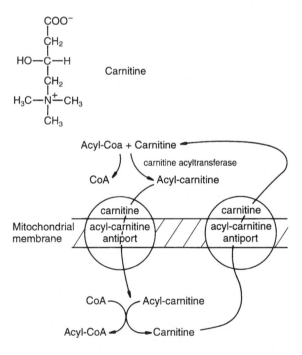

FIGURE 11.26 Acyl-CoA is unable to cross the mitochondrial membrane. Transport of fatty acids into mitochondria occurs as acyl carnitine, through the acyl carnitine antiport and resynthesis to acyl-CoA.

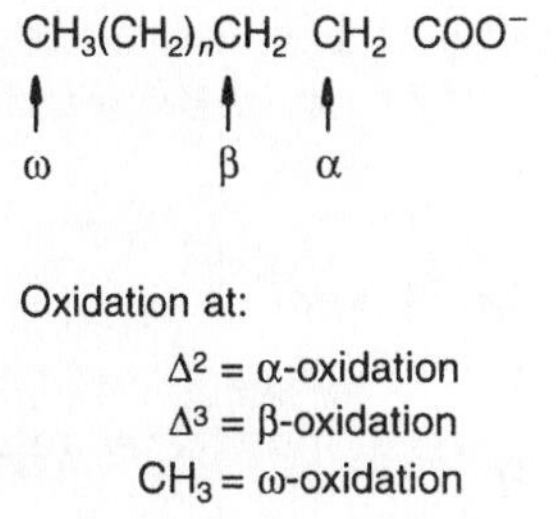

$$CH_3(CH_2)_nCH_2\ CH_2\ COO^-$$

$$\omega \qquad\qquad \beta \quad\ \alpha$$

Oxidation at:

Δ^2 = α-oxidation
Δ^3 = β-oxidation
CH_3 = ω-oxidation

FIGURE 11.27 Degradation of fatty acids. Nomenclature of oxidation: α, β, ω.

bodies oxidize long-chain fatty acids to medium-chain fatty acids which are then transported to mitochondria to complete the degradation.

Two carbon atoms in the form of acetyl-CoA are removed from the fatty acid by the successive action of four enzymes: acyl-CoA dehydrogenase, enoyl-CoA hydrolase, 3-L-hydroxyacyl-CoA dehydrogenase, and thiolase:

1. The first step in the β-oxidation cycle is the introduction of a *trans* α,β double bond to the hydrocarbon chain of the activated fatty acid by the flavoprotein enzyme acyl-CoA dehydrogenase.
2. Enoyl-CoA hydrolase then catalyses the addition of water across the *trans* double bond of the unsaturated acyl-CoA to form 3-L-hydroxy-acyl-CoA.
3. L-3-hydroxyacyl-CoA dehydrogenase converts the L-hydroxy fatty acid into 3-ketoacyl CoA.
4. Acyl-CoA : acetyl-CoA acyl transferase (or thiolase) catalyses a thiolytic cleavage of the keto acid, removing an acetyl-CoA fragment and replacing it with an –SH group from CoA.

The shorter acyl fatty acid may then progress through a repeat of this sequence of oxidation, hydration, oxidation and cleavage until the entire chain is reduced to 2-carbon lengths and enters the TCA cycle.

Acids of odd chain length may yield propionic acid, e.g. in the liver (Figure 11.29). Myocardial tissue cannot perform propionate oxidation. Branched chain fatty acids with an even number of carbon atoms may eventually yield propionate.

Many natural fatty acids are unsaturated, with

most double bonds being *cis*-orientated. When unsaturated fatty acids are β-oxidized, the unsat-

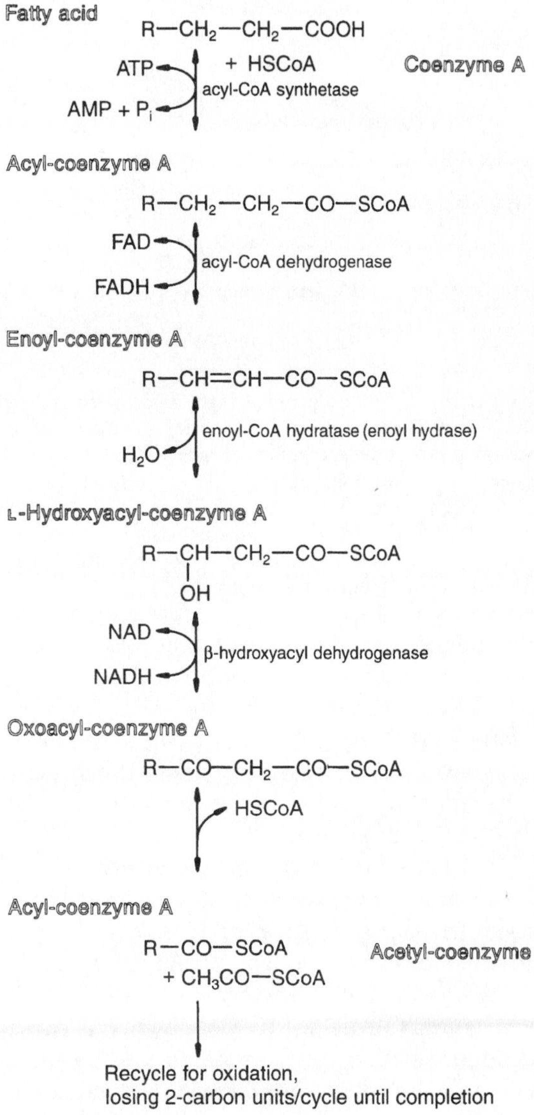

Fatty acid $\quad$ R—CH$_2$—CH$_2$—COOH

ATP $\qquad$ + HSCoA $\qquad$ Coenzyme A
$\qquad\qquad$ acyl-CoA synthetase
AMP + P$_i$

Acyl-coenzyme A

$$R—CH_2—CH_2—CO—SCoA$$

FAD
$\qquad$ acyl-CoA dehydrogenase
FADH

Enoyl-coenzyme A

$$R—CH=CH—CO—SCoA$$

$\qquad$ enoyl-CoA hydratase (enoyl hydrase)
H$_2$O

L-Hydroxyacyl-coenzyme A

$$R—CH—CH_2—CO—SCoA$$
$$\quad\ \ OH$$

NAD
$\qquad$ β-hydroxyacyl dehydrogenase
NADH

Oxoacyl-coenzyme A

$$R—CO—CH_2—CO—SCoA$$

$\qquad$ HSCoA

Acyl-coenzyme A

$$R—CO—SCoA \qquad Acetyl\text{-}coenzyme\ A$$
$$+ CH_3CO—SCoA$$

Recycle for oxidation,
losing 2-carbon units/cycle until completion

FIGURE 11.28 β-Oxidation. This is the usual form of oxidation of fatty acids as the acyl-CoA form. This is oxidized to unsaturated enoyl-CoA by a flavoprotein, followed by hydration to an alcohol group which is oxidized by NAD to a ketone. The ketoacyl-CoA is cleaved with another CoA, producing acetyl-CoA and a new acyl-CoA compound which is two carbons shorter than the original. This shorter acyl-CoA can then undergo the same reaction sequence.

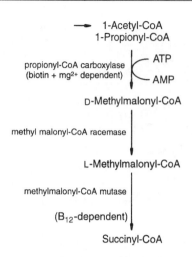

$$CH_3(CH_2)_nCH=CH(CH_2)_nCOOH$$

ATP + HSCoA

AMP + Pi

$$CH_3(CH_2)_nCH=CH(CH_2)_nCOSCoA$$
cis

FIGURE 11.29 Oxidation of fatty acids with an odd number of carbon atoms. The final products of β-oxidation are one molecule of acetyl-CoA and one molecule of propionyl-CoA. The propionyl-CoA is metabolized to D-methyl malonyl-CoA and L-methyl malonyl-CoA. The final reaction proceeds to succinyl-CoA. The enzyme (methyl malonyl-CoA mutase) has an absolute requirement for vitamin B_{12}.

urated acids have *cis* double bonds and this may prove difficult for β-oxidation. An isomerase converts the *cis*-3 compound into the necessary *trans*-2 fatty acyl-CoA. β-Oxidation then continues, removing a further two carbons and then dehydrogenation to produce a 2-*trans*,4-*cis* decadienoyl-CoA from linoleoyl-CoA (Figure 11.30).

The acetyl-CoA enters the TCA cycle. Acetoacetate and β-hydroxybutyrate may accumulate as ketone bodies. Acetoacetyl-CoA may be converted to the free acid and CoA. Alternatively, acetoacetyl-CoA may be converted into hydroxymethylglutaryl-CoA (HMG-CoA) which is subsequently cleaved to free acetoacetic acid. HMG is important in cholesterol synthesis. Ketone bodies are metabolic substrates for brain and liver metabolism (see section 11.6.15). Acetyl-CoA therefore can be involved in the citric acid cycle or ketogenesis. The particular pathway which is followed depends on the rate of β-oxidation and on the redox state of the mitochondria controlling the oxidation of malate to oxaloacetate.

The overall rate of β-oxidation is dependent on:

- the availability of free fatty acids
- the rate of utilization of β-oxidation products
- feed-back mechanisms

The concentration of plasma free fatty acids is controlled by the inhibitory effect of insulin, the stimulating effect of glucagon and the breakdown of triacylglycerols. In muscle, the rate of β-oxidation is dependent on the plasma free fatty acid

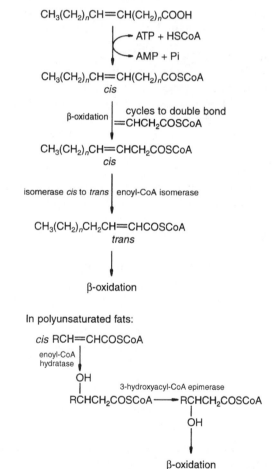

FIGURE 11.30 Oxidation of unsaturated fatty acids. Unsaturated bonds cannot act as substrates for the acyl-CoA dehydrogenase enzyme. This is overcome by an isomerization of the double bond from the *cis* to the *trans* conformation. A normal substrate for enoyl-CoA hydratase and the normal β-oxidation follows. In polyunsaturated fats, 3-hydroxyacyl-CoA epimerase hydrates and inverts the configuration of the hydroxyl at carbon-3 from the D-isomer to the L-isomer.

concentration and energy requirements. A decrease in energy demand by muscles results in increased concentrations of NADH and acetyl-CoA. An increase in NADH/NAD$^+$ ratios inhibits mitochondrial TCA cycle activity. Liver metabolism of lipids is much more complicated because of the conflicting effects of lipids, carbohydrates and ketone bodies. Malonyl-CoA inhibits carnitine palmitoyltransferase and reduces the movement of acyl groups into mitochondria for oxidation. Malonyl-CoA is produced by acetyl-CoA carboxylase which is regulated by hormones. Fatty acid synthesis and degradation are both regulated by liver hormonal concentrations.

Under normal conditions **peroxisomes** contribute up to 50% of the total fatty oxidation activity of the liver. Such oxidation is increased by hypolipidaemic drugs, high-fat diets, starvation or diabetes and is limited to medium-chain and acetyl-CoA moieties. Peroxisomal β-oxidation generates less ATP than mitochondrial β-oxidation. Peroxisomes also oxidize pristanic acid, derived from phytol, the long-chain alcohol attached to chlorophyll.

Defects in peroxisome function

There are a number of genetically determined autosomal recessive peroxisomal defects. Zellweger's syndrome is a severe condition that results in errors of development and death before 1 year. Neonatal adrenoleucodystrophy, infantile Refsum's disease and hyperpipecolic acidaemia are fatal before childhood. X-linked adrenoleucodystrophy (incidence 1 in 20–30 000) is caused by a mutation in an enzyme acting on very long-chain fatty acids. Hyperoxaluria type I results from a defect in the enzyme which transaminates glyoxalate to oxalate, which then accumulates.

α-Oxidation

α-Oxidation is so-called because only one carbon – the carboxyl carbon – is lost at each step and the α-carbon becomes oxidized to the new carboxyl group (Figure 11.31). This fatty acid oxidation is microsomal; the fatty acids do not have to be oxidized, and non-esterified fatty acids are permitted as substrates. This system is not linked to high energy production.

α-Oxidation is important in the formation of α-hydroxy fatty acids and for chain shortening. This is significant for molecules that cannot be metabolized directly by β-oxidation. For example, brain

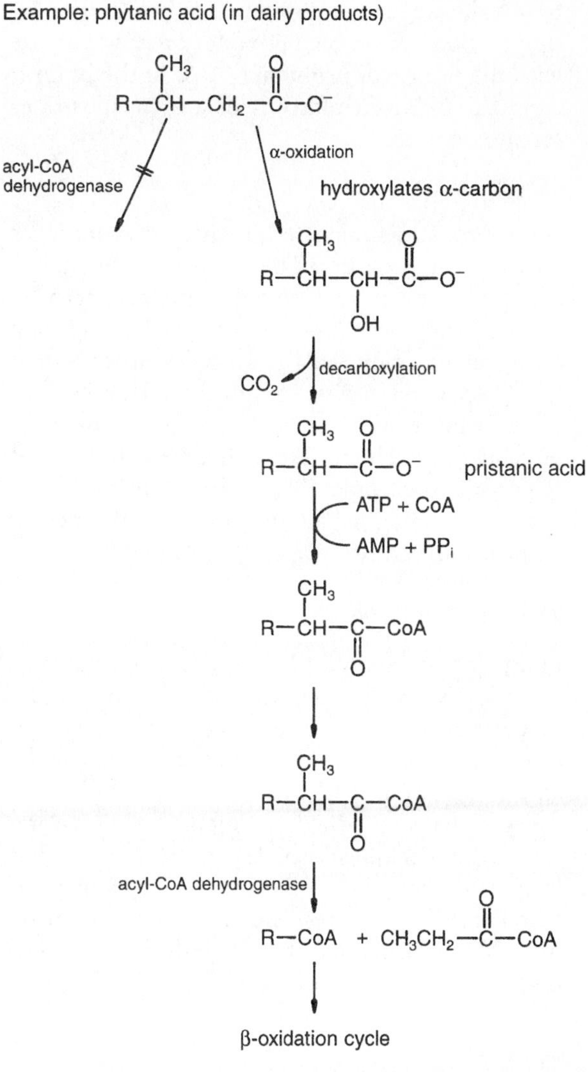

FIGURE 11.31 α-Oxidation is important in fatty acids with methyl substitutions at carbon-3. These are not substrates for acyl-CoA dehydrogenase.

cerebrosides and other sphingolipids contain α-hydroxy fatty acids. A mixed function oxidase breaks down α-hydroxy fatty acids and requires NAD, and oxygen and results in 1-carbon loss and the liberation of carbon dioxide.

An α-oxidation system in liver and kidney is important for the breakdown of branched-chain fatty acids, e.g. phytanic derived from phytol in animals. Methyl fatty acids, where the methyl group is β to a carboxyl group therefore cannot be β-oxidized, can be shortened by one carbon by α-oxidation, after which β-oxidation can continue with the release of propionyl-CoA. In the brain it is probable that free fatty acids are the substrates for α-oxidation.

ω-Oxidation

ω-Oxidation of straight-chain fatty acids produces dicarboxylic products. This is an oxidative process at the opposite end from the fatty acid carboxyls producing ω or (ω–1) hydroxy acids and then dicarboxylic acids. This is a slower process than β-oxidation. However, in substituted derivatives ω-oxidation is an important first step to the subsequent β-oxidation. An ω-hydroxy fatty acid requires cytochrome P_{450}, oxygen and NADPH and is an intermediate for the mixed-function oxidase enzyme.

11.7.6 Peroxidation

Lipids, when exposed to oxygen, form **peroxides**. If this occurs in membranes then membrane destruction occurs. Lipid peroxidation occurs in three separate processes, initiation, propagation and termination.

Peroxidation products are important, e.g. short- or medium-chain aldehydes from unsaturated fatty acids give rise to rancidity. The spoilage of frozen foods is the result of such a transformation. More pleasant tastes of fresh green leaves, oranges and cucumbers are also derived from aldehydes.

11.7.7 Lipids as energy stores

All plants use lipids as a long-term form of stored energy, particularly as triacylglycerols. Animals store the fat in adipose tissue whereas fish use their flesh or liver as a lipid store.

11.7.8 The biosynthesis of triacylglycerols

The fatty acid composition of animal acylglycerols is influenced by the diet and ultimately by the vegetable oils eaten. Such metabolism varies between species and from organ to organ within a species.

The glycerol phosphate pathway for triacylglycerol formation is the progressive esterification of glycerol 3-sn-phosphate and 1-acylglycerol 3-sn-phosphate with long-chain acyl-CoA fatty acids. Critical to this pathway is phosphatidic acid in both phospholipid and triacylglycerol biosynthesis. The diacylglycerol derived from phosphatidic acid forms the building block for triacylglycerols as well as phosphoglycerols. The transfer of acyl (fatty acids) groups from acyl-CoA to glycerol 3-phosphate requires two

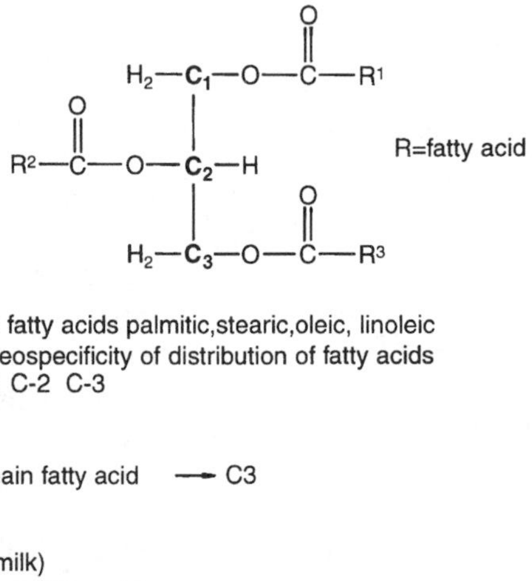

Predominant fatty acids palmitic,stearic,oleic, linoleic
There is stereospecificity of distribution of fatty acids between C-1 C-2 C-3

milk fat
    short chain fatty acid   ⟶ C3

animal fat
(not human milk)
    short chain fatty acids
    unsaturated fatty acids ⟶ C2
    polyunsaturated fatty acids
        mammals ⟶ C3
        fish ⟶ C2

seed oils
    acetate   ⟶ C3

FIGURE 11.32 Lipids as an energy store. Acyl glycerols. There is stereospecificity in the distribution of fatty acids between C1, C2 and C3.

enzymes, specific for positions 1 and 2. The enzyme responsible for position 1 exhibits marked specificity for saturated acyl-CoA-thiolesters,

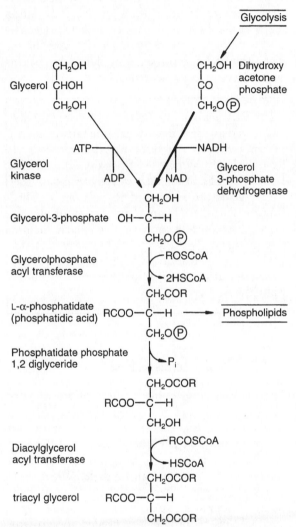

FIGURE 11.33 Synthesis of triacyl glycerol. Glycerol-3-phosphate is metabolized to L-α-phosphatidate (phosphatidic acid). L-α-phosphatidate can be further metabolized to phospholipids or 1-2-diglycerides and triacyl glycerol. Of major importance is the supply of diacyl glycerol and fatty acyl-CoA. The supply of fatty acid is increased post prandially and from biosynthesis of carbohydrate. Increased insulin post prandially stimulates the synthesis of malonyl-CoA, increasing fatty acid synthesis. Phosphatidic acid phosphatase activity may be increased by glucocorticoids.

whereas the second enzyme shows specificity towards mono- and dienoic fatty acyl-CoA-thiolesters. This explains the saturated and unsaturated difference between positions 1 and 2 (Figure 11.32).

The conversion of phosphatidic acid to diacylglycerol is a rate-limited step catalysed by phosphatidate phosphohydrolase. The final step in synthesis is the transfer of fatty acid from acyl-CoA to the diacylglycerol; the diacylglycerol acyltransferase has a wide fatty acid specificity (Figure 11.33).

An alternative precursor for 1-acylglycerol 3-sn-phosphate utilizes the dihydroxyacetone phosphate pathway; 1-acyl dihydroxyacetone phosphate acts as an intermediate. The conversion of glucose into triacylglycerol through the dihydroxyacetone phosphate (DHAP) pathway requires NADPH. The glycerol phosphate pathway consumes NADH. NADPH is associated with fatty acid biosynthesis. The activity of DHAP is enhanced under conditions of increased fatty acid synthesis and relatively reduced with starvation, or a relatively high-fat diet.

Enteric triacylglycerols

Triacylglycerols are resynthesized from monoacylglycerols which have been absorbed following small intestinal hydrolysis. The reactions are catalysed by enzymes in the enterocytes of the endoplasmic reticulum. The 2-monoacylglycerols are more readily esterified than the 1-monoacylglycerols. The rate of the initial esterification depends on which fatty acid is esterified in C2; short-chain saturated or longer-chain unsaturated fatty acids, are favoured. Diacyl transferase hasa specificity for 1,2-sn-diacylglycerols. Diacylglycerols with two unsaturated or mixed acid fatty acids are preferred as substrates to disaturated compounds.

11.7.9 Acylglycerol catabolism

Lipases are important in the splitting of the fatty acid ester from the primary hydroxyl groups of glycerol.

Fatty acids with a chain length of less than 12 carbon atoms, especially the very short chain lengths of milk fats, are cleaved more rapidly than the C14–18 chain length fatty acids. The very long-chain polyenoic acids (C20 : 5 and 22 : 6 of fish oils and marine mammals) are hydrolysed slowly.

Different tissues contain different lipases which vary in their substrate specificity. Lipoprotein lipase hydrolyses triacylglycerols associated with proteins, and is involved in the catabolism of serum lipoproteins. Adipose tissue triacylglycerol

lipase is activated by phosphorylation by hormones such as catecholamines (Figure 11.34). There is a monoacylglycerol lipase in many cells. Animals are unable to convert lipid into carbohydrate because of the irreversible decarboxylation steps in the TCA cycle (isocitrate dehydrogenase) and α-ketoglutarate dehydrogenase. Each of the two fatty acid carbons entering the Krebs cycle as acetyl-CoA is lost as carbon dioxide.

> The fatty acids at positions 1 and 3 are initially removed at equal rates. Once one fatty acid has been removed the resulting diacylglycerol is more slowly hydrolysed than the original triacylglycerol and the monoacylglycerol once the fatty acids are racemized to position 1. This means that there is an accumulation of monoacylglycerols and non-esterified fatty acids. Some lipases hydrolyse both the fatty acids in the primary position of acylglycerols and also the fatty acids esterified in position 1 of phosphoglycerides.

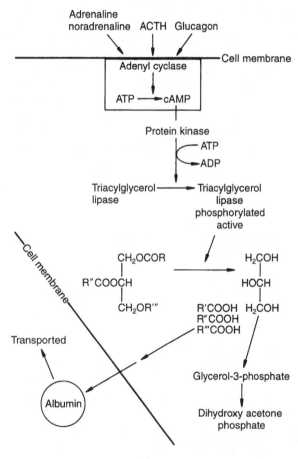

FIGURE 11.34 Acyl glycerol catabolism. The rate of hydrolysis of triacyl glycerol is dictated by lipase activity (phosphorylated form). The activity of lipase is controlled by a protein kinase which is controlled by cAMP. This is under hormone control by adrenalin, noradrenaline, ACTH, glucagon and probably other hormones.

11.7.10 Triacylglycerols as fuels

Triacylglycerols in animals are a source of fatty acids and hence act as a metabolic fuel. This is a controlled system. Blood glucose concentrations are maintained at a constant level and restored rapidly to that level. Glycogen is stored as an emergency fuel in liver and muscle. When these stores are replete, excess carbohydrate is converted to fats and stored in adipose tissue. Optimal supplies of protein are needed for growth, tissue repair and enzyme synthesis; excess protein amino acids are converted into fat. Carbohydrate and protein are interconvertible and can be converted into fat. Fat, however, is only stored or oxidized but not converted into protein or carbohydrate.

Humans require continued sources of nutrient energy, regardless of times of eating. Initially, plasma glucose concentrations are derived from the diet, glycogen breakdown and to some extent from muscle protein. Additional energy comes from adipose tissue in the form of glycerol (converted into glucose in the liver) and fatty acid oxidation, particularly in muscles (Figure 11.35). Net triacylglycerol synthesis occurs when energy demands exceed immediate requirements. When there is a preponderance of fat in the diet, fat synthesis from carbohydrate is depressed in the tissues. Absorbed triacylglycerols are converted into lipoproteins and circulate in the blood. The fatty acids are released from the acylglycerols at the endothelial surfaces of cells, catalysed by lipoprotein lipase and enter the cells. The fatty acids are then re-esterified into acylglycerols. Acylglycerol synthesis is most important in the small intestine where the resynthesis of triacylglycerols follows fat digestion.

> The liver is important in relation to the synthesis of triacylglycerols or carbohydrates. Adipose tissue is involved in longer-term storage and the mammary gland synthesizes milk fat during lactation. In the intestine, dietary fatty acids are the main source of lipids. Elsewhere in the body the glycerol phosphate pathway is all-important.

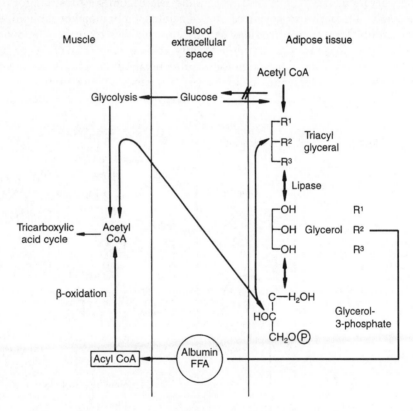

FIGURE 11.35   Glucose and fatty acids in muscle and adipose tissue. Triacyl glycerol is metabolized to glycerol and glycerol-3-phosphate which can pass into the tricarboxylic acid cycle. Triacyl glycerol fatty acids can pass from the cell and are transported bound to albumin. After transport, they enter cells. The fatty acid is converted to acyl-CoA, undergoes β-oxidation to acetyl-CoA and enters the tricarboxylic acid cycle. Glucose is oxidatively degraded to acetyl-CoA to glycerol-3-phosphate. Acetyl-CoA enters the tricarboxylic acid cycle. Glycerol is oxidized to glycerol-3-phosphate which enters the glycolysis pathway to acetyl-CoA. Acetyl-CoA, however, cannot reverse the process to synthesize glycose. There is hormone control of glucose entry into the muscle and adipose tissue and lipase activity for triacyl glycerol hydrolysis.

When the metabolic processes are stressed, tri-acylglycerols are synthesized in the liver, heart and skeletal muscle. Fatty acids are mobilized from adipose tissue as a result of stimulation by catecholamines and other hormonal changes (Figure 11.36).

The concentrations of acyltransferases in cells is influenced by nutritional status. Glycerol 3-phosphate, which is important for acylglycerol synthesis, is regulated by similar factors to those responsible for glycolysis and gluconeogenesis. Starvation reduces the glycerol 3-phosphate cell concentrations which are restored by re-feeding. The intracellular concentration of acyl-CoA is increased during starvation.

The main metabolic alternative to acylglycerol synthesis is β-oxidation. The relative activities of glycerol phosphate acyltransferase and carnitine palmitoyl transferase control the interaction of these two pathways. The underlying control mechanism is still poorly understood. The rate-limiting enzyme in mammalian acylglycerol biosynthesis is phosphatide phosphohydrolase, the activity of which increases with dietary sucrose and fat, ethanol and starvation, all of which result in high concentrations of plasma free fatty acids. This enzyme is also increased during liver regeneration following partial hepatectomy, in obese individuals, in diabetes and in association with drugs which result in reduced plasma lipid concentrations.

When there is an increased supply of saturated and monounsaturated fatty acids to the liver then phosphatide phosphohydrolase activity increases. When this enzyme activity is low, the phosphatidic acid becomes a substrate for the biosynthesis of acidic membrane phospholipid, e.g. phosphatidylinositol. This requires a supply of unsaturated fatty acids greater than could be used in acylglycerol metabolism. This occurs when there is a predominance of unsaturated fats.

The biosynthesis of phospholipids takes prece-

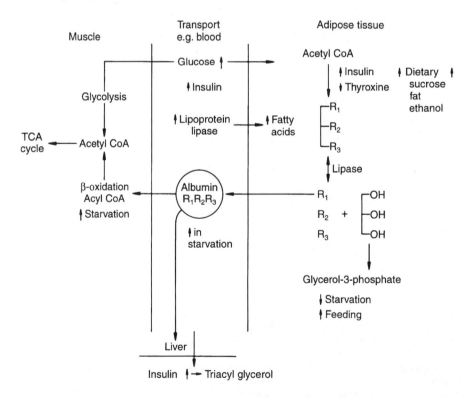

FIGURE 11.36 Controls on carbohydrate and lipid metabolism by feeding, starvation and hormones.

dence over triacylglycerols when the synthesis of diacylglycerol is relatively reduced, membrane turnover and biosecretion being biologically more important than the accumulation of storage triacylglycerol. The relatively low K_m of choline phosphotransferase for diacylglycerols is an important factor in such metabolic control.

> Phosphatidate phosphohydrolase is physiologically inactive until adsorbed on to the membrane on which the phosphatidate is being synthesized. This adsorption is regulated by both hormones and substrate. Cyclic AMP displaces the enzyme from the membranes. Increasing concentrations of non-esterified fatty acids and their CoA esters promote attachment. This has the effect of decreasing the intracellular concentrations of cyclic AMP. This ensures more effective adherence at lower fatty acid concentrations. Metabolic control, in this instance is dependent on enzyme activity, which in turn is dependent on location in the cells. Phosphorylcholine cytidyl transferase, which regulates the biosynthesis of phosphatidylcholine, is another enzyme whose activity is dependent upon location in the cell.

The ratio of circulating insulin to glucagon influences the fate of fatty acids entering the liver. High ratios favour esterification of fatty acids into acylglycerols with a low rate of β-oxidation. Insulin increases the rate of glucose transport into fat cell membrane, and also increases the synthesis of the enzyme lipoprotein lipase, thereby catalysing fatty acid and glycerol release from triacylglycerol. Within the cell insulin stimulates the synthesis of lipids by increasing the activity of lipogenic enzymes in general. Insulin also inhibits fat globule triacylglycerol breakdown and inhibits the hormone-sensitive triacylglycerol lipase activity by causing a decrease in the production of cAMP. Thyroid hormones stimulate triacylglycerol biosynthesis in the liver and suppress triacylglycerol synthesis in fat. This may be a general effect on

metabolic turnover rather than a specific enzyme effect. The control of phosphatide phosphohydrolase is also controlled by glucocorticoids.

11.7.11 Body fat

Triacylglycerol storage in humans

There are two forms of adipose tissue in the body, brown and white adipose tissue, white being the more abundant, brown adipose tissue being the more specialized. White adipose tissue is widely distributed throughout the body, largely subcutaneously and has insulating and protective functions. The fat is held in adipocytes, large single lipid globules surrounded by a ring of cytoplasm. These cells are unusual in being able to expand, and can absorb fat from circulating lipoproteins. Such a process requires hydrolytic breakdown of the triacylglycerols, through release of fatty acids by lipoprotein lipase. The fatty acids are transported into the cell before being incorporated into triacylglycerols.

> An important organ of fat storage is the breast, where the duration of lactation is important for the newborn animal. Milk consists of globules composed of triacylglycerols, small amounts of cholesterol, fat-soluble vitamins and hydrophobic complex lipids. The milk fat globule is surrounded by protein phospholipid and cholesterol and has an average diameter of 1–2 μm. In the production of milk fat globules fat droplets are formed in the mammary cells within the endoplasmic reticulum membrane and migrate to the apical regions of the mammary secretory cell to be enveloped in the plasma membrane. The neck of membrane is pinched off and the resulting vesicle is expelled into the lumen as a milk fat globule.

The amount of lipid stored within the adipose tissue is dictated by the number of adipocytes and their respective volume. Adipocyte number is a result of long-term processes of multiplication and differentiation of adipocyte precursor cells which is affected by genetic and environmental factors. The number of morphologically identified lipid-filled fat cells can increase even in adult humans. Women have larger femoral–gluteal fat cells than do men, while abdominal fat cell size is similar in men and women. Most of the differences in the size of subcutaneous fat depots between men and women at specific sites are due to differences in the number of fat cells. These sex-related differences disappear in the obese. Women are able to tolerate about 30 kg more body weight than men before experiencing comparable degrees of metabolic compromise.

Regulation of fat cell size

The synthesis of lipid by adipocytes is determined by high concentrations of lipoprotein lipase, insulin, glucose, free fatty acids and triglycerides as lipoproteins (Figure 11.36). Hydrolysis of stored triglyceride is favoured by circumstances tending to lower the availability of the above fuels and to activate the enzyme hormone-sensitive lipase which increases the breakdown of the triglyceride molecule into its constituents of free fatty acids and glycerol. The activation of hormone-sensitive lipase is controlled by phosphorylation–dephosphorylation. This is regulated by protein kinase A through adenylate cyclase-generated cAMP and insulin-activated protein phosphatase.

Brown fat

Energy expended for thermogenesis in brown adipose tissue (BAT) may be important in total energy expenditure. Thermogenesis may not only be a response to cold but also to eating. BAT thermogenesis may not occur normally in the obese and is a component of facultative thermogenesis. This is a process which may be switched on depending on need, under central and peripheral neural sympathetic control. This is in contrast to obligatory thermogenesis which occurs in all organs of the body as an essential mechanism to maintain life at a temperature above that of the surroundings. Obligatory thermogenesis is controlled to a significant extent by the thyroid hormones.

BAT cells are multilocular and contain several droplets of stored triacylglycerol and characteristically large mitochondria. When the cells are inactive they are filled with lipids and resemble white adipose tissue cells. BAT cells are innervated by sympathetic nerves and also contain mast cells. BAT deposits are found in interscapular, subscapular, axillary and intercostal regions and along major blood vessels of the abdomen and thorax. BAT mitochondria have a proton conductance mechanism that allows them to become reversibly uncoupled and to oxidize endogenous and exogenous substrates at a high rate. This is independent of the phosphorylation of ADP. The process is controlled by the intracellular concentration of fatty acids as a result of the breakdown of endogenous triacylglycerol.

Thermogenin

A specific protein, molecular weight 32 000 Da, binds purine nucleotides on the outer surface of the inner mitochondrial membrane. This protein is known as thermogenin, or uncoupling protein. This protein is found only in BAT mitochondria.

Adrenaline reacts with receptors on the cell surface and activates a series of metabolic pathways metabolizing endogenous triacylglycerol, exogenous triacylglycerol and glucose. BAT grows when stimulated by continued intense activation of the sympathetic supply. Measurement of thermogenesis and the quantity of BAT is not easy or practical in humans. Thyroid hormone is responsible for the thermogenic response of BAT to noradrenaline. Glucocorticoids, insulin, glucagon, pituitary hormones, sex hormones and melatonin may be involved in BAT suppression and modulation. It has been suggested that defective thermogenesis might contribute to obesity.

Summary

1. Liver and adipose tissue are the most important tissues for fatty acid biosynthesis. Most naturally occurring fatty acids have even numbers of carbon atoms.

2. The first step in fatty acid synthesis is the carboxylation of the 2-carbon fragment acetate as acetyl-CoA to malonate, catalysed by acetyl-CoA carboxylase, a biotin-containing enzyme. Acetyl-CoA is formed within the mitochondria, whose membrane is impermeable. As fatty acid synthesis occurs in the cytosol, acetyl-CoA is converted to citric acid, transported across the membrane; the reverse reaction occurs in the cytosol.

3. Fatty acid synthesis from malonyl-CoA is a series of condensation additions and reductions from carbonyl to methylene groups. The reactions are catalysed by a multienzyme system, fatty acid synthase. The end-product of fatty acid synthase enzymes is usually free palmitic acid.

4. Fatty acids must be converted into metabolically active thiol esters before further metabolism, whether this be anabolic or catabolic.

5. Fatty acid elongation is catalysed by Type III synthases or elongases present in the endoplasmic reticulum, with the addition of malonyl-CoA, to preformed acyl chains requiring NADPH as a reducing coenzyme. The important function of elongation is in the transformation of dietary essential fatty acids to the higher polyunsaturated fatty acids, n-3 and n-6 series.

6. The activity of the enzymes involved in fat synthesis decreases during starvation or with a high dietary fat intake. Polyunsaturated fatty acids are effective in reducing fatty acid synthesis. Sucrose ingestion, probably as a result of the fructose intake, results in an increase in lipogenesis in the liver.

7. Fatty acid synthesis is controlled by acetyl-CoA carboxylase; this requires a hydroxy tricarboxylic acid, e.g. citrate or isocitrate as substrate. Long-term regulation of acetyl-CoA carboxylase depends on nutritional status, hormonal developmental and genetic conditions.

8. Unsaturated fatty acids are important in regulating the physical properties of lipoproteins and membranes, regulation of metabolism in cells and as precursors for physiologically active compounds, e.g. eicosanoids.

9. The biosynthesis of unsaturated fatty acids includes an oxidative mechanism by which a double bond is introduced directly into a saturated long-chain fatty acid. Humans can introduce a double bond at the C-9 position (Δ^9 desaturase), and C-4, C-5 and C-6 but not beyond C-9.

10. This means that unsaturated fatty acids beyond C-9 have to be met from the diet and are essential, i.e. linoleic acid $18:2\ \Delta^{9,12}$ and α-linolenic acid $18:3\ \Delta^{9,12,15}$.

11. Fatty acid breakdown involves oxidation at defined bonds in the acyl chain or oxidation at defined double bonds in particular unsaturated fatty acids. The main forms of oxidation are termed α, β and ω, depending on the position of the carbon of the acyl chain which is oxidized. The end-product is acetyl-CoA, which passes into the TCA cycle.

12. All plants use lipids as stored energy, particularly as triacylglycerols. Animals store fat in adipose tissue whereas fish use their flesh or liver as a lipid store.

13. The glycerol phosphate pathway for triacylglycerol formation is the progressive esterification of glycerol 3-sn phosphate and 1-acylglycerol 3-sn phosphate with long-chain acyl-CoA fatty acids using phosphatidic acid in both phospholipid and triacylglycerol biosynthesis.

14. Triacylglycerols are resynthesized in the enterocyte from monoacylglycerols absorbed following small intestinal hydrolysis. Absorbed triacylglycerols are converted into lipoproteins and circulate in the blood. Fatty acids are released from the acylglycerols, enter the cells and are then re-esterified into acylglycerols.

15. Lipases are important in the splitting of the fatty acid ester from the primary hydroxyl groups of glycerol. The fatty acids at position 1 and 3 are initially removed at equal rates.
16. Triacylglycerols in humans are a source of fatty acids and hence act as a metabolic fuel, but are not converted into protein or carbohydrate. This is a controlled system. Net triacylglycerol synthesis occurs when energy demands exceed immediate requirements. When there is a preponderance of fat in the diet, fat synthesis from carbohydrate is depressed in the tissues.
17. Glycerol 3-phosphate is regulated by similar factors as for glycolysis and gluconeogenesis. Starvation reduces the glycerol 3-phosphate cell concentrations, but these are restored by re-feeding.
18. The biosynthesis of phospholipids takes precedence over triacylglycerols when the synthesis of diacylglycerol is relatively reduced, membrane turnover and biosecretion being biologically more important than accumulation of storage triacylglycerol.
19. The ratio of circulating insulin to glucagon influences the fate of fatty acids entering the liver; high ratios favour esterification of fatty acids into acylglycerols and a low rate of β-oxidation.
20. The synthesis of lipid by adipocytes is determined by high concentrations of insulin, glucose, free fatty acids and triglycerides as lipoproteins. The activation of hormone-sensitive lipase is controlled by phosphorylation–dephosphorylation. This is regulated by activities of protein kinase A through adenylate cyclase-generated cAMP and insulin-activated protein phosphatase.
21. Breast milk is an important organ of fat storage; the duration of lactation is important for the newborn animal.
22. Energy expended for thermogenesis in brown adipose tissue may be important in total energy expenditure as a component of facultative thermogenesis which is under central and peripheral neural sympathetic control.

Further reading

British Nutrition Foundation Task Force Report (1987) *Trans fatty acids*, Chapman & Hall, London.

British Nutrition Foundation Task Force Report (1992) *Saturated Fatty Acids; Nutritional and Physiological Significance*, Chapman & Hall, London.

Gurr, M.I. and Harrow, J.L. (1991) *Lipids*, Chapman & Hall, London.

Halliwell, B. (1993) Renaissance of peroxisomes. *Lancet*, 341, 92.

Himms-Hagen, J. (1985) Brown adipose tissue metabolism and thermogenesis. *Annual Review of Nutrition*, 5, 69–94.

Leibel, R.L., Edens, N. K. and Fried, S.K. (1989) Physiologic basis for the control of body fat distribution in humans. *Annual Review of Nutrition*, 9, 417–43.

MacDougald, O.A. and Lane, M.D. (1995) Transcriptional regulation of gene expression during adipocyte differentiation. *Annual Review of Biochemistry*, 64, 345–74.

Vergroesen, A.J. and Crawford, M. (1989) *The Role of Fats in Human Nutrition*, 2nd edn, Academic Press, London.

11.8 Eicosanoids

Eicosanoids are hormones derived from arachidonic acid.
- There are two families: cyclo-oxygenase-dependent pathway (prostaglandins, prostacyclins, thromboxanes); and lipoxygenase-dependent pathway (leukotrienes)
- These hormones act locally and briefly
- The source of the arachidonic acid is triacylglycerols in the cell wall.

11.8.1 Introduction

The eicosanoids are a family of important hormones and are specific oxidation products of C20 polyunsaturated fatty acids, derived from arachidonic acid. These hormones are very potent biologically, are important local hormones generated *in situ*, rapidly metabolized and briefly active in the immediate vicinity to where they are synthesized.

11.8.2 Nomenclature

The different eicosanoids all consist of a cyclopentane substituted ring. The biological differences are dictated by the structure of the substituted ring. The names are usually abbreviated: for prostaglandins to PG, followed by the letter E, F, G, H or I; and for thromboxanes, to TX, followed by the letter A or B

The subscript 1, 2 or 3 indicates the number of double bonds in the side chain structure in all series:

- 1 is *trans* Δ 13
- 2 is *trans* Δ 13, *cis* 5
- 3 is *trans* 13, *cis* 5, *cis* 17.

There are two major families dependent upon the synthetic pathway:

1. The cyclo-oxygenase-dependent pathway: prostaglandins (PG); prostacyclins (PG1); and thromboxanes (TX)
2. The lipoxygenase-dependent pathway: leukotrienes (LT)

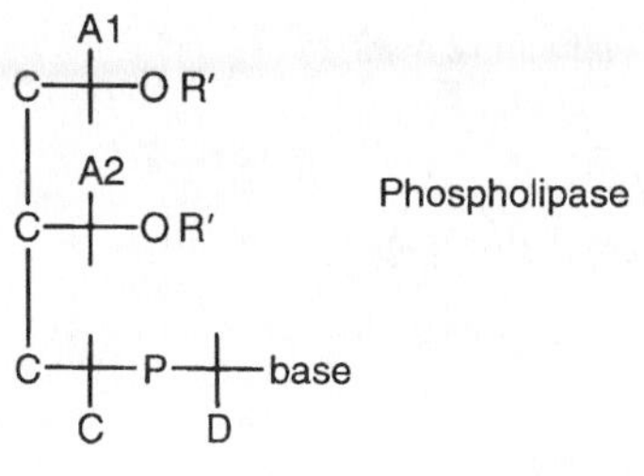

Within the cell the concentration of arachidonic acid is small. The immediate source of arachidonic acid is from cell membrane phospholipids, phosphatidylcholine and phosphatidylinositol. The release of the arachidonic
acid is catalysed by phospholipase A_2 and C hydrolysing the ester link of arachidonic acid at C-2 of the glycerol. The arachidonic acid may then be further metabolized to biologically active eicosanoids or re-esterified to membrane phospholipid (Figure 11.37).

11.8.3 The cyclo-oxygenase pathway

The synthesis of prostaglandins, prostacyclins and thromboxanes occurs by the oxidation of arachidonic acid followed by cyclization to form a cyclopentane ring C-8 to C-12. Substitutions to the cyclopentane ring R_1 and R_2 determine biological activity (Figure 11.38).

The fatty acid cyclo-oxygenase is a microsomal enzyme (Figure 11.39) which:

- adds molecular oxygen as a peroxide across C9 and C11 of the arachidonic acid
- adds a hydroperoxide (–OOH) at C15 to form the cyclic endoperoxide PGG_2. The next transformation is catalysed by the peroxidase to the 15-hydroxy PGH_2, which is the starting point for subsequent metabolism.

Free radical intermediates inactivate fatty acid cyclo-oxygenase, as do certain non-steroidal anti-inflammatory drugs, e.g. aspirin or indomethacin. These drugs compete with the fatty acid substrate for the prostaglandin endoperoxide synthetase active site. Aspirin also acetylates a serine hydroxyl at or near the active site which permanently inactivates the enzyme. The acetylenic fatty acids such as eicosa-5,8,11,14 tetraenoic have the same effect (Figure 11.40).

FIGURE 11.37 Classification of phospholipases based on the bond in the phospholipid that is hydrolysed, A1, A2, C, D. Phospholipase B hydrolyses lysophospholipids releasing a fatty acid.

The metabolites include:

- prostaglandins of the E, F, A, B, C and D types
- prostacyclin PGI_2
- thromboxane A_2

While most cells have the ability to produce all three types of metabolite there is a bias within platelets and macrophages to produce thromboxane, and for endothelial cells to produce prostacyclin. Mast cells predominantly produce PGD_2 and microvessels PGE_2.

The **nomenclature** of these compounds is based on prostanoic acid, an acid which does not exist in nature (Figure 11.41). They contain 20 carbon atoms, including a 5- or 6-carbon cyclic structure. Prostaglandins and prostacyclins have a cyclic ring of five carbons, thromboxanes a ring of six carbons. The C-8 to C-12 positions are closed to form a five-membered ring, prostanoic acid, in prostanoids. Most of the prostaglandins have an OH group at the C-15 position.

Prostaglandins have a wide range of activities. They are important in smooth muscle contraction throughout the body. PGE series, PGI_2 and PGD_2 are vasodilators whereas thromboxane is a vasoconstrictor and stimulates platelet aggregation. Some prostaglandins increase cAMP concentrations, resulting in inhibition of platelet release reactions, lysosomal enzyme release from neutro-

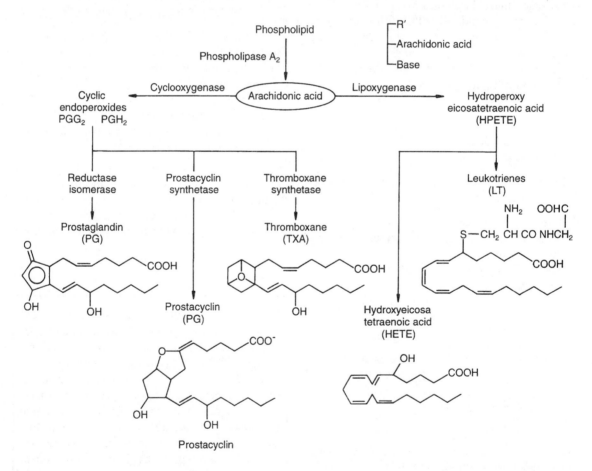

FIGURE 11.38 Arachidonic acid metabolism to eicosanoids. The eicosanoids are products of the cyclo-oxygenase or lipoxygenase pathways.

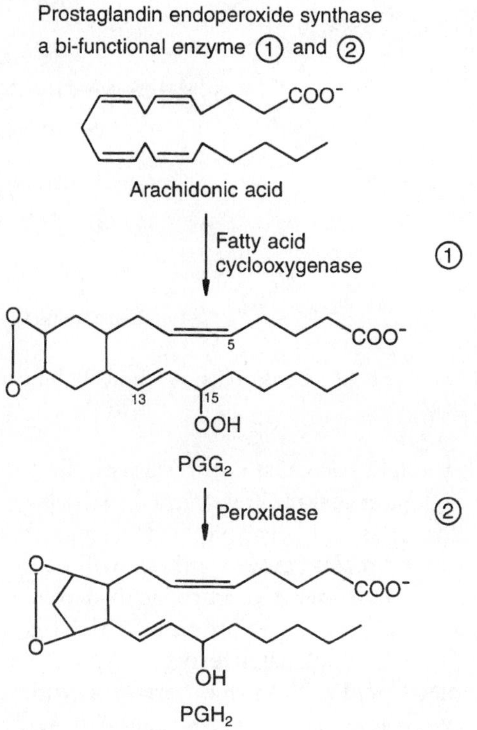

Prostaglandin endoperoxide synthase
a bi-functional enzyme ① and ②

Arachidonic acid

Fatty acid
cyclooxygenase ①

PGG₂

Peroxidase ②

PGH₂

FIGURE 11.39 Cyclo oxygenase pathway from arachidonic acid with 15 PGH₂ the starting point for subsequent metabolism (enzymes fatty acid cyclo oxygenase and peroxidase).

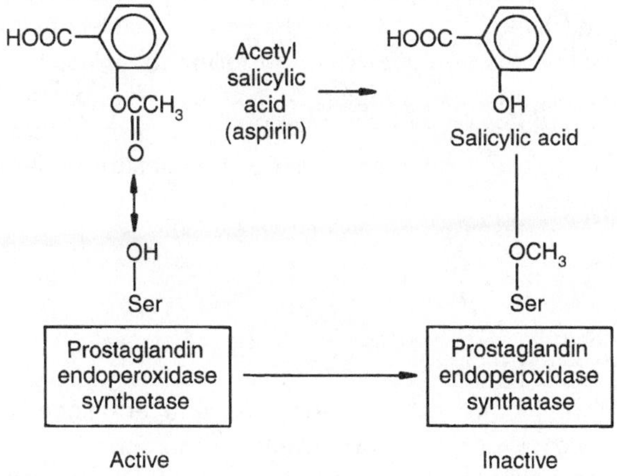

Acetyl
salicylic
acid
(aspirin)

Salicylic acid

Ser

Prostaglandin
endoperoxidase
synthetase

Prostaglandin
endoperoxidase
synthatase

Active

Inactive

FIGURE 11.40 Prostaglandins and aspirin. Inhibition of prostaglandin endoperoxidase synthetase by acetylation by acetyl salicylic acid.

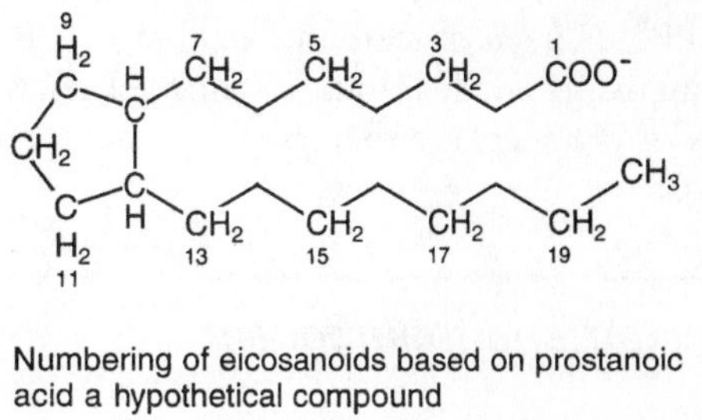

Numbering of eicosanoids based on prostanoic acid a hypothetical compound

FIGURE 11.41 Prostanoic acid, a theoretical acid. This structure is used to number the carbons of the eicosanoids.

phils and inhibition of mast cell and basophil histamine release.

Prostacyclin synthesis is largely found in the vascular and gastric tissues. They also have anti-aggregator and pulmonary vasodilator functions.

The difference between the E and F prostaglandin series lies in the keto or hydroxyl group at position 9 respectively. The E-type PGs are 11,15-dihydroxy-9-keto compounds and the F type 9,11,15-trihydroxy structures. The suffix α or β indicates the stereochemistry of the hydroxyl at C-9. The suffix 1, 2 or 3 refers to how many double bonds are contained in the prostaglandin structure. There are two side chains, one of which (R1) is attached to C-8, and carries a carboxyl group, the other (R2) is attached to C-12 and has a hydroxyl group at C-15.

Thromboxane activity is primarily to aggregate platelets and as potent vasoconstrictors.

11.8.4 Leukotrienes and mono-eicosatetraenoic acids

The leukotrienes have a role in smooth muscle contraction, mononuclear leucocytes and tumour cells. A lipoxygenase catalyses the formation of 5-hydroperoxy-6,8,11,14-eicosatetraenoic acid

(5-HPETE) from arachidonic acid which is then enzymatically dehydrated to form LTA_4. Two series of products are then produced according to needs.

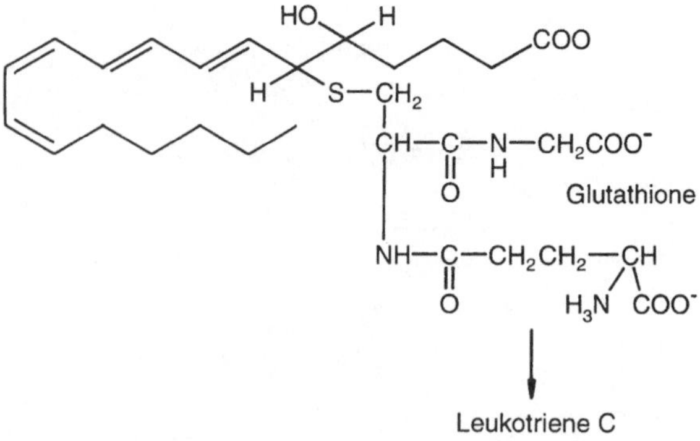

FIGURE 11.42 Leukotrienes and glutathione substitutions.

Leukotriene metabolism

- Enzymatic hydrolysis to LTB_4. This causes chemoattraction in neutrophils and subsequent aggregation. The role is to recruit circulating cells at inflammatory sites.
- Peptidoleukotriene formation. The addition of glutathione at C-6 forms LTC_4. Loss of glutamine (enzyme, γ-glutamyltransferase) forms LTD_4 and glycine residues (enzyme, dipeptidase) to form LTE_4 in sequence from these reactions. These leukotrienes participate in a number of inflammatory processes and in slow-reacting anaphylaxis. The peptidoleukotrienes have profound effects on the tone in arteries and microvascular vessels. They also have a role in microvascular leakage.

acids thought necessary contrasts with the 1 mg of prostaglandin metabolites formed in 24 hours by humans. It is unlikely that a deficiency of substrates for prostaglandin synthesis will occur.

The n-6 unsaturated fatty acids linoleic and arachidonic acid are synthesized to the 3-series and 4-series leukotrienes respectively, whereas the α-linolenic acid n-3 of fish oil origin is synthesized to the 5-series leukotrienes. Linoleic acid and arachidonic acid are synthesized to the 1 and 2 series prostaglandins respectively, whereas α-linolenic acid is synthesized to the 3-series. The relative contribution of n-3 and n-6 fatty acids in the diet will therefore dictate the spectrum of types and biological potencies of the prostanoids and leukotrienes. Other than this dietary control the production of eicosanoid formation is dependent upon phospholipase A_2 activity. This function is activated and inhibited by a number of messengers in a similar manner to the inositol cyclase system.

11.8.5 Eicosanoids and dietary fatty acids

It has been suggested that only those fatty acids capable of being converted into the $\Delta^{5,8,11,14}$-tetraenoic fatty acids of chain length C-19,C-20 and C-22 are essential because these give rise to physiologically active eicosanoids. Eicosanoids are metabolized very rapidly and are excreted in urine or bile. The dietary intake (10 g) of essential fatty

Summary

1. The eicosanoids are a family of important potent locally acting hormones and are specific oxidation products of C-20 polyunsaturated fatty acids, derived from arachidonic acid.
2. The different eicosanoids consist of a cyclopentane substituted ring; biological differences are dictated by the structure of the substituted ring.
3. There are four main groups of such cyclic compounds and other biologically active products of lipoxygenase and mono-oxygenase activity on arachidonic acid.

4. The immediate source of arachidonic acid is from cell membrane phospholipids, phosphatidylcholine and phosphatidylinositol. The release of the arachidonic acid is catalysed by phospholipase A_2 and C hydrolysing the ester link of arachidonic acid to C-2 of the glycerol.

5. There are two major families dependent upon the synthetic pathway. The cyclo-oxygenase-dependent pathway (prostaglandins, prostacyclins and thromboxanes) and the lipoxygenase-dependent pathway (leukotrienes).

6. The first step in the synthesis of prostaglandins, prostacyclins and thromboxanes is the oxidation of arachidonic acid followed by cyclization to form a cyclopentane ring C-8 to C-12. Substitutions to the cyclopentane ring R_1 and R_2 determine biological activity. The fatty acid cyclo-oxygenase is a microsomal enzyme.

7. Non-steroidal anti-inflammatory drugs, e.g. aspirin or indomethacin, compete with the fatty acid substrate for the prostaglandin endoperoxide synthetase active site. Aspirin also acetylates a serine hydroxyl at or near the active site which permanently inactivates the enzyme.

8. It has been suggested that only those fatty acids capable of being converted into the $\Delta^{5,8,11,14}$-tetraenoic fatty acids of chain length C-19, C-20 and C-22 are essential because these give rise to physiologically active eicosanoids.

Further reading

Lands, W.E.M. (1991) Biosynthesis of prostaglandins. *Annual Review of Nutrition,* **11**, 41–61.

Weatherall, D.J., Ledingham, J.G.G. and Warrell, D.A.

(1996) *Oxford Textbook of Medicine,* 3rd edn, Oxford University Press, Oxford.

11.9 Cholesterol and lipoproteins

- Cholesterol is synthesized from acetate.
- The rate-limiting step is 3-hydroxy-3-methylglutaryl-CoA (HMG-CoA) reductase.
- Cholesterol is important in cell wall structure and also as a source of hormones.
- Cholesterol is carried round the body in the form of lipoproteins.
- Cholesterol is removed from the body as bile acids following 7-α-hydroxylation in the liver and subsequent metabolism to bile acids.
- Abnormalities of the metabolism of fats are reflected in the blood concentrations of lipoproteins, chylomicrons and high, low and very low density lipoproteins (HDL, LDL and VLDL, respectively).

11.9.1 Introduction

The synthesis of sterols is closely related to growth, development and differentiation of all cells. All cholesterol is synthesized from acetate or ingested in the diet.

11.9.2 Cholesterol biosynthesis

In the synthesis of cholesterol (Figures 11.43 and 11.44) three acetates condense to form 3-hydroxy-3-methylglutaryl-coenzyme A (HMG-CoA), which are then catalysed by the enzyme HMG-CoA reductase to form mevalonic acid.

Mevalonate is converted by two kinase reactions to 5-diphosphomevalonate, losing a carbon atom en route. The enzyme has a requirement for Mg^{2+} and uses ATP as a substrate. Two isomers are formed – isopentenyl diphosphate and 3,3-dimethylallyl diphosphate – which condense to form geranyl diphosphate. There follows a

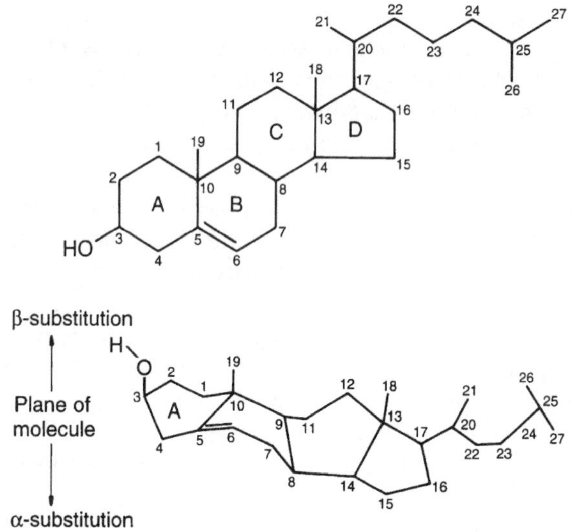

FIGURE 11.43 Structure of cholesterol including the three-dimensional structure. Substitutions above the plane of the molecule are β; substitutions below the plane of the molecule are α.

series of condensation reactions with the eventual formation of squalene, followed by lanosterol and finally a series of oxidation steps to form cholesterol.

HMG-CoA reductase

This is an irreversible reaction, a rate-limiting step, requiring two molecules of NADH. Coenzyme A is released. HMG-CoA reductase is controlled by an elaborate phosphorylation system (Figure 11.45). The enzyme is inactive when phosphorylated by HMG-CoA reductase kinase. The phosphorylation is reversed by HMG-CoA reductase phosphatase. The HMG-CoA reductase kinase is active in the phosphorylated form, this being catalysed by HMG-CoA reductase kinase. Many factors regulate cholesterol biosynthesis and many of these are involved in the activity of the HMG-CoA reductase.

There appears to be a diurnal cycle in cholesterol synthesis. This is largely due to changes with the circadian rhythm in the activity of HMG-CoA reductase, cholesterol 7-α-hydroxylase, which is involved in bile acid synthesis and lysosomal acid cholesteryl ester hydrolase, which hydrolyses cholesteryl esters derived from lipoproteins. The sterol carrier protein, which is a major regulatory protein of lipid metabolism and transport, has a diurnal cycle. It is probable that mRNA plays a major role in controlling the synthesis of HMG-CoA reductase.

The input of cholesterol to the liver is provided by two sources:

- uptake of circulating lipoprotein cholesterol
- direct *de novo* synthesis of cholesterol

The balance between input and output determines the concentration of cholesterol inside the liver cell. Excess cholesterol which is not readily disposed of, can be converted to the fatty acyl ester by the action of the liver enzyme acyl-CoA : cholesterol acyltransferase (ACAT). HMG-CoA reductase activity and LDL receptor expression can be regulated by the concentration of free cholesterol. This suggests that the pool of cholesterol may form part of a self-regulatory system.

There are wide species variations in the rate of synthesis of cholesterol, but 10 mg/day/kg body weight is typical in humans. Endogenous synthesis in humans is important in the adjustment of the cholesterol body pool. Ingestion of cholesterol increases plasma cholesterol concentration dependent upon the dietary unsaturated/saturated fatty acid content. There are wide variations in the ability to absorb dietary cholesterol; humans can absorb 2–4 mg/day/kg body weight.

Cholesterol feeding markedly inhibits hepatic cholesterogenesis, varying with species. In rats and dogs, inhibition is almost complete; in humans, it is 40%. Cellular cholesterol concentration is a balance between synthesis or uptake of preformed cholesterol and degradation.

Most of the cholesterol-rich lipoprotein is removed by the liver. Remnants of chylomicrons generated by the action of the lipoprotein lipase are cleared by the liver by a saturable receptor-mediated process. Only a fraction corresponding

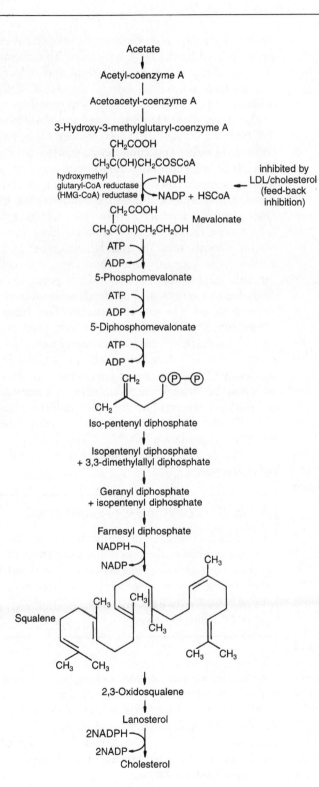

Acetate

Acetyl-coenzyme A

Acetoacetyl-coenzyme A

3-Hydroxy-3-methylglutaryl-coenzyme A

CH_2COOH

$CH_3C(OH)CH_2COSCoA$

hydroxymethyl
glutaryl-CoA reductase
(HMG-CoA) reductase

NADH

NADP + HSCoA

inhibited by
LDL/cholesterol
(feed-back
inhibition)

CH_2COOH

$CH_3C(OH)CH_2CH_2OH$ Mevalonate

ATP

ADP

5-Phosphomevalonate

ATP

ADP

5-Diphosphomevalonate

ATP

ADP

CH_2 O—P—P

CH_2

Iso-pentenyl diphosphate

Isopentenyl diphosphate
+ 3,3-dimethylallyl diphosphate

Geranyl diphosphate
+ isopentenyl diphosphate

Farnesyl diphosphate

NADPH

NADP

Squalene

CH_3

CH_3 CH_3

CH_3

CH_3

CH_3

CH_3

CH_3 CH_3

CH_3 CH_3

2,3-Oxidosqualene

Lanosterol

2NADPH

2NADP

Cholesterol

FIGURE 11.44 Cholesterol synthesis. The starting compound is acetate, which in the synthetic process passes through five isoprenoid intermediates. The conversion of 3-hydroxy-3-methyl-glutaryl-CoA to mevalonate is the rate-limiting step. This is irreversible and is catalysed by hydroxymethylglutaryl-CoA (HMG-CoA) reductase. Three molecules of isopentenyl diphosphate condense to form farnesyl diphosphate, two molecules of which condense to form squalene. There is further cyclization to form cholesterol.

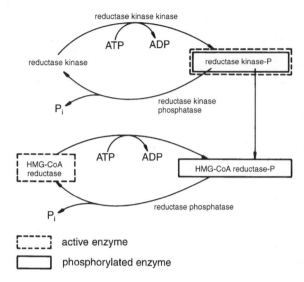

active enzyme

phosphorylated enzyme

FIGURE 11.45 The control of β-hydroxy-β-methyl-glutaryl-CoA reductase. The enzyme is activated by phosphorylation. This phosphorylated state is controlled by HMG-CoA reductase kinase which in itself is active in the phosphorylated form.

to 50–60% of VLDL remnants are directly removed by the liver in humans. The major cholesterol-carrying lipoprotein is LDL, up to 85% of these particles being taken up by the liver at apoB receptors and to a lesser extent by non-specific processes. Cholesterol carried in the HDL may be directly removed by the liver.

VLDL is the major lipoprotein secreted by the liver. Its metabolism in the blood leads to the production of intermediate density lipoproteins (IDL) and LDL (apoB-100-containing lipoproteins). The liver is the major organ which clears these modified plasma lipoproteins by having receptors for apoB-100 or apoE. Ingestion of significant amounts of dietary cholesterol and saturated fatty acids can raise the concentration of plasma lipoprotein cholesterol.

There is a general stimulatory effect of dietary lipids on cholesterogenesis. However, the effect of fats on HMG-CoA reductase activity may be related to the chemistry of the lipid. The microscopic fluidity of HDL, which are responsible for cholesterol efflux from the cell, is increased following dietary intake of unsaturated fats. This

increased fluidity may allow more cholesterol to be solubilized by the HDL molecules. This allows greater cellular efflux of cholesterol and reduces end-product inhibition effects on HMG-CoA reductase.

The synthesis of HMG-CoA reductase is inhibited by LDL. Cholesterol enters the hepatic cell as cholesterol esters and cholesterol, transported on LDL and influences sterol synthesis through a feed-back regulation on the concentration of LDL membrane receptors and on the concentration of HMG-CoA reductase. There are lipoprotein receptors on the surface membranes of liver, fibroblasts, smooth muscles and lymphocytes. LDL bind to the receptor and then enter the cell through endocytosis. The LDL protein is degraded to amino acids and the cholesterol esters are hydrolysed. The free cholesterol is then transported into the cytosolic compartment, probably on a sterol carrier protein. Cholesterol from LDL reduces HMG-CoA reductase enzyme activity by accelerating degradation and reducing synthesis. This may be through the reduction of the amount of mRNA responsible for the reductase or through phosphorylation of the enzyme. The increased degradation may be through localization of cholesterol in the membrane domain of the reductase gene.

Inhibition of HMG-CoA reductase

The active inhibitor of transcription of the reductase maybe the oxysterol derivative of cholesterol. The 25-hydroxylated cholesterol may attach to a protein which interacts with the genetic machinery, at either the transcriptional or translational level.

Cytosolic protein factors belong to a family which can influence the effect of lipids on the activity of HMG-CoA reductase; these include sterol carrier proteins, fatty acid binding protein, and z protein. These may influence the binding of the hydroxysterol metabolites of cholesterol or may bind lipid inhibitors.

11.9.3 Cholesterol catabolism to bile acids

In the degradation of cholesterol to bile acids, the first step is rate-limiting, 7-α-hydroxylation, a microsomal enzyme reaction. Cytochrome P_{450}, NADPH and molecular oxygen are required. There is then hydroxylation at C-12 for the precursor of cholic acid, following which, in sequence the C-4/C-5 bond is saturated and the carbonyl group at C-3 reduced. There is a series of oxidative steps in the C-20 to C-27 side chain in the mitochondria. This leads to the oxidation of the alcohol group at C-26 to a carboxyl.

In this manner, cholic acid, 3,7,12-trihydroxycholanoic acid and chenodeoxycholic, 3,7-dihydroxycholanoic acid are formed. They are conjugated with taurine or glycine and excreted in bile to rotate in an enterohepatic circulation.

Bile acids are a major product of the hepatic degradation of cholesterol and may exert a regulatory role on the overall hepatic sterol metabolism. Depletion of the enterohepatic circulation as with ileal bypass or as a result of cholestyramine therapy depletes the bile acid pool and alters hepatic metabolism of cholesterol. Bile acid synthesis increases, and this causes an increased rate of cholesterol synthesis and increased LDL receptor activity. This may act through a decrease in the hepatocyte regulatory pool of cholesterol. Not only the size but also the composition of the bile acid pool may contribute to the regulation of hepatic sterol metabolism. This may relate to the physicochemical characteristics of the constituents of the bile acid pool. The hydrophobic–hydrophilic balance of the pool seems to dictate most of the effects of bile acids on hepatic cholesterol metabolism.

Bile acids are amphiphilic molecules (one side is hydrophobic and the other hydrophilic) as a result of hydroxyl groups. There is increased water solubility as a result of conjugation to taurine and glycine.

The bile acid pool of humans consists of cholic acid and chenodeoxycholic acid in approximately equal amounts (40% of the total pool) and the bacterial degraded bile acids deoxycholic acid (20–30%) and trace amounts of lithocholic acid and ursodeoxycholic acid.

> Lithocholic acid is the most hydrophobic bile acid followed by deoxycholic acid, chenodeoxycholic acid and cholic acid. Hydrophobic bile acids such as deoxycholic acid and chenodeoxycholic acid inhibit HMG-CoA reductase whereas cholic acid does not appear to have this effect. The more hydrophobic the bile acid which predominates, the greater the inhibitory effect on endogenous bile acid synthesis.

11.9.4 Plasma cholesterol concentration

There is a seasonal rhythm of plasma cholesterol; in the same individuals the concentration between November and January is lower than between February and April. The timing of the trough varies, but this variation persists for women as well as men. There are similar changes in lipoprotein and apolipoprotein concentrations.

Saturated fats differ in their effects on serum cholesterol. Palmitic (C-16), myristic (C-14) and lauric acid (C-12) raise the serum cholesterol. Lauric acid and palmitic acid raise the total and LDL lipoprotein concentration, which is in contrast to the stable triglycerides and HDL cholesterol. Oleic acid has no effect on cholesterol concentrations.

> ### Key's equation
>
> This describes a change in plasma cholesterol that occurs in volunteers when alterations in the dietary intake of saturated and polyunsaturated fats (all *cis*-linoleic and α-linoleic acids) are made.
>
> Change in plasma cholesterol (mg/100 ml)
> $$= 1.3 \, (2\delta S - \delta P)$$
>
> where δS = difference in percentage energy derived from saturated fat, and δP = difference in percentage energy derived from polyunsaturated fat.

Plasma cholesterol concentrations should be under 6.5 mmol/l, between 6.5–7.8 mmol/l is regarded as moderately increased, and above this figure is severely increased. There is a general consensus that serum cholesterol concentrations have some role in the genesis of coronary heart disease.

11.9.5 Genetic control of plasma cholesterol

ApoA-I, C-I, A-IV gene cluster

The risk of developing coronary artery disease (CAD) is inversely correlated with plasma HDL cholesterol and apoA-I concentrations. There are at least 12 variants of apoA-I but in general such variants are rare, occurring in less than 0.1% of the population. As the genes for apoA-I, C-III and A-IV are clustered, polymorphism of one may also affect the others.

Apolipoprotein genes
Chromosomal location

The genes for apoA-I, apoC-II and apoA-IV are clustered on the long arm of chromosome 11. The apoC-III gene is approximately 2.6 kb downstream from the apoA-I gene but in the opposite transcriptional orientation. The apoA-IV gene is 7.5 kb downstream from the apoC-III gene. The genes for apoE, apoC-I and apoC-II are found on chromosome 19. The genes for apoB and apoA-II are on the short arm of chromosome 2 and the long arm of chromosome 1 respectively. The gene for apoD is on the long arm of chromosome 3.

In heart disease, which is multi-factorial in origin, environmental factors, diet, exercise, stress and genetic influences act on a variety of lipoproteins, apolipoproteins, enzyme and tissue specific events involved in lipoprotein metabolism. All of these are highly regulated and potentially related to vascular disease causation. Other host factors such as vascular wall biology and platelet function are important. Age, blood pressure, cigarette smoking, increased LDL cholesterol, decreased

It has been suggested that the genes for apoA-I, A-II, A-IV, C-I, C-II, C-III and E belong to a multi-gene family with a common ancestral gene. ApoB is not thought to be part of this family since its exon–intron organization is so different from that of the other apolipoprotein genes. From protein and DNA sequences a common structural element is suggested. In general there are four exons and three introns. The diversion results probably from a series of duplication events.

Apolipoprotein gene variation in human populations may be studied using biochemical genetics, DNA sequencing and genetic linkage. The genetic markers most commonly used are restriction fragment length polymorphisms (RFLPs).

Genetic polymorphism of apoE is derived from three alleles, e2, e3 and e4 at a single autosomal gene locus. These give rise to six phenotypes E2/2, E3/3, E4/4, E4/2, E4/3 and E3/2:

- E4/4 homozygous subjects have raised LDL concentrations compared with E3/3 subjects
- E4/4 subjects have an increased plasma cholesterol on an enhanced cholesterol containing diet compared with E3/3
- E4/4 subjects are characterized by IDL particles being more readily taken up by $\alpha2$-macrophages in the liver than other types

The concentration of plasma LDL is dependent upon VLDL secretion rate and conversion to LDL and its removal. The rate of VLDL to LDL conversion depends on the allelic difference in the apoE phenotype as well as VLDL particle size. ApoE binds with high affinity to the apoB receptor and hence the rate of removal of LDL may be affected by competition. The apoB receptor regulates the rate of removal of LDL as well as its synthesis from VLDL.

HDL cholesterol and diabetes are significant independent factors. Serum fibrinogen is also an independent risk factor.

ApoA-I and apoB, the major lipoproteins of HDL and LDL respectively, are markers of heart disease (CAD). In measuring such lipoproteins, cholesterol-lowering diets and β-blockers each have effects on concentrations which may distort the results, e.g. propranolol has a profound influence on lipid concentrations; β-blockers may increase triglyceride concentrations, lower HDL cholesterol and apoA-I and reduce LDL cholesterol, with little effect on apoB concentrations. This results in the formation of more dense LDL particles associated with hypertriglyceridaemia. ApoB is less directly related to risk of CAD than LDL cholesterol and even total cholesterol concentrations. For these reasons the expert panel, the National Cholesterol Education Programme of the United States, suggests that total cholesterol in HDL cholesterol is useful for screening in the non-fasting state. In the fasting state, total cholesterol, triglyceride and HDL cholesterol can be measured so that LDL cholesterol can be calculated.

Hepatic receptors to apoB and apoE are important in regulating cholesterol concentrations.

The binding capacity of apoB is determined genetically. However, hormonal factors, e.g. corticosteroids and oestrogens increase, and dietary factors including cholesterol decrease, the number of active LDL receptors. LDL concentrations in humans are more influenced by rates of LDL synthesis rather than removal as LDL receptor activity is low in humans. Linoleic acid replacing saturated acids may reduce LDL synthesis rates. The lipid composition of lipoprotein particles may affect their physical properties and hence their interactions with receptors.

Low serum cholesterol

A low serum cholesterol concentration is not necessarily beneficial. There is an inverse relationship between serum cholesterol and death from colonic cancer and non-medical deaths, most of which are violent.

Cholesterol and behaviour

Depression is associated with cholesterol concentrations under 4.14 mmol/l. This is age-dependent, especially in men. Depression may be associated with a reduction in dietary intake, weight loss and hence reduced serum cholesterol. That is, the reduction in serum cholesterol may be an effect rather than a cause of the depression.

There is an increased suicide rate in trials where the serum cholesterol has been reduced either by diet or drugs. Engelberg has suggested that changes in the central nervous neurotransmitter serotonin may be important in this curious association. A low cholesterol concentration in the brain reduces the number of serotonin receptors. The blood cholesterol is in equilibrium with the membrane cholesterol and therefore changes in one will affect the other.

A curious anomaly is that there is an enhanced rate of coronary heart disease in smokers and an association between smoking and suicide. Also the tissue adipose linoleic content is reduced in smokers.

11.9.6 Hyperlipoproteinaemia

Classification

There is a World Health Organization classification of hyperlipoproteinaemia. This is simple and allows meaningful recording and treatment regimes to be instituted. However, the more subtle features of the genetic basis of these are not brought out by this classification, and more narrow lipid bands are discernible.

Familial hypercholesterolaemia is due to defective hepatic receptors to apoB and premature development of atherosclerosis. Increased concentrations of cholesterol result from increased conversion of VLDL to LDL and a slow rate of removal of LDL from blood. This defect does not respond to diet and drug treatment is more appropriate. The common polygenic hypercholesterolaemia leads to mild increases in LDL, patients with the apoE4/4 phenotype respond better to such treatment than apoE3/3 patients. Type III

Classification of hyperlipoproteinaemia

Type	Frequency	Plasma Cholesterol	Plasma Triglyceride	Lipoprotein increases	Cause (deficiency)
I	rare	normal	markedly increased	chylomicron	(lipoprotein lipase; apoC-II)
IIa	common	increased	normal	LDL	(LDL receptor) defect
IIb	common	increased	increased	LDL VLDL	over production VLDL
III	rare	markedly increased	markedly increased	IDL chylomicrons	(apoE-2 defect), delayed VLDL remnant clearance
IV	common	slight increase	markedly increased	VLDL	overproduction VLDL (defective lipoprotein lipase lipase)
V	rare	slight increase	markedly increased	VLDL chylomicrons	Genetic or diabetes, alcohol or obesity (lipoprotein lipase, VLDL triglyceride)

Major genetic hyperlipidaemias

Type WHO type	Frequency	Cholesterol	Plasma Trigylceride	Lipoprotein increase	Cause (deficiency)
Familial hypercholesterol-aemia IIa, IIb	uncommon	markedly increased	normal	LDL	(LDL receptor activity)
Polygenic hyper-cholesterolaemia IIa	common	increased	normal	LDL	LDL overproduction, reduced catabolism, diet
Familial combined hyperlipidaemia IIa, IIb, IV	common	increased	increased	LDL VLDL	overproduction VLDL-apoB-100, impaired VLDL metabolism
Familial hyper-triglyceridaemia IV, V	uncommon	increased	markedly increased	chylomicrons VLDL	impaired VLDL triglyceride and apoB-100 metabolism

Major secondary hyperlipidaemias

	Plasma Cholesterol	Plasma Triglyceride
Hypothyroidism	increased	normal
Nephrotic syndrome	increased	increased
Diabetes mellitus	normal	increased
Alcohol excess	normal	increased
Anorexia nervosa	increased	normal

hyperlipoproteinaemia is often found in patients who are homozygous for apoE2/2 phenotype. This is characterized by increased IDL and reduced LDL concentrations The high IDL concentrations are possibly due to poor binding of the defective apoE to the hepatic B/E receptor. This condition is associated with atherosclerosis in the peripheral vessels and responds well to a reduced fat intake.

Such classifications of the population indicate that an excess of a particular item of diet in relation to another may test out the enzyme constitution of an individual. Such a constitution will be determined genetically.

The beneficial effect on plasma lipids of n-3 unsaturated fatty acids in hyperlipidaemic patients is very much dependent upon the nature of the hyperlipidaemia and genetic makeup.

In contrast, the effect on plasma lipids of n-6 unsaturated fatty acids on hyperlipidaemic patients is minimal.

Atherosclerosis

The process of the development of atherosclerosis is increased in the presence of:

- high plasma concentrations of cholesterol, more particularly the transporting lipoproteins, e.g. VLDL, IDL, LDL, apoB and Lp(a)
- reduced concentrations of HDL, especially HDL$_2$ and apoAl

High concentrations of plasma cholesterol are associated with increased concentrations of LDL, VLDL or IDL. Increased plasma triacylglycerol concentrations are associated with increased VLDL or IDL or chylomicron remnants.

Hyperlipoproteinaemias IIa, IIb, III, IV, lipoprotein apoB, LDL and VLDL and low concentrations of HDL are associated with an increased incidence of atherosclerosis.

Dietary linoleic acid and dietary intervention, e.g. reduced fat intake and increased dietary fibre may result in **atherosclerotic regression.**

The atherogenic process
If the protein fractions and lipid components of VLDL, LDL, IDL and chylomicron remnants are modified by oxidation, acetylation or glycosylation these can be ingested by monocytes which have become tissue macrophages. Glycosylation can occur in high glucose concentrations The polyunsaturated fatty acids will be particularly sensitive to oxidation. Linoleic-rich LDL particles are more readily oxidized than oleic acid-enriched particles. The lipid peroxides and breakdown products form oxidation products with cholesterol and the result is a modified apoB structure which no longer enters the normal LDL receptor but is scavenged. Antioxidants such as vitamin E and ascorbic acid prevent oxidation occurring. Lp(a) may be particularly taken up by the scavenger pathway rather than the normal apoB receptor pathway. Lp(a) inhibits tissue plasminogen activity and inhibits the breakdown of mural thrombi.

When modified lipoproteins are engorged with cholesterol they are taken up by macrophages to form **foam cells.** When native LDL is attached to a normal cell receptor the result is not atherogenic. Foam cells are trapped in the vascular intimal lining and become the fatty streak of the arterial intima. Platelet activation occurs and as a result of an interaction with the vascular wall and the production of growth factors, hyperplasia of smooth muscle occurs with resultant intimal thickening. A relaxing factor derived from the epithelium – possibly nitric oxide – is inhibited by foam cell formation and the smooth muscle of the vessels are less relaxed.

Vegetable oils (n-6-enriched) are associated with less atherosclerosis than saturated acid rich diets.

EPA (20 : 5 n-3) and DHA (22 : 6 n-3) decrease the concentrations of VLDL, IDL and chylomicron fractions but increase LDL in some patients. These effects may well be independent of the receptor system. Intimal thickening due to proliferation and migration of proliferated smooth muscle cells from deep layers is reduced by EPA and DHA.

The shape and form of the lipoproteins affects how they are taken up by the cell receptors. This can be modified and altered by genetic and dietary influences.

Summary

1. The synthesis of sterols is closely related to growth, development and differentiation of all cells.
2. All cholesterol is synthesized from acetate. The irreversible, rate-limiting reaction in cholesterol synthesis is catalysed by HMG-CoA reductase. This enzyme is inactive when phosphorylated by HMG-CoA reductase kinase. Many of the factors that regulate cholesterol biosynthesis are involved in HMG-CoA reductase activity.
3. Plasma cholesterol enters the hepatic cell as cholesterol esters and cholesterol, transported on low-density lipoproteins (LDL) and influences sterol synthesis through a feed-back regulation on the concentration of LDL membrane receptors and of HMG-CoA reductase.
4. There appears to be diurnal and seasonal cycles in cholesterol synthesis, due largely to changes in the activities of HMG-CoA reductase and cholesterol 7-α-hydroxylase, which is involved in bile acid synthesis and lysosomal acid cholesteryl ester hydrolase.
5. Excess cholesterol is esterified to fatty acids by the liver enzyme acyl-CoA : cholesterol acyltransferase (ACAT).
6. There is a general stimulatory effect of dietary lipids on cholesterol synthesis.
7. The first step in the degradation of cholesterol to bile acids is a rate limiting, 7-α-hydroxylation, a microsomal enzyme reaction. Cytochrome P_{450}, NADPH and molecular oxygen are required. There is then hydroxylation at C-12 for the precursor of cholic acid. Cholic acid 3,7,12-trihydroxycholanoic acid and chenodeoxycholic-3,7-dihydroxy cholanoic acid are formed and conjugated with taurine or glycine and excreted in bile; these rotate in an enterohepatic circulation.
8. Bile acids have a major regulatory role on the overall hepatic sterol metabolism. Not only the size but also the composition of the bile acid pool may contribute to the regulation of hepatic sterol metabolism.
9. The risk of developing coronary artery disease is inversely correlated with plasma HDL cholesterol and apoA-I concentrations. As the genes for apoA-I, C-III and A-IV are clustered, polymorphism of one of these genes may also affect the others.
10. The hyperlipoproteinaemias are classified by the changes in chylomicron, HDL, LDL and VLDL concentration. The type of abnormality determines outcome and treatment.

Further reading

Carulli, N., Paola, L., Bertolotti, M., Caruddi, F., Tripodi, A., Abate, N. and Dilengite, M. (1990) The effects of bile acid pool composition on hepatic metabolism of cholesterol in man. *Italian Journal of Gastroenterology*, **22**, 88–96.

Engelberg, H. (1992) Low serum cholesterol and suicide. *Lancet*, **339**, 727–9.

Fisher, E.A., Coates, P.M. and Cortner, J.A. (1989) Gene polymorphisms and variability of human apolipoproteins. *Annual Review of Nutrition*, **9**, 139–60.

Kritchevsky, D. (1992) Variation in plasma cholesterol levels. *Nutrition Today*, **27**, 21–3.

McGee, J. O'D., Isaacson, P.G. and Wright, N.A, (1992) *Oxford Textbook of Pathology*, Oxford University Press, Oxford.

Muldoon, M.F. *et al.* (1990) Lowering serum cholesterol concentrations and mortality: a quantitative review of primary prevention trials. *British Medical Journal*, **301**, 309–14.

Rudney, H. and Sexton, R.C. (1986) Regulation of cholesterol synthesis. *Annual Review of Nutrition*, **6**, 245–72.

Small, D.M. (1988) Progression and regression of atherosclerotic lesions. Insights from lipid physical biochemistry. *Arteriosclerosis*, **8**, 103–29.

Spady, D.K., Woollett, L.A. and Dietschy, J.M. (1993) Regulation of plasma LDL-cholesterol by dietary cholesterol and fatty acids. *Annual Review of Nutrition*, **13**, 355–81.

Weatherall, D.J., Ledingham, J.G.G. and Warrell, D.A. (1996) *Oxford Textbook of Medicine*, 3rd edn, Oxford University Press, Oxford.

11.10 Amino acid metabolism

- Amino acids are essential or non-essential.
- Only L-amino acids are involved in protein synthesis.
- The necessary amount and types of L-amino acids must be present in order for any particular protein to be synthesized.
- Non-essential amino acids are synthesized by transamination.
- Glutamate is an important source of amino groups for transamination.
- Following deamination the carbon skeletons of amino acids can pass into the metabolic pool, usually the TCA cycle.
- Ammonia is produced following deamination of amino acids.
- Ammonia is toxic but is converted to urea through the urea cycle and excreted in urine.

11.10.1 Introduction

Amino acids can be classified as essential or non-essential. An essential amino acid is one that has to be supplied in the diet in order to maintain a positive nitrogen balance.

Fewer than half of the protein amino acids can be synthesized by de novo pathways. The remainder must be supplied by nutrients.

11.10.2 Amino acid synthesis

The pathway to amino acid synthesis does not arise from a particular carbon source, but rather from a few key intermediates in the central metabolic pathways of all cells. This is independent of the carbon source, the glycolytic pathway, pentose phosphate pathway or the TCA cycle.

The synthesis of amino acids, with the exception of cysteine and tyrosine, is linked to the glycolytic TCA cycle by transamination or ammonia fixation. The α-amino group is central to all amino acid synthesis and is derived from ammonia from the amino groups of L-glutamate. From this glutamine, proline and arginine are synthesized.

Glutamic dehydrogenase facilitates a reversible reaction between glutamic acid and oxaloacetic acid. Glutamic acid is the key source of amino groups for transamination. The transaminase aspartate aminotransferase activity is also important and is found in tissue cytoplasm and mitochondria.

Amino acids in adult humans

Essential	*Non-essential*
Isoleucine	Alanine
Leucine	Arginine
Lysine	Asparagine
Methionine	Aspartate
Phenylalanine	Cysteine
Threonine	Glutamate
Tryptophan	Glutamine
Valine	Glycine
	Histidine
	Proline
	Serine
	Tyrosine

---

glutamate + oxaloacetate ↔
α-ketoglutarate + aspartate

11.10.3 Catabolism of amino acids

The catabolism of dietary proteins from amino acids is important in higher animals, while the catabolism of storage proteins is important in seeds such as beans or peas. Amino acids that are catabolized come from three different sources:

- dietary proteins
- storage proteins
- metabolic turnover of endogenous proteins

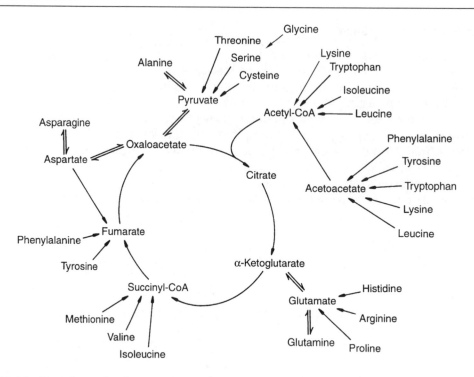

FIGURE 11.46   The relationship between amino acids, deamination and the tricarboyxlic acid (TCA) cycle.

All cells are involved in such metabolic turnover. This means that protein-containing structures and their constituent amino acids can be recycled into other proteins or derivatives.

Protein catabolism starts with the hydrolysis of the covalent peptide bonds linking amino acid residues in a polypeptide chain. This proteolytic process yields free amino acids and peptides by the action of peptidase. Endopeptidases hydrolyse peptide bonds remote from the ends of the molecule whereas peptidases which remove amino acids from the end of the molecule are called exopeptidases. Such peptidases are either aminopeptidases or carboxypeptidases according to the end of the peptide at which digestion takes place. The degradation of amino acids requires the removal of the α-amino nitrogen through two forms of deamination, i.e. transamination or oxidative deamination.

Transamination occurs when an amino acid gives its amino group to α-ketoglutarate produc-ing α-keto acid and glutamate. Most transaminases contain pyridoxal-5'-phosphate as a coenzyme. In order that α-ketoglutarate is regenerated for further transamination, oxidative deamination is necessary using NAD-linked glutamate dehydrogenase. This allows the net conversion of amino acid groups to ammonia.

Deamination products of amino acids (Figure 11.47)

Amino acid(s)	Product
Ile, Leu, Lys	→ Acetyl-CoA
Tyr, Phe	→ Acetoacetate
Gln, Pro, Arg	→ Glu → α-ketoglutarate
His	→ Glu → α-ketoglutarate
Thr, Met, Val	→ Succinyl-CoA
Tyr, Phe, Asp	→ Fumarate
Asp, Asn	→ Oxaloacetate
Ser, Gly, Cys	→ Pyruvate
Trp	→ Alanine → Pyruvate

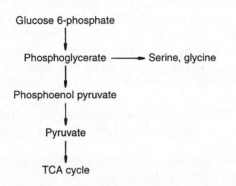

An interesting phenomenon is the presence of widely distributed and highly active D amino acid oxidase. D Amino acids are seldom encountered by animals except as breakdown products of bacterial cell walls. It is just possible that this is a perpetual protection against bacterial amino acid.

Central to the catabolism of amino acids is the formation of a dicarboxylic acid intermediate of pyruvate or acetyl-CoA in the TCA cycle. The 4-carbon dicarboxylic acids can stimulate tricarboxylic acid function. They leave the cycle either by:

FIGURE 11.47 The contribution of glycolysis to amino acids.

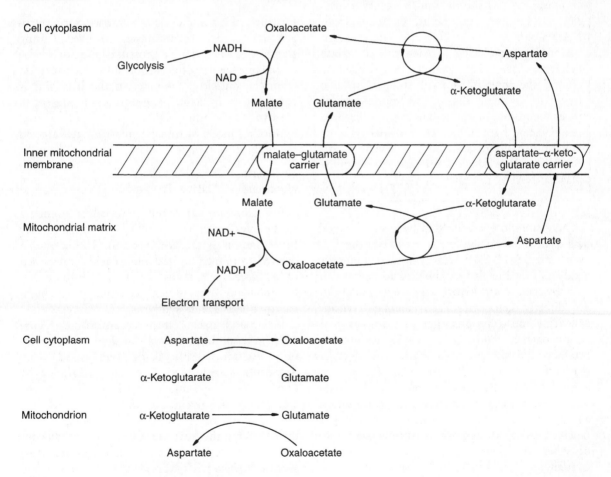

FIGURE 11.48 Because NADH cannot cross the mitochondrial membrane the electrons are carried by a malate–aspartate shuttle.

- the conversion by gluconeogenesis of oxaloacetate to phosphoenolpyruvate; such amino acids are called **glycogenic**
- pyruvate can also be formed and converted to acetyl-CoA and oxidized completely to carbon dioxide and water

---

The malate shuttle

This involves malate dehydrogenase. Oxaloacetate cannot cross the inner mitochondrial membrane. Aspartate crosses from the mitochondrion and is deaminated, yielding oxaloacetate; malate then returns to the mitochondria. In this shuttle, which occurs primarily in the liver and heart, electrons from NADH outside the mitochondria reduce oxaloacetate to malate using an isoenzyme of malate dehydrogenase. The malate dehydrogenase crosses the inner membrane using a specific transport system. Inside the mitochondrial inner membrane space malate is re-oxidized to oxaloacetate by a mitochondrial isoenzyme of malate dehydrogenase. Both isoenzymes are NAD-linked, so electrons are moved from cytoplasmic NADH to matrix NADH.

There is a close relationship between the products of deamination of amino acids and the TCA cycle (Figure 11.46). Acids whose deamination products are directly metabolized to acetyl-CoA or acetoacetate are **ketogenic**; leucine and lysine are directly ketogenic. Phenylalanine, tyrosine, isoleucine and tryptophan are partially ketogenic or glucogenic. Amino acids whose deamination products directly enter the citric acid cycle as pyruvate or citric acid are **glucogenic** and produce a net synthesis of glucose. Eighteen of the amino acids yield glucose and therefore are important as energy sources as well as having structural importance. This is of significance during starvation or when the diet is reduced in carbohydrate content.

The transaminase **aspartate aminotransferase**, as well as catalysing the synthesis of aspartate, is important in the transport of reducing equivalents between mitochondria and cytoplasm.

The **malate shuttle** (Figure 11.48) is important in re-oxidizing NAD to NADH, which is generated in abundance during glycolysis in the cytoplasm. NADH cannot cross the mitochondrial membrane. Reactions which require NADH may produce reaction products which after oxidizing NADH in the cytoplasm cross into the mitochondria and are re-oxidized and then return to the cytoplasm.

The transaminase **alanine aminotransferase**, which catalyses the reaction

pyruvate + glutamate $\leftrightarrow$
alanine + α-ketoglutarate

provides an alanine cycle which transports nitrogen and carbon from muscle to liver. During starvation, when muscle protein is a major source of gluconeogenesis, the carbon substrate is carried from the muscle as alanine. In the liver this is converted to pyruvate. The nitrogen is present in the liver as glutamic acid.

Another mode of transporting nitrogen around the body is glutamine

glutamate + NH_3 + ATP $\rightarrow$
glutamine + ADP + P_i

In this manner glutamine can deliver ammonia to the liver for urea production, or to the kidney for excretion as the NH^+ cation. The ammonia derived from metabolized amino acids is disposed of as **urea** in humans. Urea is non-toxic but metabolically expensive, requiring four high-energy bonds for its synthesis.

Serine and **glycine** are interconvertible (Figure 11.49). Serine undergoes a transhydroxymethylation reaction, serine being synthesized from 3-phosphoglyceric acid. Glycine is degraded by two routes. The route to pyruvate involves the conversion of glycine to serine and involves the addition of hydroxymethyl through 5,10-methylenetetrahydrofolate. The serine is then converted to pyruvate by serine dehydratase. Most glycine however is oxidized to CO_2, NH_4^+ and methylene ($-CH_2-$) which is accepted by tetrahydrofolate in a reversible reaction.

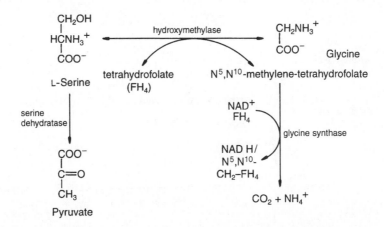

FIGURE 11.49 Serine and glycine interrelationship and catabolism. This transmethylation reaction involves tetrahydrofolate and N^5, N^{10}-methylene tetrahydrofolate.

Alanine is involved in a reversible transaminase reaction to pyruvate as part of the glucose alanine cycle.

Tyrosine catabolism involves **tyrosine–glutamate transaminase**. Tyrosine can also be catabolized to the skin colour pigment, melanin. The enzyme involved, tyrosinase, is found solely in melanosomes which are specialized pigment-producing cells in the skin and other tissues. Dopa (3,4-dihydroxyphenylalanine) is synthesized from tyrosine by tyrosine hydroxylase and is a precursor for noradrenaline and adrenaline. This takes place entirely in the adrenal glands.

Phenylalanine degradation occurs through the intermediary tyrosine by the action of phenyl-alanine-4-mono-oxygenase (Figure 11.50). This requires tetrahydrobiopterin as a co-substrate. This biopterin remains in the reduced form through the action of NADPH. Tyrosine is not an essential amino acid unless phenylalanine is present in large amounts. Where there is an absence or deficiency of the hydroxylation to tyrosine, high concentrations of phenylalanine are to be found in the blood. This is known as **phenylketonuria**.

The major pathway for **tryptophan** catabolism in the liver is through kynurenine, which is metabolized in the liver through α-ketoadipate (also an intermediate in lysine degradation) (Figure 11.51). Kynurenine is also involved in the synthesis of the coenzyme nicotinamide.

Arginine, histidine, proline, glutamic acid and **glutamine** are readily converted to α-ketoglutarate (Figure 11.52). Arginine is not transaminated but is a precursor of various essential polyamines and is important in the urea cycle.

Methionine, isoleucine and **valine** are degraded to succinyl-CoA. The three keto-acids produced by deamination of valine, isoleucine and leucine are decarboxylated by the same enzyme complex. This enzyme also acts on pyruvate and α-keto-butyrate which are products of both threonine and methionine metabolism.

Aspartate and **asparagine** are deaminated to oxaloacetate, and thence to the metabolic pathway into the TCA pool.

Regulation of amino acid catabolism

Amino acid catabolic enzymes are under hormonal control, though the activity of some is influenced by diet (Figure 11.53).

The degradation of arginine (and its precursor ornithine), serine and tryptophan is affected by the protein content of the diet. High-protein diets stimulate the enzymes involved, whether they be ornithine–glutamate transaminase, or urea cycle enzymes or deaminases. Threonine deaminase is stimulated by a high-protein diet. In contrast, tryptophan oxygenase and tyrosine–glutamate transaminase are stimulated by glucocorticoids.

471

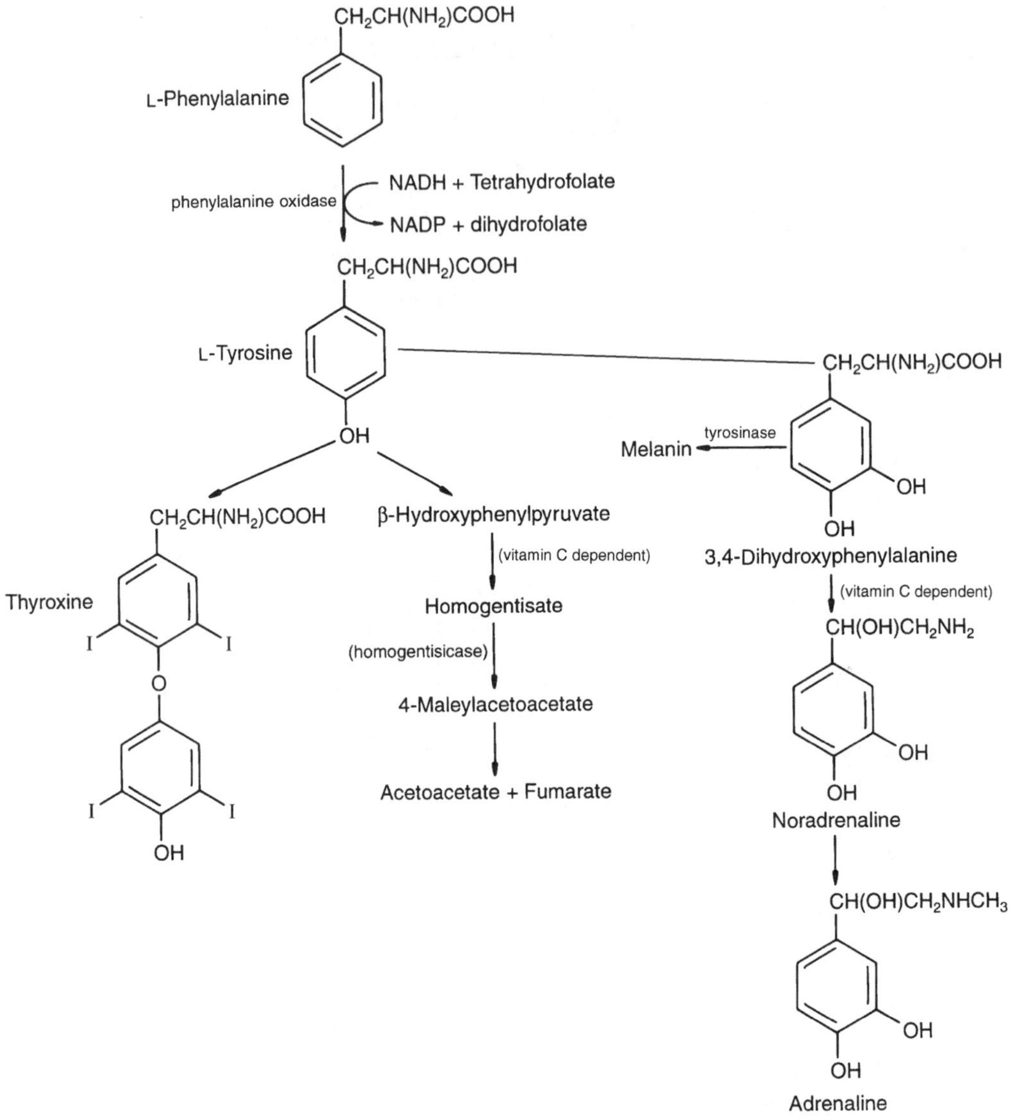

FIGURE 11.50 Aromatic amino acids. Phenylalanine, an essential amino acid, is an important source of tyrosine, which may be catabolized to fumarate and acetoacetate or to 3,4-dihydroxyphenylalanine and thence to melanin, which is a pigmented cutaneous polymer controlled by tyrosinase. 3,4-Dihydroxyphenylalanine may also be synthesized to noradrenaline and adrenaline. L-tyrosine may also be iodinated to thyroxine or degraded to acetoacetate and fumarate. (Note that tyrosine is not an essential amino acid unless phenylalanine is deficient.)

These catabolic enzymes may vary in activity with age. Tryptophan oxygenase activity is low in the newborn rat but increases after 12 days, an increase associated with a rise in adrenal activity. Glucocorticoids induce the formation of the enzyme in young rats and stimulate enzyme production in adults. Ornithine transcarbamoylase and the other urea-forming enzymes are synthesized after birth and are induced by a high-protein diet.

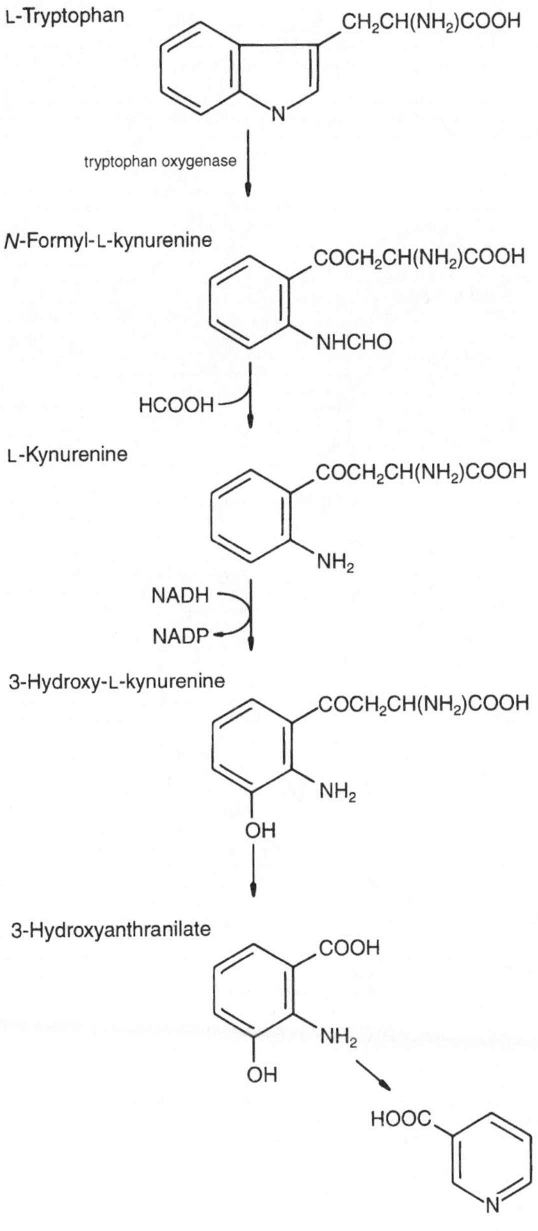

L-Tryptophan $CH_2CH(NH_2)COOH$

tryptophan oxygenase

N-Formyl-L-kynurenine $COCH_2CH(NH_2)COOH$

NHCHO

HCOOH

L-Kynurenine $COCH_2CH(NH_2)COOH$

NH_2

NADH

NADP

3-Hydroxy-L-kynurenine $COCH_2CH(NH_2)COOH$

NH_2

OH

3-Hydroxyanthranilate COOH

NH_2

OH

HOOC

Nicotinate

FIGURE 11.51 Tryptophan is the only amino acid with an indole ring and is an essential amino acid. It can be catabolized to kynurenine and synthesized to the coenzyme NAD.

Many amino acids are taken into cells by sodium ion-dependent transport systems (Figure 11.54) which convert the energy of the electrochemical sodium gradient across the plasma membrane into osmotically active amino acid gradients with intracellular/extra-cellular concentration ratios of up to 30. Such gradients cause water to move into the cell and lead to cell swelling. Liver cells swell by as much as 12% within 2 minutes under the influence of glutamine and increased cellular hydration is maintained as long as the amino acid is present. Hormones also change the cellular hydration state, that is, the cell volume, by modulating the activity of the ion transport system in the plasma membrane. Insulin increases cellular hydration by causing Na^+, K^+ and Cl^- to accumulate within the cell due to activation of the Na^+-H^+ antiporter, Na-K-2 Cl co-transport and the Na^+-K^+ ATPase. Glucagon on the other hand induces cell shrinkage.

Cellular hydration status is an important determinant of protein catabolism in health and disease.

In the liver, **cell swelling** inhibits the breakdown of glycogen, glucose, RNA and protein and at the same time stimulates synthesis of glycogen, RNA, DNA and protein. The opposite metabolic pathway is triggered by cell shrinkage. Cell swelling is a proliferative anabolic signal whereas cell shrinkage is antiproliferative and catabolic. Hormone-induced changes in cellular hydration are 'second messengers' of hormone action. It is possible that the antiproteolytic effects of some amino acids and insulin and the proteolytic action of glucagon may be explained by the influences of these on cellular hydration.

11.10.4 Glutamine

Glutamine is a non-essential amino acid and is the most abundant free amino acid in the body. It has the highest plasma concentration of any amino acid and provides approximately 5% of the whole body free amino acid pool. Glutamine is an oblig-

473

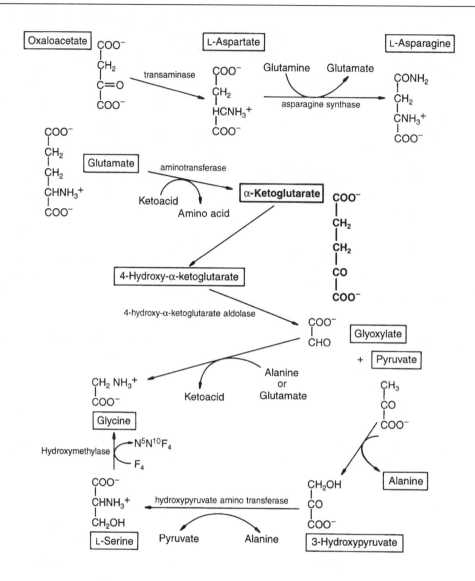

FIGURE 11.52   Inter-relationships of amino acids through α-ketoglutarate.

atory fuel for intestinal cells and other rapidly dividing cells, e.g. the cells of the immune system and in its role in regulation of acid–base balance by providing the most important substrate for renal ammoniagenesis. Glutamine provides precursors for nucleic acid biosynthesis and protein synthesis and its regulation. Glutamine is important in various sites in the body.

Intestines

Intestines contain high glutaminase activity and very little glutamine synthetase activity. The intestines therefore break down glutamine, which provides an important energy source in addition to glucose, short-chain fatty acids and ketone bodies. The glutamine is derived either from the blood

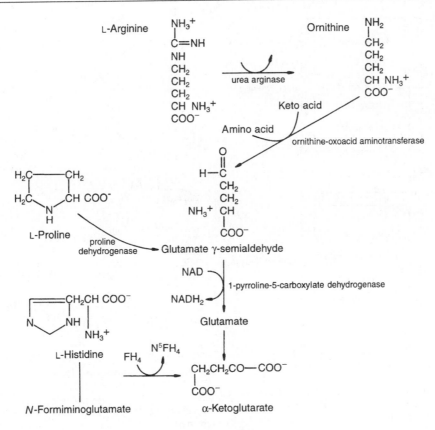

FIGURE 11.53 Regulation of diet and amino acid catabolism. Arginine, ornithine, serine and tryptophan degradation is increased by a high protein content of the diet.

stream or the intestinal lumen. Most of the glutamine is utilized by the intestinal mucosa. The large bowel uses much less glutamine than the small intestine.

Intestinal glutamine uptake and metabolism accounts for approximately 60% of ammonia released from the viscera drained by the portal vein. Glutamine is converted to ammonia and glutamate, which is then transaminated to alanine and to a lesser extent to other amino acids and organic acids. Colonic bacteria in the gut lumen produce ammonia by splitting urea, creating another source of ammonia. Glutamine breakdown and urea splitting make the intestines a major ammonia-producing organ. Ammonia and amino acids are carried in the portal vein to the liver and then to the urea cycle. The alanine carbon skeleton is utilized in hepatic gluconeogenesis.

Muscle

Glutamine synthetase activity in skeletal muscle is low. However, because muscle occurs in such substantial quantities, it is one of the principal glutamine-synthesizing organs. In addition, skeletal muscle contains glutaminase activity and is an organ of net glutamine release in the physiological situation. Skeletal muscle contains 70–80% of the total body free amino acid pool and glutamine forms 60% of the muscle pool. Glutamine release from skeletal muscle increases with changes in muscle protein turnover or release from the free glutamine pool. Glutamine forms approximately 30% of the amino acid released from skeletal

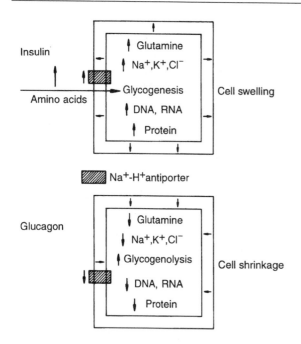

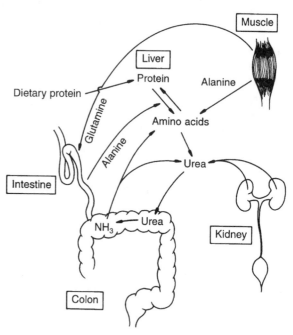

FIGURE 11.54 The entrance of amino acids into the cell through the sodium-dependent transport system increases the osmotic active pressure within the cell, which swells. Insulin increases the activity of the Na^+, H^+ antiporter, increases glycogenesis and DNA, RNA and protein synthesis. Glucagon has the opposite effect.

FIGURE 11.55 Urea is produced in the liver from dietary and endogenous protein amino acids. Some 70% of the urea is excreted in the urine, with 30% being retained in the body. Most tissues release nitrogen as alanine or glutamine. Alanine results from the transamination of glutamate to pyruvate. Glutamine utilizes ammonia (enzyme glutamine synthetase).

muscle but constitutes only 5% of muscle protein amino acids. Glutamine is a carbon and nitrogen carrier from skeletal muscle to splanchnic organs.

The cellular tissue concentration of glutamine falls during severe illness. Glutamine is considered to be an essential fuel for small intestinal function. This leads some clinicians to regard glutamine to be conditionally indispensable in critical illness.

11.10.5 Ammonia and urea

Ammonia is converted to urea (Figure 11.55), a soluble end-product which is readily excreted in urine at approximately 30 g/day.

Ammonia is a toxic substance whereas urea is relatively harmless to tissue function and well-being. All the nitrogen in urea is derived from two precursors, the ammonium ion and aspartate.

Glutamine dehydrogenase and ammonia

Glutamine is an abundant amino acid and is readily transaminated by other amino acids. Glutamate in a reversible reaction is oxidatively deaminated to α-ketoglutarate and ammonia by the mitochondrial enzyme, glutamate dehydrogenase.

> The transfer of nitrogen from an amino group is to a keto acid; the most active enzyme is aspartate aminotransferase
>
> L-glutamate + oxaloacetate → α-ketoglutarate + L-aspartate
>
> The aminotransferase contains pyridoxal phosphate.

Glutamine synthetase

Ammonia can be carried to the liver as glutamine, formed in a reaction catalysed by glutamine synthetase in which NH_3 is added to glutamate to form glutamine. Once the glutamine reaches the liver or small intestine the enzyme glutaminase releases the ammonia

Skeletal muscle transports NH_3 to the liver in the form of alanine. This is derived from a transamination reaction between pyruvate and glutamate. In the liver alanine reacts with α-ketoglutarate to form pyruvate and glutamate, a reaction catalysed by alanine transaminase. If the blood glucose concentration is low, then pyruvate is converted to glucose via gluconeogenesis. This glucose can be returned to the skeletal muscle for energy purposes. This is known as the glucose alanine cycle and is important in the muscular activity of the organism.

Urea cycle

Urea is derived from the hydrolysis of arginine – catalysed by arginase – leaving L-ornithine which is recycled, enabling a continuous production of urea.

The urea cycle (Figure 11.56), which is confined to the liver, is a metabolic pathway for the disposal of ammonia. In each cycle, two nitrogens are eliminated; one from the oxidative deamination of glutamate and the other from the α-amino group of aspartate. Urea is excreted into the blood stream and removed through the kidneys in the urine.

Each enzyme is a single polypeptide chain produced by a single-copy nuclear gene. The first two enzymes of the urea cycle are to be found within the mitochondrial matrix and the other three are cytosolic (Figure 11.57). The substrates sequentially move from one enzyme to another by channelling. In the liver the urea cycle enzymes are found predominantly in the periportal hepatocytes. This spatial distribution may be modulated by diet and hormones. In addition to the removal of ammonia the urea cycle is important in pH homeostasis by regulating bicarbonate con-

centrations. Some of the enzymes of the urea cycle are to be found in the small intestine and liver forming an independent arginine biosynthetic pathway.

The urea cycle

Five enzymes are involved:

1. Carbamyl phosphate synthetase I (carbamoyl phosphate synthetase)

   carbon dioxide + ammonia → carbamoyl phosphate

2. Ornithine transcarbamylase (L-ornithine carbamoyltransferase)

   carbomyl phosphate + ornithine → citrulline

citrulline is transported from the mitochondria.

3. Argininosuccinate synthetase (L-citrulline : L-aspartate ligase)

   citrulline + aspartate + ATP → arginosuccinate + AMP

4. Argininosuccinate lyase (L-argininosuccinate arginine-lyase)

   arginosuccinate → arginine + fumarate

5. Arginase (L-arginine ureohydrolase)

   arginine → ornithine + urea

Fumarate is in equilibrium with malate. In this way urea production is tied into other metabolic pathways. Malate is converted to pyruvate in fatty acid synthesis or gluconeogenesis as required.

The importance of endogenous arginine synthesis is very much dependent on age, physiological state and species of animal. Young animals require dietary arginine for optimal growth whereas endogenous arginine synthesis meets the arginine requirements of most adult omnivores.

The potential ureagenic capacity of the liver is

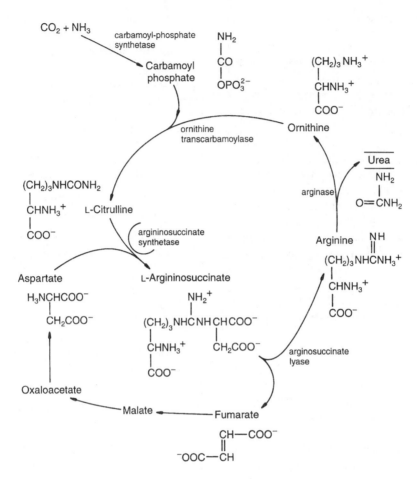

FIGURE 11.56 The urea cycle. Ammonia reacts with carbon dioxide and ATP to produce carbamoylphosphate within the mitochondria. The urea cycle produces urea which diffuses into the blood, is carried to the kidney and excreted.

limited by argininosuccinate synthetase activity. As the urea cycle normally does not function at full potential, the rate of ureagenesis is controlled by substrate availability.

Urea production in the adult human varies with the dietary protein intake. The metabolic regulation of urea cycle enzymatic activity is very much species-specific, perhaps depending on the type of diet to which an animal is normally adapted. Urea cycle enzyme activity is greatest in response to starvation and high-protein diets, but is reduced in response to low-protein or protein-free diets. The amino acid composition of the proteins has little or no effect on enzyme activity. Diet-dependent changes in urea cycle activity are the consequence of changes in enzyme mass and altered enzyme synthesis rates. Therefore, dietary regulation of enzyme level is acting at a pre-translational step. There is a co-ordination in the regulation of the five mRNAs involved. Sepsis, trauma, uraemia or cancer alter urea synthesis and levels of urea cycle enzymes.

Arginine biosynthesis
The small intestine is the principal source of citrulline in adult mammals, the synthesis being

catalysed by carbamyl phosphate synthetase I and ornithine transcarbamylase in mucosal epithelial cells. The kidney has a significant capacity for converting citrulline to arginine, by argininosuccinate synthetase and argininosuccinate lyase, the third and fourth enzymes of the urea cycle in the proximal tubules of the kidney. The kidney is a major site of arginine biosynthesis and renal production of arginine is limited by the availability of citrulline. Renal argininosuccinate synthetase and argininosuccinate lyase activities and the responsible mRNA increase as dietary protein intake increases.

Regulation of carbamyl phosphate synthetase I (CPS-1) is dependent on the concentration of its essential cofactor N-acetylglutamate. Arginase is a rate-limiting enzyme at a junction point in the process wherein arginine may be used for ureagenesis, for protein synthesis or polyamine synthesis.

Free ammonia formed by oxidation and the deamination of glutamine is converted into carbamoyl phosphate. The carbamoyl group is trans-ferred to the terminal amino group of ornithine to form L-citrulline and then into the urea cycle.

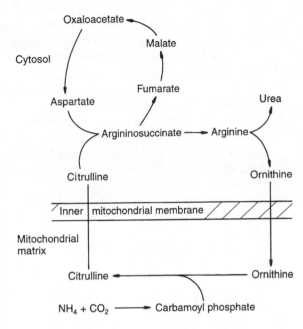

FIGURE 11.57 Urea cycle compartmentalism. Part of the urea cycle takes place within the mitochondria. Ornithine must cross the mitochondrial membrane to be available for the formation of citrulline. The transport of ornithine into the mitochondria uses the same transporter system that carries citrulline out of the organelles.

Summary

1. Dietary amino acids are either essential or non-essential. Non-essential amino acids can be synthesized from key intermediates in the glycolytic pathway, pentose phosphate pathway or the TCA cycle.
2. The synthesis of amino acids involves transamination or ammonia fixation. The α-amino group is generally derived from the amino groups of L-glutamate.
3. The catabolism of dietary proteins yields amino acids for recycling. Other sources of amino acids for catabolism include storage proteins and the metabolic turnover of endogenous proteins.
4. Protein catabolism involves the hydrolysis of covalent peptide linkages. The degradation of amino acids requires the removal of the α-amino nitrogen through deamination in the form of transamination or oxidative deamination. Transamination is the donation of the amino group to α-ketoglutarate, which is then regenerated by oxidative deamination.

5. There is a close link between the deamination products of amino acids and the TCA cycle. Amino acids directly converted to acetyl-CoA are ketogenic, but deamination products directly entering the TCA cycle are glucogenic.

6. The malate shuttle is important in maintaining a required balance between NAD and NADH across the mitochondrial membrane. This involves NAD-linked enzymes and the movement of aspartate and malate across the mitochondrial membrane, producing oxaloacetate on either side; the process is then repeated.

7. The transaminase alanine aminotransferase reaction allows the transportation of nitrogen from muscle to liver in the form of alanine.

8. Amino acid catabolic enzymes are under hormonal and dietary control.

9. Glutamine, while being a non-essential amino acid, is important as the most abundant free amino acid. It is an obligatory fuel for intestinal and immune cells, has a role in acid–base balance, provides α-amino groupings for renal ammoniagenesis, and is a precursor in nucleic acid biosynthesis.

10. The end-products of nitrogen catabolism are ammonia and urea. Ammonia is toxic and is carried to the liver as glutamine. Muscle transports ammonia to the liver in the form of alanine.

11. In the urea cycle, urea is produced by the sequential removal of nitrogen atoms. Urea production varies as a function of dietary protein intake, being reduced in starvation.

Further reading

Fuller, M.F. and Garlick, P.J. (1994) Human amino acid requirements. *Annual Review of Nutrition*, **14**, 217–42.

Kilberg, M.S., Stevens, B.R. and Novak, D.A. (1993) Recent advances in mammalian amino acid transport. *Annual Review of Nutrition*, **13**, 137–65.

Lacey, J.M. and Wilmore, D.W. (1990) Is glutamine a conditionally essential amino acid? *Nutrition Reviews*, **48**, 297–309.

Morris, S.M. Jr (1992) Regulation of enzymes of urea and arginine synthesis. *Annual Review of Nutrition*, **12**, 81–101.

Souba, W.W. (1991) Glutamine: a key substrate for the splanchnic bed. *Annual Review of Nutrition*, **11**, 285–308.

Waterlow, J.C. (1995) Whole body protein turnover in humans – past, present and future. *Annual Review of Nutrition*, **15**, 57–92.

Waterlow, J.C. and Stephen, J.M.L. (1981) *Nitrogen Metabolism in Man*, Applied Sciences Publishers, London.

Windmueller, H.G. (1982) Glutamine utilization by the small intestine. *Advances in Enzymology*, **53**, 210–37.

Zubay, G. (1992) **Biochemistry**, W.C. Brown, USA.

11.11 Amino acid neurotransmitters

• Nerve cells transmit messages from one cell to another via synapses.
• Neuroamines are important neurotransmitters.

11.11.1 Introduction

Nerve cells are the basic units of the nervous system which incorporates the brain, spinal cord and nerves. Nerves in the periphery may be motor or sensory, that is, stimulating or transmitting information to organs or tissues. A nerve cell receives, conducts and transmits signals over large distances. Though nerve cells are made in a wide variety of forms, the form of the signal is always the same: changes in the electrical potential across the nerve cells' plasma membrane.

11.11.2 Amino acid neurotransmitters

Stimuli are received in a variety of ways including directly onto the nerve cell surface. Neuronal cells transmit from one to another via **synapses** (Figure 11.58). The small chemical neurotransmitters are very varied, and pass from one synapse to the next at great speed. The important property of the nerve cell is its **excitability**. Whatever the signal, voltage-gated cation channels generate the action potentials (Figure 11.59).

An action potential is set off by a depolarization of the plasma membrane, to a less negative value. The proteins of these voltage-gated channels are very similar in their amino acid structure among a wide range of species. This is in accord with the evolution of these systems from quite primitive to complex organisms.

There are also GTP-binding protein (G-protein)-linked receptors which respond to signals evoked by neuropeptides. These are more complex and longer acting. Unlike the G-protein-coupled receptors, the ligand-gated channels do not require second messenger systems for signal transduction.

Metabolically driven pumps are effective in establishing concentration differences across neuronal and glial cell membranes. Membrane gates or channels can be classified into two types, controlled by chemical molecules or ligands on transmembrane voltage. Opening and closing the gates allows ions to flow across the membrane in the direction required, creating the electrical neuronal activity.

Excitatory neurotransmitters

The most studied is the nicotinic acetylcholine receptor. The brain also contains important excitatory amino acids, glutamate and aspartate. These are widely distributed through the spinal cord and brain and are active at concentrations of 10^{-15} M. They induce rapid membrane depolarization, in a manner resembling nicotinic acetylcholine receptor activity.

Once aspartate and glutamate are released into the synapse they are rapidly taken up by a high-affinity transport system in the nerve endings.

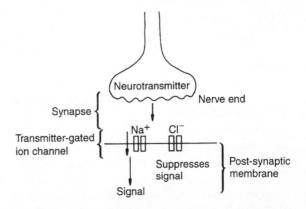

FIGURE 11.58 Signals pass from nerve-end to nerve-end at specialized sites called **synapses**. A presynaptic cell is separated from the postsynaptic cell. The neurotransmitter crosses the synaptic space and binds to a transmitter-gated ion channel. Opening the channel gate provokes an electrical change and a neurological signal is transmitted.

Glutamate is transformed to glutamine in the glial cells and once in contact with the glutaminergic nerve endings, is deaminated to glutamate. A similar transport system applies to aspartate.

Inhibitory neurotransmitters

These include γ-aminobutyric acid (GABA$_A$), an inhibitory transmitter in the brain, and glycine, which is effective in the spinal cord and brain stem. They operate by controlling a channel specific to small anions, e.g. chloride. The inhibitory

Excitatory
  Acetylcholine
  Glutamate
  Serotonin
  Open cation channel
  Influx Na$^+$
  Depolarizes post-synaptic membrane
    to fire action potential

Inhibitory
  γ-Aminobutyric acid (GABA)
  Glycine
  Open Cl$^-$ channel
  Keep post-synaptic membrane polarized

FIGURE 11.59 Excitatory neurotransmitters open cation channels and fire a nervous impulse. Inhibitory neurotransmitters open Cl$^-$ channels which suppress the signal.

action of $GABA_A$ is potentiated by the psycho-active drugs benzodiazepines and barbiturates. The glycine-activated channels are found on post-synaptic membranes in the brain stem and spinal cords of mammals. They are very selective for small anions, e.g. chloride and show marked affinity for strychnine.

Most neurotransmitters use two types of receptors and neurotransmitters; $GABA_A$ and nicotinic acetylcholine act by directly opening ion channels. $GABA_A$, $5\text{-}HT_1$, $5HT_2$ and muscarinic acetylcholine and glutamate are linked by G-proteins to a variety of effector molecules which include ion channels and enzymes that generate diffusible second messengers. Glutamate is the major excitatory neurotransmitter in the brain. In the spinal cord and brain stem glycine is the dominant inhibitory neurotransmitter, whereas in higher brain regions GABA is the dominant inhibitor.

Action potentials

In nerves, action potentials are produced by voltage-gated sodium channels which allow Na^+ to enter the cell along an electrochemical gradient. Another system is the Ca^+ channel. The membrane has an automatic inactivating system. A voltage-gated K^+ reverses the charge by an efflux of K^+ which reverses the polarity changes. These channels may be present in one of three forms which reflect the energy state.

- membrane at rest   highly polarized gates closed
- membrane energy low   depolarized   gate open
- membrane inactivated   inactivated

There are five distinct subgroups of glutamate receptors which are important in neuronal activity. Glutamate receptors exist as a diverse family which comprise three subtypes. The $GABA_A$ receptor is the major molecular site of the inhibitory activities of the brain. The $GABA_A$ brain synapse receptor is the site of action for:

- GABA agonist/antagonist site
- benzodiazepine site
- picrotoxin site
- a depressant site which includes the barbiturates

$GABA_A$ receptors

The receptor is a heterotetrameric protein which has an array of sites which span the channel. There appear to be two subunit types, α (M_r 53 000 Da) and β (M_r 57 000 Da). The GABA sites are on two β-subunits and the benzodiazepine receptor consists of two α-subunits. The receptor contains a proline moiety which may give the receptors flexibility with a bend of 20–25° and a helix which is hydroxy-rich on one side and has aliphatic side chains on the other. The latter may interact with lipid hydrocarbon chains. Clusters of arginine and lysine molecules at the channel mouth act in both $GABA_A$ and glycine receptors as an anion-concentrating device and to increase the driving force for anion flow upon opening the channel.

Nitric oxide

Nitric oxide (NO) is a neurotransmitter, produced enzymatically in post-synaptic structures in response to activation of amino acid receptors. NO is derived from arginine by enzymatic removal of one of the terminal guanidino nitrogens of the amino acid L-arginine by nitrogen oxide synthase which is found in six isoforms. NO synthase is found in all tissues of the brain, the highest in the cerebellum and the lowest in the medulla. The granule cell in the neurone appears to be particularly active in NO synthesis. Haemoglobin is an important antagonist of NO activity. Arginine metabolism in the brain is tied in with urea metabolism. Some citrulline is formed during NO synthase activity. The brain urea cycle resynthesizes arginine from co-product ornithine.

NO diffuses to act on other cellular pre-synaptic nerve endings and astrocyte receptors. The major action of NO is to activate soluble guanylate cyclase and to increase cGMP levels in target cells. NO is highly reactive and an unstable free radical species with a half-life of 4 seconds.

A major role of NO is to stimulate cGMP synthesis, which directly regulates cation channels and inhibits inositol phospholipid hydrolysis. Arachidonic acid and its metabolites mimic or oppose NO at the guanylate cyclase level. NO binds to the iron–sulphur centres of enzymes, e.g. those involved in the mitochondrial electron transport chain, citric cycle and DNA synthesis.

NO acts as a neurotransmitter in the brain and peripheral nervous tissues and may be the neurotransmitter in the non-adrenergic, non-cholinergic neurotransmission system. This is a possible role for NO in angiotensin and oxytocin release and the stomach receptive relaxation reflex.

NO regulates blood flow by acting on the endothelial cells lining the vessels, binding to iron in a haem moiety attached to guanylyl cyclase, activating the enzyme to form cGMP. Such increases cause smooth muscle to relax. Through the action of cGMP, NO inhibits platelet aggregation and may function as an endogenous antithrombotic agent in endothelial cells.

Neuronal PrP protein and prions

The PrP protein is found in many tissues but has an important role in maintaining neural cell function. There are two normal variants of the protein in Caucasian populations. One has a methionine, the other a valine at the 129th amino acid. There are two copies of PrP in each cell. An individual may be homozygous (two PrP_{val} or PrP_{meth}) or heterozygous (only PrP_{val} or PrP_{meth}).

A prion is an infective protein similar in structure to PrP which may interfere with or substitute for the PrP protein. The prion causing scrapie in sheep is dissimilar to the somewhat similar prions responsible for bovine spongiform encephalopathy (BSE) and Creutzfeldt–Jacob disease (CJD) in man. Susceptibility to infection with prions may be enhanced in individuals homozygous for PrP_{val}.

Summary

1. Nerve cells, regardless of anatomy or function, signal to each other from one synapse to the other using chemical neurotransmitters, thus inducing an action potential.
2. The signals are received at voltage-gated channel receptors.
3. There are both excitatory and inhibitory neurotransmitters, which include amino acids.
4. Nitric oxide, which is widely distributed throughout the body and has many functions, is a rapidly acting neurotransmitter derived enzymatically from arginine.
5. The PrP protein important in nerve function exists in two forms. This is relevant to bovine prion infection.

Further reading

Barnard, E.A. *et al.* (1991) Molecular biology of GABA$_A$ receptors. Joint Meeting of the British Pharmacological Society and the Association Francaise des Pharmacologistes, Lyon, April 25–27. *Fundamentals of Clinical Pharmacology*, 5, 381.

Edmunds, B., Gibb, A.J. and Colquhon, D. (1995) Mechanisms of activation of glutamate receptors. *Annual Review of Physiology*, 57, 495–519.

Garthwaite, J. (1991) Glutamate, nitric oxide and cell-cell signalling in the nervous system. *Trends in Neurological Sciences*, 14, 60–7.

Gasic, G.P. and Heinemann, A. (1991) Receptor coupled to ionic channels: the glutamate receptor family. *Current Opinion in Neurobiology*, 1, 20–6.

Smith, C.U.M. (1989) *Elements of Molecular Neurobiology*, John Wiley & Sons, Chichester.

Snyder, S.H. and Bredt, D.S. (1991) Nitric oxide as a neuronal messenger *Trends in Neurological Sciences*, 12, 125–7.

Chapter 12

Specific nutritional requirements

12.1 Introduction

The science of nutrition aims to define the overall quantity and individual constituents of nutrition for all of the human population.

Humans deal uniformly badly with a deficiency of a nutrient whether this be the very short term (oxygen), the short term (water) or longer term, e.g. survival times depend upon body stores at the onset of deprivation. The consequence of a deficiency of single or several nutrients from the diet may be lethal over a period of time.

Modern western populations are faced with an excess of food. This may lead to an over-generous intake of nutrients. The manner in which each individual deals with this is in part dependent upon their genetic and isoenzymic constitution. The basal metabolic rate and level of physical activity are also important. The excess intake may be in all nutrients or be confined to a few specific nutrients. The stage of life and health status are also important.

12.2 Growth

- Growth is an increase in physical size, and development of form and function.
- Growth includes whole body, individual organ and cell growth, tissue turnover, and the synthesis of chemicals.
- These activities are controlled by nutrient intake and by hormone activity.

12.2.1 Introduction

Growth is frequently used as a non-specific term to describe the changes associated with the development of form and function. Usually this is an increase in height, weight and composition of body and size of various organs.

Growth requires nutrient utilization to be directed to energy storage, cell multiplication and skeletal utilization. These metabolic fuels are coordinated in their role in growth by systemic and local regulatory factors. Growth includes:

- whole body growth
- individual organ growth
- cellular growth and replication
- tissue turnover and repair

All of these may respond to external stimuli, for example nutrient intake and the hormone environment within the body, locally and systemically.

Ways in which growth is studied are:

- dimensional
- compositional

• developmental or functional

These overlap and are all involved in the development of functional maturity. Growth may readily be measured as length, weight and chemical composition. By such measurements, abnormalities of growth can be identified.

The amount of dietary energy and protein necessary for growth varies from species to species and is influenced by the time taken for the growth phase. Differences between species and the genetically determined rate of proliferation will influence cell number and body mass.

Maternal nutrient supply to the foetus in the latter part of gestation will affect the rate of growth and the composition of the infant. As the foetus develops, the fractional rate of weight gain and the rate of cell proliferation slow markedly. Hypertrophy is the reversible increase in the size of a cell through the accumulation of more structural components. Hypertrophy becomes more important in the more mature foetus, as does the differential growth of different cells, tissue and body protein mass and organs.

Hypertrophy, especially in cell size is also important during postnatal life. **Hyperplasia** is an increase in the number of cells in a tissue or organ, and an increase in the organ size. As the infant grows there is a steady slowing of the fractional rate of weight gain and protein deposition.

The increase in weight from birth to adult maturity is determined by the proportion of the life span spent in maturation and growth. The development of reproductive maturity usually means the end of growth. Humans are somewhat different from other mammals in that body weight increases 15-fold from the birth weight. Protein deposition continues for a quarter of the time between birth and death.

12.2.2 Phases of organ growth

There are three phases of organ growth:

1. Formation of the endodermal, ectodermal and mesodermal layers; these cell lines are then committed to a specific lineage of cell specialization and structure.

2. Differentiation of cell lines into primary differential states.
3. Maturation to full physiological and metabolic function.

An example of the phases of organ growth

In the development of skeletal muscle, which is derived from a population of myoblasts, the formation of cells involves transcriptional activation. Two genes – *myd* (*my*ogenic *d*etermination gene) and *MyoD*$_\lambda$ (*my*oblast *d*etermination gene number *one*) – are activated sequentially at the beginning of muscle growth.

Skeletal muscle differentiation

Cell differentiation continues after the withdrawal of the myoblast nuclei from the cell cycle. The cells differentiate into multinucleated myotubes. Protein products of the differential genes are important in such change, along with promotor–enhancer regions of other genes, e.g. muscle type creatine kinase and α-actin. In the foetus, interactions between fibroblasts, myotubes and growing skeleton lead to the formation of tendon attachments. At birth the skeletal muscles are close to completing the primary stages of differentiation. The muscle then develops into three myosin fibre types. The maturing motor sensory system innervates the muscle to produce externally regulated contractions.

12.2.3 Transmission mechanisms in growth

Communications between cells are important for coordinated behaviour in multicellular organisms. There are numerous signalling molecules. Some regulators, e.g. steroids, retinoids and thyroid hormones, circulate in the blood and act on distant targets. They pass through the plasma membrane and interact with cytoplasmic receptor molecules. The vast majority of extracellular signals, e.g. neurotransmitters, nucleotides, leukotrienes, peptides and prostaglandins, act in a localized manner. These molecules act by combining with specific cell surface receptors.

Insulin-like growth factor (IGF) resembles insulin in hormone structure, receptor structure and action. There are two IGFs named 1 and 2. The gene for IGF-1 is located on chromosome 12 while that for IGF-2 is on chromosome 11. Both are made of a single chain of amino acids with interchain disulphide bridges connected by a C-peptide and a terminal D extension. Both IGFs stimulate the multiplication of a wide variety of cells, and promote processes related to energy storage. This includes amino acid and protein synthesis in muscle, and fatty acid synthesis in the liver. They stimulate processes which are important in skeletal elongation, including synthesis of RNA, DNA, protein and proteoglycan in chondrocytes and cartilage. IGF-2 is important in foetal life, whereas circulating and tissue concentrations are low in the adult except in the brain. IGF-2 may act as a neurotransmitter. The concentration of IGF-1 is low in infants and increases in adults.

Binding of IGF-1 to its receptor activates a whole cascade of cellular events including phosphoinositide-derived messenger molecules. This affects glucose transport and the phosphorylation of substrate proteins which mediate slower processes, e.g. the activation and inactivation of genes. IGF-1 may also activate phospholipase C which hydrolyses cellular membrane phosphoinositides that are important in phosphate energy metabolism. IGF-1 may, through phosphorylation, be involved in the activation of transcription. This is involved in the formation of messenger RNA, movement of mRNA to the cytosol, ribosomal translation of mRNA and the synthesis of protein and the intracellular processing of protein, including glycosylation.

Synthesis of IGF-1 in the liver is a switch mechanism which results in the utilization of nutrients for growth. IGF-1 may amplify the action of endocrine hormones. The level of dietary protein is important in sustaining circulating IGF-1. There is also local production of IGF-1 by many tissues.

Peptide regulatory factor (PRF) has a low molecular weight (< 80 kDa), a short or intermediate range of action, is very specific and has a high affinity for cell surface receptors. The peptide also has the ability to affect cellular differentiation and proliferation. Peptide regulatory factor acts locally on adjacent cells or on the secreting cell. The PRF is often glycosylated, and the glycosylated form has an equivalent action to the free form, e.g. granulocyte macrophage colony stimulating factor. Free forms of erythopoietin do not act equally with the glycosylated form. The free form of follicle stimulating hormone acts as an antagonist or antihormone to the glycosylated form. PRF is now identified to be a group of multifunctional molecules with a range of biological effects.

Transforming growth factor (TGF) was originally described as being capable of inducing phenotypic transformation of untransformed cell lines. TGF-α is now known to be expressed during embryonic development and in adult pituitary cells, epithelial cells and macrophages.

PRF binds to specific high-affinity cell surface receptors; the effect is:

- a post-receptor signalling mechanism which alters the pattern of gene expression
- ligand–receptor complexes are internalized and increase activity or affect down-regulation that may act as a negative feed-back loop.

The effect of a particular PRF on a single cell type is not necessarily the same on all occasions. The effect depends on the background set by other signalling molecules which communicate by a signalling language read by the cell.

Some PRFs may down-regulate other unrelated receptors, which is called **transmodulation**. A cascade of PRF may amplify the original stimulus or lead to a feed-back inhibition of the primary mediator. The mechanism is partially understood for muscle contraction and glycogen metabolism. The main regulatory mechanism is through the phosphorylation and dephosphorylation of proteins. The initial transmembrane signalling process is transmitted to long-term patterns of gene expression. This regulation leads to cell proliferation and differentiation.

Phosphorylation of tyrosine residues on proteins is less frequently the control mechanism on the protein than the phosphorylation of serine or threonine. This means that in the regulation of enzymes, phosphorylation of only a limited fraction of the total molecule may significantly change the catalytic activity and so initiate a regulatory cascade.

One signalling pathway in the regulation of cell growth is hydrolysis of the membrane lipid phosphatidylinositol 4,5-bisphosphate, a lipid kinase which phosphorylates phosphatidylinositol.

Serine kinase is a cell cycle regulator and is essential to the start of the cell cycle at G1 and/or the entry of cells into mitosis. Phosphorylation of the tyrosine residues of the enzyme results in increased protein kinase activity, reaching a maximum as cells progress from S phase towards mitosis.

Some receptors transmit their signals through GTP-dependent coupling, e.g. control of cyclic GMP phosphodiesterase, phospholipase A_2 and membrane channels.

Hormone activation of cyclic AMP formation involves sequential action of three proteins:

- receptor
- a guanine-nucleotide-dependent coupling protein: G-protein, α, β and γ
- adenylate cyclase

The receptor is a transmembrane glycoprotein in which, when a hormone is bound, a conformational change takes place. This is transmitted through the receptor protein to the signalling domain. This system can transmit a large amount of information to the cell during regulation. Each stimulus has its own receptor but the receptors respond to each other. All the receptors that interact with G proteins have a common amino acid pattern. Each domain represents one transit of the folded polypeptide chain through the plasma membrane.

12.2.4 Organ growth controls

Increased physiological activity leads to organ enlargement, while disuse leads to atrophy. Trophic hormones are secreted when physiological demand is increased. These stimulate target organs to greater functional activity, trigger cell hypertrophy and hyperplasia. Some organs react to the physiological demands to which they are committed. The parathyroid hormone responds directly to serum calcium, islets of Langerhans to plasma glucose, and zona glomerulosa of the adrenal cortex to sodium. Skeletal muscle and bone similarly respond to need. Other organs are very complex in the regulatory mechanisms of growth, e.g. the liver

Differences in growth control

There are differences in growth control between somatic and visceral organs and tissues. Somatic tissues – muscle, heart, blood vessels, bone, skin and connective tissue – undertake the mechanical functions of the body. Their growth is controlled by local influences. Visceral organs are found in the body cavities surrounded by somatic tissue. These are involved in secretions and biochemical processing, and include blood cells, endocrine organs and some exocrine organs including the liver and lungs. Their growth is regulated by more systemic hormonal and nutritional influences.

12.2.5 Overall body growth

Growth is only possible if energy intake exceeds energy expenditure. Waterlow (1961) has shown that the rate of weight gain depends more closely on the intake of calories than of protein (within a range of 100–200 kcal/kg/day and protein intake of 2–7 g/kg/day; in children, 150 kcal/kg/day and a protein intake of 3–4 g/kg/day is adequate). The requirement for weight gain is less in adults. Weight gain of lean tissue is quite different from weight gain of adipose tissue.

Body and tissue chemical composition

As the organ and body mass increase during development, so the hydration of the body decreases. The distribution of water through the tissues alters, with more being found in intracellular compartments. The nitrogen : weight ratio is a measure of relative chemical maturity. As maturity approaches the protein deposition slows and fat storage increases in importance. Fat stores require little accompanying deposition of water.

Height

There are many important factors affecting height which include ethnic, genetic, dietary and environmental aspects. What is not known is whether the achievement of maximal height is healthy. Does maximum height and maximum growth make for longevity? Does such realization of potential affect the propensity to obesity, hypertension, diabetes, coronary artery disease and diet related cancers? Obviously, short stature due to bad nutrition or illness is undesirable. This does not mean that feeding children for maximum growth and physical development is advantageous to health and longevity, e.g. it has been shown that energy restriction of as little as 10% subsequently reduces tumour incidence in rodents.

Growth and function

Throughout life, organs, e.g. the skeleton, liver and kidney and small intestine, respond to the needs of the body in relation to environment and workload. Both hyperplasia (increase in cell number) and hypertrophy (increase in cell size), are stimulated. The heart and liver are relatively larger in relation to total body protein cell mass at birth than at later stages of development. The intestine and skeletal muscle grow more after birth than *in utero*. The long bones appear to reach a genetically determined length irrespective of growth in other organs.

Growth regulation is divided between genetic–temporal and functional aspects. During foetal life, protein disposition is less sensitive to maternal nutritional deprivation than after birth and lactation. At weaning, when the change to an adult expression of gene expression is virtually complete, organ growth, e.g. muscle, is very sensitive to undernutrition.

From ancient times it has been believed that the span of the arms and the height of a normal man are equal. Vitruvius, the Roman architect, claimed that the perfect proportion of a man was for his span and height to fit within a square or a circle. Such human proportions were central to the artistic and architectural traditions of classical art. Height is greater than span up to age 4, equal at that age and the span is greater than height thereafter. This is the case in two-thirds of Caucasian adults. Another measurement of functional value is the belief that the child is ready to go to school when able to touch the left ear by placing the right arm over the head.

The child who is not growing to expected potential is identified from the use of centiles and height and weight charts over a defined period of months or years. Reduced growth can occur for social and economic reasons, familial trends, malnutrition, chronic inflammatory disease, renal insufficiency, cardiac causes and endocrine insufficiency.

Growth hormone deficiency as a consequence of pituitary deficiency can be treated by recombinant DNA-produced hormone. When the illness is overcome, the child may to some extent catch up the deficient growth (catch-up growth). The catch-up is dependent upon the age of the child and the length of the illness. The earlier in life and the shorter the illness the better.

Catch-up growth

This is accelerated weight gain and is primarily seen in three areas:

1. Preterm infants where the object is to achieve rates of growth similar to those which would have normally occurred *in utero*.
2. Children recovering from severe undernutrition.
3. Adults recovering from the stress of trauma, infection or surgery.

The diet for catch-up will reflect the needs of depleted tissue and varies for different tissues. There is relative preservation of visceral tissue at the expense of muscle and adipose tissue.

Immature animals will seek to obtain and use nutrients to enable a rate of protein disposition to achieve maximum growth. After birth the rate of protein development will depend upon the quantity and biological quality of the essential amino acids and dietary proteins. These enable the genetic programme to be performed.

There are a series of genetically and developmentally regulated priorities for growth. Nutrition is important in the achievement of the genetic potential of the body as a whole and for individual organs. Nutritional deprivation depletes the body of initially fat, then protein. These differences are reflected in changes in the intracellular and extracellular water spaces.

Total body water is used to measure lean tissue mass. During the early phase of rapid weight gain there is a disproportionate increase in body hydration which returns to normal once recovery is established.

Some amino acids, e.g. infused arginine, are claimed to increase growth hormone secretion. Oral ornithine is believed to be twice as effective as oral arginine.

Hormones and growth

Farm animals can be induced to grow faster and with more lean meat content with the use of oestrogenic hormones, oestradiol benzoate and progesterone or testosterone. These increase growth rate by 8–15% and feed conversion by 5–10%. Hormonal implants include testosterone analogues and oestrogenic-like substances. The animals have to be adequately fed to achieve the full benefits of the treatment. Such stimulant effects of hormones has also been noted by certain sportsmen and women.

Summary

1. Growth is an increase in weight, height, composition of body organs, cells and tissue repair. This requires appropriate nutrient intake of energy and protein.
2. Hypertrophy is a reversible increase in cell size. Hyperplasia is an increase in the number of cells. Organ growth is in three phases, the development of cell lines, differentiation and maturation to mature function.
3. Growth at cellular and organ level is a coordinated process, controlled by various specific signalling molecules, many of which are hormones. The growth process at maturity is in response to need. Disuse leads to atrophy.
4. Growth is only possible if energy intake exceeds expenditure. Height is dependent upon sex and ethnic, genetic, dietary and environmental factors
5. Catch-up growth requires a high quality protein and energy intake. This growth spurt is seen in children who, following failure to achieve their age growth norms, are subsequently well and eating sufficient for growth.

Further reading

Burns, H.J.G. (1990) Growth promotors in humans. *Proceedings of the Nutrition Society,* **49**, 467–72.

Buttery, P.J. and Dawson, J.M. (1990) Growth promotion in farm animals. *Proceedings of the Nutrition Society,* **49**, 459–66.

Carter-Su, C., Schwarz, J. and Smit, L.S. (1996) Molecular mechanism of growth hormone action. *Annual Review of Physiology,* **57**, 187–208.

Goss, R.J. (1990) Similarities and differences between mechanisms of organ and tissue growth regulation. *Proceedings of the Nutrition Society,* **49**, 437–42.

Jackson, A.A. (1990) Protein requirements for catch-up growth. *Proceedings of the Nutrition Society,* **49**, 507–16.

Loveridge, N., Farquharson, C. and Scheven, B.A.A. (1990) Endogenous mediators of growth. *Proceedings of the Nutrition Society,* **49**, 443–50.

Phillips, L.S., Harp, J.B., Goldstein, S., Klein, J. and Pao, C.-I. (1990) Regulation and action of insulin-like growth factors at the cellular level. *Proceedings of the Nutrition Society,* **49**, 451–8.

Reeds, P.J. and Fiorotto, M.L. (1990) Growth in perspective. *Proceedings of the Nutrition Society,* **49**, 411–20.

Schott, G.D. (1992) The extent of man from Vitruvius to Marfan. *Lancet,* **340**, 1518–20.

Symonds, M.E. (1996) Regulatory factor in the control of development and maturity. *Proceedings of the Nutrition Society,* **55**, 519–70.

Walker, A.R.P., Walker, B.F., Glatthaar, I.I. and Vorster, H.H. (1994) Maximum genetic potential for adult stature that is aimed as desirable. *Nutrition Review,* **52**, 208–15.

Waterlow, J.C. (1961) The rate of recovery of malnourished infants in relation to the protein and calorie levels of the diet. *Journal of Tropical Pediatrics,* **7**, 16–22.

Wright, N.A. (1992) in *Oxford Textbook of Pathology* (eds J.O'D. McGee, P.G. Isaacson and N.A. Wright), Oxford University Press, Oxford.

12.3 Bone structure

12.3.1 Introduction

The skeleton provides scaffolding which enables a mobile terrestrial life. The skeleton consists of bones, joints, tendon sheaths and fascia. The bony skeleton forms the attachment for tendons which enable muscles to exert their movements.

Bones consist of 35% mineral salts (largely calcium and phosphorus), 20% organic matter (mostly collagen) and 45% water; 99% of the body calcium is found in bone. The bones of the skeleton are responsible for:

- support to the body as a whole
- support to muscles
- providing distances between muscle attachments to enable coordinated movement
- movement at joints
- protection
- a reserve of calcium and other minerals
- bone marrow to produce blood constituents

Every bone will have some role in each of these functions but some will be more conspicuous in a particular function than others:

- The vertebrae support the whole body.
- The long bones of the limbs, humerus, radius, ulna, tibia, fibula and femur, support muscles of movement and give height to the individual; their shape is designed for an upright stance.

Almost all bones have **joints**. In the long bones these are at either end. The bones of the wrist and ankle are small and largely articular surfaces. Some bones have only one articular surface, e.g. the scapula.

The protective function of the bones is best illustrated by the pelvic bone, skull, vertebrae and sternum and ribs. The pelvic bones protect the ovaries, uterus, bladder and rectosigmoid region of the colon. The skull and vertebrae protect the brain and spinal column and the sternum and ribs, the heart and lungs. Each bone has a blood supply, which has a nutrient role

A long bone (Figure 12.1) consists of:

- a shaft or diaphysis
- an epiphysis, which is the end of the bone and includes the articular surface

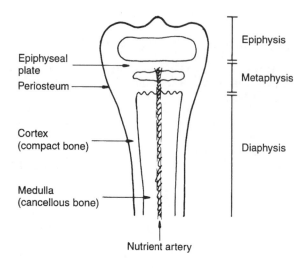

Epiphyseal plate

Periosteum

Cortex (compact bone)

Medulla (cancellous bone)

Nutrient artery

Epiphysis

Metaphysis

Diaphysis

FIGURE 12.1 Cross-section through the head of a long bone. The structure is initially cartilage, which is replaced by bone. Osteoclasts erode cartilage and bone matrix; osteoblasts secrete bone matrix.

- a metaphysis, between the diaphysis and the epiphysis, a junctional zone
- an epiphyseal plate, cartilage between the epiphysis and the metaphysis; this is important in growing bone

12.3.2 Bone tissue and cells

Bone is formed in three layers (Figure 12.1):

- The **periosteum** is a fibrous membrane covering the bone.
- The **cortex** is the hard sheath of compact bone.
- The **medulla** contains the marrow and spongy (cancellous) bone.

Microscopic structure of bone

Mature bone

In both compact and cancellous bone the bone is lamellar. Collagen fibres are arranged in parallel sheets (lamellae). In cortical bone the lamellae are wrapped around the blood vessels in a concentric manner. These cylinders (Haversian systems or osteons) lie in the long axis of the bone. Between each osteon are interstitial lamellae. Cancellous

bone is made of lamellar bone separated by cement lines.

In immature bone, growing bone, repaired fracture and pathological conditions, woven or immature bone is found. There is random packing of the collagen and the matrix is rich in ground substance. This woven bone is not strong and is replaced by lamellar bone.

Bone cells

There are three types of bone cell:

- **Osteoblasts**: these line the external surface of the bone trabeculae, the inner aspect of the periosteum and the surface of bone lining the osteons. They produce collagen and proteins forming the organic bone matrix. They deposit and remove calcium.
- **Osteocytes**: these are osteoblasts trapped in mature bone, which cannot synthesize protein but have a role in mature bone maintenance and are the sensors for the changes in mechanical demand.
- **Osteoclasts**: these cells are responsible for bone resorption.

The roles of osteoblasts and osteoclasts are closely related and act in equilibrium.

12.3.3 Bone structure

Bone consists of a lamellae and a mineral matrix of calcium hydroxyapatites deposited on the protein collagen matrix of bone. This gives bone rigidity. The organic lamellae is 90% collagen (largely type I) produced by the osteoblasts. The synthesis of collagen is under the control of several genes. Collagen undergoes a series of post-translational modifications before being deposited in the growing bone. This collagen is then calcified between the collagen molecules. Bone contains other proteins, phosphoproteins and lipids.

Many of the trace elements (Zn, Mn and Cu) are required for the growth, development and maintenance of healthy bone. Lead and cadmium have a toxic effect on bone. Fluoride increases bone density in advanced osteoporosis. Fluorine readily substitutes for the hydroxyl ion in bone hydroxyapatite, creating a less acid-soluble, more stable crystal. Fluorine does not diffuse into bone but may be incorporated during growth. Hydroxyapatite crystals run parallel to collagen fibres in bone, whereas fluoroapatite crystals run perpendicularly to the fibres. Fluoride increases osteoblast number and bone formation.

12.3.4 Bone formation and growth

Bone formation

Foetal bone

Connective tissue is ossified, i.e. calcified by intra-membranous or endochondrial ossification. The resultant woven bone is remodelled by osteoclasts to form the mature skeleton. In flat bony structures, as in the skull and scapula, the centres of ossification are not on articular surfaces but expand to form a continuous mass of woven bone.

Endochondrial ossification occurs in the epiphyseal growth plate and is how long bone, vertebrae and the pelvis grow. Cartilage is laid down and then calcified. This is a growth process and when growth stops at maturity, the epiphysis closes. Completion of calcification occurs in different bones at different intrauterine ages.

Bone ossification after birth
The baby must be flaccid to pass through the birth canal without harm to the baby or the mother. The timing of the completion of ossification in a bone enables the bone age as compared with the chronological age of a person to be measured. This is important in nutritional assessments.

For epiphyseal closure in bones with epiphyses at either end, the first epiphysis to appear is the last to close. Growth continues longest at the shoulder, knee and wrist. Different parts of the bone complete ossification at different ages.

Ossification is best determined by X-ray examination. Full details of the expected age of ossification are to be found in atlases of bone ossification, available in radiology departments.

Bone remodelling is a continuous process during adult life. There is a 25% turnover of cancellous bone and 2–3% of cortical bone. The rate varies with age and the particular bone. Bones exposed to stress, e.g. weight-bearing and exercise, will remodel at a different rate to less pressured bones.

Ossification of the ulna and radius

Age of ossification in sections of bones

Primary centre in shaft		7 weeks
Secondary centre	lower end	year 2
	upper end	year 5
	closure upper epiphysis	year 18
	closure lower epiphysis	year 20

Bone remodelling

This begins with the retraction of the lining cells, derived from osteoblasts that cover bone surfaces. Mononucleated osteoclast precursors fuse on the denuded surface to form differentiated osteoclasts. Once activation has started, a volume of bone is replaced over a 2–3-week period. Then the osteoclasts are replaced by mononucleated cells preparatory to new bone formation on lacunar surfaces and the summoning of osteoblast precursors to the resorption lacunae. This is a phase of reversal, converting to formation controlled by poorly understood systems. Osteoblasts fill the cavity with new organic matrix or osteoid which after an interval of 25–35 days becomes mineralized. This cycle takes several months and results in a new area of cancellous bone and a new Haversian system in cortical bone.

Bone growth

Skeletal growth and metabolism are dependent upon a number of factors:

- **Calcium control**: factors include parathyroid hormone, calcitonin and 1,25-dihydroxycholecalciferol
- **Control of skeletal growth**: factors include growth hormone; insulin-like growth factor-1; thyroid hormone; adrenal corticosteroids; androgens; oestrogens; vitamin C; vitamin A
- **Control mechanisms** (the overall rate of remodelling): factors include parathyroid hormone activity; 1,25-dihydroxyvitamin D; calcitonin

Long bone

At birth, the long bone columnar structure of the epiphyseal growth plate is incompletely formed. Longitudinal growth results from the increased production of cells (hypertrophic chondrocytes) originating in the germinal layer which then becomes the proliferative layer of the bone. These cells produce the cartilaginous matrix which then becomes calcified, resorbed and replaced by osteoblast-mediated bone formation. Bone longitudinal growth is dependent upon the increased production of cells and their subsequent expansion. After birth, bone growth is activated by growth hormone.

Growth hormone and insulin-like growth factor are found either free or bound to protein in the circulation. There are two binding proteins (> 100 kDa and 80–85kDa) The 80–85 kDa protein is responsible for 80–90% of the growth hormone binding. The concentrations of these two proteins vary from subject to subject and in sickness and in health. The levels are low in neonates but increase at 1 year. The biological significance of these proteins is not known but they may have a protective role and prolong the half-life of the hormones.

Vitamin D influences the availability of calcium with effects on bone metabolism. The degree of mechanical loading also affects bone modelling. Vitamin C is essential for the synthesis of connective tissue, through the hydroxylation of pro-line and lysine. Vitamin A stimulates osteoclast bone resorption.

12.3.5 Bone mass and age

Bone mass increases rapidly during the first year of life and during the rapid growth at the time of puberty. Pubertal growth is greater in the male than the female. The most critical needs during the growth phase are dietary proteins. Vitamin D and calcium provision are less important, but of significance.

The pubertal spurt occurs for 2 years after the menarche in girls with a maximum velocity at 12.5 years and for 2 years in the male with a maximum at 14 years. During this period calcium accumulates in the bone of girls at about half the rate of boys. By the age of 16–18 years calcium retention falls. Dietary supplementation of calcium during the pubertal spurt in children results in significant increases in bone mass which may well be of very long standing value, even until after the menopause.

Peak bone mass – the amount of bone at the completion of linear growth – is under genetic control as well as the dietary provision of calcium during growth and mechanical stress (Figure 12.2). There does not appear to be a genetic component to bone loss rate. Bone loss is greater in the female than the male of the same age.

Peak bone mass is on average achieved at the ages 35–44 years for cortical bone and somewhat

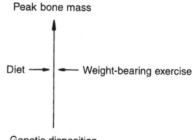

FIGURE 12.2 Peak bone mass is dependent upon diet and weight-bearing exercise, but most of all the genetic disposition, possibly mediated through the 1,25-dihydroxy vitamin D-specific receptors.

earlier for trabecular bone. White males have a 5–30% greater peak bone mass than females, and Afro-Asians 10% greater than Caucasians. Age-related bone loss results from gradual thinning of trabeculae through declining osteoblastic function.

Bone strength and the menopause

Rapid postmenopausal bone loss is osteoclast in origin, through increased activity as oestrogen activity declines. Once osteoclastic resorption has breached the trabecular plate then new bone formation is not possible and tissue is rapidly lost. Trabeculae are separated and intertrabecular connective tissue is reduced. The rapid disruption of trabecular microarchitecture after the menopause seriously reduces bone strength.

Each year, 0.3% cortical bone loss occurs in males and females up to the menopause, when there is a decrease in oestrogen production and an accelerated loss in females. The periosteal circumference increases slowly but there is also an expansion of the medullary cavity. Osteoclastic activity is important in such change. The postmenopausal bone loss is independent of calcium intake, but this is a somewhat controversial area.

Bone loss begins in the fourth or fifth decades of life varying from one skeletal site to another. The mass of the skeleton shrinks and in advanced old age has fallen to approximately 70% of the value in youth. This is reflected in a deterioration of the cortical bone area to the total bone area in bones such as the metacarpal bone.

During confinement to bed and during space travel (see p. 524) there is a negative calcium balance. A reduction in bone density can be modified by exercise. The relationship between bone density and dietary calcium intake is not immediate or direct. Calcium supplementation in the mature individual only benefits those with a particularly pronounced life-long deficiency in dietary calcium.

With age, the control of calcium intake and output is less efficient, and intestinal calcium absorption and the renal ability to hydroxylate 25-hydroxyvitamin D declines. Secondary hyperparathyroidism may then occur. This may stimulate bone remodelling through activation of osteoclast activity and with reduced available calcium the result is bone thinning. In contrast, in the postmenopausal state, parathyroid concentrations fall with a resultant fall in 1,25-dihydroxyvitamin D concentration followed by reduced calcium absorption. The rate of bone loss may be accelerated by several factors such as smoking and loss of oestrogen production. Low bone density will also coincide with ageing and frailty, especially in the female – who has an increased risk of bone fractures. The low bone density, in addition to increasing the risk of fracture, is likely to be a marker of other variables, including diet, which influence mortality, for example old women with a low bone density are at risk of dying of non-traumatic causes, e.g. stroke. Declining bone structure is exaggerated in osteoporosis where there is a reduction in bone density.

12.3.6 Osteoporosis

Clinical osteoporosis is the thinning of the bone with normal mineralization, severe enough to lead to fracture with minimal trauma. For a fracture to occur, the trauma must be greater than the strength of the bone. The factors which increase the chance of fracture include reduced bone mass, loss of trabeculae in cancellous bone and skeletal fragility, trauma and the type of fall. Protective muscle mass, which declines in the elderly is protective from the effects of trauma.

Ovarian function (menopause, surgical removal, athletic pursuits, stress-related amenorrhoea, anorexia nervosa) is important in retaining bone homeostasis, the rationale for hormone replacement therapy (HRT).

Corticosteroid osteoporosis in patients treated with long-term corticosteroids, Crohn's disease, ulcerative colitis, and autoimmune chronic hepatitis may be due to inhibition of osteoblast activity and a decrease in intestinal calcium absorption.

Bone density of all regions of the hip is strongly related to risk of hip fracture. There is a 2.6-fold increase in risk for each standard deviation reduction in bone density. A woman whose femoral neck bone density is at the 10th percentile would have a 25% lifetime risk compared with an 8% risk for a woman at the 90th percentile. The decline in bone density with age only partially accounts for the increasing risk of hip fracture with age. There is a doubling of the risk of fracture with age, but falls directly increase bone fractures. It has been suggested that present-day women have a reduced bone density compared with previous generations.

A study of proximal femur bone density in 100- to 200-year-old human skeletal material of known age at death and from a population whose way of life is well documented

A church in London was being restored, which required the temporary removal from a crypt of bodies buried between 1729 to 1852. Dual energy X-ray absorptiometry showed the bone density to be greater in the previous generation than the present in both premenopausal and postmenopausal women. The physical activity of the 18th and 19th century women was hard, many working as weavers.

Exercise – especially of the weight-bearing type – is important in maintaining bone density. Parity conserves bone. Smoking, alcohol and poor diet are predisposing factors in reducing bone density.

The incidence of vertebral fractures is up to six times as common in women than men. Men are protected for several reasons. Age for age the bone density is higher, in part because the bones are larger and in part because androgens have a direct effect on chondrocytes of the growth plate and on osteoblasts. The anabolic effect of androgens on muscles increases the stress on muscles during exercise and increased bone deposition. The production of testosterone and oestradiol continues well into old age in the male. Men die on average five years earlier than women, which reduces the risk of developing an age-related condition. When men do present it is with vertebral crush fractures. Predisposing illnesses include treatment with steroids, and malabsorption, e.g. following gastric surgery. Heavy tobacco smoking, excessive drinking, a low calcium diet and a slothful lifestyle are other factors. Curiously enough, obesity is protective in men. Steroids inhibit bone formation whereas osteoclastic activity continues. Hypogonadism leads to trabecular perforation rather than thinning.

Treatment with fluoride does not influence the fracture rate compared with controls. Fluoride therapy in osteoporosis increases unmineralized osteoid and delays mineralization. New bone formation occurs in the axial skeleton, which is mostly trabecular bone. The new bone matrix has a woven rather than lamellar appearance. The therapeutic band is narrow at 30–50 mg NaF/day. The treated bone looks similar to those in osteomalacia but calcium and vitamin D supplements do not alter the picture.

A 5% increase in peak bone mass would lead to a 40–50% reduction in fracture rate. The calcium intake during childhood and adolescence may be important in preventing fractures in the elderly. Increased dietary intake of milk in childhood results in increased bone mass in adult life. Early exposure to a good calcium intake can influence bone density as an adult. It is possible that the present USA RDA for calcium intake for children to young adults is half that required, at the present recommendation of 800 mg/day.

12.3.7 Osteomalacia

This is a failure of mineralization of bone. Rickets is a condition which develops in the growing child affecting both bone and growth cartilage leading to distorted bones. This is due to vitamin D deficiency (see p. 230).

Osteomalacia in the adult results in thin bony cortex and pathological fractures which are often incomplete (Looser's zones). These fractures do not unite until the nutritional problem is resolved. An important feature is osteoid mineralization failure. However, other complicating bony changes may take place as there may be multiple associated deficiencies.

Osteomalacia and osteoporosis maybe complicated by chronic gastrointestinal and liver diseases including coeliac disease, Crohn's disease, partial gastrectomy and pancreatic insufficiency.

Primary biliary cirrhosis and other chronic conditions of bile insufficiency, e.g. biliary atresia and hypoplasia are important causes of osteomalacia and bone demineralization. This is because insufficient calcium phosphate and vitamin D absorption will result from maldigestion and malabsorption. Symptoms and signs of osteomalacia include bone pain, muscle weakness, hypophosphataemia, increased serum alkaline phosphatase and parathyroid hormone concentrations.

12.3.8 Diagnosis of osteoporosis and osteomalacia

Bone demineralization is diagnosed by a number of methods. A 'rough and ready' screening test for the development of osteoporosis is reduction in height. Women's height peaks at 27 years and men's at 21 years. Men of predominantly north-ern European stock will be 2.6 cm shorter at age 60 and 6 cm shorter at age 80; women will be 2.6 cm shorter at 60 and 6.7 cm shorter at 80. Assessments of bone include bone mineral density, total lumbar spine (dual energy X-ray absorption), plasma alkaline phosphatase, plasma osteocalcin and serum parathyroid hormone.

The only method to distinguish between osteomalacia and osteoporosis is with a full-thickness iliac crest bone biopsy.

Early bone demineralization can be detected by radiogrammetry, photonabsorptiometry and tomography. Radiometry provides magnified views of bone, particularly metacarpal bones, and indicates the thickness of cortical bone. However, it does not indicate total skeletal mass. [125]Iodine photon absorptiometry is typically used for measurements on the radius of the non-dominant arm. This precise accurate method correlates with total skeletal mass, but not trabecular bone in the iliac crest or vertebral bodies. It does not distinguish between osteoporosis and osteomalacia. A complicating factor is that peripheral bones, arm and leg bones, axial bones and the pelvis and vertebrae may react differently to deficiencies. Radiogrammetry and ^{125}I-photon absorptiometry may be insensitive in the early detection of osteoporosis where the axial skeleton may be more involved.

Summary

1. The skeleton consists of bones, joints, tendon sheaths and fascia. The bony skeleton forms the attachment for tendons which enable muscles to exert their movements. Bones consist of 35% mineral salts (largely calcium and phosphorus), 20% organic matter (mostly collagen) and 45% water; 99% of the body calcium is found in bone.
2. Each bone has a blood supply, which has a nutrient role. A long bone consists of a shaft or diaphysis; an epiphysis with an articular surface; a metaphysis; a junctional zone and an epiphyseal plate; cartilage between the epiphysis and the metaphysis is important in growing bone.
3. Bone is formed in three layers, periosteum, cortex and medulla. Periosteum is a fibrous membrane covering the bone. The cortex is the hard sheath of compact bone. The medulla contains the marrow and spongy (cancellous) bone.
4. There are three types of bone cells. Osteoblasts line the external surface of the bone trabeculae, the inner aspect of the periosteum and the surface of bone lining the osteons; they produce collagen and proteins forming the organic bone matrix, and deposit and remove calcium. Osteocytes are osteoblasts trapped in mature bone; they have a role in mature bone main-

tenance and are the sensors for the changes in mechanical demand. Osteoclasts are responsible for bone resorption.

5. Bone consists of lamellae and a mineral matrix of calcium hydroxyapatites deposited on the protein collagen matrix of bone; this gives bone rigidity. The organic lamellae is 90% collagen (largely type I) produced by the osteoblast. This collagen is then calcified between the collagen molecules. Bone contains other proteins, phosphoproteins and lipids.

6. Bone is being continuously remodelled throughout life. The rate varies with age and the particular bone. Bones exposed to stress, e.g. weight-bearing and exercise, will remodel at a different rate to less-pressured bones.

7. With age, bone develops osteoporotic changes wherein bone structure and density decline with a continued calcium content.

8. Bone calcium content declines in osteomalacia.

Further reading

Anderson, D. (1992) Osteoporosis in men. *British Medical Journal,* 305, 489–90.

Anderson, J.J.B. (1992) The role of nutrition in the functioning of skeletal tissue. *Nutrition Review,* 50, 388–94.

Beattie, J.H. and Avenell, A. (1992) Trace element nutrition and bone metabolism. *Nutrition Research Reviews,* 5, 167–88.

Branca, F., Robins, S.P., Ferro-Luzzi, A. and Golden, M.H.N. (1992) Bone turnover in malnourished children. *Lancet,* 340, 1493–6.

Browner, W.S., Seeley, D.G., Vogt, T.M. and Cummings, S.R. (1991) Non-trauma mortality in elderly women with low bone density. *Lancet,* 338, 355–8.

Cullen, K. (1992) Motivating people to attend screening for osteoporosis. *British Medical Journal,* 305, 521–2.

Cummings, S.R., Black, D.M., Nevitt, M.C., Browner, W., Cauley, J., Ensrud, K., Genant, H.K., Palermo, L., Scott, J. and Vogt, T.M. (1993) Bone density at various sites for prediction of hip fractures. *Lancet,* 341, 72–5.

Dempster, D.W. and Lindsay, R. (1993) Pathogenesis of osteoporosis. *Lancet,* 341, 797–801.

Johnston, C.C. and Slemenda, C.W. (1992) Changes in skeletal tissue during the ageing process. *Nutrition Reviews,* 50, 385–7.

Leader (1992) Maximizing peak bone mass: calcium supplementation increases bone mineral density in children. *Nutrition Reviews,* 50, 335–7.

Lees, B., Molleson, T., Arnett, R.A. and Stevenson, J.C. (1993) Differences in proximal femur bone density over two centuries. *Lancet,* 341, 673–5.

Loveridge, N., Farquharson, C. and Scheven, B.A.A. (1990) Endogenous mediators of growth. *Proceedings of the Nutrition Society,* 49, 443–50.

McGee, J. O'D., Isaacson, P.G. and Wright, N.A. (1992) *Oxford Textbook of Pathology,* Oxford University Press, Oxford.

Reeve, J., Green, J.R., Hesp, R. *et al.* (1990) Determinants of axial bone loss in the early post menopause: The Harrow postmenopausal bone loss study, in *Osteoporosis 1990* (eds C. Christiansen and K. Overgaard), Osteopress, Copenhagen, pp. 101–3.

Schneider, V.S., LeBlanc, A. and Rambaut, P.C. (1989) Bone and mineral metabolism, in *Space Physiology and Medicine* (eds A.E. Nicogossian, C.L. Huntoon and S.L. Pool), Lea & Febiger, Philadelphia, pp. 214–21.

Weatherall, D.J., Ledingham, J.G.G. and Warrell, D.A. (1996) *Oxford Textbook of Medicine,* 3rd edn, Oxford University Press, Oxford.

12.4 Pregnancy

- Good maternal nutrition before and during pregnancy is important both for the developing baby and the mother.
- Malnourished teenage mothers are particularly at risk of failing to meet good nutritional standards.
- Pica, a pathological craving for unconventional or curious foods, may occur in pregnancy.

12.4.1 Introduction

The chances of pregnancy are reduced if there is a reduction in energy intake to less than 6.3 MJ/day. Such restriction, as in anorexia nervosa, inhibits ovulation. The lactational amenorrhoea method of contraception is based on the belief that a mother who is breast feeding, or almost fully breast feeding, her infant and who remains amenorrhoeic has a less than 2% chance of pregnancy during the first 6 months after childbirth.

12.4.2 Mother

An expectant mother needs a good mixed diet, both before and during pregnancy, adequate in total number of calories, protein, vitamins and minerals. Her nutrition well before pregnancy is most important as this will establish the long-term prospects for her baby both as a child and into middle and even old age. During pregnancy weight and diet should be monitored. Iron, calcium and folic acid supplements (as milk) are important.

A woman should gain weight during the first two trimesters and thereafter gain 0.5 kg/week until term, with a total gain of 12.5 kg, made up of 4.8 kg foetus, placenta and liquor, 1.3 kg uterus and breasts, 1.25 kg blood, 1.2 kg extracellular water and 4 kg fat. When the expectant mother is obese, hypertension is a dangerous complication. Women who are under weight at the beginning of pregnancy should gain weight at a greater rate than normal-weight women, i.e. 0.58 kg/week. Adolescent mothers should also put on weight at a rate similar to the underweight group to allow for normal growth of the mother.

Under-nutrition, particularly leading to rickets during childhood and adolescence, may lead to stunted growth with a resulting small pelvis. The passage of the baby through this distorted pelvis may lead to a difficult and prolonged labour. This is now a rarity in Western obstetric practice.

Mortality during pregnancy is considerable in many developing countries, where between 20–45% of deaths among women of child-bearing age are related to pregnancy. This compares with less than 1% in the UK. Women in rural Africa have a 1 in 21 chance of dying of pregnancy-related causes, in South Asia and Latin America 1 in 38 and 1 in 73 respectively, which compares with 1 in 6366 for the United States and 1 in 9850 for Northern Europe. The reasons for this are poor education, poverty, insanitary conditions and malnutrition. World-wide, 500 000 women die of pregnancy-related causes annually, 99% in the developing countries. Bangladesh, Bhutan, India, Nepal, Pakistan and Sri Lanka account for 43% of this total.

12.4.3 Foetus

The weight of the baby is influenced by the protein–energy intake of the mother. In the second trimester the mother should be laying down fat in preparation for later foetal requirements. Protein-calorie malnutrition at this stage may have consequences for later infant development. In the third trimester the mother's weight gain affects the infant's final weight. This underlines the significance of the timing of maternal dietary supplementation, i.e. first, second or third trimester.

If a developmental process be restricted of its fastest rate by any agency at any time, not only will this delay the process, its ultimate value will also be restricted. This applies even when the restricting influence is removed and the fullest rehabilitation obtained.

The 'vulnerable period hypothesis' states that when there is stress to the pregnancy, then the older the foetus the better it can cope with that stress. This is reflected in the potential for development, e.g. height, number of cells in certain organs, as well as the rate of growth in height or number of cells. Some developments occur at only one period, others during two separate periods, e.g. increase in height and weight occur both during infancy and at puberty. If a stress occurs at an early stage in the growth phase then a catch-up is possible.

Foetal brain development

The development of the baby's brain and neurological system, both *in utero* and during the early stages of extrauterine life are at risk during the vulnerable periods. The brain undergoes neuroblast development during the first trimester and nerve glial development occurs during the third trimester continuing into the first year of life. Nutritional deficiency can lead to psychomotor retardation. The later period of gestation has been likened to a culture medium. Excess or deficient availability of nutritional growth factors may have long-term consequences which extend into even middle life. Premature babies may require specialized feeding in order that growth and development should be normal.

12.4.4 Birth weight

The birth weights of infants of well-fed mothers are on average higher than those of poorly fed women. The baby is particularly at risk where the mother is herself a young girl 12–14 years old, and who has herself not completed growth and development before pregnancy. The reduced weight of the baby may be due to prematurity or to retarded intrauterine growth. If the mother's weight is less than 40 kg then the baby is particularly at risk of low birth weight. When a baby's weight is less than 2500 g the chances of survival are reduced. Infant deaths are nine times higher in the developing countries than in industrialized countries. Other causes of low birth weight babies are smoking and excess alcohol consumption; these may also influence the infant's mental development.

12.4.5 Pica

Pica is a pathological craving for normal food constituents or for substances not commonly regarded as food by the local culture. The name 'pica' is derived from a mediaeval Latin word meaning 'magpie'. The magpie was known to pick up a range of objects to satisfy hunger or curiosity. The diagnosis of pica depends on cultural attitudes as well as amounts ingested and degrees of craving.

Pica has been described since the time of ancient Greece and sometimes can have quite devastating effects. On the plantations of the United States and the Caribbean slaves sometimes showed a craving for clay. When pica is superimposed on a diet which is conducive to pellagra, that is a diet of salt pork, corn bread and molasses, then a condition which is known as *cachexia africane* develops: weakness, pallor, oedema, enlargement of the liver, spleen and lymph nodes, anorexia, tachycardia, ulceration of the skin and death.

Conditions in which pica occurs

The craving for particular food substances during pregnancy is a widely recognized phenomenon. In a survey in pregnant women, cravings for fruit and other sweet, sour or sharp-tasting foods were not uncommon. Pregnant women may eat clay and starch, soil and refrigerator frost. It is debatable whether chewing tobacco is a form of pica. Pica is prevalent among children, particularly psychotic, mentally retarded children.

Causes of pica

Pica has been suggested to be a craving produced by nutrient or mineral deficiency. The salt lick of wild animals is widely considered a response to a deficiency. On the other hand, pica has been suggested to be due to deeply ingrained cultural customs within societies. However, pica may not be limited to any age, sex or racial grouping.

Consequences of pica

These depend on what is being eaten, so that lead toxicity may occur if lead is a feature of the material being eaten. Nutritional status may be affected by pica; the consumption of excessive amounts of a single substance may reduce the intake of normal dietary sources of nutrient. For example excessive ingestion of laundry starch may depress intakes of required food. If the amylophagia does not interfere with appetite then obesity may result. Pica may reduce the bioavailability of minerals. Alternatively, pica may not matter.

Summary

1. The expectant mother needs a good, mixed diet before and during the pregnancy. Breast feeding and anorexia nervosa reduce the chances of pregnancy.
2. The weight of the baby is influenced by the protein–energy intake of the mother. The provision of nutrients for brain development is important. If the mother weighs under 40 kg then the baby is liable to a reduced birth weight. Young, growing mothers are another group whose babies are vulnerable.
3. Pica is a craving for a normal food constituent or unusual substances. This may be secondary to some explicable or inexplicable nutritional deficiency. Unless the substance is harmful or there is an underlying deficiency then the phenomenon is harmless.

Further reading

Barker, D.J.P. (1992) in *Infant Origins of Common Diseases in Human Nutrition; A Continuing Debate* (eds M.A. Eastwood, C. Edwards and D. Parry), Chapman & Hall, London, pp. 17–30.

Danford, D.E. (1982) Pica and nutrition. *Annual Review of Nutrition,* 2, 303–22.

Dobbing, J. (1968) Undernutrition and the developing brain, in *Applied Neurochemistry* (eds A.N. Davison and J. Dobbing), Blackwell Scientific Publications, Oxford, pp. 289–94.

Freinkel, N. (1980) Of pregnancy and progeny. *Diabetes,* 29, 1023–39.

Harries, J.N, and Hughes, D.F. (1957) An enumeration of the 'cravings' of some pregnant women. *Proceedings of the Nutrition Society,* 16, 20–1.

Rees, J.M. *et al.* (1992) Weight gain in adolescents during pregnancy: rate related to birth weight outcome. *American Journal of Clinical Nutrition,* 56, 868–73.

Stein, Z., Sussex, M., Sanger, G. and Marolla, F. (1975) Mental performance after prenatal exposure to famine, in *Famine and Human Development; The Dutch Hunger Winter of 1944–45,* Oxford University Press, New York, pp. 197–214.

Wharton, B. (1992) Food and biological clocks. *Proceedings of the Nutrition Society,* 51, 145–53.

12.5 Lactation

- Human breast milk is the most suitable food for human babies.
- The amount and constituents of milk alter with time after the birth of the baby.
- Milk constituents are to an extent dependent upon the mother's nutrition.

12.5.1 Introduction

Human breast milk is the best feed for human babies. Mothers in developing countries breast feed their babies. Many supplement their milk with water or teas from the first week of life in the belief that it will relieve pain, colic and earache, prevent and treat colds and constipation, soothe fretfulness and quench thirst. Human milk contains approximately 70 kcal gross energy, 1.0 g protein, 4 g fat, and 7 g lactose per 100 g. For the first 5 days after birth, milk is rich in **colostrum** and immunoglobulins. The nitrogen content of human milk decreases rapidly during the early weeks of lactation and slowly thereafter. The nutrient content of human milk is very adequate for growth except for iron (0.6 mg/l at 2 weeks) at parturition.

12.5.2 Breast feeding

Healthy infants at 2, 4 and 6 months need about 800, 900 and 1000 ml of milk daily, respectively. Most mothers can produce sufficient milk to meet the needs of their baby at 4 months but only a minority can at 6 months. Supplementary feeding starts when the baby is aged between 4 and 6 months.

Supplementing breast milk in the first 4 months of life is unnecessary, even in hot climates and may be harmful. Breast milk has an osmolarity which rarely exceeds 700 mmol/l. The osmolarity of breast milk in all environments, even in high temperatures and varying degrees of humidity, is within the normal concentrating ability of the child's kidneys. Breast feeding on demand particularly in developing countries is important in giving the infant only breast milk, ensuring a regime that is sterile in the first 4–6 months of life.

12.5.3 Lactational capacity

An important influence on the amount and nutrient content of maternal milk production is the maternal lactational capacity. The amount of milk that the child is able to drink will affect the infant's milk nutrient intake and subsequent growth. The maternal physical activity, the thermic effect of food or maternal and infant illness may also affect the amount and content of milk produced. Maternal dietary intake before and during pregnancy affects maternal adipose tissue and nutrient stores at parturition, lactational capacity and infant size at birth. Whether the child is solely fed on the breast or whether there is supplementary feeding will affect the infant's growth.

Lactational capacity is probably a function of genetic heritage, age, parity, breast enlargement during pregnancy and nutritional history. It may respond to improved dietary intake and increased infant demand. Lactational capacity is usually measured by milk production – infant milk intake

plus residual milk. For well-nourished women lactational capacity is greater than milk production. The amount of milk a woman produces corresponds to infant milk intake, which corresponds to infant size and growth rate and that is affected by infant vigour in suckling.

Among well-nourished women dietary intake accounted for approximately 13% of the variability of infant milk intake. The size of the mother was not associated with lactational performance during the first 1–4 months postpartum. If there is prolonged lactation, i.e. 3–12 months, some association has been found between maternal nutritional status and milk composition. In general, however the mother is able to maintain milk production, even in poor circumstances.

Seasonal food shortages and reduced dietary intakes are associated with a decrease in milk production. In undernourished women given a dietary supplement, infant milk intake increased by 100 ml/day, though there was no effect on the protein content. Infants of the food-supplemented mothers did not gain significantly more weight during the supplemented period, though there was an increase in maternal body fat content. There is an association between maternal dietary intake and infant milk intake.

The availability of sufficient nutrients for milk biosynthesis is a necessary but not absolute condition for increasing milk production. Milk production requires a combination of adequate lactational capacity, adequate infant demand, adequate nutrients and good suckling vigour

The relationship between maternal dietary intake, maternal nutritional status, milk production and infant growth will vary during lactation. Maternal dietary intake during lactation influences the maternal adipose nutrient stores, and hence the amount of nutrients available for milk biosynthesis. The proportion of ingested nutrients available for milk biosynthesis is influenced by maternal nutrient stores. Such nutrient stores will contribute to the nutrients available for milk biosynthesis and this will depend on maternal dietary intake.

For the mother, lactation requires markedly increased food intake. This may deplete stores in some parts of the body structure. Fat synthesis is reduced and there is loss of adipose tissue.

An alternative to breast feeding is modified cows' milk, boiled and then cooled for drinking. This should be given only when the baby is approaching 6–12 months of age. Disadvantages in supplementary feeding are those of hygiene and infection.

12.5.4 Composition of infant feeds (including breast milk)

If modified milks are incorrectly made up, there may be contamination of the milk or a high sodium intake. Cows' milk has a high sodium content – between 20–25 mmol/l (human milk has 7 mmol/l).

Na content of milks (mmol/150 ml):

Cow	3.3
Human	0.9
Whey-based	1.0
Infant formula (AMSS recommendation)	0.9–2.0

Constituents of human milk

Protein: caseins; α-lactalbumin; β-lactalbumin
Non-protein nitrogen: urea; creatine; creatinine; uric acid; glucosamines; α-amino nitrogen; nucleic acids; nucleotides
Immunological factors: secretory IgA; immunoglobulins; lactoferrin; lysozyme
Enzymes: α-amylase; bile salt-stimulated lipase; glutathione peroxidase; γ-glutamyl transferase; lipoprotein lipase; trypsin
Hormones and growth factors: pituitary hormones; brain gut peptides; growth factors; steroids; non-steroid hormones

---

The **bioavailability of nutrients** in human maternal milk is high. Such favourable bioavailability is maximal when the milk and baby belong to the same species. The infant is, however, born with substantial iron stores. Iron concentrations in breast milk decline to 0.3 mg/l at 20 weeks. The absorption efficiency of iron and zinc is high.

Milk proteins are very digestible and are a ready source of amino acids to the infant. α-Lactalbumin forms 25–30% of milk protein and is a major nitrogen source. This protein also has a role in the breast synthesis of lactose and also chelates calcium and zinc.

12.5.5 Nutritional value of breast feeding

Human milk protects the baby against infections of the gastrointestinal tract or respiratory system. There may also be long-term protective influences. The compounds which are protective are present in milk in highest concentrations during the early phase of lactation.

During the first year of life endocrine function is not fully developed in the infant. Milk-borne hormones may be of importance in supplementing the endogenous secretions of the child. Such hormones are not present in formula feeds. Bile salt-stimulated lipase augments the modest pancreatic function of the infant and ensures good lipid digestion during the period of lactation.

Essential fatty acids (mean ± s.e.m.) in human breast milk by diet

	Omnivores	Vegetarians
Linoleic acid	10.9 (1.0)	22.4 (1.26)
α-Linolenic acid	0.49 (0.06)	0.70 (0.91)
Docosahexaenoic acid (DHA)	0.31 (0.07)	0.11 (0.08)

Human milk is particularly suited to ensure human brain development. The linoleic acid, α-linolenic acid, arachidonic acid and docosahexaenoic acid content of milk is particularly important. The fatty acid composition of human milk is very variable, depending upon the diet of the mother. A topic of interest is whether vegetarian and omnivorous mothers produce milks of equivalent essential fatty acid content.

Breast milk from vegan mothers is enriched in linoleic acid and reduced in DHA. Breast milk has a role in the adaptation of the gastrointestinal tract of the newborn infant to oral feeding. Milk contains both nutrients and non-nutrient factors which are important in adaptation.

> The type of milk is very species-dependent. The baby seal drinks a milk rich in fat (490 g/l). The shrew doubles its body weight in 24 hours drinking milk with a protein content of 10 g/l. These milks contain double the equivalent nutrient values to human milk.

The neonatal gastrointestinal tract:

- is an organ of nutrition, with digestive, absorptive, secretory and motile functions
- is part of the immune system containing both humoral and cellular elements of the gut-associated lymphoid tissue
- is a large and diffuse endocrine organ secreting locally acting gut hormones
- plays a role in conservation of water and electrolytes

The newborn receive immunoglobulins, trophic factors, digestive enzymes, other physiologically active peptides and oligosaccharides in the milk. The newborn baby's gastrointestinal tract is quite underdeveloped and the infant has no teeth. Breast milk is rich in secretory immunoglobulin (sIgA), which resists gastric hydrolysis. Lactoferrin and sIgA provide 30% of the milk protein, giving passive protection at the gastrointestinal epithelium to exclude foreign antigens. Antigens to which the mother has been exposed in her diet are recognized at M cells within Peyer's patches in the maternal small intestine. Primed plasma cells pass to the lymphatic system and to the breasts where specific sIgA is synthesized and secreted in

milk. The neonatal infant's gastrointestinal tract becomes adapted to postnatal requirements.

The trophic factors secreted in milk include epidermal growth factor, a small polypeptide with mitogenic, anti-secretory and cytoprotective properties. This peptide is also present in amniotic fluid and colostrum. Their role is to activate mucosal function, to reduce gastric hydrolysis of milk macromolecules and to protect the gastrointestinal epithelium. Epidermal growth factor receptors (EGF) are found on enterocytes from the 19th week of gestation increasing slowly during the first two trimesters and then rising steeply thereafter. EGF has a role in inducing intestinal lactase and depresses the activity of sucrase.

The newborn baby has an immediate requirement for fat for energy, insulation, neural tissue and membrane synthesis. As the pancreatic (10% of adult) and hepatobiliary function (50% of adult) is not developed, milk lipase is important. The benefit of such depressed hydrolytic systems is that immunoglobulins and enzymes in milk are not hydrolysed in the gastrointestinal tract. The immature infant intestine readily absorbs macro and micromolecules.

12.5.6 Potential disadvantages of breast feeding

The linear growth and weight gain decreases after 4–6 months of breast feeding. At this same period the zinc content of the breast milk declines from 40 µmol/l at 1 month to 10–15 µmol/l at 6 months. If children of poorly fed mothers, breast-fed over a prolonged period of time are to maintain growth, then 5 mg of zinc restores weight gain to anticipated average values. Supplementation with iron and ergocalciferol is also important during prolonged breast feeding.

Breast milk from mothers who smoke contains cotinine a stable metabolite of nicotine (ten times that of controls). Cotinine was found, attributed to passive smoking, in the milk of 10% of mothers who did not smoke. By 1 year there appears to be no measurable effect of maternal smoking on infantile growth.

There is concern over the transmission of the virus associated with HIV in breast milk from infected mothers, though studies in Italy indicate breast feeding not to be a major route of HIV-1 transmission. The HIV virus has been found in human milk, concentrations appearing highest in milk from recently infected women and those with the advanced disease. The viruses appear to be shed into milk intermittently so that while any sample may appear to be free from the virus, this may not rule out the possibility of virus excretion. Where there are adequate supplies of properly prepared infant formula milk, women infected with the HIV virus are advised not to breast feed. In the Third World the benefits of breast feeding exceed the potential risk of HIV transmission.

Summary

1. Breast milk is the best food for the baby and does not require supplementation in the first months of life. The nutritional constituents of milk reduce with time.
2. The maternal lactational capacity is important, and is determined by genetic heritage, age, parity and nutritional history. Seasonal food shortages and dietary intake may reduce milk production.
3. Supplementary feeds introduce the risk of infection and harmful constituents, e.g. excess sodium.
4. The bioavailability of nutrients in human breast milk is high. Human milk protects the baby against infections. Milk-borne hormones may be of importance to the baby. The human milk content of essential and long-chain fatty acids is of paramount importance in brain development. There is no significant difference between the milks of adequately nourished vegetarian and omnivorous mothers.
5. Viruses can be intermittently transmitted in human milk.

Further reading

Ellis, L.A. and Picciano, M.F. (1992) Milk borne hormones: regulators of development in neonates. *Nutrition Today*, 27, 6–13.

Martino, M. de *et al.* (1992) HIV-1 transmission through breast milk: appraisal of risk according to duration of feeding. *AIDS*, 6, 991–7.

Rasmussen, K.M. (1992) The influence of maternal nutrition on lactation. *Annual Review of Nutrition*, 12, 103–19.

Ruff, A.J. (1992) Breast feeding and maternal infant transmission of human immunodeficiency virus type 1. *Journal of Pedriatrics*, 121, 325–9.

Sanders, T.A.B. and Reddy, S. (1992) Infant brains and diet. *British Medical Journal*, 340, 1093–4.

Schulte-Hobein, B. *et al.* (1992) Cigarette smoke exposure and development of infants throughout the first year of life: influence of passive smoking and nursing on cotinine levels in breast milk and infants' urine. *Acta Paediatrica*, 81, 550–7.

Vernon, R.G. (1992) The effects of diet on lipolysis and its regulation. *Proceedings of the Nutrition Society*, 51, 397–408.

Walravens, P.A., Chakar, A., Mokni, R., Denise, J. and Lemonennier, D. (1992) Zinc supplementation in breast-fed infants. *Lancet*, 340, 683–5.

Weaver, L.T. (1992) Breast and gut: the relationship between lactating mammary function and neonatal gastrointestinal function. *Proceedings of the Nutrition Society*, 51, 155–63.

12.6 Weaning

- Weaning is the gradual withdrawal of breast or formula milk and introduction of foods suitable for the infant.
- The timing of weaning is important.
- The food onto which the baby is weaned must provide adequate energy, protein, vitamins and minerals.

12.6.1 Introduction

Weaning is the gradual withdrawal of breast milk and introduction of other foods including suitably prepared adult food and the milk of other animals. The child is weaned from the breast or formula milk and introduced to semi-solid or solid food. These new foods become the source of energy and nutrient intake. It is becoming evident that the nutritional status of the child during the first year of life and early development has a major effect on the child's well-being as a middle-aged adult.

12.6.2 Age of weaning

The usual practice in Western countries is to wean before 6 months. An earlier weaning pattern is found in urban areas. This may not be unrelated to the need for women to return to paid employment. In Europe and North America more than 90% of children receive some semi-solid food by the age of 9 months, supplemented by breast milk or a modern infant formula. The age of weaning in developing countries may be quite different from developed countries.

The usual recommendation is that the majority of infants should receive a milk-only diet until 12 weeks. Weaning should be a gradual process thereafter and babies offered a mixed diet not later than the age of 6 months. Energy intake per kg body weight declines rapidly from birth to the age of 6 months and then rises again, as the baby becomes more active. Babies who are not weaned until age 3–6 months grow differently from very early weaners. Predominantly breast-fed babies grow more slowly, but have a greater head circumference than formula-fed babies.

12.6.3 Foods and methods

Cereals are the most common first weaning food. In developing countries the cereal is determined by availability. The energy density and fat content is low. The baby has to eat a considerable bulk to meet energy needs. Adding milk increases protein–energy ratio and the overall protein quality. Cereal or milk alone would be inadequate. Water content and hence intake is important, otherwise the gruel becomes very viscous. Germination, malting and roasting reduces viscosity and increases the energy density.

Europe and North America

In the early weaning months breast milk, or more commonly a formula, provides a substantial portion of the total energy and nutrient intake. In later infancy and beyond, cows' milk may provide one-quarter of the total energy and one-third of the protein intake. Strained foods are introduced later and are frequently based on full adult meals. Parents are able to apply their knowledge of a reasonable mixture of food for adults to their children's diet.

Developing countries

In these countries weaning is undertaken by stressed, possibly pregnant, hardworking rural mothers. Sometimes if the mother is working in the fields, the responsibility falls upon slightly older sisters.

Breast milk is the source of nutrition for the first 6 months of life. Weaning from maternal milk onto food comes towards the middle of the first year. An energy gap may develop from that stage.

Traditional weaning foods may be contaminated with enteropathic bacteria. The result is diarrhoea, and undernutrition which may compound the nutritional inadequacies of a child who has been receiving insufficient milk. More than half of the developing world's children are undernourished with retardation of growth and development. Lactation is sometimes continued in part to reduce the chances of the next pregnancy.

The weaning foods are gruels based on cereal flours, maize, rice, millet, sorghum and wheat, the starch being the important energy source. The boiled starch gelatinizes and produces a viscous paste which may contain resistant starch (p. 349). The calorie content may be increased by oils, fat or sugar. Malting may help partially digest the starch.

> Cereals are germinated by soaking in the dark for 48–72 hours, dried, toasted and after removal of the sprouting bits, ground and milled to flour. The starches are partially hydrolysed. There may be some solubilization of proteins, amino acids and vitamins. Their concentrations may increase and trypsin inhibitors and phytic acid content be reduced. The energy content may be doubled, by such processes. Fermentation and amylase-enrichment may improve both the nutritional content and also the hygiene of the feed.

Composition of weaning foods

There are many differences in approach to weaning in different countries. The substances recommended are based on energy and protein content per unit weight or per unit energy. There is often fortification of natural foods.

France has its own detailed recommendations. In **Thailand**, there is a recommendation for protein of not less than 2.5 g/100 kcal, amino acid score not less than 70% of the FAO/WHO reference pattern, fat 2.6 g/100 kcal and linoleic acid not less than 300 mg/100 kcal. Supplementary food mixtures containing rice and soya bean, groundnuts, sesame or mung beans are also used. This mixes a starch (rice)-containing, relatively low-protein source with a vegetable with a high protein content. This has had profound effects on reducing moderate and severe protein energy malnutrition.

It is not impossible that the occurrence of coeliac disease is a result of premature weaning onto gluten-containing foods, e.g. wheat.

12.6.4 Level of nutrition

Overnutrition

During the first year of life the deposition of fat increases from approximately 11% of body weight at birth to 25% at 6 months, being steady from ages 6–12 months and slowly decreases to 16% during the toddler years.

Solid foods, particularly those high in protein and electrolytes, i.e. added salts, increase the solute loads. The kidney must be able to respond by producing a more concentrated urine, perhaps of twice the plasma osmolality. This should result in increased thirst with the baby drinking more milk or formula to increase weight.

Undernutrition

Weaning may be forced on a mother and her baby through insufficient breast milk for the baby's hunger and nutritional requirements. There is only a limited relationship between a mother's food intake and her ability to breast feed. A strong, healthy baby who suckles frequently and well appears to be the most important stimulus to maternal prolactin production and hence milk secretion.

Infant milk made up in unclean bottles is a source of **infection** to babies. There is a contrast between baby health in Europe or North America and the developing world. Babies in Europe or North America who are healthy at 4 weeks old will grow into healthy adults. In the developing world, the early postnatal months and toddler years are very dangerous periods.

The most significant risks of deficiency conditions to the weaning infant and requirements for supplementation are identified in the earlier sections on specific nutrients, particularly essential fatty acids and amino acids, calcium and vitamin D, vitamin A, iron and zinc.

Summary

1. Weaning is the process of withdrawal from dependence upon breast milk to prepared semi-solid and then solid food. The age of weaning varies from culture to culture, country to country and economic status.

2. Cereals are the most common first weaning food with added milk and other protein sources. The weaning food must be adequate in energy, protein and essential dietary constituents.
3. The mode of preparation will vary, with the major risks being bacterial contamination of the feed.
4. The correct amount of nutrients given is critical to the baby's future well-being. Depending upon the economic status in which the infant lives there is the potential for under- or over-provision.

Further reading

Barker, D.J.P. (1992) *Infant Origins of Common Diseases in Human Nutrition; A Continuing Debate* (eds M.A. Eastwood, C. Edwards and D. Parry D), Chapman & Hall, London, pp. 17–30.

Leader (1991) Solving the weanling's dilemma: power-flour to fuel the gruel. *Lancet,* 338, 604–5.

Wharton,B. (1989) Weaning and child health. *Annual Review of Nutrition,* 9, 377–94

12.7 Childhood and youth

- Childhood and youth are the periods of transition of physical, hormonal, reproductive and emotional status, between the baby and the adult.
- Good nutrition is reflected in growth, providing height, weight and appropriate bodily function and freedom from illnesses associated with malnutrition.

12.7.1 Introduction

Height is believed to be a good indicator of nutritional status, though why is not clear, since diet is not the only environmental factor that influences height. Other factors include genetic make-up, childhood illnesses, and sleep patterns.

An important measure is **height velocity** rather than height as a single measurement. The short child in the 25th percentile or growing less than 4 cm per year should always be considered for investigation. However, some children grow slowly before achieving their final height. Height is an indicator of past growth rather than present health.

Normal growth takes place in distinct spurts in a cyclical manner with a periodicity of about 2 years. This cyclical pattern is maintained even during growth lags.

Food which satisfies the energy gap, micro-nutrient and nitrogen needs is essential for growth and high activity. However, access to effective primary health care is also essential for good growth. The progress of growth can be followed by regular weight and height measurements and compared with tables for the populations. During **adolescence** there are increased nutritional requirements of adequate protein, vitamins and minerals and calorie intake.

Growth is dependent upon a series of factors including: (i) the inherited genetic potential; (ii) available nutrition to enable genetic potential to find expression; and (iii) good health.

12.7.2 Growth patterns

The prepubertal growth spurt is a qualitative as well as a quantitative growth. Body shape and composition also change. Boys enter puberty with one-sixth of the body as fat and finish puberty developing muscle so that the fat content falls to one-tenth. Girls' bodies over the same period change from one-sixth fat to one-quarter fat. Menarche appears to depend upon the weight-for-height ratio, and a minimum fat of 17% is a requirement.

At the pubertal growth spurt, boys grow 20 cm in height and 20 kg in weight – a greater growth than that seen in girls at the same stage.

Indian, Japanese and US children have similar growth patterns up to the age of 9 years. Thereafter a growth gap develops during the remaining years of growth between the first two nationals and the US adolescents. It is not clear whether this difference is environmental or genetic. The Indian and Japanese diet is largely cereal-derived and there may be dietary deficiency or dietary antinutrients which cause the lack of growth. A survey by the Japanese Ministry of Education in 1992 of 700 000 children aged between 4 and 18 years showed they had grown significantly in height compared with their parents. Boys aged 13–14 years old were 9.5 cm taller and 9 kg heavier than their parents at the same age. At 17–18 years these Japanese in 1992 were 5 cm taller than 30 years ago. Much of this difference was in leg length.

In the Netherlands, there is a tradition of monitoring children's height that goes back to the 19th century. During the last century and the early part of this century differences of 5–11 cm at various ages were noted between those children whose fathers had a 'low' as compared with a 'high' occupational status. This gap is now reduced to differences of 1–3 cm. Similar closure of height differentials between socially different children, economically disadvantaged and rich children have been observed in India. A British tradition, unproven except in folklore, is that the child's height at 2 years 6 months is half that of the final growth achievement.

Since the 1980s began a number of people in western countries have developed alternative lifestyles. One such style is the macrobiotic diet, consisting of organically grown cereals, vegetables and pulses and small amounts of seaweed, fermented foods, nuts and seeds and occasional seafood. There has been anxiety that such a diet may result in health risks. In a study of 4- to 18-month-old macrobiotic-fed children, a lack of energy and protein containing foods led to muscle wasting and intellectual retardation. However, a longer-term study found that long-standing mild to moderate malnutrition did not lead to a reduction in mental development. Another dictate of intellectual and emotional development is the social and family environment in which the child is reared. A child that is given attention will do better in intellectual and emotional progress than the neglected infant receiving the same nutritional support.

A curious feature of the development of the stunted child is the anatomy of the deficient height. This may be the sitting height, that is, coccyx to top of head height or standing height where leg length is important. In stunted USA blacks and Australian Aborigines the sitting height is deficient; in Japanese, standing height is deficient and in Indians the deficit is symmetrical. Possibly some complicating micronutrient is deficient or the timing of the deficiency at some vulnerable period occurs primarily affecting trunk or leg growth.

A study of children reared in London on vegan diets showed that such children developed normally, provided that care is taken with vitamin B_{12} replenishment and that the problems of bulky diets and nutrient dilution are recognized. These children are frequently not immunized and problems can arise in susceptibility to infectious diseases.

The stunted child may be helped by nutritional replenishment for example the child with iron-deficiency anaemia may partially benefit from iron replenishment. On the other hand, full restoration of health may depend upon other deficient nutrients being replenished. Stunting of growth due to insufficient nutrition may also be aggravated by coincidental infection.

12.7.3 Attitudes to growth and height

In development and normal growth, there is an instinctive feeling that tall equates with capability. Height may protect against myocardial infarction but not against cancer, and the converse applies to short stature. It still has to be shown that final growth height is advantageous in terms of health.

> Height may be a cosmetic attribute useful in certain sporting activities but not necessarily the ultimate measure of nutritional success. The taller of the two candidates in the USA Presidential elections, has with one exception always won. Comparable statistics for other countries are not readily available. Some quite short men, e.g. Napoleon Bonaparte, Alexander the Great, Attila the Hun, Byron, Cervantes, Cromwell, Sir Frances Drake, Louis XIV, Admiral Lord Nelson, Shakespeare, Socrates and St Paul have been extremely successful. However, tallness in women does not necessarily carry the same cache in Western society.

12.7.4 Nutritional recommendations in childhood and growth

A number of nutrition experts have recommended that all children consume diets containing no more than 30% of calories as fat. It is important to ensure that a sufficiency of calories are provided for growth and energy requirements and that a sufficiency of PUFAs are provided for essential functions.

Children of less than 2 years old should not take foods rich in non-starch polysaccharides at the expense of more energy-rich foods which they require for adequate growth. It is not at the present time possible to define a non-starch polysaccharide intake for children. Children need energy derived from carbohydrates in all forms. For convenience, their nourishment is facilitated if their simple energy requirements are met at the same time as other needs, e.g. protein and essential nutrients.

12.7.5 Nutrition and intellect in childhood and youth

There has been considerable debate on whether supplementing the diet with vitamins improves children's performance in intelligence testing. When an improvement is reported, the improvement is in non-verbal rather than in verbal measures. Verbal intelligence reflects educational and other experiences, whereas the non-verbal measures basic biological functioning. Improvement may also reflect increased working capacity and attention capacity. A poor diet may be associated with a poorer performance on non-verbal intelligence tests. The response to the trials is greater in schools with socially deprived children.

What is not known is the degree of undernutrition which results in a reduced intellectual performance. There may be a subgroup of children who are undernourished and whose optimal intellectual development could be ensured by dietary supplements. Such children are found more frequently in developing countries than in the developed world.

Summary

1. Development in growing children is usually measured as height or rate of increase in height over a defined period of time. Height is determined both by genetic background and nutrition.
2. There is a prepubertal growth spurt in which height, body shape and composition change. The characteristic body shapes of males and females are formed at this time. The prepubertal growth pattern varies in different racial and economic groups. Young people on defined diets should ensure that specific nutrient requirements are met, e.g. vitamin B_{12}.
3. The correct dietary intake is critical during childhood and growth.

Further reading

Benton, D. (1992) Vitamin–mineral supplements and intelligence. *Proceedings of the Nutrition Society*, **51**, 295–302.

Butler, G.E., McKie, M. and Ratcliffe, S.G. (1990) The cyclical nature of prepubertal growth. *Annals of Human Biology*, **17**, 177–98.

Evans, D., Bowie, M.O., Hansen, J.L.O, Moodie, A.D. and Spicy, H.I.J. (1980) Intellectual development and nutrition. *Pediatrics, 97*, 358–363.

Frisch, R.E. (1977) Food intake, fatness and reproductive ability, in *Anorexia Nervosa* (ed. R.A. Vigersky), Raven Press, New York, pp. 149–61.

Geissler, C. (1996) Adolescent nutrition: are we doing enough? *Proceedings of the Nutrition Society*, **55**, 321–67.

Herens, M.C., Dagnelie, P.C., Kleber, R.J., Mol, M.C.J. and van Staveren, W.A. (1992) Nutrition and mental development of 4–5-year-old children on macrobiotic diets. *Journal of Human Nutrition and Dietetics, 5*, 1–9.

Leader (1992) Too tall? *Lancet, 339*, 339–40.

Mackenbach, J.P. (1991) Narrowing inequalities in children's height (letter). *Lancet, 338*, 764.

Nelson, M. (1992) Vitamin and mineral supplementation and academic performance in school children. *Proceedings of the Nutrition Society*, **51**, 303–13.

NNMB (1991) Report of repeat survey (1988–90). National Institute of Nutrition, ICMR.

Sanders, T.A.B. and Manning, J. (1992) The growth and development of vegan children. *Journal of Human Nutrition and Dietetics, 5*, 11–21.

12.8 Middle age

- Middle age is the established group at the peak of social and professional achievement. A healthy middle age is dependent upon good nutrition in the formative years.
- Food intake should maintain the body in health and allow tissue replacement to take place.
- Middle age is the time when excess nutrition especially fat and alcohol occurs with adverse consequences for weight and disease.

Middle age might be defined as the age group between 35–59 years. This is the establishment group, at the peak of social and professional achievement, beset by a number of problems and responsibilities:

- the empty nest (children leaving home)
- the middle life crisis
- change of life, the menopause, etc.
- the period of transition and change before retirement
- the period when 'health and illness' problems begin to become more frequent

Growth that is of functional value has stopped. During middle age, fat stores usually increase and there is a decline in organ function – especially of pulmonary and renal capacity, which reduce by half between the ages of 30 and 90. The decline in organs and systems is gradual except for the abrupt cessation of ovarian function at the menopause. At the menopause the female is released from the constant threat of iron deficiency through menstruation and pregnancy. This is a period of life when if there is a sufficiency of food available, food intake, storage and metabolism should be balanced.

The middle age group is where individuals begin to die in increasing numbers. Overall it is believed that the middle aged die of diseases associated with excess of the wrong foods, too much alcohol, and smoking – or all three. Those middle aged who die or who suffer from these conditions are disadvantaged by their genetic make-up and consequently by a vulnerability to dietary constituents, alcohol and tobacco. The individual's weight, especially if male, and the amount of exercise taken are of importance. An individual should be of acceptable weight (BMI 20–25) and take 70 or more minutes of exercise a week which causes tachycardia.

A broadly based diet is to be commended. It is not yet clear if it is too late to alter the diet of the middle aged. The early years may be more important in establishing the individual's constitu tion.

The most important factors in longevity are heredity, the nutritional status of one's mother during uterine life, nutrition during the first year of life, and luck. It is not possible to choose parents, or culture or exposure to war or fatal accidents. Perhaps the next generation will benefit from improving social and nutritional influences.

It is important when prescribing healthy diets to see the whole person coping with the varied problems identified and who will eat, drink and smoke in an individual manner. The health problems which develop in this age group are dependent upon a number of factors, only some of which are avoidable:

- vulnerabilities which are dependent upon the genetic make-up of the individual
- exposure to factors which are controllable to a varying extent, including smoking and employment hazards
- exposure to dangerous infections
- excess food challenging the genetic make-up
- overall or individual factors, including shortage of food, alcohol, exercise and body weight.

Summary

1. Middle age is a period of equilibrium in nutrition. There is no further growth except the laying down of undesirable fat.
2. The decline of body organs is beginning, and tissue repair is important.
3. The prospect of premature death is in part dictated by genetic make-up and early nutritional patterns but may be reduced by attention to weight, exercise, smoking habits and an adequate diet.

Further reading

Abel, T., McQueen, D.V., Backett, K. and Currie, C. (1992) Patterns of unhealthy eating behaviours in a middle aged Scottish population. *Scottish Medical Journal*, 37, 170–4.

Eastwood, M.A., Edwards, C. and Parry, D. (1992) *Human Nutrition: A Continuing Debate*, Chapman & Hall, London.

Garrow, J.S. (1988) *Obesity and Related Diseases*, Churchill Livingstone, London.

12.9 Old age

- Life expectancy varies from country to country; this is dictated by nutrition and genetic make-up.
- Old age is complicated by disease with consequences for nutrition.

12.9.1 Introduction

Ageing may be defined as regression of physiological functions accompanied by advancing age. Malnourished populations do not have a long expectation of life. In developed countries it is possible that the maximum expectation of life for the human of approximately 85 years is being achieved.

12.9.2 Epidemiology

Longevity has increased quite markedly this century with the most marked increase being in Japan

TABLE 12.1 Changes in longevity, in Japan 1890–1990

Year	Life expectancy (years)	
	Males	Females
1890*	36	37
1947⁺	50	50
1990++	76	82

*New Zealand, Australia and Sweden highest
⁺Sweden and Western Europe highest
++Japan highest

TABLE 12.2 Changes in the Japanese diet, 1910–1989 (daily intake per capita)

Dietary component	1910	1989
Carbohydrate (g)	430	190
Fat (g)	13	59
Animal protein (g)	3	42

(Table 12.1), where the diet has changed over the century (Table 12.2). The Japanese diet is still based on rice with a balance of animal, mostly fish and vegetable foods.

Longevity statistics

Part of the problem in assessing longevity is how and why age is registered in any population. The accuracy of the registers of birth is not uniform through out the world. In some countries it was the practice to take one's father's birth certificate during late teens to avoid military service and hence appear older. Another important element is prosperity and hence mode of living. So that when the Gross National Product exceeds $7000 per capita then the life expectancy of the inhabitants of the country exceeds 70 years.

A major factor in life expectation is the protein content of the diet. Intake over 50 g/day is an important divider between short and long life expectations.

A group of the elderly who are particularly at risk are the unsupervized, either living alone, disabled or in an unfriendly environment, e.g. poor-quality nursing home or even within an uncaring family environment.

12.9.3 The ageing process

It is not known what causes ageing, and whether the process occurs at an organ or a cellular level. Perhaps the organ is slowly damaged or compromised by the environment. The brain shrinks with age, most probably due to cell death. The concentrations of most hormones decline with age, affecting thermoregulatory function and blood glucose concentrations. Imahori (1992) has proposed that there are supervising organs, e.g. brain and thymus which decay before the other organs. The other organs follow an inevitable down hill path. Alternatively, cells may be genetically programmed to die or wear out. Cell death has been linked to a lethal or senescent gene. The only factors which modify the effects of ageing are nutrition, exercise and blood pressure control.

Physiology

As the individual ages, physical activity declines and so less dietary energy is required. Energy expenditure in the young and old is very different. All individuals have to meet the basal needs of the body in both body and mental activity. It is the spare energy which varies. The young have abounding energy, well in excess of need. This gives an amplitude of activity which is physical, social and intellectual. As the years accumulate this excess to basal energy requirement reduces. Less free energy is available for physical, social and intellectual activity. The basal energy devoted to basal activity declines less markedly, but assumes a larger proportion of the reducing energy and activity. This is a source of great frustration to the elderly as the body tends to decline in strength and stamina in advance of the mind.

Many individuals abandon physical activity after the age of 30 years. The Canada Fitness Survey suggests that the average male over 30 years of age finds walking uphill at 3 miles/hour severe exertion.

As the sedentary way of life becomes more ingrained, the capacity to change decreases, and the degree of unfitness increases. When people with a sedentary habit try to exercise they draw on available energy stores which are limited and they readily tire. Endurance is therefore limited.

Stamina is age-related. Some of the deterioration is unavoidable; the remainder is amenable to intervention. There is a reduction in muscle mass of 10–15% between 40 and 70 years. This means that less muscle is responsible for an increasing work load and hence more readily fatigued.

Assessing fitness

Fitness, in general is measured in the measurement of VO_{2max}. A more day-to-day assessment is the point at which the person is distressed by a degree of exercise and the speed of restoration of well-being is slow.

In the elderly, a small reduction in activity can result in loss of mobility. Physical activity in the elderly is important in maintaining cardiovascular well-being, maintaining muscle mass, reducing falls, osteoblast formation and maintaining bone density. In one study, 1 hour of walking twice a week for 8 months increased bone density by 3.5%, compared with a 2.7% decrease in the controls. Flexibility and joint movements benefit.

Pathology

Body composition

During adult life there is a slow decrease in lean body mass, and total body potassium. By 70 years, 40% of skeletal muscle has been lost compared with young adult life. Women retain their lean body mass up to the age of 50, whereas men begin to lose lean body mass from the age of 30. There is an increase in body fat which accumulates throughout life up to the eighth decade. In older people, fat tissue accumulates on the trunk,

in the abdominal region and from subcutaneous tissue to fat surrounding organs.

Total body water decreases with age. This decline begins in middle-age in men and in women after the age of 60 years. Bone mass diminishes with age from about the age of 30. Bone mineral and matrix disappear more rapidly than deposition of bone tissue. Trabecular bone is lost at an earlier age than cortical bone, which is significant as bone mass is important in bone strength.

It is not obvious that there are age-related effects on the **gastrointestinal tract**. A distinction has to be made between the elderly fit living in their familiar surroundings and the old and frail who require care. Morphologically, the changes observed with age, include a reduction in gastric acid, mucus, gastric enzyme, pancreatic enzyme excretion and bile with age, and a reduction in intestinal wall strength. The latter occurs in part because of cross-linkage changes in the collagen of the intestinal wall. **Constipation** is said to be more common in the elderly but again only in the old and frail.

There are increases in **blood pressure** and also cardiac enlargement and decreased contractility of the heart muscle and cardiac output. Circulating concentrations of testosterone and oestrogen, parathyroid hormone, triiodothyronine and aldosterone are reduced. There is a reduction in **renal function** and a reduction in kidney size so that the kidney mass is 30% less at the age of 80 than at 30. **Brain activity** changes with an increase in slow-wave activity, slowing of the alpha-wave frequency and an increased beta-wave activity. The cells of the **immune system** are affected by age and also the T-cell lymphocytes are reduced. There is a reduction in serum IgG concentrations and an increase in IgA.

Metabolism

The energy requirements of an individual can be described by an energy balance equation:

Energy stored = energy intake − energy expenditure

Basal metabolic rate (BMR) is affected by familial and genetic influences and nutritional metabolic and disease conditions. BMR is reduced in the elderly. This reduction is due to the age-related fall in lean body mass and the loss of the muscle mass. The value is unchanged from younger adults when expressed in relation to lean body, cell or fat-free mass.

The thermic effect of food

This is the energy required for the ingestion, digestion, absorption process and storage of the energy-yielding nutrients. The energy cost of the thermic effect of food varies according to the immediate metabolic fate of these nutrients. There are consequent changes in the metabolic cycle and activation of the sympathetic nervous system, the responsiveness of which declines with age. The size, frequency and composition of meals affect the thermic effect of food. The less that is eaten by the elderly, the less pronounced will this effect be.

The capacity to dissipate excess energy as heat may be different between the elderly and the young adult. However, it is possible that dietary-induced thermogenesis is quantitatively similar or leads to the same approximate degree in young and elderly subjects in the short term, although it may not be the case in the long term.

While **immunocompetence** declines with age, such loss is very varied between different individuals. Nutrition is an important determinant of immunocompetence.

Glucose metabolism is altered in elderly people. Insulin deficiency is a contributor to diabetes in the elderly. Total body protein decreases with age as a result of a declining skeletal muscle mass. Whole-body protein synthesis and breakdown and muscle protein breakdown are significantly lower in elderly people compared with the young. Urinary creatinine excretion, which is an index of muscle mass, is greater in the elderly compared with young subjects. It may be that the reduced muscle protein metabolism in the elderly relates to the metabolism of the amino acid glutamine. There is a reduction in the maximal capacity to utilize oxygen during exercise with increasing age.

The total daily energy intake decreases progressively from approximately 11.3 MJ (2700 kcal) at 30 years to 8.8 MJ (2100 kcal) at 80 years.

Alzheimer's disease

Alzheimer's disease is a progressive dementia which may afflict individuals from late middle age, though it is not an inevitable feature of the ageing process. The cerebral changes include cerebral amyloid angiopathy, neuritic amyloid plaques and neurofibrillary tangles. This condition is the cause of 75% of dementias.

Potential contributors to nutritional problems in elderly people

Physical factors
Reduced total energy needs
Declining absorptive and metabolic capacities
Chronic diseases, restrictive diets
Loss of appetite, anorexia
Changes in taste or odour perception
Poor dentition, reduced salivary flow
Lack of exercise
Physical disability (restricting the capacity to purchase, cook, or eat a varied diet)
Drug–nutrient interactions
Side effects of drugs (nausea, altered taste)
Alcoholism
Sociopsychological factors
Depression
Loneliness
Social isolation
Bereavement
Loss of interest in food or cooking
Mental disorders
Food faddism
Lack of self-worth
Inadequate diets caused by cultural and religious influences
Socioeconomic factors
Low income
Inadequate cooking or storage facilities
Poor nutrition knowledge
Lack of transportation
Shopping difficulties
Cooking practices resulting in nutrient losses
Inadequate cooking skills

Apolipoprotein E (apoE) is a plasma protein involved in cholesterol transport. It is also produced and secreted in the central nervous system by astrocytes. The brain contains large amounts of apoE mRNA in amounts second only to the liver. In Alzheimer's disease apoE is bound to extracellular senile plaques, to intracellular neurofibrillary tangles, and at sites of cerebral vessels congophilic angiopathy. ApoE may be involved in the pathogenesis of late onset or familial Alzheimer's disease.

Once old age has been reached, alteration of the diet, e.g. to reduce obesity, has little effect on longevity, unless the diet is deficient. Osteomalacia and other resultants of long-standing malnutrition may be corrected, without necessarily benefiting the elderly.

Elderly individuals with fractured neck of femur, do badly if underweight. These patients mobilize more readily and without morbidity if they are fed supplementary food during the convalescent period.

12.9.4 Food intake and metabolism

Food intake declines with age, as does appetite. The reasons for this will include reduced energy expenditure, ability to shop and a reduced income. There may be changes in the satisfaction and pleasure in eating particular foods or nutrients. The result may be a somewhat monotonous diet, especially in the socially isolated.

The source of energy nutrients in the elderly may be important. There is a deterioration of glucose tolerance with ageing due to impaired peripheral tissue insulin sensitivity, late insulin secretion and altered hepatic glucose output. There are positive epidemiological connections between fat intake and body composition, suggesting that the composition of the diet is a factor in determining body energy balance and composition.

Dietary carbohydrate and fat have unequal effects on energy substrate metabolism and body energy balance. The conversion of carbohydrate to fat is an important pathway of carbohydrate disposal in the individual who is not close to body energy balance or whose glycogen stores are not saturated.

Dietary carbohydrate promotes carbohydrate oxidation and reduces lipid oxidation whereas dietary fat does not increase fat oxidation or influence carbohydrate oxidation.

Imbalances between intake and oxidation are more likely to occur for fat than for carbohydrate. Carbohydrate balance is under a strict metabolic and body fat balance. Consequently:

- carbohydrate and fat balances are under different regulatory controls
- an appreciable storage of glycogen occurs before a significant lipogenesis occurs from dietary carbohydrate
- adjustment of fat oxidation to altered dietary fat intake occurs eventually, but only after a significant increase of adipose tissue

Thus overall the balance of fat to carbohydrate intake may be very important. The source of the energy intake is apparently as important as the level of energy intake with a voluntary restriction of fat intake.

Summary

1. The ageing process is a generalized decline in physiological function and stamina, which is a feature of the number of years lived, rather than of a disease process.
2. Longevity is more likely with a good diet and regular exercise, though there is a genetic contribution.
3. There is a slow reduction in body mass, mineral and water with age. Fat may accumulate in the centre of the body. BMR and other measures of metabolism are reduced with age.
4. Alzheimer's disease is a progressive dementia with characteristic clinical and pathological lesions in the brain.

5. Once old age has been reached the general debility may alter nutritional intake for physical and/ or intellectual reasons. However, once old age has been achieved, changes in diet have little effect on longevity, unless the diet is deficient.

Further reading

Chandra, R.K. (1992) Nutrition and immunity in the elderly. *Nutrition Reviews,* 50, 367–71.

Fries, J.F. (1980) Aging, natural death and the compression of morbidity. *New England Journal of Medicine,* 303, 130–5.

Hosoda, S., Bamba, T., Nakago, S., Fujiyma, Y., Senda, S. and Hirata, M. (1992) Age-related changes in the gastrointestinal tract. *Nutrition Reviews,* 50, 374–7.

Imahori, K. (1992) How I understand aging. *Nutrition Review,* 50, 351–2.

Matsuzaki, T. (1992) Longevity, diet and nutrition in Japan: epidemiological studies. *Nutrition Reviews,* 50, 355–9.

Rolls, B.J. (1992) Aging and appetite. *Nutrition Reviews,* 50, 422–6.

Stephens, T., Craig, C.L. and Ferris, B.F. (1986) Adult physical fitness in Canada: findings from the Canada Fitness Survey. *Canadian Journal of Public Health,*77, 285–90.

Young, V.R. (1992) Energy requirements for the elderly. *Nutrition Reviews,* 50, 95–101.

12.10 Sport

- The nutritional requirements for an athlete are very dependent upon the type of sport.
- The physical make-up of endurance athletes and short intense activity athletes differs with consequences for nutrition.
- The metabolic needs of endurance athletes in whom stamina is vital include a high carbohydrate intake to replenish body glycogen stores.

12.10.1 Introduction

A good diet for an athlete is dependent upon the sport, age of the athlete, fitness and freedom from injury. Modern sport, football, rugby, athletics, cycling, American football, tennis and hockey are now so competitive that previously unsuspected demands on physical fitness and strength are expected.

12.10.2 Metabolic needs

The metabolic needs of endurance athletes (long-distance runners, distance swimmers and cyclists) are different from intermittent activity athletes (football, hockey, cricket and golf) and sports of short but intense duration, e.g. sprinters and wing three-quarters at rugby. Body-building sportsman such as weight lifters have yet different needs (Figure 12.3).

The requirements of endurance athletes are to

	Endurance sports	Ball games	Strength sports
Requirements	Stamina	Stamina, coordination	Strength, coordination
Muscle types	1 (slow-twitch)	1 or 2	2 (fast-twitch)
Dietary emphasis	Carbohydrates	Mixed diet	Protein
	Glycogen loading		Muscle mass

FIGURE 12.3 General characteristics of requirements in different sports: muscle types and dietary emphasis in endurance sports, ball games and strength sports. All these dietary emphases are in addition to a balanced diet of carbohydrates, protein, fat, fibre, minerals and vitamins.

establish a store of readily retrievable energy in a frame which has to be light in weight in order to perform well. The typical long-distance runner or cyclist is a lean individual packed with large stores of glycogen. A first-class football player may run 10 miles during a game, yet a goalkeeper may be required to perform in a gymnastic manner intermittently. Sprinters over the range of 100–800 metres will rapidly expend energy over short periods of 10 seconds to less than 2 minutes. Other sports such as cricket, baseball, golf – which are coordination sports – require a feeling of well-being to concentrate. The nutritional needs of a fast bowler at cricket will be different from those of a golfer. It is unlikely that the nutritional needs of this group are likely to differ from the prudent advice offered to the population of their age. The strength sports, e.g. weight lifters, prop-forwards in rugby, heavy-weight boxers and American footballers, build massive muscle structure to perform deeds of strength.

There is therefore a range of requirements including:

1. The provision of energy over an extended period from a modest muscle mass meeting the need to sustain speed for long periods.
2. The development of a strong physique able to push and pull massive loads and, during the same game, to run distances up to 10 miles over the 90 minutes of the game.
3. The requirement for bursts of energy over very brief periods of time, e.g. sprinters, whether these are athletes or wingers in rugby.

A well-balanced diet should contain 50–60% of the calorie intake as carbohydrate, fats no more than 35% and protein approximately 15%. In addition, there should be sufficient amounts of fibre, vitamins and minerals (as suggested for the population in general) but possibly increased pro rata to energy intake. That is, athletes who are using large amounts of calories will eat large amounts of energy with the nutrients in the above proportions. Some sports will require modest increases of a particular nutrient, e.g. carbohydrate for endurance athletes and protein for strength sport. It is important that such diets

are sensible and within the overall concept of a balanced diet.

Alcohol is a continuing problem in sport. Athletes are strong vigorous young people sometimes with time on their hands. Victories are to be celebrated. Many athletes have taken an undue amount of alcohol and paid the price.

12.10.3 Efficiency of physical work

Work efficiency may be seen as biochemical or as mechanical efficiency:

- **biochemical efficiency**: the efficiency of muscle contraction is the product of the coupling efficiency, i.e. the work performed per unit ATP hydrolysed and the efficiency of energy transduction to ATP by oxidation of substrate. Such biochemical efficiency is tied into the availability of substrate largely provided from glucose.
- **mechanical efficiency**: this is the ratio between mechanical work and the energy expended in doing this. To determine the energy cost of the work, energy expenditure at zero load has to be subtracted. This may not be the same as basal metabolic rate because of position and movement of limbs. A further factor is the speed of work as efficiency is reduced with increasing speed.

While the body mass index (BMI) may range from 20.5 in middle distance runners to 23.6 in pentathletes, the cost of work in these people (in kJ/kg/m) varies from 3.6 to 3.95, and the mechanical efficiency (as a percentage) from 34.1% to 22.6% respectively. In contrast, untrained subjects will have a cost of work of 4 kJ/kg/m, and a mechanical efficiency of 19–20%. Energy cost in moving 1 kg over 1 metre is approximately the same for all athletes.

Training does not produce any change in the metabolic efficiency of the muscles.

Skeletal muscle is not uniform in type; rather, there are at least two distinct muscle types. **Type 1 fibres**, which are slender and contain an abundance of mitochondria, oxidative enzymes and fat, have a slow twitch speed and are resistant to fatigue. **Type 2 muscle fibres** (fast twitch) are broader and more coarse than type 1, have less mitochondria and fat. The enzymes are those for anaerobic metabolism and fast twitch muscles and have a high glycogen content. Type 2 muscle fibres subdivide into types 2a and 2b, dependent upon their ATPase enzymes and content. The chemistries of the muscles in types 1 and 2 are different, as also is the neurone supply. The provision and distribution of these types of muscles may well decide the type and excellence of an athlete. There is a positive correlation between the proportion of type 1 slow-twitch fibres and VO_2 at the anaerobic threshold, a measure of maximum working capacity. Slow-twitch fibres are more efficient than fast-twitch fibres in terms of mechanical force development per unit ATP.

12.10.4 Stamina

Endurance athletes

Endurance running can be defined as regular physically demanding exercise including frequent sessions of 90 minutes or more, wherein significant demands are made on body stores of energy.

The limitation of stamina for endurance runners has been shown to include the depletion of glycogen in muscle and hepatic stores.

Energy and glycogen

The glycogen content of resting muscle is 1.5 g/ 100 g wet tissue, which is approximately 300–500 g in 28 kg of muscle. To this should be added 70–100 g of hepatic glycogen. In total, this has the potential of 8.4 MJ (2000 kcal) energy, which may last 100 minutes in really vigorous exercise.

The energy expenditure of an individual can be estimated from a determination of the oxygen uptake, which increases linearly with exercise intensity until a maximum oxygen uptake VO_{2max} is achieved. This is a reflection of the individual's cardiovascular capacity for oxygen transport and an indication of the levels of exercise an individual might tolerate. The oxygen cost of exercise expressed as a percentage of the maximum oxygen uptake (% VO_{2max}) gives an indication of the physiological stress on the individual.

During endurance running there is a substantial contribution from carbohydrate to energy metabolism. After an hour or more, there is a shift towards fat catabolism. Fatigue comes with depletion of muscle glycogen stores, the amount of which is increased by endurance training.

The **type of exercise** is important; cyclists exercise in a consistent pattern of action on their machines, therefore they will expend all the glycogen in a patterned manner. In running, stride length and running pattern change with the undulations of the course and also can be altered consciously by experienced runners. The load is spread over a wider mass of muscles and hence the depletion of glycogen is less profound. As exercise progresses, free fatty acids are mobilized from adipose tissue. This metabolic source does not prevent fatigue as the fatty acid oxidation is associated with a slower rate of resynthesis compared with carbohydrate. Training increases the aerobic capacity of the muscles and allows them to use fatty acids more effectively. Increased blood fatty acid concentrations as a result of anxiety at the start of a run may conserve glycogen metabolism for later in the race.

Adequate muscle glycogen stores can be achieved by a diet rich in carbohydrates eaten with an endurance training programme. In a study of individuals running a 30-km race, glycogen-loading diets did not increase speed but enabled individuals to hold their optimal speed longer. This led to reductions in times of 3.2% in experienced distance runners and 7.6% in active physical education students. The more highly trained individuals utilize glycogen and to some extent lipids more effectively than the average fit individual. A reduction in performance time of 3.2% would improve a 2 hour 30 minute marathon time to 2 hours 25 minutes personal best time and would justify the regime to some – especially if the profound feeling of exhaustion in the latter part of such a run, known colloquially as 'the wall' is delayed.

The accumulation of glycogen in muscle is dependent upon an initial glycogen depletion by exercise and a carbohydrate-rich diet thereafter. The rate of muscle glycogen restoration is faster than liver glycogen repletion, but is dependent upon training status. The more glycogen to be restored, the longer time period of feeding necessary, for example a week for marathon runners. The rate of repletion of glycogen may depend on the amount of carbohydrate eaten. The carbohydrate intake should be of the order of 500 g/day, which is 70 % of the calorie intake.

Other distance runners

The energy expenditure for different intensities of exercise for a lightly built woman and man will be different from that of a plump and moderately enthusiastic runner. Ten minutes' mild jogging a day will consume 2.1–2.5 MJ (500–600 kcal) in a week. Such activity will be better for personal feeling of well-being than actual weight loss. Determination is an important factor in extending exercise capability, especially as weight loss and improvements in serum lipid concentrations appear to be dependent upon the intensity and duration of the exercise programme. There is a relationship between the amount of exercise performed and the physiological change with exercise. To obtain significant alterations in weight and serum lipids, it is necessary to run at least 9 miles and 12 miles a week respectively for a year. In another study walking and running for 9–15 miles a week for 8 weeks resulted in weight loss of an average of 1.5 kg, an average increase of maximal oxygen consumption of 0.25 l/min, an average decrease in serum cholesterol from 5.34 to 5.15 mmol/l with increases in HDL.

These studies involved middle-aged men and women but the benefits of exercise are not confined to this age group. In a further study elderly men and women (aged on average 63 years) joined in a 12-month endurance programme, 4.6 miles of vigorous walking a week for 6 months, followed by 6 months of 100 minutes of jogging a week. At the end of both periods there were significant improvements in insulin response to a glucose tolerance test and maximal oxygen uptake. Improved serum lipids occurred after the second period. The problem with all of these studies is that there were significant muscular skeletal injuries limiting exercise tolerance.

The majority of runners of all ages have modest ambitions, health, weight control and the occasional marathon, half-marathon or 10-km run. The intense interest in nutrition and exercise comes from long-distance competitive runners. This is an uncompromising pastime with personal best time the omnipresent goal. Improvements in times of a few percent may make the difference between regional or national representation, a place in the Olympics final or unplaced in a local race. In the realization of the marginal differences between success and failure, athletes look to nutritionists for a boost in performance as much as to their inherent natural ability and training programme. Usually the chosen diet has no scientific basis but owes much to the confidence that the athlete feels towards his trainer or whoever suggests the diet or supplement. At the end of the day an important factor in endurance running is personal resolve, an attribute which no diet or dietary supplement so far has claimed to improve.

12.10.5 Water requirements

Perhaps the most important single influence on performance over distance is sweating, and the extent to which lost water is replaced during the run. Fluid loss and need for replacement depend very much on the temperature and humidity, varying from 3–4 kg at 100°F (38°C) and 80–100% humidity to 0.5–1 kg loss at 60°F or less (15°C) and less than 40% humidity.

It is well established that a decrease in body weight of more that 2% by exercise-induced sweating places severe demands on the cardiovascular and thermal regulatory systems. Even in a temperate climate the cumulative sweat loss in a marathon is often 3–4 litres, with an associated loss of electrolytes. The problem of fluid loss becomes even more important. The most consistent item of nutrition which is ignored in sport is water intake. Repeatedly in races as diverse as local to Olympic marathon runs, some hero sets out to run the fastest marathon ever run and in order to achieve this ignores the watering stations. The great and dramatic failures in marathon history have been caused by dehydration. Acute renal failure has been reported in participants of ultramarathons in South Africa some 24–48 hours after completing the run.

> During competition in a warm environment, a marathon runner running at 240 m/min (that is, a marathon run in 2 h 46 min) will use an estimated 12–14 kcal/min and lose 0.6 l/min, which may mean losses in excess of 6 l overall. Despite drinking water during the race the runner may lose 8% of body weight and 13–14% of body water – a hazardous loss, especially in an individual who continues to run.

During a long run, water reserves are maintained by the regular drinking of water and the metabolic water from glycogen metabolism. The water balance of a distance runner is improved if 500 ml of fluid is drunk 15–30 minutes before the race. Further small draughts of fluid should be drunk at regular intervals during the race. The balance is important. Too much fluid and the runner has to break rhythm to stop and pass urine; too little fluid and dehydration is a prospect. Drinks stations should be sited every 3 miles in a marathon, with sponges available in between. The drinks should provide electrolytes or just water according to the runners' choice. Runners, especially the inexperienced, should drink at every station. It is also important that the water is absorbed. The osmolality of the drink determines gastric emptying time. It is desirable that gastric

emptying time is such to avoid bloating and allow for comfortable running, and so a drink of 250 ml of water or dilute glucose for taste are recommended. The shortest gastric emptying time is achieved with an osmolality of 250 mosmol/l. A suitable drink for endurance athletes contains 25 g/l or less of glucose and 10 mM/l sodium and 5 mM/l potassium. These recommendations apply to any vigorous extended sport in hot conditions, e.g. soccer or rugby played at speed in the heat.

Sometimes after a long run, thirst is insatiable and lots of fluid must be drunk. Urine output is a good guide to need. After demanding runs, the urine may be sparse and concentrated. One rough rule of thumb is that for every 500 g lost, that two 250-ml glasses of water should be drunk.

12.10.6 Dietary requirements

Energy

Energy needs vary from athlete to athlete and sport to sport. The muscle mass and therefore energy utilization increases with training. The proportion of energy expenditure as resting metabolic rate is reduced. In the Tour de France cycle run the energy turnover is 3.5–5.5 times BMR, the energy intake being about 35 MJ/day. Ingesting this is a real problem.

Amenorrhoeic female athletes tend to have a lower calorie intake than menstruating athletes with similar fitness and sporting requirements.

Protein

The protein needs for endurance athletes (4–18 hours/week) are increased above those of the average population at 1.2–1.4 g/kg/day. The athlete will meet all nitrogen needs with a protein intake which is 15% of total energy intake.

Carbohydrates

There are many theories as to the best way to increase the muscle glycogen content before endurance races. Basically, a running work load, which regularly makes demands on the muscle glycogen content is necessary, accompanied by a diet which includes at least 500 g of carbohydrate/

day with the fat content correspondingly decreased.

> The USA and UK DRV for protein is 0.8g/kg/day. In a large energy intake the additional demands are easily reached. Sports in which lifting strength is important, e.g. body-builders and rugby and American football players require between 1.2 and 1.7 g/kg/day. The moderately active athlete will have a protein turnover of 1–2% of total protein per day (125–250%/day); 75% of this is recycled.

During the two to three days before a race athletes reduce training and replenish their carbohydrate stores with 10 g/kg/day of carbohydrates. The carbohydrates with a high glycaemic index are the preferred foods, e.g. bread, potato, pasta, rice or glucose. During the long events over several hours then instantly available carbohydrate as glucose is taken to meet the immediate needs. Glycogen resynthesis is most rapid in the 4–5 hours after prolonged exercise. Maximum rates of resynthesis of glycogen are achieved when the equivalent of 1 g/kg of carbohydrate is eaten every 2 hours in the first 5 hours after the exercise.

Fat

Stored fat, muscle triglyceride, adipose tissue fat and low density lipoproteins are important energy sources during moderately intense exercise. Training facilitates the breakdown of fats during exercise, through amounts of tissue lipase, and release and response to insulin. A diet rich in fat, unlike carbohydrates, is not conducive to better performance. It is not known if when fat is lost during exercise it is necessary to replenish tissue fat n-6 and n-3 polyunsaturated fatty acids and vitamin E.

Other nutrients

Despite claims which are sometimes strident there is no evidence that any other nutrient whether mineral, vitamin, protein, fat or whole food influences performance, though in the growing athlete a balanced – often massive energy input may be necessary. Iron-deficiency anaemia could be a problem in the menstruating female athlete but iron therapy should be based on biochemical and haematological measurements.

Specific needs

Dancers and gymnasts

These athletes are figure, as well as performance, conscious. Nutrition restriction is a feature of their programmes with consequent energy and mineral deficiencies; this may be accompanied by eating disorders, endocrine problems and dangers of fractures from thinned bone structure.

Wrestlers

These sports are dominated by concepts of nutrition which are ancient in concept and bereft of modern scientific evidence. Bouts last about 10 minutes and require muscle power, endurance and utilize aerobic and anaerobic pathways. The competitors often achieve the weight for a fight by fluid and food deprivation. Wrestlers can lose 2–5 kg in weight during a fight and fluid and electrolyte replacement is important. A diet of 12% protein, 58% carbohydrates and 30% fat is recommended for their sport.

12.10.7 Drugs and sport

It is possible artificially, dangerously and illegally to increase athletic performance with amphetamine- and steroid-type drugs. Some athletes use such drugs in this way.

Amphetamines

These stimulate the nervous system by bringing about the release of dopamine into the synaptic cleft. Amphetamine also facilitates the release of noradrenaline. Amphetamine is usually taken as a tablet and as a consequence of increased circulating adrenaline there is an increase in heart rate, breathing and alertness. Such a drug is therefore believed to increase endurance, speed and power, postpone tiredness and hunger, and is the choice of cyclists. The risk is an increase in heart rate and respiration leading to cardiac failure and death.

Overdosage is also a danger. In the long term, hypertension, strokes, addiction, delusions, hallucinations and mental illness are all risks.

Anabolic steroids

These can be taken as tablets or by injection. The effect is to mimic the male hormone testosterone and increase muscle bulk, body hair and aggression. This is believed to result in increased strength and speed and is the choice of some sprinters, weight-lifters and throwers (e.g. shot, discus).

There is an increased risk of liver cancer, hyperlipidaemia and fluid retention in both men and women. Men are at risk of impotence and acne; women of developing male body composition, acne, infertility and increased body hair. Injecting with re-used needles carries the usual risk of viral transmission, hepatitis and AIDS.

Drug testing

In championships – and occasionally randomly – athletes are chosen for urine testing. The samples are divided into two containers which are sealed. One is sent for analysis at an approved laboratory; the other is retained as a back-up sample. If the test is negative then both are destroyed. If the first is positive then the second sample is tested in the presence of the athlete or representative. If negative then all is well. If positive then disqualification results.

Summary

1. The diet of athletes must meet their energy requirements and the demands posed by the speed and length of time of exercise. Sports requiring coordination have no special dietary needs. Increased protein calorie intake may be required in the body-building sports whereas a high carbohydrate diet is required by endurance runners.
2. Sport makes demands on biochemical and mechanical efficiency. Mechanical efficiency declines with the amount of work involved in the sport.
3. Skeletal muscle consists of two types of fibre: type 1 is for aerobic metabolism and is resistant to fatigue; type 2 is for anaerobic metabolism.
4. Endurance runners require stores of glycogen to complete their prolonged activity. The glycogen is stored in muscles, the storage capacity being increased by training. Increased carbohydrate intake is required during such training periods.
5. In endurance sports, water is lost and must be replaced during the run or severe and life-threatening consequences result.
6. Energy requirements vary from sport to sport; the most demanding is the Tour de France cycle race. Protein requirements are modestly and carbohydrate massively increased in endurance sports. Fat has no merit in stamina provision.
7. There is little evidence that other nutrients are required in amounts in excess of the non-exercising population of the same age.
8. It is possible artificially, but both dangerously and illegally, to increase athletic performance with amphetamine- and steroid-type drugs.

Further reading

Costill, D.L. (1977) in *The Marathon: Physiological, Medical, Epidemiological and Psychological Studies* (ed. P. Milvy), New York Academy of Sciences.

Durnin, J.V.G.A. (1985) The cost of exercise. *Proceedings of the Nutrition Society*, **44**, 273–82.

Fentem, P.H. (1985) Exercise and the promotion of health. *Proceedings of the Nutrition Society*, **44**, 297–302.

Nutrition for Sport (1984) Joint publication of the United States National Association for Sport and Physical Education, The Nutrition Foundation, The Swanson Center for Nutrition and the United States Olympic Committee.

Rowe, W.J. (1992) Extraordinary unremitting endurance exercise and permanent injury to the normal heart. *Lancet*, **340**, 712–14.

Shephard, R.J. (1982) *Physiology and Biochemistry of Exercise*, Praeger Publishers, New York.

Wasserman, D.H. (1995) Regulation of glucose fluxes during exercise. *Annual Review of Physiology*, **57**, 191–218.

Williams, C. and Devlin, J.T. (1992) *Foods, Nutrition and Sports Performance*, E & F.N. Spon, London.

Williams, P.T., Wood, P.D., Haskell, W.L. and Vranizan, K. (1982) The effects of running mileage and duration on plasma lipoprotein levels. *Journal of the American Medical Association*, **247**, 2674–9.

12.11 Nutrition in space

- Energy requirements in space are the same as on earth.
- Fluid balance, bone and muscle status are affected by space flight with demineralization of bones and muscle atrophy.

12.11.1 Introduction

The first space flight in 1961 by Yuri Gagarin lasted 108 minutes. Since then, teams of nutritional scientists involved in the care of the American and Russian astronauts have devised meals suitable for space flight. There is no refrigeration in space and food must therefore be storable at ambient temperature. Flights now may extend over 12 months. The available space in the capsules is limited and has precluded studies of metabolic responses. However, it is known that flight can induce persistent negative energy, alter nitrogen and potassium balance, and cause loss of body mass. These probably reflect changes in body composition, energy utilization and endocrine status.

12.11.2 Body composition changes

Fluid balance, bone and muscle status are affected at different rates in space. Some of the mass lost is body fluid, of the order of 500–900 ml. Plasma volume decreases by an order of 6–13% at the end of the flight.

There is also bone demineralization, which is a gradual process dependent upon the length of mission. A flight of 184 days resulted in a loss of 20% in bone. Both compact and trabecular bone structure is lost. There is increased excretion of calcium and phosphorus during space flight.

Muscles, particularly of the leg, atrophy in space though the cause may be secondary to inadequate exercise or insufficient food or the effect of altered gravity. As the muscles atrophy urinary potassium, nitrogen and uric acid increase.

The energy requirements in space are the same as those on Earth, though carbohydrate utilization may be increased. Mean respiratory quotients may increase from a mean of 0.887 ± 0.09 before flight to 1.041 ± 0.09 during flight and returning to approximately pre-flight figures on return. The total energy requirement for work in space is of the order of 30–45 kcal/kg/day depending upon body mass, length of flight and activities. The loss of lean body mass is puzzling and increasing protein intake may increase the risk of renal stones.

Further reading

Lane, H.W. (1992) Nutrition in space: evidence from the US and the USSR. *Nutrition Reviews,* 50, 3–6.

12.12 Tobacco and smoking

- Tobacco smoking is an important cause of disease and may influence nutrient intake.

12.12.1 Introduction

There are a number of conditions in which nutrition is regarded as having an aetiological role, but in which smoking is also implicated.

12.12.2 Association between smoking and disease

Smoking is generally recognized as an aetiological factor in a number of diseases:

- **Cardiovascular disease,** including atherosclerosis, coronary heart disease, peripheral vascular occlusive disease and cerebrovascular disease.
- **Pulmonary disease,** including lung cancer, chronic obstructive lung disease and chronic mucus hypersecretion.

There are also a number of other conditions which have been associated with smoking, though smoking does not absolutely lead to disease. Yet less than 0.5% of the smoking population ever contracts the major diseases that have been linked with cigarette smoking. There may be either a genetic vulnerability or resistance to these tobacco-related diseases.

It is estimated that 50–55 million people are cigarette smokers in the United States. Each year there are approximately 115 000 people who die from coronary heart disease, 126 000 from lung cancer, 57 000 from chronic obstructive airways disease, and 27 000 people from stroke and whose deaths are directly attributed to cigarette smoking. It can be shown that a section of the non-smoking population will also develop and die from such disorders. This may possibly – but not necessarily – be due to secondary or **passive smoking,** from smokers' fumes. Another paradox is that in Japan, which has one of the highest *per capita* consumption of cigarettes, the incidence of lung cancer is one of the lowest in the world. Diet might be one of the factors which modify the risk of developing cardiovascular or pulmonary disease in people who smoke cigarettes.

Free radicals and singlet molecular oxygen react with biological membranes to cause membrane lipid oxidation and/or peroxidation of polyunsaturated phospholipids. This reaction produces membrane damage or molecular changes in cellular DNA, carbohydrates or proteins which results in initiation or worsening of chronic cardiovascular or pulmonary disease. The combustion of tobacco during cigarette smoking promotes the oxidation of polycyclic aromatic hydrocarbons and produces free radicals. It has been estimated that there are 10^{14} free radicals per inhalation in the tar phase.

Long-term cigarette smoking results in a decrease in plasma levels of HDL and HDL apoproteins, and an increase in LDL. The free radicals from cigarette smoke could be the first step in the lipid peroxidation of unsaturated fatty acids in the membranes of LDL particles. Lipid peroxidation of LDL begins only after the LDL particle has been depleted of intrinsic lipophilic antioxidants, e.g. α-tocopherol (vitamin E) and β-carotene. Membrane oxidation of the LDL results in its uptake by macrophages in the subendothelial space of the vascular wall and development of fatty streaks and atherosclerosis. Smokers also have reduced plasma levels of β-carotene and vitamins C and E. A variety of nutritional agents may reduce or inhibit carcinogenesis. Vitamin A, retinoids and retinoic acid can in some studies suppress malignancy and even reverse premalignant phenotypes. A diet that has a reduced vitamin C, carotene, fibre, fruit, vegetable, whole-grain cereals and bread content, and substantial fat, cholesterol, energy, red meats and processed meats, increases the likelihood of tobacco-related diseases.

Smoking is recognized as one of the major risk factors for coronary heart disease. Diet may be one of the risk factors which alters with smoking cessation. Smoking affects taste preferences and those who stop tend to have an improved health consciousness. For men, their diet appears to improve with duration of ex-smoking and reaches approximately the level of nutrient intake of the never-smokers by 10 years after stopping smoking. There is an initial increase in sugar and energy consumption which is short-lived. This may be the result of immediate compensatory eating of sweets instead of smoking, followed by a combination of altered taste preference and the avoidance of sugar to resist weight gain. Ex-smokers eat a diet with an increased polyunsaturated fatty acid, fibre and vitamin intake which might suggest increasing health awareness and long-term changing of dietary habits.

Further reading

Diana, J.N. and Pryor, W.A. (1993) Tobacco smoking and nutrition; the influence of nutrition on tobacco and associated risks. *Annals of the New York Academy of Sciences*, 686, 1–11.

US Office on Smoking and Health (1979) *Smoking and Health. A Report of the Surgeon General*, US Government Printing Office, Washington, DC.

Chapter 13

Dietary deficiency

- Famine is a failure to provide adequate nutrition to sustain a population
- Famine is found where there is a failure of crops or supply of food due to adverse weather, war, social deprivation or pestilence.

13.1 Famine

13.1.1 Introduction

A definition of health should include the concept of sustainability or the ability of the ecosystem to support life in quantity and quality. Entrapment occurs when a community exceeds the food-carrying capacity of the land, when it lacks the ability to obtain food to sustain its population, or when its people are forced to migrate from a bad to a worse situation. Entrapment leads to dependence on outside aid, forced migration, starvation or civil war. Reduced mortality rate, continuing high birth rates, increased longevity, poor uptake of contraception and precarious food sources raise the question of whether populations are in danger of entrapment.

13.1.2 Geographical factors

There is much starvation throughout the world. A terrible example of the problem is Southern Africa where the changing weather patterns are resulting in reduced rainfall, resulting in rivers, wells and water holes drying up. Harvests deteriorate and widespread death from starvation becomes inevitable. A state of famine exists and an inevitable death from starvation is the fate of millions of Africans. Countries, e.g. Zimbabwe – formerly seen as the regional source of food – are now faced with drought and crop failure. Crops are reduced to less than half or even a quarter of past or expected crops. While war and pestilence are factors, the main problem is lack of rain. All countries have been affected regardless of their political or economic status.

13.1.3 Ecological factors

Trees are chopped down to provide fuel to cook the food and this weakens the stability of the soil and allows erosion to occur. This may be overcome by the planting of fodder trees and grasses. Half of the world's population depends upon wood for fuel, used for cooking and keeping warm at night. As trees disappear there is a reduced uptake of carbon dioxide by the trees. They are burned and the result is an increased production of carbon dioxide. Trees are a natural sink for carbon dioxide.

Ozone is an important shield from ultraviolet irradiation and may be lost in the presence of gases such as chlorofluorocarbons (CFCs) which are used in refrigerators.

Nearly three-quarters of the Earth's surface is covered in salt water and unavailable for human needs. Some 1.5% of children in the Third World die of diarrhoea from contaminated water before the age of five.

Soil is a delicate balance between water, clay, humus and sand. An undue loss of one constituent can lead to a desert of useless soil. Organic waste is needed to replenish the active principles in the soil.

As the earth warms then there will be the thawing of the polar ice and the seas will rise. This will affect such countries as Bangladesh where more than a million of the population live on ground which would be covered by a 1-metre rise in the water level. Cyclones are particularly dangerous to such populations and occur when the surface temperature of the sea rises over 29°C.

War and **civil strife** are important creators of famine. Peace can often only be achieved or continued if both sides feel that aid is reaching them equally. The food must be of the correct type. Malnourished people need a balanced diet. The available crops are largely starch and low in protein content, e.g. derere, a boiled wild okra and mahewa, a fermented millet. Local maize is more likely to grow than an imported variety. Maize is cheap and plentiful but beans and oil and fresh fruit and vegetables are also necessary. A clean water supply is also critical. To preserve the population, it is vital that seeds and tools are provided at the correct time for planting.

Summary

1. Famine occurs when a community's food needs exceed its supplies. This a common problem throughout the world. War and pestilence are important contributors to the problem but the fundamental cause in many areas is lack of rain.
2. The problem is compounded by the felling of trees for fuel, the loss of ozone in the atmosphere, and the consequent effects on the weather, the polar ice mass and sea levels.

Further reading

Smith, R. (1993) Over population and over consumption. *British Medical Journal,* 306, 1285–6.

Weaver, L.T. and Beckerleg, S. (1993) Is health a sustainable state a village study in the Gambia? *Lancet,* 341, 1327–30.

13.2 Starvation

- Starvation may result from shortcomings in total energy intake or of specific nutrients.
- The body may partially adapt to starvation, deficiency of total energy intake or deficiency of specific nutrients, e.g. protein.
- A person's physical activity may adapt to reduced nutritional and energy intake.

13.2.1 Introduction

There are many causes of starvation:

- insufficiency of food
- digestive tract malfunction and malabsorption of specific and total nutrients
- impaired appetite due to disease, e.g. cancer, or psychological, e.g. anorexia nervosa
- abnormal tissue metabolism, e.g. renal disease, hepatic disease or hyperthyroidism
- severe long-standing infection
- voluntary political hunger strike
- oppressed groups

Oppressed groups are found everywhere and in every country. One such group are the begging eunuchs of Bombay, the Hijras or Chakkas, who were sold by poverty-stricken parents, subjected to castration and penectomy without anaesthesia and thereafter left to beg.

13.2.2 Adaptation

When starvation is encountered, the body attempts to **adapt** to the problem. The definition of adaptation to starvation is the process of change of a defined entity, e.g. cell, organ, body or society in response to a defined cause, e.g. infection, starvation. Alternatively in an otherwise fit individual successful adaptation is achieved when the body or a function is retained within an acceptable range. Adaptations may be classified as:

1. **Biological and genetic**; this is the response which is dictated by the genetic make-up of the individual.
2. **Physiological and metabolic**; the body switches metabolic processes from synthetic to conserving, from storage to the organized release of carbohydrates and fats and then proteins, the object being to protect vital organs.
3. **Behavioural and social**; in societies where starvation is experienced parents will protect their children. Mothers will give their own food to their offspring. The breadwinner will also be given preference. The level of activity of starving people may decline to match intake. In prolonged starvation these orderly systems break down and individuals become more aware of self-survival rather that the protection of their society and even family.

Variations in response to starvation

An individual with substantial lipid stores will cope with starvation better than an individual who is thin or enfeebled by illness.

There are great variations between individuals with regard to energy intake, just as there are large differences in recorded intake between similar individuals engaged in similar activities. There are also intra-individual variations, the coefficient of variation in the same individual over 3 weeks perhaps being of the order of 20%. The variability in energy expenditure appears to be less than that in intake.

The coefficient of variability of expenditure per m² surface area while lying, sitting and standing is about 16%. The coefficient of variability for BMR (adjusted for age, sex and body weight) is 7–10%. The average amount of protein needed for zero nitrogen balance may have a coefficient of variability of 12.5%. There are substantial interindividual differences in protein turnover. The 'P ratio' is the ratio of protein stored to total energy stored during weight gain. Conversely, the ratio of protein loss to energy loss during weight loss may be characteristic and fixed for each person.

Another variable is the nutritional intake (energy and protein) and existing body weight. A balance is needed to maintain body weight at a required energy expenditure.

A lower BMI limit compatible with acceptable functional capacity is also an important milestone in a starving population. In many Third World countries the average BMI is 18–19 with a range of 15–23. Hard work is possible with a BMI of 15–16 and an estimated body fat content of 6%. It may be that a BMI of 13 should be regarded as the absolute lower limit of what is acceptable.

The rate of weight loss or BMI status is all-important. A rapid loss of weight is very dangerous whereas a slow loss or failure to achieve acceptable weight may be much more compatible with a healthy life. Death is likely to occur when 40% of body weight has been lost, i.e. a BMI of about 13 in a person whose initial BMI was 22.

Children and starvation

Children are very sensitive to deficient intakes of energy and protein, because of their growth needs, when retardation of growth can be very conspicuous. Many children in Third World and semi-starving situations are severely stunted in height. Height deficits in older children and adults are probably established in the first years of their life. Stunting of growth in children is an adaptation which requires less food and therefore the child is more likely to survive. However, survival without an acceptable functional capacity is not a favourable adaptation. It is assumed that everyone has a right to develop their full genetic potential. In terms of food intake this aspiration requires that there is total equality of access to a broadly based diet.

Physiological efficiency

Efficiency is the ratio of

$$\frac{\text{Energy expended during work} - \text{energy expended at rest}}{\text{Total energy expended while working}}$$

Energy cost of activities and occupations might also be expressed as multiples of the BMR. In normal subjects the cost per kg is independent of body weight. Gross energy expenditure per kg is probably the most important useful measurement for comparing individuals or groups. There may well be significant differences in work from a low efficiency in individuals with a high dietary energy intake compared with those with a high efficiency on a low intake.

In real life, individuals vary in their mode of activity and how they work, play and rest. Energy intake may be achieved with considerable economy of effort and few superfluous movements. Short alternating periods of work and rest allow the same amount of work in a given time with a much lower heart rate and lactic acid level than if the work and rest periods are longer. The greatest rate at which a person can use oxygen during continuous exercise is VO_{2max}, which is low in undernourished people. Indices of physical fitness when corrected for body weight may be equal or even greater. The most important adaptation to a low energy intake is to have a low body weight.

Diet-induced thermogenesis

Individuals who eat substantial amounts of food may have a larger thermic response than modest eaters. An inverse relationship between body fat content and the thermic response to eating fat has been shown. This has not been shown for protein or carbohydrate. The energy content of the meal – and not its composition – may be important, suggesting that thermogenesis is unlikely to be important in causing or maintaining obesity. Diet-induced thermogenesis is not a catabolic response but represents the energy cost of protein synthesis and the conversion of carbohydrate to fat. There may be a reduction in diet-induced thermogenesis where protein turnover is depressed. Diet-induced thermogenesis uses approximately 10% of the energy value of a meal. Over a whole day this could account for approximately 200 kcal.

13.2.3 Effects of starvation

Fasting

During fasting, blood glucose concentrations need to be maintained to ensure cerebral function. Calories are drawn from fat but not protein, in a system controlled by hormonal and metabolic changes.

The sequence of metabolic changes (and time at which such changes occur) after the onset of starvation is:

- effects on gastrointestinal absorption (1–6 hours)
- glycogenolysis (1–2 days)
- gluconeogenesis (12 hours to 1 week)
- ketosis (from 3 days onwards)

The first phase of fasting depends on the carbohydrate concentration of the preceding meal. If the preceding meal is large and predominantly carbohydrate in content, then the liver subsequently removes glucose from the blood in response to increased insulin and decreased glucagon secretion. The glucose is incorporated into glycogen and later metabolized to pyruvate and lactate. The body stores of glycogen are modest (70–100 g in liver and 300–400 g in muscle) and last for approximately 12–24 hours.

After 12–16 hours, gluconeogenesis starts, as a result of glucagon excess over insulin, increased hepatic cyclic AMP and an increased concentration of free fatty acids. This leads to increased fat oxidation. Over the next period of time the utilization of ketoacids becomes more efficient. The measurement of the rate at which dependence upon glucose or gluconeogenesis occurs is somewhat dependent upon the technique used. These differences are quantitative rather than qualitative in their importance.

When food is unavailable the body has to rely on its own stores. The crucial provision of glucose for the brain and elsewhere, depends on liver glycogen and subsequently on the synthesis of glucose by both the liver and kidneys, initially from muscle protein amino acids. Collagen proteins – which represent 25% of muscle – are preserved. The liver, intestine, skin, brain and adipose tissue, can contribute 1 kg of protein amino acid. Muscle glycogen falls progressively during the first 5 days of starvation and may contribute 140 g of glucose to the brain after hepatic glycogen reserves are exhausted and circulating blood ketones are insufficiently concentrated to supply the brain.

Muscle protein catabolism releases mainly alanine and glutamine. Alanine is the preferred substrate for gluconeogenesis in the liver and glutamine contributes to gluconeogenesis in the kidneys. In prolonged starvation the flow of alanine from muscle falls and this is reflected in a steady decline in urea synthesis and excretion. After 10 days of fasting, ammonia becomes the main urinary nitrogen product.

There is an increase of ketone bodies from fatty acids in the liver, to be used by most tissues, including the brain. Initially, ketone production is small, as fasting continues they increase progressively to become the dominant substrate. Two main ketone bodies – acetoacetate and 3-hydroxybutyrate – are formed and found in urine, generated from acetyl-CoA in liver. Atrophy of tissues is the most characteristic feature of starvation. The wasting and loss of weight is initially rapid, but gradually slows down. The actively metabolizing cell mass is reduced and requires less energy to maintain activity. Unnecessary voluntary movements are curtailed.

A healthy, non-obese subject can lose 25% of weight without endangering life. During starvation, in a 65 kg man, 3 kg of protein, 6.5 kg of fat, 200 g of carbohydrate, 6 kg of intracellular water and 70% of fat are lost. These represent 300 MJ reserves of nutrition and last for approximately 50 days. The major weight loss in the first week is 1.5 kg of body water, when water-bound glycogen is released and excreted; 1 g of glycogen binds 3–5 g of water.

Protein deficiency in starvation causes a fall in concentration in plasma albumin which also contributes to oedema.

Most people with primary under nutrition recover rapidly with access to food. Over 20 MJ/day may be consumed when free food is available.

Adaptation to low energy intakes

The energy expenditure of subjects engaged in minimum physical activity is 1.4 × basal metabolic rate. The four major components of energy expenditure are:

- basal metabolic rate
- physical activity
- growth in children
- pregnancy and lactation

There are four adaptation processes to starvation:

- weight loss
- reduction in voluntary and conscious activity
- unconscious economy of activity
- true metabolic adaptation

Energy expenditure is related to weight or lean body mass. Total energy expenditure rarely exceeds twice the BMR.

> The relationship of BMR (as kcal per person per day) to body weight is not linear. In healthy subjects the BMR/kg rises as body weight falls and this is independent of height, e.g. a young adult will have an expected BMR of 25 kcal/ kg/day whereas at 55 kg the BMR is 27.6 kcal/ kg/day.

The effect of weight loss

The effects are variable. There is a rapid initial fall in BMR in response to reduced energy intake followed by a smaller fall in parallel with loss of weight. Tissues such as muscle and skin with low metabolic rate are preferentially lost while the visceral tissues and brain with high metabolic rate tend to be preserved. A person with a low body weight has from a physiological point of view reduced metabolic rate per unit lean body mass.

BMR appears to be the same for large and small eaters. Indians have a BMR that is significantly lower by approximately 9% than European or North Americans. This may be due to climate, ethnic group or diet content or adequacy. The climate may have only a very small effect on BMR. Very underweight Indian labourers may have adapted to weight loss as they are able to continue to be active and fit.

In the muscles of malnourished patients there is an increase in the ratio of slow to fast fibres in the muscles. This is a result of reduction of fast fibres (type 1), the slow (type 2) being better preserved. Similar effects on muscle fibres have been shown in hypothyroidism. Hence, it is possible that adaptation is a relative preservation of slow-twitch fibres adapting for a new lifestyle.

13.2.4 Dietary protein deficiency

Nitrogen balance

Nitrogen balance is obtained from obligatory nitrogen loss and efficiency of restoration of nitrogen by food protein.

Obligatory losses

In well-nourished young men the average obligatory nitrogen loss is 60 mg/kg/day; 35–40 mg is excreted in urine, 15–20 mg in faeces, and 5 mg from the skin. The faecal component depends on the diet and may be increased with a high-fibre diet. Nitrogen is lost from the skin as urea and is trivial. Urinary nitrogen loss is the most important. Even on minimal protein intake 50% of urinary nitrogen is urea. Ammonia excretion is determined by the need to maintain acid–base balance and is lower on vegetable rather than animal protein diets. Uric acid, creatinine and free amino acids appear to be a consistent loss.

Efficiency of utilization

This is obtained from the slope of the line comparing nitrogen balance to intake. Nitrogen intake is required to replace obligatory losses at a constant weight. The protein requirement of different groups is approximately 0.6 g protein/kg/day. These figures are based on short-term studies and longer-term studies may give different results. There are day-to-day variations in urinary nitrogen excretion. Changes from a high to a low protein intake result in the loss of so-called **labile protein,** the source of which is unknown.

On a normal diet 30% of the urea produced in the liver passes into the colon to be split into ammonia by bacterial urease. This may be recycled to urea and part taken up as non-essential amino acids. On a low protein intake there is an increase in the proportion of the urea produced utilized for amino acid synthesis but no increase in the absolute amount taken up. Urea may be incorporated into amino acids and hence into proteins as a result of transamination of NH_3.

Energy balance affects nitrogen intake. Each extra kilocalorie reduces urinary nitrogen loss by about 1.5 mg. Not infrequently, low protein intake and low energy intake go together, so that there are small stores of body fat and adaptation to starvation is difficult.

Nitrogen metabolic balances

There are two nitrogen cycles: *input/output* and *synthesis/breakdown*, which connect through the free amino acid pool. When one is not in balance the other must also be out of balance, since alterations in the size of the free pool are small in relation to flux. The two cycles are not inter-related when they are in balance but they are related indirectly when there is a reduction in the rate of protein turnover.

If there is a small reduction in energy intake: protein turnover is 15% of the BMR so that a reduction in rate of protein turnover saves calories. There is less net protein synthesis in response to the ingestion of food and hence smaller requirements for essential amino acids. Survival would be possible on a dietary protein mixture of reduced biological value. The increase in protein breakdown in the post-absorptive state is reduced, influx of amino acids into the free pool is reduced and hence less loss by oxidation.

The time-course of reduction in the urea cycle enzymes closely parallels changes in urinary nitrogen output. The rate limiting enzyme is probably arginosuccinate synthetase. Another adaptation to a reduced protein intake is the rate of input of amino nitrogen to the urea cycle. The enzymes of much of the amino acid metabolic pathways are modified by dietary protein intake. When urea production is reduced, though it may be a result of toxic accumulation of ammonia, this may be removed by recycling to amino acids and hence into protein. However, protein synthesis requires sufficient essential amino acids available in appropriate amounts. The limiting factor is the rate of oxidation of the carbon skeleton of the essential amino acids. It is possible that the important amino acids in this situation are the branched-chain amino acids. These form 20% of the amino acid residues of most proteins but their concentration in the amino acid pool is relatively low and therefore supply becomes rate limiting. The first step in their catabolism is transamination in muscle followed by irreversible decarboxylation of the ketoacids by the branched-chain amino acid dehydrogenase complex.

Branched-chain amino acid dehydrogenase is widely distributed in the body. The K_m of the enzyme is close to the concentration of the substrate in the free pool so that activity is concentration-dependent. The enzyme exists in an inactive phosphorylated form which is activated by dephosphorylation. There is also an activator protein that reactivates the enzyme without dephosphorylation. These result in a reduction in dehydrogenase activity when dietary protein is restricted.

Adaptation to low protein intakes

Achieving nitrogen balance over a wide range of dietary protein intakes is important. When the capacity to economize on nitrogen metabolism is exhausted then there is a reduction in lean body mass. Protein requirement in an individual at a constant height is directly related to lean body mass. In children, there is a reduction in the rate of growth in weight with resultant stunting in height. Pregnancy and lactation are different physiological states, wherein the priority is the foetus and its growth is at the expense of the mother's own tissues.

Sub-maintenance intakes

At sub-maintenance intakes, rates of turnover of protein are reduced. At the same intake the turnover rate is lower in children with kwashiorkor. Such children are unable to increase their turnover rate in response to an infection and excrete less nitrogen than infected children who are normally nourished. Muscle mass is important in the response of whole-body protein turnover to low protein intakes. The diurnal changes in whole-body protein turnover in response to fasting and feeding run in parallel with rate of change of muscle protein synthesis. In children, when the protein intake is reduced there is an immediate fall in the rate of albumin synthesis. Thereafter, there is a fall in the rate of breakdown and a shift of albumin from the extravascular to intravascular compartment. This maintains the intravascular circulating albumin mass.

When amino acid supplies are limited the rate of protein turnover falls in many tissues and in the whole body. Protein turnover is relatively less reduced in the liver and visceral tissues than in muscle. The reduction for the body as a whole is small. The thyroid hormones are important in regulating protein turnover.

The body is very efficient with a rapid response to variations in amino acid supply, economizing when they are in short supply and disposing when in excess. There are, however, limits to this adaptive capacity.

Summary

1. The causes of starvation are many, but lack of food, especially energy and protein are critical deficiencies.
2. During starvation there is adaptation to the new dietary restrictions. The adaptation is both genetically and biologically determined, with physiological, metabolic, behavioural and social changes. There is considerable inter-individual variation in adaptation; children are particularly vulnerable.
3. Physiological efficiency is important in determining successful adaptation.
4. The effects of starvation are weight loss, changes in metabolism, and increasing dependence upon available nutrient stores. The specificity of such release of stores causes some metabolic imbalances, e.g. ketone body production.
5. Adaptations to low energy and low protein intakes are different. Low energy intake leads to weight loss and reduced energy. Dietary protein reduction has a profound effect on protein structure in all forms throughout the body.

Further reading

Allahbadia, G.N. and Shah, N. (1992) Begging eunuchs of Bombay. *Lancet,* 339, 48–9.

Burges, R.C. (1956) Deficiency disease in prisoners of war at Changi, Singapore. *Lancet,* ii, 411–18.

Leader (1992) Insights into fasting. *Lancet,* 339, 152–3.

Olubodon, J.O.B., Jaiyesimi, A.E.A., Fakoya, E.A. and Olasoda, O.A. (1991) Malnutrition in prisoners admitted to a medical ward in a developing country. *British Medical Journal,* 303, 693–4.

Payne, P.R. and Dugdale, A.E. (1977) Model for the prediction of energy, balance and bodyweight. *Annals of Human Biology,* 4, 425–35.

Waterlow, J.C. (1986) Metabolic adaptation to low intakes of energy and protein. *Annual Review of Nutrition,* 6, 495–526.

Waterlow, J.C. (1994) Childhood malnutrition in developing nations. *Annual Review of Nutrition,* 14, 1–20.

13.3 Protein energy malnutrition

- Protein energy malnutrition is a spectrum of nutritional inadequacies from insufficient total energy or sufficient energy but with inadequate protein intake.

13.3.1 Introduction

The term protein energy malnutrition (PEM) is applied to a group of clinical conditions of both adults and children, kwashiorkor, famine oedema, marasmus and cachexia. Severe under-nutrition in adults is found in famine or is secondary to illness, such as anorexia nervosa or cancer.

PEM is the most important social health problem in developing countries and is an important factor in the development and mortality of over half the children in such countries who do not survive for more than 5 years.

PEM is not, however, confined to poor countries. This condition may also occur in the families of the poor in the Western world and also as a result of neglectful parenting, child abuse and ignorance. In the afflicted children there is a failure of growth.

13.3.2 Clinical presentation

Marasmus is caused by a lack of dietary energy, protein and other nutrients in the growing infant. It occurs in infants under 1 year, in populations where infants have insufficient food. There are inadequate amounts of all required dietary components. The common background is of a family with frequent pregnancies, early and abrupt weaning followed by dirty and dilute artificial feeding of the infant. Repeated infections develop and the child may be treated with water, rice water and other non-nutritious foods.

Marasmus also occurs in total starvation. This condition is found in the young baby leading to severely compromised growth. The baby is cachectic, alert and ravenous.

The term **kwashiorkor** was introduced into modern medicine by Cecily Williams in 1931 and the Ghanaian word means 'the sickness which the second child gets when the next baby is born'. Kwashiorkor occurs due to quantitative and qualitative deficiencies of dietary proteins with an otherwise adequate energy intake. It occurs in the second year of life after a prolonged period of breast feeding. The child is weaned to a traditional diet which because of poverty is deficient in protein. There is no supplementation of milk.

Pitting oedema is always found in kwashiorkor; the oedema is both dependent and periorbital. The reason for the oedema is the low serum albumin and the consequent reduced hydrostatic pressure created by the plasma proteins. Because of this there is a failure to bind water by hydrophilic proteins. In addition, there is, possibly due to increased permeability a leakage of fluid from the vascular compartment into the extracellular tissues.

> In kwashiorkor, sodium and water retention in the extracellular fluid may reach 50% in severe cases. There is inappropriate distribution of sodium and water throughout the compartments of the extracellular fluids. The intravascular volume may be depleted. Fluid may accumulate in the peritoneum, pleural cavity or pericardium. In severe cases the entire body and internal organs become oedematous.

The combination of loss of soluble proteins and the excess of sodium and water is responsible for the oedema. There are frequent deficiencies of other intracellular ions, e.g. magnesium, zinc, phosphorus, iron and copper with consequent effects on metabolism.

Children with PEM fail to grow, are apathetic while resting, but cry when nursed. Adults become introspective and apathetic. The cerebral functions that are most affected are the higher cerebral functions with consequent intellectual impairment.

Not all organs of the body are uniformly affected by PEM, those most affected being the least essential to life. There is a reduction in the gastrointestinal tract mucosa, reduced salivary glands, fat stores, muscle mass, heart, liver, pancreas, reproductive organs, and thymus. The baby may have a mild normochromic anaemia due to reduced haematopoietic activity. Brain size is unaffected, though it is difficult to know if the changes in size of an organ mean changes in function. Does a smaller organ function as well as previously? Does a starved brain of constant size function as well as when adequately fed? The marasmic child is more sensitive to bacterial, viral and fungal infections, due in part to a reduced thymus size and circulating T-cell population. The spleen and adrenals may be increased in size. The patients readily succumb to infective diarrhoea and hence malabsorption occurs. As a result, kwashiorkor may develop in addition to the marasmus.

In the child there is a range of clinical signs which, at the one extreme can be called marasmus, and at the other, kwashiorkor. All gradations between these two are seen clinically and can occur at all ages. Each of the clinical signs has a biochemical and metabolic basis.

Advanced protein malnutrition in adults is associated with progressive weight loss, thirst, craving for food, weakness, lax, pale and dry skin with pigmented patches and loss of turgor; thinned hair and pedal (famine) oedema. There is an increased risk of infections probably due to reduced serum immunoglobulins. The serum albumin is reduced to under 30 g/l. Skin infections are common, including tinea versicolor and scabies. This condition has been described in neglected prisoners in overcrowded war camps and civil prisons in developing countries.

13.3.3 Response to treatment

Feeding usually results in a restoration of function and growth. Cerebral function may, however, make only a partial recovery particularly in the infant. It is important not to overload the intestine during the early stages of re-feeding. The enzyme systems and body proteins are present in insufficient amounts to cope. The intestinal mucosa may be very thin and unable to tolerate too much food in the lumen. After the liberation of the concentration camps in World War II, some inmates died from intestinal perforation due to the enthusiastic feeding of steaks and other foods for which the intestine was ill-prepared. The important early remedies should include the eradication of infection, the provision of vitamins and trace elements and then a slow increment of protein and energy intake. A feature of the starved when re-fed is that the intake of water exceeds the hydrophilic capacity of the body, e.g. proteins. Therefore in the previously compensated, oedema develops. This will only resolve when protein synthesis is seriously underway. Many adults make a full recovery.

Summary

1. Protein energy malnutrition (PEM) is a group of deficiency conditions where all combinations and degrees of energy and protein deficiency are represented.
2. Marasmus is a lack of dietary energy, protein and other nutrients, and is found in total starvation. Kwashiorkor is a deficiency of dietary protein with sufficient calorie intake.
3. In PEM the organs essential for life are conserved if possible at the cost of the expendable tissues, e.g. muscle mass.
4. Feeding can restore growth in the young, but intellectual recovery may not be complete.

Further reading

McCarrison, R. (1921) *Studies in Deficiency Disease*, Oxford Medical Publications, Oxford.

Waterlow, J.C. and Stephen, J.M.L. (eds) (1981) *Nitrogen Metabolism in Man*, Applied Science Publishers, London.

Williams, C.D. (1933) A nutritional disease of childhood associated with a maize diet. *Archives of Disease in Childhood*, 8, 423–33.

World Health Organization (1981) *The Treatment and Management of Severe Protein Energy Malnutrition*, WHO, Geneva.

Chapter 14

Nutrition in the aetiology of diseases

- The response of an individual to excess nutrition, whether total energy or particular nutrients, is individual and dependent upon genetic make-up.
- Nutrition and genetic make-up have an important aetiological role in the development of some diseases, e.g. cancer of the colon and coronary heart disease.

14.1 Genetic and environmental aspects

The aetiology of many human diseases includes the consequence of environmental factors. Until recently, infection was the major cause of premature mortality in the developing countries, but this is changing. When the gross national product per capita in a country becomes greater than $1200 per annum then the proportion of deaths from cardiovascular disease increases sharply. Equally, the proportion of deaths from cancer increases progressively as the gross national product per capita increases. The proportion of animal fat in the diet also increases progressively with increasing gross national product.

Diet, smoking and industrial pollution are important factors. Populations living in the Western world constantly seek the various elements in the environment which can be changed in order to prevent premature death. Hence an interest in diet in relation to coronary artery disease, and cancer such as breast and colon. In addition, human immunodeficiency virus (HIV) is causing significant numbers of deaths, particularly in Africa.

The interplay between diet, environment and genetic predisposition is important in most diseases. Two problems – cancer and coronary heart disease – are discussed in this section. Diet is a recognized aetiological factor. The individual's genetic predisposition determines an individual response, which is dependent upon the isoenzyme constitution.

14.2 Cancer

14.2.1 Aetiology

Both epidemiological and experimental evidence indicates that cancer results from an accumulation of several distinct molecular events. Cancer is a prime example of the influence of environment and external agencies on the development of disease, e.g. smoking and lung cancer and certain industrial processes, e.g. tyre manufacture and cancer of the bladder. The individual's genetic make-up predisposes to cancer, an important example being cancer of the colon. Cancer of the breast and other cancers may be less well studied but are equally important.

14.2.2 Incidence

In 1980, there were approximately 6.3 million new cases of cancer throughout the developed and the developing countries. The mortality rates for cancer are significantly lower in the developing countries than in the industrialized countries. In Thailand, cancer mortality rates per 100 000 per annum are 54 for males and 36 for females. The figures for Mexico are 77 and 78 and < 100 per 1000 for most South American countries. In contrast, the cancer mortality rates are > 100 for females and 150 for males in most developed countries. Lung cancer accounts for 25–33% of cancer-related mortality in men.

In the United States, one-third of human cancer cases have been attributed to dietary and nutritional factors. These include excess calories in the diet, the type and amount of fat intake, dietary contaminants, e.g. moulds, cooking-related mutagens, naturally occurring toxins and lack of fruit and vegetables in the diet.

Any diet may contain a number of carcinogenic materials which can be classified into naturally occurring chemicals, synthetic compounds and compounds produced by cooking. The naturally occurring group includes mycotoxins and plant alkaloids, the second group food additives and

pesticides, and the third group polycyclic aromatic hydrocarbons and heterocyclic amines.

Some constituents of moulds are carcinogenic. Aflatoxin B_1 is associated with liver cancer. Other suggested carcinogens include hydrazines in some mushroom species, allylisothiocyanate in brown mustard, oestragole in basil leaves and safrole in natural root beer. Biogenetically engineered food constituents are untested in their potential for carcinogenesis.

Cancers in which infective agents have been suggested to be important in the pathogenesis are cancer of the uterine cervix, hepatoma, nasopharyngeal cancer and cancer of the bladder infected with bilharzia. Of the cancers in which cigarette smoking and possibly diet are important, lung, breast and colon are most common.

Lung cancer rates increase at 0.5% per year, while stomach cancer rates are falling by about 2% per year. As cigarette smoking becomes more common in Asia, China and Africa, so lung cancer is beginning to be a problem there. Colorectal cancer is almost twice as common in the industrialized countries as in the developing countries. Cancers of the upper digestive system account for one-third of cancer in the world and are associated with tobacco chewing and smoking habits and diet.

14.2.3 Geographical variations

Africa and the Middle East

In Africa, the important cancers in men are hepatoma, lymphoma, Kaposi's sarcoma, melanoma and prostatic, bladder, stomach, oesophageal and penile cancer. Cancer of the lung is rare. In women, cancer of the cervix, followed by breast, liver and stomach are the most common. There are differences in different parts of equatorial Africa in the frequency of these cancers. In Middle Eastern countries bladder cancer caused by endemic schistosomiasis is common.

Asia

There are regional differences and variations between different religious groups. In women, cancer of the cervix predominates, particularly in southern India, followed by cancer of the breast, cancer of the mouth, oesophagus and stomach. In affluent Indians breast cancer is as common as in industrial countries. In men, cancer of the stomach and oesophagus predominate, followed by cancer of the mouth and pharynx. Oral cancer is associated with chewing 'pan' (areca-lime and tobacco wrapped in betel leaves). In Pakistani women breast cancer is more common than cervical cancer. In south-east Asia, liver, lung, oesophageal, stomach and nasopharyngeal cancers are among the most prominent cancers in men. Again, there are differences between different racial groups within the same geographical area.

Latin America

The differences depend on socioeconomic developments. The types and rates vary between the developed countries (such as Argentina and Chile) and very poor countries (such as Bolivia). The prevalent disease also depends on the relative numbers of migrant European, native Indian or African populations. There is a high incidence of gallbladder cancer in Bolivia and southern Mexico, oesophageal cancer in southern Brazil, northern Argentina, Paraguay and Uruguay and gastric cancer in Chile, Columbia and Costa Rica. The highest incidence of cervical cancer in the world is found in Brazil.

14.2.4 Colonic cancer

Colonic cancer is a good example of the relationship between genetic predisposition and environment. There are community differences in the incidence of colonic cancer, Scotland having a 2- to 3-fold higher prevalence than the south of England. A person raised in Scotland who moves south retains the enhanced risk. A move north increases the risk, though the time scale is long.

Most cancers of the colon arises spontaneously but there is a distinct familial predisposition. There are individuals who are very much at risk of cancer of the colon, polyposis coli and rare proliferative adenomatous states. Family history makes their surveillance more defined and necessary. As the number of relatives who have had cancer of the colon increases so does the risk. One relative imposes a trivial increased risk, but with two or three affected relatives then surveillance for polyps by regular testing of faeces for occult blood and by flexible sigmoidoscopy is mandatory. If a polyp is found in the left side of the colon or a defined hereditary risk is identified then examination of the entire colon by colonoscopy is required.

The aetiology of colorectal cancer and the adenomatous polyp to carcinoma progression is multifactorial. Colonic cancer is characterized by a well-defined premalignant phase in the adenomatous polyp. The first change may take place in one cell in the crypt. This may occur frequently and there may well be processes which respond to such a conversion. Sometimes a changed cell survives with the prospect of growth, polyp formation and malignancy.

Colonic cancer may develop as a sporadic event in 75% of the population, or be related less commonly to a genetic predisposition, wherein the individual is born with a gene mutation. Such gene mutations may be one of several required for malignant transformation. These at risk individuals will develop cancer at an earlier age than the overall population and at multiple sites within the gastrointestinal tract. They include familial adenomatous polyposis coli and hereditary non-polyposis coli family syndromes (Lynch types I and II). The same accumulation of molecular events occurs in all types of colonic cancer but in the sporadic older age group at a later stage. The genes which mutate include oncogenes which induce and maintain cell transformation and tumour suppressor genes. There can also be mutation of genes involved in DNA mismatch repair, resulting in a characteristic pattern of DNA damage in some tumours.

During the change from polyp to cancer there is an increasing *ras* mutation and loss of specific chromosomal regions containing tumour suppression genes (allelic loss). The sequence is an accumulation of change initiated by mutation in a tumour suppressor gene in chromosome 5. Loss of methyl groups from DNA is an early contribution to altered cell mitosis, adenoma formation with *ras* mutation and loss of genetic material in the long arm of chromosome 18 and the short arm of chromosome 17. In colonic adenomas only 20% showed abnormalities such as 18q-allelic deletions or −17p deletions while in carcinomas these abnormalities were seen in 70% or more of the tumours examined.

14.2.5 Dietary influences on development of cancer

There is evidence that **dietary factors** are important. The complex role of diet includes excessive energy intake and possibly a high consumption of saturated fat and protein and a low consumption of dietary fibre and micronutrients. Epidemiological studies have shown a relationship between colorectal cancer and the per capita consumption of meat and fat. There is an inverse correlation between fibre intake and the prevalence of colorectal carcinoma; the influence of various types of dietary fibre on the risk of colorectal cancer has not been investigated. It is difficult to separate the risks of a deficiency of dietary fibre from that of anti-oxidants present in the fruit and vegetables, as opposed to cereal fibres. Epidemiological studies on the influences of the individual dietary constituents and the risk of colorectal cancer must control for energy balance, micronutrients, i.e. calcium, selenium, vitamins D, A, C and E, and also the type as well as the amount of dietary fibre.

Epidemiologists now believe that dietary factors play a key role in the causation and prevention of large bowel cancer.

Kritchevsky (1993) has shown that caloric (i.e. energy) restriction, will inhibit the growth of spontaneous or experimentally induced tumours. Exercise reduces the risk of chemically induced tumours in rats and vigorous occupational activity has been shown to reduce colon cancer risk.

The rates of colorectal cancer occurrence in various countries increase with the consumption of red meat and animal fat and with low fibre consumption. Saturated fat consumption has been shown to be positively associated with risk of colorectal adenoma and **dietary fibre** was inversely associated with the risk of adenoma. All sources of fibre, whether vegetable, fruit or grain were associated with a degree of protection against developing adenomata. For subjects on a high saturated fat, low fibre diet the relative risk of developing an adenomatous polyp was 3.7 times greater when compared with individuals on a low saturated fat, high fibre diet.

There are international differences in the effect of red meat on colon cancer rates. This raises the possibility that there are differences in meat production techniques; an important factor which is not discussed is the addition of growth factors, anabolic steroids, or antibiotics to the feed of animal stock.

Dietary fibre has a complex relationship to the causation of colon carcinoma. There should be differences between fermentable and non-fermentable fibres in their protective potential against colon cancer. Fermentable fibres produce short-chain fatty acids through fermentation, while poorly fermentable fibres dilute intestinal contents. Fibre has complex effects on the colon, altering intestinal microfloral activity, aqueous phase bile acids, mutagenicity of intestinal contents, alterations in bacterial enzymes, response to hormones and other peptide growth hormones, whether these be local or systemic. Dietary fibre affects the enterohepatic circulation of hormones, intestinal transit time, colonic pH, as well as delaying the absorption of dietary energy.

The average stool weight for a range of countries varies from 70 to 470 g/day and appears to be inversely related to colon cancer risk. There is a significant correlation between fibre intake and mean daily stool weight. Diets providing more than 18 g of non-starch polysaccharides per day would result in daily stool weights of 140 g or more and might reduce the risk of bowel cancer.

14.2.6 The biology of cancer

Despite epidemiological and experimental evidence that cancer requires a number of quite distinct molecular alterations, there is still hope that a single cause, e.g. the control of cell division, lies behind the genesis of the cancer. The cell cycle is regulated in a similar manner in all higher organisms. There are two major control points in the cycle, one at the G_1/S transition, i.e. where DNA replication starts and the other at entry to mitosis. These are determined by protein kinases, enzymes regulated by protein cofactors (cyclins), the concentrations of which increase and fall during the cell cycle. Cyclin-dependent kinases (Cdks) and cyclins have many subfractions. The gene involved in cyclin D1 is to be found on human chromosomes 11q13 and is involved in a number of tumours including β-cell lymphoma and parathyroid. Amplification and/or over expression of cyclin D1 has been shown to be involved in breast and oesophageal cancers.

Resistance to chemotherapy

Biochemical changes in malignant cells give rise to resistance to chemotherapy, which may be either intrinsic or acquired. Some tumours such as the colon are inherently resistant to chemotherapy. Goldie and Coldman have suggested that the more cells present in a tumour, the greater the risk of resistance. A small mature colonic tumour may go through many divisions but remain small by shedding cells. There are a number of steps in the sensitivity of cells to drugs and these include:

1. **Multiple drug resistance** associated with p-glycoprotein 170 found in cancer cells. Drugs against which resistance readily develops have a common origin in that they are all obtained from plants, fungi and bacteria. Resistance appears to be related to the amount of p-glycoprotein 170 in the tumour cell. The function of p-glycoprotein is to pump toxic substances (particularly of plant origin) from the cell. This perhaps accounts for why the oesophagus, stomach and colon are so resistant to chemotherapy. These tissues are exposed to and may provide protection from noxious roots and berries. The multi-drug resistance-1 gene associated with p-glycoprotein 170 is amplified in patients receiving drugs of plant or bacterial origin. Placement of this gene into a previously drug sensitive cell makes that cell resistant to chemotherapy.

2. **Gene amplification**, which results in an increased enzyme production; for example, the gene controlling the production of dihydrofolate reductase which is specifically inhibited by methotrexate. Cells become resistant to methotrexate in various ways. Methotrexate passes through the cell membrane by simple diffusion or an energy-dependent mechanism. On passing through the cell it is polyglutamated and cannot diffuse out. Mutations leading to impaired active transport or a failure of polyglutamation lead to a decreased uptake and hence the cell becomes resistant. The cell may become resistant to methotrexate when the drug, having passed into the cell, meets a variant of the enzyme dihydrofolate reductase. This affects the binding of methotrexate to the enzyme. The most common mechanism of resistance is by amplification of the gene for dihydrofolate reductase.

3. **The presence of topoisomerase II**, which is important in the action of the chemotherapeutic agents doxorubicin, amsacrine and etoposide. These agents bind to topoisomerase II, preventing the binding of the DNA strands broken as a result of therapy. If topoisomerase II is reduced, or absent, the drugs cannot bind to DNA, and DNA strand re-ligation occurs and the cells can then divide.

These approaches to the biology of cancer can only be encouraging news for the ultimate treatment of the condition – particularly those of the gastrointestinal tract – which are so refractory to chemotherapy and radiotherapy.

14.2.7 Metabolic consequences of cancer

Cachexia

Neoplastic disease is not infrequently complicated by wasting, weakness, anorexia and anaemia. As the tumour increases in size, muscle mass and adipose tissue diminish. It is curious that the liver, kidney, adrenal glands and spleen are spared and may even enlarge. Early in the process, total body protein may be unchanged, though redistributed from muscle to tumour. At a later stage, total body protein declines as anorexia is more pronounced. The mechanism responsible is unknown.

Cachectic cancer patients may not only have protein calorie under-nutrition, but may also be deficient of vitamins and minerals. The amount of weight loss varies and may in fact precede clinical presentation of the cancer in a good proportion of patients. The frequency of weight loss varies between 31% with non-Hodgkin's lymphoma, 14% of patients with breast cancer, and 87% of patients with gastric carcinoma. There may also be reduced creatinine : height indices, serum albumin and vitamins A and C. Patients with visceral protein and lean body mass depletion have a worse prognosis than patients in whom weight is retained. However, while there is a general association between protein calorie malnutrition and survival, a cause-and-effect relationship has not been established. Weight loss may be because of reduced food intake, gastrointestinal malabsorption or endogenous metabolic abnormalities leading to combinations of impaired protein synthesis, breakdown or hypermetabolism. Mechanical factors, dysphagia, intestinal obstruction or ascites are important reasons for tissue weight loss. Chemotherapeutic drugs which reduce appetite are also important factors in weight loss.

The mechanism by which tumours depress appetite is unknown. Other factors decreasing appetite include pain, drugs for pain relief, radia-tion enteritis and depression. Intestinal absorption appears to remain intact.

Metabolic abnormalities

There is a variable metabolic response to cancer. There is no consistency between tumour type and increases and decreases in metabolic rate. There is variable production of hormones such as cytokines which may be due to differing tumour tissue types and degree of differentiation. The progressive wasting of host tissue contrasts with the vigorous growth of tumour tissue. Tumour cells may divide under conditions where host cells atrophy.

The hexoses taken up by a malignant cell may, in addition to being involved in a glycolytic pathway, be shunted to the pentose phosphate pathway for the synthesis of both DNA and RNA. Hexoses in malignant cells are also used in glycosylation of membrane structures, proteins and lipids. The cancer cell is not infrequently a high consumer of glucose and a producer of lactate. Hexoses used by many malignant cells may support proliferation in ways other than the production of energy, ultimately contributing to the growth advantage of these cells. The ability of tumours to synthesize fatty acid varies widely, though the synthesis rate is inadequate for replication. Consequently, tumour cells may obtain required fatty acids from host tissues.

There may be differences in the uptake of amino acids by malignant cells and normal cells. Malignant cells may use disproportionately large amounts of certain amino acids. There is a marked loss of weight and content of protein and glycogen in the liver. Hepatic protein synthesis is reduced in tumour-containing livers. There are marked changes in the activity of many liver enzymes with abundant production of foetal isoenzymes. Enzymes important in carbohydrate and amino acid metabolism are altered, such that there is increased glycolysis and disproportionate metabolism of some amino acids.

Skeletal muscle

Skeletal muscle mass declines as the tumour grows. This appears to be due to decreased incorporation of amino acids into muscle protein, which is probably the result of a tumour-specific defect in protein synthesis. On the other hand, these changes may be those of starvation. Glucose metabolism is also reduced in muscles of patients with carcinomatosis. The rate of endogenous glucose production and turnover is increased in under-nourished compared with normally nourished cancer patients. Lactate production is increased in patients with metastatic cancer but the range of production rates is broad. Adipose tissue loss is very marked in patients with cancer.

Tumour cells also produce hormones and hormone-like factors. It is possible that these tumour-secreted metabolically active products have a deleterious effect on host metabolism. However, all studies comparing tumour growth rates with and without nutritional supplementation show an acceleration in tumour growth rate when nutrient intake is increased.

Summary

1. Diet, smoking and industrial pollution are important aetiological factors in the disease process in humans. The impact of these on any individual will be determined by exposure, the genetic make-up, gender and geographical location. There is a relationship between socioeconomic status and the prevalence of types of disease.
2. Cancer and coronary heart disease are good examples of this interplay.
3. Some cancers have been attributed to dietary causes, e.g. liver cancer and aflatoxin B_1.
4. Colonic cancer has both genetic and dietary components to its aetiology. Most cancers of the colon arise spontaneously, but there is a distinct familial predisposition.
5. Colonic cancer develops as a polyp which undergoes malignant transformation; with such changes are accompanying gene changes. The genes which mutate include oncogenes that induce and maintain cell transformation and tumour suppressor genes. There can also be mutation of genes involved in DNA mismatch repair, resulting in a characteristic pattern of DNA damage in some tumours. During the change from polyp to cancer there is an increasing *ras* mutation and loss of specific chromosomal regions containing tumour suppression genes (allelic loss).
6. The dietary contributions to this cancer include a deficiency of dietary fibre, particularly fruit and vegetables, and excess of red meat, protein and fat in the diet. Excessive energy intake may be a factor.
7. Once the malignant transformation has occurred, there are complex metabolic changes which have implications for treatment, including chemotherapy. A damaging change is the very complicated process of cachexia.

Further reading

Bond, J.H. (1993) Polyp guideline: diagnosis, treatment and surveillance for patients with non-familial colorectal polyps. *Annals of Internal Medicine*, 119, 836–43.

Burnstein, M.J. (1993) Dietary factors related to colorectal neoplasms. *Surgical Clinics of North America*, 73, 13–29.

Cummings, J.H., Bingham, S.A., Heaton, K.W. *et al.* (1992) Fecal weight, colon cancer risk and dietary intake of non-starch polysaccharides (dietary fiber). *Gastroenterology*, 103, 1783–9.

Fearon, E.R. and Vogelstein, B. (1990) A genetic model for colorectal tumorigenesis. *Cell*, 61, 759–67.

Giovannucci, E., Stampfer, M.J., Colditz, G. *et al.* (1992) Relationship of diet to risk of colorectal adenoma in men. *Journal of the National Cancer Institute*, 84, 91–8.

Goldie, J.H. and Coldman, A.J. (19979) A mathematical model for relating the drug sensitivity of tumors to the spontaneous mutation rate. *Cancer Treatment Reports*, 63, 1727–33.

Gottesman, M.M. and Pastan, I. (1993) Biochemistry

of multidrug resistance mediated by the multidrug transporter. *Annual Review of Biochemistry*, **62**, 385–427.

Hayes, J.D. and Wolf, C.R. (1990) Molecular mechanisms of drug resistance. *Biochemical Journal*, **272**, 281–95.

Kritchevsky, D. (1993) Colorectal cancer: the role of dietary fat and caloric restriction. *Mutation Research*, **290**, 63–70.

Lawson, D.H., Richmond, A., Nixon, D.W. and Redman, D. (1982) Metabolic approaches to cancer cachexia. *Annual Review of Nutrition*, **2**, 277–301.

Malpas, J.S. (1994) Some aspects of cancer medicine.

Journal of the Royal College of Physicians of London, **28**, 136–42.

Mcgraff, I. and Lidvike, G. (1993) Cancer in developing countries: opportunities and challenges. *Journal of the National Cancer Institute*, **85**, 862–73.

Steel, M. (1994) Cyclins and cancer: wheels within wheels. *Lancet*, **343**, 931–2.

Vogelstein, B., Fearon, E.R., Hamilton, S.R. *et al.* (1988) Genetic alteration during colorectal tumor development. *New England Journal of Medicine*, **319**, 525–32.

Wynder, E.L., Reddy, B.S. and Weisburger, J.H. (1992) Environmental dietary factors in colorectal cancer. Some unresolved issues. *Cancer*, **70**, 1222–8.

14.3 Atheroma

Atherosclerosis involves the intima of medium and large arteries.

14.3.1 Epidemiology

Epidemiological studies have shown that those populations which have a high coronary heart disease (CHD) rate have a substantial dietary intake of saturated fat. The consensus, in population studies, is that the higher the average serum cholesterol the greater the risk of CHD. Worldwide cross-sectional surveys are suggestive that low-fat, low-cholesterol diets result in low lipid concentrations and reduced atherosclerotic heart disease. The development of atherosclerosis is complicated by lack of exercise, total calorie intake, obesity, smoking and stress. In an international comparison the French and Japanese have the lowest incidence of CHD (Figure 14.1) and associated mortality rate (Figure 14.2). It is difficult to explain the relative freedom of the French from the CHD problem in the light of their diet and serum cholesterol concentrations. The French cuisine is not based on a low-fat diet. One suggested explanation is that the drinking of red wine provides protective anti-oxidants.

Normal subjects develop higher serum cholesterol concentrations when fed saturated fat compared with concentrated polyunsaturated fat diets. In general, polyunsaturated fats lower plasma cholesterol concentrations. Fatty acids from fatty fish, such as mackerel, salmon and trout contain unsaturated fatty acids of the n-3

type. Vegetable oils contain n-6 fatty acids. Adding fish oils to the diet of normal subjects or hyperlipidaemic subjects results in a decrease in total triglycerides, VLDL and LDL cholesterol concentration and prolonged clotting times.

When the percentage of fat in the diet is lowered there is a consequent increase in dietary carbohydrates preferably as starch. A high carbohydrate diet of sucrose results in increased triglycerides and VLDL triglycerides. Hepatic secretion of VLDL increases and VLDL contains more triglyceride. There is a fall in HDL concentrations.

There are various accounts of the effect of **dietary fibre** on plasma lipids. The mechanism whereby fibre reduces serum cholesterol has generally been seen to be an increase in the excretion of sterols in the form of bile acids in the stool. The bile acids are bound to fibre, or lignin within the fibre. However those fibres which are most successful in reducing the serum cholesterol are fermented in the caecum, e.g. pectin. This suggests that the effect is through some other mechanism which includes the activity of bacteria in the colon. The bile acids may be bound to bacteria.

The recommendation by COMA and other authorities is to decrease the overall intake of fat in the diet to less than 35% of total calorie intake and that the fat should include poly- and mono-

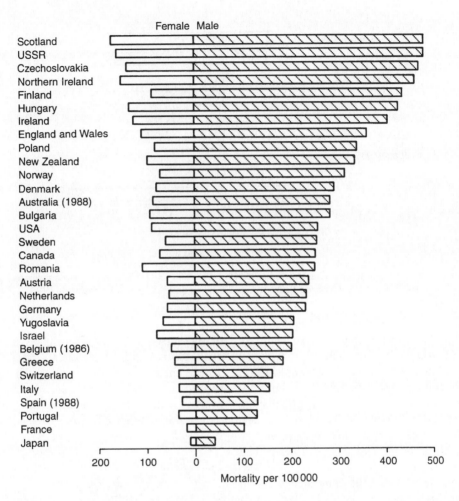

FIGURE 14.1 Coronary heart disease worldwide, 1989. Mortality per 100 000, age 40–69 years (WHO data, 1992). Reproduced from *Scotland's Health – A Challenge to Us All*, Report of a Working Party to the Chief Medical Officer for Scotland, 1993, with permission of the Scottish Office.

unsaturated fats. The energy intake should be maintained by an increase in starch-containing foods and the daily diet should contain fruit and vegetables.

14.3.2 Pathology

There are diffuse thickenings of the musculo-elastic intima of the vessels. This abnormal process may start in childhood. Three main lesions are described: the fatty streak, the fibrous plaque, and a complicated lesion. The **fatty streaks** are an

accumulation of intimal smooth muscle cells surrounded by fat. Lipid-laden macrophages are a feature of the early lesion. The lipid in the fatty streak is largely cholesterol oleate in foam cells. The cholesterol deposited in the plaque is from circulating lipoproteins, especially LDL cholesterol. This uptake is not to LDL receptors. A modification of the LDL particle may be involved in its uptake by the vascular epithelium by a scavenger receptor which is independent of the LDL receptor. The requirements to be taken up include lipid oxidation of the unsaturated fats in

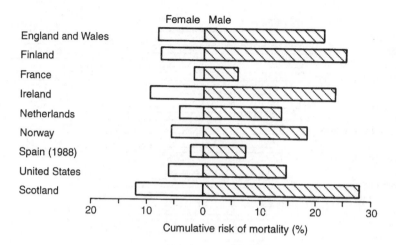

FIGURE 14.2 Coronary heart disease, 1989. Cumulative risk of mortality expressed as a percentage of the population aged 35–74 years in a number of countries (WHO data, 1992). Reproduced from *Scotland's Health – A Challenge to Us All*, Report of a Working Party to the Chief Medical Officer for Scotland, 1993, with permission of the Scottish Office.

the LDL lipid. The size of the LDL particle may determine the oxidation potential.

The **fibrous plaque** consists of fat-laden smooth muscle cells and macrophages, the fat being cholesterol linoleate. There are deposits of collagen, elastic fibres and proteoglycan. This lesion undergoes further change with calcification, cell necrosis, mural thrombosis and haemorrhage. The result is the **complicated lesion**.

The fibrous plaque and the complicated lesion are associated with vascular occlusion which is a feature of coronary heart disease (Figure 14.3). The lesions develop at the junction of vessels, more frequently in the arteries of the legs, head and heart and the rarely the arms. These lesions are more common when there is hypertension (cerebrovascular disease), or smoking (peripheral vascular disease).

The occlusion of an artery may result from:

- haemorrhage into an atherosclerotic plaque
- spasm in a diseased vessel
- thrombosis on an atheromatous and stenosed vessel

Thrombosis is the most common cause of vessel occlusion and occurs as a result of altered blood flow, altered blood constituents or an abnormality

of the vessel wall. An abnormal vessel wall can alter the binding of platelets – which are negatively charged – to this surface. The vessel wall endothelium is protective against this aggregation process. Metabolic processes in the endothelium reduce the adherence of the platelets.

14.3.3 Familial hypercholesterolaemia

This is characterized clinically by raised plasma LDL concentrations, xanthomas and early CHD due to atherosclerosis. Familial hypercholesterolaemia is an autosomal dominant trait and homozygotes are more severely affected than heterozygotes. It is a commonly occurring inherited metabolic condition with a population frequency of 1 in 500 for heterozygotes and 1 per 10^6 for homozygotes. There are populations, however, where the frequency is higher, e.g. Lebanese, French Canadians, Afrikaaners and Jews of Lithuanian origin. Raised concentrations of plasma cholesterol in familial hypercholesterolaemia result from mutations in the LDL-receptor gene. Plasma cholesterol and LDL cholesterol concentrations may be increased substantially. It is probable that differences in lifestyle have no

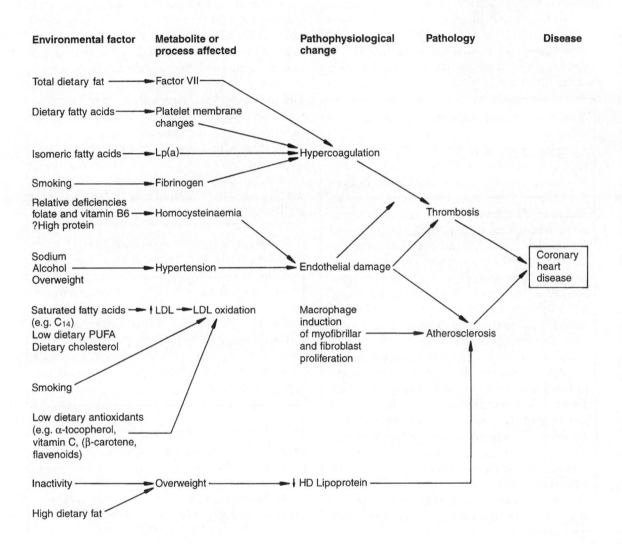

Environmental factor	Metabolite or process affected	Pathophysiological change	Pathology	Disease

FIGURE 14.3 A modern view of the development of coronary heart disease (CHD). Reproduced from *Scotland's Health – A Challenge to Us All*, Report of a Working Party to the Chief Medical Officer for Scotland, 1993, with permission of the Scottish Office. Note: all arrows indicate a stimulatory (i.e. aggravating) effect except ↑LDL = raised LDL; (↓HD Lipoprotein = lowered HDL. The three well-characterized risk factors are smoking, hypertension and LDL. These are highly interactive but may only explain half the regional or group variation in CHD. The metabolite response to dietary change shows variation between people, these differences being increasingly recognized as of genetic origin. Physical activity has many effects apart from increasing HDL concentrations, including physical dilatation of coronary vessels, making them less likely to undergo occlusion. There are also behavioural and metabolic confounders: smokers eat a diet richer in sugar and salt and with fewer vegetables and fruit. Smoking also accelerates the use of vitamin C which, in this scheme, is shown as a protective factor. Not included in this diagram is the potential programming of one or more metabolic processes in foetal and infant life which may then predispose the adult to CHD. The role of breast feeding is also not shown. PUFA, polyunsaturated fatty acids.

effect on homozygotes with familial hypercholesterolaemia, though there are individual variables in plasma cholesterol concentration. When the homozygotes are subdivided according to LDL-receptor gene mutations there is less variability among the groups.

Variations among subjects with a $>10\,\text{kb}$ deletion is two-fold, and among the exon 3 mutation group it is 1.5-fold. Plasma cholesterol concentrations are higher in subjects with a $>10\,\text{kb}$ deletion than in those with the exon 3 mutation. The $>10\,\text{kb}$ deletion is a *null mutation* (functional class I) or receptor-negative, whereas exon mutation is LDL-receptor (functional class III) and therefore receptor-defective. CHD and coronary deaths are more frequent in homozygotes in whom receptor activity is less than 2% of normal compared with subjects in whom receptor activity is 20–30% of normal. In homozygotes, mutations that impair rather than abolish receptor function tend to produce lower plasma concentrations of cholesterol, to be more responsive to treatment and to result in less severe CHD. There is a large difference in effect of null (deletion of the promotor and exon 3) versus missense alleles (exon 3, $Trp_6 \rightarrow Gly$) on cholesterol concentration in homozygotes. Heterozygotes with these mutations have very little difference in blood cholesterol concentrations. This may be due to interference of the missense allele (wherein an amino acid is changed) with the function of the normal allele.

There are modifying effects of other genes on cholesterol concentration. Homozygous lipoprotein lipase deficiency in familial hypercholesterolaemia heterozygotes reduces LDL cholesterol concentrations to normal. Other genes can effect the risk of coronary disease.

In liver cells LDL receptor activity is not subject to the normal suppressive effects of cholesterol. This is because of expression of 7-α-hydroxylase in liver where this enzyme metabolizes hydroxy-lated sterols as well as cholesterol to bile acids. Placing the LDL receptor gene into non-hepatic cells causes them to take on the hepatic phenotype of resistance of the LDL receptor to repression by cholesterol. Bile acid sequestrants work by increasing oxysterol flux through this pathway and increasing LDL receptor gene expression.

A diet rich in cholesterol and saturated fats, nevertheless has the greatest adverse effect. In a population with a low fat and cholesterol intake a genetic propensity to an increased blood cholesterol does not exhibit itself. Nature and nurture each have a large role in familial hypercholesterolaemia.

Homozygosity for the apoE2 allele which does not bind to the LDL receptor has severe effects on the familial hypercholesterolaemia phenotype. Inheritance of high concentrations of Lp(a) reduces the risk of coronary disease in subjects with familial hypercholesterolaemia. Transcription of the LDL receptor is negatively regulated by intracellular cholesterol. This process is mediated by hydroxylated cholesterol metabolites.

Hyperhomocysteinaemia is readily corrected by supplementation with folate or betaine. Homocysteine is metabolized by *trans*-sulphuration to cysteine through cystathionine or by methylation to methionine. Homocysteinuria is due to an inborn metabolic defect in the *trans*-sulphuration or remethylation of homocysteine.

There is a **polymorphism** in the gene for the angiotensin-converting enzyme (ACE). A deletion/insertion has been found in intron 16 of the *ACE* gene. Deletion polymorphism has been found more frequently in patients who have suffered a myocardial infarction. Individuals who are homozygous for the deletion allele have plasma ACE concentrations which are twice as high as individuals homozygous for the insertion allele.

Many retrospective studies have shown a positive relationship between moderate hyperhomocysteinaemia and occlusive arterial disease and myocardial infarction. The accumulation of excessive homocysteine causes damage to endothelial and smooth muscle cells and alters the activity of coagulation factors.

In defined populations at risk of CHD, there is evidence of an individual variation in response to changes in dietary fat, some individuals showing a large and others a small response to dietary change. This is not due to variable dietary compliance and may be determined by individual lipid metabolic activities, apolipoprotein concentrations and other polygenic factors.

Summary

1. Coronary heart disease (CHD) is associated with populations who eat a high saturated fat intake and in whom the serum lipids are increased. The CHD in such populations is, to a large extent, concentrated within genetically vulnerable families.
2. Dietary constituents which increase protection from CHD include n-3- and n-6- containing fatty acids and oils, fruit and vegetables, and possibly red wine.
3. The atheromatous plaque formation which is the basis of CHD progresses through a fatty streak, a fibrous plaque and a complicated arterial wall lesion. These are associated with the deposition of cholesterol in the form of lipoproteins, e.g. LDL lipoprotein.
4. Familial hypercholesterolaemia is an autosomal dominant trait with varying frequency in the population. The LDL-receptor gene mutations are well described and the mutation dictates the metabolic response.
5. Other gene mutations, e.g. apoE2 allele have severe effects on blood cholesterol concentrations.

Further reading

Cox, C., Mann, J., Sutherland, W. and Ball, M. (1995) Individual variations in plasma cholesterol response to dietary saturated fat. *British Medical Journal,* **311**, 1260–4.

Goldberg, A.C. and Schonfeld, G. (1985) Effects of diet on lipoprotein metabolism; *Annual Review of Nutrition,* **5**, 195–212.

Kang, S.-S., Wong, P.W.K. and Malino, W. (1992) Hyperhomocysteinaemia as a risk factor for occlusive vascular disease. *Annual Review of Nutrition,* **12**, 279–98.

McGee, J.O'D., Isaacson, P.G. and Wright, N.A. (1992) *Oxford Textbook of Pathology,* Oxford Medical Publications, Oxford.

Parthasarathy, S., Steinberg, D. and Witztum, J.L. (1992) The role of oxidized low-density lipoproteins in the pathogenesis of atherosclerosis. *Annual Review of Medicine,* **43**, 219–25.

Weatherall, D.J., Ledingham, J.G.G. and Warrell, D.A. (1996) *Oxford Textbook of Medicine,* 3rd edn, Oxford Medical Publications, Oxford.

Index

Page numbers appearing in **bold** refer to figures and tables.